A Primer, and Beyond, on Composition Operators on the Unit Disk

This is a contemporary treatment of composition operators on Banach spaces of analytic functions in one complex variable. It provides a step-by-step introduction, starting with a review (including full proofs) of the key tools needed and building the theory with a focus on Hardy and Bergman spaces.

Several proofs of operator boundedness (Littlewood's principle) are given, and the authors discuss approaches to compactness issues and essential norm estimates (Shapiro's theorem) using different tools such as Carleson measures and Nevanlinna counting functions. Membership of composition operators in various ideal classes (Schatten classes for instance) and their singular numbers are studied.

This framework is extended to Hardy–Orlicz and Bergman–Orlicz spaces and, finally, weighted Hardy spaces are introduced, with a full characterization of those weights for which all composition operators are bounded.

This will be a valuable resource for researchers and graduate students working in functional analysis, operator theory, or complex analysis.

PASCAL LEFÈVRE is a professor at the University of Artois, France. After a PhD thesis on harmonic analysis, Banach spaces, and interplay with basic number theory, he has worked for twenty years on composition operators.

HERVÉ QUEFFÉLEC is an emeritus professor at the University of Lille, France. He completed his PhD thesis on harmonic analysis, Banach spaces, and Dirichlet series and has worked on composition operators for the past twenty-five years.

NEW MATHEMATICAL MONOGRAPHS

All the titles listed below can be obtained from good booksellers or from Cambridge University Press. For a complete series listing visit www.cambridge.org/mathematics

18. T. Downarowicz *Entropy in Dynamical Systems*
19. C. Simpson *Homotopy Theory of Higher Categories*
20. E. Fricain and J. Mashreghi *The Theory of H(b) Spaces I*
21. E. Fricain and J. Mashreghi *The Theory of H(b) Spaces II*
22. J. Goubault-Larrecq *Non-Hausdorff Topology and Domain Theory*
23. J. Śniatycki *Differential Geometry of Singular Spaces and Reduction of Symmetry*
24. E. Riehl *Categorical Homotopy Theory*
25. B. A. Munson and I. Volić *Cubical Homotopy Theory*
26. B. Conrad, O. Gabber and G. Prasad *Pseudo-reductive Groups (2nd Edition)*
27. J. Heinonen, P. Koskela, N. Shanmugalingam and J. T. Tyson Sobolev *Spaces on Metric Measure Spaces*
28. Y.-G. Oh *Symplectic Topology and Floer Homology I*
29. Y.-G. Oh *Symplectic Topology and Floer Homology II*
30. A. Bobrowski *Convergence of One-Parameter Operator Semigroups*
31. K. Costello and O. Gwilliam *Factorization Algebras in Quantum Field Theory I*
32. J.-H. Evertse and K. Győry *Discriminant Equations in Diophantine Number Theory*
33. G. Friedman *Singular Intersection Homology*
34. S. Schwede *Global Homotopy Theory*
35. M. Dickmann, N. Schwartz and M. Tressl *Spectral Spaces*
36. A. Baernstein II *Symmetrization in Analysis*
37. A. Defant, D. García, M. Maestre and P. Sevilla-Peris *Dirichlet Series and Holomorphic Functions in High Dimensions*
38. N. Th. Varopoulos *Potential Theory and Geometry on Lie Groups*
39. D. Arnal and B. Currey *Representations of Solvable Lie Groups*
40. M. A. Hill, M. J. Hopkins and D. C. Ravenel *Equivariant Stable Homotopy Theory and the Kervaire Invariant Problem*
41. K. Costello and O. Gwilliam *Factorization Algebras in Quantum Field Theory II*
42. S. Kumar *Conformal Blocks, Generalized Theta Functions and the Verlinde Formula*
43. P. F. X. Müller *Hardy Martingales*
44. T. Kaletha and G. Prasad *Bruhat–Tits Theory*
45. J. Schwermer *Reduction Theory and Arithmetic Groups*
46. J. Kotus and M. Urbański *Meromorphic Dynamics I*
47. J. Kotus and M. Urbański *Meromorphic Dynamics II*
48. S. Garibaldi, H. P. Petersson and M. L. Racine *Albert Algebras over Commutative Rings*
49. D. Pellicer *Abstract Chiral Polytopes*
50. C. Xu *K-stability of Fano Varieties*
51. D. Denisov, D. Korshunov and V. Wachtel *Markov Chains with Asymptotically Zero Drift*
52. B. Simon *Phase Transitions in the Theory of Lattice Gases*
53. P. Lefèvre and H. Queffélec *A Primer, and Beyond, on Composition Operators on the Unit Disk*
54. T. Saito *Ramification Groups of Local Fields: with Geometric Applications*

A Primer, and Beyond, on Composition Operators on the Unit Disk

PASCAL LEFÈVRE
Université d'Artois

HERVÉ QUEFFÉLEC
Université de Lille

CAMBRIDGE
UNIVERSITY PRESS

CAMBRIDGE
UNIVERSITY PRESS

Shaftesbury Road, Cambridge CB2 8EA, United Kingdom

One Liberty Plaza, 20th Floor, New York, NY 10006, USA

477 Williamstown Road, Port Melbourne, VIC 3207, Australia

314-321, 3rd Floor, Plot 3, Splendor Forum, Jasola District Centre, New Delhi – 110025, India

Cambridge University Press is part of Cambridge University Press & Assessment, a department of the University of Cambridge.

We share the University's mission to contribute to society through the pursuit of education, learning and research at the highest international levels of excellence.

www.cambridge.org
Information on this title: www.cambridge.org/9781009388894
DOI: 10.1017/9781009388917

When citing this work, please include a reference to the DOI 10.1017/9781009388917

First published 2026

A catalogue record for this publication is available from the British Library

A Cataloging-in-Publication data record for this book is available from the Library of Congress

ISBN 978-1-009-38889-4 Hardback

To Anna and Aleksandra

To Clémence, Yann, and Jérémy

Contents

Foreword

Mathematics is often described as the "queen of the sciences," but it is within the realm of complex analysis that one finds some of its most profound and exquisite expressions. The beauty of the subject comes from its intricate interplay between algebra, geometry, and topology.

From the graceful behavior of holomorphic functions to the deep insights provided by the Riemann Mapping Theorem, complex analysis is filled with results that are not only mathematically profound but also aesthetically pleasing.

This book explores these themes in depth, focusing on composition operators on various spaces of holomorphic functions. These operators, while technical, are imbued with the same beauty that mathematicians such as Poincaré and von Neumann recognized. Their study not only opens up new frontiers in operator theory but also brings to light the delicate balance of mathematical rigor and aesthetic harmony that is characteristic of complex analysis.

The study of composition operators on spaces of holomorphic functions represents a rich and dynamic area of complex analysis with deep connections to operator theory, functional analysis, and the broader field of mathematical analysis. The goal of this book is to provide a comprehensive exploration of these operators, offering both theoretical foundations and contemporary advancements in the field.

Composition operators are central objects of study due to their ability to combine structural properties of holomorphic functions with the intricacies of operator theory. This book examines the fundamental properties of these operators, their spectral theory, and their behavior on various function spaces. By delving into both classical and modern results, it aims to provide readers with a solid understanding of the interplay between composition operators and holomorphic function spaces.

After a chapter specifying the prerequisites, boundedness of composition operators is explored, emphasizing this highly non-trivial question. Then, compactness theorems are presented using the classical Carleson embedding point of view as well as Nevanlinna counting functions. The membership of composition operators to Schatten classes is also detailed, before moving to approximation numbers on Hardy and Bergman spaces. The last part is mainly devoted to composition operators on Hardy–Orlicz, Bergman–Orlicz, and weighted Hardy spaces.

The book is designed for researchers and advanced graduate students in complex analysis, operator theory, and functional analysis. It assumes a basic knowledge of complex function theory, but the material is presented in a way that makes it accessible to anyone with a fundamental understanding of these areas. Throughout the chapters, readers will find both theorems and proofs that not only shed light on the nature of composition operators but also highlight their connections to other important topics in mathematical research.

The approach taken in this book is both rigorous and intuitive, with a focus on clarity and detailed exposition. Each chapter builds upon the previous one, guiding the reader through a progressive exploration of the subject. Special attention is given to the most recent advancements, with discussions on open problems and future directions for research in the very last part of the book.

This book does not intend to give an exhaustive exposition of the huge number of results on composition operators but to present, in as self-contained a manner as possible, the most elegant tools and a flavor of remarkable results by the richness of the ideas involved in their proof.

I believe that this book serves as both a valuable resource and a source of inspiration for anyone working in the fascinating field of composition operators and holomorphic function spaces.

Isabelle Chalendar
Université Gustave Eiffel, France

Preface

The birth act of composition operators can be considered as the Littlewood subordination principle (around 1924) that proves the following: given an analytic self-map φ of the unit disk, vanishing at the origin, and a function $f \in H^2$, the Hardy space, the function $f \circ \varphi$ is in H^2 as well, with smaller H^2-norm. In modern terms, the (linear) composition operator C_φ with symbol φ, given by $f \mapsto f \circ \varphi$, is a contraction on H^2. But Littlewood does not use the term operator, defines $f \circ \varphi$ as being *subordinate* to f, and his result proves that subordination decreases the H^2-norm (as well as that of partial sums of f). The assumption $\varphi(0) = 0$ was later dropped.

The second step can be seen as the work of harmonic analysts (Beurling, Kahane, Katznelson, Leibenson, Rudin) on the characterization of symbols φ such that, given a compact abelian group G and its Wiener algebra $W(G)$:

(i) The (linear) composition operator C_φ, $f \mapsto f \circ \varphi$, maps $W(G)$ to itself (φ has to be piecewise affine and is called a homomorphism).
(ii) The (non-linear) superposition operator $f \mapsto \varphi \circ f$ maps $W(G)$ to itself (φ has to be analytic and is said to operate on $W(G)$).

Once more, boundedness is the only concern and the terminology is different.

In the late sixties, the connection between functions (the symbol φ) and operators (the operator C_φ) becomes more pregnant with the works of Nordgren, Ryff, H. Schwarz, and in the early seventies with J. Shapiro and P. Taylor, or Power. For example, the compactness of $C_\varphi : H^2 \to H^2$ is characterized in terms of a Carleson measure m_φ, which has to be "vanishing Carleson," or in terms of a Nevanlinna counting function N_φ, which has to be "vanishing Littlewood," with even a simple closed formula for the essential norm of C_φ in terms of N_φ. Other Banach spaces (Hilbertian or not), like Hardy spaces H^p or Bergman spaces $\mathcal{B}^p$, are considered.

Now, with a blink to the title "A primer, and beyond ..." the purpose of this book is twofold:

(i) First, we present some basic facts of function theory (Lindelöf, Denjoy-Wolff, Julia-Carathéodory, Carleson, etc.) and of operator theory (Schmidt decomposition, various singular numbers, etc.) needed here. We then enter into the theory of composition operators from X to X where X can be classical Hardy spaces, or Bergman spaces (with classical weights, or radial weights).

Our aim here is to enable PhD students or researchers to enter into the theory of composition operators with detailed proofs, and different viewpoints on basics.

(ii) Second, we try to present some of the progresses done in the theory during the past twenty years or so. The literature on composition operators extremely increased since the nineties, and we had to make some choices: without entering into details, here is a short, non-exhaustive list:

(a) A contraction property (kind of homogeneity) of the Carleson measures m_φ or σ_φ, which will play an important role in Chapter 11 on Orlicz spaces.

(b) An intrinsic comparison between m_φ and N_φ, independent of any Banach space of analytic functions, which helps unifying the theory.

(c) Luecking characterized the Schattenness (or the membership of the sequence $(a_n(C_\varphi))$ in ℓ^p). We study the individual behavior of those approximation numbers $a_n(C_\varphi)$, which can tend to 0 arbitrarily slowly for C_φ compact, and we give a closed formula for

$$\lim_n \left(a_n(C_\varphi)\right)^{1/n},$$

as well as estimates on specific examples: lens maps, cusp map, Shapiro-Taylor maps, and so on.

(d) The compactness of $C_\varphi : H^\infty \to H^\infty$ is quite simple to study, and takes place if and only if $\|\varphi\|_\infty < 1$. This leads to coining new Banach spaces, which interpolate more closely between H^2 and H^∞, which we call Hardy–Orlicz spaces (we also consider Bergman–Orlicz spaces), present quite new features, and highly depend on the implied Orlicz function ψ. In particular, we show that the equivalence pointed previously on H^∞ is quite far from taking place in the "Orlicz" framework.

(e) In the final chapter, going back (essentially) to the Hilbertian case, we give a full characterization of those weights $\beta = (\beta_n)$ for which *all symbols* induce a bounded composition operator $C_\varphi : H^2(\beta) \to H^2(\beta)$.

We wish to thank Frédéric Bayart, Isabelle Chalendar, Emmanuel Fricain, and Etienne Matheron for helping us to improve the redaction of this book.

We also would like to take the opportunity to address here our warmest thanks to our colleagues and friends Daniel Li and Luis Rodríguez-Piazza for more than twenty-five years of intense collaboration, with a great part dedicated to the study of composition operators. This book owes much to them.

1
Introduction: Toolbox

In the present book, we are interested in studying composition operators

$$f \longmapsto C_\varphi(f) = f \circ \varphi,$$

where the *symbol* φ is an analytic self-map of $\mathbb{D} = \{z \in \mathbb{C} : |z| < 1\}$, the open unit disk of $\mathbb{C}$, and f belongs to various Banach spaces of holomorphic functions on $\mathbb{D}$. More generally, for w analytic on $\mathbb{D}$, the weighted composition operator with symbols φ and w is defined by $f \longmapsto w.(f \circ \varphi)$.

Let us fix some elementary notations and the framework: the boundary of $\mathbb{D}$, the unit circle (the 1-torus), is denoted $\mathbb{T} = \partial\mathbb{D} = \{z \in \mathbb{C} : |z| = 1\}$.

The normalized arclength measure on $\mathbb{T}$ is denoted by m. It is actually the Haar measure on $\mathbb{T}$ viewed as a compact abelian group. The normalized area measure on $\overline{\mathbb{D}}$ is denoted $d\sigma$, but most often we shall work on $\mathbb{D}$ and keep the same notation since obviously $\sigma(\overline{\mathbb{D}} \setminus \mathbb{D}) = \sigma(\mathbb{T}) = 0$. We can write for every Borel subset $E \subset \overline{\mathbb{D}}$:

$$\sigma(E) = \int_{\mathbb{T}} \int_0^1 \mathbb{1}_E \, 2r dr \, dm,$$

where $\mathbb{1}_E$ denotes the characteristic function of the set E.

We denote by $\mathscr{H}(\mathbb{D})$ the class of holomorphic functions on $\mathbb{D}$. In the present monograph, we shall focus on classical spaces: Hardy spaces, the disk algebra, Bergman and Dirichlet spaces, all included in $\mathscr{H}(\mathbb{D})$.

The set $\mathbb{N}_0$ stands for the set of non-negative integers: $0, 1, 2, \ldots$; and $\mathbb{N}$ for the set of positive integers: $1, 2, \ldots$.

The notation $A \lesssim B$ means that $A \leqslant CB$ for some constant C. Same with $\gtrsim$, and $\approx$ means both $\lesssim$ and $\gtrsim$.

Finally a usual notation concerns conjugate exponents: for $p \in [1, +\infty]$, the conjugate exponent of p is $p' \in [1, +\infty]$ defined by $\dfrac{1}{p} + \dfrac{1}{p'} = 1$.

1.1 Reminders on Holomorphic Functions

In this section, we recall some classical results on holomorphic (or subharmonic) functions, offering sometimes proofs for the sake of completeness. The letter Ω will always denote a domain of the complex plane, and $\mathcal{H}(\Omega)$ the class of holomorphic functions on Ω. If (f_n) is a sequence in $\mathcal{H}(\Omega)$ converging uniformly to some function $f \in \mathcal{H}(\Omega)$ on all compact subsets of Ω, it will be convenient to write in short

$$f_n \xrightarrow{uc} f.$$

Later, we shall specialize to $\Omega = \mathbb{D}$, the unit disk.

Theorem 1.1 (Hurwitz) *Let (f_n) be a sequence of $\mathcal{H}(\Omega)$ such that*

$$f_n \xrightarrow{uc} f$$

with f a non-zero function.

If $z \in \Omega$ and $f(z) = 0$, there exists a sequence (z_n) of points of Ω such that $z_n \to z$ and $f_n(z_n) = 0$ for each n.

From the Hurwitz theorem, we get (iii) in the next theorem.

Theorem 1.2 (Montel) *Let (f_n) be a* normal *sequence of $\mathcal{H}(\Omega)$, that is, a sequence whose terms are uniformly bounded on compact subsets of Ω. Then:*

(i) *(f_n) contains a subsequence (f_{n_k}) such that $f_{n_k} \xrightarrow{uc} f$ with $f \in \mathcal{H}(\Omega)$.*
(ii) *If all convergent subsequences have the same limit f, then $f_n \xrightarrow{uc} f$.*
(iii) *Let $E \subset \mathbb{C}$. If $f_n \xrightarrow{uc} f$ and $f_n(\Omega) \subset E$, then either $f(\Omega) \subset E$ or f is a constant c with $c \in \partial E$.*

Let us also recall two basic results.

Theorem 1.3 (Riemann) *Let Ω be a simply connected domain with $\Omega \neq \mathbb{C}$, and $a \in \Omega$. Then, there exists a (unique) conformal map $h : \Omega \to \mathbb{D}$ such that*

$$h(a) = 0 \qquad \text{and} \qquad h'(a) > 0.$$

Moreover, if Ω is the interior of a Jordan curve, h extends to a homeomorphism from $\overline{\Omega}$ to $\overline{\mathbb{D}}$.

Next, the classical Schwarz lemma shows:

Lemma 1.4 (Schwarz) *Let $\varphi : \mathbb{D} \to \mathbb{D}$ be an analytic map fixing the origin, that is, $\varphi(0) = 0$. Then*

$$\forall z \in \mathbb{D}, \quad |\varphi(z)| \leqslant |z| \quad \text{and} \quad |\varphi'(0)| \leqslant 1, \tag{1.1}$$

with equality in (1.1) *for some* $z \neq 0$, *or for the derivative, if and only if* φ *is a rotation that is,* $\forall z \in \mathbb{D}$, $\varphi(z) = \lambda z$, *where* λ *is a unimodular constant.*

Another key result is also Jensen's formula.

Let f be a non-zero analytic function on $\mathbb{D}$. This formula (see (1.2) below), which relates the growth of f and the number $N(t)$ of its zeros (counted with their multiplicity) in the disk $\overline{D}(0,t)$, for $t \in (0,1)$, is of fundamental importance.

Theorem 1.5 (Jensen) *Let* $f \in \mathscr{H}(\mathbb{D})$ *with* $f(0) \neq 0$ *and* $r \in (0,1)$. *Let* $z_1, \dots, z_p$ *be the zeros of* f *inside* $\overline{D}(0,r)$, *counted with multiplicity. Then*

$$m_0(r) := \int_{\mathbb{T}} \log(|f(r\xi)|)\, dm(\xi) = \log(|f(0)|) + \sum_{j=1}^{p} \log\left(\frac{r}{|z_j|}\right). \tag{1.2}$$

Equivalently,

$$\int_0^r \frac{N(t)}{t} dt = \int_{\mathbb{T}} \log\left|\frac{f(r\xi)}{f(0)}\right| dm(\xi). \tag{1.3}$$

Proof: We refer to [153, p. 307]. We just comment on the (quite useful) reformulation (1.3): set $r_j = |z_j|$, and

$$\chi_j(t) = 1 \ \text{ if } r_j \leqslant t, \qquad \chi_j(t) = 0 \ \text{ if } r_j > t,$$

so that $N(t) = \sum_{j=1}^{p} \chi_j(t)$. Then, according to (1.2):

$$\begin{aligned}\int_0^r \frac{N(t)}{t} dt &= \sum_{j=1}^{p} \int_0^r \frac{\chi_j(t)}{t} dt = \sum_{j=1}^{p} \int_{r_j}^r \frac{dt}{t} \\ &= \sum_{j=1}^{p} \log \frac{r}{r_j} = \int_{\mathbb{T}} \log\left|\frac{f(r\xi)}{f(0)}\right| dm(\xi).\end{aligned}$$ □

An application of Jensen's formula is the following set with the notations of (1.2):

$$m_0^+(f,r) = \int_{\mathbb{T}} \log^+(|f(r\xi)|)\, dm(\xi) \geqslant \log(|f(0)|) + \sum_{j=1}^{p} \log\left(\frac{r}{|z_j|}\right),$$

where $\log^+(x) = \max\{\log(x); 0\}$.

Define the *Nevanlinna class* $\mathcal{N}$ as the set of holomorphic functions f on $\mathbb{D}$ such that $\sup_{r<1} m_0^+(f,r) < \infty$. With this definition, we can state the following proposition:

Proposition 1.6 *Assume that $f \in \mathcal{N}$ with $f \neq 0$. Let (z_j) be the set of its zeros counted with multiplicity. Then*

$$\sum_{j=1}^{\infty}(1 - |z_j|) < \infty.$$

Proof: We can assume $f(0) = 1$ and drop the dependence in f. Let $M_0 = \sup_{r<1} m_0^+(r)$ (note that, by subharmonicity (see below), $m_0^+(r)$ increases with r). Now (1.2) and Fatou's lemma give

$$\sum_{j=1}^{\infty} \log(1/|z_j|) = \sum_{j=1}^{\infty} \varliminf_{r\to 1} (\log^+(r/|z_j|)) \leqslant \varliminf_{r\to 1} \sum_{j=1}^{\infty}(\log^+(r/|z_j|))$$
$$\leqslant \varliminf_{r\to 1} m_0^+(r) = M_0 < \infty.$$

The result follows because $\log(1/|z_j|) \sim 1 - |z_j|$ as $j \to \infty$. □

1.2 Reminders on Subharmonic Functions

Definition

Let Ω be an open subset of the plane. A function $u : \Omega \to \mathbb{R} \cup \{-\infty\}$ is said to be subharmonic if

(i) u is upper semi-continuous ("usc" for short).
(ii) For any $a \in \Omega$, there exists $r = r_a > 0$ such that $\overline{D}(a, r_a) \subset \Omega$ and

$$u(a) \leqslant \int_{\mathbb{T}} u(a + r\xi)\, dm(\xi) \qquad \text{for all } r \leqslant r_a. \tag{1.4}$$

We will denote sh(Ω) the set of subharmonic functions on Ω. The following proposition gathers some basic stability properties of sh(Ω) and examples of such functions.

Proposition 1.7 *We have*

(i) *$sh(\Omega)$ is a convex cone.*
(ii) *If $u, v \in sh(\Omega)$, so does* $\max(u, v)$.
(iii) *If $u \in sh(\Omega)$ and if $\psi : \mathbb{R} \to \mathbb{R}$ is convex and non-decreasing, $\psi(u) \in sh(\Omega)$.*
(iv) *If $f \in \mathscr{H}(\Omega)$, then* $\log|f| \in sh(\Omega)$.
(v) *If $u \in sh(\Omega)$ and $a \in \Omega$, for any $0 < r \leqslant r_a$, it holds*

$$u(a) \leqslant \int_B u\, dV,$$

where $B = \overline{D}(a, r)$ and dV is the normalized area measure on B.

with equality in (1.1) *for some* $z \neq 0$*, or for the derivative, if and only if* φ *is a rotation that is,* $\forall z \in \mathbb{D}$, $\varphi(z) = \lambda z$*, where* λ *is a unimodular constant.*

Another key result is also Jensen's formula.

Let f be a non-zero analytic function on $\mathbb{D}$. This formula (see (1.2) below), which relates the growth of f and the number $N(t)$ of its zeros (counted with their multiplicity) in the disk $\overline{D}(0,t)$, for $t \in (0,1)$, is of fundamental importance.

Theorem 1.5 (Jensen) *Let* $f \in \mathscr{H}(\mathbb{D})$ *with* $f(0) \neq 0$ *and* $r \in (0,1)$*. Let* $z_1, \ldots, z_p$ *be the zeros of* f *inside* $\overline{D}(0,r)$*, counted with multiplicity. Then*

$$m_0(r) := \int_{\mathbb{T}} \log(|f(r\xi)|)\, dm(\xi) = \log(|f(0)|) + \sum_{j=1}^{p} \log\left(\frac{r}{|z_j|}\right). \tag{1.2}$$

Equivalently,

$$\int_0^r \frac{N(t)}{t} dt = \int_{\mathbb{T}} \log\left|\frac{f(r\xi)}{f(0)}\right| dm(\xi). \tag{1.3}$$

Proof: We refer to [153, p. 307]. We just comment on the (quite useful) reformulation (1.3): set $r_j = |z_j|$, and

$$\chi_j(t) = 1 \ \text{ if } r_j \leqslant t, \qquad \chi_j(t) = 0 \ \text{ if } r_j > t,$$

so that $N(t) = \sum_{j=1}^{p} \chi_j(t)$. Then, according to (1.2):

$$\begin{aligned} \int_0^r \frac{N(t)}{t} dt &= \sum_{j=1}^{p} \int_0^r \frac{\chi_j(t)}{t} dt = \sum_{j=1}^{p} \int_{r_j}^r \frac{dt}{t} \\ &= \sum_{j=1}^{p} \log \frac{r}{r_j} = \int_{\mathbb{T}} \log\left|\frac{f(r\xi)}{f(0)}\right| dm(\xi). \end{aligned}$$ □

An application of Jensen's formula is the following set with the notations of (1.2):

$$m_0^+(f,r) = \int_{\mathbb{T}} \log^+(|f(r\xi)|)\, dm(\xi) \geqslant \log(|f(0)|) + \sum_{j=1}^{p} \log\left(\frac{r}{|z_j|}\right),$$

where $\log^+(x) = \max\{\log(x); 0\}$.

Define the *Nevanlinna class* $\mathcal{N}$ as the set of holomorphic functions f on $\mathbb{D}$ such that $\sup_{r<1} m_0^+(f,r) < \infty$. With this definition, we can state the following proposition:

Proposition 1.6 *Assume that $f \in \mathcal{N}$ with $f \neq 0$. Let (z_j) be the set of its zeros counted with multiplicity. Then*

$$\sum_{j=1}^{\infty}(1 - |z_j|) < \infty.$$

Proof: We can assume $f(0) = 1$ and drop the dependence in f. Let $M_0 = \sup_{r<1} m_0^+(r)$ (note that, by subharmonicity (see below), $m_0^+(r)$ increases with r). Now (1.2) and Fatou's lemma give

$$\sum_{j=1}^{\infty} \log(1/|z_j|) = \sum_{j=1}^{\infty} \underline{\lim}_{r\to 1} (\log^+(r/|z_j|)) \leqslant \underline{\lim}_{r\to 1} \sum_{j=1}^{\infty}(\log^+(r/|z_j|))$$
$$\leqslant \underline{\lim}_{r\to 1} m_0^+(r) = M_0 < \infty.$$

The result follows because $\log(1/|z_j|) \sim 1 - |z_j|$ as $j \to \infty$. □

1.2 Reminders on Subharmonic Functions

Definition

Let Ω be an open subset of the plane. A function $u : \Omega \to \mathbb{R} \cup \{-\infty\}$ is said to be subharmonic if

(i) u is upper semi-continuous ("usc" for short).
(ii) For any $a \in \Omega$, there exists $r = r_a > 0$ such that $\overline{D}(a, r_a) \subset \Omega$ and

$$u(a) \leqslant \int_{\mathbb{T}} u(a + r\xi)\, dm(\xi) \qquad \text{for all } r \leqslant r_a. \tag{1.4}$$

We will denote sh(Ω) the set of subharmonic functions on Ω. The following proposition gathers some basic stability properties of sh(Ω) and examples of such functions.

Proposition 1.7 *We have*

(i) *$sh(\Omega)$ is a convex cone.*
(ii) *If $u, v \in sh(\Omega)$, so does* $\max(u, v)$.
(iii) *If $u \in sh(\Omega)$ and if $\psi : \mathbb{R} \to \mathbb{R}$ is convex and non-decreasing, $\psi(u) \in sh(\Omega)$.*
(iv) *If $f \in \mathscr{H}(\Omega)$, then* $\log|f| \in sh(\Omega)$.
(v) *If $u \in sh(\Omega)$ and $a \in \Omega$, for any $0 < r \leqslant r_a$, it holds*

$$u(a) \leqslant \int_B u\, dV,$$

where $B = \overline{D}(a, r)$ and dV is the normalized area measure on B.

Proof: (i) and (ii) are obvious. Next, (iii) follows from the classical (convexity) Jensen's inequality. Finally, we set $u = \log|f|$ and take $a \in \Omega$ with $\overline{D}(a, r_a) \subset \Omega$; let $0 \leqslant r \leqslant r_a$. We must prove that

$$u(a) \leqslant \int_{\mathbb{T}} u(a + r\xi)\, dm(\xi).$$

If $f(a) = 0$, $u(a) = -\infty$ and there is nothing to prove. If $f(a) \neq 0$, we use Jensen's formula

$$\int_{\mathbb{T}} \log\left|\frac{f(a + r\xi)}{f(a)}\right| dm(\xi) = \int_0^r \frac{N(t)}{t} dt,$$

where $N(t)$ denotes the number of zeros of f in $\overline{D}(a, t)$. The right-hand side is non-negative and this ends the proof. Note that (iii) and (iv) together give

$$f \in \mathscr{H}(\Omega) \Longrightarrow |f|^p \in \mathrm{sh}(\Omega)$$

for all $p > 0$. Finally, (v) is obvious from (1.4) by integration in polar coordinates. □

For an exhaustive presentation of subharmonic functions, we refer to [71], notably pp. 46–48. We will need more specifically the following maximum principles.

Maximum Principle

Theorem 1.8 (Maximum principle 1) *Let $\Omega \subset \mathbb{C}$ be a bounded open set, and m a real number. Let $u : \Omega \to \mathbb{R} \cup \{-\infty\}$ be bounded above and subharmonic on Ω and satisfying*

$$\overline{\lim_{\substack{z \to \xi \\ z \in \Omega}}}\; u(z) \leqslant m, \quad \textit{for any } \xi \in \partial\Omega.$$

Then $u \leqslant m$ on Ω.

Proof: Since $\partial\omega \subset \partial\Omega$ when ω is a connected component of Ω, we can assume that Ω is connected. Let $M := \sup_{z \in \Omega} u(z) < \infty$ and (z_n) a sequence of points in Ω such that $u(z_n) \to M$. Since Ω is bounded, taking a subsequence if necessary, we can assume that $z_n \to \xi \in \overline{\Omega}$ and we distinguish two cases.

- $\xi \in \partial\Omega$. Then, $M = \lim_{n \to \infty} u(z_n) \leqslant m$ by hypothesis, and we are done.
- $\xi \in \Omega$. Consider the set E defined by

$$E = \{a \in \Omega\,;\, u(a) = M\} = \{a \in \Omega\,;\, u(a) \geqslant M\}.$$

The set E is closed in Ω because u is usc, and E is non-empty since $\xi \in E$. Let us show that E is open in Ω. We will need the following lemma.

Lemma 1.9 *Let X be a compact metric space, μ a probability measure on X, charging any non-empty open set, and $v : X \to \mathbb{R} \cup \{-\infty\}$ a usc non-positive function such that $\int_X v\, d\mu = 0$. Then, $v \equiv 0$.*

Indeed, if $v(x_0) =: -2\varepsilon < 0$ for some $x_0 \in X$, let W be an open neighborhood of x_0 such that $v \leqslant -\varepsilon$ on W. Then

$$\int_X v\, d\mu \leqslant \int_W v\, d\mu \leqslant -\varepsilon\mu(W) < 0,$$

a contradiction, and Lemma 1.9 is proved.

Now, we come back to the proof of Theorem 1.8. When $a \in E$, let $B = \overline{D}(a, r_a) \subset \Omega$ where r_a is as in (1.4), and dV the normalized area measure on B. Using (v) in Proposition 1.7, we have $u(a) \leqslant \int_B u(z)\, dV(z)$. Hence

$$0 = u(a) - M \leqslant \int_B (u(z) - M)\, dV(z) \leqslant 0,$$

so that $\int_B (u(z) - M)\, dV(z) = 0$. But then, Lemma 1.9 gives $u(z) - M = 0$ on B, and $B \subset E$. Since Ω is connected, this proves that $E = \Omega$ and that $u \equiv M$ on Ω. Once more, we get $M \leqslant m$, ending the proof. □

Here is a first application of Theorem 1.8.

Theorem 1.10 *Let $u : \Omega \to \mathbb{R} \cup \{-\infty\}$ be a subharmonic function and $a \in \Omega$. Then the function*

$$I(r) = \int_{\mathbb{T}} u(a + r\xi)\, dm(\xi), \quad 0 < r < d(a, \Omega^c),$$

is non-decreasing.

Proof: Let $0 \leqslant r_1 < r_2 < d(a, \Omega^c)$. Denote $\tilde{u}$ the restriction of u to the circle $C(a, r_2)$; let v be a continuous function on that circle, with $v \geqslant \tilde{u}$; still denote by v its harmonic extension to $D(a, r_2)$ (so that v is defined on $\overline{D}(a, r_2)$). By Theorem 1.8 applied to $u - v$, one has $u \leqslant v$ in $\overline{D}(a, r_2)$, so that

$$I(r_1) \leqslant \int_{\mathbb{T}} v(a + r_1\xi)\, dm(\xi) = v(a) = \int_{\mathbb{T}} v(a + r_2\xi)\, dm(\xi).$$

Now, since $\tilde{u}$ is scs, there is a sequence (v_n) of continuous functions on $C(a, r_2)$ that decreases pointwise to $\tilde{u}$. Passing to the monotone limit in the inequality

$$I(r_1) \leqslant \int_{\mathbb{T}} v_n(a + r_2\xi)\, dm(\xi)$$

gives the result $I(r_1) \leqslant I(r_2)$. □

As shown by Theorem 1.8, subharmonic functions satisfy a maximum principle as moduli of analytic functions do, when Ω is bounded. This is one of their main interests. They also satisfy a kind of Phragmén–Lindelöf property (or better said here a maximum principle) for *unbounded* Ω. Since this fact is perhaps less familiar to the potential reader (see [71, theorem 2.3, or theorem 5.16]), we detail a proof below.

Theorem 1.11 (Maximum principle 2) *Let $\Omega \subset \mathbb{C}$ be open with $\Omega \neq \mathbb{C}$. Let $u : \Omega \to \mathbb{R}\cup\{-\infty\}$ be subharmonic on Ω,* bounded above *on Ω, and satisfying, for some real number m,*

$$\overline{\lim_{\substack{z\to\xi \\ z\in\Omega}}}\ u(z) \leqslant m, \quad \text{for any } \xi \in \partial\Omega.$$

Then, $u(z) \leqslant m$ on Ω. This applies in particular when $u = |f|$ where f is holomorphic and bounded on Ω.

Proof: Since $\Omega \neq \mathbb{C}$, we can find some point $a \in \partial\Omega$. Fix a point $z_0 \in \Omega$ and let $\varepsilon, \delta > 0$. We select $r = r_\varepsilon > 0$ such that, setting $C(a,r) = \{z \in \mathbb{C} : |z-a| = r\}$,

$$|z_0 - a| > r \quad \text{and} \quad u \leqslant m + \varepsilon \text{ on } C(a,r) \cap \Omega.$$

Let then $R > r$ to be adjusted, $A(r,R)$ the annulus $r < |z-a| < R$, and U the *bounded* open set $U = \Omega \cap A(r,R)$. We will apply the first maximum principle to U and the subharmonic function (note that $\log \frac{|z-a|}{r}$ is harmonic outside a !):

$$v(z) = u(z) - \delta \log\left(\frac{|z-a|}{r}\right).$$

We observe that

$$\partial U \subset \big(\partial\Omega \cap \{|z-a| \geqslant r\}\big) \cup \big(C(a,r)\cap\Omega\big) \cup \big(C(a,R)\cap\Omega\big) =: F_1 \cup F_2 \cup F_3.$$

Set in general $L_w(\xi) := \overline{\lim\limits_{\substack{z\to\xi \\ z\in U}}}\ w(z)$ and $M = \sup\limits_{z\in\Omega} u(z)$, to save notation. We next observe that

- If $\xi \in F_1$, $L_v(\xi) \leqslant L_u(\xi) \leqslant m$.
- If $\xi \in F_2$, $L_v(\xi) = L_u(\xi) \leqslant m + \varepsilon$.
- If $\xi \in F_3$, $L_v(\xi) \leqslant M - \delta \log \frac{R}{r}$.

We now adjust R so as to have $z_0 \in U$ and $M - \delta \log \frac{R}{r} \leqslant m$. The maximum principle for the bounded open set U and the function v, subharmonic on U, gives us

$$u(z_0) = v(z_0) + \delta \log\left(\frac{|z_0 - a|}{r}\right) \leqslant m + \delta \log\left(\frac{|z_0-a|}{r}\right) + \varepsilon.$$

We successively let δ and ε tend to 0 to get $u(z_0) \leqslant m + \varepsilon$, and $u(z_0) \leqslant m$, as claimed. □

1.3 The Disk Case

We now switch to the case of the unit disk $\mathbb{D}$, for which classes of analytic functions (Hardy, Nevanlinna) appear. It is first convenient to coin the notions of Stolz angles and Stolz regions.

1.3.1 Stolz Sets

Definition 1.12 The *Stolz angle* $S_\alpha(u)$ is associated with a point $u \in \partial\mathbb{D}$ and an angle $0 < \alpha < \pi/2$ is by definition the interior of the convex hull of the point u and the disk $D(0, \sin\alpha)$; in other terms

$$S_\alpha(u) = \bigcup_{|\theta|<\alpha}]u, u(1 - \lambda(\theta)e^{i\theta})[\quad \text{with } \lambda(\theta) = \cos\theta + \sqrt{\sin^2\alpha - \sin^2\theta}$$

(see Figure 1.1).

The *Stolz region* $\Gamma_a(u)$ is defined ([184, p. 312] or [34, pp. 50–51]) for $a > 1$ as

$$\Gamma_a(u) = \{z \in \mathbb{D} : |z - u| < a(1 - |z|)\}$$

(see Figure 1.2).

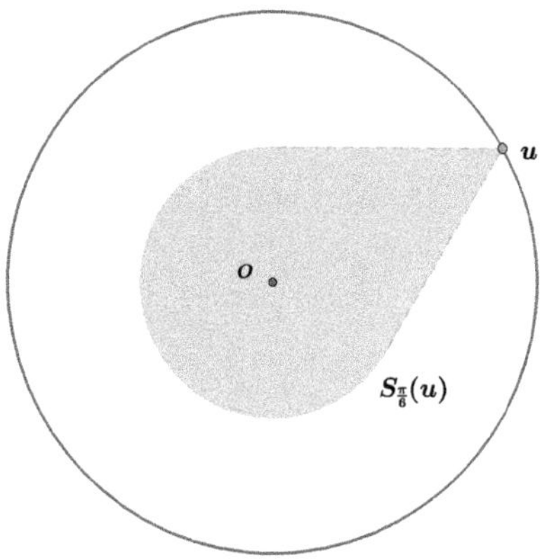

Figure 1.1 Stolz angle.

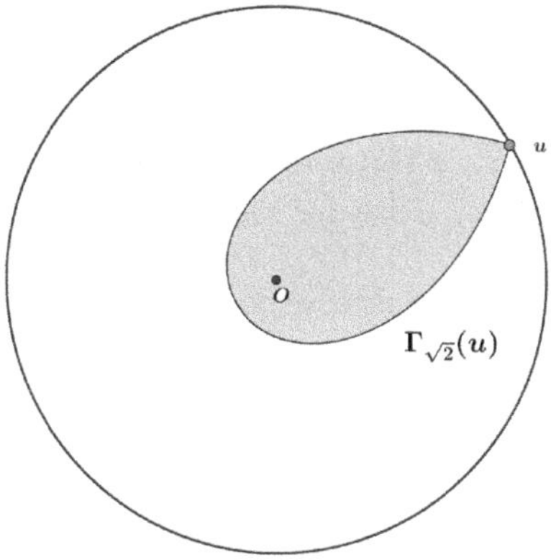

Figure 1.2 Stolz region.

The sets $S_\alpha(u)$ and $\Gamma_a(u)$ increase with α and a respectively; they are convex, look (more or less) like ice-cream cones, and are "equivalent" in the sense of Proposition 1.13:

Proposition 1.13 *The following inclusions hold for $a > 1$ and $|\alpha| < \pi/2$.*

(i) *One has $S_\alpha(u) \subset \Gamma_a(u)$ for $a = \dfrac{1+\sin\alpha}{1-\sin\alpha}$.*

(ii) *One has $\Gamma_a(u) \subset S_\alpha(u)$ for $\cos\alpha = \dfrac{1}{a}$.*

Proof: Up to a rotation, we can assume that $u = 1$.

Let $z \in S_\alpha(u)$. By definition, there exist $\delta \in (0,1)$ and $w \in D(0, \sin\alpha)$ such that $z = (1-\delta)\cdot 1 + \delta \cdot w$. In other words, $z = 1-\delta+\delta w = 1-\delta(1-w)$ with $|w| < \sin\alpha$. Then $|z-1| = \delta|1-w| < \delta(1+\sin\alpha)$ while $1-|z| > \delta(1-\sin\alpha)$, giving the first inclusion.

Let now $z = 1 - re^{i\theta} \in \Gamma_a(u)$ with $r > 0$ and $|\theta| < \pi/2$.

We have $|z-1| < a(1-|z|)$, that is, $a|z| < a-r$. Squaring and rearranging gives $r < \frac{2a}{a^2-1}(a\cos\theta - 1)$, in particular $\cos\theta > 1/a = \cos\alpha$ and $|\theta| < \alpha$. Moreover, for $|\theta| < \alpha$,

$$\frac{2a}{a^2-1}(a\cos\theta - 1) < \cos\theta + \sqrt{\cos^2\theta - \cos^2\alpha} = \lambda(\theta),$$

because (with $x = \cos\theta$) the function

$$f(x) = \frac{a^2+1}{a^2-1}x - \sqrt{x^2 - a^{-2}} - \frac{2a}{a^2-1}$$

is negative, since it is convex on $[1/a, 1]$ (we have $f'' \geqslant 0$) and satisfies $f(1/a) < 0$ and $f(1) < 0$. □

These particular subsets of the unit disk play a key role in the following situation:

Definition 1.14 Let $f \in \mathscr{H}(\mathbb{D}), u \in \partial\mathbb{D}$, and $l \in \mathbb{C}$.

When $\lim_{\substack{z \to u \\ z \in S_\alpha(u)}} f(z) = l$ exists for every $0 < \alpha < \pi/2$, we say that f has the *angular limit* l at u and we shall write

$$\angle \lim_{z \to u} f(z) = l.$$

1.3.2 Limits at the Boundary

In the following we recall some important properties on the boundary behavior of functions belonging to Hardy spaces. Even though we shall give definitions and more properties of the Hardy spaces in Sections 1.11.2 and 1.11.4, let us recall here quickly how they are defined:

$$H^\infty = \left\{ f \in \mathscr{H}(\mathbb{D}) : \|f\|_\infty = \sup_{z \in \mathbb{D}} |f(z)| < \infty \right\},$$

and, for $p > 0$,

$$H^p = \left\{ f \in \mathscr{H}(\mathbb{D}) : \sup_{r \in (0,1)} \int_{\mathbb{T}} |f(rz)|^p \, dm(z) < \infty \right\},$$

equipped with the quasi-norm (this is a norm when $p \geqslant 1$)

$$\|f\|_p = \sup_{r<1} \left(\int_{\mathbb{T}} |f(rz)|^p \, dm(z) \right)^{1/p} = \sup_{r<1} \|f_r\|_{L^p(\mathbb{T})},$$

where, for $r \in (0, 1)$ and $z \in \overline{\mathbb{D}}$,

$$f_r(z) = f(rz).$$

We can now recall a basic theorem of Fatou.

Theorem 1.15 (Fatou) *Let $f \in H^\infty$. Then, the radial limit*

$$f^*(u) := \lim_{r \to 1^-} f(ru) \quad \textit{exists for m-almost every } u \in \mathbb{T}.$$

As a corollary,

$$\angle \lim_{z \to u} f(z) = f^*(u) \quad \textit{for almost all } u \in \mathbb{T}.$$

Moreover, if $f \not\equiv 0$, then $f^*(u) \neq 0$ for almost all $u \in \mathbb{T}$.

Note that the corollary part is a nice application (see [45] p. 6 or [152] p. 168 for detailed proofs) of the forthcoming Theorem 1.18 to follow, and the last assertion follows from Jensen's formula.

Throughout this book, it will be convenient to extend the definition of f^* to the whole closed unit disk $\overline{\mathbb{D}}$:

Definition 1.16 Let f satisfy the conclusion of Fatou's theorem (see Theorem 1.15 above). We set

- $f^*(z) = f(z)$ when $z \in \mathbb{D}$.
- $f^*(u) = \lim\limits_{r \to 1^-} f(ru)$ when $u \in \mathbb{T}$ is like in Fatou's theorem (i.e., the radial limit does exist).
- $f^*(u) = 0$ (for instance) for other $u \in \mathbb{T}$ (hence lying in a subset of $\mathbb{T}$ with m-measure 0).

Of course, in the latter case, we can choose any value other than 0; it will be irrelevant anyway.

When $f \in H^\infty$, it is clear that we get $f^* \in L^\infty(\overline{\mathbb{D}})$.

In Theorem 1.17, denote again by $\mathcal{N}$ the Nevanlinna class, that is, the set of analytic functions f on $\mathbb{D}$ such that

$$\sup_{0<r<1} \int_{\mathbb{T}} \log^+(|f(ru)|)\, dm(u) < \infty.$$

Obviously, $H^p \subset \mathcal{N}$ for any $0 < p \leqslant \infty$.

Theorem 1.17 (Beurling) *Any function $f \in \mathcal{N}$ can be written as a quotient U/V of bounded analytic functions on $\mathbb{D}$, with V zero-free. Hence, Fatou's theorem extends to $\mathcal{N}$.*

As a consequence, if $f \in \mathcal{N}$ then $f^*(u)$ exists almost everywhere on $\mathbb{T}$, and $m(\{f^* = 0\}) = 0$, unless f is the zero function.

In order to ease the further presentation of the delicate Julia–Carathéodory's theorem, the following application of Montel's theorem is useful.

Theorem 1.18 (Lindelöf 1) *Let $u \in \partial\mathbb{D}$ and f be an analytic function on $\mathbb{D}$,* bounded on each Stolz angle $S_\beta(u)$.

Assume that the radial limit $\lim_{r \to 1^-} f(ru) = l \in \mathbb{C}$ exists. Then,

$$\angle \lim_{w \to u} f(w) = l.$$

Proof: Fix $0 < \alpha < \pi/2$. Let $\alpha < \beta < \pi/2$ and consider the sequence of functions defined by

$$f_n(z) = f\big((1 - 1/n)u + z/n\big), \text{ where } z \in S_\beta(u),\ n = 1, 2, \ldots$$

By our assumption, the functions f_n are bounded on $S_\beta(u)$, so they form a normal family on that open set. Moreover, if a subsequence of (f_n) converges to g, we have $g(ru) = l$ for $0 < r < 1$. By Montel's theorem, f_n converges to l uniformly on compact subsets of $S_\beta(u)$, in particular on the set

$$K = \left\{ z = u - re^{i\theta} \; ; \; |\theta| \leqslant \alpha \, , \; \frac{\cos\alpha}{2} \leqslant r \leqslant \cos\alpha \right\}.$$

Let now $w = u - \rho e^{i\theta} \in S_\alpha(u)$ with $|\theta| < \alpha$ and $0 < \rho \leqslant \frac{\cos\alpha}{2} \cdot$ Take a positive integer n such that $\frac{\cos\alpha}{2\rho} \leqslant n \leqslant \frac{\cos\alpha}{\rho} \cdot$ Then w can be written as

$$w = (1 - 1/n)u + (1/n)z \quad \text{with } z = u - n\rho e^{i\theta} \in K.$$

This shows that $f(w) = f_n(z) \to l$ as $w \in S_\alpha(u) \to u$. □

In Theorem 1.18, the point u is approached from $S_\alpha(u)$ or $\Gamma_a(u)$ in a *non-tangential way.* A more general version of this theorem is sometimes desirable. This version notably helps to show that the composition of two inner functions is again an inner function. We refer to ([152], p. 168). Here, we give a detailed proof of this fairly delicate theorem.

Theorem 1.19 (Lindelöf 2) *Let f be a* bounded *analytic function on the unit disk $\mathbb{D}$ and $u \in \partial\mathbb{D}$. Assume that the radial limit $\lim_{r\to 1^-} f(\gamma(r)) = l$ exists for* some *continuous curve $\gamma : [0,1) \to \mathbb{D}$, which terminates at u, that is, such that $\lim_{r\to 1^-} \gamma(r) = u$. Then, one has*

$$\angle \lim_{z\to u} f(z) = l.$$

Proof: We can assume $l = 0$, $u = 1$, and $\|f\|_\infty \leqslant 1$. Let S be the vertical strip $\{z : |\mathrm{Re}(z)| < 1\}$, with boundary lines $\Delta_{\pm 1} = \{z : \mathrm{Re}(z) = \pm 1\}$ and let $\psi : \mathbb{D} \to S$ be the conformal map given by

$$\psi(z) = \frac{2i}{\pi} \log \frac{1+z}{1-z} = \frac{2i}{\pi} \log \left|\frac{1+z}{1-z}\right| - \frac{2}{\pi}\theta(z),$$

with $\theta(z) = \arctan \frac{2\mathrm{Im}\, z}{1-|z|^2} \cdot$ This map satisfies $\psi(0) = 0$ and $\psi([0,1)) = i\mathbb{R}^+$.

We set

$$F = f \circ \psi^{-1}, \; \Gamma = \psi \circ \gamma, \text{ with } F \circ \Gamma = f \circ \gamma \text{ and } \lim_{r\to 1^-} \mathrm{Im}\,\Gamma(r) = +\infty.$$

Of course, since f is bounded, F is bounded on S as well and $\|F\|_\infty \leqslant 1$.

By Theorem 1.18, we only have to show that $f(r) \to 0$ as $r \to 1^-$, or that $F(iy) \to 0$ as $y \to +\infty$. Fix $0 < \varepsilon < 1$. Choose $y \geqslant \mathrm{Im}\,\Gamma(0)$ large enough to ensure that $|F(\Gamma(r))| \leqslant \varepsilon$ as soon as $\mathrm{Im}\,\Gamma(r) \geqslant y$. We will show that

$$|F(iy)| \leqslant \varepsilon^{1/4}. \tag{1.5}$$

This will give us the result. To that effect, let $r_0 < 1$ be the "last time" for which $\operatorname{Im}\Gamma(r_0) = y$. We put

$$\Gamma(r_0) = x_0 + iy, \qquad E = \big\{\Gamma(r) : r_0 \leqslant r < 1\big\},$$

$$\widetilde{E} = \{\overline{z} + 2iy : z \in E\}, \quad \text{and} \quad \Omega = S \setminus (E \cup \widetilde{E}).$$

The set $E \cup \widetilde{E}$ represents a kind of a closed barrier inside S between Δ_1 and Δ_{-1}, and which can be described as a continuous curve $C : \mathbb{R} \to \mathbb{C}$ with $\operatorname{Im} C(\mathbb{R}) = \mathbb{R}$, parametrized so as to satisfy $\lim\limits_{t\to\pm\infty} \operatorname{Im} C(t) = \pm\infty$.

We also define $G(z) = F(z)\overline{F(\overline{z} + 2iy)}\,\varepsilon^{\frac{1+z}{2}}$ and clearly $G \in \mathscr{H}(S)$ and is bounded.

Assume w.l.o.g. that $x_0 < 0$, so that $iy \in \Omega$ (if $x_0 = 0$, there is nothing to prove, and we would proceed similarly if $x_0 > 0$). We first need a basic topological lemma (the proof is postponed).

Lemma 1.20 *Let ω be the connected component of Ω containing iy. Then*

(i) $[iy, iy+1) \subset \omega$.
(ii) $\partial\omega \subset E \cup \widetilde{E} \cup \Delta_1$.

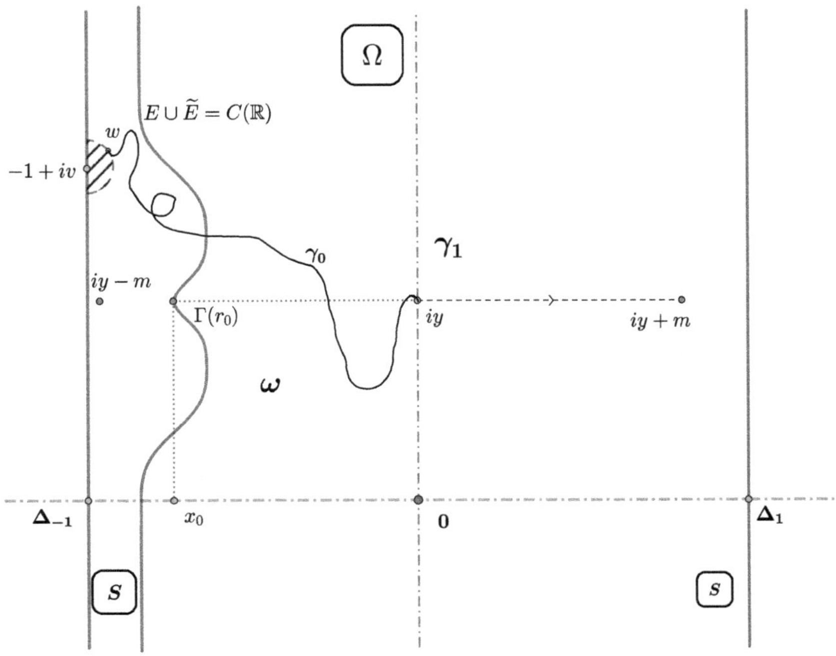

Figure 1.3 The barrier $C(\mathbb{R})$, and the curve γ_1.

Let us first show that this lemma gives (1.5).

Let $\xi \in \partial\omega$ and set $L(\xi) = \overline{\lim}_{\substack{z \to \xi \\ z \in \omega}} |G(z)|$. Three cases occur:

- $\xi \in E$ so that $\xi = \Gamma(r)$ with $r \geqslant r_0$. Then, since $|G(z)| \leqslant |F(z)|$, we get $L(\xi) \leqslant |F(\xi)| \leqslant \varepsilon$.
- $\xi \in \widetilde{E}$. Then, $\overline{\xi} + 2iy \in E$, $|G(z)| \leqslant |F(\overline{z} + 2iy)|$, and again $L(\xi) \leqslant \varepsilon$.
- $\xi \in \Delta_1$. Then, $|G(z)| \leqslant |\varepsilon^{\frac{1+z}{2}}|$ and $L(\xi) \leqslant \varepsilon$.

By the lemma, we get $L(\xi) \leqslant \varepsilon$ for all $\xi \in \partial\omega$. By Theorem 1.11 (maximum principle for the bounded analytic function G and the unbounded open set ω), it follows that $|F(iy)|^2\varepsilon^{1/2} = |G(iy)| \leqslant \varepsilon$. This proves (1.5).

Now let us prove Lemma 1.20. The first assertion follows from the definition of r_0. Next, we obviously have

$$\partial\omega \subset \partial\Omega \subset \partial S \cup E \cup \widetilde{E},$$

so it only remains to prove that $\partial\omega \cap \Delta_{-1} = \emptyset$. Suppose that $-1 + iv \in \partial\omega$; let $0 < \delta < 1$ be small enough to ensure that $\overline{D}(-1+iv,\delta) \cap (E \cup \widetilde{E}) = \emptyset$, so that $\overline{D}(-1+iv,\delta) \cap S \subset \omega$, as a connected set contained in Ω and intersecting ω. Since ω is arcwise connected, there is a curve $\gamma_0 : [0,1] \to \omega$ connecting $w = -1 + \delta + iv$ and iy: $\gamma_0(0) = w, \gamma_0(1) = iy$. Set

$$a = \min_{0 \leqslant \lambda \leqslant 1} \operatorname{Im}[\gamma_0(\lambda)], \quad b = \max_{0 \leqslant \lambda \leqslant 1} \operatorname{Im}[\gamma_0(\lambda)], \quad c = \max_{0 \leqslant \lambda \leqslant 1} \left|\operatorname{Re}[\gamma_0(\lambda)]\right| < 1.$$

We can find t_a, t_b and m satisfying $\max(c, 1-\delta) < m < 1$ such that

$$\operatorname{Im} C(t_a) = a,\ \operatorname{Im} C(t_b) = b,\ \operatorname{Im} C(t) \in [a,b],\ |\operatorname{Re} C(t)| \leqslant m, \text{ for all } t \in [t_a, t_b]$$

(take t_a the first time when $\operatorname{Im} C(t) = a$ and t_b the first time $> t_a$ when $\operatorname{Im} C(t) = b$). Let $\alpha = -m + iv \in D(-1 + iv, \delta)$. Define now γ_1 as the curve obtained by concatenation of the segment $[\alpha, w]$, the curve γ_0, and then the segment $[iy, iy + m]$. Finally, set $\gamma_2(\lambda) = C(t_a + \lambda(t_b - t_a))$, $0 \leqslant \lambda \leqslant 1$. We see that both continuous curves γ_1 and γ_2 join the opposite sides of the rectangle $R = [-m, m] \times [a, b]$ and have images included in that rectangle.

But then those curves must intersect (a consequence of Brouwer's fixed point theorem in dimension 2, see, for example, [144, p. 159 and 175]), which is impossible, since $\omega \cap (E \cup \widetilde{E}) = \emptyset$. Indeed, the image of $\gamma_1 \subset \omega$ and the image of $\gamma_2 \subset C(R) = E \cup \widetilde{E}$. This contradiction ends the proof. □

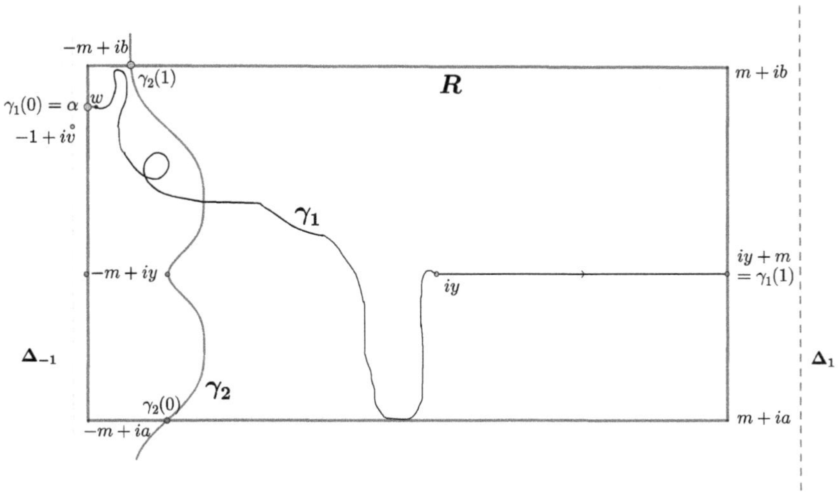

Figure 1.4 A zoom-in on Figure 1.3 to focus on R, γ_1, γ_2.

Here is an important corollary of Lindelöf's second theorem concerning the composition with a symbol φ.

Corollary 1.21 *Let $f \in \mathcal{N}$ and $\varphi : \mathbb{D} \to \mathbb{D}$ analytic. Then,*

$$(f \circ \varphi)^* = f^* \circ \varphi^* \quad m\text{-almost everywhere on } \mathbb{T}.$$

Proof: We treat only the case when φ is not constant, else it is obvious.

Assume first that $f \in H^\infty$. By Fatou's theorem for φ and $f\circ\varphi$, there is a set $E \subset \mathbb{T}$, with $m(E) = 1$, such that the two radial limits $\varphi^*(u) = v$, $(f \circ \varphi)^*(u)$ exist if $u \in E$. For such an u, pass to the limit, as $r \to 1^-$, in

$$(f \circ \varphi)(ru) = f[\varphi(ru)].$$

The left-hand side tends to $(f \circ \varphi)^*(u)$. Hence, by Theorem 1.19 applied to f and the curve $\gamma(r) = \varphi(ru)$, $f^*(v)$ exists and equals $(f \circ \varphi)^*(u)$.

Next, write $f = U/V$ with $U, V \in H^\infty$ and V zero-free. Let again $E \subset \mathbb{T}$, with $m(E) = 1$, such that $\varphi^*(u) = v$, $(U \circ \varphi)^*(u)$, $(V \circ \varphi)^*(u)$ exist for $u \in E$ with moreover $(V \circ \varphi)^*(u) \neq 0$ (observe that Fatou's theorem applies to $V \circ \varphi \in H^\infty$).

We get from the preceding that, for such u, $U^*(v)$ and $V^*(v)$ exist, and moreover

$$(f \circ \varphi)^*(u) = \frac{(U \circ \varphi)^*(u)}{(V \circ \varphi)^*(u)} = \frac{U^*(v)}{V^*(v)} = \lim_{r\to 1^-} \frac{U(rv)}{V(rv)},$$

so that $f^*(v) = f^*(\varphi^*(u))$ exists and is equal to $(f \circ \varphi)^*(u)$. This ends the proof. □

The results to follow belong more specifically to the theory of analytic self-maps of $\mathbb{D}$.

1.3.3 Pseudo-hyperbolic Distance

There exists a metric on $\mathbb{D}$ that is much better adapted to the study of analytic self-maps of $\mathbb{D}$ than the usual metric. It is denoted by ρ and called the *pseudo-hyperbolic metric*. It helps particularly in the study of iterations of self-maps of $\mathbb{D}$, and of interpolating sequences. We follow the general presentation of [4, pp. 128–129], adding some details. See also the quite interesting paper [178]. This presentation is very analytic. We will see another one, slightly more geometric, in Section 1.3.4 devoted to the Schwarz–Pick lemma.

We begin by recalling a general fact on the geodesic distance on the unit sphere S of a complex Hilbert space H, equipped with its scalar product $\langle .,. \rangle$.

Proposition 1.22 *For $x, y \in S$, put $|\langle x, y\rangle| = \cos\eta(x, y)$ with $0 \leqslant \eta(x, y) \leqslant \frac{\pi}{2}\cdot$ Then, $\eta(x, y)$ satisfies the triangle inequality on S.*

Proof: To show that

$$\eta(x,z) \leqslant \eta(x,y) + \eta(y,z),$$

we assume that $H = \mathbb{C}^3$ and that, on the canonical basis of $\mathbb{C}^3$, all angles lie between 0 and $\pi/2$, we have:

(i) $z = (0,0,1)$.
(ii) $y = (0,a,b)$ with $|a| = \sin\beta$, $|b| = \cos\beta$.
(iii) $x = (u,v,w)$ with $|u| = \sin\alpha\cos\theta$, $|v| = \sin\alpha\sin\theta$, $|w| = \cos\alpha$.

Then, $\eta(x,z) = \alpha$, $\eta(y,z) = \beta$, and we must show that

$$\alpha \leqslant \beta + \eta \quad \text{where } \eta = \eta(x,y).$$

We can assume that $\alpha \geqslant \beta$. We then observe that

$$\cos\eta = |\langle x,y\rangle| = |a\overline{v} + b\overline{w}| \leqslant \sin\beta\sin\alpha\sin\theta + \cos\beta\cos\alpha \leqslant \cos(\alpha - \beta),$$

so that $\eta \geqslant \alpha - \beta$. □

Let now H be a Hilbert space of functions on a set X, such that the pointwise evaluations δ_a, $\delta_a(f) = f(a)$, $a \in X$ are continuous linear forms on H, so that for some $K_a \in H$ (the so-called reproducing kernel of H at an arbitrary point $a \in X$),

$$f(a) = \langle f, K_a\rangle \quad \text{for all } f \in H. \tag{1.6}$$

We will use the notation $K(z,a) = K_a(z)$. We moreover assume that H is strongly separating, namely that K_a and K_b are linearly independent for $a \neq b$. In particular, the functionals δ_a separate points of X: for every $f \neq g \in H$, there exists some $a \in \mathbb{D}$ such that $f(a) \neq g(a)$. We note $u_a = \frac{K_a}{\|K_a\|}$ the normalized reproducing kernel at a. The following holds.

Theorem 1.23 *Let H be strongly separating on X and*

$$\rho(a,b) = \sqrt{1 - |\langle u_a, u_b\rangle|^2}.$$

Then, ρ is a metric on X, called the pseudo-hyperbolic metric induced by H.

Proof: Let $\alpha = \eta(u_a,u_c), \beta = \eta(u_b,u_c), \gamma = \eta(u_a,u_b)$ be the geodesic distances between (u_a,u_c), (u_b,u_c), and (u_a,u_b) respectively. We observe that $\rho(x,y) = \sin\eta(u_x,u_y)$ in general, and we have to show that $\sin\alpha \leqslant \sin\beta + \sin\gamma$, knowing from Proposition 1.22 that $\alpha \leqslant \beta + \gamma$.

1. If $\beta + \gamma \leqslant \pi/2$, we get $\sin\alpha \leqslant \sin(\beta+\gamma) \leqslant \sin\beta + \sin\gamma$.
2. If $\beta + \gamma \geqslant \pi/2$, we get $\sin\gamma \geqslant \sin(\pi/2 - \beta) = \cos\beta$, and

$$\sin\beta + \sin\gamma \geqslant \sin\beta + \cos\beta \geqslant \sqrt{\sin^2\beta + \cos^2\beta} = 1 \geqslant \sin\alpha.$$

The separation axiom for ρ follows from our strong separation assumption. □

If we specialize to $X = \mathbb{D}$, the unit disk, and $H = H^2$, the Hardy space of the disk, we know (see Section 1.11.4) that

$$K_a(z) = \frac{1}{1-\bar{a}z} \quad \text{and} \quad \|K_a\|^2 = K_a(a) = \frac{1}{1-|a|^2},$$

so that strong separation holds and that

$$\rho(a,b) = \sqrt{1 - \frac{(1-|a|^2)(1-|b|^2)}{|1-\bar{a}b|^2}} = \left|\frac{a-b}{1-\bar{a}b}\right|, \tag{1.7}$$

where the last equality is immediate but quite useful.

We will denote by $\Delta(a,r)$ the open *pseudo-hyperbolic disk* of center $a \in \mathbb{D}$ and radius $0 < r < 1$. It is useful to note the following proposition.

Proposition 1.24 *Hyperbolic disks are euclidean disks, and more precisely, we have*

$$\Delta(a,r) = D(a^*,r^*) \quad \textit{with} \quad a^* = \frac{a(1-r^2)}{1-r^2|a|^2}, \; r^* = \frac{r(1-|a|^2)}{1-r^2|a|^2}.$$

In this book, we shall make an extensive use of the maps (*Möbius functions*)

$$\varphi_a(z) = \frac{a-z}{1-\bar{a}z} \qquad \text{for } z \in \overline{\mathbb{D}}, \tag{1.8}$$

where $a \in \mathbb{D}$.

Proof: We can assume $0 < a < 1$. By definition, $\Delta(a,r) = \varphi_a(D(0,r))$. Hence, by conformal mapping, $\Delta(a,r)$ is the euclidean disk with diameter $[\varphi_a(r), \varphi_a(-r)]$, that is, $\Delta(a,r) = D(a^*, r^*)$ with (noting that φ_a decreases on $(-1,1)$)

$$a^* = \frac{\varphi_a(r) + \varphi_a(-r)}{2}, \qquad r^* = \frac{\varphi_a(-r) - \varphi_a(r)}{2}.$$

This gives the result. □

We mention here a result to be used several times in the book (see Chapters 9 and 10).

Lemma 1.25 *Suppose that $a, b \in \mathbb{D}$ satisfy $|a-b| \leqslant M \min(1-|a|, 1-|b|)$. Then,*

$$\rho(a,b) \leqslant \frac{M}{\sqrt{M^2+1}} =: \chi < 1.$$

Similarly, if $|a-b| \geqslant m \max(1-|a|, 1-|b|)$,

$$\rho(a,b) \geqslant \frac{m}{\sqrt{m^2+4}}.$$

Proof: Set $\delta = \min(1-|a|, 1-|b|)$ and just observe that

$$\frac{1}{\rho^2(a,b)} - 1 = \frac{(1-|a|^2)(1-|b|^2)}{|a-b|^2} \geqslant \frac{(1-|a|)(1-|b|)}{|a-b|^2} \geqslant \frac{\delta^2}{\delta^2 M^2} = \frac{1}{M^2}.$$

The second assertion is proved similarly. □

1.3.4 The Schwarz–Pick Lemma

We shall need more flexible forms of the Schwarz lemma. We first recall the following:

Theorem 1.26 *The group $G = \mathrm{Aut}\,\mathbb{D}$ of holomorphic automorphisms of $\mathbb{D}$ is the group of the following rational linear maps:*

$$\varphi(z) = \lambda \frac{a-z}{1-\bar{a}z} = \lambda \varphi_a(z), \quad |\lambda| = 1,\ a \in \mathbb{D}. \tag{1.9}$$

Proof: One easily checks the relations, valid for $a \in \mathbb{D}$ and $z \in \overline{\mathbb{D}}$:

$$\varphi_a(0) = a; \quad \varphi_a(a) = 0; \quad \varphi_a'(z) = -\frac{1 - |a|^2}{(1 - \overline{a}z)^2}. \tag{1.10}$$

$$1 - |\varphi_a(z)|^2 = \frac{(1 - |a|^2)(1 - |z|^2)}{|1 - \overline{a}z|^2}, \tag{1.11}$$

and hence

$$\frac{|\varphi_a'(z)|}{1 - |\varphi_a(z)|^2} = \frac{1}{1 - |z|^2}.$$

In particular, $\varphi_a(\mathbb{D}) \subset \mathbb{D}$ and $\varphi_a(\partial\mathbb{D}) \subset \partial\mathbb{D}$. One also checks that $\varphi_a(z) = w$ if and only if $z = \varphi_a(w)$, that is, $\varphi_a \circ \varphi_a = I$ where I is the identity of $\mathbb{D}$. This shows that $\varphi_a \in G$ and $\varphi_a^{-1} = \varphi_a$, and also that $\lambda\varphi_a \in G$ for $|\lambda| = 1$. Conversely, let $\varphi \in G$ satisfy $\varphi(0) = 0$. The Schwarz lemma for φ and φ^{-1} gives $|\varphi(z)| \leqslant |z|$ and $|\varphi^{-1}(w)| \leqslant |w|$, so that $|\varphi(z)| \geqslant |z|$ by taking $w = \varphi(z)$. Hence, $|\varphi(z)| = |z|$ and the Schwarz lemma again gives $\varphi = \lambda I$ with $|\lambda| = 1$. In the general case, let $a = \varphi^{-1}(0)$. Then, $\varphi \circ \varphi_a \in G$ and $\varphi \circ \varphi_a(0) = 0$ so that $\varphi \circ \varphi_a = \lambda I$ and finally $\varphi = \lambda\varphi_a$ with $a \in \mathbb{D}$ and $\lambda \in \partial\mathbb{D}$. □

Note the important identity

$$\rho(a,b) = |\varphi_a(b)| = |\varphi_b(a)|. \tag{1.12}$$

Note also that (1.11) and the description of G imply

$$\frac{|\varphi'(z)|}{1 - |\varphi(z)|^2} = \frac{1}{1 - |z|^2} \quad \text{for all } \varphi \in G. \tag{1.13}$$

The group G is the analog for the unit disk of the group of translations for the complex plane; in other terms:

Proposition 1.27 *The group G acts transitively on $\mathbb{D}$, namely, if $a,b \in \mathbb{D}$, there exists $\varphi \in G$ such that $\varphi(a) = b$.*

Since the group G acts transitively on $\mathbb{D}$, the origin plays no privileged role in the Schwarz lemma, and its extension, called the *Schwarz–Pick lemma*, is very well adapted to the iteration of analytic functions $\varphi : \mathbb{D} \to \mathbb{D}$, which reveal contractions for the pseudo-hyperbolic metric, whereas automorphisms are isometries for this metric.

Lemma 1.28 (Schwarz–Pick) *Let $\varphi : \mathbb{D} \to \mathbb{D}$ be holomorphic and $a,z \in \mathbb{D}$. Then*:

(i) *One has $\rho\big(\varphi(a), \varphi(z)\big) \leqslant \rho(a,z)$, with equality for a pair of distinct points a and z if and only if $\varphi \in G$.*

(ii) *It also holds*

$$\frac{|1-\overline{\varphi(a)}\varphi(z)|^2}{1-|\varphi(z)|^2} \leqslant \frac{1-|\varphi(a)|^2}{1-|a|^2} \times \frac{|1-\overline{a}z|^2}{1-|z|^2};$$

moreover, we have an equality if and only if $\varphi \in G$.

(iii) *One has* $\dfrac{|\varphi'(a)|}{1-|\varphi(a)|^2} \leqslant \dfrac{1}{1-|a|^2}$, *with equality if and only if* $\varphi \in G$.

(iv) *The function* $|.|$ *is a contraction for* ρ*:*

$$\forall a,b \in \mathbb{D}, \quad \rho(|a|,|b|) \leqslant \rho(a,b). \tag{1.14}$$

(v) *For every* $z \in \mathbb{D}$, *we have*

$$\Delta(z) := \frac{1-|\varphi(z)|}{1-|z|} \geqslant \frac{1-|\varphi(0)|}{1+|\varphi(0)|}. \tag{1.15}$$

Let us mention an immediate useful consequence of (1.15):

$$\lim_{r\to 1^-} \inf\{|z|\,;\ z \in \mathbb{D} \text{ and } |\varphi(z)| \geqslant r\} = 1. \tag{1.16}$$

Proof:

(i) Set $c = \varphi(a)$, $\psi = \varphi_c \circ \varphi \circ \varphi_a$, and consider the diagram

$$0 \xrightarrow{\varphi_a} a \xrightarrow{\varphi} c \xrightarrow{\varphi_c} 0,$$

which indicates that $\psi(0) = 0$. By the Schwarz lemma, we have $|\psi(z)| \leqslant |z|$ for $z \in \mathbb{D}$, or again $|\varphi_c(\varphi(z))| \leqslant |\varphi_a(z)|$ after changing z into $\varphi_a(z)$. This reads as $\rho(\varphi(z),c) \leqslant \rho(z,a)$ and gives the stated inequality, since $c = \varphi(a)$.

Moreover, we have an equality if and only if we have an equality when we applied the Schwarz lemma, so that $\psi(z) = \lambda z$ for some $\lambda \in \mathbb{T}$, hence $\varphi(z) = \varphi_c \circ \varphi_a(\lambda z)$.

(ii) This other form of Schwarz–Pick's inequality follows from (i) via (1.11), which implies

$$\begin{aligned}\frac{(1-|\varphi(a)|^2)(1-|\varphi(z)|^2)}{|1-\overline{\varphi(a)}\varphi(z)|^2} &= 1-|\varphi_{\varphi(a)}(\varphi(z))|^2 = 1-\rho(\varphi(a),\varphi(z))^2\\ &\geqslant 1-\rho(a,z)^2 = \frac{(1-|a|^2)(1-|z|^2)}{|1-\overline{a}z|^2},\end{aligned}$$

and one takes inverses to get the result.

The equality corresponds exactly to the equality $\rho(\varphi(a),\varphi(z)) = \rho(a,z)$, which is equivalent to $\varphi \in G$.

(iii) Next, the chain rule and (1.10) give us

$$\psi'(0) = \varphi_c'(c)\varphi'(a)\varphi_a'(0) = \frac{1}{1-|\varphi(a)|^2}\varphi'(a)(1-|a|^2).$$

The Schwarz lemma $|\psi'(0)| \leqslant 1$ with its equality cases gives the statement.

(iv) For every $a, b \in \mathbb{D}$, we have

$$1-\rho^2(a,b) = \frac{(1-|a|^2)(1-|b|^2)}{|1-\overline{a}b|^2} \leqslant \frac{(1-|a|^2)(1-|b|^2)}{(1-|a||b|)^2} = 1-\rho^2(|a|,|b|),$$

and the conclusion follows.

(v) Finally, by the preceding property (1.14) and the Schwarz–Pick lemma:

$$\frac{|\varphi(z)|-|\varphi(0)|}{1-|\varphi(z)||\varphi(0)|} \leqslant \rho(|\varphi(z)|,|\varphi(0)|) \leqslant \rho(\varphi(z),\varphi(0)) \leqslant |z|.$$

This gives $|\varphi(z)| \leqslant \dfrac{|z|+|\varphi(0)|}{1+|z||\varphi(0)|}$ and $1-|\varphi(z)| \geqslant \dfrac{(1-|z|)(1-|\varphi(0)|)}{1+|\varphi(0)|}$, as claimed. □

The important inequality (1.15) saying that $\Delta(z)$ is always bounded below is useful in the theory. It will reappear when studying the set $\mathcal{F}$ of finite Blaschke products: Δ is bounded *above* exactly when $\varphi \in \mathcal{F}$.

We now come back to the metric ρ (for more applications and geometric interpretations, we refer to [87, pp. 31–42] or [184, pp. 66–68]). Since (recall (1.13)), for $\varphi \in G$,

$$\frac{|\varphi'(z)|}{1-|\varphi(z)|^2} = \frac{1}{1-|z|^2} \quad \text{for all } z \in \mathbb{D},$$

Poincaré was led to define the hyperbolic distance d on $\mathbb{D}$ as follows.

One first defines the Poincaré length of the piecewise C^1-curve $\gamma : [0,1] \to \mathbb{D}$ by the formula

$$L(\gamma) = \int_\gamma \frac{|dz|}{1-|z|^2} := \int_0^1 \frac{|\gamma'(t)|}{1-|\gamma(t)|^2}\, dt.$$

Then, the "true" hyperbolic distance $d(a,b)$ between two points $a, b \in \mathbb{D}$ is defined as the geodesic distance of a and b for the Poincaré length, namely

$$d(a,b) = \inf\{L(\gamma)\},$$

where the infimum is taken over piecewise C^1-curves γ in $\mathbb{D}$, with origin $\gamma(0) = a$ and extremity $\gamma(1) = b$. By definition, d satisfies the triangle inequality. Indeed, if γ connects a to b and δ connects b to c, the curve η defined by

$$\eta(t) = \gamma(2t) \text{ if } 0 \leqslant t \leqslant 1/2; \quad \eta(t) = \delta(2t-1) \text{ if } 1/2 \leqslant t \leqslant 1$$

is piecewise-C^1, connects a to c, and satisfies

$$L(\eta) = \int_0^{1/2} \frac{2|\gamma'(2t)|}{1-|\gamma(2t)|^2}\, dt + \int_{1/2}^1 \frac{2|\delta'(2t-1)|}{1-|\delta(2t-1)|^2}\, dt = L(\gamma) + L(\delta).$$

Thanks to (1.13), one has $L(\varphi \circ \gamma) = L(\gamma)$ for all $\varphi \in G$, so that d is invariant under G:

$$d[\varphi(a), \varphi(b)] = d(a,b) \quad \text{for all } \varphi \in G. \tag{1.17}$$

The invariance of d under G makes it possible to compute explicitly $d(a,b)$ in the general case ([184], pp. 66–68).

Lemma 1.29 *For all $a, b \in \mathbb{D}$, it holds*

$$d(a,b) = \frac{1}{2}\log\frac{1+\rho(a,b)}{1-\rho(a,b)}, \quad \textit{or else} \quad \rho(a,b) = \tanh d(a,b). \tag{1.18}$$

Proof: By the invariance of d under G, we can assume that $a = 0$ and $b \geqslant 0$. If $\gamma = u + iv : [0,1] \to \mathbb{D}$ is a piecewise C^1-curve with origin $a = 0$ and end b, we note that $|\gamma'| \geqslant |u'|$ and that $\frac{1}{1-|\gamma|^2} \geqslant \frac{1}{1-u^2}$, so that

$$\begin{aligned} L(\gamma) &= \int_0^1 \frac{|\gamma'(t)|}{1-|\gamma(t)|^2}\,dt \geqslant \int_0^1 \frac{|u'(t)|}{1-u(t)^2}\,dt \geqslant \left|\int_0^1 \frac{u'(t)}{1-u(t)^2}\,dt\right| \\ &= \int_{u(0)}^{u(1)} \frac{dv}{1-v^2} = \int_0^b \frac{dv}{1-v^2} = \frac{1}{2}\log\frac{1+b}{1-b}\,. \end{aligned}$$

Equality is achieved with $\gamma(t) = tb$, a parametrization of the segment $[0,b]$. □

It easily follows that ρ is a metric as well and is also invariant under G (which could be checked by hand). We can be more specific.

Lemma 1.30 *The metric ρ satisfies the strong triangle inequality*

$$\rho(a,c) \leqslant \frac{\rho(a,b)+\rho(b,c)}{1+\rho(a,b)\rho(b,c)}.$$

Proof: We first indicate a proof using the addition formula for the function tanh:

$$\tanh(u+v) = \frac{\tanh u + \tanh v}{1 + \tanh u \tanh v}, \quad u, v \geqslant 0.$$

Now, we see that

$$\rho(a,c) = \tanh d(a,c) \leqslant \tanh[d(a,b)+d(b,c)] = \frac{\tanh d(a,b) + \tanh d(b,c)}{1 + \tanh d(a,b)\tanh d(b,c)};$$

hence

$$\rho(a,c) \leqslant \frac{\rho(a,b)+\rho(b,c)}{1+\rho(a,b)\rho(b,c)}\,.$$

Here is a second, direct proof. Since ρ is G-invariant, we can assume that $b = 0$ without loss of generality. It follows from (1.7) that

$$1-\rho^2(a,c) = \frac{(1-|a|^2)(1-|c|^2)}{|1-\bar{a}c|^2} \geqslant \frac{(1-|a|^2)(1-|c|^2)}{(1+|a||c|)^2} = 1-\rho^2(|a|,-|c|),$$

whence

$$\rho(a,c) \leqslant \rho(|a|,\,-|c|) = \frac{|a|+|c|}{1+|a||c|},$$

which was to be proved. □

1.4 The Julia–Carathéodory Theorem

The Julia–Carathéodory theorem is a basic result in the study of analytic self-maps φ of $\mathbb{D}$ and can be first informally phrased as follows.
"Let u be a point of $\partial\mathbb{D}$. If the *unrestricted liminf* (written $\underline{\lim}$) of $\frac{1-|\varphi(z)|}{1-|z|}$ as $z \to u$ is finite, then the *restricted lim* of the same quantity exists as well and is finite, and moreover φ has a restricted derivative at u."

Before stating the theorem, it is convenient to mention and prove the following lemma, which can be seen as a Schwarz–Pick lemma "pushed to the boundary." To appreciate this lemma, we first coin a definition.

Definition 1.31 Let $u \in \partial\mathbb{D}$ and $\lambda > 0$. The *horodisk* $H(u,\lambda)$ is the open euclidean disk

$$H(u,\lambda) := \left\{ z : \left| z - \frac{u}{1+\lambda} \right| < \frac{\lambda}{1+\lambda} \right\} = D\left(\frac{u}{1+\lambda}, \frac{\lambda}{1+\lambda} \right).$$

It is internally tangent to the unit disk at u.

Here is now an important preliminary lemma.

Lemma 1.32 (Julia) *Let $\varphi : \mathbb{D} \to \mathbb{D}$ be analytic and $u \in \partial\mathbb{D}$. Set*

$$\underline{\lim}_{z\to u} \frac{1-|\varphi(z)|}{1-|z|} =: d(u).$$

Then:

(i) $d(u) > 0$ *for any* $u \in \partial\mathbb{D}$.
(ii) *If* $d(u) < \infty$, *there exists* $v \in \partial\mathbb{D}$ *such that*

$$\frac{|v-\varphi(z)|^2}{1-|\varphi(z)|^2} \leqslant d(u)\,\frac{|u-z|^2}{1-|z|^2} \qquad \textit{for all } z \in \mathbb{D}.$$

(iii) *For any* $\lambda > 0$, *one has*

$$\varphi[H(u,\lambda)] \subset H\big(v, \lambda d(u)\big).$$

Proof: We have $d(u) > 0$ since $\dfrac{1-|\varphi(z)|}{1-|z|} \geqslant \dfrac{1-|\varphi(0)|}{1+|\varphi(0)|}$, in view of Lemma 1.28.

Assume that $d(u) < \infty$. Let (a_n) be a sequence of points of $\mathbb{D}$ converging to u, such that $\dfrac{1-|\varphi(a_n)|}{1-|a_n|} \to d(u)$ and $\varphi(a_n) \to v$, with $|v| = 1$ since $d(u) < \infty$. We pass to the limit in the Schwarz–Pick inequality

$$\frac{|1-\overline{\varphi(a_n)}\varphi(z)|^2}{1-|\varphi(z)|^2} \leqslant \frac{1-|\varphi(a_n)|^2}{1-|a_n|^2} \times \frac{|1-\overline{a_n}z|^2}{1-|z|^2}$$

to obtain

$$\frac{|1-\overline{v}\varphi(z)|^2}{1-|\varphi(z)|^2} \leqslant d(u) \times \frac{|1-\overline{u}z|^2}{1-|z|^2},$$

or else

$$\frac{|v-\varphi(z)|^2}{1-|\varphi(z)|^2} \leqslant d(u) \times \frac{|u-z|^2}{1-|z|^2},$$

Now, a simple calculation gives, for any $w \in \partial\mathbb{D}, z \in \mathbb{D}$, and $\lambda > 0$, the equivalence

$$\frac{|w-z|^2}{1-|z|^2} < \lambda \iff \left|z - \frac{w}{1+\lambda}\right|^2 < \Big(\frac{\lambda}{1+\lambda}\Big)^2.$$

Applying this to the pairs (u, z) and $(v, \varphi(z))$, we see that the final part of the lemma is just a reformulation of the second one. □

We can now state and prove the full Julia–Carathéodory theorem, following more or less closely ([34] p. 51).

Theorem 1.33 (Julia–Carathéodory) *Let* $\varphi : \mathbb{D} \to \mathbb{D}$ *be analytic and* $u \in \partial\mathbb{D}$. *The three following statements are equivalent.*

(i) *In the following, the liminf is taken* unrestrictedly *as* $z \to u, z \in \mathbb{D}$*:*

$$\varliminf_{z\to u} \frac{1-|\varphi(z)|}{1-|z|} =: d(u) < \infty.$$

(ii) *The angular derivative of* φ *exists at* u, *meaning that there exists* $v \in \partial\mathbb{D}$ *such that the angular* (restricted) *limits*

$$\varphi(u) := \angle \lim_{z\to u} \varphi(z), \quad \varphi'(u) := \angle \lim_{z\to u} \frac{v-\varphi(z)}{u-z}$$

exist, with respective values v *and* $d(u)\overline{u}v$.

(iii) *Functions* φ *and* φ' *have at point* u *angular limits* v *and* l *with* $|v| = 1$ *and* $l = d(u)\overline{u}v$.

Moreover, if those conditions hold,

$$\angle \lim_{z \to u} \frac{1 - |\varphi(z)|}{1 - |z|} = d(u).$$

In particular, for any $a > 1$,

$$\sup_{z \in \Gamma_a(u)} \frac{1 - |\varphi(z)|}{1 - |z|} < \infty.$$

Proof: The proof of the most delicate implication (i) $\Longrightarrow$ (ii) will be done in several steps.
(a) We first show that the following radial limit exists:

$$\lim_{r \to 1} \varphi(ru) = v. \tag{1.19}$$

Indeed, Julia's lemma gives for some $v \in \partial\mathbb{D}$

$$|v - \varphi(ru)|^2 \leqslant \frac{|v - \varphi(ru)|^2}{1 - |\varphi(ru)|^2} \leqslant d(u)\,\frac{|u - ru|^2}{1 - r^2} = d(u)\,\frac{1 - r}{1 + r}\cdot$$

By Lindelöf's theorem (see Theorem 1.18), this implies $\angle \lim_{z \to u} \varphi(z) = v$ (we will take the liberty to write $v = \varphi(u)$).
(b) We next show that

$$\lim_{r \to 1} \frac{1 - |\varphi(ru)|}{1 - r} = d(u). \tag{1.20}$$

Indeed, if $T(r) = \frac{1 - |\varphi(ru)|}{1 - r}$, one has by definition $\underline{\lim}_{r \to 1} T(r) \geqslant d(u)$. Julia's lemma gives as above

$$(1 - |\varphi(ru)|)^2 \leqslant |v - \varphi(ru)|^2 \leqslant d(u)(1 - |\varphi(ru)|^2)\frac{1 - r}{1 + r}\cdot \tag{1.21}$$

After simplification

$$T(r) \leqslant d(u)\frac{1 + |\varphi(ru)|}{1 + r}\cdot$$

This gives, in view of (1.19), $\overline{\lim}_{r \to 1} T(r) \leqslant d(u)$, proving (1.20).

(c) We now show that

$$\lim_{r \to 1} \frac{|v - \varphi(ru)|}{1 - r} = d(u). \tag{1.22}$$

Indeed, $\underline{\lim}_{r\to 1} \frac{|v-\varphi(ru)|}{1-r} \geqslant \underline{\lim}_{r\to 1} T(r) = d(u)$, and (1.21) gives

$$\left(\frac{|v-\varphi(ru)|}{1-r}\right)^2 \leqslant d(u)\frac{1-|\varphi(ru)|^2}{1-r^2},$$

so that

$$\overline{\lim_{r\to 1}} \left(\frac{|v-\varphi(ru)|}{1-r}\right)^2 \leqslant d^2(u)$$

by (1.20), which proves (1.22).

(d) In this step, we show that

$$\lim_{r\to 1} \frac{\varphi(u)-\varphi(ru)}{u-ru} = \lim_{r\to 1} \frac{v-\varphi(ru)}{u-ru} = d(u)\,\overline{u}v \tag{1.23}$$

(we will then write $\varphi'(u) = d(u)\,\overline{u}v$). To this effect, we first observe that

$$\frac{1-\operatorname{Re}\overline{v}\varphi(ru)}{1-r} \to d(u).$$

Indeed,

$$2\frac{(1-\operatorname{Re}\overline{v}\varphi(ru))}{1-r} = \frac{1-|\varphi(ru)|^2}{1-r} + \frac{|v-\varphi(ru)|^2}{1-r}$$

and the RHS tends to $2d(u)+0 = 2d(u)$ by (1.20) and (1.22). But now

$$\left|\frac{v-\varphi(ru)}{1-r} - d(u)v\right|^2 = \left|\frac{v-\varphi(ru)}{1-r}\right|^2 + d^2(u) - 2d(u)\frac{(1-\operatorname{Re}\overline{v}\varphi(ru))}{1-r}$$
$$\longrightarrow 2d^2(u) - 2d^2(u) = 0,$$

and we are done.

(e) The final step consists in applying Lindelöf's theorem 1.19 to pass from radial convergence to angular one. Let $\Gamma_a(u)$ be a Stolz region. Then the function $f(z) = \frac{v-\varphi(z)}{u-z}$, holomorphic in $\mathbb{D}$, is bounded in $\Gamma_a(u)$. Indeed,

$$1-|\varphi(z)|^2 \leqslant 2(1-|\varphi(z)|) \leqslant 2|v-\varphi(z)|.$$

Julia's lemma then gives, when $z \in \Gamma_a(u)$,

$$|v-\varphi(z)|^2 \leqslant d(u)(1-|\varphi(z)|^2)\frac{|z-u|^2}{1-|z|^2} \leqslant 2d(u)|v-\varphi(z)|\,|z-u|\,a\frac{1-|z|}{1-|z|^2}\cdot$$

After simplification by $|v-\varphi(z)|$, we get $|f(z)| \leqslant 2ad(u)$ for $z \in \Gamma_a(u)$. It follows from (1.23) that $f(ru)$ has the limit $\varphi'(u)$ as $r \to 1^-$. Hence, f

converges to $\varphi'(u)$ sectorially by Lindelöf's theorem 1, whose assumptions are met. Finally, for any $a > 1$, we have

$$\sup_{z\in\Gamma_a(u)} \frac{1-|\varphi(z)|}{1-|z|} \leqslant a \sup_{z\in\Gamma_a(u)} \left|\frac{v-|\varphi(z)}{z-u}\right| < \infty.$$

For the implication (ii) $\Longrightarrow$ (iii), it is convenient to note the following lemma.

Lemma 1.34 *Let $u \in \partial\mathbb{D}$, $a > 1$, and $\delta = \frac{1}{1+2a}$ · Then*

$$z \in \Gamma_a(u) \text{ and } |\lambda| \leqslant \delta|z-u| \Longrightarrow z+\lambda \in \Gamma_{2a}(u).$$

Proof: We see that

$$|z+\lambda-u| \leqslant |z-u| + |\lambda| \leqslant (1+\delta)|z-u| \leqslant (1+\delta)a(1-|z|).$$

Now

$$1-|z+\lambda| \geqslant 1-|z|-|\lambda| \geqslant 1-|z|-\delta|z-u| \geqslant (1-|z|)(1-a\delta).$$

The result follows since $\dfrac{(1+\delta)a}{1-a\delta} = 2a$. □

Now, we show that φ' is angularly bounded at u.

Fix $a > 1$. Set $M_a = \sup\limits_{s\in\Gamma_{2a}(u)} \left|\dfrac{\varphi(s)-v}{u-s}\right|$.

If $z \in \Gamma_a(u)$ and $\rho = \delta|z-u|$, then $z+\rho w \in \Gamma_{2a}(u)$ for $w \in \mathbb{T}$ by Lemma 1.34, and the Cauchy formula gives

$$\begin{aligned}\varphi'(z) &= (\varphi-v)'(z)\\ &= \int_{\mathbb{T}} \frac{\varphi(z+\rho w)-v}{\rho w}\, dm(w)\\ &= \int_{\mathbb{T}} \left(\frac{\varphi(z+\rho w)-v}{z+\rho w-u}\right)\left(\frac{z+\rho w-u}{\rho w}\right) dm(w).\end{aligned}$$

Since $\left|\dfrac{z+\rho w-u}{\rho w}\right| \leqslant 1+\dfrac{|z-u|}{\rho} = 1+1/\delta$, the integrand is bounded by $M_a(1+1/\delta)$ so that $|\varphi'(z)| \leqslant M_a(1+1/\delta)$ for $z \in \Gamma_a(u)$.

If now $z = ru$ with $0 < r < 1$, we see that

$$\frac{z+\rho w-u}{\rho w} = 1-\frac{u}{\delta w},$$

and Lebesgue's dominated convergence theorem when applied in the integral above shows that

$$\lim_{r\to 1^-} \varphi'(ru) = \int_{\mathbb{T}} \varphi'(u)\Big(1-\frac{u}{\delta w}\Big)\, dm(w) = \varphi'(u).$$

So (iii) holds by Lindelöf's theorem 1 again.

Finally, to show that (iii) implies (i), let $M = \sup_{0<r<1} |\varphi'(ru)| < \infty$. Then

$$|v - \varphi(ru)| = \left| \int_r^1 \varphi'(tu)\, dt \right| \leqslant M(1-r).$$

Hence, by definition,

$$d(u) \leqslant \varliminf_{r \to 1^-} \frac{|v - \varphi(ru)|}{1 - |ru|} \leqslant M. \qquad \square$$

1.5 The Denjoy–Wolff Theorem

Let $\varphi : \mathbb{D} \to \mathbb{D}$ be a holomorphic map. We study here the behavior of the iterates $\varphi^{[n]}$ of φ, with $\varphi^{[n]} = \varphi \circ \cdots \circ \varphi$ (n times). We designate by $I : \mathbb{D} \to \mathbb{D}$ the identity map and by $N(\varphi)$ the number of fixed points of φ in $\mathbb{D}$, counted with multiplicity. Here is a first, simple proposition, which can be essentially attributed to H. Cartan. We provide a proof.

In the sequel, we recall that the notation $\xrightarrow{uc}$ means that we have uniform convergence on compact subsets, equivalently convergence in $\mathscr{H}(\mathbb{D})$.

Proposition 1.35 *Let $\varphi : \mathbb{D} \to \mathbb{D}$ be analytic. Then:*

(i) *If $N(\varphi) \geqslant 2$, we have $\varphi = I$.*
(ii) *If $N(\varphi) = 1$ and if $\varphi(a) = a$ with $a \in \mathbb{D}$, two exclusive cases occur:*
  (a) *$|\varphi'(a)| < 1$, and $\varphi^{[n]} \xrightarrow{uc} a$.*
  (b) *$|\varphi'(a)| = 1$, and φ is an elliptic automorphism of $\mathbb{D}$, holomorphically conjugate to the rotation $z \mapsto \varphi'(a)z$.*

For the definition of *elliptic automorphism*, see the proof of Theorem 1.36.

Proof: (i) Assume $a \in \mathbb{D}$ is a fixed point. The map $\psi = \varphi_a \circ \varphi \circ \varphi_a : \mathbb{D} \to \mathbb{D}$ satisfies $\psi(0) = 0$. By the Schwarz lemma, we know that $|\psi(z)| \leqslant |z|$ and $|\psi'(0)| \leqslant 1$. If a is a double fixed point, that is, $\varphi'(a) = 1$, then we have $\psi'(0) = 1$ and the case of equality in the Schwarz lemma gives $\psi(z) = \lambda z$ for every $z \in \mathbb{D}$, with a unimodular constant λ. Since $\lambda = \psi'(0) = 1$, $\varphi = I$. In the case of two distinct fixed points $a, b \in \mathbb{D}$ of φ, then $b' = \varphi_a(b) \neq 0$ satisfies $\psi(b') = b'$. The case of equality in the Schwarz lemma gives again $\psi(z) = \lambda z$ for every $z \in \mathbb{D}$, with a unimodular constant λ, and testing $b' \neq 0$ gives also $\lambda = 1$, so that $\varphi = I$.

For (ii), set $\lambda = \varphi'(a)$, so that $|\lambda| \leqslant \frac{1-|\varphi(a)|^2}{1-|a|^2} = 1$ by the Schwarz–Pick lemma. If $|\lambda| < 1$, then $\varphi^{[n]} \xrightarrow{uc} a$ by a result of Cartan, which is proved as follows: the sequence $(\varphi^{[n]})$ is a normal family, and since a is attractive for φ, there is an open neighbourhood $U \subset \mathbb{D}$ such that $\varphi^{[n]}(z) \to a$ when $z \in U$.

But then, all limits of convergent subsequences of $\varphi^{[n]}$ are equal to a on U and hence on $\mathbb{D}$. Now, Montel's theorem gives the result.

If $|\lambda| = 1$, it holds with the previous notations:

$$\psi'(0) = \varphi_a'(a)\varphi'(a)\varphi_a'(0) = \frac{1}{1-|a|^2}\lambda(1-|a|^2) = \lambda;$$

hence $\psi(z) = \lambda z$ and $\varphi = \varphi_a \circ \psi \circ \varphi_a$ is conjugate to the rotation $z \mapsto \lambda z$. □

The case $N(\varphi) = 0$ is handled by the following theorem.

Theorem 1.36 (Denjoy–Wolff) *Let $\varphi : \mathbb{D} \to \mathbb{D}$ be analytic with no fixed point inside $\mathbb{D}$. Then there exists a point $\omega \in \partial\mathbb{D}$, called the Denjoy–Wolff point of φ, such that*

$$\varphi^{[n]} \xrightarrow{uc} \omega.$$

Moreover, the angular derivative $\varphi'(\omega)$ exists, and

$$0 < \varphi'(\omega) \leqslant 1.$$

Remark. By misuse of language, we shall also call *Denjoy–Wolff* point the point a in Proposition 1.35(ii).

Proof: We distinguish two cases.

Case A. Assume that $\varphi \in G = \operatorname{Aut}\mathbb{D}$, that is, $\varphi = e^{i\theta}\varphi_p$, with $p \in \mathbb{D}$ and $-\pi \leqslant \theta < \pi$. Then, we content ourselves with describing the situation in this elementary case of fractional linear mappings (see [147, chapters 15] with θ changed into $\theta + \pi$).

1. (*Hyperbolic case*) $|p| > \cos(\theta/2)$. Then, φ has two distinct fixed points $a, b \in \partial\mathbb{D}$ with $0 < \varphi'(a) < 1$ and $\varphi'(a)\varphi'(b) = 1$. Hence, $N(\varphi) = 0$. The point a is attractive and is the Denjoy–Wolff point; b is repelling and the relation

$$\frac{\varphi^{[n]}(z) - a}{\varphi^{[n]}(z) - b} = \big(\varphi'(a)\big)^n \frac{z-a}{z-b}$$

shows that $\varphi^{[n]} \xrightarrow{uc} a$. A typical case is $a = 1$, $b = -1$ ($p = -r$ and $\theta = \pi$) and

$$\varphi(z) = \frac{z+r}{1+rz}, \quad \varphi'(1) = \frac{1-r}{1+r}, \quad \text{with } 0 < r < 1.$$

2. (*Parabolic case*) $|p| = \cos(\theta/2)$. Then φ has a double fixed point $a \in \partial\mathbb{D}$ with $\varphi'(a) = 1$, and again $N(\varphi) = 0$. One has the relation

$$\frac{1}{\varphi^{[n]}(z) - a} - \frac{1}{z-a} = nc$$

with $c \neq 0$ since $\varphi \neq I$. Hence $\varphi^{[n]} \xrightarrow{uc} a$, and $a \in \partial\mathbb{D}$ is the Denjoy–Wolff point of φ.

3. (*Elliptic case*) $|p| < \cos(\theta/2)$. This case does not occur if $N(\varphi) = 0$. Indeed, if $p = 0$, one has $\varphi(0) = 0$ and $N(\varphi) = 1$. And if $p \neq 0$, one of the fixed points a of φ belongs to $\mathbb{D}$, again a contradiction.

The case $\varphi \in G$ is achieved.

Case B. We assume that $\varphi \notin G$. There are two steps.

1. We claim that $|\varphi^{[n]}(0)| \to 1$. Indeed, by Schwarz–Pick, the sequence $\rho(\varphi^{[n+1]}(0), \varphi^{[n]}(0))$ is non-increasing and tends to a limit l. If the conclusion fails, there is $a \in \mathbb{D}$ and some subsequence (n_k) such that $\varphi^{[n_k]}(0) \to a$. Set $b = \varphi(a)$. Then, $\rho(\varphi^{[n_k+1]}(0), \varphi^{[n_k]}(0)) \to \rho(a,b)$ and $\rho(\varphi^{[n_k+2]}(0), \varphi^{[n_k+1]}(0)) \to \rho(\varphi(a), \varphi(b))$, so that

$$\rho(\varphi(a), \varphi(b)) = \rho(a,b) = l.$$

Since $\varphi \notin G$, Schwarz–Pick gives us $a = b$, contradicting the assumption $N(\varphi) = 0$.

2. There exists $\omega \in \partial\mathbb{D}$ such that $\varphi^{[n]} \xrightarrow{uc} \omega$ and $0 < d(\omega) \leqslant 1$. Indeed, let $0 < r_n < 1$ with $\lim_{n\to\infty} r_n = 1$. By Rouché's theorem, one can find a sequence (b_n) in $\mathbb{D}$ such that

$$r_n \varphi(b_n) = b_n, \tag{1.24}$$

and actually $|b_n| < r_n$.

We can assume that $b_n \to \omega \in \overline{\mathbb{D}}$. If $\omega \in \mathbb{D}$, we get $\varphi(\omega) = \omega$, contradicting our assumption $N(\varphi) = 0$. Hence $\omega \in \partial\mathbb{D}$. Moreover, $1 - |\varphi(b_n)| \leqslant 1 - |b_n|$ by (1.24), so that $d(\omega) \leqslant 1$, and $d(\omega) > 0$ by Julia's lemma. Assume now that some subsequence $(\varphi^{[n_k]})$ of the normal family $(\varphi^{[n]})$ satisfies

$$\varphi^{[n_k]} \xrightarrow{uc} g \in \mathscr{H}(\mathbb{D}).$$

We see that $|g(0)| = \lim_{k\to\infty} |\varphi^{[n_k]}(0)| = 1$ by the first step. By Montel's theorem, g is a unimodular constant w. Now, by Julia's lemma, we have for fixed $\lambda > 0$

$$\varphi^{[n]}[H(\omega,\lambda)] \subset H(\omega,\lambda),$$

which implies that $w \in \overline{H(\omega,\lambda)} \cap \mathbb{T}$. But $\overline{H(\omega,\lambda)} \cap \mathbb{T} = \{\omega\}$ so that $w = \omega$ and that, by Montel's theorem once more,

$$\varphi^{[n]} \xrightarrow{uc} \omega.$$

The conclusion follows from Julia–Carathéodory's theorem (with $u = v = \omega$), which gives $\varphi'(\omega) = d(\omega)\omega\overline{\omega} = d(\omega)$. □

Remark. Speaking of automorphisms, a natural question emerges: does an analytic self-map $\varphi : \mathbb{D} \to \mathbb{D}$ exist that is both surjective and locally invertible, that is, $\varphi'(z) \neq 0$ for all $z \in \mathbb{D}$, but which is not an automorphism of $\mathbb{D}$?

For entire functions, the answer is yes (take for φ a primitive of e^P with P a polynomial of degree larger than 2), but for the disk? The answer is *again yes*, as J. F. Burnol (private communication) showed us: let $g(z) = \dfrac{z+1}{z-1}$, then we have $f(z) = z e^{g(z)} : \mathbb{C} \setminus \{1\} \to \mathbb{C}$ with

$$|f(z)| = |z| \exp\left(\frac{|z|^2 - 1}{|z-1|^2}\right), \tag{1.25}$$

so that by (1.25) $f(\mathbb{D}) \subset \mathbb{D}$, and it defines a symbol (still denoted by f for short) with $f(0) = 0$ and a derivative vanishing at the single point $z_0 = 2 - \sqrt{3}$ (easy computation), and with

$$f(\mathbb{D} \setminus \{z_0\}) = \mathbb{D}$$

by Picard's great theorem, for example (and (1.25) again since $|f(z)| \geqslant 1$ when $|z| \geqslant 1$ and $z \neq 1$).

Now take

$$F = f \circ \varphi_{z_0} \circ e^g$$

with the (surjective) arrows

$$\mathbb{D} \xrightarrow{e^g} \mathbb{D} \setminus \{0\} \xrightarrow{\varphi_{z_0}} \mathbb{D} \setminus \{z_0\} \xrightarrow{f} \mathbb{D}.$$

Moreover, F' never vanishes on $\mathbb{D}$ since $\varphi_{z_0}(e^g) \neq z_0$ on $\mathbb{D}$.

Much more on this question, with a broader and more conceptual view, is to be found in a paper in preparation on *uniformization principles* by Burnol and Touzé.

1.6 Nevanlinna's Counting Function and Littlewood's Inequality

Here is an extension of the Schwarz lemma, quite useful in the study of composition operators (notably in connection with Littlewood's identity), and due to... Littlewood. We first coin the important Nevanlinna counting function N_φ defined by

$$N_\varphi(w) = \begin{cases} \displaystyle\sum_{\varphi(z)=w} \log 1/|z| & \text{if } w \neq \varphi(0),\ w \in \varphi(\mathbb{D}), \\ +\infty & \text{if } w = \varphi(0), \\ 0 & \text{if } w \notin \varphi(\mathbb{D}). \end{cases} \tag{1.26}$$

Theorem 1.37 (Littlewood) *Let $\varphi : \mathbb{D} \to \mathbb{D}$ be analytic. Then, for any $w \in \mathbb{D} \setminus \{\varphi(0)\}$, it holds*

$$N_\varphi(w) \leqslant \log \left| \frac{1 - \overline{w}\varphi(0)}{w - \varphi(0)} \right| .$$

Let us mention that this very nice inequality is a "super Schwarz lemma": it means, when $\varphi(0) = 0$,

$$N_\varphi(w) \leqslant \log \left(\frac{1}{|w|} \right) .$$

In other words, we have in this case

$$|\varphi(z)| \leqslant \prod_{\varphi(a)=\varphi(z)} |a| \quad (\leqslant |z|). \tag{1.27}$$

Proof: We can assume $w \neq \varphi(0)$ and $w \in \varphi(\mathbb{D})$. Fix $\rho \in (0,1)$ and let $z_1, \ldots, z_p$ be zeros of $\varphi - w$ in $\rho\mathbb{D}$ (not necessarily distinct, taking into account the multiplicity). We have a factorization

$$\varphi_w \circ \varphi = \prod_{j=1}^{p} \varphi_{z_j} \times \psi,$$

where the analytic function ψ satisfies $\|\psi\|_\infty \leqslant 1$ by the maximum modulus principle, and the fact that, for any $a \in \mathbb{D}$, $\lim_{|z| \to 1^-} |\varphi_a(z)| = 1$. Taking $z = 0$, we obtain, since $|\psi(0)| \leqslant 1$,

$$|\varphi_w(\varphi(0))| \leqslant \prod_{j=1}^{p} |\varphi_{z_j}(0)| = \prod_{j=1}^{p} |z_j|.$$

Or again, rearranging,

$$\sum_{j=1}^{p} \log(1/|z_j|) \leqslant \log(1/|\varphi_w(\varphi(0))|);$$

hence, since ρ is arbitrary,

$$N_\varphi(w) \leqslant \log \left| \frac{1 - \overline{w}\varphi(0)}{w - \varphi(0)} \right| .$$

□

The Nevanlinna counting function is defined with the help of the function log that is a harmonic function and hence satisfies the mean value theorem. A part of this property is inherited by N_φ:

Theorem 1.38 *Let $\varphi : \mathbb{D} \to \mathbb{D}$ be an analytic function and $a \in \mathbb{D}$.*

(i) *Let $r \in (0, |\varphi(0) - a|)$. We have*

$$\int_{\mathbb{T}} N_\varphi(a + r\xi)\, dm(\xi) \geqslant N_\varphi(a).$$

(ii) *For every disk $\Delta \subset \mathbb{D}$ centered in a and with radius R, not containing $\varphi(0)$,*

$$\int_{\Delta} N_{\varphi}(w)\, d\sigma(w) \geqslant R^2 N_{\varphi}(a).$$

Proof: Let $u, v \in \mathbb{D}$. Assume first that $|u| < |v|$ in the following, so that the function $z \in \overline{\mathbb{D}} \mapsto \log \big||v| - uz\big|$ is harmonic as the real part of an analytic function on $\mathbb{D}$, continuous on $\overline{\mathbb{D}}$, so

$$\int_{\mathbb{T}} \log |u\xi - v|\, dm(\xi) = \int_{\mathbb{T}} \log \big||v| - uz\big|\, dm(z) = \log |v|.$$

Thanks to the rotation invariance of the measure m, we get for any $u, v \in \mathbb{D}$ with $|u| \neq |v|$,

$$\int_{\mathbb{T}} \log |u\xi - v|\, dm(\xi) = \log \big(\max(|u|, |v|)\big), \tag{1.28}$$

and actually (1.28) is still valid when $|u| = |v|$ by a standard limit argument.

Fix $\rho \in (0, 1)$ and $w \in \mathbb{D}$ with $|w| = r$. From the Jensen formula (1.2), applied to $z \in \mathbb{D} \mapsto \varphi(z) - w$, we have

$$\log |\varphi(0) - w| + \sum_{j=1}^{m} \log \frac{\rho}{|z_j(\rho)|} = \int_{\mathbb{T}} \log |\varphi(\rho z) - w|\, dm(z), \tag{1.29}$$

where $z_1(\rho), \ldots, z_m(\rho)$ are the zeros of $z \mapsto \varphi(z) - w$ with modulus less than ρ, counted with multiplicities.

Let us denote (with the previous notation)

$$N_{\varphi}(w, \rho) = \sum_{j=1}^{m} \log \frac{\rho}{|z_j(\rho)|}, \tag{1.30}$$

so that

$$N_{\varphi}(w, \rho) = \int_{\mathbb{T}} \log |\varphi(\rho z) - w|\, dm(z) - \log |\varphi(0) - w|. \tag{1.31}$$

Note that, by its definition, $N_{\varphi}(w, \rho)$ grows to $N_{\varphi}(w)$ when ρ grows to 1. Now we apply (1.31) with $w = a + r\xi$: thanks to Fubini's theorem, we get using (1.28), and since $r < |a - \varphi(0)|$,

$$\begin{aligned}
&\int_{\mathbb{T}} N_\varphi(a + r\xi, \rho)\, dm(\xi) \\
&\quad = \int_{\mathbb{T}}\int_{\mathbb{T}} \log|\varphi(\rho z) - a - r\xi|\, dm(\xi)dm(z) - \log|\varphi(0) - a| \\
&\quad = \int_{\mathbb{T}} \log\max(|\varphi(\rho z) - a|, r)\, dm(z) - \log|\varphi(0) - a| \\
&\quad \geqslant \int_{\mathbb{T}} \log|\varphi(\rho z) - a|\, dm(z) - \log|\varphi(0) - a| = N_\varphi(a, \rho).
\end{aligned}$$

Finaly, taking the limit when $\rho \to 1^-$, we get

$$\int_{\mathbb{T}} N_\varphi(a + r\xi)\, dm(\xi) \geqslant N_\varphi(a).$$

Now, integrating over $r \in (0, R)$ we get

$$\int_{\Delta} N_\varphi(w)\, d\sigma(w) = \int_0^R \left(\int_{\mathbb{T}} N_\varphi(a + r\xi)\, dm(\xi) \right) 2r dr \geqslant R^2 N_\varphi(a).$$

□

1.7 Interpolating Sequences

1.7.1 Lagrange's Interpolation

Given n distinct complex numbers $z_1, \ldots, z_n$ (the knots) and n complex numbers $w_1, \ldots, w_n$ (the data), Lagrange's interpolation problem consists in finding $f \in \mathcal{P}_{n-1}$, the vector space spanned by the monomials $1, \ldots, z^{n-1}$, such that

$$f(z_j) = w_j, \quad \text{for } 1 \leqslant j \leqslant n.$$

Lagrange's idea (remember also the Chinese remainder theorem) consists in starting with $l_j \in \mathcal{P}_{n-1}$ satisfying $l_j(z_k) = \delta_{j,k}$, or equivalently $l_j(z) = \dfrac{l(z)}{(z - z_j)l'(z_j)}$ where $l(z) = \prod_{1 \leqslant j \leqslant n} (z - z_j)$; then take

$$f = \sum_{j=1}^{n} w_j l_j.$$

What now with an *infinite* sequence (z_j) of distinct knots in the unit disk $\mathbb{D}$, and an infinite sequence (w_j) of data, if f is looked for inside the set $\mathscr{H}(\mathbb{D})$ of all holomorphic functions in $\mathbb{D}$? This is still possible, as shown by the following theorem:

Theorem 1.39 *Given an infinite sequence (z_j) of distinct points in $\mathbb{D}$, with no accumulation point in $\mathbb{D}$, and an infinite sequence (w_j) of data, there exists $f \in \mathscr{H}(\mathbb{D})$ such that*

$$f(z_j) = w_j, \quad \textit{for every } j = 1, 2, \ldots$$

Proof: Let $f_j \in \mathscr{H}(\mathbb{D})$ be functions satisfying $f_j(z_k) = \delta_{j,k}$:

$$f_j(z) = \frac{P(z)}{(z - z_j)P'(z_j)},$$

where P is an infinite product of the Weierstrass type. We would like to take $f = \sum_{j=1}^{\infty} w_j f_j$, which works formally. As the convergence of the series is not guaranteed, one "corrects" in replacing f_j by $g_j = f_j e_j$ where $e_j(z) = \exp(c_j(z - z_j))$, c_j being complex constants.

We still have $g_j(z_k) = \delta_{j,k}$, and we can adjust the c_j's with $|c_j|$ large and $c_j z_j = |c_j|\,|z_j|$ [35, pp. 422–424] in such a way that the series $\sum_{j=1}^{\infty} w_j g_j$ converges uniformly on every compact subset of $\mathbb{D}$. Then, we take

$$f = \sum_{j=1}^{\infty} w_j g_j,$$

and we have $f(z_j) = w_j$ for every $j \geqslant 1$. □

This method (formal series + correction) is reminiscent of one proof of the Borel C^∞ theorem ($D^j f(0) = w_j$, Hermite interpolation). But as soon as a quantitative version is required, things become much more delicate, for example, in Borel theorem (if the w_j's have controlled growth, the function f can be taken in some good class of functions $C(\{M_n\})$; see [135]). A similar phenomenon occurs with bounded holomorphic functions in $\mathbb{D}$.

1.7.2 Pick's Interpolation

If we look for an interpolating function f satisfying $\|f\|_\infty \leqslant 1$ if $\sup |w_j| < 1$, we find it too ambitious. For example, if $n = 2$, the Schwarz–Pick inequality implies

$$\rho(w_1, w_2) = \rho(f(z_1), f(z_2)) \leqslant \rho(z_1, z_2),$$

or else the matrix $A = (a_{i,j}) = \left(\dfrac{1 - w_i \overline{w_j}}{1 - z_i \overline{z_j}}\right)$ must be non-negative (see Theorem 1.40). In other terms, our sup-norm preserving interpolation can take place only under strong constraints on the data and the knots.

A remarkable fact due to Pick is that the positivity of this matrix A is a necessary and sufficient condition in the case of Hardy spaces.

It is first convenient to slightly generalize the context: let H be a Hilbert space of analytic functions on $\mathbb{D}$ with reproducing kernel K, and denote

$$K_a(b) = \langle K_b, K_a \rangle = K(b,a).$$

Let next $\mathcal{M}(H)$ be the space of *multipliers* of H, namely the space of functions f analytic on $\mathbb{D}$ such that $fh \in H$ for all $h \in H$, equipped with the multiplier norm

$$\|f\|_{\mathcal{M}(H)} := \sup_{\|h\|_H \leqslant 1} \|fh\|_H.$$

For example, $\mathcal{M}(H)$ is (isometrically) H^∞ when $H = H^2$.

The following was proved by Pick [58, p. 7] when $H = H^2$.

Theorem 1.40 (Pick) *Let $(z_1, \dots, z_n)$ and $(w_1, \dots, w_n)$ be two n-tuples of points of $\mathbb{D}$. Then, the following are equivalent:*

(i) *The matrix* $A = (a_{i,j}) = \left(\dfrac{1 - w_i\overline{w_j}}{1 - z_i\overline{z_j}}\right)_{i,j} = \big((1 - w_i\overline{w_j})K(z_i, z_j)\big)_{i,j}$ *is non-negative.*

(ii) *There exists $f \in B_{H^\infty}$ such that $f(z_k) = w_k$, where $k = 1, \dots, n$.*

That (ii) implies (i) is rather formal and holds in general. Let $M_f : H^2 \to H^2$ be the operator of multiplication by f, which satisfies

$$\|M_f\| = \|f\|_{\mathcal{M}(H^2)} \leqslant 1 \quad \text{and} \quad M_f^*(K_z) = \overline{f(z)}\, K_z \quad \text{for all } z \in \mathbb{D}. \tag{1.32}$$

If $(\lambda_1, \dots, \lambda_n) \in \mathbb{C}^n$, then by setting $g = \sum_j \overline{\lambda_j} K_{z_j}$, we see that, thanks to (1.32),

$$\begin{aligned}\sum a_{i,j}\lambda_i\overline{\lambda_j} &= \sum (1 - w_i\overline{w_j})\langle K_{z_j}, K_{z_i}\rangle \lambda_i\overline{\lambda_j} \\ &= \left\|\sum \overline{\lambda_j} K_{z_j}\right\|^2 - \left\|\sum \overline{w_j\lambda_j} K_{z_j}\right\|^2 \\ &= \|g\|^2 - \left\|M_f^*(g)\right\|^2 \geqslant 0.\end{aligned}$$

The reverse implication will not be proved here (see [58, p. 7] or [4, chapter 6] for a proof). Indeed, we make no further use of Pick's theorem (a motivation for Carleson's interpolation theorem to follow) in this book, even though we will briefly return to it with Carleson measures.

Let us however mention that Theorem 1.40 gave rise to the following definition:

Definition 1.41 A Hilbert space H of analytic functions on $\mathbb{D}$ is said to have the *Pick property* if Theorem 1.40 holds replacing B_{H^∞} by $B_{\mathcal{M}(H)}$.

The Hardy space H^2, the Dirichlet space $\mathscr{D}$ with the norm

$$\left\| \sum_{n=0}^{\infty} a_n z^n \right\|^2 = \sum_{n=0}^{\infty} (n+1)|a_n|^2$$

have this property, the Bergman space does not.

The space H is said to have the *complete Pick property* if, for all integers n, s, t and all $s \times t$ matrices $W_1, \ldots, W_n$, the positivity of the Pick matrix

$$\left[(I_{\mathbb{C}^t} - W_i W_j^*) K(z_i, z_j) \right]$$

implies the existence of a function F in the unit ball of $\mathscr{M}(H \otimes \mathbb{C}^s, H \otimes \mathbb{C}^t)$ such that

$$F(z_i) = W_i,\ i = 1, \ldots, n.$$

We refer to [4, chapter 5] for more details. In particular, the complete Pick property amounts to the fact that $1 - 1/K$ is positive definite.

Before proceeding, it is convenient to define Blaschke products.

1.7.3 Blaschke Products

A Blaschke product is a holomorphic function on $\mathbb{D}$ of the form

$$B(z) = \prod_{n=1}^{\infty} \frac{|z_n|}{z_n} \frac{z_n - z}{1 - \overline{z_n} z} =: \prod_{n=1}^{\infty} b_{z_n}(z) \tag{1.33}$$

(agreeing that $b_0(z) = z$) where the points (distinct or not) $z_n \in \overline{\mathbb{D}}$ satisfy the so-called Blaschke condition

$$\sum_{n=1}^{\infty} (1 - |z_n|) < \infty. \tag{1.34}$$

Point out that $b_\xi(z) = 1$ when $\xi \in \mathbb{T}$, so that this definition covers the case of finite Blaschke products (taking in this case only finitely many z_n's in $\mathbb{D}$).

We shall denote $\mathcal{F}$ the *set of finite Blaschke products*.

Condition (1.34) ensures that the series $\sum_{n=1}^{\infty} \left(1 - b_{z_n}(z)\right)$ converges normally on compact subsets of $\mathbb{D}$, hence the product (1.33) converges uniformly on compact subsets of $\mathbb{D}$ and defines a holomorphic function on $\mathbb{D}$ whose zeros are exactly the z_n belonging to $\mathbb{D}$, with multiplicity. Clearly, $|B(z)| \leqslant 1$ on $\mathbb{D}$. A more detailed study of Blaschke products and applications is to follow (see for instance Section 1.9 dedicated to inner functions); nevertheless, we mention here a first meeting with a very useful factorization principle.

Proposition 1.42 *Let $f \neq 0$ be a bounded analytic function on $\mathbb{D}$ and B be the Blaschke product formed by the zeros of f (with multiplicities).*

Then we can write $f = FB$ where F is a bounded analytic function on $\mathbb{D}$, which is zero-free and satisfies $\|F\|_\infty = \|f\|_\infty$.

Before proving this result, let us mention that the (countable) set $Z(f)$ of zeros of $f \in H^\infty$ satisfies the condition (1.34) (recall Proposition 1.6) so that B is indeed defined.

We shall prove a corresponding result for functions belonging to H^p (see Proposition 1.75).

Proof: The function $F = \dfrac{f}{B}$ is *a priori* defined and holomorphic on $\mathbb{D} \setminus Z(f)$. But writing B_n the partial Blaschke product (of order n) associated to B, the function

$$F_n = \frac{f}{B_n}$$

is well defined on $\mathbb{D}$, and analytic, by definition of the zeros and multiplicity. Moreover, $\lim F_n(z)$ exists for every $z \in \mathbb{D}$, and this allows to extend F to $\mathbb{D}$ holomorphically. Clearly $f = FB$ on $\mathbb{D}$ and F is zero-free.

On the other hand, for every $n \geqslant 1$, the function B_n is actually analytic on a disk containing $\overline{\mathbb{D}}$. Therefore, for every $r \in (0,1)$, close enough to 1, we have

$$\sup_{|z| \leqslant r} |f(z)| \leqslant \sup_{|z| \leqslant r} |F_n(z)| = \sup_{|z|=r} \frac{|f(z)|}{|B_n(z)|} \leqslant \frac{\|f\|_\infty}{\min_{|z|=r} |B_n(z)|},$$

and since $\lim\limits_{r \to 1^-} \min\limits_{|z|=r} |B_n(z)| = 1$, we get $\|F_n\|_\infty = \|f\|_\infty$.

We conclude that, for every $z \in \mathbb{D}$, we have

$$|f(z)| \leqslant |F(z)| = \lim_n |F_n(z)| \leqslant \|f\|_\infty;$$

hence $F \in H^\infty$ with $\|F\|_\infty = \|f\|_\infty$. □

1.7.4 Carleson's Interpolation

The problem is now as follows: if the sequence $w = (w_j)$ of data is *bounded* ($w \in \ell^\infty$), can one take some interpolating function f bounded in $\mathbb{D}$ ($f \in H^\infty$) as well that is, $f \in H^\infty$? The full answer was given by Carleson in 1958 [17]. Note that Carleson's paper begins with comments on the Pick theorem and its slightly too rigid aspect. Carleson's work is all but formal and involves the pseudo-hyperbolic distance ρ on $\mathbb{D}$. This distance is well adapted here, because it makes the analytic self-maps f of $\mathbb{D}$ contractions for ρ, that is,

$\rho(f(z), f(w)) \leqslant \rho(z,w)$. The main reason why we are concerned here with interpolating sequences is the following:

In the study of analytic Hilbert (or Banach) spaces, reproducing kernels play a basic role. And, for example, if (z_j) is a sequence of distinct points of $\mathbb{D}$, the Riesz (nearly orthogonality) character of the associated sequence (K_{z_j}) of reproducing kernels is closely, and quite quantitatively, related to the interpolating character of (z_j), in particular in connection with Bernstein numbers of composition operators. We will come back to that in detail later.

The Necessary and Sufficient Condition of Carleson

Here we state and comment Carleson's theorem. The second part is due to Earl.

Theorem 1.43 (Carleson) *Given an infinite sequence $Z = (z_j)$ of distinct points of $\mathbb{D}$, a necessary and sufficient condition in order that every bounded sequence $w = (w_j)_{j\geqslant 1}$ can be written as $w_j = f(z_j)$ with $f \in H^\infty$ is the following:*

$$\textit{if} \quad \delta_n = \prod_{j\neq n} \rho(z_n, z_j), \quad \textit{then} \quad \inf_{n\geqslant 1} \delta_n = \delta =: \delta(Z) > 0. \tag{1.35}$$

If this condition is met, there exists a best constant $I_Z < \infty$ and a constant $C_\delta > 0$, depending only on δ, satisfying: if $w = (w_j)_{j\geqslant 1} \in \ell^\infty$ with $\|w\|_\infty \leqslant 1$, there exists an interpolating function f such that $\|f\|_\infty \leqslant I_Z \leqslant C_\delta$. The following estimates hold:

$$\frac{1}{\delta} \leqslant I_Z \leqslant C_\delta \leqslant \frac{C}{\delta} \log\Big(1 + \frac{1}{\delta}\Big), \tag{1.36}$$

where C is a numerical constant.

Set also

$$\tau = \frac{\delta}{1 + \sqrt{1-\delta^2}} \qquad \textit{with } \frac{2\tau}{1+\tau^2} = \delta.$$

Then

$$I_Z \leqslant K_\delta := \tau^{-2} = \frac{2 - \delta^2 + 2\sqrt{1-\delta^2}}{\delta^2}\,. \tag{1.37}$$

We begin with a comment on Carleson's condition. Let (z_n) be a sequence of distinct points of $\mathbb{D}$ satisfying Blaschke's condition $\sum_{n=1}^\infty (1 - |z_n|) < \infty$. The infinite product

$$B(z) = \prod_{k\geqslant 1} \frac{|z_k|}{z_k} \frac{z_k - z}{1 - \overline{z_k} z} = \prod_{k\geqslant 1} b_{z_k}(z)$$

converges uniformly on compacta of $\mathbb{D}$, and the Blaschke product

$$B_n = \frac{B}{b_{z_n}} = \prod_{k \neq n} b_{z_k} \tag{1.38}$$

plays a privileged role, since, if (z_n) moreover satisfies (1.35), then

$$|B_n(z_n)| = \prod_{k \neq n} \rho(z_n, z_k) \text{ and } \inf_n |B_n(z_n)| = \delta, \tag{1.39}$$

the Carleson constant of the sequence (z_n). Here is a useful reformulation.

Proposition 1.44 *Let (z_n) be an interpolating sequence of distinct points of $\mathbb{D}$, with Carleson constant δ. Then*

$$(1 - |z_n|^2)|B'(z_n)| \geqslant \delta \qquad \textit{for each} \quad n \geqslant 1. \tag{1.40}$$

Proof: We have $B = B_n b_{z_n}$, so that

$$|B'(z_n)| = |B_n(z_n) b'_{z_n}(z_n)| = \frac{|B_n(z_n)|}{1 - |z_n|^2} \geqslant \frac{\delta}{1 - |z_n|^2}.$$

□

Let us come back to Theorem 1.43. The quantities $\delta = \delta(Z)$ and the optimal I_Z are respectively called *the Carleson separation constant* and *the interpolation constant* of Z. If $\delta > 0$, the sequence Z is said to be interpolating. Observe that such a sequence Z must satisfy Blaschke's condition $\sum_{n=1}^{\infty}(1 - |z_n|) < \infty$ (interpolate $w_j = \delta_{j,1}$), but this will not be sufficient in general. The very existence of interpolating sequences is an issue to start.

The minoration of C_δ is easy (as soon as one interpolates at least two points). For, if $w_j = \delta_{n,j}$, any interpolating function f_n writes $f_n = B_n g_n$ with

$$\|f_n\|_\infty = \|g_n\|_\infty \quad \text{and} \quad \|g_n\|_\infty \geqslant |g_n(z_n)| = 1/|B_n(z_n)| = 1/\delta_n,$$

whence

$$I_Z \geqslant \frac{1}{\delta} > 1.$$

This is no surprise when one thinks of the Schwarz–Pick theorem ([35, p. 368] or [4]). Observe that (1.36) is interesting when δ is close to 0, and that (1.37) becomes more interesting when δ is close to 1.

This theorem of Carleson is rather deep. The full proof of (1.36) can be found in [58, chapter VII, pp. 278–285]. We also recommend the fairly old proof of (1.37) due to J. Earl in the 1970s [47], not so well known, but which has a double advantage:

- The approach is original, in the sense that it is based on a topological argument on *topology of euclidean spaces* $\mathbb{R}^p$ (Invariance of Domain Theorem).
- Instead of using linear combinations of finite Blaschke products, the proof uses *a single finite Blaschke product*, whose zeros will be chosen with extreme care, thanks to the topological theorem mentioned. As a consequence, it has a quasi-isometric character *when* δ *is close to* 1, which is in our opinion one of its most interesting features. Several other proofs will be mentioned in Chapter 5.

Remark 1.45 Let us stress that Carleson's condition of Theorem 1.43 can be efficiently checked. Here are a few examples.

(i) $z_n = 1 - n^{-2}$. Blaschke's condition holds, but

$$\delta_n \leqslant \rho(z_n, z_{n+1}) = \frac{(n+1)^2 - n^2}{(n+1)^2 + n^2 - 1} \sim \frac{1}{n},$$

hence $\inf_n \delta_n = 0$, and the sequence is not interpolating. In fact, it is well known, and easy to check, that an increasing sequence (z_n) of $[0,1)$ is interpolating if and only if it is *Newman*:

$$1 - z_{n+1} \leqslant \beta(1 - z_n), \quad \text{where } 0 < \beta < 1.$$

(ii) $z_n = \dfrac{n}{n+i} = \dfrac{n^2 - in}{n^2+1} \in C(1/2, 1/2)$, $z_n \to 1$. Then, for $j \neq n$,

$$\rho(z_n, z_j) = \left|\frac{n-j}{1+i(n-j)}\right| = \left[1 + (n-j)^{-2}\right]^{-1/2}$$

so that

$$\delta_n \geqslant \prod_{k \neq n} \left[1 + k^{-2}\right]^{-1/2} \geqslant \prod_{k=1}^{\infty} \left[1 + k^{-2}\right]^{-1},$$

and the sequence is interpolating. Nevertheless, $1 - |z_n| \sim \frac{1}{2n^2}$ tends slowly to zero.

(iii) $z_n = 1 - r^n$, $0 < r < 1$. A similar computation shows that

$$\delta_n \geqslant \prod_{k=1}^{\infty} \left(\frac{1 - r^k}{1 + r^k}\right)^2, \tag{1.41}$$

and the sequence is interpolating.

(iv) $z_n = a + r\omega^n$ where $1 \leqslant n \leqslant N$ and $\omega = e^{\frac{2i\pi}{N}}$ and $r \in (0,1)$ with $|a| + r < 1$.

We have (the proof is left as exercise: see Exercise I.1)

$$I_Z \leqslant C^N \qquad \text{with } C = C(a,r). \tag{1.42}$$

Remark 1.46 The estimate of I_Z given by (1.36) in Theorem 1.43 is sharp when δ is close to 1, but not when it is close to 0. We have $I_Z \leqslant C_\delta \leqslant C\,\delta^{-1}(1 + \log \delta^{-1})$, and this is optimal ([35, pp. 381–382] for example). Whereas (1.37) only gives $I_Z \lesssim \delta^{-2}$ as $\delta \to 0$. But observe that $\tau \to 1$ as $\delta \to 1$, hence $I_Z \to 1$ as well, since $I_Z \leqslant K_\delta$.

Therefore (1.37) provides us with the interesting corollary:

Corollary 1.47 *Let $\varepsilon > 0$. Then there exists an interpolating sequence $Z = (z_n)$ whose interpolation constant I_Z satisfies*

$$I_Z \leqslant 1 + \varepsilon.$$

Remark 1.48 When $\|w\|_\infty \leqslant \tau^2$, the interpolating function f of the second part of Theorem 1.43 has the form $f(z) = uB(z)$ where u is a unimodular constant and B a Blaschke product [58, p. 300]. And in the general case, it has the form $\lambda B(z)$, where λ is a constant depending on the size of $w = (w_j)$.

1.7.5 Applications

1. An interesting and rather unexpected application of interpolating sequences can be found in the paper [79], where the authors give several counterexamples to a conjecture of Peller: if T is a power-bounded operator on a Hilbert space H, then for all $\varepsilon > 0$, T is similar to an operator S with $\sup_{n\in\mathbb{N}} \|S^n\| \leqslant 1 + \varepsilon$ (it is known that $\varepsilon = 0$ is impossible in general).
2. Generally, interpolating sequences have a wealth of applications to Operator Theory on a Hilbert space and to Functional Analysis. Many of them will appear in this book (see, e.g., [115]). Other examples can be found in [61] or [144].

1.8 Outer Functions

Here is an important class of analytic functions on the disk, the so-called *outer* functions. Instead of dropping their definition out of the blue, we first mention a theorem of Szegö.

Theorem 1.49 (Szegö) *Let $p > 0$ and h be a non-negative and p-integrable function on $\mathbb{T}$. The following are equivalent:*

(i) *There exists some $F \in H^p$ such that $|F^*| = h \quad m$-a.e.*
(ii) $\displaystyle\int_{\mathbb{T}} \log h \, dm > -\infty$.

When this condition is met, one can take

$$F(z) = \exp\left[\int_{\mathbb{T}} \frac{w+z}{w-z} \log h(w)\, dm(w)\right] \tag{1.43}$$

and one has

$$\log|F(0)| = \int_{\mathbb{T}} \log|F|\, dm = \int_{\mathbb{T}} \log h \, dm.$$

Observe that if (ii) holds, $\log h$ is automatically integrable since $\log^+(h) \leqslant h^p/p$. In any case, this motivates a closer study of the functions given by (1.43).

Definition 1.50 A function $F \in H^1$ is called *outer* if it satisfies one of the following equivalent definitions [74, p. 63].

(i) If $f \in H^1$ and $|f(u)| \leqslant |F(u)|$, a.e. on $\mathbb{T}$, then $|f(z)| \leqslant |F(z)|$, for all $z \in \mathbb{D}$.
(ii) $\log|F(0)| = \displaystyle\int_{\mathbb{T}} \log|F(u)|\, dm(u)$.
(iii) It has the form given by (1.43).

Another basic property of outer functions is the following [153, p. 350]:

Theorem 1.51 *A function $F \in H^2$ is* cyclic *for the forward shift S (i.e., FH^2 is dense in H^2) if and only if F is an outer function of H^2.*

Here are two other important classes of functions.

1.9 Inner Functions

Definition 1.52 An inner function is a holomorphic function f on $\mathbb{D}$ that is *bounded* and moreover satisfies

$$|f^*(u)| = 1 \text{ m-almost everywhere on } \mathbb{T},$$

where $f^*(u) = \lim_{r \to 1^-} f(ru)$.

We first show that Blaschke products are inner.

Theorem 1.53 *A Blaschke product B is an inner function.*

Proof: Consider the partial products $\pi_n(z) = \prod_{k=1}^{n} \frac{|z_k|}{z_k} \frac{z_k - z}{1 - \overline{z_k} z}$ associated with B. They are clearly inner (and in the disk algebra).

Point out that, for every $z \in \mathbb{D}$, we have $|B(z)| = \lim |\pi_n(z)| \leqslant 1$.

A simple computation gives for $u < v$ positive integers

$$\int_{\mathbb{T}} |\pi_v - \pi_u|^2 dm = 2\left[1 - \operatorname{Re} \int_{\mathbb{T}} \frac{\pi_v}{\pi_u} dm\right] = 2\left[1 - \operatorname{Re} \frac{\pi_v(0)}{\pi_u(0)}\right]$$
$$= 2\left[1 - \prod_{u<k\leqslant v} |z_k|\right].$$

This shows that π_n is a Cauchy sequence in H^2 of the circle, and hence in H^2 of the disk, with a H^2-limit necessarily equal to B, thanks to the continuity of the point evaluation at every point of $\mathbb{D}$. Since $\pi_n^* \to B^*$ in H^2, passing to a subsequence if necessary, we get $\pi_n^*(\xi) \to B^*(\xi)$ m-almost everywhere. But $|\pi_n^*(z)| = 1$ for every $n \geqslant 1$ and every $z \in \mathbb{D}$. This gives $|B^*(\xi)| = 1$ m-almost everywhere and $B \in H^\infty$, so that B is inner. □

Beurling [74, p. 63], completing Szegö's theorem, proved the following.

Theorem 1.54 (Beurling) *Any non-zero function $f \in H^1$ has an essentially (i.e., up to a unimodular constant) unique decomposition $f = IF$ where I is inner and F outer. We can always take*

$$F(z) = \exp\left[\int_{\mathbb{T}} \frac{w+z}{w-z} \log|f(w)|\, dm(w)\right].$$

Moreover, for any positive function h on the circle such that $\int_{\mathbb{T}} \log h\, dm > -\infty$, the function

$$F(z) = \exp\left[\int_{\mathbb{T}} \frac{w+z}{w-z} \log h(w)\, dm(w)\right]$$

is outer and satisfies $|F(u)| = h(u)$ a.e. on $\mathbb{T}$.

We will denote by $\mathcal{I}$ the set of all inner functions, so that $\mathcal{I} \subset B_{H^\infty}$, the unit ball of H^∞. As seems reasonable, inner functions are extreme points of B_{H^∞}. There are others [74, p. 138]. For further use, we mention the following theorem:

Theorem 1.55 (i) *A function* $f \in B_{H^\infty}$ *is an* extreme point *of* B_{H^∞} *if and only if*

$$\int_{\mathbb{T}} \log(1 - |f(u)|)\, dm(u) = -\infty.$$

(ii) *A function* $f \in B_{H^\infty}$ *is an* exposed point *of* B_{H^∞} *if and only if*

$$m(|f| = 1) > 0.$$

The second part is due to Amar and Lederer [2]. Recall that a point a of a convex subset K of a complex Banach space X is said to be exposed (i.e., super-extreme) if there exists a linear functional $L \in X^*$ such that

$$\mathrm{Re}(L(a)) > \mathrm{Re}(L(x)) \quad \text{for all } x \in K \setminus \{a\}.$$

1.9.1 Singular Inner Functions

Blaschke products have by definition a wealth of zeros. Here is an example of zero-free inner function.

• The "singular" function $\psi_\alpha(z) = \exp\left(-\alpha \dfrac{1+z}{1-z}\right)$, where $\alpha > 0$, satisfies

$$|\psi_\alpha(z)| = \exp\left(-\alpha \frac{1-|z|^2}{|1-z|^2}\right) \leqslant 1 \text{ and } \psi_\alpha(\mathrm{e}^{it}) = \exp\big(-i\alpha \cot t/2\big).$$

This function ψ_α can be written as $\psi_\alpha(z) = \exp\big(-\alpha\chi(z)\big)$, where $\chi(z) = \dfrac{1+z}{1-z}$ is the Cayley transform from $\mathbb{D}$ onto $\mathbb{C}_+$, the right half of the complex plane. We have

$$\|\psi_\alpha\|_\infty = \sup_{w \in \mathbb{C}_+} |\exp(-\alpha w)| = 1.$$

Actually, for every $\alpha > 0$, the function ψ_α can be analytically extended to $\mathbb{C} \setminus \{1\}$. Moreover for every $z \in \mathbb{T} \setminus \{1\}$, we have $|\psi_\alpha(z)| = 1$ so that $|\psi_\alpha^*| = 1$, m-a.e. on $\mathbb{T}$, and ψ_α is an inner function.

It should be observed that the function $f(z) = (\psi_\alpha(z))^{-1} = \exp\left(\alpha \frac{1+z}{1-z}\right)$ satisfies $|f^*| = 1$, m-a.e. on $\mathbb{T}$, but is *not* an inner function (it is not bounded on $\mathbb{D}$).

More generally, if μ is a finite, non-negative, singular measure on $\mathbb{T}$, the zero-free function

$$S(z) = \exp\left(-\int_{\mathbb{T}} \frac{w+z}{w-z}\, d\mu(w)\right)$$

is inner (the previous example corresponds to $\mu = a\delta_1$ with $a > 0$), as we shall see, and hence any product $I = BS$, with B a Blaschke product, and S a singular inner function, is inner. There are no others, as shown by the following theorem:

Theorem 1.56 (Beurling) *Any inner function I can be written in a "unique" way as*

$$I = BS, \quad \text{where } S(z) = \exp\left(-\int_{\mathbb{T}} \frac{w+z}{w-z}\, d\mu(w)\right),$$

where B is a Blaschke product and μ a positive, singular (w.r.t. m) measure on circle $\mathbb{T}$. The inner function S is called the singular part of I, and the factorization $I = BS$ the Beurling decomposition of I.

Proof: We will only give the ideas. Since $I \in H^\infty$, the Blaschke product B formed from the zeros of I is convergent (recall Proposition 1.6). The quotient $S = I/B$ is again bounded by 1 and zero-free (recall Proposition 1.42), hence can be written $S = \mathrm{e}^{-g}$ with $g = u + iv$ and $u > 0$ throughout $\mathbb{D}$. We can also assume $v(0) = 0$. By Herglotz's theorem, with P_z denoting the Poisson kernel at z:

$$u(z) = \int_{\mathbb{T}} P_z(w)\, d\mu(w),$$

where μ is a positive, finite measure on $\mathbb{T}$. So that

$$g(z) = \int_{\mathbb{T}} C_z(w)\, d\mu(w),$$

where $C_z(w) = \frac{w+z}{w-z}$ is the Cauchy kernel at z. Since S is again inner, the radial limit $\lim_{r\to 1} g(rv)$ is 0 for m-almost $v \in \mathbb{T}$. But $\lim_{r\to 1} u(rv) = D\mu(v)$, the derivative of μ at v, for m-almost $v \in \mathbb{T}$ [153, pp. 242–243]. Hence $D\mu(v) = 0$ for m almost such v, implying that μ is singular w.r.t. m, and the converse holds. □

The following characterization of Blaschke products among inner functions is useful.

Theorem 1.57 (Frostman 1) *Let f be an inner function. For $0 < r < 1$, we set*

$$J(r, f) = \int_{\mathbb{T}} \log|f(ru)|\, dm(u) \quad \text{and} \quad J(f) = \lim_{r\to 1^-} J(r, f).$$

Then, $J(f) \leqslant 0$ and $J(f) = 0$ if and only if f is a Blaschke product.

Proof: Observe first that $J(r, f)$ increases with r since $\log|f|$ is a subharmonic function. Hence, the limit $J(f) \leqslant 0$ does exist (remember that $|f(z)| \leqslant 1$). Next, consider the Beurling decomposition $f = BS$ with B a Blaschke product, and

$$S(z) = \exp\left[-\int_{\mathbb{T}} \frac{w+z}{w-z}\, d\mu(w)\right],$$

with μ a positive singular measure. Thus,

$$J(r, f) = J(r, B) + J(r, S), \quad J(f) = J(B) + J(S),$$

and we will prove that

$$J(B) = 0,\ J(S) = -\int_{\mathbb{T}} d\mu.$$

This will give the result.

The second relation is clear: since $\log|S(z)| = -\int_{\mathbb{T}} P_z(v) d\mu(v)$,

$$\begin{aligned} J(r, S) &= -\int_{\mathbb{T}}\int_{\mathbb{T}} P_{ru}(v)\, dm(u)\, d\mu(v) = -\int_{\mathbb{T}}\int_{\mathbb{T}} P_r(v\overline{u})\, dm(u)\, d\mu(v) \\ &= -\int_{\mathbb{T}} d\mu(v). \end{aligned}$$

The first relation goes as follows. We may assume that $B(0) \neq 0$ and denote the zeros of B by z_n where $n \geqslant 1$, arranged in increasing order. Fix an integer N. For $r > |z_N|$, Jensen's formula gives us

$$\begin{aligned} J(r, B) &\geqslant \sum_{n=1}^{N} \log\frac{r}{|z_n|} + \log|B(0)| = \sum_{n=1}^{N} \log\frac{r}{|z_n|} + \sum_{n=1}^{\infty} \log|z_n| \\ &= N\log r + \sum_{n=N+1}^{\infty} \log|z_n|. \end{aligned}$$

Hence, $J(B) \geqslant \sum_{n=N+1}^{\infty} \log|z_n|$ and finally $J(B) \geqslant 0$, by letting N tend to ∞. Since trivially $J(B) \leqslant 0$, we are done. □

We can now explain why Blaschke products play a central role among all inner functions, with the help of the following theorem:

Theorem 1.58 (Frostman 2) *For every inner function I, the Frostman shift*

$$\varphi_a \circ I = \frac{a - I}{1 - \overline{a}I}$$

is a Blaschke product for quasi-all (in the sense of logarithmic capacity), in particular for σ-almost all, points a of $\mathbb{D}$.

Proof: We consider, for $a \in \mathbb{D}$, the shifted function $g_a = \varphi_a \circ I$. This is again an inner function.

We first prove that g_a is a Blaschke product for σ-almost every $a \in \mathbb{D}$. To that effect, we will show that

$$\int_{\mathbb{D}} J(g_a)\,d\sigma(a) = 0. \tag{1.44}$$

Since clearly $J(g_a) \leqslant 0$ for each a, this will imply that

$$J(g_a) = 0 \quad \sigma\text{-almost everywhere,}$$

giving the result in view of Theorem 1.57.

We first consider $J(r, g_a)$, $0 < r < 1$. From (1.28), we have for any $\rho \in (0, 1)$

$$\int_{\mathbb{T}} \log \left| \frac{I(z) - \rho v}{1 - \overline{v}\rho I(z)} \right| dm(v) = \max(\log \rho, \log |I(z|) \quad \text{for all } z \in \mathbb{D}.$$

Integrating in polar coordinates ($a = \rho v$) then gives

$$\int_{\mathbb{D}} \log \left| \frac{I(z) - a}{1 - \overline{a}I(z)} \right| d\sigma(a) = \int_0^1 2\rho \max(\log \rho, \log |I(z)|)\,d\rho.$$

So that, by Fubini's theorem (taking $z = ru$),

$$\begin{aligned}\int_{\mathbb{D}} J(r, g_a)\,d\sigma(a) &= \int_{\mathbb{T}} \int_{\mathbb{D}} \log \left| \frac{I(ru) - a}{1 - \overline{a}I(ru)} \right| d\sigma(a)\,dm(u) \\ &= \int_{\mathbb{T}} \int_0^1 2\rho \max(\log \rho, \log |I(ru)|)\,d\rho\,dm(u). \end{aligned} \tag{1.45}$$

Now, by monotone convergence (recall that $J(r, g_a) \leqslant 0$), the LHS of (1.45) tends to $\int_{\mathbb{D}} J(g_a)\,d\sigma(a)$ as $r \to 1^-$. And the function in the RHS is greater than $2\rho \log \rho$ and non-positive, and moreover tends to 0 almost everywhere, since I is inner. We can thus apply the dominated convergence theorem to conclude that the RHS converges to 0 as $r \to 1^-$, proving (1.44) and the m-almost everywhere result.

Next, assume that the general conclusion fails. Then, there exist a compact set $K \subset \mathbb{D}$ and a positive non-zero measure μ supported in K such that $J(g_a) < 0$ for all $a \in K$ and such that the logarithmic potential

$$\lambda \longmapsto \int_K \log^+ \left(1/|\lambda - a|\right) d\mu(a)$$

is a continuous and bounded function on $\mathbb{C}$ [50, chapter 2]. It ensures that the opposite of the Green potential

$$G_\mu(\lambda) := \int_K \log \left(\left| \frac{\lambda - a}{1 - \overline{a}\lambda} \right| \right) d\mu(a)$$

is a continuous and bounded function: $\overline{\mathbb{D}} \to \mathbb{R}^-$, vanishing on $\partial\mathbb{D}$. Now, fix $r < 1$. Jensen's formula and Fubini's theorem again give

$$\int_K J(r, g_a)\, d\mu(a) = \int_{\mathbb{T}} G_\mu\big(I(ru)\big)\, dm(u). \tag{1.46}$$

Finally, pass to the limit in (1.46) as $r \to 1^-$. We are allowed to do so by the monotone convergence theorem for the LHS, and by the Lebesgue dominated convergence for the RHS, since G_μ is continuous and bounded on $\overline{\mathbb{D}}$. We get, since $|I(u)| = 1$ m-almost everywhere,

$$\int_K J(g_a)\, d\mu(a) = \int_{\mathbb{T}} G_\mu\big(I(u)\big)\, dm(u) = 0.$$

This is a contradiction. □

1.9.2 Examples and Applications of Inner Functions

Here are examples, borrowed from [130], with some additional details. Recall that $\mathcal{F}$ denotes the set of finite Blaschke products. We shall first prove here the following theorem:

Theorem 1.59 *Let $f(z) = \sum_{k=0}^{\infty} c_k z^k$ be an inner function, $n \in \mathbb{N}$, and let A_n be the infinite Hankel matrix with entries $a_{j,k}^{(n)}$, where*

$$a_{j,k}^{(n)} = |c_{n+j+k}|.$$

If $\|A_n\| < 1$, then $f \in \mathcal{F}$.

Proof: It is convenient to display the following lemma:

Lemma 1.60 *Let f be inner, with $f \notin \mathcal{F}$. Then there exists a sequence $(z_k)_{k\geqslant 1}$ of points of $\mathbb{D}$ such that*

$$|z_k| \to 1 \text{ and } f(z_k) \to 0.$$

Indeed, if such a sequence does not exist, we can find $0 < r_0 < 1$ and $\delta > 0$ such that

$$|z| \geqslant r_0 \Rightarrow |f(z)| \geqslant \delta. \tag{1.47}$$

Then, f only has finitely many zeros in $\mathbb{D}$; let $B \in \mathcal{F}$ be the corresponding Blaschke product. Owing to (1.47), both functions f/B and B/f belong to H^∞ and have modulus 1 a.e. on $\mathbb{T}$, since $f \in \mathcal{I}$. Using the maximum modulus principle, this leads to $|f| = |B|$ on $\mathbb{D}$, hence $f = cB$ with c a unimodular constant, so that $f \in \mathcal{F}$.

We also recall a classical lemma of Banach space theory.

Lemma 1.61 *Let X be a Banach space with dual space X^*, and S a closed subspace of X with orthogonal $S^\perp$. Then for any $x^* \in X^*$, it holds*

$$d(x^*, S^\perp) = \sup_{\substack{x\in S\\ \|x\|=1}} |x^*(x)|.$$

As a consequence, we have for each function $F \in L^\infty$

$$d(F, H^\infty) = \sup_{\substack{h\in H_0^1\\ \|h\|_1=1}} \left| \int_{\mathbb{T}} Fh\, dm \right|.$$

For the consequence, we take $X = L^1$, $S = H_0^1$ the set of functions in H^1 that vanish at the origin and $X^* = L^\infty$ with the duality

$$\langle F, h\rangle = \int_{\mathbb{T}} Fh\, dm.$$

Then, $S^\perp = H^\infty$.

Coming back to the proof of Theorem 1.59, suppose that $f \notin \mathcal{F}$. We claim that, for any integer $n \geqslant 0$ and any function $g \in H^\infty$, we have

$$\|z^n - f(z)g(z)\|_\infty \geqslant 1. \tag{1.48}$$

Indeed, if (z_k) is as in Lemma 1.60, we get

$$\|z^n - f(z)g(z)\|_\infty \geqslant |z_k|^n - |f(z_k)g(z_k)| \geqslant |z_k|^n - |f(z_k)|\, \|g\|_\infty,$$

which gives (1.48) by letting k tend to infinity. But since $f \in \mathcal{I}$, (1.48) becomes $d(z^n\overline{f}, H^\infty) \geqslant 1$, and Lemma 1.61 when applied to $F = z^n\overline{f}$ shows that, for any $n \geqslant 0$ and any $0 < \varepsilon < 1$, there exists $h(z) = \sum_{l=0}^\infty b_l z^l \in H^1$ satisfying (consider $zh \in H_0^1$)

$$\|h\|_1 = 1, \quad \text{and} \quad \left| \int_{\mathbb{T}} \xi^{n+1}\overline{f}(\xi)h(\xi)\, dm(\xi) \right| > 1 - \varepsilon;$$

equivalently

$$\|h\|_1 = 1, \quad \text{and} \quad \left| \sum_{l=0}^{\infty} \overline{c_{n+l+1}}\, b_l \right| > 1 - \varepsilon. \tag{1.49}$$

Now, by Hardy's theorem $H^1 = H^2 \times H^2$, we can write

$$b_l = \sum_{p+q=l} x_p y_q \text{ with } \sum_{k\geqslant 0} |x_k|^2 = \sum_{k\geqslant 0} |y_k|^2 = 1,$$

so that (1.49) becomes, with $x = (x_p)$ and $y = (y_q)$,

$$\|x\|_2 = \|y\|_2 = 1, \text{ and} \left| \sum_{p,q\geqslant 0} \overline{c_{n+p+q+1}}\, x_p y_q \right| > 1 - \varepsilon. \tag{1.50}$$

This implies, setting $u = (|x_p|)$ and $v = (|y_q|)$, that

$$\langle A_{n+1}(u), v\rangle = \sum_{p,q\geqslant 0} \left|\overline{c_{n+p+q+1}}\, x_p y_q\right| \geqslant \left|\sum_{p,q\geqslant 0} \overline{c_{n+p+q+1}}\, x_p y_q\right| > 1-\varepsilon;$$

hence $\|A_{n+1}\| > 1-\varepsilon$ and $\|A_{n+1}\| \geqslant 1$ for each integer $n \geqslant 0$, proving our theorem. □

As a nice corollary, we get the following theorem:

Theorem 1.62 *If $f(z) = \sum_{k=0}^{\infty} c_k z^k$ is an inner function satisfying the condition $\overline{\lim}_{n\to\infty} n|c_n| < 1/\pi$, in particular if $c_n = o(1/n)$, then $f \in \mathcal{F}$.*

Proof: Let $n \geqslant 1$ be large enough to ensure $(n+j+k)|c_{n+j+k}| \leqslant \alpha/\pi$, where $0 < \alpha < 1$. With obvious notations, it clearly holds that

$$|a_{j,k}| \leqslant b_{j,k} \Rightarrow \|A\| \leqslant \|B\|.$$

Hence, if $B = (b_{j,k})$ is the matrix with entries $b_{j,k} = \frac{\alpha}{\pi}\frac{1}{j+k+1} =: \frac{\alpha}{\pi} h_{j,k}$, we have $|a_{j,k}^{(n)}| \leqslant b_{j,k}$. It follows that $\|A_n\| \leqslant \|B\| \leqslant \alpha < 1$ by the properties of the Hilbert matrix $H = (h_{j,k})$, which notably satisfies $\|H\| = \pi$. An appeal to Theorem 1.59 ends the proof. □

A useful complement, well in the spirit of composition operators, is the following theorem:

Theorem 1.63 *If $f(z) = \sum_{k=0}^{\infty} a_k z^k$ is a non-constant* zero-free *inner function, then*

$$\sum_{k>n} |a_k|^2 \geqslant \delta n^{-1/2},$$

where δ is a positive constant depending only on a_0. In particular, one cannot have $a_n = o(n^{-3/4})$ and this is sharp.

Proof: The idea is the following:
• If $f(z) = \psi_a(z) = \sum_{k=0}^{\infty} c_k z^k$, we have an "explicit" expression of $c_n = c_n(a)$'s in terms of generalized Laguerre polynomials, giving the result.
• In the general case, f is strongly subordinate to some ψ_a, $a > 0$, namely $f = \psi_a \circ \varphi$ with a symbol φ such that $\varphi(0) = 0$. So by denoting $\psi_a(z) = \sum_{k=0}^{\infty} c_k z^k$, Littlewood's strong subordination principle (see Theorem 2.16) gives

$$\sum_{k=0}^{n} |a_k|^2 \leqslant \sum_{k=0}^{n} |c_k|^2.$$

Since f and ψ_a are inner, this amounts to

$$\sum_{k>n} |a_k|^2 \geqslant \sum_{k>n} |c_k|^2.$$

As we already know that $\sum_{k>n} |c_k|^2 \geqslant \delta n^{-1/2}$, we get the same with a_k. Now, suppose that $a_k = o(k^{-3/4})$. Then $\sum_{k>n} |a_k|^2 = o(n^{-1/2})$, and we get a contradiction. The details are as follows:

• Since $c_n = \mathrm{e}^{-a} L_n^{(-1)}(2a)$ where $L_n^{(-1)}$ is the Laguerre polynomial of parameter -1, the analysis of those polynomials [130, p. 253] implies that there are positive constants c, d depending only on a such that

$$c_k = c\, k^{-3/4} \cos(d\sqrt{k} + \pi/4) + O\big(k^{-5/4}\big). \tag{1.51}$$

Next, by Kuzmin–Landau's lemma or Euler–Maclaurin's summation formula [35, p. 23], we know that, for some constant C and every integer N,

$$\left| \sum_{k=1}^{N} \mathrm{e}^{i(2d\sqrt{k}+\pi/2)} \right| \leqslant C\sqrt{N}.$$

This and an Abel summation by parts imply

$$\sum_{k>n} k^{-3/2} \cos(2d\sqrt{k} + \pi/2) = O\left(\sum_{k>n} \sqrt{k} \times k^{-3/2-1}\right) = O(n^{-1}).$$

Then, using (1.51) and linearizing $\cos^2(d\sqrt{k} + \pi/4)$, we obtain

$$\sum_{k>n} |c_k|^2 \geqslant 2cn^{-1/2} + O(n^{-1}) \geqslant \delta n^{-1/2},$$

which is our first step.

• Let $\mathbb{C}_+$ be the right half-plane. Without loss of generality, we can assume that $f(0) > 0$ and we put $f(0) = \mathrm{e}^{-a}$ with $a > 0$. Since f is zero-free and $|f| < 1$ in $\mathbb{D}$, we can write $f = \mathrm{e}^{-ag}$ with $g : \mathbb{D} \to \mathbb{C}_+$ holomorphic and $g(0) = 1$. Let

$$\varphi = \frac{g-1}{g+1} : \mathbb{D} \to \mathbb{D}.$$

Then, φ is a symbol, $\varphi(0) = 0$, and $f = \psi_a \circ \varphi$ as claimed. □

A small detour via Blaschke products: we saw that an inner function $f \notin \mathcal{F}$ cannot have Taylor coefficients decreasing as $o(1/n)$, but $O(1/n)$ is allowed as shown by the following theorem:

Theorem 1.64 *There exists an infinite Blaschke product* $B(z) = \sum_{n=0}^{\infty} c_n z^n$ *satisfying* $c_n = O(1/n)$.

Proof: Let (z_n) be a *Newman interpolating sequence* of distinct points of $\mathbb{D}$, namely

$$1 - |z_{k+1}| \leqslant \alpha(1 - |z_k|) \quad \text{where } \alpha < 1.$$

Let $B(z) = \sum_{n=0}^{\infty} c_n z^n$ be the corresponding infinite Blaschke product. By the residue theorem and the unimodularity of B, it holds

$$\overline{c_n} = \frac{1}{2i\pi} \int_{\mathbb{T}} \frac{z^{n-1}}{B(z)}\, dz = \sum_{k=1}^{\infty} \frac{z_k^{n-1}}{B'(z_k)}.$$

Whence, using Proposition 1.44,

$$|c_n| \leqslant C \sum_{k=1}^{\infty} (1 - |z_k|)|z_k^{n-1}| \quad \text{with } C = 2\delta^{-1}.$$

But Newman's condition implies $(1 - |z_k|) \leqslant \dfrac{1}{1-\alpha}(|z_{k+1}| - |z_k|)$, so that with $C' = \frac{C}{1-\alpha}$,

$$\begin{aligned} |c_n| &\leqslant C' \sum_{k=1}^{\infty} (|z_{k+1}| - |z_k|)|z_k^{n-1}| \leqslant C' \sum_{k=1}^{\infty} \int_{|z_k|}^{|z_{k+1}|} t^{n-1} dt \\ &\leqslant C' \int_0^1 t^{n-1} dt = \frac{C'}{n}. \end{aligned}$$

This ends the proof. □

1.9.3 Finite Blaschke Products

Contrary to what one might think at first glance, finite Blaschke products (the set denoted by $\mathcal{F}$) are a very rich topic (see the nice book [57]) and have unexpected applications. We first give a simple characterization of $\mathcal{F}$, to be used later on.

Theorem 1.65 *Let $\varphi : \mathbb{D} \to \mathbb{D}$ be analytic. Then*

$$\inf_{z \in \mathbb{D}} \frac{1 - |\varphi(z)|}{1 - |z|} \geqslant \frac{1 - |\varphi(0)|}{1 + |\varphi(0)|} > 0.$$

Moreover, the following are equivalent:

(i) $\varphi \in \mathcal{F}$.
(ii) $\displaystyle\sup_{z \in \mathbb{D}} \frac{1 - |\varphi(z)|}{1 - |z|} < \infty$.
(iii) $\displaystyle\lim_{|z| \to 1} |\varphi(z)| = 1$.

Proof: The first inequality is a reminder of Lemma 1.28(iv).

Next, if $\varphi(z) = c \prod_{k=1}^{n} \frac{z - z_k}{1 - \overline{z_k} z} \in \mathcal{F}$, with $|c| = 1$, the obvious inequality

$$\left| \prod_{k=1}^{n} u_k - \prod_{k=1}^{n} v_k \right| \leqslant \sum_{k=1}^{n} |u_k - v_k|$$

for complex numbers in $\overline{\mathbb{D}}$ gives

$$1 - |\varphi(z)|^2 \leqslant \sum_{k=1}^{n} (1 - |b_{z_k}(z)|^2) = \sum_{k=1}^{n} \frac{(1 - |z|^2)(1 - |z_k|^2)}{|1 - \overline{z_k} z|^2},$$

and it follows that, for every $z \in \mathbb{D}$,

$$\frac{1 - |\varphi(z)|^2}{1 - |z|^2} \leqslant \sum_{k=1}^{n} \frac{1 + |z_k|}{1 - |z_k|} =: C.$$

Whence $\frac{1 - |\varphi(z)|}{1 - |z|} \leqslant 2C$.

The second condition obviously implies the third. Assume that the latter holds. Then, the condition of Lemma 1.60 fails, and $\varphi \in \mathcal{F}$. □

Remark 1.66 There are *infinite* Blaschke products B nevertheless satisfying

$$\lim_{r \to 1^-} \frac{1 - |B(r e^{i\theta})|}{1 - r} < \infty \qquad \text{for each } \theta.$$

We use the following (see [57, p. 55] or [165, p. 183]):

Theorem 1.67 (Frostman) *Let B be a Blaschke product with zeros denoted (z_n), and $\xi \in \mathbb{T}$. Then, B has an angular derivative (in particular a radial limit with modulus* 1*) at ξ if and only if*

$$\sum_{n=1}^{\infty} \frac{1 - |z_n|}{|\xi - z_n|^2} < \infty. \tag{1.52}$$

Now, take $z_n = (1 - \varepsilon_n)\, e^{i\theta_n}$, where the sequences (ε_n) and (θ_n) satisfy

$$0 < \varepsilon_n < 1/2,\ 0 < \theta_n < \pi/2, \quad \text{and} \quad \sum_{n=1}^{\infty} \frac{\varepsilon_n}{\theta_n^2} < \infty.$$

For example, $\theta_n = 1/n$, $\varepsilon_n = 1/(2n^4)$. We see that (1.52) is satisfied at $\xi = 1$, because

$$|1 - z_n| \geqslant |\mathrm{Im}\,(1 - z_n)| = (1 - \varepsilon_n) \sin\theta_n \geqslant 1/2 \times 2/\pi \times \theta_n = (1/\pi)\theta_n.$$

It is obviously satisfied at $\xi \in \mathbb{T}$, $\xi \neq 1$ since $z_n \to 1$. Thus, Frostman's condition in Theorem 1.67 is satisfied everywhere, and we are done.

In another direction, we can construct so-called *slow Blaschke products B* having the property that

$$\lim_{|z|\to 1} \frac{1-|B(z)|}{1-|z|} = \infty.$$

See Section 4.6 for details and applications to composition operators.

1.10 Change of Variable Formula

We will repeatedly be in need in this book of a change of variable formula for non-injective mappings, which is a special case of the so-called *coarea formula*, but will be sufficient for our purposes. A good reference is [54, pp. 244–245].

Theorem 1.68 *Let Ω and Ω' be two open subsets of $\mathbb{R}^n$ and $\varphi : \Omega \to \Omega'$ be a $\mathscr{C}^1$ map. Let $u : \Omega \to \mathbb{R}^+ \cup \{+\infty\}$ be a measurable map, with associated weighted counting function*

$$N_{\varphi,u}(y) = \sum_{\varphi(x)=y} u(x) \leqslant +\infty, \quad \textit{for } y \in \Omega'$$

(agreeing that $N_{\varphi,u}(y) = 0$ if $y \notin \varphi(\Omega)$). Then, for any $f \in L^1(\Omega', N_{\varphi,u}.dy)$, it holds that

$$\int_{\Omega'} f(y) N_{\varphi,u}(y)\, dy = \int_{\Omega} f(\varphi(x)) J_\varphi(x) u(x)\, dx,$$

where $J_\varphi(x)$ denotes the jacobian *of φ at x.*

Note that with $u = 1$ and φ injective, we recover the usual change of variable formula. When $\Omega = \mathbb{D}$ and $\varphi : \mathbb{D} \to \mathbb{D}$ is an analytic map, Theorem 1.68 takes the following form, also called the Stanton formula:

Theorem 1.69 (Stanton) *Let $\varphi : \mathbb{D} \to \mathbb{D}$ be analytic, and $u : \mathbb{D} \to \mathbb{R}^+ \cup \{+\infty\}$ a measurable function, with associated weighted counting function*

$$N_{\varphi,u}(w) = \sum_{\varphi(z)=w} u(z) \leqslant +\infty, \quad \textit{for } w \in \mathbb{D}.$$

Then, for any $f \in L^1(\mathbb{D}, N_{\varphi,u}.d\sigma)$, it holds that

$$\int_{\mathbb{D}} f(w) N_{\varphi,u}(w)\, d\sigma(w) = \int_{\mathbb{D}} f(\varphi(z))\, |\varphi'(z)|^2 u(z)\, d\sigma(z).$$

Proof: Indeed, one takes $n = 2$ and it suffices to note that $J_\varphi(z) = |\varphi'(z)|^2$. □

1.11 Banach Spaces of Analytic Functions

In the sequel, we denote, as usual, $C(K)$ the space of continuous functions on the compact set K, equipped with the sup norm.

1.11.1 Banach Subspaces of $\mathscr{H}(\mathbb{D})$

We recall that the topology on the Fréchet space $\mathscr{H}(\mathbb{D})$ is the topology of uniform convergence on compact subsets of $\mathbb{D}$.

We say that $X \subset \mathscr{H}(\mathbb{D})$ is a Banach space of analytic functions on $\mathbb{D}$ if it is continuously embedded in $\mathscr{H}(\mathbb{D})$, meaning that

$$\forall r \in (0,1), \exists M(r) > 0\,,\, \forall f \in X\,, \qquad \sup_{|z|\leqslant r} |f(z)| \leqslant M(r)\|f\|_X.$$

With this definition it is clear that point evaluations are bounded on X.

More generally, the norm of point evaluation of the derivatives is uniformly bounded on X when a stays in a compact set of $\mathbb{D}$. We have indeed, by Cauchy's formula, for every $r \in (0,1)$, $f \in X$ and $j \in \mathbb{N}_0$:

$$\sup_{|a|\leqslant r} \left|f^{(j)}(a)\right| \leqslant \frac{j!\,2^j}{(1-r)^j} M\left(\frac{1+r}{2}\right) \|f\|_X. \tag{1.53}$$

In particular, when H is a Hilbert space, writing as usual K_a the reproducing kernel on H at point a, we have, for every compact set $Q \subset \mathbb{D}$,

$$\sup_{a\in Q} \|K_a\|_H < \infty. \tag{1.54}$$

1.11.2 The Space H^∞

We have already seen in this book the Hardy space of bounded analytic functions on $\mathbb{D}$ defined by

$$H^\infty = \left\{f \in \mathscr{H}(\mathbb{D}) : \|f\|_\infty = \sup_{z\in\mathbb{D}} |f(z)| < \infty\right\}.$$

This space plays a key role in many respects. Of course it is a very natural and common Banach space of analytic functions. Moreover in this book, we are interested in composition operators associated to a symbol $\varphi : \mathbb{D} \to \mathbb{D}$, and the function φ is therefore analytic and bounded with $\|\varphi\|_\infty \leqslant 1$.

We already introduced inner functions (see 1.9): such a function I belongs to H^∞ with $|I^*| = 1$ a.e. on $\mathbb{T}$. Möbius functions and Blaschke products are therefore examples of functions in H^∞.

We also already mentioned (see Section 1.9.1) another class of examples of inner functions given by the family

$$\forall z \in \mathbb{D}, \quad \psi_\alpha(z) = \exp\left(-\alpha \frac{1+z}{1-z}\right), \text{ where } \alpha > 0. \tag{1.55}$$

Let us mention the following result whose proof uses this family of functions.

Proposition 1.70 *The space H^∞ is not separable.*

In particular, (analytic) polynomials are not dense in H^∞.

Proof: We use the family of functions defined by (1.55).

If there were a countable family $(v_n)_{n\geqslant 1}$, dense in H^∞, then, for every $\alpha > 0$, there would exist an integer n_α such that $\|v_{n_\alpha} - \psi_\alpha\|_\infty < 1$. The key fact is that for every $\alpha, \beta > 0$ with $\alpha \neq \beta$, we have

$$\begin{aligned} \|\psi_\alpha - \psi_\beta\|_\infty &= \sup_{w\in\mathbb{C}_+} \left|\exp(-\alpha w) - \exp(-\beta w)\right| \\ &= \sup_{w\in\overline{\mathbb{C}_+}} \left|\exp(-\alpha w) - \exp(-\beta w)\right| = 2. \end{aligned}$$

Therefore the map $\alpha \in (0, +\infty) \longmapsto n_\alpha \in \mathbb{N}$ is one to one, but this would imply that $(0, +\infty)$ is countable, which is false. □

Of course the pointwise evaluation $f \in H^\infty \longmapsto f(z)$ is bounded with norm 1 for any $z \in \mathbb{D}$. For the pointwise evaluation of the derivative, we have:

Lemma 1.71 *Given $z \in \mathbb{D}$, the functional $f \in H^\infty \mapsto f'(z)$ is bounded with norm $\dfrac{1}{1-|z|^2}$.*

Proof: We first justify that the norm of this functional is at most $\dfrac{1}{1-|z|^2}$.

Argument 1. Remembering the Schwarz–Pick inequality (see Lemma 1.28), we have the result from (iii). Indeed when $\|f\|_\infty < 1$, we have

$$|f'(z)| \leqslant \frac{1-|f(z)|^2}{1-|z|^2} \leqslant \frac{1}{1-|z|^2}.$$

Argument 2. It is clearly true for $z = 0$ since, for instance, for every $r \in (0,1)$

$$|f'(0)| = \left|\frac{1}{2i\pi}\int_{\partial r\mathbb{D}} \frac{f(w)}{w^2}\, dw\right| \leqslant \frac{1}{r}\|f\|_\infty.$$

Since r is arbitrary, we get $|f'(0)| \leqslant \|f\|_\infty$.

Applying this result to $F = f \circ \varphi_z$, we get the upper estimate. Indeed, recall that $\varphi_z(w) = \dfrac{z - w}{1 - \bar{z}w}$ for $w \in \mathbb{D}$, so $(1-|z|^2)|f'(z)| = |F'(0)| \leqslant \|F\|_\infty = \|f\|_\infty$.

The equality for the norm of the functional is seen testing φ_z since $\|\varphi_z\|_\infty = 1$ and $|\varphi_z'(z)| = \dfrac{1}{1 - |z|^2}$. □

1.11.3 The Disk Algebra

The disk algebra $A(\mathbb{D})$ is defined as $C\big(\overline{\mathbb{D}}\big) \cap \mathscr{H}(\mathbb{D}) \subset H^\infty$ and is equipped with the sup norm

$$\|f\|_\infty = \sup_{z \in \mathbb{D}} |f(z)|.$$

It is straightforward to check that it is a Banach space.

With the viewpoint of harmonic analysis, there is another description of $A(\mathbb{D})$. Indeed, using the Poisson transform, we can easily check that $A(\mathbb{D})$ is isometrically isomorphic to the space $\{f \in C(\mathbb{T});\ \forall n < 0,\ \widehat{f}(n) = 0\}$.

Let us give some details: actually we claim that the map

$$T: \quad \begin{array}{ccc} C\big(\overline{\mathbb{D}}\big) \cap \mathscr{H}(\mathbb{D}) & \longrightarrow & \big\{f \in C(\mathbb{T});\ \forall n < 0,\ \widehat{f}(n) = 0\big\} \\ f & \longmapsto & f_{|\mathbb{T}} \end{array}$$

is an isometric isomorphism.

It is well defined: for any $f \in C\big(\overline{\mathbb{D}}\big) \cap \mathscr{H}(\mathbb{D})$, we have for every $n < 0$

$$\widehat{f}(n) = \int_{\mathbb{T}} f(z)\bar{z}^n dm(z) = \frac{1}{2i\pi} \int_{\mathbb{T}} f(z)\bar{z}^{n+1} dz = \frac{1}{2i\pi} \int_{\mathbb{T}} f(z) z^{|n|-1}\, dz = 0,$$

since $z \in \overline{\mathbb{D}} \mapsto f(z)z^{|n|-1}$ is continuous, and it is analytic on $\mathbb{D}$. Clearly T is an isometry thanks to the maximum principle.

Now it suffices to justify that T is onto: given $g \in C(\mathbb{T})$ with $\widehat{g}(n) = 0$ for every $n < 0$, we consider the Poisson transform f of g:

$$f(r\xi) = P_r * g(\xi) \qquad \text{for } \xi \in \mathbb{T} \text{ and } r \in [0, 1),$$

and we define $f = g$ on $\mathbb{T}$.

It turns out that f is continuous on $\overline{\mathbb{D}}$. Such a function f, without other assumption than continuity of g, is harmonic on $\mathbb{D}$ since by definition

$$\forall z \in \mathbb{D}, \qquad f(z) = \int_{\mathbb{T}} \frac{1 - |z|^2}{|1 - \overline{w}z|^2} g(w)\, dm(w).$$

But we can rewrite it. For every $z \in \mathbb{D}$

$$f(z) = \sum_{n=0}^{+\infty} \widehat{g}(n) z^n,$$

so that f is analytic on $\mathbb{D}$. By definition $T(f) = g$.

Let us mention some of the usual properties of $A(\mathbb{D})$:

- It is easily seen that (analytic) polynomials are dense in $A(\mathbb{D})$:

$$A(\mathbb{D}) = \overline{span\{z^n \,;\, n \in \mathbb{N}_0\}}^{\|.\|_\infty}.$$

- $A(\mathbb{D})$ is therefore a separable space, but there is a stronger result: $A(\mathbb{D})$ has a Schauder basis. This is far from obvious and due to Bočkarev [10].
- We can interpolate any continuous function on a compact subset of $\mathbb{T}$ with m-measure zero by a function of $A(\mathbb{D})$ (see the Rudin–Carleson theorem below).

Let us prove this statement. We reproduce here an argument due to Danielyan [37]. First we need a result due to Fatou:

Lemma 1.72 (Fatou) *Let E be a closed subset of $\mathbb{T}$ with m-measure 0. Then there exists a function $\chi_E \in A(\mathbb{D})$ such that $\chi_E(z) = 1$ for every $z \in E$ and $|\chi_E(z)| < 1$ for every $z \in \mathbb{T} \setminus E$.*

Sketch of proof: We do not give all the details, but we can define χ_E with an explicit formula. First, since E has measure 0, there exists a function $\omega : \mathbb{T} \to [-\infty, -1]$ such that $\omega = -\infty$ on E only, and $\omega(u) \to -\infty$ when $u \in \mathbb{T}$ approaches E. The function ω is continuously differentiable on $\mathbb{T} \setminus E$ and $\omega \in L^1(\mathbb{T})$.

Then consider the function defined by

$$\forall z \in \mathbb{D}, \quad f_E(z) = \exp\left[\left(\int_{\mathbb{T}} \frac{1 + \bar{u}z}{1 - \bar{u}z}\, \omega(u)\, dm(u)\right)^{-1}\right].$$

Here one can extend continuously f_E on the closed unit disk to a function χ_E (belonging to the disk algebra), which fulfills the required properties. For a detailed proof, see [74, p. 80]. □

Theorem 1.73 (Rudin–Carleson Theorem) *Let K be a closed subset of $\mathbb{T}$ with m-measure zero and f be a continuous function on K. Then there exists some $F \in A(\mathbb{D})$ such that $F = f$ on K. Moreover, given $\delta > 0$, one can find such a function F with $\|F\|_\infty \leqslant (1 + \delta) \sup_{t \in K} |f(t)|$.*

Actually it is also possible (up to more work) to construct such a function satisfying $\|F\|_\infty = \sup_{t\in K} |f(t)|$.

Proof: Since f is uniformly continuous, given $\varepsilon \in (0, 1/2)$, we can cover K by a finite number of disjoint open intervals $I_1, \dots, I_n$ of the torus satisfying

$$\forall k \in \{1, \dots, n\}, \forall z, z' \in K \cap I_k, \quad \left|f(z) - f(z')\right| < \varepsilon.$$

Actually, since these intervals are open and K is closed, the endpoints cannot belong to K. In particular, $E_k = K \cap I_k$ is closed for every $k \in \{1, \dots, n\}$ and we can associate a function $\chi_{E_k} \in A(\mathbb{D})$ by Lemma 1.72.

Since for every $k \in \{1, \dots, n\}$, the set $\mathbb{T} \setminus I_k$ is compact, we can find an integer $N \geqslant 1$ such that $|\chi_{E_k}|^N < \dfrac{\varepsilon}{n}$ on $\mathbb{T} \setminus I_k$. Choosing any $t_k \in E_k$, we can define the function

$$h = \sum_{k=1}^{n} f(t_k)\chi_{E_k}^N,$$

which belongs to $A(\mathbb{D})$ and satisfies $\|h\|_\infty \leqslant (1+\varepsilon)\sup_{t\in K}|f(t)|$. Moreover, for every $z \in K$, $|h(z) - f(z)| < \varepsilon\big(1 + \sup_{t\in K}|f(t)|\big)$.

Introducing the restriction operator $T : g \in A(\mathbb{D}) \mapsto g_{|K} \in C(K)$, the previous result actually expresses that for every $f \in C(K)$ with $\|f\|_\infty < 1$, there exists some function $g \in A(\mathbb{D})$ with norm less than $1 + \varepsilon$, such that $\|T(g) - f\| \leqslant 2\varepsilon$. Mimicking the proof of the open mapping theorem, we get that T is onto and there exists some $F \in A(\mathbb{D})$ such that $T(F) = f$ and $\|F\|_\infty \leqslant \dfrac{1+\varepsilon}{1-2\varepsilon}$. Finally, choose ε small enough to satisfy $\dfrac{1+\varepsilon}{1-2\varepsilon} \leqslant 1 + \delta$. □

1.11.4 Hardy Spaces

We focus on Hardy spaces H^p for finite p. Let us quickly recall how classical Hardy spaces are defined (see [45] for more details).

When $p > 0$:

$$H^p = \left\{ f \in \mathscr{H}(\mathbb{D}) : \sup_{r<1} \int_{\mathbb{T}} |f(rz)|^p \, dm(z) < \infty \right\},$$

and the associated (quasi-)norm is defined by

$$\|f\|_{H^p} = \|f\|_p = \sup_{r<1} \left(\int_{\mathbb{T}} |f(rz)|^p \, dm(z) \right)^{1/p} = \sup_{r<1} \|f_r\|_{L^p(\mathbb{T})}.$$

Here we recall that m stands for the (normalized) Lebesgue measure on the torus, and here $f_r(z) = f(rz)$ with $r \in (0, 1)$ and $z \in \overline{\mathbb{D}}$.

In the particular case $p = 2$ (the Hilbertian case), thanks to the Parseval formula, we can express the norm with the Taylor coefficients of the function. Indeed, let $f(z) = \sum_{n=0}^{+\infty} a_n z^n$ be analytic on $\mathbb{D}$:

$$\|f\|_{H^2} = \|f\|_2 = \sup_{r<1} \left(\sum_{n=0}^{+\infty} |a_n|^2 r^{2n} \right)^{1/2} = \left(\sum_{n=0}^{+\infty} |a_n|^2 \right)^{1/2}.$$

We are mainly interested in the case $p \geqslant 1$, and it is easy to check that, in this case, H^p spaces are Banach spaces.

Moreover, when $p = 2$, the Hardy space H^2 is a Hilbert space associated to the scalar product

$$\langle f, g \rangle_{H^2} = \int_{\mathbb{T}} f \overline{g} \, dm = \sum_{n=0}^{+\infty} a_n \overline{b_n}$$

when $f(z) = \sum_{n=0}^{+\infty} a_n z^n$ and $g(z) = \sum_{n=0}^{+\infty} b_n z^n$.

Let us mention some (very useful) properties of analytic functions belonging to the Hardy spaces.

Theorem 1.74 (Fatou-Nevanlinna) *Let $p \geqslant 1$ and $f \in H^p$.*

(i) *The function f admits almost everywhere radial limits on the torus, that is, $f^*(\xi) = \lim_{r \to 1^-} f(r\xi)$ is defined for almost every ξ of the torus.*
(ii) *The function f^* belongs to $L^p(\mathbb{T})$ and $\|f\|_{H^p} = \|f^*\|_{L^p(\mathbb{T})}$.*
Actually f^* belongs to the space $\{g \in L^p(\mathbb{T}) \mid \hat{g}(n) = 0 \text{ for every } n < 0\}$.
(iii) *If $f^* = 0$ on a subset of $\mathbb{T}$ with positive measure, then $f = 0$.*

Here $\hat{g}(n)$ stands for the usual nth Fourier coefficient of g.

For a proof see [153, 17.11].

Conversely, if we consider any $g \in L^p(\mathbb{T})$, with $\hat{g}(n) = 0$ for every $n < 0$, the Poisson integral of g at point $z = r\mathrm{e}^{2i\pi\theta}$,

$$P[g](z) = \int_0^1 P_r\left(\mathrm{e}^{2i\pi(\theta - t)}\right) g\left(\mathrm{e}^{2i\pi t}\right) dt,$$

belongs to H^p. Moreover $(P[g])^* = g$.

Therefore, let us emphasize that H^p is naturally identified with the (closed) subspace of $L^p(\mathbb{T})$:

$$\{g \in L^p(\mathbb{T}) \mid \hat{g}(n) = 0 \text{ for every } n < 0\}.$$

These remarks allow to adopt a double viewpoint on Hardy spaces: we can have the viewpoint of functions on the open unit disk, and the viewpoint of functions on the unit circle. Gathering all this, we can consider that $f \in H^p$ is defined not only on $\mathbb{D}$ but on the whole $\overline{\mathbb{D}} = \mathbb{D} \cup \mathbb{T}$ (even if it means defining f as 0 on a subset of $\mathbb{T}$ with measure zero). In the sequel, we shall sometimes omit the star, that is, identifying f and f^*, but only when this identification does not induce any ambiguity.

We shall need in the sequel the following weak form of the Riesz factorization theorem (see [45, theorem 2.8] or [153, 17.17] for the full statement).

This result will be crucial to bypass issues with taking powers of an arbitrary holomorphic function. With the following factorization, we shall be able to legally consider powers of our functions in H^p up to an inner factor.

Proposition 1.75 *Let $p \geqslant 1$ and $f \in H^p$ with $f \neq 0$.*

(i) *The function f can be written as $f = Bg$, where B is an inner function (even a Blaschke product) and g does not vanish on $\mathbb{D}$.*
(ii) *The function g belongs to H^p with $\|f\|_{H^p} = \|g\|_{H^p}$.*
(iii) *For every $q \geqslant 1$, there exists $G \in H^q$ such that $|G^*| = |f^*|^{\frac{p}{q}}$ a.e.*
(iv) *The function f can be written as $f = g_1.g_2$ where g_1, g_2 in H^{2p} with $\|g_1\|_{H^{2p}} = \|g_2\|_{H^{2p}} = \|f\|_{H^p}^{\frac{1}{2}}$.*
(v) *Any (non-zero) $f \in H^p$ can be written as the sum of two functions g_1, $g_2 \in H^p$, not vanishing on $\mathbb{D}$, and satisfying $\|g_k\|_{H^p} \leqslant k\|f\|_{H^p}$ for $k \in \{1;2\}$.*

Proof: (i) Since $f \neq 0$ and $f \in H^p$, the zero set $Z(f)$ of f is countable and so can be described as a sequence $(a_n)_{n\in\mathbb{N}}$ (each zero is counted with multiplicity) satisfying the Blaschke condition $\sum(1 - |a_n|) < +\infty$ (recall Proposition 1.6). Then we can consider the associated Blaschke product B. The function $g = f/B$ is then well defined, does not vanish, and is holomorphic on $\mathbb{D} \setminus Z(f)$. Actually g has a non-zero limit at each a_{n_0} by definition (taking into account the multiplicity), and hence can be extended analytically in a_{n_0} with a non-zero value. This defines a non-vanishing function g on $\mathbb{D}$, which is holomorphic and satisfies $f = Bg$ on $\mathbb{D}$.

(ii) Point out that $f^* = B^*.g^*$ a.e. on $\mathbb{T}$, hence $|f^*| = |g^*|$ a.e. on $\mathbb{T}$, since B is inner. Nevertheless, this remark is not sufficient to conclude directly.

Let us consider the partial Blaschke product B_n (of order $n \geqslant 1$) of B. For every $n \geqslant 1$, by definition of B with zeros (with multiplicity) of f, there exists $f_n \in \mathscr{H}(\mathbb{D})$ such that, on $\mathbb{D}$, $f = f_n.B_n$. For $\rho \in (0,1)$ close enough to 1 (so that $\mathbb{D} \setminus \rho\mathbb{D}$ does not contain any zero of B_n), we have

$$\int_{\mathbb{T}} |f_n(\rho\xi)|^p dm(\xi) = \int_{\mathbb{T}} \frac{|f(\rho\xi)|^p}{|B_n(\rho\xi)|^p} dm(\xi) \leqslant \frac{1}{\min_{\rho\mathbb{T}} |B_n|^p} \int_{\mathbb{T}} |f(\rho\xi)|^p dm(\xi)$$
$$\leqslant \frac{\|f\|_{H^p}^p}{\min_{\rho\mathbb{T}} |B_n|^p}.$$

But for every fixed $r \in (0,1)$, we have

$$\int_{\mathbb{T}} \frac{|f(r\xi)|^p}{|B_n(r\xi)|^p}\, dm(\xi) = \int_{\mathbb{T}} |f_n(r\xi)|^p\, dm(\xi) \leqslant \int_{\mathbb{T}} |f_n(\rho\xi)|^p\, dm(\xi)$$

for every $\rho > r$, in particular when $\rho \in (0,1)$ is close enough to 1 as above. We get

$$\int_{\mathbb{T}} \frac{|f(r\xi)|^p}{|B_n(r\xi)|^p}\, dm(\xi) \leqslant \lim_{\rho\to 1^-} \frac{\|f\|_{H^p}^p}{\min_{\rho\mathbb{T}} |B_n|^p} = \|f\|_{H^p}^p.$$

Hence, by Fatou,

$$\begin{aligned}\int_{\mathbb{T}} |g(r\xi)|^p\, dm(\xi) = \int_{\mathbb{T}} \frac{|f(r\xi)|^p}{|B(r\xi)|^p}\, dm(\xi) &= \int_{\mathbb{T}} \underline{\lim} \frac{|f(r\xi)|^p}{|B_n(r\xi)|^p}\, dm(\xi)\\ &\leqslant \underline{\lim} \int_{\mathbb{T}} \frac{|f(r\xi)|^p}{|B_n(r\xi)|^p}\, dm(\xi) \leqslant \|f\|_{H^p}^p;\end{aligned}$$

hence

$$\|g\|_{H^p} \leqslant \|f\|_{H^p}.$$

On the other hand, for every $r \in (0,1)$, we clearly have

$$\int_{\mathbb{T}} |f(r\xi)|^p\, dm(\xi) \leqslant \int_{\mathbb{T}} |g(r\xi)|^p\, dm(\xi),$$

since $|B(r\xi)| \leqslant 1$ for every $\xi \in \mathbb{T}$.

We conclude that $\|f\|_{H^p} = \|g\|_{H^p}$.

Let us mention too that the proof above actually shows also that, for every $n \geqslant 1$, we have $\|f_n\|_{H^p} = \|f\|_{H^p}$.

(iii) We use the previous decomposition obtained in (i) and define $G = g^{\frac{p}{q}}$, which is well defined since g does not vanish on the simply connected domain $\mathbb{D}$. Then

$$\|G\|_{H^q}^q = \|g\|_{H^p}^p = \|f\|_{H^p}^p < \infty.$$

Therefore, $G \in H^q$ and $|G^*| = \left|\left(g^{\frac{p}{q}}\right)^*\right| = |g^*|^{\frac{p}{q}}$ a.e., so $|G^*| = |f^*|^{\frac{p}{q}}$ a.e.

(iv) From (i) and (ii), we know that the function f can be written as $f = Bg$ where g does not vanish on $\mathbb{D}$ and $\|g\|_{H^p} = \|f\|_{H^p}$. Now we can define the analytic function $g_1 = g^{\frac{1}{2}}$ in a natural way. It satisfies $\|g_1\|_{H^{2p}}^2 = \|g\|_{H^p} = \|f\|_{H^p}$. Let $g_2 = Bg_1$. We also have $\|g_2\|_{H^{2p}} = \|f\|_{H^p}^{\frac{1}{2}}$. Finally $g_1.g_2 = Bg = f$.

(v) This is obvious when f does not vanish on $\mathbb{D}$. Else the decomposition obtained in (i) provides B and g with a non-trivial B, that is, B is not equal to 1. Actually since B is inner and non-constant, it is never equal to 1 on $\mathbb{D}$. Write $g_1 = g$ and $g_2 = f - g = (B-1)g$. Then the conclusion follows. □

Point Evaluation on Hardy Spaces

The pointwise evaluation at $z \in \mathbb{D}$ on the Hardy space H^p

$$\delta_z : \begin{array}{ccc} H^p & \longrightarrow & \mathbb{C} \\ f & \longmapsto & f(z) \end{array}$$

is an important tool (as for any space of analytic functions), and we know the exact value of its norm:

Proposition 1.76 *We have*

(i) *On the Hilbert space H^2, the functional δ_z is associated to the reproducing kernel $w \in \overline{\mathbb{D}} \longmapsto K_z(w) = \dfrac{1}{1-\bar{z}w}$.*

(ii) *For every $p \geqslant 1$, we have $\|\delta_z\|_{(H^p)^*} = \left(\dfrac{1}{1-|z|^2}\right)^{1/p}$.*

Proof: We propose two proofs.

First proof. We have the Taylor expansion

$$\delta_z(f) = f(z) = \sum_{n \geqslant 0} a_n z^n = \langle f, K_z \rangle_{H^2},$$

where $K_z(w) = \dfrac{1}{1-\bar{z}w} = \sum_{n\geqslant 0} \bar{z}^n w^n$. In particular,

$$\|\delta_z\|^2_{(H^2)^*} = \|K_z\|^2_{H^2} = \delta_z(K_z) = \frac{1}{1-|z|^2}.$$

Now, thanks to Proposition 1.75(iii), we may write $f \in H^p$ as $f = Bg = BG^{2/p}$ on $\mathbb{D}$, with $G \in H^2$ and B inner. Thus,

$$|\delta_z(f)|^p = |B(z)|^p |G(z)|^2 \leqslant |G(z)|^2 \leqslant \|\delta_z\|^2_{(H^2)^*} \|G\|^2_{H^2} = \frac{1}{1-|z|^2} \|f\|^p_{H^p}.$$

So

$$\|\delta_z\|_{(H^p)^*} \leqslant \left(\frac{1}{1-|z|^2}\right)^{1/p}.$$

Testing with $F = K_z^{2/p}$, we get the reverse inequality. Indeed we point out that

$$\|F\|_p = \|K_z\|_2^{\frac{2}{p}} = \frac{1}{(1-|z|^2)^{\frac{1}{p}}}.$$

By definition

$$\|\delta_z\|_{(H^p)^*} \|F\|_p \geqslant F(z) = \frac{1}{(1-|z|^2)^{\frac{2}{p}}}.$$

Rearranging we get the conclusion.

Another proof for the upper estimate. This second proof gives a first flavor of composition operator techniques.

Since $f(0) = \dfrac{1}{2i\pi}\displaystyle\int_{\mathbb{T}} \frac{f(w)}{w}\,dw$, we have $|f(0)| \leqslant \|f\|_1 \leqslant \|f\|_p$.

Now given $f \in H^p$ and $z \in \mathbb{D}$, we consider $g(w) = \dfrac{(1-|z|^2)^{\frac{1}{p}}}{(1-\bar{z}w)^{\frac{2}{p}}} f \circ \varphi_z(w)$ where φ_z is the Möbius function

$$\varphi_z(w) = \frac{z-w}{1-\bar{z}w}\cdot$$

We have $g \in H^p$ and

$$\|g\|_p^p = \int_{\mathbb{T}} |f \circ \varphi_z|^p \, |\varphi_z'| \, dm = \|f\|_p^p$$

after a change of variable.

We already know that $|g(0)| \leqslant \|g\|_p = \|f\|_p$. But $g(0) = (1-|z|^2)^{\frac{1}{p}} f(z)$. Rearranging we get the upper estimate. □

The following result gives some useful estimates for the derivative.

Proposition 1.77 *For every $p \geqslant 1$ and $z \in \mathbb{D}$, the pointwise evaluation $\delta_z^{(1)}$ of the derivative*

$$\begin{array}{ccc} H^p & \longrightarrow & \mathbb{C} \\ f & \longmapsto & f'(z) \end{array}$$

is bounded on H^p with norm satisfying

$$\frac{2}{p} \cdot \frac{|z|}{(1-|z|^2)^{1+\frac{1}{p}}} \leqslant \left\|\delta_z^{(1)}\right\|_{(H^p)^*} \leqslant \frac{(1+|z|)^{\frac{2}{p}}}{(1-|z|^2)^{1+\frac{1}{p}}}\cdot$$

In particular, when z approaches the boundary of $\mathbb{D}$, that is, $|z| \to 1^-$, we have $\left\|\delta_z^{(1)}\right\|_{(H^p)^*} \approx \dfrac{1}{(1-|z|^2)^{1+\frac{1}{p}}} \longrightarrow +\infty\cdot$

On the other hand, it is clear that the lower estimate is not sharp at all when z is close to the origin. For instance, it is very easy to see that $\left\|\delta_0^{(1)}\right\|_{(H^p)^*} = 1$ and

$$\left\|\delta_z^{(1)}\right\|_{(H^2)^*} = \frac{(1+|z|^2)^{\frac{1}{2}}}{(1-|z|^2)^{\frac{3}{2}}}\,\cdot \tag{1.56}$$

From this we also could deduce sharper estimates for any value of p, but we shall not need this in the following.

Proof: Thanks to the Cauchy formula, we can write

$$f'(z) = \frac{1}{2i\pi} \int_{\mathbb{T}} \frac{f(w)}{(z-w)^2}\, dw.$$

Then taking the modulus and using Hölder inequality, we get

$$|f'(z)| \leqslant \|f\|_p \left(\int_{\mathbb{T}} \frac{1}{|z-w|^{2p'}}\, dm(w) \right)^{\frac{1}{p'}}$$

(of course the integral is replaced by the sup norm for $p = 1$, that is, $p' = \infty$).

For $p = 1$, $\sup\limits_{|w|=1} \dfrac{1}{|z-w|^2} = \dfrac{1}{(1-|z|)^2}$, which gives the right upper estimate.

For $p > 1$, we have

$$\left(\int_{\mathbb{T}} \frac{1}{|(z-w)^2|^{p'}}\, dm(w) \right)^{\frac{1}{p'}} \leqslant \left(\int_{\mathbb{T}} \frac{1}{|(z-w)^2|}\, dm(w) \right)^{\frac{1}{p'}} \left(\sup_{|w|=1} \frac{1}{|z-w|^2} \right)^{\frac{1}{p}},$$

but, thanks to Parseval identity, we have

$$\int_{\mathbb{T}} \frac{1}{|(z-w)^2|}\, dm(w) = \left\| \frac{1}{(1-\overline{z}\cdot)} \right\|_{H^2}^2 = \frac{1}{1-|z|^2}.$$

Gathering everything, we conclude again for the upper estimate, and rearranging we get

$$|f'(z)| \leqslant \frac{(1+|z|)^{\frac{2}{p}}}{(1-|z|^2)^{1+\frac{1}{p}}} \|f\|_p .$$

For the lower estimate, it suffices to test again the function $K_z^{2/p}$. □

Duality

H^p spaces are dual spaces: for $p \in (1, \infty)$, H^p is the dual of $H^{p'}$ and the duality is described by the duality bracket

$$\langle f, g \rangle = \int_{\mathbb{T}} f(z) g(\overline{z})\, dm(z),$$

where $f \in H^p$ and $g \in H^{p'}$.

This is a classical result: we refer to [45] or [153] for details.

For $p = 1$, H^1 is also a dual space. This can be seen quickly with the help of harmonic analysis with the description of H^1 as the space of functions in $L^1(\mathbb{T})$ whose spectrum is included in the set of non-negative integers. Let us introduce the following notation, where $X \subset M(\mathbb{T})$ and $\Lambda \subset \mathbb{Z}$:

$$X_\Lambda = \left\{ f \in X \mid \forall n \notin \Lambda,\ \widehat{f}(n) = 0 \right\}.$$

Then, the classical Riesz brothers theorem states that

$$H^1(\mathbb{T}) = M_{\mathbb{N}}(\mathbb{T}) = \{\mu \in M(\mathbb{T}) \mid \forall n < 0,\ \widehat{\mu}(n) = 0\},$$

so that $H^1(\mathbb{T})$ appears as the dual of the quotient space $C(\mathbb{T})/C_{\mathbb{Z}^-}(\mathbb{T})$ for the duality described above. The space H^∞ can also be described as the dual of the quotient space $L^1(\mathbb{T})/H^1_0$.

Let us emphasize the following proposition.

Proposition 1.78 *Let $p \geqslant 1$. On the unit ball of H^p, the weak-star topology coincides with the topology of uniform convergence on compact subsets of $\mathbb{D}$.*

Of course, by reflexivity, when $p > 1$, the weak-star topology is actually the weak topology on H^p.

Proof: First let us point out that the weak-star topology on the unit ball is metrizable since H^p is the dual of a separable space. Therefore, we shall work with sequences.

Assume that the sequence (f_n), in the unit ball of H^p, is weak star convergent to 0. Fix a compact subset K of $\mathbb{D}$. There exists $\rho \in (0,1)$ such that every $z \in K$ satisfies $|z| \leqslant \rho$. We claim that $\sup\{|f_n(z)|\,;\ z \in K\} \longrightarrow 0$. If it were not true, up to an extraction we would find $\delta > 0$ and (z_n) in K such that $|f_n(z_n)| \geqslant \delta$. By compactness of K, and up to another extraction $z_n \longrightarrow c \in K$.

Using Proposition 1.77, we have

$$\big|f_n(z_n) - f_n(c)\big| \leqslant \big|z_n - c\big| \sup_{|w|\leqslant\rho} |f_n'(w)| \leqslant 4\frac{\big|z_n - c\big|}{(1-\rho^2)^2}, \tag{1.57}$$

since $\|f_n\|_p \leqslant 1$.

The functional δ_c is continuous on H^p (see 1.76) so $f_n(c) \longrightarrow 0$ by weak convergence when $p > 1$. For $p = 1$, the functional $\delta_c \in (H^1)^*$ is actually inherited from the continuous function $z \mapsto \frac{1}{1-\bar{c}z}$, which lies in the predual of $H^1(\mathbb{T})$, so that $f_n(c) \longrightarrow 0$ as well by weak-star convergence. Finally passing to the limit in (1.57), we get a contradiction.

Now assume that the sequence (f_n), in the unit ball of H^p, is convergent to 0 uniformly on compact subsets of $\mathbb{D}$.

Choose any ξ in the predual of H^p, which can be represented by some g, which lies in $H^{p'}$ when $p > 1$, and lies in $C(\mathbb{T})$ when $p = 1$. For every $r \in (0,1)$, using $g_r = P_r * g$, we have

$$\overline{\lim}|\langle f_n, g\rangle| \leqslant \overline{\lim}|\langle f_n, g_r\rangle| + \|g - g_r\|_{p'} . \sup \|f_n\|_p.$$

We point out that $\langle f_n, g_r\rangle = f_n * (P_r * g) = (f_n * P_r) * g$; therefore, we have $|\langle f_n, g_r\rangle| \leqslant \|f_n * P_r\|_p \|g\|_{p'}$. But $\|f_n * P_r\|_p \leqslant \|f_n * P_r\|_\infty = \sup\{|f_n(z)|;\ |z| \leqslant r\} \to 0$ by hypothesis, so

$$\overline{\lim}|\langle f_n, g\rangle| \leqslant \|g - g_r\|_{p'}. \sup \|f_n\|_p .$$

We know that $g_r \to g$ as $r \to 1^-$. Therefore, $\overline{\lim}|\langle f_n, g\rangle| = 0$. □

In the Hilbertian case, we already mentioned that the H^2 norm can be expressed in a very simple way with the Taylor coefficients only. Let us mention now another expression of the H^2 norm with help of the derivative. This result is extremely useful in the theory of composition operators.

It can be viewed as a Green formula in H^2, but we give in this Hilbertian case an elementary proof. A similar result holds in H^p (see Theorem 1.80).

Theorem 1.79 (Littlewood–Paley formula) *Let $f \in H^2$. We have*

$$\|f\|_{H^2}^2 = |f(0)|^2 + 2\int_{\mathbb{D}} |f'(w)|^2 \log\frac{1}{|w|}\, d\sigma(w)\,.$$

In particular,

$$\|f\|_{H^2}^2 - |f(0)|^2 = \|f - f(0)\|_{H^2}^2 \approx \int_{\mathbb{D}} |f'(w)|^2 (1 - |w|)\, d\sigma(w).$$

Proof: Let us start from the integral in the right-hand side term. Write

$$f(z) = \sum_{n=0}^{\infty} a_n z^n \quad \text{for every } z \in \mathbb{D}.$$

We rewrite the integral in polar coordinates as

$$\int_{\mathbb{D}} |f'(w)|^2 \log\frac{1}{|w|}\, d\sigma(w) = \int_0^1 \left(\int_{\mathbb{T}} |f'(rz)|^2\, dm(z)\right)\left(\log\frac{1}{r}\right) 2r dr,$$

and we compute

$$\int_{\mathbb{T}} |f'(rz)|^2\, dm(z) = \sum_{n=0}^{\infty} (n+1)^2 |a_{n+1}|^2 r^{2n}$$

by a simple application of the Parseval formula.

Now, via Fubini, we get

$$2\int_{\mathbb{D}} |f'(w)|^2 \log\frac{1}{|w|}\, d\sigma(w) = \sum_{n=0}^{\infty} 4(n+1)^2 |a_{n+1}|^2 \int_0^1 r^{2n+1} \log\frac{1}{r}\, dr,$$

but via the change of variable $x = r^{2(n+1)}$,

$$\int_0^1 4(n+1)^2 r^{2n+1} \log\frac{1}{r}\,dr = \int_0^1 \log\frac{1}{x}\,dx = 1,$$

so we get

$$2\int_{\mathbb{D}} |f'(w)|^2 \log\frac{1}{|w|}\,d\sigma(w) = \sum_{n=0}^{\infty} |a_{n+1}|^2 = \sum_{n=1}^{\infty} |a_n|^2 = \|f\|_2^2 - |f(0)|^2.$$

Concerning the last (approximation) statement, the first equality is obvious, and the last estimate follows from the fact that

$$1 - |w| \leqslant \log\frac{1}{|w|} \leqslant \frac{1}{|w|} - 1 = \frac{1-|w|}{|w|} \tag{1.58}$$

for every $w \in \mathbb{D}\setminus\{0\}$. Then we easily have a lower estimate, with the help of the right-hand side of (1.58):

$$\int_{\mathbb{D}} |f'(w)|^2 (1-|w|)\,d\sigma(w) \leqslant \int_{\mathbb{D}} |f'(w)|^2 \log\frac{1}{|w|}\,d\sigma(w).$$

For the upper estimate, fix $\rho \in (0, 1/2)$. We have on one hand

$$\int_{\{|w|\geqslant\rho\}} |f'(w)|^2 \log\frac{1}{|w|}\,d\sigma(w) \leqslant \frac{1}{\rho}\int_{\mathbb{D}} |f'(w)|^2 (1-|w|)\,d\sigma(w).$$

On the other hand, from (1.56), we know that, for every w in the disk centered at 0 with radius $\rho \leqslant \frac{1}{2}$,

$$|f'(w)| \leqslant \frac{4\sqrt{5}}{3\sqrt{3}}\|f - f(0)\|_2 \leqslant \sqrt{3}\|f - f(0)\|_2.$$

So

$$\int_{\{|w|\leqslant\rho\}} |f'(w)|^2 \log\frac{1}{|w|}\,d\sigma(w) \leqslant 3\|f - f(0)\|_2^2 \int_{\{|w|\leqslant\rho\}} \log\frac{1}{|w|}\,d\sigma(w)\,.$$

But, using (1.58) again,

$$\int_{\{|w|\leqslant\rho\}} \log\frac{1}{|w|}\,d\sigma(w) = 2\int_0^{\rho} r\log(1/r)\,dr \leqslant 2\rho,$$

so that we get

$$\|f - f(0)\|_2^2 \leqslant \frac{2}{\rho}\int_{\mathbb{D}} |f'(w)|^2 (1-|w|)\,d\sigma(w) + 12\rho\|f - f(0)\|_2^2\,.$$

Choosing $\rho = \frac{1}{13}$ (for instance) and rearranging, we get the result. □

Though we shall not use it in the following when $p \neq 2$, there is a corresponding formula for any value of p.

Theorem 1.80 (Green formula in Hardy spaces) *Let $p \geqslant 1$ and $f \in H^p$. We have*

$$\|f\|_p^p = |f(0)|^p + \frac{p^2}{2}\int_{\mathbb{D}} |f'(z)|^2 |f(z)|^{p-2} \log\frac{1}{|z|}\, d\sigma(z).$$

In particular, $$\|f - f(0)\|_p^p \approx \int_{\mathbb{D}} |f'(z)|^2 |f(z)|^{p-2}(1-|z|)\, d\sigma(z).$$

Proof: Recall that, from Green's classical formula, for $g \in \mathscr{C}^2(\mathbb{D})$, we have for every $r \in (0,1)$ (since $z \mapsto g(rz)$ belongs to $\mathscr{C}^2(\overline{\mathbb{D}})$)

$$\int_{\mathbb{T}} g(r\xi)\, dm(\xi) = g(0) + \frac{1}{2}\int_{D(0,r)} \Delta g(z) \log\left(\frac{r}{|z|}\right) d\sigma(z), \tag{1.59}$$

where Δ is the usual Laplacian operator.

Now fix f a non-zero analytic function on $\mathbb{D}$.

We know that we can write $|f|^p = f^{\frac{p}{2}}\overline{f}^{\frac{p}{2}}$ on a suitable (convex) neighborhood of any point z where f does not vanish. Therefore, for any such point $z \in \mathbb{D}$, we have at point z

$$\begin{aligned}\Delta|f|^p &= 4\frac{\partial^2}{\partial z \partial \bar{z}}|f|^p = 4\frac{\partial}{\partial z} f^{\frac{p}{2}}\left(\frac{p}{2}\overline{f'}\,\overline{f}^{\frac{p}{2}-1}\right)\\ &= 4\left(\frac{p}{2}f'\, f^{\frac{p}{2}-1}\right)\left(\frac{p}{2}\overline{f'}\,\overline{f}^{\frac{p}{2}-1}\right) = p^2|f'|^2|f|^{p-2}.\end{aligned} \tag{1.60}$$

From (1.59), since the zero set of f has σ-measure 0, we get

$$\int_{\mathbb{T}} |f|^p(r\xi) dm(\xi) = |f|^p(0) + \frac{p^2}{2}\int_{D(0,r)} |f'(z)|^2|f(z)|^{p-2} \log\left(\frac{r}{|z|}\right) d\sigma(z).$$

Now let r grow to 1 and we get the conclusion.

The second part of the statement can be proved exactly like the corresponding formula for $p = 2$ in Theorem 1.79. □

1.11.5 Bergman Spaces

In this section, we focus on Bergman spaces. Let us recall how they are defined (see [46] and [73] for more details). We shall first introduce the classical (unweighted) Bergman spaces $\mathscr{B}^p$ for $p \geqslant 1$, which are very natural spaces of analytic functions. Then we shall extend this definition to the weighted versions $\mathscr{B}^p_w$.

When $p \in [1, +\infty)$:

$$\mathscr{B}^p = \left\{ f \in \mathscr{H}(\mathbb{D}) : \int_{\mathbb{D}} |f|^p \, d\sigma < \infty \right\} = \mathscr{H}(\mathbb{D}) \cap L^p(\mathbb{D}, d\sigma)$$

and

$$\|f\|_{\mathscr{B}^p} = \left(\int_{\mathbb{D}} |f|^p \, d\sigma \right)^{\frac{1}{p}}.$$

Using polar coordinates, it is sometimes useful to write

$$\|f\|_{\mathscr{B}^p} = \left(\int_0^1 \int_{\mathbb{T}} |f(r\xi)|^p \, 2r dm(\xi) \, dr \right)^{\frac{1}{p}} = \left(\int_0^1 \|f_r\|_{H^p}^p \, 2r dr \right)^{\frac{1}{p}},$$

which implies in particular that $H^p \subset \mathscr{B}^p$ with $\|f\|_{\mathscr{B}^p} \leqslant \|f\|_{H^p}$.

In the particular case $p = 2$ (the Hilbertian case), we can express the norm with the Taylor coefficients of the function. Indeed, let $f(z) = \sum_{n=0}^{+\infty} a_n z^n$ be analytic on $\mathbb{D}$. For every $r \in (0, 1)$, we have

$$\|f_r\|_{H^2}^2 = \sum_{n=0}^{+\infty} |a_n|^2 r^{2n},$$

so

$$\|f\|_{\mathscr{B}^2}^2 = \int_0^1 \sum_{n=0}^{+\infty} 2|a_n|^2 r^{2n+1} \, dr.$$

Therefore,

$$\|f\|_{\mathscr{B}^2} = \left(\sum_{n=0}^{+\infty} \frac{|a_n|^2}{n+1} \right)^{1/2}. \tag{1.61}$$

It is easy to check that all these spaces are Banach spaces. Moreover, when $p = 2$, the Bergman space $\mathscr{B}^2$ is a Hilbert space associated to the scalar product

$$\langle f, g \rangle_{\mathscr{B}^2} = \int_{\mathbb{D}} f \overline{g} \, d\sigma = \sum_{n=0}^{+\infty} \frac{1}{n+1} a_n \overline{b_n}$$

when $f(z) = \sum_{n=0}^{+\infty} a_n z^n$ and $g(z) = \sum_{n=0}^{+\infty} b_n z^n$.

In the spirit of Theorem 1.79 let us mention now another expression of the $\mathscr{B}^2$ norm with the help of the derivative. Again this result will be useful in the theory of composition operators.

Theorem 1.81 (Classical Littlewood–Paley formula for the Hilbertian Bergman space) *Let $f \in \mathcal{B}^2$. We have*

$$|f(0)|^2 + \frac{1}{2}\int_{\mathbb{D}} |f'(z)|^2(1-|z|^2)^2\, d\sigma(z) = |f(0)|^2 + \sum_{n=1}^{+\infty} \frac{n}{(n+2)} \cdot \frac{|a_n|^2}{(n+1)}$$
$$\approx \|f\|^2_{\mathcal{B}^2}. \qquad (1.62)$$

Proof: Write $f(z) = \sum_{n=0}^{\infty} a_n z^n$ for every $z \in \mathbb{D}$. We rewrite the integral in the left-hand side in polar coordinates as

$$\int_{\mathbb{D}} |f'(z)|^2(1-|z|^2)^2\, d\sigma(z) = \int_0^1 \Big(\int_{\mathbb{T}} |f'(r\xi)|^2\, dm(\xi)\Big)\big(1-r^2\big)^2\, 2r dr,$$

and we have

$$\int_{\mathbb{T}} |f'(r\xi)|^2\, dm(\xi) = \sum_{n=0}^{\infty} (n+1)^2 |a_{n+1}|^2 r^{2n}.$$

Now, via Fubini, we get

$$\int_{\mathbb{D}} |f'(z)|^2(1-|z|^2)^2\, d\sigma(z) = \sum_{n=0}^{\infty} 2(n+1)^2 |a_{n+1}|^2 \int_0^1 r^{2n+1}\big(1-r^2\big)^2\, dr,$$

but

$$2\int_0^1 r^{2n+1}\big(1-r^2\big)^2\, dr = \frac{1}{n+1} - \frac{2}{n+2} + \frac{1}{n+3} = \frac{2}{(n+1)(n+2)(n+3)}.$$

Gathering everything we get the result. □

Actually there is another elementary approach that leads to an exact version of the Littlewood–Paley type formula in the framework of Bergman spaces. Nevertheless, though less sharp, and following the tradition, we shall often use (1.62) in the sequel when we do not need exact estimates.

Theorem 1.82 (Exact Littlewood–Paley formula for $\mathcal{B}^2$) *Let $f \in \mathcal{B}^2$. We have*

$$\|f\|^2_{\mathcal{B}^2} = |f(0)|^2 + \int_{\mathbb{D}} |f'(z)|^2 \Big(\log\frac{1}{|z|^2} - (1-|z|^2)\Big)\, d\sigma(z). \qquad (1.63)$$

Note that, when $|z| \to 1$, we have

$$\log\frac{1}{|z|^2} - (1-|z|^2) = -\log\big(1-(1-|z|^2)\big) - (1-|z|^2) \sim \frac{1}{2}(1-|z|^2)^2$$

and compare with (1.62).

Proof: One way to prove (1.63) is to start from the integral, and the computations follow the lines of the proof of (1.62). Nevertheless, we choose to start from the Littlewood–Paley formula for H^2, and this explains where the formula comes from in an elementary way.

Let $f \in \mathscr{B}^2$. Then for every $r \in (0,1)$, we have $f_r \in H^\infty \subset H^2$ (here $f_r(z) = f(rz)$), and we can apply Theorem 1.79 to the function f_r:

$$\|f_r\|_{H^2}^2 = |f(0)|^2 + 2\int_{\mathbb{D}} \left|rf'(rz)\right|^2 \log\left(\frac{1}{|z|}\right) d\sigma(z).$$

Now we take the integral over $(0,1)$ (with respect to the probability measure $2r dr$) and use Fubini:

$$\|f\|_{\mathscr{B}^2}^2 = |f(0)|^2 + 4\int_{\mathbb{D}}\int_0^1 \left|f'(rz)\right|^2 r^3 \log\left(\frac{1}{|z|}\right) dr\, d\sigma(z),$$

but, using polar coordinates, we get

$$\begin{aligned}
&\int_{\mathbb{D}}\int_0^1 \left|f'(rz)\right|^2 r^3 \log\left(\frac{1}{|z|}\right) dr\, d\sigma(z) \\
&\quad = 2\int_{\mathbb{T}}\int_0^1\int_0^1 \left|f'(r\rho\xi)\right|^2 r^3 \rho \log\left(\frac{1}{\rho}\right) dr\, d\rho\, dm(\xi).
\end{aligned}$$

Consider the change of variable $(r,\rho) \in (0,1)^2 \longmapsto (t,s) = (r\rho, r/\rho) \in \Omega$ with $\Omega = \{(u,v) \mid u \in (0,1),\ v \in (u, 1/u)\}$. Its inverse is $(t,s) \longmapsto (\sqrt{ts}, \sqrt{t/s})$, whose jacobian at point (t,s) is equal to $1/(2s)$. We get

$$\begin{aligned}
&\int_{\mathbb{D}}\int_0^1 \left|f'(rz)\right|^2 r^3 \log\left(\frac{1}{|z|}\right) dr\, d\sigma \\
&\quad = \frac{1}{2}\int_{\mathbb{T}}\int_0^1\int_t^{\frac{1}{t}} \left|f'(t\xi)\right|^2 t^2 \log\left(\frac{s}{t}\right) ds\, dt\, dm \\
&\quad = \frac{1}{4}\int_{\mathbb{D}} \left|f'(w)\right|^2 I(w)\, d\sigma(w),
\end{aligned}$$

where, writing $t = |w|$,

$$I(w) = t\int_t^{\frac{1}{t}} \log\left(\frac{s}{t}\right) ds = t^2\int_1^{\frac{1}{t^2}} \log(x)\, dx = \log\frac{1}{|w|^2} - (1 - |w|^2).$$

We get the result. □

The following result extends Theorem 1.81 to other values of p and is due to Hardy and Littlewood.

Proposition 1.83 *Let $p \geqslant 1$ and f be an analytic function on $\mathbb{D}$. Then f belongs to $\mathscr{B}^p$ if and only if $(1 - |z|^2)|f'(z)| \in L^p(\mathbb{D}, \sigma)$.*

Moreover,

$$\|f - f(0)\|_{\mathscr{B}^p} \approx \left(\int_{\mathbb{D}} |f'(z)|^p (1 - |z|^2)^p \, d\sigma(z) \right)^{1/p}. \tag{1.64}$$

More precisely, we have

$$c_p \, \|f - f(0)\|_{\mathscr{B}^p} \leqslant \left(\int_{\mathbb{D}} |f'(z)|^p (1 - |z|^2)^p \, d\sigma(z) \right)^{1/p} \leqslant 6 \|f - f(0)\|_{\mathscr{B}^p} \tag{1.65}$$

for some constant $c_p > 0$.

As an immediate corollary, we can state the following:

Corollary 1.84 *The Bloch space is included in every Bergman space $\mathscr{B}^p$ where $p \geqslant 1$.*

We recall here that an analytic function $f : \mathbb{D} \to \mathbb{C}$ is a Bloch function if $\sup_{z \in \mathbb{D}} (1 - |z|^2)|f'(z)| < \infty$. Equivalently f belongs to the *Bloch space* if its Bloch norm is finite, that is, $|f(0)| + \sup_{z \in \mathbb{D}} \left(1 - |z|^2\right) |f'(z)| < \infty$.

A function f belongs to the *Little Bloch space* if it belongs to the Bloch space and satisfies $\lim_{|z| \to 1} \left(1 - |z|^2\right) |f'(z)| = 0$.

We shall not study extensively Bloch spaces and their composition operators in the present book (nevertheless see Exercise I.6).

Let us point out that the constant on the left-hand side in Proposition 1.83 must depend on p. Indeed, taking some analytic Bloch function f and letting p go to the infinity, the central term tends to the Bloch norm of $f - f(0)$, whereas the norm in the left-hand side tends to $\|f - f(0)\|_\infty$. Since there exist some unbounded Bloch functions, the constant c_p cannot be uniform.

For the proof of Proposition 1.83, we shall need the following lemma. We give here a simple version in the space $\mathscr{B}_1^p = \mathscr{H}(\mathbb{D}) \cap L^p(\mathbb{D}, 2(1 - |z|^2)\, d\sigma)$ equipped with its natural norm (we shall meet again the probability measure $\sigma_1 = 2(1 - |z|^2)\, \sigma$ in the framework of weighted Bergman spaces) and leave the corresponding result for other general weighted Bergman spaces as an exercise.

Lemma 1.85 *Let $p \geqslant 1$ and h be an analytic function on $\mathbb{D}$ such that $h(0) = 0$. Then h belongs to $\mathscr{B}_1^p$ if and only if $z \mapsto Q(z) = \dfrac{h(z)}{z}$ belongs to $\mathscr{B}_1^p$. Moreover*

$$\|h\|_{\mathscr{B}_1^p} \approx \|Q\|_{\mathscr{B}_1^p}.$$

Proof: Though different, the proof of this lemma shares some idea with the proof of Proposition 1.92.

Keeping in mind that $h(0) = 0$, we define Q at the origin by $Q(0) = h'(0)$, so that Q is analytic. Since, for every $z \in \mathbb{D}$, we have $|Q(z)| \geqslant |h(z)|$, we concretely only have to control the norm of Q and we claim that

$$\int_{\mathbb{D}} |Q|^p \, d\sigma_1 \leqslant \frac{7.4^p}{3} \int_{\mathbb{D}} |h|^p \, d\sigma_1.$$

Indeed, we write

$$\int_{\mathbb{D}} |Q|^p \, d\sigma_1 = \int_{\{|z| \leqslant \frac{1}{4}\}} |Q(z)|^p \, d\sigma_1(z) + \int_{\{|z| > \frac{1}{4}\}} |Q(z)|^p \, d\sigma_1(z).$$

On one hand

$$\int_{\{|z| > \frac{1}{4}\}} |Q(z)|^p \, d\sigma_1(z) \leqslant 4^p \int_{\mathbb{D}} |h|^p \, d\sigma_1. \tag{1.66}$$

On the other hand, with $Q_r(z) = Q(rz)$,

$$\begin{aligned}\int_{\{|z| \leqslant \frac{1}{4}\}} |Q(z)|^p \, d\sigma_1(z) &= 4 \int_0^{\frac{1}{4}} r(1 - r^2) \, \|Q_r\|_{H^p}^p \, dr \\ &\leqslant 4 \int_0^{\frac{1}{4}} \left(r + \frac{1}{4}\right) \|Q_r\|_{H^p}^p \, dr,\end{aligned}$$

but, by Fatou's lemma ($r \to \|Q_r\|_{H^p}$ is non-decreasing), $\|Q_r\|_{H^p} \leqslant \|Q_{r+\frac{1}{4}}\|_{H^p}$.

Hence, since $1 - r^2 \geqslant \dfrac{3}{4}$ when $r \leqslant \dfrac{1}{2}$,

$$\begin{aligned}\int_{\{|z| \leqslant \frac{1}{4}\}} |Q(z)|^p \, d\sigma_1(z) &\leqslant 4 \int_0^{\frac{1}{4}} \left(r + \frac{1}{4}\right) \|Q_{r+\frac{1}{4}}\|_{H^p}^p \, dr = 4 \int_{\frac{1}{4}}^{\frac{1}{2}} \|Q_r\|_{H^p}^p \, r dr \\ &\leqslant \frac{4}{3} \int_{\{|z| > \frac{1}{4}\}} |Q(z)|^p \, d\sigma_1(z).\end{aligned}$$

Using (1.66), the claim is proved and it finishes the proof of the lemma. □

Proof of Proposition 1.83: First point out that

$$Tg(z) = (1 - |z|^2) \int_{\mathbb{D}} \frac{6\bar{w}(1 - |w|^2)}{(1 - \bar{w}z)^4} g(w) \, d\sigma(w)$$

is well defined for every $z \in \mathbb{D}$ when $g \in L^1(\mathbb{D}, \sigma)$. Moreover, when f is analytic on $\mathbb{D}$, we claim that $Tf(z) = (1 - |z|^2) f'(z)$. Indeed, it suffices to see that it is true for $f(z) = \beta_n(z) = z^n$ for any $n \in \mathbb{N}_0$, but

$$\int_{\mathbb{D}} \frac{6\bar{w}(1-|w|^2)}{(1-\bar{w}z)^4} w^n \, d\sigma(w) = 2\int_0^1 (1-r^2) r^{n+2} \left(\int_{\mathbb{T}} \frac{6\bar{\xi}\xi^n}{(1-r\bar{\xi}z)^4} \, dm(\xi) \right) dr,$$

and, thanks to the Cauchy formula,

$$\begin{aligned} \int_{\mathbb{T}} \frac{6\bar{\xi}\xi^n}{(1-r\bar{\xi}z)^4} \, dm(\xi) &= \frac{3!}{2i\pi} \int_{\mathbb{T}} \frac{\xi^{n+2}}{(\xi - rz)^4} \, d\xi = \beta_{n+2}^{(3)}(rz) \\ &= \begin{cases} 0 & \text{if } n = 0, \\ n(n+1)(n+2)(rz)^{n-1} & \text{if } n \geqslant 1, \end{cases} \end{aligned}$$

so that $\int_{\mathbb{D}} \frac{6\bar{w}(1-|w|^2)}{(1-\bar{w}z)^4} w^n \, d\sigma(w) = \begin{cases} 0 & \text{if } n = 0, \\ nz^{n-1} & \text{if } n \geqslant 1. \end{cases}$ Therefore, we proved the claim $T\beta_n(z) = (1-|z|^2)\beta_n'(z)$ for every $n \geqslant 0$ and $Tf(z) = (1-|z|^2) f'(z)$.

For every $g \in L^1(\mathbb{D}, \sigma)$, we have $Tg \in L^1(\mathbb{D}, \sigma)$ with

$$\|Tg\|_1 \leqslant 6\|g\|_1. \tag{1.67}$$

Indeed

$$\|Tg\|_1 \leqslant 6 \int_{\mathbb{D}} |g(w)| \left(\int_{\mathbb{D}} \frac{1-|z|^2}{|1-\bar{w}z|^4} \, d\sigma(z) \right) (1-|w|^2) \, d\sigma(w)$$

thanks to Fubini. But, for any $w \in \mathbb{D}$, we have

$$\begin{aligned} \int_{\mathbb{D}} \frac{1-|z|^2}{|1-\bar{w}z|^4} \, d\sigma(z) &= \int_{\mathbb{D}} \frac{|\varphi_w'(z)|^2 (1-|z|^2)}{(1-|w|^2)^2} \, d\sigma(z) \\ &= \frac{1}{(1-|w|^2)^2} \int_{\mathbb{D}} (1-|\varphi_w(z)|^2) \, d\sigma(z), \end{aligned}$$

where $\varphi_w(z) = \dfrac{w-z}{1-\bar{w}z}$, after a change of variable.

But we know that $1 - |\varphi_w(z)|^2 = \dfrac{(1-|z|^2)(1-|w|^2)}{|1-\bar{w}z|^2}$.

So we get

$$\begin{aligned} \int_{\mathbb{D}} \frac{1-|z|^2}{|1-\bar{w}z|^4} \, d\sigma(z) &= \frac{1}{1-|w|^2} \int_{\mathbb{D}} \frac{1-|z|^2}{|1-\bar{w}z|^2} \, d\sigma(z) \\ &= \frac{1}{1-|w|^2} \int_0^1 \frac{2(1-r^2)r}{1-|w|^2 r^2} \, dr \\ &= \frac{1}{1-|w|^2} \int_0^1 \frac{1-x}{1-|w|^2 x} \, dx \leqslant \frac{1}{1-|w|^2}. \end{aligned}$$

This proves (1.67).

On the other hand, for every $z \in \mathbb{D}$, thanks to the estimate of the same previous integral, we have $|Tg(z)| \leqslant 6\|g\|_\infty$.

By interpolation (we could also simply use the Schur test lemma (see Theorem A.1) with the test function $\mathbb{1}$), T is bounded from $L^p(\mathbb{D})$ into itself with norm less than 6. In particular, when f is analytic,

$$\left(\int_{\mathbb{D}} |f'(z)|^p (1-|z|^2)^p \, d\sigma(z)\right)^{1/p} \leqslant 6\|f - f(0)\|_{\mathscr{B}^p}.$$

For the reverse inequality, assume that f is analytic on $\mathbb{D}$ and such that the map $w \in \mathbb{D} \mapsto (1-|w|^2)f'(w)$ belongs to $L^p(\mathbb{D},\sigma)$. Actually, we rather consider

$$\tilde{f}(z) = f(z) - f(0) - zf'(0),$$

so that we have $w \in \mathbb{D} \mapsto F(w) = \frac{1}{\bar{w}}(1-|w|^2)\tilde{f}'(w) \in L^p(\mathbb{D},\sigma)$ as well, thanks to Lemma 1.85 with an equivalent norm.

We define the analytic function g on $\mathbb{D}$ by

$$\forall z \in \mathbb{D}, \quad g(z) = \int_{\mathbb{D}} \frac{(1-|w|^2)^2}{\bar{w}(1-\bar{w}z)^3} \tilde{f}'(w) \, d\sigma(w) = \int_{\mathbb{D}} \frac{(1-|w|^2)}{(1-\bar{w}z)^3} F(w) \, d\sigma(w).$$

For every $r \in (0,1)$, we have $\displaystyle\int_{\mathbb{T}} \xi \tilde{f}'(r\xi) \, dm(\xi) = 0$, hence $g(0) = 0$.

In the spirit of the first part of the proof, we claim that the operator $h \mapsto S(h)$, defined by

$$S(h)(z) = \int_{\mathbb{D}} \frac{(1-|w|^2)}{|1-\bar{w}z|^3} h(w) \, d\sigma(w),$$

is bounded on $L^p(\mathbb{D},\sigma)$ for $p \geqslant 1$.

It is straightforward for $p = 1$ (using Fubini's theorem).

To see this for $p > 1$, simply use the Schur test lemma (Theorem A.1) with the test function $w \in \mathbb{D} \mapsto (1-|w|^2)^{-\delta}$, where $0 < \delta < \min(1/p; 2/p')$. Since $|g| \leqslant S(|F|)$, we get $g \in L^p(\mathbb{D},\sigma)$ with

$$\|g\|_{\mathscr{B}^p} \leqslant m_p \|F\|_{L^p(\mathbb{D},\sigma)},$$

where m_p depends on p only.

In other words, g belongs to $\mathscr{B}^p$ and its derivative satisfies

$$\forall z \in \mathbb{D}, \qquad g'(z) = \int_{\mathbb{D}} \frac{3(1-|w|^2)^2}{(1-\bar{w}z)^4} \tilde{f}'(w) \, d\sigma(w) = \tilde{f}'(z).$$

Therefore, since both $\tilde{f}$ and g vanish at 0, we have $g = \tilde{f} = f - f(0) - zf'(0)$ and

$$\|f - f(0) - zf'(0)\|_{\mathscr{B}^p} \leqslant m_p \|F\|_{L^p(\mathbb{D},\sigma)} \leqslant M_p \|(1-|w|^2)(f' - f'(0))\|_{L^p(\mathbb{D},\sigma)},$$

where M_p depends on p only, thanks to Lemma 1.85.

Finally, we point out that $f'(0) = 2\int_{\mathbb{D}} f'(z)\,(1-|z|^2)\,d\sigma(z)$; hence

$$|f'(0)| \leqslant 2\|(1-|w|^2)f'\|_{L^1(\mathbb{D},\sigma)} \leqslant 2\|(1-|w|^2)f'\|_{L^p(\mathbb{D},\sigma)}.$$

We get $\|f - f(0)\|_{\mathscr{B}^p} \leqslant (3M_p + 2)\|(1-|w|^2)f'\|_{L^p(\mathbb{D},\sigma)}$. □

Analytic functions belonging to Bergman spaces do not have so nice boundary properties as those belonging to Hardy spaces. More specifically, a function $f \in \mathscr{B}^p$ may fail to have some radial limit almost everywhere. Actually, it is even worse:

Proposition 1.86 *There exist functions belonging to every Bergman space $\mathscr{B}^p$ (for every $p \geqslant 1$) yet having radial limits almost nowhere on $\mathbb{T}$.*

Proof: Actually we are going to construct such a function f belonging to the Bloch space: this function then belongs to every classical Bergman space thanks to Corollary 1.84.

Namely, consider some random choices of sign $\{\theta_n\}$, that is, independent, equidistributed, centered random variables taking their values in $\{\pm 1\}$ on some probability space $(\Omega, \mathbb{P})$. Actually, since we are going to use a Fourier-type argument, we can also realize Ω as a compact abelian group. Concretely, we may assume that $\Omega = \{\pm 1\}^{\mathbb{N}_0}$ so that the Rademacher functions appear as coordinate functions.

We claim that the function

$$F_\omega(z) = \sum_{n=0}^{\infty} \theta_n(\omega) z^{2^n}, \quad \text{where } z \in \mathbb{D},$$

belongs to the Bloch space for any $\omega \in \Omega$.

Indeed, for every $z \in \mathbb{D}$, gathering powers $k \in \left\{2^{n-1}, \ldots, 2^n - 1\right\}$ (when $n \geqslant 1$), we have

$$\left|F'_\omega(z)\right| \leqslant \sum_{n=0}^{\infty} 2^n |z|^{2^n - 1} \leqslant 1 + 2\sum_{n=1}^{\infty} \sum_{k=2^{n-1}}^{2^n-1} |z|^k \leqslant 2\sum_{k=0}^{\infty} |z|^k = \frac{2}{1-|z|}. \tag{1.68}$$

Hence $(1-|z|^2)\left|F'_\omega(z)\right| \leqslant 4$, which proves the claim.

In the sequel, we denote $R_N F_\omega(z) = \sum_{n=N}^{\infty} \theta_n(\omega) z^{2^n}$, where $z \in \mathbb{D}$, $\omega \in \Omega$, and $N \in \mathbb{N}$.

Now our argument mimics the one in [45, theorem A.4]. We are going to show that *for almost every* ω, there is no $E \subset \mathbb{T}$ with positive measure such that the function F_ω has a radial limit on E.

We introduce the sets

$$\mathscr{E} = \left\{(\xi,\omega) \in \mathbb{T} \times \Omega \mid \lim_{r\to 1} F_\omega(r\xi) \text{ does not exist}\right\}$$

and, for $\xi \in \mathbb{T}$,

$$\mathscr{E}_\xi = \left\{\omega \in \Omega \mid (\xi,\omega) \in \mathscr{E}\right\}.$$

By a standard argument based on Fubini's theorem, it suffices to show that for every $\xi \in \mathbb{T}$, $\mathbb{P}(\mathscr{E}_\xi) = 1$.

Let us assume that for some (now fixed) $\xi \in \mathbb{T}$, we have $\mathbb{P}(\Omega \setminus \mathscr{E}_\xi) > 0$.

For every integer q, the set $E_q = \left\{\omega \notin \mathscr{E}_\xi \mid \forall r \in (0,1), |F_\omega(r\xi)| \leqslant q\right\}$ defines a non-decreasing sequence of measurable sets whose union contains $\Omega \setminus \mathscr{E}_\xi$, which has a positive measure. Therefore, there exists some integer q_0 such that $\mathbb{P}(E_{q_0}) \geqslant \frac{1}{2}\mathbb{P}(\Omega \setminus \mathscr{E}_\xi) > 0$. Hence we get a set $A = E_{q_0}$ with positive measure, satisfying $A \subset \Omega \setminus \mathscr{E}_\xi$ and

$$\forall \omega \in A, \forall r \in (0,1), \quad |F_\omega(r\xi)| \leqslant q_0.$$

Moreover, for every N, we have

$$\forall \omega \in A, \forall r \in (0,1), \qquad |R_N F_\omega(r\xi)| \leqslant q_0 + N.$$

On one hand, we have the upper estimate

$$\int_A |R_N F_\omega(r\xi)|^2 \, d\mathbb{P} \leqslant (q_0 + N)^2 \mathbb{P}(A).$$

On the other hand, we have

$$\int_A |R_N F_\omega(r\xi)|^2 \, d\mathbb{P} = \mathbb{P}(A) \sum_{i=N}^{\infty} r^{2^{i+1}} + \sum_{\substack{i,j \geqslant N \\ i \neq j}} r^{2^i} r^{2^j} \int_A \varepsilon_i \varepsilon_j \, d\mathbb{P},$$

but $\int_A \varepsilon_i \varepsilon_j \, d\mathbb{P}$ is actually the Fourier coefficient at $\varepsilon_i\varepsilon_j$ of the function $\mathbb{1}_A$ (recall that $\varepsilon_i\varepsilon_j$ is a particular function of the Walsh system).

By the Bessel inequality,

$$\sum_{i \neq j} \left|\widehat{\mathbb{1}_A}(\varepsilon_i\varepsilon_j)\right|^2 \leqslant \|\mathbb{1}_A\|_2^2 = \mathbb{P}(A);$$

hence we can choose N large enough (from now on fixed) to ensure

$$\sum_{\substack{i,j\geqslant N\\ i\neq j}} \left|\widehat{\mathbb{1}_A}(\varepsilon_i\varepsilon_j)\right|^2 \leqslant \frac{\mathbb{P}(A)^2}{4}.$$

From the Cauchy–Schwarz inequality, we get

$$\left|\sum_{\substack{i,j\geqslant N\\ i\neq j}} r^{2^i} r^{2^j} \int_A \varepsilon_i\varepsilon_j \, d\mathbb{P}\right| \leqslant \frac{\mathbb{P}(A)}{2}\left(\sum_{\substack{i,j\geqslant N\\ i\neq j}} r^{2^{i+1}} r^{2^{j+1}}\right)^{\frac{1}{2}} \leqslant \frac{\mathbb{P}(A)}{2}\sum_{i=N}^{\infty} r^{2^{i+1}}.$$

Finally, we get the inequality

$$\frac{\mathbb{P}(A)}{2}\sum_{i=N}^{\infty} r^{2^{i+1}} \leqslant (q_0+N)^2\,\mathbb{P}(A),$$

where $\mathbb{P}(A) > 0$ and $\sum_{i=N}^{\infty} r^{2^{i+1}} \longrightarrow +\infty$ when $r \to 1$.

This contradiction proves that $\mathbb{P}(\mathscr{E}_\xi) = 1$ for any $\xi \in \mathbb{T}$, as claimed. □

Point out that if one wants to check only that $F_\omega \in \mathscr{B}^2$ (which is weaker than what we proved), then we do not need to use Proposition 1.83. Indeed, for any $\omega \in \Omega$, we immediately see that

$$\|F_\omega\|^2_{\mathscr{B}^2} = \sum_{\substack{k=2^n\\ n\geqslant 0}} \frac{1}{k+1} = \sum_{n=0}^{\infty} \frac{1}{2^n+1} < \infty.$$

A variant of the estimate (1.68) also avoids Proposition 1.83: we can conclude directly that $F_\omega \in \mathscr{B}^p$ for every $p \geqslant 1$ and for any ω, just pointing out that we have $|F_\omega(z)| \leqslant 2\log\frac{1}{1-|z|}$. Nevertheless, keep in mind that the intersection of all Bergman spaces is larger than the Bloch space.

In another direction, the only function f belonging to some Hardy space, vanishing on a subset of $\mathbb{T}$ with positive m-measure, is $f = 0$ (see Theorem 1.74). This is not true anymore in the Bergman framework. Actually a striking result of Kahane–Katznelson [78] states the following:

Theorem 1.87 (Kahane–Katznelson) *Let u be a positive non-decreasing unbounded function on* $(0,1)$. *Let ϕ be a positive non-decreasing function on* $[0, +\infty)$ *such that $\phi(t) \to 0$ when $t \to 0$. Let $(\delta_n)_n$ be a positive sequence belonging to the space c_0 of vanishing sequences. Let h and g be two measurable functions taking their values in* $[-\infty, +\infty]$.

Then there exists an analytic function $F(z) = \sum a_n z^n$ *on* $\mathbb{D}$ *such that*

- $\|F_r\|_\infty = o\big(u(r)\big)$ *(when* $r \to 1$*).*
- $\sum_{n\geqslant 0} |a_n|^2 \phi(|a_n|) < \infty.$
- $\sum_{n\geqslant 0} \delta_n |a_n|^2 < \infty.$
- $\lim_{r\to 1} Re(F(r\xi)) = h(\xi)$ *and* $\lim_{r\to 1} Im(F(r\xi)) = g(\xi)$ *a.e. on* $\mathbb{T}$.

Choosing a suitable function u, for example, $u(r) = \sqrt{\log(1/(1-r))}$, and $h = -\infty$, Theorem 1.87 provides an analytic function F. Then it is easy to check that the function $f = \exp(F)$ belongs to every Bergman space $\mathscr{B}^p$ (where $p \geqslant 1$), never vanishes, and satisfies $\lim_{r\to 1} |f(r\xi)| = \lim_{r\to 1} \exp(Re(F(r\xi))) = 0$, a.e.

Point Evaluation on Bergman Spaces

We consider the pointwise evaluation at $z \in \mathbb{D}$ on the Bergman space $\mathscr{B}^p$,

$$\delta_z : \begin{array}{ccc} \mathscr{B}^p & \longrightarrow & \mathbb{C} \\ f & \longmapsto & f(z), \end{array}$$

and estimate its norm.

Proposition 1.88 *We have*

- *On the Hilbert space* $\mathscr{B}^2$*, the functional* δ_z *is associated to the reproducing kernel* $w \in \overline{\mathbb{D}} \longmapsto K_z(w) = \dfrac{1}{(1-\bar{z}w)^2}$.
- *For every* $p \geqslant 1$*, we have* $\big\|\delta_z\big\|_{(\mathscr{B}^p)^*} = \dfrac{1}{\big(1-|z|^2\big)^{\frac{2}{p}}}$.

Proof: We have the Taylor expansion

$$\delta_z(f) = f(z) = \sum_{n\geqslant 0} a_n z^n = \langle f, K_z \rangle_{\mathscr{B}^2},$$

where $K_z(w) = \sum_{n\geqslant 0}(n+1)\bar{z}^n w^n = \dfrac{1}{(1-\bar{z}w)^2}$ · In particular,

$$\big\|\delta_z\big\|^2_{(\mathscr{B}^2)^*} = \big\|K_z\big\|^2_{\mathscr{B}^2} = \delta_z(K_z) = \frac{1}{(1-|z|^2)^2}.$$

Now for any $p \geqslant 1$, we first notice that $f(0) = \displaystyle\int_{\mathbb{D}} f(w)\, d\sigma(w)$. Therefore, we have

$$|f(0)| \leqslant \|f\|_{\mathscr{B}^1} \leqslant \|f\|_{\mathscr{B}^p}.$$

Now given $f \in \mathscr{B}^p$ and $z \in \mathbb{D}$, we consider $g(w) = \dfrac{(1-|z|^2)^{\frac{2}{p}}}{(1-\bar{z}w)^{\frac{4}{p}}} f \circ \varphi_z(w)$, where φ_z is the Möbius function associated to z (see (1.8)).

We have $g \in \mathscr{B}^p$ and

$$\|g\|_{\mathscr{B}^p}^p = \int_{\mathbb{D}} |f \circ \varphi_z|^p \, |\varphi_z'|^2 \, d\sigma(z) = \|f\|_{\mathscr{B}^p}^p$$

after a change of variable.

We saw that $|g(0)| \leqslant \|g\|_{\mathscr{B}^p} = \|f\|_{\mathscr{B}^p}$, but $g(0) = (1-|z|^2)^{\frac{2}{p}} f(z)$. Rearranging, we get the upper estimate.

With $F = K_z^{2/p}$, we get the reverse inequality since

$$\|F\|_{\mathscr{B}^p} = \|K_z\|_{\mathscr{B}^2}^{\frac{2}{p}} = \frac{1}{(1-|z|^2)^{\frac{2}{p}}}.$$

By definition

$$\left\|\delta_z\right\|_{(\mathscr{B}^p)^*} \|F\|_{\mathscr{B}^p} \geqslant F(z) = \frac{1}{(1-|z|^2)^{\frac{2}{p}}}.$$

Rearranging, we get the conclusion. □

Duality

$\mathscr{B}^p$ spaces are dual spaces: for $p > 1$, $\mathscr{B}^p$ is the dual of $\mathscr{B}^{p'}$ and the duality is described by the duality bracket

$$\langle f, g\rangle = \int_{\mathbb{D}} f(z) g(\bar{z}) \, d\sigma(z),$$

where $f \in \mathscr{B}^p$ and $g \in \mathscr{B}^{p'}$.

This is a classical result: we refer to [46] or [73] for details.

For $p = 1$, $\mathscr{B}^1$ is also a dual space: it is the dual of the little Bloch space, whereas the dual of $\mathscr{B}^1$ is the Bloch space, but we shall not detail this aspect here.

Let us mention the following result that is similar to Proposition 1.78 in the framework of Bergman spaces.

Proposition 1.89 *Let $p \geqslant 1$. On the unit ball of $\mathscr{B}^p$, the weak-star topology coincides with the topology of uniform convergence on compact subsets of $\mathbb{D}$.*

Of course, when $p > 1$, the weak-star topology and the weak topology coincide. We leave the proof as an exercise.

Finally, let us mention that, from the point of view of geometry of Banach spaces, Bergman spaces are like classical spaces of sequences.

Theorem 1.90 *Let $p \geqslant 1$.*
The Bergman space $\mathcal{B}^p$ is isomorphic to the sequence space ℓ^p.

We do not give the proof here: see [118] or [180, theorem 11].

1.11.6 Weighted Bergman Spaces

The weighted versions of the classical Bergman spaces introduced in the present section are often naturally invited in investigations about operator theory on spaces of analytic functions. *We shall focus only on radial weights here.* The most classical one is the weight $\omega_\alpha(z) = (\alpha+1)(1-|z|^2)^\alpha$ where $\alpha > -1$. Point out that the measure $\sigma_\alpha = \omega_\alpha.\sigma$ is a probability measure on $\mathbb{D}$.

Given a continuous function $\omega : (0,1) \to (0,\infty)$ such that $\omega \in L^1(0,1)$, we extend it by $\omega(z) = \omega(|z|)$ for every $z \in \mathbb{D} \setminus \{0\}$.

We can consider the weighted Bergman space

$$\mathcal{B}^p_\omega := \mathcal{H}(\mathbb{D}) \cap L^p(\mathbb{D}, \omega\, d\sigma)$$

consisting of functions analytic in $\mathbb{D}$ and L^p-integrable with respect to the weight ω. The space $\mathcal{B}^p_\omega$ is equipped with the norm

$$\|f\|_{\mathcal{B}^p_\omega} = \left(\int_{\mathbb{D}} |f(z)|^p\, \omega(z)\, d\sigma(z)\right)^{\frac{1}{p}} = \left(\int_0^1 \|f_r\|^p_{H^p}\, 2r\omega(r)\, dr\right)^{\frac{1}{p}}.$$

In particular, since $\|f_r\|_{H^p} \leqslant \|f\|_{H^p}$, we have $H^p \subset \mathcal{B}^p_\omega$ with $\|f\|_{\mathcal{B}^p_\omega} \leqslant \|f\|_{H^p}$.

We shall simply write $\mathcal{B}^p_\alpha$ instead of $\mathcal{B}^p_{\omega_\alpha}$.

Since the weight is radial, we can easily give an expression of the norm with the Taylor coefficient when $p = 2$. Indeed, let $f(z) = \sum_{n=0}^{+\infty} a_n z^n$ be analytic on $\mathbb{D}$. For every $r \in (0,1)$, we have

$$\|f_r\|^2_{H^2} = \sum_{n=0}^{+\infty} |a_n|^2 r^{2n};$$

so

$$\|f\|^2_{\mathcal{B}^2_\omega} = \sum_{n=0}^{+\infty} \omega_n |a_n|^2 \quad \text{where } \omega_n = 2\int_0^1 r^{2n+1}\omega(r)\, dr.$$

In particular, for every $\alpha > -1$, we have $\|f\|^2_{\mathcal{B}^2_\alpha} = \sum_{n=0}^{+\infty} c_n(\alpha)|a_n|^2$,

where

$$c_n(\alpha) = \frac{\Gamma(\alpha+2)\, n!}{\Gamma(\alpha+n+2)} \sim \frac{\Gamma(\alpha+2)}{(n+1)^{\alpha+1}}, \tag{1.69}$$

and the reproducing kernel at $z \in \mathbb{D}$ is given by

$$w \longmapsto K_{z,\alpha}(w) = K_\alpha(w,z) := \sum_{n=0}^{+\infty} \frac{1}{c_n(\alpha)} \bar{z}^n w^n = \frac{1}{(1-\bar{z}w)^{\alpha+2}}. \tag{1.70}$$

The pointwise evaluation at $z \in \mathbb{D}$ on the Bergman space $\mathcal{B}^p_\omega$ is difficult to estimate precisely in the case of general weights (nevertheless see Theorem 1.96 for an accurate upper estimate), but we can give an exact estimate in the case of classical weights.

Proposition 1.91 *For every $p \geqslant 1$ and $\alpha > -1$, the point evaluation is bounded on $\mathcal{B}^p_\alpha$ and we have*

$$\left\|\delta_z\right\|_{(\mathcal{B}^p_\alpha)^*} = \frac{1}{\left(1-|z|^2\right)^{\frac{2+\alpha}{p}}}.$$

In particular, for $p = 2$, we have

$$\|K_{z,\alpha}\|_{\mathcal{B}^2_\alpha} = \left\|\delta_z\right\|_{(\mathcal{B}^2_\alpha)^*} = \frac{1}{\left(1-|z|^2\right)^{\frac{2+\alpha}{2}}}.$$

Proof: The case $p = 2$ follows directly from the previous value of the reproducing kernel. The proof for an arbitrary $p \geqslant 1$ mimics that of Proposition 1.88, so we just sketch it.

Since $f(0) = \int_{\mathbb{D}} f(w)\, d\sigma_\alpha(w)$, we have $|f(0)| \leqslant \|f\|_{\mathcal{B}^p_\alpha}$.

Now given $f \in \mathcal{B}^p_\alpha$ and $z \in \mathbb{D}$, we consider $g(w) = \dfrac{\left(1-|z|^2\right)^{\frac{2+\alpha}{p}}}{(1-\bar{z}w)^{\frac{4+2\alpha}{p}}} f \circ \varphi_z(w)$, where φ_z is the Möbius function (1.8). We remark that

$$\|g\|^p_{\mathcal{B}^p_\alpha} = (\alpha+1)\int_{\mathbb{D}} |f\circ\varphi_z|^p\, |\varphi_z'|^2 (1-|\varphi_z|^2)^\alpha\, d\sigma(z) = \|f\|^p_{\mathcal{B}^p_\alpha}$$

after a change of variable. Hence

$$\left(1-|z|^2\right)^{\frac{2+\alpha}{p}} |f(z)| = |g(0)| \leqslant \|g\|_{\mathcal{B}^p_\alpha} = \|f\|_{\mathcal{B}^p_\alpha},$$

and we get the upper estimate.

The inequality is seen to be sharp testing the function $z \mapsto \dfrac{\left(1-|z|^2\right)^{\frac{2+\alpha}{p}}}{(1-\bar{z}w)^{\frac{4+2\alpha}{p}}}$. □

The control of the norm of the point evaluation allows to give some other equivalent expressions for the norm in $\mathcal{B}^p_\alpha$.

Proposition 1.92 *Let $p \geqslant 1$, $\alpha > -1$, and $\gamma \in (0,2)$. Let g be an analytic function. Then $g \in L^p(\mathbb{D},\sigma_\alpha)$ if and only if $g \in L^p(\mathbb{D},|z|^{-\gamma}\sigma_\alpha)$. Moreover,*

$$\int_{\mathbb{D}} |g(z)|^p \, d\sigma_\alpha(z) \approx \int_{\mathbb{D}} |g(z)|^p \frac{1}{|z|^\gamma} \, d\sigma_\alpha(z).$$

In other words, $\mathscr{B}_\alpha^p = \mathscr{H}(\mathbb{D}) \cap L^p(\mathbb{D},|z|^{-\gamma}\sigma_\alpha)$ with equivalence of norms.

Proof: We claim that we have

$$\int_{\mathbb{D}} |g(z)|^p \frac{1}{|z|^\gamma} \, d\sigma_\alpha(z) \lesssim \int_{\mathbb{D}} |g(z)|^p \, d\sigma_\alpha(z)$$

for every $g \in \mathscr{B}_\alpha^p$.

Indeed, we first point out that

$$\begin{aligned}&\int_{\mathbb{D}} |g(z)|^p \frac{1}{|z|^\gamma} \, d\sigma_\alpha(z)\\ &\quad= \int_{\{|z|\leqslant \frac{1}{2}\}} |g(z)|^p \frac{1}{|z|^\gamma} \, d\sigma_\alpha(z) + \int_{\{|z|> \frac{1}{2}\}} |g(z)|^p \frac{1}{|z|^\gamma} \, d\sigma_\alpha(z).\end{aligned}$$

But

$$\int_{\{|z|> \frac{1}{2}\}} |g(z)|^p \frac{1}{|z|^\gamma} \, d\sigma_\alpha(z) \leqslant 2^\gamma \int_{\mathbb{D}} |g(z)|^p \, d\sigma_\alpha(z)$$

and

$$\int_{\{|z|\leqslant \frac{1}{2}\}} |g(z)|^p \frac{1}{|z|^\gamma} \, d\sigma_\alpha(z) \leqslant \int_{\{|z|\leqslant \frac{1}{2}\}} \frac{\|\delta_z\|^p_{(\mathscr{B}_\alpha^p)^*}}{|z|^\gamma} \|g\|^p_{\mathscr{B}_\alpha^p} \, d\sigma_\alpha(z) \leqslant I \, \|g\|^p_{\mathscr{B}_\alpha^p},$$

where

$$\begin{aligned} I &= \int_{\{|z|\leqslant \frac{1}{2}\}} \frac{1}{(1-|z|^2)^{\alpha+2}|z|^\gamma} \, d\sigma_\alpha(z)\\ &= \int_0^{\frac{1}{2}} \frac{2(\alpha+1)}{r^\gamma(1-r^2)^2} \, r dr \leqslant \frac{8(\alpha+1)2^\gamma}{9(2-\gamma)} < \infty. \end{aligned}$$

The claim is proved.

The reverse inequality is obvious in this case. □

We extend here the result of Theorem 1.82 to the weighted framework and will specify the results for standard weights.

Theorem 1.93 (Littlewood–Paley formula for $\mathscr{B}_\omega^2$) *Let $f \in \mathscr{B}_\omega^2$, where ω is a weight satisfying $\int_{\mathbb{D}} \omega \, d\sigma = 1$. We have*

$$\|f\|^2_{\mathcal{B}^2_\omega} = |f(0)|^2 + \int_{\mathbb{D}} \left|f'(z)\right|^2 \widetilde{\omega}(z)\, d\sigma(z), \tag{1.71}$$

where $\widetilde{\omega}(z) = 4 \int_{|z|}^1 x\,\omega(x) \log\left(\frac{x}{|z|}\right) dx.$

Remark 1.94 Given a weight ω, we emphasize the role played in the preceding result by the *associated weight* $\widetilde{\omega}$ defined on $(0, 1)$ by

$$\widetilde{\omega}(s) = 4 \int_s^1 x\,\omega(x) \log\left(\frac{x}{s}\right) dx, \tag{1.72}$$

and the formula $\widetilde{\omega}(z) = \widetilde{\omega}(|z|)$ extends the definition for $z \neq 0$.

We easily check that $\widetilde{\omega}$ is non-increasing (for instance, computing the derivative, which does not require here an extra regularity for ω) and convex on $(0, 1)$. Point out too that $\widetilde{\omega}(s) = O(|\log(s)|)$ when $s \to 0$ (testing $\omega(x) = 1/\sqrt{x}$ for instance; this is optimal in general). Therefore, $\widetilde{\omega} \in L^1(0, 1)$ and we also have

$$\int_0^1 \widetilde{\omega}(s)\, ds = 4 \int_0^1 x^2 \omega(x)\, dx.$$

Moreover $\lim_{s\to 1^-} \widetilde{\omega}(s) = 0$ and we can again extend its definition by $\widetilde{\omega}_{|\mathbb{T}} = 0$.

In the case of the standard weights (with $\alpha > -1$), when $|z| \to 1$, we have

$$\widetilde{\omega_\alpha}(z) \sim \frac{1}{\alpha + 2}(1 - |z|^2)^{\alpha+2} \approx \omega_{\alpha+2}(z)\,. \tag{1.73}$$

Indeed, using the change of variable $s = (1 - x^2)^{\alpha+1}$,

$$\begin{aligned}\widetilde{\omega_\alpha}(z) &= 4(\alpha + 1) \int_{|z|}^1 \left(1 - x^2\right)^\alpha \log\left(\frac{x}{|z|}\right) x dx \\ &= \int_0^{(1-|z|^2)^{\alpha+1}} \log\left(1 - s^{\frac{1}{\alpha+1}}\right) ds - \log\left(|z|^2\right)(1 - |z|^2)^{\alpha+1}.\end{aligned}$$

But $\log\left(|z|^2\right)(1 - |z|^2)^{\alpha+1} = -(1 - |z|^2)^{\alpha+2} + o\left((1 - |z|^2)^{\alpha+2}\right)$ and

$$\begin{aligned}&\int_0^{(1-|z|^2)^{\alpha+1}} \log\left(1 - s^{\frac{1}{\alpha+1}}\right) ds \\ &= -\int_0^{(1-|z|^2)^{\alpha+1}} s^{\frac{1}{\alpha+1}}\, ds + o\left(\int_0^{(1-|z|^2)^{\alpha+1}} s^{\frac{1}{\alpha+1}}\, ds\right).\end{aligned}$$

We get

$$\widetilde{\omega_\alpha}(z) = -\frac{\alpha+1}{\alpha+2}(1-|z|^2)^{\alpha+2} + (1-|z|^2)^{\alpha+2} + o\left((1-|z|^2)^{\alpha+2}\right)$$
$$\sim \frac{1}{\alpha+2}(1-|z|^2)^{\alpha+2} \tag{1.74}$$

as claimed.

We mention an immediate consequence of this discussion:

$$\|f - f(0)\|^2_{\mathscr{B}^2_\alpha} \approx \int_{\mathbb{D}} |f'(z)|^2 (1-|z|^2)^{\alpha+2}\, d\sigma(z) \tag{1.75}$$

for any $\alpha > -1$.

In other words, an analytic function f on $\mathbb{D}$ belongs to $\mathscr{B}^2_\alpha$ if and only if $f' \in \mathscr{B}^2_{\alpha+2}$.

Proof of Theorem 1.93: We shall follow the same strategy as for the proof of (1.63).

Let $f \in \mathscr{B}^2_\omega$. Then for every $r \in (0,1)$, we have $f_r \in H^\infty \subset H^2$ and from Theorem 1.79 we know that

$$\|f_r\|^2_{H^2} = |f(0)|^2 + 2\int_{\mathbb{D}} |rf'(rz)|^2 \log\left(\frac{1}{|z|}\right) d\sigma(z).$$

Now we take the integral over $(0,1)$ with respect to the probability measure $2r\omega(r)dr$ and use Fubini:

$$\|f\|^2_{\mathscr{B}^2_\omega} = |f(0)|^2 + 4\int_{\mathbb{D}}\int_0^1 |f'(rz)|^2 r^3 \log\left(\frac{1}{|z|}\right) \omega(r)\, dr\, d\sigma(z),$$

but, using polar coordinates,

$$\int_{\mathbb{D}}\int_0^1 |f'(rz)|^2 r^3 \log\left(\frac{1}{|z|}\right) \omega(r)\, dr\, d\sigma(z)$$
$$= 2\int_{\mathbb{T}}\int_0^1\int_0^1 |f'(r\rho\xi)|^2 r^3 \rho \log\left(\frac{1}{\rho}\right)\omega(r)\, dr\, d\rho\, dm(\xi).$$

Considering again the map $(r,\rho) \in (0,1)^2 \mapsto (t,s) = (r\rho, r/\rho) \in \Omega$ with $\Omega = \{(u,v) \mid u \in (0,1),\ v \in (u, 1/u)\}$, we get

$$\int_{\mathbb{D}}\int_0^1 |f'(rz)|^2 r^3 \log\left(\frac{1}{|z|}\right) \omega(r)\, dr\, d\sigma(z) = \frac{1}{4}\int_{\mathbb{D}} |f'(z)|^2\, \widetilde{\omega}(z)\, d\sigma(z),$$

where $\widetilde{\omega}(z) = |z| \displaystyle\int_{|z|}^{\frac{1}{|z|}} \log\left(\frac{s}{|z|}\right) \omega\left(\sqrt{s|z|}\right) ds$.

We get the result thanks to the change of variable $x = \sqrt{s|z|}$. □

In the same spirit, we can extend Proposition 1.83 to the following weighted version.

Proposition 1.95 *Let $p \geqslant 1$ and $\alpha > -1$. Let f be an analytic function on $\mathbb{D}$. Then f belongs to $\mathscr{B}_\alpha^p$ if and only if $(1-|z|^2)|f'(z)| \in L^p(\mathbb{D}, \sigma_\alpha)$. Moreover,*

$$\|f - f(0)\|_{\mathscr{B}_\alpha^p} \approx \left(\int_{\mathbb{D}} |f'(z)|^p (1-|z|^2)^p \, d\sigma_\alpha(z) \right)^{1/p}. \tag{1.76}$$

The proof is left as an exercise: the same ideas as in the proof of Proposition 1.83 work.

We now come back to the issue of estimating point evaluation in full generality. In the literature, some estimates appear but most often (always?) some regularity condition is required. We give below an upper estimate *without any regularity or growth condition* on the weight, showing that the point evaluations are always bounded.

Theorem 1.96 (Point evaluation on weighted Bergman spaces) *Let ω be a radial weight, $p \geqslant 1$, and $z \in \mathbb{D}$. Then:*

(i) *When $z = 0$, we have*

$$\left\|\delta_0\right\|_{(\mathscr{B}_\omega^p)^*} = \left\|a_0^*\right\|_{(\mathscr{B}_\omega^p)^*} = \|\omega\|_{L^1(\mathbb{D},\sigma)}^{-\frac{1}{p}}.$$

(ii) *When $z \neq 0$, we have*

$$\left\|\delta_z\right\|_{(\mathscr{B}_\omega^1)^*} \leqslant \|\omega\|_{L^1(\mathbb{D},\sigma)}^{-1} + \frac{2}{\widetilde{\omega}(z)}.$$

(iii) *For every $p > 1$ and $z \neq 0$, we have*

$$\left\|\delta_z\right\|_{(\mathscr{B}_\omega^p)^*} \leqslant \|\omega\|_{L^1(\mathbb{D},\sigma)}^{-\frac{1}{p}} + \frac{2^{1+\frac{1}{p}}}{\widetilde{\omega}(z)^{\frac{1}{p}}}.$$

In particular

$$\left\|\delta_z\right\|_{(\mathscr{B}_\omega^p)^*} = O\left(\frac{1}{\widetilde{\omega}(z)^{\frac{1}{p}}}\right) \quad \textit{when } |z| \to 1^-.$$

Actually in (ii) and (iii) the upper bounds can be extended to $z = 0$, once $1/\widetilde{\omega}(0)$ can be potentially understood as 0 if $\widetilde{\omega}(z) \to +\infty$ when $z \to 0$. Anyway having in mind that $\widetilde{\omega}$ is non-increasing on $(0,1)$, $\widetilde{\omega}(0^+) = \lim_{0^+} \widetilde{\omega}$ does exist in $(0, +\infty]$.

Let us observe that, up to constants, these upper estimates have a behavior near the boundary that coincides with the one for classical weights (see Proposition 1.91 and (1.74)).

Another consequence of Theorem 1.96 is that the evaluations on $\mathscr{B}^p_\omega$ are uniformly bounded on compact subsets of $\mathbb{D}$ (still keep in mind that $\widetilde{\omega}$ is non-increasing on $(0,1)$). Therefore, we have the following immediate corollary:

Corollary 1.97 *Let ω be a radial weight, $p \geqslant 1$, and $z \in \mathbb{D}$. Then the space $\mathscr{B}^p_\omega$ is a Banach space of analytic functions on $\mathbb{D}$ in the sense of Section 1.11.1.*

Before giving the proof, we first observe the following lemma that provides the exact value of the norm of the linear functional associating the nth Taylor coefficient:

$$f \in \mathscr{B}^p_\omega \longmapsto a_n^*(f) \qquad \text{with } f(z) = \sum_{n=0}^{+\infty} a_n^*(f)\, z^n \text{ for } z \in \mathbb{D}, \tag{1.77}$$

and where $n \geqslant 0$ is an integer.

Lemma 1.98 *Let ω be a radial weight, $p \geqslant 1$, and $n \geqslant 0$ be an integer. Then*

$$\left\|a_n^*\right\|_{(\mathscr{B}^p_\omega)^*} = \left[\int_0^1 2\omega(r) r^{np+1}\, dr\right]^{-\frac{1}{p}}.$$

In particular $\left\|a_0^*\right\|_{(\mathscr{B}^p_\omega)^*} = \|\omega\|_{L^1(\mathbb{D},\sigma)}^{-\frac{1}{p}}$.

Proof: For every $f \in \mathscr{B}^p_\omega$ and $r \in (0,1)$, we have

$$a_n^*(f) r^n = \int_{\mathbb{T}} f(r\xi)\, \bar{\xi}^n\, dm(\xi),$$

so that, for every positive $g \in L^{p'}((0,1), 2r\omega(r)\, dr)$, we have

$$\begin{aligned}\left|a_n^*(f)\right| \int_0^1 g(r) r^n\, 2r\omega(r) dr &\leqslant \int_0^1 g(r) \|f_r\|_{H^1}\, 2r\omega(r) dr \\ &\leqslant \int_0^1 g(r) \|f_r\|_{H^p}\, 2r\omega(r) dr,\end{aligned}$$

where $f_r(z) = f(rz)$; hence, taking the supremum over positive g in the unit ball of $L^{p'}((0,1), 2r\omega(r)\, dr)$ and then over f in the unit ball of $\mathscr{B}^p_\omega$, we get

$$\left\|a_n^*\right\|_{(\mathscr{B}^p_\omega)^*} \left[\int_0^1 2r^{np+1}\omega(r)\, dr\right]^{\frac{1}{p}} \leqslant 1.$$

The upper estimate is proved.

Now we prove the lower estimate, simply testing the monomial z^n:

$$1 = |a_n^*(z^n)| \leqslant \left\| a_n^* \right\|_{(\mathscr{B}_\omega^p)^*} \left\| z^n \right\|_{\mathscr{B}_\omega^p} = \left\| a_n^* \right\|_{(\mathscr{B}_\omega^p)^*} \left[\int_0^1 2r^{np+1} \omega(r)\, dr \right]^{\frac{1}{p}} .$$

□

Proof of Theorem 1.96: Actually the proof of (iii) is also valid for $p = 1$; nevertheless, the proof in (ii) is a bit different (and gives a slightly better constant).

(i) follows directly from Lemma 1.98.

(ii) We first observe that the relation $f(z) = \sum_{n=0}^{+\infty} a_n^*(f) z^n$ obviously implies

$$\left\| \delta_z \right\|_{(\mathscr{B}_\omega^1)^*} \leqslant \sum_{n=0}^{+\infty} \| a_n^* \|_{(\mathscr{B}_\omega^1)^*} |z|^n . \tag{1.78}$$

What's more: from the Cauchy–Schwarz inequality, we have for every $n \geqslant 1$

$$\begin{aligned} \widetilde{\omega}(z)^2 &= 16 \left(\int_{|z|}^1 r\, \omega(r) \log\left(\frac{r}{|z|}\right) dr \right)^2 \\ &\leqslant 8 \left(2 \int_0^1 r^{n+1}\, \omega(r)\, dr \right) \left(\int_{|z|}^1 r^{-n+1}\, \omega(r) \log^2\left(\frac{r}{|z|}\right) dr \right), \end{aligned}$$

and this implies, thanks to Lemma 1.98,

$$\| a_n^* \|_{(\mathscr{B}_\omega^1)^*} \leqslant \frac{8}{\widetilde{\omega}(z)^2} \int_{|z|}^1 r^{-n+1}\, \omega(r) \log^2\left(\frac{r}{|z|}\right) dr .$$

Relation (1.78) becomes

$$\begin{aligned} \left\| \delta_z \right\|_{(\mathscr{B}_\omega^1)^*} &\leqslant \|\omega\|_{L^1(\mathbb{D},\sigma)}^{-1} + \frac{8}{\widetilde{\omega}(z)^2} \int_{|z|}^1 \sum_{n=1}^{+\infty} \left(\frac{|z|}{r}\right)^n r\omega(r) \log^2\left(\frac{r}{|z|}\right) dr \\ &\leqslant \|\omega\|_{L^1(\mathbb{D},\sigma)}^{-1} + \frac{8}{\widetilde{\omega}(z)^2} \int_{|z|}^1 \frac{\log^2\left(\frac{r}{|z|}\right)}{\frac{r}{|z|} - 1} r\omega(r)\, dr, \end{aligned}$$

and we obtain, since $\log(x) \leqslant x - 1$ (for $x > 0$),

$$\left\| \delta_z \right\|_{(\mathscr{B}_\omega^1)^*} \leqslant \|\omega\|_{L^1(\mathbb{D},\sigma)}^{-1} + \frac{8}{\widetilde{\omega}(z)^2} \int_{|z|}^1 \log\left(\frac{r}{|z|}\right) r\omega(r)\, dr \leqslant \|\omega\|_{L^1(\mathbb{D},\sigma)}^{-1} + \frac{2}{\widetilde{\omega}(z)} \cdot$$

(iii) For any $r \in (|z|, 1)$, we have $\dfrac{z}{r} \in \mathbb{D}$ and from the well-known result on H^p we know that

$$|f(z)| = \left|f_r\left(\frac{z}{r}\right)\right| \leqslant \frac{\|f_r\|_{H^p}}{\left(1 - |\frac{z}{r}|^2\right)^{\frac{1}{p}}} \leqslant \frac{\|f_r\|_{H^p}}{\left(1 - |\frac{z}{r}|\right)^{\frac{1}{p}}},$$

where $f_r(z) = f(rz)$, but it will be more accurate to apply this to the function $w \mapsto (f(w) - f(0))/w$ so that

$$|f(z) - f(0)| \leqslant \left|\frac{z}{r}\right| \frac{\|f_r - f(0)\|_{H^p}}{\left(1 - |\frac{z}{r}|\right)^{\frac{1}{p}}} \leqslant \frac{2\|f_r\|_{H^p}}{\left(|\frac{r}{z}| - 1\right)^{\frac{1}{p}}}.$$

Multiplying by $4r\,\omega(r)\log\left(\dfrac{r}{|z|}\right)$ and integrating over $(|z|, 1)$, we obtain

$$|f(z) - f(0)|\,\widetilde{\omega}(|z|) \leqslant 4\int_{|z|}^{1} \frac{2r\|f_r\|_{H^p}\,\omega(r)\log\left(\frac{r}{|z|}\right)}{\left(|\frac{r}{z}| - 1\right)^{\frac{1}{p}}}\,dr.$$

Use Hölder's inequality (in spaces with associated measure $2r\omega(r)\,dr$) to get

$$|f(z) - f(0)|\,\widetilde{\omega}(|z|) \leqslant 4\|f\|_{\mathscr{B}^p_\omega}\left(\int_{|z|}^{1} \frac{\left(\log\left(\frac{r}{|z|}\right)\right)^{p'}}{\left(|\frac{r}{z}| - 1\right)^{\frac{p'}{p}}}\,2r\omega(r)\,dr\right)^{\frac{1}{p'}}.$$

We observe that $\dfrac{p'}{p} = p' - 1 > 0$ and we use that $\log(x) \leqslant x - 1$ to get

$$\begin{aligned}|f(z) - f(0)|\,\widetilde{\omega}(|z|) &\leqslant 4\|f\|_{\mathscr{B}^p_\omega}\left(\int_{|z|}^{1} \log\left(\frac{r}{|z|}\right) 2r\omega(r)\,dr\right)^{\frac{1}{p'}} \\ &= 2^{1+\frac{1}{p}}\|f\|_{\mathscr{B}^p_\omega}\left(\widetilde{\omega}(|z|)\right)^{\frac{1}{p'}}.\end{aligned}$$

In other words, since $|f(0)| = |a_0^*(f)| \leqslant \|\omega\|_{L^1(\mathbb{D},\sigma)}^{-\frac{1}{p}}\|f\|_{\mathscr{B}^p_\omega}$,

$$|f(z)| \leqslant |f(0)| + \frac{2^{1+\frac{1}{p}}}{\left(\widetilde{\omega}(|z|)\right)^{\frac{1}{p}}}\|f\|_{\mathscr{B}^p_\omega} \leqslant \left[\|\omega\|_{L^1(\mathbb{D},\sigma)}^{-\frac{1}{p}} + \frac{2^{1+\frac{1}{p}}}{\left(\widetilde{\omega}(|z|)\right)^{\frac{1}{p}}}\right]\|f\|_{\mathscr{B}^p_\omega},$$

and this gives the result. □

1.12 Operator Theory

1.12.1 Compact and Weakly Compact Operators

For Banach spaces X and Y, we denote $B(X,Y)$ the set of bounded operators $T : X \to Y$, and by B_X the closed unit ball of X.

The spectrum $\sigma(T)$ of an operator T is $\{\lambda \in \mathbb{C} \mid \lambda I - T \text{ is not invertible}\}$, and the point spectrum $\sigma_p(T)$ is the set of all the eigenvalues of T.

Definition 1.99 (Compact and weakly compact operators) We recall

- A bounded operator T between X and Y is called compact if $T(B_X)$ is relatively compact in Y. The set of which is denoted $\mathscr{K}(X,Y)$, and $\mathscr{K}(X)$ when $Y = X$.
- A bounded operator T between X and Y is called weakly compact if $T(B_X)$ is relatively weakly compact in Y. The set of which is denoted $\mathscr{W}(X,Y)$, and $\mathscr{W}(X)$ when $Y = X$.

We first focus on *compactness*, which can alternatively be formulated as: for any bounded sequence (x_n) of X, some subsequence $(u(x_{n_k}))$ is norm-convergent in Y.

A third formulation goes as follows: denoting in general by $N(A,B) \leqslant +\infty$ the smallest number of translates of B needed to cover A, we have

$$N(T(B_X), \varepsilon B_Y) < +\infty \text{ for each } \varepsilon > 0.$$

Then, for every integer $n \geqslant 1$, we denote $\mathfrak{e}_n(T)$ the nth *entropy number* of T, namely

$$\mathfrak{e}_n(T) = \inf\left\{\varepsilon > 0\,;\; N(TB_X,\ \varepsilon B_Y) \leqslant 2^{n-1}\right\}$$

(the map $n \mapsto \mathfrak{e}_n(T)$ is a kind of an inverse function to $\varepsilon \mapsto N(TB_X,\ \varepsilon B_Y)$). Clearly, whatever the Banach spaces X, Y are, T is compact if and only if $\mathfrak{e}_n(T)$ tends to 0. Those numbers are studied in greater detail in Appendix G.

The space X is said to have the *approximation property* if the identity of X can be approximated by finite rank operators uniformly on compact subsets of X. This is the case when X has a Schauder basis ([108, chapter 9]).

We now recall some basic properties of those compact operators ([108], [180]).

Theorem 1.100 (i) *$\mathscr{K}(X,Y)$ is a norm-closed subspace of $B(X,Y)$ containing the set $\mathscr{A}(X,Y)$ of all operators approximable by finite rank operators.*

(ii) $\mathscr{A}(X,Y) = \mathscr{K}(X,Y)$ *for all* Y *if and only if* X *has the approximation property.*

(iii) *Compact operators form* a two-sided ideal, *meaning that if*

$$A \in B(V,X), T \in \mathscr{K}(X,Y),\ B \in B(Y,Z),$$

then $BTA \in \mathscr{K}(V,Z)$.

(iv) *(Schauder's theorem) An operator* $T : X \to Y$ *is compact if and only if its adjoint* $T^* : Y^* \to X^*$ *is.*

A basic second theorem (Riesz theory) goes as follows:

Theorem 1.101 *Let* $T \in \mathscr{K}(X)$ *and* I *be the identity on* X. *The following are equivalent:*

(i) $I - T$ *is injective.*
(ii) $I - T$ *is surjective.*
(iii) $I - T$ *is invertible in* $B(X)$.

As a consequence, $\sigma(T)\backslash\{0\} = \sigma_p(T)\backslash\{0\}$. *Moreover,* $\sigma_p(T)\backslash\{0\}$ *is countable and consists of a sequence* (λ_n) *with limit* 0.

This theorem makes the spectral theory of compact operators very similar to that of operators in finite dimension. All the more as the point spectrum $\sigma_p(T)$ will be at our disposal for compact (weighted) composition operators, thanks to the König and Gunatillake theorems (see Theorems 8.2 and 8.6).

It is worth mentioning the special case of Hilbert spaces H.

Theorem 1.102 *(Polar decomposition) Let* $T \in \mathscr{K}(H)$. *Then there exist two orthonormal sequences* $(a_n)_{n\geqslant 1}$ *and* $(b_n)_{n\geqslant 1}$, *and a sequence* $(s_n)_{n\geqslant 1}$ *of non-negative numbers with limit zero such that*

$$T(x) = \sum_{n=1}^{\infty} s_n \langle x, b_n\rangle a_n. \tag{1.79}$$

Symbolically

$$T = \sum_{n=1}^{\infty} s_n\, a_n \otimes b_n,$$

or $T = U|T|$ *with* $|T| = (T^*T)^{\frac{1}{2}}$ *and* U *is a partial isometry.*

Observe the order in which a_n and b_n are put. In writing $T = \sum_{n=1}^{\infty} s_n a_n \otimes b_n$, the dependence is linear w.r.t. a_n (the left variable) and antilinear w.r.t. b_n (the right variable). The numbers s_n are none other than the eigenvalues of $|T|$

in decreasing order and are called the *singular* values (or numbers) of T. They will be revisited in Appendix G.

For non-compact operators, it is interesting to know how far from compact they are. There are many ways to express it, but one is to estimate the distance to the space of compact operators:

Definition 1.103 (Essential norm) Let $T : X \to Y$ be a bounded operator between two Banach spaces. The *essential norm* of T is its distance to the space of compact operators:

$$\|T\|_e = d\big(T, \mathscr{K}(X,Y)\big) = \inf\big\{\|T - K\| \,:\, K \in \mathscr{K}(X,Y)\big\}\,. \tag{1.80}$$

This is the norm of the class of T in the quotient space $B(X,Y)/\mathscr{K}(X,Y)$. When $X = Y$, the algebra $B(X)/\mathscr{K}(X)$ is called the Calkin algebra.

Since $\mathscr{K}(X,Y)$ is closed in $B(X,Y)$, it is clear that T is compact if and only if $\|T\|_e = 0$.

When several spaces come into play, we shall write sometimes $\|T\|_{e,X}$ (instead of $\|T\|_e$) to insist that this is the essential norm of T, viewed as an operator from X to X.

We now switch to *weakly compact* operators $X \to Y$ for which the following theorem turns out to be basic ([180, p. 49] or [108, vol 1, p. 2]).

Theorem 1.104 *(Eberlein–Smulian) A set $A \subset X$ is relatively weakly compact if and only if every sequence $(a_n) \subset A$ has a weakly convergent subsequence.*

First, an analogue of Theorem 1.100 holds.

Theorem 1.105 *Let X, Y be Banach spaces.*

(i) *$\mathscr{W}(X,Y)$ is a norm-closed subspace of $B(X,Y)$.*
(ii) *Weakly compact operators form a* two-sided ideal, *meaning that if*

$$A \in B(V,X), T \in \mathscr{W}(X,Y),\ B \in B(Y,Z),$$

then $BTA \in \mathscr{W}(V,Z)$.
(iii) *(Gantmacher's theorem) An operator $T : X \to Y$ is weakly compact if and only if its adjoint $T^* : Y^* \to X^*$ is.*

A key result on weakly compact operators ([180, p. 51]) is now the following:

Theorem 1.106 *An operator $T : X \to Y$ is weakly compact if and only if T factors through a* reflexive *space R: $T = VU$ where $U : X \to R$ and $V : R \to Y$ are bounded.*

In particular, $B(X,Y) = \mathscr{W}(X,Y)$ as soon as X or Y is reflexive.

This theorem indicates that there are many more weakly compact than compact operators. In our setting with $X = Y = H^p$, the most important cases will accordingly be H^1 and H^∞.

1.12.2 Hilbert–Schmidt Operators and Schatten Classes

Let H be a Hilbert space. We later abbreviate orthonormal basis to ONB.

One of the simplest set of compact operators showing up is the following.

Definition 1.107 The operator $T \in B(H)$ is called *Hilbert–Schmidt* if, for some ONB $(e_n)_{n\geqslant 0}$ of H, it holds

$$\sum_{n=0}^{\infty} \|T(e_n)\|^2 < \infty. \tag{1.81}$$

The set of those operators is denoted $HS(H) = HS$, and we define

$$\|T\|_{HS} = \left(\sum_{n=0}^{\infty} \|T(e_n)\|^2\right)^{1/2}.$$

Then, we have the following theorem.

Theorem 1.108 (i) *If $\sum_{n=0}^{\infty} \|T(e_n)\|^2 < \infty$ for some ONB (e_n), then we have $\sum_{n=0}^{\infty} \|T(f_n)\|^2 < \infty$ for all ONB (f_n).*
Moreover

$$\sum_{n=0}^{\infty} \|T(e_n)\|^2 = \sum_{n=0}^{\infty} \|T(f_n)\|^2 = \sum_{n=0}^{\infty} \|T^*(e_n)\|^2.$$

(ii) *The inclusion $HS \subset \mathcal{K}(H)$ holds.*
(iii) *$T \in HS$ if and only if its singular numbers s_n are square-summable. In that case,*

$$\sum_{n=0}^{\infty} \|T(e_n)\|^2 = \sum_{n=1}^{\infty} s_n^2 =: \|T\|_{HS}^2.$$

(iv) *$T \mapsto \|T\|_{HS}$ is a Hilbertian norm on the two-sided ideal HS.*

Since membership in HS amounts to membership of $(T(e_n))$ in $\ell^2(H)$, it is tempting to replace the exponent 2 by an arbitrary exponent $p \in (0, \infty)$, which naturally leads to the following definition.

Definition 1.109 Let $0 < p < \infty$. The Schatten class $\mathscr{S}^p = \mathscr{S}^p(H)$ is formed by those compact operators T such that the sequence $(s_n) = (s_n(T)) \in \ell^p$, equipped with

$$\mathfrak{s}_p(T) = \left(\sum_{n=1}^{\infty}(s_n(T))^p\right)^{1/p}. \tag{1.82}$$

With this definition, the following holds.

Theorem 1.110 *We have the following properties.*

(i) *$\mathscr{S}^p$ is a two-sided ideal of $B(H)$.*
(ii) *The map $T \mapsto \mathfrak{s}_p(T)$ is a Banach norm on $\mathscr{S}^p$ for $1 \leqslant p \leqslant \infty$, and a quasi-norm on $\mathscr{S}^p$ for $0 < p < 1$.*
(iii) *If $0 < p < q$, then $\mathscr{S}^p \subset \mathscr{S}^q$ and $\mathfrak{s}_p(T) \geqslant \mathfrak{s}_q(T)$.*
(iv) *$T \in \mathscr{S}^2$ if and only if $T \in HS$. Moreover*

$$\|T\|_{HS} = \mathfrak{s}_2(T) = \left(\sum_{n=1}^{\infty} s_n^2\right)^{1/2}.$$

(v) *$T \mapsto \mathfrak{s}_2(T)$ is a* Hilbertian norm *on the two-sided ideal $\mathscr{S}^2$.*

For $p = 2$, the spaces HS and $\mathscr{S}^2$ coincide, isometrically.

It is natural to ask what remains when $p \neq 2$ and the following will reveal useful (see [84, p. 40] and [39, theorem 4.6.b]; see also Lemma F.6 for a proof.

Theorem 1.111 *Let $T \in \mathscr{K}(H)$ and (e_n) an ONB. Then:*

(i) *If $0 < p \leqslant 2$ and $\sum_{n=0}^{\infty} \|T(e_n)\|^p < \infty$, then $T \in \mathscr{S}^p$ and*

$$\mathfrak{s}_p(T) \leqslant \left(\sum_{n=0}^{\infty} \|T(e_n)\|^p\right)^{1/p}.$$

(ii) *If $2 \leqslant p < \infty$ and $T \in \mathscr{S}^p$, then*

$$\left(\sum_{n=0}^{\infty} \|T(e_n)\|^p\right)^{1/p} \leqslant \mathfrak{s}_p(T).$$

(iii) *For every $p \geqslant 1$,*

$$\mathfrak{s}_p(T) = \sup\left(\sum_{n\geqslant 1} \left|\langle T(h_n), k_n\rangle\right|^p\right)^{\frac{1}{p}},$$

where the supremum runs over all the orthonormal sequences $\left(h_n\right)_{n\geqslant 1}$ and $\left(k_n\right)_{n\geqslant 1}$ in H.

A basic tool for establishing the assertions of Theorem 1.110 are the Horn and Weyl inequalities, to be revisited and exploited, which say the following:

Theorem 1.112 *(Horn) Let $A, B \in \mathscr{K}(H)$. Then, for all integers $n \geqslant 1$,*

$$\prod_{j=1}^{n} s_j(AB) \leqslant \prod_{j=1}^{n} s_j(A)s_j(B).$$

One will observe that these inequalities do not compare pointwise $s_n(AB)$ and $s_n(A)s_n(B)$ but their partial products. As long as one deals with those "matrix inequalities," this is intrinsic to the theory (pointwise inequalities fail) and will motivate the notion of subordination (and log-subordination) of sequences to be used again in Chapter 12 (see Appendix F for reminders).

Theorem 1.113 *(Weyl) Let $T \in \mathscr{K}(H)$, with eigenvalues (λ_n) and singular values (s_n). Then, for all integers $n \geqslant 1$,*

$$\prod_{j=1}^{n} |\lambda_j| \leqslant \prod_{j=1}^{n} s_j.$$

In the preceding satement, the eigenvalues are enumerated in the non-increasing order of their modulus.

The same observation as for Horn inequalities holds.

Even if we will not make detailed use of those facts, it is worth mentioning that the Banach space properties of Schatten classes $\mathscr{S}^p$ are quite similar to those of the spaces L^p, even if the latter are never isomorphic to $\mathscr{S}^p$! Indeed, we have the following:

Theorem 1.114 *Fix $1 < p < \infty$. Then:*

(i) *The dual space of $\mathscr{S}^p$ is $\mathscr{S}^{p'}$, with $1/p + 1/p' = 1$.*
(ii) *The space $\mathscr{S}^p$ is uniformly convex and satisfies Clarkson-type inequalities called Mc Carthy's inequalities.*
(iii) *The space $\mathscr{S}^p$ has type $\inf(p, 2)$ and cotype $\sup(p, 2)$.*

For (i) and (ii), we refer to [168, p. 20]. The third item is due to Nicole Tomczak–Jaegermann (see [176] for instance).

1.12.3 Trace Class Operators and Nuclear Operators

We recall that the simplest (non-trivial) operators that we can easily produce on general Banach spaces X and Y are rank 1 operators:

$$x \longmapsto (x^* \otimes y)(x) = x^*(x)y,$$

where $x^* \in X^*$ and $y \in Y$ (not equal to 0).

Then we can produce finite sums of such operators and obtain finite rank operators. We can go slighly further.

Definition 1.115 Let X and Y be Banach spaces. A nuclear operator from X to Y is an operator T such that

$$\forall x \in X, \qquad T(x) = \sum_{n \geqslant 0} x_n^*(x) y_n, \qquad \text{where} \sum_{n \geqslant 0} \|x_n^*\|_{X^*} \|y_n\|_Y < \infty.$$

We write $T \in \mathscr{N}(X,Y)$ (simply $\mathscr{N}(X)$ when $X = Y$).

In other words, it is an absolutely converging sum of rank 1 operators with the previous notations:

$$T = \sum_{n \geqslant 0} x_n^* \otimes y_n \qquad \text{with} \sum_{n \geqslant 0} \|x_n^*\|_{X^*} \|y_n\|_Y < \infty. \tag{1.83}$$

It is plain that any nuclear operator is a compact operator: $\mathscr{N}(X,Y) \subset \mathscr{K}(X,Y)$.

For any nuclear operator T, we may define its nuclear norm $\|T\|_{\mathscr{N}}$ as the lower bound of the sums $\sum \|x_n^*\|_{X^*} \|y_n\|_Y < \infty$ over all the decompositions as in (1.83). It is clear that it defines a norm and that $\big(\mathscr{N}(X,Y), \|.\|_{\mathscr{N}}\big)$ is a Banach space. Moreover, $\|T\| \leqslant \|T\|_{\mathscr{N}}$ for every nuclear operator.

Let us mention a link between approximation numbers and the nuclear operators. We recall the definition of *approximation numbers* (cf. Definition G.1); where $T : X \to Y$ in an operator between Banach spaces X and Y,

$$\forall n \geqslant 1, \qquad a_n(T) = \inf_{\operatorname{rank} R < n} \|T - R\|.$$

In particular $a_1(T) = \|T\|$.

We emphasize the following link between the nuclear norm and the operator nom for finite rank operators.

Lemma 1.116 *Let $T : X \to Y$ be an operator between Banach spaces X and Y with finite rank:* $\dim T(X) = d \geqslant 1$.

Then T is a nuclear operator with norm

$$\|T\|_{\mathscr{N}} \leqslant d \|T\|. \tag{1.84}$$

Proof: We can choose an Auerbach system in $T(X)$, that is, vectors $u_1, \dots, u_d$ in the unit ball of $T(X)$ and $\xi_1, \dots, \xi_d$ in the unit ball of the dual of $T(X)$, with $\xi_i(u_j) = \delta_{i,j}$ (the Kronecker symbol). We can write for every $y \in T(X)$:

$$y = \sum_{k=1}^{d} \xi_k(y) u_k$$

hence, for every $x \in X$, we have $T(x) = \sum_{k=1}^{d} \psi_k(x) u_k$ where $\psi_k = \xi_k \circ T \in X^*$.

Since, for each k, we have $\|\psi_k\| \leqslant \|\xi_k\| \, \|T\| = \|T\|$, we conclude that

$$\|T\|_{\mathscr{N}} \leqslant \sum_{k=1}^{d} \|\psi_k\|.\|u_k\| \leqslant d\|T\| . \qquad \square$$

Proposition 1.117 *Let $T : X \to Y$ be an operator between Banach spaces X and Y such that $\sum a_n(T) < \infty$.*

Then T is a nuclear operator.

Proof: In this proof, we write a_m instead of $a_m(T)$. By definition, for every $n \geqslant 0$, there exists an operator T_n, with rank strictly less than 2^n such that

$$\|T - T_n\| \leqslant a_{2^n} + \frac{1}{3^n} \cdot \tag{1.85}$$

It is clear that $T_0 = 0$.

Therefore, for every $n \geqslant 1$,

$$\|T_n - T_{n-1}\| \leqslant a_{2^n} + a_{2^{n-1}} + 4.3^{-n} \leqslant 2a_{2^{n-1}} + 4.3^{-n}. \tag{1.86}$$

On the other hand, write $S_n = T_n - T_{n-1}$. Since $S_n(X)$ has finite dimension $d < 3.2^{n-1}$, we get from Lemma 1.116 that

$$\|T - T_n\|_{\mathscr{N}} \leqslant 3.2^{n-1} \|T_n - T_{n-1}\|.$$

We clearly have $T = \sum_{n \geqslant 1} \left(T_n - T_{n-1}\right)$ thanks to (1.85) (recall $T_0 = 0$) since $a_n \to 0$. Moreover, using (1.86), we obtain

$$\sum_{n \geqslant 1} \left\|T_n - T_{n-1}\right\|_{\mathscr{N}} \lesssim 1 + \sum_{m \geqslant 0} 2^m a_{2^m} \leqslant 1 + 2 \sum_{n \geqslant 1} a_n < \infty$$

since (a_n) is non-increasing.

Therefore, the series $\sum_{n \geqslant 1} \left(T_n - T_{n-1}\right)$ converges in the Banach space $\mathscr{N}(X,Y)$ to a nuclear operator, which must be T because it is also its limit in the space $B(X,Y)$ (recall the comparison of the norms). We conclude that T is a nuclear operator. $\square$

Nuclear Operators on Hilbert Spaces

In this framework, thanks to the Riesz representation theorem, the definition of a nuclear operator can be rewritten as $T \in \mathscr{N}(H)$ if and only if

$$\forall x \in H, \qquad T(x) = \sum_{n \geqslant 0} \langle x, v_n \rangle y_n, \qquad \text{where} \sum_{n \geqslant 0} \|v_n\|_H \|y_n\|_H < \infty .$$

Let us produce some natural examples in this framework: we fix an infinite dimensional separable (complex) Hilbert space H and $(\mathrm{e}_n)_{n \geqslant 0}$ an orthonormal basis.

Given $\lambda = (\lambda_n)_{n \geqslant 0}$ a bounded sequence in $\mathbb{C}$, we naturally define a bounded operator by the relations $D_\lambda(\mathrm{e}_n) = \lambda_n \mathrm{e}_n$ for every $n \geqslant 0$. Actually, for every $x \in H$,

$$D_\lambda(x) = \sum_{n \geqslant 0} x_n \lambda_n \mathrm{e}_n \qquad \text{where } x = \sum_{n \geqslant 0} x_n \mathrm{e}_n,$$

and

$$\|D_\lambda(x)\|_H^2 = \sum_{n \geqslant 0} |x_n \lambda_n|^2 \leqslant \|\lambda\|_\infty^2 \|x\|_H^2 .$$

Moreover

$$\|D_\lambda\| \geqslant \sup_{n \geqslant 0} \|T(\mathrm{e}_n)\|_H = \sup_{n \geqslant 0} |\lambda_n|,$$

so that $\|D_\lambda\| = \|\lambda\|_\infty$.

One easily checks that D_λ is compact if and only if $\lambda \in c_0$. By definition of the Hilbert–Schmidt class, it is also clear that D_λ is a Hilbert-Schmidt operator if and only if $\lambda \in \ell^2$.

Proposition 1.118 *With the previous notations, we fix $\lambda \in \ell^\infty$.*
The operator $D_\lambda \in \mathscr{N}(H)$ if and only if $\lambda \in \ell^1$.

We could use Proposition 1.117, but we give a simple self-contained proof.

Proof: Observe that the functional $x \mapsto \mathrm{e}_n^*(x) = \langle x, \mathrm{e}_n \rangle = x_n$, where $x = \sum_{n \geqslant 0} x_n \mathrm{e}_n$, is bounded with norm 1.

Assume that $\lambda \in \ell^1$. Then $\sum_{n \geqslant 0} \lambda_n \, \mathrm{e}_n^* \otimes \mathrm{e}_n$ is a bounded operator that coincides with D_λ on an orthonormal basis; hence is equal to D_λ.

For the converse, assume that

$$\forall x \in H, \qquad D_\lambda(x) = \sum_{n \geqslant 0} \langle x, v_n \rangle y_n, \qquad \text{where} \sum_{n \geqslant 0} \|v_n\|_H \|y_n\|_H < \infty .$$

Then

$$\sum_{n\geqslant 0}|\lambda_n| = \sum_{n\geqslant 0}\big|\langle \mathrm{e}_n, D_\lambda(\mathrm{e}_n)\rangle\big| \leqslant \sum_{k\geqslant 0}\sum_{n\geqslant 0}\big|\langle v_k, \mathrm{e}_n\rangle\langle \mathrm{e}_n, y_k\rangle\big|$$
$$\leqslant \sum_{k\geqslant 0}\|v_k\|_H\,\|y_k\|_H < \infty$$

using the Cauchy–Schwarz inequality (in ℓ^2, not H !). □

The following proposition clarifies the link between nuclear and Hilbert–Schmidt operators on Hilbert spaces.

Proposition 1.119 *On a Hilbert space H, nuclear operators are exactly the set of products of two Hilbert–Schmidt operators.*

Proof: Fix an ONB $(b_n)_{n\geqslant 0}$.

If $T = AB$ with $A, B \in \mathscr{S}^2$, then for every $x \in H$, we have

$$T(x) = A\big(B(x)\big) = \sum_{n\geqslant 0}\langle B(x), b_n\rangle A(b_n)$$

with $\sum \|A(b_n)\|^2<\infty$ and $\sum \|B(b_n)\|^2 = \sum \|B^*(b_n)\|^2<\infty$. Therefore, the map $x \mapsto \langle B(x), b_n\rangle A(b_n)$ is a rank 1 operator with norm $\|A(b_n)\|.\|B^*(b_n)\|$, and this sequence is summable.

Now assume that T is nuclear: for every $x \in H$,

$$T(x) = \sum_{n\geqslant 0}\langle x, u_n\rangle y_n \quad \text{with} \quad \sum \|u_n\|.\|y_n\| < \infty\,.$$

Inspired by the previous argument, we define the operator A by $A(b_n) = \sqrt{\frac{\|u_n\|}{\|y_n\|}}\,y_n$ (it is 0 when $y_n = 0$) and the operator B as the adjoint of the operator V defined by $V(b_n) = \sqrt{\frac{\|y_n\|}{\|u_n\|}}.u_n$ (it is 0 when $u_n = 0$). Clearly A and B are Hilbert–Schmidt operators (in particular bounded on H). We easily check that $AB = T$. □

It turns out that we can define a trace for nuclear operators. More precisely, we have the following theorem:

Theorem 1.120 (Trace class operators) *Let $T \in B(H)$. We have $T \in \mathscr{N}(H)$ if and only if, for any orthonormal basis (b_n), $\sum_n \langle |T|(b_n), b_n\rangle < \infty$.*

For such an operator, we define the trace of the operator T:

$$\mathrm{Tr}(T) = \sum_{n\geqslant 0}\langle T(b_n), b_n\rangle,$$

and this quantity does not depend on the choice of the orthonormal basis.

Let us mention that in the trace formula, the sum does not depend on the order of summation (since it is true for any orthonormal basis), so the convergence is absolute:

$$\sum_{n\geqslant 0} \left|\langle T(b_n), b_n\rangle\right| < \infty .$$

Moreover, let us mention that, for such an operator T, there exists an orthonormal basis (b_n) such that $\sum_{n\geqslant 0} \left\|T(b_n)\right\| < \infty$. Indeed, using (1.87) in the proof below, we have (with the same notations) $T(\beta_n) = U|T|(\beta_n) = \lambda_n U\beta_n$ with $(\lambda_n) \in \ell^1$ and $\|T(\beta_n)\| = \lambda_n$.

Observe that, for any orthonormal sequence $(b_n)_{n\geqslant 0}$, $\langle |T|(b_n), b_n\rangle \geqslant 0$; hence $\mathrm{Tr}(|T|)$ makes sense with values in $[0, +\infty]$. In particular, $T \in \mathcal{N}(H)$ if and only if $\mathrm{Tr}(|T|) < \infty$.

Proof: Assuming $T \in \mathcal{N}(H)$, we can write

$$\forall x \in H, \qquad T(x) = \sum_{n\geqslant 0} \langle x, v_n\rangle y_n, \qquad \text{where } \sum_{n\geqslant 0} \|v_n\|_H \|y_n\|_H < \infty.$$

Point out that

$$\forall x \in H, \qquad T^*(x) = \sum_{n\geqslant 0} \langle x, y_n\rangle v_n.$$

Since $T = U|T|$ where U is a partial isometry satisfying $U^*T = U^*U|T| = |T|$, we have, for any orthonormal basis $(b_n)_{n\geqslant 0}$,

$$\langle |T|(b_n), b_n\rangle = \langle U^*T(b_n), b_n\rangle.$$

Then

$$\sum_{n\geqslant 0} \langle |T|(b_n), b_n\rangle \leqslant \sum_{k\geqslant 0}\sum_{n\geqslant 0} \left|\langle b_n, v_k\rangle\langle U^*y_k, b_n\rangle\right|,$$

and we conclude via Cauchy–Schwarz like in the proof of Proposition 1.118.

For the converse, we claim that $|T|$ is a compact operator. Indeed, it turns out that $|T|^{\frac{1}{2}}$ is a Hilbert–Schmidt operator:

$$\sum_{n\geqslant 0} \||T|^{\frac{1}{2}}(b_n)\|^2 = \sum_{n\geqslant 0} \langle |T|(b_n), b_n\rangle < \infty;$$

hence it is compact. A fortiori, $|T|$ itself is compact.

Since $|T|$ is a positive (self-adjoint) compact operator, we can write a Schmidt decomposition:

$$|T|(x) = \sum_{n\geqslant 0} \lambda_n \langle x, \beta_n\rangle \beta_n, \tag{1.87}$$

where $(\beta_n)_{n\geqslant 0}$ is an orthonormal basis and λ_n's are non-negative.

The hypothesis now reads as $\sum_{n\geqslant 0} \lambda_n < \infty$. The conclusion follows.

Finally, let us show that the quantity $\sum_{n\geqslant 0} \langle T(b_n), b_n\rangle$ does not depend on the choice of the orthonormal basis $(b_n)_{n\geqslant 0}$.

Let us fix two orthonormal bases $(b_n)_{n\geqslant 0}$ and $(b'_n)_{n\geqslant 0}$.

$$\sum_{n\geqslant 0}\langle T(b_n), b_n\rangle = \sum_{n\geqslant 0}\sum_{k\geqslant 0}\overline{\langle b_n, b'_k\rangle}\langle T(b_n), b'_k\rangle = \sum_{k\geqslant 0}\sum_{n\geqslant 0}\langle b'_k, b_n\rangle\langle b_n, T^*(b'_k)\rangle,$$

so

$$\sum_{n\geqslant 0}\langle T(b_n), b_n\rangle = \sum_{k\geqslant 0}\langle b'_k, T^*(b'_k)\rangle = \sum_{k\geqslant 0}\langle T(b'_k), b'_k\rangle. \qquad \square$$

Finally, we mention that in this framework, nuclear operators coincide with operators in the Schatten class $\mathscr{S}^1$.

Theorem 1.121 $\mathscr{N}(H) = \mathscr{S}^1$.

Proof: Assume that $T \in \mathscr{N}(H)$. From the proof of Theorem 1.120 (see (1.87)), we know that

$$\forall x \in H, \qquad |T|(x) = \sum_{n\geqslant 0}\lambda_n\langle x, \beta_n\rangle\beta_n,$$

where $\lambda \in \ell^1$ and (β_n) is orthonormal. It means that $T \in \mathscr{S}^1$.

Now assume that $T \in \mathscr{S}^1$. We can write $T = U|T|$ where U is a partial isometry and

$$\forall x \in H, \qquad |T|(x) = \sum_{n\geqslant 0}\lambda_n\langle x, \beta_n\rangle\beta_n,$$

where (β_n) is an orthonormal basis and $\lambda \in \ell^1$.

$$\forall x \in H, \qquad T(x) = U|T|(x) = \sum_{n\geqslant 0}\lambda_n\langle x, \beta_n\rangle U\beta_n;$$

so T is a nuclear operator. $\square$

2

Boundedness of Composition Operators

2.1 First Meeting with C_φ

Given an analytic self-map $\varphi : \mathbb{D} \to \mathbb{D}$, called the *symbol* of the composition operator, we can always define

$$f \in \mathscr{H}(\mathbb{D}) \longmapsto C_\varphi(f) = f \circ \varphi \in \mathscr{H}(\mathbb{D}) .$$

Point out that for two symbols φ and ψ, we have $C_\varphi \circ C_\psi = C_{\psi\circ\varphi}$.

Many questions naturally then arise, in many directions. For functional analysts, the main viewpoint is the following: fix X and Y two spaces of holomorphic functions on $\mathbb{D}$ and focus on the properties of the composition operator

$$C_\varphi : f \in X \longmapsto f \circ \varphi \in Y .$$

There are immediately many natural questions: *when is C_φ defined, bounded? When is it compact? When is it "very" compact? More generally, when does C_φ belong to one of the operator ideals of our toolbox?*

Ideally, we aim to understand the link

$$\textbf{“operator } C_\varphi\textbf{”} \overset{?}{\longleftrightarrow} \textbf{“symbol } \varphi\textbf{”}$$

In other words, the aim is to understand how the properties of the *operator* C_φ (which involves *theory of operators*) are connected to the properties of the *symbol* φ (which involves *theory of functions*). Even though the pioneer work of Littlewood [119] (and then Riesz [149]) appeared in 1925 (but rather focused on functions and geometry), the very first such considerations seem to come back to Nordgren, Ryff, Schwartz in the 1960s [131, 155, 158] and Shapiro–Taylor [166] a bit later. Many others followed since then.

Potentially, there are as many questions as there are some 3-uplets (operator property, space X, space Y). Of course, there are also some specific results when we impose some restrictions on the symbol: it can be some geometric restriction (the shape of $\varphi(\mathbb{D})$ for instance), or some other kind of constraints: the valence, fixing the origin, being inner, ...

2.2 H^∞ and the Disk Algebra

Composition operators are clearly well defined and bounded on H^∞. Indeed, given a symbol φ, it is obvious that $f \circ \varphi \in H^\infty$ for every $f \in H^\infty$, with norm less than $\|f\|_\infty$ and $\|C_\varphi\|_{B(H^\infty)} \leqslant 1$. Since the constant function $\mathbb{1}$ satisfies $\mathbb{1} \circ \varphi = \mathbb{1}$, we have $\|C_\varphi\|_{B(H^\infty)} = 1$.

The same occurs for the disk algebra except that we have to add the extra assumption $\varphi \in A(\mathbb{D})$. Indeed if C_φ is well defined on $A(\mathbb{D})$ then $\varphi = C_\varphi(z)$ must belong to $A(\mathbb{D})$; and once we have $\varphi \in A(\mathbb{D})$, we have $f \circ \varphi \in A(\mathbb{D})$ for every $f \in A(\mathbb{D})$.

We summarize these results in the following proposition.

Proposition 2.1 *Let $\varphi : \mathbb{D} \to \mathbb{D}$ be analytic. Then*

- *C_φ is bounded on H^∞ and $\|C_\varphi\|_{B(H^\infty)} = 1$.*
- *C_φ is bounded on $A(\mathbb{D})$ if and only if $\varphi \in A(\mathbb{D})$. In that case, $\|C_\varphi\|_{B(A(\mathbb{D}))} = 1$.*

The next step is to determine whether the same situation occurs for other spaces of analytic functions. Our first interest is for Hardy spaces.

2.3 Boundedness on Hardy Spaces

2.3.1 Inner Symbols

We shall present in this section several proofs of the boundedness on Hardy spaces of the operator C_φ for inner functions. The framework may vary from a result to another.

Let us begin with a proof valid in the very special case of the Hilbertian framework and inner function.

Proposition 2.2 *Let $\varphi : \mathbb{D} \to \mathbb{D}$ be an inner function. Then C_φ is bounded on H^2 and*

$$\|C_\varphi\|_{B(H^2)} = \sqrt{\frac{1 + |\varphi(0)|}{1 - |\varphi(0)|}}\,.$$

Proof: Since, on one hand, the polynomials are dense in H^2 and, on the other hand, finitely supported sequences are dense in ℓ^2, it suffices to consider finite sums. Let (a_n) be a finitely supported sequence and write $f(z) = \sum_{n\geqslant 0} a_n z^n$. We have $C_\varphi(f) = \sum_{n\geqslant 0} a_n \varphi^n$ and we can estimate its norm (writing φ instead of φ^*):

$$\|C_\varphi(f)\|_2^2 = \sum_{k,l\geqslant 0} a_k \overline{a_l} \int_{\mathbb{T}} \varphi^k \overline{\varphi}^l \, dm.$$

But, for $k \geqslant l$, we have, since φ is inner (i.e., has modulus 1 a.e. on $\mathbb{T}$):

$$\int_{\mathbb{T}} \varphi^k \overline{\varphi}^l \, dm = \int_{\mathbb{T}} |\varphi|^{2l} \varphi^{k-l} \, dm = \int_{\mathbb{T}} \varphi^{k-l} \, dm.$$

Therefore, $\int_{\mathbb{T}} \varphi^k \overline{\varphi}^l \, dm = \varphi^{k-l}(0)$ when $k \geqslant l$.

In the same way $\int_{\mathbb{T}} \varphi^k \overline{\varphi}^l \, dm = \overline{\varphi^{l-k}(0)}$ when $k \leqslant l$.

We obtain

$$\|C_\varphi(f)\|_2^2 = \sum_{k,l\geqslant 0} a_k \overline{a_l} \widehat{\psi}(l-k) = \widehat{\psi |f|^2}(0) = \langle \psi f, f \rangle_{L^2(\mathbb{T})},$$

where ψ is nothing but the Poisson kernel P_a with $a = \varphi(0)$, that is,

$$P_a(\xi) = 1 + \sum_{n=1}^{+\infty} \bar{a}^n \xi^n + \sum_{n=1}^{+\infty} a^n \bar{\xi}^n = \frac{1-|a|^2}{|1-\bar{a}\xi|^2}\cdot$$

Since ψ is real valued and positive, the Toeplitz operator $T_\psi : H^2 \to H^2$ (see Appendix C) is self-adjoint and positive, hence

$$\sup_{\|f\|_2=1} \langle f, \psi f \rangle_{L^2(\mathbb{T})} = \sup_{\|f\|_2=1} \langle f, T_\psi(f) \rangle_{H^2} = \|T_\psi\| = \|\psi\|_\infty,$$

and we conclude that $\|C_\varphi\|_{B(H^2)}^2 = \|P_a\|_\infty = \dfrac{1+|\varphi(0)|}{1-|\varphi(0)|}\cdot$ □

We give another proof (actually a variant since the ideas are similar) but for general Hardy spaces H^p for any $p \geqslant 1$.

Theorem 2.3 *Let $\varphi : \mathbb{D} \to \mathbb{D}$ be an inner function. Then C_φ is bounded on H^p and*

$$\|C_\varphi\|_{B(H^p)} = \left(\frac{1+|\varphi(0)|}{1-|\varphi(0)|}\right)^{\frac{1}{p}}.$$

We first need the following simple lemma (see Definition 5.24).

Lemma 2.4 *Let $\varphi : \mathbb{D} \to \mathbb{D}$ be an inner function. Then the pull-back measure m_φ is actually the measure $P_{\varphi(0)}\, m$.*

Proof: For any continuous function $h \in C(\mathbb{T})$, since φ is inner, we have that $\varphi(z) \in \mathbb{T}$ for a.e. $z \in \mathbb{T}$, and $z \mapsto h \circ \varphi(z)$ is defined for a.e. $z \in \mathbb{T}$, and measurable. Moreover we have

$$\int_{\mathbb{T}} h \circ \varphi \, dm = \int_{\mathbb{T}} h \, dm_\varphi.$$

Applying this relation for every $n \in \mathbb{N}$ with $h(z) = z^n$, we get

$$\widehat{m_\varphi}(-n) = \int_{\mathbb{T}} \varphi^n \, dm = \varphi^n(0) = \widehat{P_{\varphi(0)}}(-n).$$

We can also test the function $h(z) = \bar{z}^n$ to get the equality for positive indices, but since m_φ and $P_{\varphi(0)} m$ are real-valued (finite) measures, we directly have equality of Fourier coefficients with positive index. We get the conclusion. □

Proof of Theorem 2.3.: For every polynomial f, we have

$$\|f \circ \varphi\|_{H^p}^p = \int_{\mathbb{T}} |f|^p \, dm_\varphi = \int_{\mathbb{T}} |f(z)|^p \frac{1 - |a|^2}{|1 - \bar{a}z|^2} \, dm(z), \qquad (2.1)$$

where $a = \varphi(0)$, thanks to Lemma 2.4.

Therefore the norm of C_φ appears clearly as the norm of the multiplier operator on H^p, associated to the weight $M(z) = \dfrac{(1 - |a|^2)^{\frac{1}{p}}}{(1 - \bar{a}z)^{\frac{2}{p}}}$ (see Appendix B.2). We get

$$\|C_\varphi\|_{B(H^p)} = \|M\|_\infty = \left(\frac{1 + |a|}{1 - |a|}\right)^{1/p}.$$

□

Let us mention that an automorphism θ of the disk can be written as $\theta = c\varphi_a$ (for some $a \in \mathbb{D}$ and some $c \in \mathbb{T}$), where φ_a is the Möbius transformation $\varphi_a(z) = \dfrac{a - z}{1 - \bar{a}z} \cdot$ We get

$$\|C_\theta\|_{B(H^p)} = \left(\frac{1 + |\theta(0)|}{1 - |\theta(0)|}\right)^{1/p}.$$

Let us mention too the following immediate consequence of (2.1), which shows that a composition operator associated to an inner symbol behaves almost like an isometry.

Proposition 2.5 *Let φ be an inner symbol and $p \geqslant 1$.*
For every f in H^p, we have

$$\|f \circ \varphi\|_{H^p} \geqslant \left(\frac{1 - |\varphi(0)|}{1 + |\varphi(0)|}\right)^{\frac{1}{p}} \|f\|_{H^p}. \tag{2.2}$$

In particular, when φ is inner and fixes the origin, C_φ is an isometry on H^p.

This implies that C_φ cannot be compact on H^p when φ is inner.

2.3.2 The "Dévissage" Principle

We explain here a general principle that allows to often focus on symbols fixing the origin.

Assume that we already know that every composition operator C_φ associated to a symbol φ fixing the origin is bounded on H^p (this will be proved in the sequel). Then we can conclude that every composition operator C_φ is bounded on H^p.

Indeed, let $\varphi : \mathbb{D} \to \mathbb{D}$ and denote $a = \varphi(0) \in \mathbb{D}$. We define $\widetilde{\varphi} = \varphi_a \circ \varphi$. This symbol fixes the origin. Point out that $\varphi = \varphi_a \circ \widetilde{\varphi}$. Our assumption implies that $C_{\widetilde{\varphi}}$ is bounded on H^p. Therefore $C_\varphi = C_{\widetilde{\varphi}} \circ C_{\varphi_a}$ is bounded on H^p. Moreover

$$\left\|C_\varphi\right\|_{B(H^p)} \leqslant \left(\frac{1 + |a|}{1 - |a|}\right)^{1/p} \left\|C_{\widetilde{\varphi}}\right\|_{B(H^p)}. \tag{2.3}$$

A General *Dévissage* Principle

We will use this trick several times in similar situations, for boundedness, not only on H^p but on many *other* Banach spaces $X \subset \mathscr{H}(\mathbb{D})$.

Assume that we are able to prove that a composition operator $C_\psi : X \to X$ associated to a symbol ψ belongs to some operator ideal $\mathcal{I}$ when ψ fixes the origin under some condition on the symbol ψ. Take any symbol φ and consider $\widetilde{\varphi} = \varphi_a \circ \varphi$ where $a = \varphi(0) \in \mathbb{D}$; then $\widetilde{\varphi}$ fixes the origin and the composition operator $C_{\widetilde{\varphi}}$ will belong to $\mathcal{I}$ as well (as soon as φ satisfies the right condition required for $\widetilde{\varphi}$). Assume also that we already proved the boundedness of $C_\theta : X \to X$ when the symbol θ is an automorphism. We can then conclude that $C_\varphi = C_{\widetilde{\varphi}} \circ C_{\varphi_a}$ belongs to $\mathcal{I}$.

2.3.3 The Subordination Principle when $p = 2$

Boundedness via the Backwardshift Operator

This is the original argument given by Littlewood in 1925 [119] (although Littlewood assumed that the symbol φ is univalent and $\varphi(0) = 0$).

We consider a symbol φ fixing the origin. The *dévissage* principle 2.3.2 will then give the final conclusion.

We introduce the backwardshift operator B on H^2 defined by

$$B(f)(z) = \sum_{n\geqslant 0} a_{n+1}z^n = \frac{f(z) - f(0)}{z}, \quad \text{where } f(z) = \sum_{n\geqslant 0} a_n z^n.$$

The operator is clearly well defined on H^2 and for $f \in H^2$, we have

$$\|B(f)\|_2^2 = \sum_{n=0}^{+\infty} |a_{n+1}|^2 = \|f\|_2^2 - |f(0)|^2 \leqslant \|f\|_2^2.$$

We mention here two key observations:

(i) $f(z) = f(0) + zB(f)(z)$, for every $z \in \mathbb{D}$.
(ii) For every $n \geqslant 0$, $B^n(f)(0) = a_n$

- Step 1: We work with a polynomial f. We have $f \circ \varphi \in H^\infty \subset H^2$. We can write, from (i), for every $z \in \mathbb{D}$:

$$f\big(\varphi(z)\big) = f(0) + \varphi(z)B(f)\big(\varphi(z)\big).$$

In other words,

$$C_\varphi(f) = f(0) + \big(M_\varphi \circ C_\varphi\big)\big(B(f)\big),$$

where M_φ is the multiplier operator on H^2, that is, $g \in H^2 \mapsto \varphi g \in H^2$, which has norm less than 1.

Since $\varphi(0) = 0$, the function φ belongs to zH^2, and $\varphi g \in zH^2$ as well, therefore is orthogonal to $\mathbb{C}.1$ for every $g \in H^2$. Hence we have

$$\|C_\varphi(f)\|_2^2 = |f(0)|^2 + \|M_\varphi C_\varphi B(f)\|_2^2 \leqslant |f(0)|^2 + \|C_\varphi B(f)\|_2^2. \tag{2.4}$$

Applying this relation (2.4) to $B(f)$, and then $B^2(f)$, and so on, we obtain by induction:

$$\begin{aligned}\big\|C_\varphi(f)\big\|_2^2 &\leqslant \big|f(0)\big|^2 + \big\|C_\varphi B(f)\big\|_2^2.\\ &\leqslant \big|f(0)\big|^2 + \big|B(f)(0)\big|^2 + \big\|C_\varphi B^2(f)\big\|_2^2.\\ &\ \vdots\\ &\ \vdots\\ &\leqslant \big|f(0)\big|^2 + \big|B(f)(0)\big|^2 + \cdots + \big|B^d(f)(0)\big|^2 + \big\|C_\varphi B^{d+1}(f)\big\|_2^2,\end{aligned}$$

where $d \geqslant 1$ is the degree of the polynomial f.

Observe that $B^{d+1}(f) = 0$ so, by (ii), we get

$$\|C_\varphi(f)\|_2^2 \leqslant \sum_{n=0}^{d} |a_n|^2 = \|f\|_2^2. \tag{2.5}$$

- Step 2: Let $f \in H^2$ and $S_N(f)$ be its Nth partial sum, where $N \geqslant 1$:

$$S_N(f)(z) = \sum_{n=0}^{N} a_n z^n\,, \quad \text{where } f(z) = \sum_{n\geqslant 0} a_n z^n.$$

Observe that

(iii) $\|S_N(f)\|_2 \leqslant \|f\|_2$.
(iv) For $g \in \mathscr{H}(\mathbb{D})$ satisfying $M = \sup_{N\geqslant 1} \|S_N(g)\|_2 < \infty$, we have $g \in H^2$ and $\|g\|_2 = M$.
(v) $S_N(C_\varphi(f)) = S_N\big(C_\varphi(S_N(f))\big)$.

Actually (v) follows from the fact that $S_N(\varphi^k) = 0$ when $k > N$ since $\varphi(0) = 0$.

Write $g = C_\varphi(f) \in \mathscr{H}(\mathbb{D})$. Since $S_N(f)$ and $S_N(g)$ are polynomials, we know from (v) that

$$\big\|S_N(g)\big\|_2 = \big\|S_N(C_\varphi(S_N(f)))\big\|_2\,.$$

But $C_\varphi(S_N(f)) \in H^2$ thanks to (2.5) and we have

$$\big\|S_N\big(C_\varphi(S_N(f))\big)\big\|_2 \leqslant \big\|C_\varphi(S_N(f))\big\|_2 \leqslant \big\|S_N(f)\big\|_2 \leqslant \big\|f\big\|_2.$$

Hence we proved that C_φ is bounded on H^2 with norm less than 1.

We would like to emphasize the stronger inequality obtained during the proof: $\|S_N(g)\|_2 \leqslant \|S_N(f)\|_2$. In other words:

$$\forall N \geqslant 0, \quad \big\|S_N(f\circ\varphi)\big\|_2 \leqslant \big\|S_N(f)\big\|_2\,. \tag{2.6}$$

Moreover, testing C_φ on the function $\mathbb{1}$, we have, when φ fixes the origin,

$$\big\|C_\varphi\big\| = 1.$$

An application of the *dévissage* principle 2.3.2 leads to

$$\big\|C_\varphi\big\|_{B(H^2)} \leqslant \sqrt{\frac{1+|\varphi(0)|}{1-|\varphi(0)|}} \qquad \text{for every symbol } \varphi. \tag{2.7}$$

Boundedness in the Hilbertian Case via the Nevanlinna Function

We already introduced the Nevanlinna function (see 1.26), but let us recall how it is defined

$$N_\varphi(w) = \begin{cases} +\infty & \text{if } w = \varphi(0). \\ 0 & \text{if } w \notin \varphi(\mathbb{D}). \\ \displaystyle\sum_{\varphi(z)=w} \log 1/|z| & \text{if } w \neq \varphi(0),\ w \in \varphi(\mathbb{D}) \end{cases}$$

(in the sum, every z occurs as many times as its multiplicity as a solution of $\varphi(z) = w$).

We already know (recall Theorem 1.37) that for every $w \neq \varphi(0)$ in the unit disk:

$$N_\varphi(w) \leqslant \log \left| \frac{1 - \overline{\varphi(0)} w}{\varphi(0) - w} \right| = O\big(1 - |w|\big) \quad \text{when } |w| \to 1^-.$$

We want to establish the boundedness of C_φ. We recall the Littlewood–Paley formula (1.79):

$$\|f\|_2^2 = |f(0)|^2 + 2 \int_{\mathbb{D}} |f'(z)|^2 \log \frac{1}{|z|} \, d\sigma(z).$$

Let us see how it implies the boundedness of C_φ on H^2:

$$\begin{aligned} \|f \circ \varphi\|_2^2 &= |f \circ \varphi(0)|^2 + 2 \int_{\mathbb{D}} |(f \circ \varphi)'(z)|^2 \log \frac{1}{|z|} \, d\sigma(z) \\ &= |f \circ \varphi(0)|^2 + 2 \int_{\mathbb{D}} \big(|f'(z)|^2 \circ \varphi\big) . \big|\varphi'(z)\big|^2 \log \frac{1}{|z|} \, d\sigma(z). \end{aligned}$$

Using the generalized (non-injective) change of variable formula (see Theorem 1.69), we get a very useful formula

$$\|f \circ \varphi\|_2^2 = |f \circ \varphi(0)|^2 + 2 \int_{\mathbb{D}} |f'(w)|^2 N_\varphi(w) \, d\sigma(w). \tag{2.8}$$

Now, since we assumed that $\varphi(0) = 0$, using the majorization $N_\varphi(w) \leqslant \log \dfrac{1}{|w|}$, we get

$$\|f \circ \varphi\|_2^2 \leqslant |f(0)|^2 + 2 \int_{\mathbb{D}} |f'(w)|^2 \log \frac{1}{|w|} \, d\sigma(w) = \|f\|_2^2,$$

which proves (again) the boundedness of C_φ on H^2 when φ fixes the origin, and also proves that its operator norm is 1 in this case.

The general case follows from the dévissage principle 2.3.2 and leads to the same upper estimate of the norm (see (2.7)).

Sarason's Approach

This approach follows the spirit of the proof of Theorem 2.2 given in the specific case of inner functions.

Let again, for $a \in \mathbb{D}$ and $w \in \mathbb{T}$, the Poisson kernel P_a be defined by

$$P_a(w) = \frac{1 - |a|^2}{|w - a|^2} = \frac{1 - |a|^2}{|1 - a\overline{w}|^2} = \operatorname{Re} \frac{1 + a\overline{w}}{1 - a\overline{w}} \cdot$$

If φ is a symbol with, say, $\|\varphi\|_\infty < 1$, and if $f \in H^2$, with boundary values still denoted by f, we can write for a.e. $z \in \mathbb{T}$:

$$C_\varphi(f)(z) = f[\varphi(z)] = \int_{\mathbb{T}} P_{\varphi(z)}(w) f(w)\, dm(w) =: \int_{\mathbb{T}} K(z, w) f(w)\, dm(w). \tag{2.9}$$

The operator C_φ on H^2 appears as the restriction to H^2 of the kernel operator T_K on $L^2 = L^2(m)$ given by the kernel K, with

$$K(z, w) = P_{\varphi(z)}(w).$$

All this makes sense because $\|\varphi\|_\infty < 1$, hence $|\varphi(z)| < 1$ for almost every $z \in \mathbb{T}$.

Now, we apply the L^2-version of the Schur test lemma (see Theorem A.1) and we only have to check the assumptions with our specific kernel K. We note that K is non-negative. Moreover

$$\int_{\mathbb{T}} P_{\varphi(z)}(w)\, dm(w) = 1 \quad \text{for almost all } z \in \mathbb{T}, \tag{2.10}$$

which is a usual property of the Poisson kernel (it can also be recovered using the same argument as below).

$$\int_{\mathbb{T}} P_{\varphi(z)}(w)\, dm(z) = \operatorname{Re} \frac{1 + \varphi(0)\overline{w}}{1 - \varphi(0)\overline{w}} \leqslant \frac{1 + |\varphi(0)|}{1 - |\varphi(0)|} \quad \text{for a.e. } w \in \mathbb{T}. \tag{2.11}$$

Indeed, for fixed $w \in \mathbb{T}$, the function

$$h(z) = \operatorname{Re} \frac{1 + \varphi(z)\overline{w}}{1 - \varphi(z)\overline{w}} = P_{\varphi(z)}(w)$$

is harmonic in $\mathbb{D}$ as the real part of an analytic function (which actually belongs to $H^\infty \subset H^1$), hence

$$h(0) = \int_{\mathbb{T}} h(z)\, dm(z),$$

and this gives (2.11). Then, Schur's test with $A = 1$ and $B = \dfrac{1 + |\varphi(0)|}{1 - |\varphi(0)|}$ gives the result.

In the general case $\|\varphi\|_\infty \leqslant 1$, we apply the preceding to the symbol defined by $z \in \mathbb{D} \mapsto \varphi(rz)$, with $0 < r < 1$, whose sup-norm is strictly less than 1. We get for any $f \in H^2$

$$\int_{\mathbb{T}} |f(\varphi(rz))|^2 \, dm(z) \leqslant \frac{1 + |\varphi(0)|}{1 - |\varphi(0)|} \cdot \|f\|^2.$$

By definition of the H^2 norm, taking the supremum over $r \in (0, 1)$, we get in one stroke (i.e., without separating the cases $\varphi(0) = 0$ and $\varphi(0) \neq 0$) that $f \circ \varphi \in H^2$.

Actually, we proved that for any symbol φ, the operator C_φ is bounded on H^2 with

$$\|C_\varphi\|_{B(H^2)} \leqslant \sqrt{\frac{1 + |\varphi(0)|}{1 - |\varphi(0)|}}. \tag{2.12}$$

2.3.4 The Subordination Principle when $p \geqslant 1$

We present here also several approaches to prove the boundedness of composition operators on H^p for any value of $p \geqslant 1$. Actually, we restrict our framework to the Banach space point of view, hence $p \geqslant 1$, but the arguments below could be extended obviously to the range $p > 0$.

From the Hilbertian Case

The idea here is to use that we already proved the boundedness on H^2.

Fix $p \geqslant 1$. For any $f \in H^p$, with $f \neq 0$, we can write $f = Bg$ where B is a Blaschke product and g does not vanish on $\mathbb{D}$ and $|f| = |g|$ a.e. on $\mathbb{T}$ (see Proposition 1.75). The function $h = g^{\frac{p}{2}}$ defines an analytic function belonging to H^2 and $\|h\|_2^2 = \|f\|_p^p$.

For every $r \in (0, 1)$, we have

$$\begin{aligned} \int_{\mathbb{T}} \left|f \circ \varphi(rz)\right|^p dm &= \int_{\mathbb{T}} \left|B \circ \varphi(rz).g \circ \varphi(rz)\right|^p dm \\ &\leqslant \int_{\mathbb{T}} \left|g \circ \varphi(rz)\right|^p dm = \int_{\mathbb{T}} \left|h \circ \varphi(rz)\right|^2 dm, \end{aligned}$$

since B is inner.

From the result on H^2, we get

$$\int_{\mathbb{T}} \left|h \circ \varphi(rz)\right|^2 dm \leqslant \left\|C_\varphi(h)\right\|_2^2 \leqslant \left\|C_\varphi\right\|_{B(H^2)}^2 \left\|h\right\|_2^2 = \left\|C_\varphi\right\|_{B(H^2)}^2 \left\|f\right\|_p^p.$$

Since this is valid for any $r \in (0, 1)$ we obtain

$$\left\|C_\varphi(f)\right\|_p \leqslant \left\|C_\varphi\right\|_{B(H^2)}^{\frac{2}{p}} \left\|f\right\|_p$$

hence C_φ is bounded on H^p and

$$\|C_\varphi\|_{B(H^p)} \leqslant \|C_\varphi\|_{B(H^2)}^{\frac{2}{p}}.$$

We could prove the reverse inequality in the same way. Actually we have proved the following proposition:

Proposition 2.6 *Let $\varphi : \mathbb{D} \to \mathbb{D}$ be analytic. Let $p, q \geqslant 1$. Then*

$$\|C_\varphi\|_{B(H^q)}^q = \|C_\varphi\|_{B(H^p)}^p. \tag{2.13}$$

In particular

$$\|C_\varphi\|_{B(H^p)} \leqslant \left(\frac{1+|\varphi(0)|}{1-|\varphi(0)|}\right)^{\frac{1}{p}}. \tag{2.14}$$

Let us mention that for a direct proof of the preceding result, it suffices to prove $\|C_\varphi\|_{B(H^q)}^q \leqslant \|C_\varphi\|_{B(H^p)}^p$ for arbitrary $p, q \geqslant 1$ (adapt the case $q = 2$). The symmetry in this inequality leads directly to an equality.

The Harmonic Majorant Approach

The argument below goes back to F. Riesz [149].

Let $\varphi\colon \mathbb{D} \to \mathbb{D}$ be analytic with $\varphi(0) = 0$, and let $g\colon \mathbb{D} \to [0, +\infty)$ be a subharmonic function. For every $r \in (0, 1)$, we claim that

$$\int_{\mathbb{T}} g\big(\varphi(rz)\big)\, dm \leqslant \int_{\mathbb{T}} g(rz)\, dm\,. \tag{2.15}$$

Indeed, let G be the harmonic function on $r\mathbb{D}$, continuous on $\overline{r\mathbb{D}}$, such that $G = g$ on $r\mathbb{T}$. Thus, $g \le G$ on $r\mathbb{D}$.

Thanks to the Schwarz lemma, $\varphi(rz)$ belongs to $\overline{r\mathbb{D}}$ for every $z \in \mathbb{T}$. So, not forgetting that $G \circ \varphi$ is harmonic, we have

$$\begin{aligned}\int_{\mathbb{T}} g\big(\varphi(rz)\big)\, dm &\le \int_{\mathbb{T}} G\big(\varphi(rz)\big)\, dm = G \circ \varphi(0)\\ &= G(0) = \int_{\mathbb{T}} G(rz)\, dm = \int_{\mathbb{T}} g(rz)\, dm,\end{aligned}$$

which was our claim.

Now we apply this inequality to the function $g(z) = |f(z)|^p$, where $f \in H^p$. Letting $r \nearrow 1^-$, we get

$$\|f \circ \varphi\|_{H^p}^p \le \|f\|_{H^p}^p, \qquad \text{hence} \qquad \|C_\varphi\|_{B(H^p)} = 1.$$

The boundedness is proved again when φ fixes the origin and we may conclude in the general case using once again the "dévissage" principle 2.3.2.

A Variant of Sarason's Approach on H^p

We give here a variant of Sarason's approach (proved in H^2) to prove directly (without the "dévissage" principle, but also without the Schur test) that C_φ is bounded on H^p.

Fix $r \in (0,1)$ and denote for a while $\phi_r(z) = \varphi(rz)$, which is defined for $z \in \overline{\mathbb{D}}$ and satisfies $\|\phi_r\|_\infty < 1$. We have

$$C_{\phi_r}(f)(z) = f[\varphi(rz)] = \int_{\mathbb{T}} P_{\varphi(rz)}(w) f(w)\, dm(w). \tag{2.16}$$

From (2.10), the measure $P_{\varphi(rz)}\, dm$ (on $\mathbb{T}$) is a probability measure for every $z \in \mathbb{T}$. From Jensen's inequality and (2.11), we get that $\|C_{\phi_r}\| \leqslant \left(\frac{1+|\varphi(0)|}{1-|\varphi(0)|}\right)^{\frac{1}{p}}$ as an operator on H^p. Finally, by definition of the H^p norm, we obtain again

$$\|C_\varphi\|_{B(H^p)} \leqslant \left(\frac{1+|\varphi(0)|}{1-|\varphi(0)|}\right)^{\frac{1}{p}}.$$

This viewpoint will be used in Theorem 11.13 to prove in detail boundedness of composition operators on Hardy–Orlicz spaces.

2.3.5 Lower Estimate of the Norm

We now know that every composition operator C_φ is bounded on H^p and already got the following upper estimate for its norm (see (2.14) for instance):

$$\left\|C_\varphi\right\|_{B(H^p)} \leqslant \left(\frac{1+|\varphi(0)|}{1-|\varphi(0)|}\right)^{\frac{1}{p}}.$$

We shall now give a lower estimate for the norm of C_φ.

Let us fix $p \geqslant 1$. Since C_φ is bounded, its adjoint is bounded as well on $(H^p)^*$ and has the same operator norm.

We are going to use a special functional on H^p: the point evaluation (recall Proposition 1.76); for every $z \in \mathbb{D}$,

$$\left\|C_\varphi^*(\delta_z)\right\|_{(H^p)^*} \leqslant \left\|C_\varphi^*\right\|_{B((H^p)^*)} \cdot \left\|\delta_z\right\|_{(H^p)^*} = \left\|C_\varphi\right\|_{B(H^p)} \cdot \left\|\delta_z\right\|_{(H^p)^*},$$

but

$$C_\varphi^*(\delta_z) = \delta_{\varphi(z)}, \tag{2.17}$$

since for every $f \in H^p$ (using the duality bracket)

$$\langle C_\varphi^*(\delta_z), f\rangle = \langle \delta_z, C_\varphi(f)\rangle = f \circ \varphi(z) = \langle \delta_{\varphi(z)}, f\rangle.$$

Therefore

$$\|C_\varphi\|_{B(H^p)} \geqslant \frac{\|\delta_{\varphi(z)}\|_{(H^p)^*}}{\|\delta_z\|_{(H^p)^*}};$$

so

$$\|C_\varphi\|_{B(H^p)} \geqslant \sup_{z\in\mathbb{D}} \frac{(1-|z|^2)^{\frac{1}{p}}}{(1-|\varphi(z)|^2)^{\frac{1}{p}}}, \tag{2.18}$$

and in particular we have the estimates

$$\frac{1}{(1-|\varphi(0)|^2)^{\frac{1}{p}}} \leqslant \|C_\varphi\|_{B(H^p)} \leqslant \left(\frac{1+|\varphi(0)|}{1-|\varphi(0)|}\right)^{\frac{1}{p}}. \tag{2.19}$$

Point out that this implies that $\|C_\varphi\|_{B(H^p)} = 1$ if and only if $\varphi(0) = 0$.

Let us mention here that one of the most striking open problems about composition operators is the exact value of the norm of C_φ:

Question: What is the exact value of the norm of $C_\varphi : H^2 \to H^2$?

The problem is equivalent to the question of the exact value of the operator norm on H^p (see (2.13)) for some $p \geqslant 1$. Besides knowing the exact value, solving this question could, maybe, give more information about the link between a composition operator and its symbol.

2.3.6 More Examples

We are interested here in the exact norm of some specific composition operators acting on H^p. We shall recover in a very elementary way the exact norm in the case of inner functions, but also estimate the norm associated to an affine symbol. We actually exploit the ideas of the proof of Proposition 2.2.

Lemma 2.7 *Let $\varphi : \mathbb{D} \to \mathbb{D}$ be analytic. On H^2, we have*

$$\|C_\varphi\|^2_{B(H^2)} \leqslant \inf_{\rho>0} \sup_{k\geqslant 0} \frac{1}{\rho^k} \sum_{l\geqslant 0} \rho^l \left| \int_{\mathbb{T}} \varphi^k \overline{\varphi}^l \, dm \right| .$$

Proof: For $f(z) = \sum_{n\geqslant 0} a_n z^n$, we have $\|C_\varphi(f)\|_2^2 = \sum_{k,l\geqslant 0} a_k \overline{a_l} \int_{\mathbb{T}} \varphi^k \overline{\varphi}^l \, dm$.

Actually, $\|C_\varphi^* C_\varphi\| = \|C_\varphi\|^2$ on H^2. The entries of the matrix of $C_\varphi^* C_\varphi$ in the canonical (orthonormal) basis of H^2 are given by

$$Q_{l,k} = \int_{\mathbb{T}} \overline{\varphi}^l \varphi^k \, dm , \qquad \text{where } k,l \geqslant 0;$$

hence we have

$$\|C_\varphi(f)\|_2^2 = \langle a, Q(a)\rangle \quad \text{and} \quad \|C_\varphi\|^2 = \|Q\|,$$

where $Q = \big(Q_{l,k}\big)_{l,k\geqslant 0}$ is viewed acting on $\ell^2 = \ell^2(\mathbb{N}_0)$.

We observe that for any $x, y \geqslant 0$, we have $xy \leqslant \frac{1}{2}(x^2 + y^2)$; hence

$$\|C_\varphi(f)\|_2^2 = \sum_{k,l\geqslant 0} Q_{l,k} a_k \overline{a_l} \leqslant \sum_{k,l\geqslant 0} \frac{1}{2} t^{k+l} |Q_{l,k}| \left(\frac{1}{t^{2k}}|a_k|^2 + \frac{1}{t^{2l}}|a_l|^2\right),$$

so that, by symmetry, we get

$$\|C_\varphi(f)\|_2^2 \leqslant \sum_{k\geqslant 0} \frac{1}{t^k}|a_k|^2 \sum_{l\geqslant 0} t^l |Q_{k,l}| \leqslant \left(\sum_{k\geqslant 0} |a_k|^2\right)\left(\sup_{k\geqslant 0} \frac{1}{t^k} \sum_{l\geqslant 0} t^l |Q_{k,l}|\right)$$

(the sums may be infinite for some choices of t).

Taking the supremum over normalized sequences $(a_n)_{n\geqslant 0}$, we conclude

$$\|C_\varphi\|_{B(H^2)}^2 \leqslant \inf_{\rho>0} \sup_k \frac{1}{\rho^k} \sum_{l\geqslant 0} \rho^l |Q_{l,k}|. \qquad \square$$

As a first application, we can recover immediately the estimate for inner functions:

$$\|C_\varphi\|_{B(H^p)} = \left(\frac{1+|\varphi(0)|}{1-|\varphi(0)|}\right)^{\frac{1}{p}}.$$

This statement is already proved twice in this book: either as an immediate corollary of Proposition 2.2 and (2.13), or in Theorem 2.3. Nevertheless, we shall present another alternative elementary proof with the help of Lemma 2.7.

Another Proof of Theorem 2.3: The estimate at the beginning of the proof of Proposition 2.2 reads as

$$\sum_{l\geqslant 0} \left|\int_{\mathbb{T}} \varphi^k \overline{\varphi}^l \, dm\right| = \sum_{l\geqslant 0} \left|\varphi^{|k-l|}(0)\right| = \frac{1+|\varphi(0)| - |\varphi(0)|^{k+1}}{1-|\varphi(0)|}.$$

Taking the supremum over k, Lemma 2.7 applied with $\rho = 1$ leads immediately to the upper estimate when $p = 2$.

For the lower estimate, consider the function $f(z) = \dfrac{\sqrt{1-r^2}}{1-r\bar{\xi}z}$, where $\xi \in \mathbb{T}$ satisfies $\varphi(0) = \xi|\varphi(0)|$, and $r \in (0,1)$.

Then we have

$$\|C_\varphi\|^2_{B(H^2)} \geqslant \|C_\varphi(f)\|_2^2 = \int_{\mathbb{T}} \frac{1-r^2}{|1-r\bar{\xi}\varphi|^2}\, dm\,.$$

But, since $|\varphi| = 1$ a.e. on $\mathbb{T}$ (φ is inner), we have

$$\frac{1-r^2}{|1-r\bar{\xi}\varphi|^2} = \mathrm{Re}\left(\frac{1+r\bar{\xi}\varphi}{1-r\bar{\xi}\varphi}\right)\cdot$$

Now, we get:

$$\begin{aligned}\int_{\mathbb{T}} \frac{1-r^2}{|1-r\bar{\xi}\varphi|^2}\, dm &= \mathrm{Re}\left(\int_{\mathbb{T}} \frac{1+r\bar{\xi}\varphi}{1-r\bar{\xi}\varphi}\, dm\right) \\ &= \mathrm{Re}\left(\frac{1+r\bar{\xi}\varphi(0)}{1-r\bar{\xi}\varphi(0)}\right) = \frac{1+r|\varphi(0)|}{1-r|\varphi(0)|}\cdot\end{aligned}$$

Therefore $\|C_\varphi\|^2_{B(H^2)} \geqslant \dfrac{1+r|\varphi(0)|}{1-r|\varphi(0)|}$ and letting $r \to 1^-$, the result follows for $p = 2$.

Finally, (2.13) (see Proposition 2.6) gives the conclusion for any $p \geqslant 1$. □

We give another interesting application of this method and we estimate the exact norm for an affine symbol (see [14, 32, 69] where the result appears with different techniques, or [142] with similar ideas).

Proposition 2.8 *Let $a \in \mathbb{C}$ and $b \in \mathbb{D}$ satisfying $|a| + |b| \leqslant 1$ and the affine symbol defined by $\varphi(z) = az + b$ for $z \in \mathbb{D}$.*

Then, on H^p, we have

$$\|C_\varphi\|_{B(H^p)} = \left(\frac{2}{1+|a|^2-|b|^2+\sqrt{\Delta}}\right)^{\frac{1}{p}},$$

where $\Delta = \big[1-(|a|-|b|)^2\big]\big[1-(|a|+|b|)^2\big]$.

In particular when $|a| + |b| = 1$, $\|C_\varphi\|_{B(H^p)} = \dfrac{1}{|a|^{\frac{1}{p}}} = \left(\dfrac{1}{1-|\varphi(0)|}\right)^{\frac{1}{p}}$.

Point out, having in mind the estimate for the essential norm in H^2 (see Chapter 6, in particular Theorem 6.1), that the norm of the affine symbol touching the boundary is exactly its essential norm on H^2.

Point out too that we can gather the two cases in the following formula when $a, b \neq 0$:

$$\|C_\varphi\|_{B(H^p)} = \left(\frac{1}{1-\rho_0|b|}\right)^{\frac{1}{p}} = \left(\frac{1}{1-\rho_0|\varphi(0)|}\right)^{\frac{1}{p}}, \tag{2.20}$$

where ρ_0 is the root belonging to $(0,1]$ of $P(\rho)=|b|\rho^2+(-1+|a|^2-|b|^2)\rho+|b|$.

Proof: If a or b is equal to 0 then the result is already known so we now assume that $a,b\in\mathbb{C}^*$.

- We first focus on the case $p=2$.

We first have to estimate $\displaystyle\int_{\mathbb{T}}\varphi^k\overline{\varphi}^l\,dm$ where $k,l\geqslant 0$:

$$\int_{\mathbb{T}}(az+b)^k\overline{(az+b)}^l\,dm=\sum_{r\geqslant 0}\binom{k}{r}\binom{l}{r}|a|^{2r}b^{k-r}\bar{b}^{l-r},$$

thanks to the binomial formula.

Therefore, for every $\rho>0$ (a suitable one will be chosen later), and every integer k:

$$\sum_{l\geqslant 0}\rho^l\left|\int_{\mathbb{T}}\varphi^k\overline{\varphi}^l\,dm\right|\leqslant\sum_{r\geqslant 0}\binom{k}{r}|a|^{2r}|b|^{k-r}\sum_{l\geqslant 0}\binom{l}{r}\rho^l|b|^{l-r}$$

(the right-hand side may be infinite).

But

$$\sum_{l\geqslant 0}\binom{l}{r}\rho^{l-r}|b|^{l-r}=\sum_{l\geqslant r}\binom{l}{r}\rho^{l-r}|b|^{l-r}=\frac{1}{(1-|b|\rho)^{r+1}},$$

so

$$\begin{aligned}\frac{1}{\rho^k}\sum_{l\geqslant 0}\rho^l\left|\int_{\mathbb{T}}\varphi^k\overline{\varphi}^l\,dm\right|&\leqslant\sum_{r\geq 0}\binom{k}{r}|a|^{2r}\left(\frac{|b|}{\rho}\right)^{k-r}\frac{1}{(1-|b|\rho)^{r+1}}\\&=\frac{1}{1-|b|\rho}\left(\frac{|b|}{\rho}+\frac{|a|^2}{1-|b|\rho}\right)^k.\end{aligned}$$

We may choose now $\rho_0\in(0,1]$ such that $\dfrac{|b|}{\rho_0}+\dfrac{|a|^2}{1-|b|\rho_0}=1$. This is possible since it is equivalent to the fact that $P(\rho)=|b|\rho^2+(-1+|a|^2-|b|^2)\rho+|b|$ vanishes somewhere in $(0,1]$ and we clearly have $P(0)>0$ whereas $P(1)\leqslant 0$. Actually, $\Delta\geqslant 0$ is its discriminant and

$$\rho_0=\frac{(1-|a|^2+|b|^2)-\sqrt{\Delta}}{2|b|}$$

is the desired root. Lemma 2.7 gives the upper estimate.

Finally, for the lower estimate, we test the function $f(z)=\dfrac{\sqrt{1-\rho^2}}{1-\rho\xi z}$ where $\rho\in(0,1)$ and $\xi\in\mathbb{T}$ satisfies $\xi b=|b|$.

Clearly, f is normalized in H^2 and

$$\|C_\varphi\|^2_{B(H^2)} \geqslant \|C_\varphi(f)\|_2^2 = \frac{1-\rho^2}{(1-\rho|b|)^2}\int_{\mathbb{T}} \frac{1}{|1-rz|^2}\,dm = \frac{1-\rho^2}{(1-\rho|b|)^2(1-r^2)},$$

where $r = \dfrac{|a|\rho}{1-|b|\rho} \in (0,1)$.

Now, we have to split the discussion into two cases.

When $|a|+|b| < 1$, we choose $\rho = \rho_0 \in (0,1)$. Since $\dfrac{1-\rho_0^2}{1-r^2} = (1-\rho_0|b|)$, we get the conclusion in this case.

When $|a|+|b| = 1$, we have

$$\|C_\varphi\|^2_{B(H^2)} \geqslant \lim_{\rho\to 1^-} \frac{1-\rho^2}{(1-\rho|b|)^2 - |a|^2\rho^2} = \frac{1}{|a|}.$$

- Now let $p \geqslant 1$. The conclusion follows directly from (2.13). □

2.4 Boundedness on Bergman Spaces

In this section, we focus on the boundedness of composition operators on Bergman spaces with radial weights. We shall see that for such weights, any composition operator associated to a symbol fixing the origin is a contraction. The boundedness of C_φ for a general symbol φ is then only conditioned by the boundedness of composition operators associated to an automorphism of the unit disk. That is why we first focus on composition operators associated to a Möbius map. A special sufficient regularity condition comes into play:

Definition 2.9 (Property (W)) Given a radial weight ω, we shall say that ω satisfies (W) when $\breve{\omega}(x) = x^2\omega\big(\sqrt{1-x}\big)$ satisfies: for every $\lambda \geqslant 1$, there exists some $C_\lambda > 0$ such that

$$y \leqslant \lambda x \implies \breve{\omega}(y) \leqslant C_\lambda \breve{\omega}(x)\,, \text{ for every } x, y \in (0,1). \qquad \text{(W)}$$

Point out that $\breve{\omega}$ and ω are linked by the relation $\breve{\omega}(1-|z|^2) = (1-|z|^2)^2\omega(|z|)$.

In the case of standard weights $\omega_\alpha(x) = (\alpha+1)\big(1-x^2\big)^\alpha$, where $\alpha > -1$, this condition (W) is clearly fulfilled since $\alpha + 2 > 1$ and

$$\breve{\omega_\alpha}(s) = (\alpha+1)s^{\alpha+2}.$$

Proposition 2.10 *Let $p \geqslant 1$, $a \in \mathbb{D}$ and ω satisfy the condition (W). Then C_{φ_a} is bounded on $\mathscr{B}^p_\omega$.*

In the particular case of standard weights, for every $\alpha > -1$, the composition operator C_{φ_a} is bounded on $\mathscr{B}^p_\alpha$ and

$$\|C_{\varphi_a}\|_{B(\mathscr{B}^p_\alpha)} = \left(\frac{1+|a|}{1-|a|}\right)^{\frac{\alpha+2}{p}}.$$

Proof: Let $f \in \mathscr{B}^p_\omega$. By definition,

$$\|f\circ\varphi_a\|^p_{\mathscr{B}^p_\omega} = \int_{\mathbb{D}} |f\circ\varphi_a(z)|^p\,\omega(z)\,d\sigma(z) = \int_{\mathbb{D}} |f\circ\varphi_a(z)|^p\,\omega(\varphi_a\circ\varphi_a(z))\,d\sigma(z),$$

and we can now use a standard change of variable: $w = \varphi_a(z)$, and we get

$$\|f\circ\varphi_a\|^p_{\mathscr{B}^p_\omega} = \int_{\mathbb{D}} |f(w)|^p \left(\frac{1-|a|^2}{|1-\bar{a}w|^2}\right)^2 \omega(\varphi_a(w))\,d\sigma(w). \tag{2.21}$$

But, we have the key relation

$$1-|\varphi_a(w)|^2 = \frac{(1-|w|^2)(1-|a|^2)}{|1-\bar{a}w|^2}; \tag{2.22}$$

so, on one hand,

$$\|f\circ\varphi_a\|^p_{\mathscr{B}^p_\omega} = \int_{\mathbb{D}} |f(w)|^p \left(\frac{1-|\varphi_a(w)|^2}{1-|w|^2}\right)^2 \omega(\varphi_a(w))\,d\sigma(w); \tag{2.23}$$

hence

$$\|f\circ\varphi_a\|^p_{\mathscr{B}^p_\omega} = \int_{\mathbb{D}} |f(w)|^p\,\frac{\breve{\omega}(1-|\varphi_a(w)|^2)}{(1-|w|^2)^2}\,d\sigma(w).$$

On the other hand, from (2.22), we have

$$1-|\varphi_a(w)|^2 \leqslant \frac{1+|a|}{1-|a|}\,(1-|w|^2),$$

and condition (W) gives

$$\breve{\omega}(1-|\varphi_a(w)|^2) \leqslant c(a)\,\breve{\omega}(1-|w|^2) = c(a)\,(1-|w|^2)^2\,\omega(w),$$

for some real $c(a) > 0$.

We conclude that

$$\|f\circ\varphi_a\|^p_{\mathscr{B}^p_\omega} \leqslant c(a)\int_{\mathbb{D}} |f(w)|^p\,\omega(w)\,d\sigma(w),$$

and that C_{φ_a} is bounded.

When $\omega(z) = \omega_\alpha(z) = (\alpha+1)(1-|z|^2)^\alpha$, we can write more accurate estimates.

Indeed since $1 - |\varphi_a(w)|^2 \leqslant \dfrac{1+|a|}{1-|a|}(1-|w|^2)$, we have

$$\frac{(1-|\varphi_a(w)|^2)^{\alpha+2}}{(1-|w|^2)^2} \leqslant \left(\frac{1+|a|}{1-|a|}\right)^{\alpha+2}(1-|w|^2)^{\alpha}.$$

Therefore, from (2.23), we get

$$\|f \circ \varphi_a\|_{\mathfrak{B}_\alpha^p}^p \leqslant \left(\frac{1+|a|}{1-|a|}\right)^{\alpha+2} \int_{\mathbb{D}} |f(w)|^p \, d\sigma_\alpha(w).$$

But actually we can estimate exactly the norm of the composition operator in this case. Indeed, we know from (2.21) that

$$\|f \circ \varphi_a\|_{\mathfrak{B}_\alpha^p}^p = \int_{\mathbb{D}} |f(w)|^p \left(\frac{1-|a|^2}{|1-\bar{a}w|^2}\right)^{\alpha+2} d\sigma_\alpha(w) = \int_{\mathbb{D}} |f(w)g(w)|^p \, d\sigma_\alpha(w),$$

where g is the analytic function

$$g(w) = \left(\frac{1-|a|^2}{(1-\bar{a}w)^2}\right)^{\frac{\alpha+2}{p}}.$$

Therefore, the norm of C_{φ_a} is the norm of the multiplier by g acting on $\mathfrak{B}_\alpha^p$ and we know from Appendix B.2 that this norm is equal to

$$\|g\|_\infty = \left(\frac{1-|a|^2}{(1-|a|)^2}\right)^{\frac{\alpha+2}{p}} = \left(\frac{1+|a|}{1-|a|}\right)^{\frac{\alpha+2}{p}}.$$

□

We can now treat the case of general symbols. Once again we shall present several approaches.

2.4.1 Hilbertian Bergman Spaces

In the spirit of the proof given for H^2 with the help of the Littlewood–Paley formula and the Nevanlinna counting function, we can prove the boundedness in Bergman spaces. We treat here directly the case of weighted Bergman spaces, but, of course, the unweighted case $\omega = 1$, which corresponds to $\widetilde{\omega}(z) = \log(1/|z|^2) - (1-|z|^2) \approx (1-|z|^2)^2$, appears as a special case.

We recall the definition of the weighted Nevanlinna counting function, associated to some radial weights ω:

$$N_{\varphi,\omega}(z) = \begin{cases} \sum_{\varphi(a)=z} \omega(a) & \text{if} z \in \varphi(\mathbb{D}) \setminus \{\varphi(0)\}, \\ +\infty & \text{if} z = \varphi(0). \\ 0 & \text{otherwise.} \end{cases}$$

We are going to prove the boundedness of composition operators on weighted Bergman spaces, but in the special case of a symbol fixing the origin, the following key property is important: it is an extension of the Littlewood inequality.

We recall the notations of Theorem 1.93: for any weight ω, we denote, for any $z \in \mathbb{D} \setminus \{0\}$:

$$\widetilde{\omega}(z) = 4 \int_{|z|}^{1} x\, \omega(x) \log\left(\frac{x}{|z|}\right) dx.$$

Lemma 2.11 (Weighted Littlewood inequality) *Let φ be a symbol and let $z \in \varphi(\mathbb{D}) \setminus \{\varphi(0)\}$.*

(i) *Let ω be a radial weight. Then*

(a) $N_{\varphi,\widetilde{\omega}}(z) = 4 \int_0^1 r\omega(r)\, N_{\phi_r}(z)\, dr,$ *where $\phi_r(w) = \varphi(rw)$.*

(b) *When $\varphi(0) = 0$, we have*

$$N_{\varphi,\widetilde{\omega}}(z) \leq \widetilde{\omega}(z).$$

(ii) *In particular, for every $\beta \geqslant 1$,*

$$\sum_{\substack{\varphi(a)=z\\ a\in\mathbb{D}}} (1-|a|^2)^\beta = O\left((1-|z|^2)^\beta\right) \quad \textit{when } |z| \to 1^-.$$

Proof: (i) (a). This is a simple application of the definition: for $z \in \varphi(\mathbb{D}) \setminus \{\varphi(0)\}$,

$$\begin{aligned} N_{\varphi,\widetilde{\omega}}(z) &= \sum_{\varphi(u)=z} \widetilde{\omega}(u) = 4 \sum_{\varphi(u)=z} \int_{|u|}^{1} r\omega(r) \log\left(\frac{r}{|u|}\right) dr \\ &= 4 \int_0^1 r\omega(r) \sum_{\varphi(u)=z} \mathbb{1}_{(|u|,1)}(r) \log\left(\frac{r}{|u|}\right) dr. \end{aligned}$$

But for every $r \in (0,1)$, we have $\mathbb{1}_{(|u|,1)}(r) \neq 0$ if and only if $a = \dfrac{u}{r} \in \mathbb{D}$; so

$$\sum_{\substack{u\in\mathbb{D}\\ \varphi(u)=z}} \mathbb{1}_{(|u|,1)}(r) \log\left(\frac{r}{|u|}\right) = \sum_{\substack{\phi_r(a)=z\\ a\in\mathbb{D}}} \log\left(\frac{1}{|a|}\right) = N_{\phi_r}(z).$$

The result follows for (a).

To prove (b), first point out that for a fixed $r \in (0,1)$, the function $\psi(z) = \dfrac{1}{r}\varphi(rz)$ is analytic on $\mathbb{D}$ and $\psi(\mathbb{D}) \subset \mathbb{D}$ thanks to the Schwarz lemma. Hence

ψ is a symbol, which satisfies $\psi(0) = 0$. Therefore, applying the Littlewood inequality (see Theorem 1.37) to ψ, we get

$$N_{\phi_r}(z) = \sum_{\substack{\psi(a)=z/r \\ a\in\mathbb{D}}} \log\left(\frac{1}{|a|}\right) \leqslant \log\left(\frac{r}{|z|}\right).$$

Thanks to (a), this implies (b).

(ii). The very particular case $\beta = 1$ comes from Theorem 1.37 and the fact that $2\log(1/|u|) \sim (1 - |u|^2)$ near the boundary.

The case $\beta > 1$ can be deduced directly from the case $\beta = 1$ using that $(1 - |a|^2)^\beta \leqslant (1 - |a|^2)\big(\delta_0(1 - |z|^2)\big)^{\beta-1}$ when $\varphi(a) = z$, for some $\delta_0 > 0$ (recall (1.15)), which is 1 when φ fixes the origin, or using that $\|\cdot\|_{\ell^\beta} \leqslant \|\cdot\|_{\ell^1}$. It can also be viewed as a consequence of (b) for $\omega = \omega_\alpha$ with $\alpha = \beta - 2 > -1$, when φ fixes the origin. Then the general case follows from the "dévissage" principle. □

Now we easily get the following theorem:

Theorem 2.12 *Let ω be a weight satisfying $\int_{\mathbb{D}} \omega\, d\sigma = 1$, and φ be a symbol.*

(i) *Assume that $\varphi(0) = 0$. Then C_φ is a contraction on $\mathscr{B}^2_\omega$.*
(ii) *Let ω be a weight satisfying (W). Then C_φ is bounded on $\mathscr{B}^2_\omega$.*
(iii) *In the particular case of standard weights, for every $\alpha > -1$, the composition operator C_φ is bounded on $\mathscr{B}^2_\alpha$ and*

$$\|C_\varphi\|_{B(\mathscr{B}^2_\alpha)} \leqslant \left(\frac{1 + |\varphi(0)|}{1 - |\varphi(0)|}\right)^{\frac{\alpha+2}{2}}.$$

Proof: (i) Assume first that φ fixes the origin. Let $f \in \mathscr{B}^2_\omega$. From (1.71), we know that

$$\|f\|^2_{\mathscr{B}^2_\omega} = |f(0)|^2 + \int_{\mathbb{D}} |f'(z)|^2\, \widetilde{\omega}(z)\, d\sigma(z).$$

Therefore, by the generalized change of variable formula,

$$\begin{aligned}\|f \circ \varphi\|^2_{\mathscr{B}^2_\omega} &= |f(0)|^2 + \int_{\mathbb{D}} |f' \circ \varphi(z)|^2\, |\varphi'(z)|^2\, \widetilde{\omega}(z)\, d\sigma(z) \\ &= |f(0)|^2 + \int_{\mathbb{D}} |f'(z)|^2\, N_{\varphi,\widetilde{\omega}}(z)\, d\sigma(z).\end{aligned}$$

From Lemma 2.11, we get

$$\|f \circ \varphi\|^2_{\mathscr{B}^2_\omega} \leqslant |f(0)|^2 + \int_{\mathbb{D}} |f'(z)|^2\, \widetilde{\omega}(z)\, d\sigma(z) = \|f\|^2_{\mathscr{B}^2_\omega}.$$

(ii) Now, we wish to prove the boundedness for a general symbol under condition (W). From Proposition 2.10, we know that C_{φ_a} is bounded where $a = \varphi(0)$. Therefore, the conclusion follows directly from the "dévissage" principle.

Point (iii) follows from the same idea but using the special case of standard weights in Proposition 2.10. □

2.4.2 From H^p to $\mathscr{B}^p_\omega$

In this section, our approach is based on the fact that we already proved the boundedness for Hardy spaces. Moreover we are not restricted here to the case of Hilbertian Bergman spaces.

Theorem 2.13 *Let $p \geqslant 1$, ω be a radial weight on $\mathbb{D}$ and $\varphi\colon \mathbb{D} \to \mathbb{D}$ be analytic. Then:*

(i) *When φ fixes the origin, C_φ is a contraction on $\mathscr{B}^p_\omega$.*
(ii) *When ω satisfies the condition (W), the composition operator C_φ is bounded on $\mathscr{B}^p_\omega$.*
(iii) *In the particular case of standard weights, for every $\alpha > -1$, the composition operator C_φ is bounded on $\mathscr{B}^p_\alpha$ and*

$$\frac{1}{\left(1-|\varphi(0)|^2\right)^{\frac{\alpha+2}{p}}} \leqslant \sup_{a\in\mathbb{D}} \left(\frac{1-|a|^2}{1-|\varphi(a)|^2}\right)^{\frac{\alpha+2}{p}} \leqslant \left\|C_\varphi\right\|_{B(\mathscr{B}^p_\alpha)} \leqslant \left(\frac{1+|\varphi(0)|}{1-|\varphi(0)|}\right)^{\frac{\alpha+2}{p}}.$$

Proof: From (2.19), we know that, on H^p,

$$\left\|C_\varphi\right\|_{B(H^p)} \leqslant \left(\frac{1+|\varphi(0)|}{1-|\varphi(0)|}\right)^{\frac{1}{p}}.$$

Let us begin with the special case of a symbol fixing the origin. Then from the Schwarz lemma we know that $|\varphi(rz)| \leqslant r|z|$ for every $z \in \mathbb{D}$ and every $r \in (0,1)$. In particular, the analytic function defined by

$$\forall z \in \mathbb{D}, \qquad \Phi_r(z) = \frac{1}{r}\varphi(rz)$$

is a symbol, fixing also the origin.

It induces a composition operator on H^p, which is a contraction, so that

$$\|f \circ \varphi\|^p_{\mathscr{B}^p_\omega} = \int_0^1 \|f_r \circ \Phi_r\|^p_{H^p} 2r\omega(r)\, dr \leqslant \int_0^1 \|f_r\|^p_{H^p} 2r\omega(r)\, dr,$$

where $f_r(z) = f(rz)$.

We get $\|f \circ \varphi\|_{\mathscr{B}^p_\omega} \leqslant \|f\|_{\mathscr{B}^p_\omega}$ and we conclude that C_φ is a contraction.

(ii) and the upper estimate in (iii) follow again from the "dévissage" principle and Proposition 2.10. For the lower estimate in (iii), we follow the lines of the corresponding lower estimate in H^p (see 2.3.5), using Proposition 1.91. See also (2.28) in Proposition 2.15. □

2.5 Complements

2.5.1 Aleksandrov–Clark's Measures

This section gives a short introduction to Aleksandrov–Clark's measures and is inspired by the expositions [150] and [156]. It is very far from being exhaustive and we refer to the mentioned expositions (for instance) for many more informations. It is closely linked to Sarason's approach of boundedness of composition operators.

Let φ be a symbol. Fix any $\alpha \in \mathbb{T}$, and we point out that, on $\mathbb{D}$,

$$\operatorname{Re}\left(\frac{\alpha + \varphi}{\alpha - \varphi}\right) = \frac{1 - |\varphi|^2}{|\alpha - \varphi|^2} \geqslant 0\,.$$

Therefore, from Herglotz theorem (see [153, theorem 11.30]), there exists a non-negative (finite) measure μ_α on $\mathbb{T}$ such that, for every $z \in \mathbb{D}$,

$$\begin{aligned}\operatorname{Re}\left(\frac{\alpha + \varphi(z)}{\alpha - \varphi(z)}\right) &= \int_{\mathbb{T}} \frac{1 - |z|^2}{|\xi - z|^2}\, d\mu_\alpha(\xi)\\ &= \int_{\mathbb{T}} \operatorname{Re}\left(\frac{\xi + z}{\xi - z}\right) d\mu_\alpha(\xi).\end{aligned} \tag{2.24}$$

The family of measure $(\mu_\alpha)_{\alpha\in\mathbb{T}}$ has many names, but we shall call them the *Aleksandrov–Clark measures* associated to φ. They naturally appear in function theory or operator theory, but we shall focus only on the link with C_φ.

From (2.24), since an analytic function on $\mathbb{D}$ taking its values in $i\mathbb{R}$ is constant, we get

$$\forall z \in \mathbb{D}, \quad \frac{\alpha + \varphi(z)}{\alpha - \varphi(z)} = \int_{\mathbb{T}} \frac{\xi + z}{\xi - z}\, d\mu_\alpha(\xi) + i \operatorname{Im}\left(\frac{\alpha + \varphi(0)}{\alpha - \varphi(0)}\right). \tag{2.25}$$

Moreover, taking simply $z = 0$ in (2.24),

$$\forall \alpha \in \mathbb{T}, \qquad \|\mu_\alpha\| = \mu_\alpha(\mathbb{T}) = \operatorname{Re}\left(\frac{\alpha + \varphi(0)}{\alpha - \varphi(0)}\right) = \frac{1 - |\varphi(0)|^2}{|\alpha - \varphi(0)|^2}. \tag{2.26}$$

In particular, when φ fixes the origin, $\|\mu_\alpha\| = 1$.

The *Aleksandrov operator* on $L^p(\mathbb{T})$ is (formally) defined by

$$f \longmapsto A_\varphi(f)(\alpha) = \int_{\mathbb{T}} f \, d\mu_\alpha.$$

Proposition 2.14 will show that A_φ is in fact well defined.

In the sequel, $P[g]$ stands for the Poisson extension of $g \in L^p(\mathbb{T})$ to $\mathbb{D}$.

Proposition 2.14 *Let φ be a symbol fixing the origin. Then*

(i) *A_φ is a contraction on $L^q(\mathbb{T})$ for every $q \geqslant 1$.*
(ii) *The adjoint of A_φ is a contraction as well on $L^p(\mathbb{T})$ for every $p > 1$ and is the operator $g \mapsto P[g] \circ \varphi$.*
Its restriction to H^p coincides with C_φ.

Proof: We first point out that for every $z \in \mathbb{D}$, we have

$$\int_{\mathbb{T}} \left(\int_{\mathbb{T}} P_z(\xi) \, d\mu_\alpha(\xi) \right) dm(\alpha) = \int_{\mathbb{T}} \operatorname{Re} \frac{\alpha + \varphi(z)}{\alpha - \varphi(z)} \, dm(\alpha) = 1 = \int_{\mathbb{T}} P_z \, dm,$$

where P_z is the Poisson kernel: $P_z(\xi) = \dfrac{1 - |z|^2}{|\xi - z|^2}$.

But the vector space spanned by the Poisson kernels $(P_z)_{z \in \mathbb{D}}$ is dense in $C(\mathbb{T})$. Therefore, by a standard approximation argument, for every $g \in C(\mathbb{T})$, we have

$$\int_{\mathbb{T}} \left(\int_{\mathbb{T}} g(\xi) \, d\mu_\alpha(\xi) \right) dm(\alpha) = \int_{\mathbb{T}} g \, dm\,. \tag{2.27}$$

Actually this relation extends to every function $g \in L^1(\mathbb{T})$.

Indeed, given $g \in L^1(\mathbb{T})$, we can find a sequence (g_k) in $C(\mathbb{T})$ converging to $|g|$ in $L^1(\mathbb{T})$, and thanks to the converse of the Lebesgue domination theorem, we may assume that the functions g_k are non-negative and satisfy $g_k \leqslant h$ a.e. with $h \in L^1(\mathbb{T})$. Therefore, from Fatou's lemma,

$$\begin{aligned}
\int_{\mathbb{T}} \left(\int_{\mathbb{T}} |g|(\xi) \, d\mu_\alpha(\xi) \right) dm(\alpha) &= \int_{\mathbb{T}} \left(\int_{\mathbb{T}} \underline{\lim}\, g_k(\xi) \, d\mu_\alpha(\xi) \right) dm(\alpha) \\
&\leqslant \underline{\lim} \int_{\mathbb{T}} \left(\int_{\mathbb{T}} g_k(\xi) \, d\mu_\alpha(\xi) \right) dm(\alpha) \\
&= \underline{\lim} \int_{\mathbb{T}} g_k \, dm = \int_{\mathbb{T}} |g| \, dm,
\end{aligned}$$

applying (2.27) to g_k and the Lebesgue domination theorem.

Therefore, by Fubini–Tonelli, $A_\varphi(g)$ is well defined a.e. and A_φ is a contraction on $L^1(\mathbb{T})$. By continuity, (2.27) is also valid on $L^1(\mathbb{T})$.

Since A_φ is obviously a contraction on $L^\infty(\mathbb{T})$, we get by interpolation that A_φ is bounded, with norm less than 1, on $L^q(\mathbb{T})$.

Then the adjoint A_φ^* is well defined on L^p for any $p > 1$. Moreover, for any $f \in L^{p'}(\mathbb{T})$ and $g \in L^p(\mathbb{T})$, we have

$$\int_{\mathbb{T}} A_\varphi(f)(\alpha)\, g(\alpha)\, dm(\alpha) = \int_{\mathbb{T}} f(\alpha)\, A_\varphi^*(g)(\alpha)\, dm(\alpha),$$

and in particular, for every $z \in \mathbb{D}$, we have (with $f = P_z$):

$$\int_{\mathbb{T}} P_{\varphi(z)}(\alpha)\, g(\alpha)\, dm(\alpha) = \int_{\mathbb{T}} P_z(\alpha)\, A_\varphi^*(g)(\alpha)\, dm(\alpha),$$

but on the other hand, for every trigonometric polynomial $g \in \mathbb{C}[\bar{z}, z]$,

$$\int_{\mathbb{T}} P_{\varphi(z)}(\alpha)\, g(\alpha)\, dm(\alpha) = g\big(\varphi(z)\big) = g \circ \varphi(z) = \int_{\mathbb{T}} P_z(\alpha)\, g \circ \varphi(\alpha)\, dm(\alpha).$$

Since trigonometric polynomials are dense in $L^p(\mathbb{T})$ on one hand, and thanks to the density of the vector space spanned by Poisson kernels on the other hand, we get the conclusion. □

2.5.2 Adjoint of a Composition Operator

The following easy general fact turns out to be very useful. We already mentioned it in H^p spaces (see (2.17)), but we would like to emphasize this equation in a more general framework.

In the following statement, $X \subset \mathscr{H}(\mathbb{D})$ means that X is a Banach space of analytic functions on $\mathbb{D}$, *continuously* included in $\mathscr{H}(\mathbb{D})$ (see Section 1.11.1). In particular, the point evaluations $\delta_a : f \in X \mapsto f(a)$ are continuous on X.

Proposition 2.15 (Mapping equation) *Let $X \subset \mathscr{H}(\mathbb{D})$ be a Banach space. Let φ be a symbol such that $C_\varphi : X \to X$ is bounded.*

Then, for every $a \in \mathbb{D}$, $C_\varphi^*(\delta_a) = \delta_{\varphi(a)}$ and

$$\|C_\varphi\| \geqslant \sup_{a \in \mathbb{D}} \frac{\|\delta_{\varphi(a)}\|}{\|\delta_a\|} \cdot \tag{2.28}$$

In particular when X is a Hilbert space, then $C_\varphi^*(K_a) = K_{\varphi(a)}$.

Proof: For every $f \in X$, we have by definition

$$C_\varphi^*(\delta_a)(f) = \delta_a\big(C_\varphi(f)\big) = f \circ \varphi(a) = f\big(\varphi(a)\big).$$

Hence $C_\varphi^*(\delta_a)(f) = \delta_{\varphi(a)}(f)$ for every $f \in X$. Finally using this relation and the fact that C_φ and C_φ^* have the same norm, the result follows.

The specific case of Hilbert spaces is clear: the reproducing kernel K_a represents the functional δ_a. □

2.5.3 The Strong Subordination Principle

Here is a refinement of Littlewood's subordination principle, which we will refer to as the *strong subordination principle*.

Let S_n be the nth partial sum operator. It is an orthogonal projection on every $H^2(\beta)$ (in particular on H^2 and every Bergman space $\mathscr{B}_\omega^2$); see Chapter 12.

Theorem 2.16 (Littlewood–Rogosinski) *Let φ be a symbol fixing the origin. For every $f \in H^2$ and every integer $n \geqslant 0$,*

$$\|S_n(f \circ \varphi)\|_2 \leqslant \|S_n(f)\|_2. \tag{2.29}$$

Point out that the boundedness of C_φ on H^2 (*the subordination principle*) expresses precisely that $\lim \|S_n(f \circ \varphi)\|_2 \leqslant \lim \|S_n(f)\|_2$, so (2.29) appears as a more general result.

Actually this inequality is nothing but (2.6) in the proof of the boundedness via the backwardshift operator. Nevertheless, we show here that it can also be obtained only knowing that C_φ is a contraction. Therefore, Theorem 2.16 follows from the general principle below.

In the following statement, $X \subset \mathscr{H}(\mathbb{D})$ has the same meaning as in 2.5.2.

Theorem 2.17 (Strong subordination principle) *Let φ be a symbol such that $\varphi(0) = 0$. Let $X \subset \mathscr{H}(\mathbb{D})$ be a Banach space such that $(z^n)_{n\in\mathbb{N}_0}$ is a Schauder basis of X and $C_\varphi : X \to X$ is bounded.*

Then, for every $f \in X$ and every integer $n \geqslant 0$,

$$\|S_n(f \circ \varphi)\|_X \leqslant K\|S_n(f)\|_X, \tag{2.30}$$

with $K = \sup\limits_{n\in\mathbb{N}_0} \|S_n\|.\|C_\varphi\|.$

This applies to H^2 or $\mathscr{B}_\omega^2$, for any (radial) weight ω, with $K = 1$, but also to every H^p or $\mathscr{B}_\omega^p$, where $p > 1$, with $K = \sup_{n\in\mathbb{N}_0} \|S_n\|$. On the other hand this also applies to the weighted Hardy spaces $H^2(\beta)$ with $K = \|C_\varphi\|$.

Proof: Let $f \in X$ with a Taylor expansion $f(z) = \sum_{k=0}^\infty a_k z^k$ for every $z \in \mathbb{D}$. But actually this is also an equality in X because we assumed that $(z^k)_{k\in\mathbb{N}_0}$ is a Schauder basis of X; hence $f = \sum_{k=0}^\infty a_k z^k$ with convergence in X.

For every $n \in \mathbb{N}_0$, the operator $S_n C_\varphi : X \to X$ is bounded and we get $S_n C_\varphi(f) = \sum_{k=0}^{\infty} a_k S_n(\varphi^k)$ in the space X. But, as in (v) in Section 2.3.3, we have, since $\varphi(0) = 0$,

$$S_n(f \circ \varphi) = S_n C_\varphi(f) = \sum_{k \leqslant n} a_k S_n(\varphi^k) = S_n \left(\sum_{k \leqslant n} a_k \varphi^k \right) = S_n C_\varphi S_n(f).$$

We conclude $\|S_n(f \circ \varphi)\|_X \leqslant K \|S_n(f)\|_X$. □

3

A First Contact with Compactness Issues

3.1 A General Criterion

We begin this chapter with a general principle. Originally, it was used in the framework of some specific spaces of analytic functions, often Hilbertian, mainly the Hardy space H^2 or the Bergman space $\mathscr{B}^2$, but we shall apply it in many spaces, so it is worth stating as a general criterion.

We shall say that a Banach space X is a Banach space of holomorphic functions on $\mathbb{D}$ if X is continuously embedded (through the canonical injection) in $\mathscr{H}(\mathbb{D})$, equipped with its natural topology: convergence of a sequence means uniform convergence on compact subsets of $\mathbb{D}$. For short, most often, we shall write $f_n \to f$ in $\mathscr{H}(\mathbb{D})$, when (f_n) converges to f uniformly on compact subsets of $\mathbb{D}$.

Definition 3.1 (Fatou property) We shall say that a Banach space X of holomorphic functions on $\mathbb{D}$ has the *Fatou property* if for every bounded sequence $(f_n)_n$ in X, which converges to f in $\mathscr{H}(\mathbb{D})$, one has $f \in X$ and $\|f\|_X \leqslant \underline{\lim} \|f_n\|_X$.

For example, it is easy to check that Hardy spaces H^p, weighted Bergman spaces, weighted Hardy spaces, and H^∞ share the Fatou property. Nevertheless, not all classical spaces share it. For instance, the disk algebra does not share this property: indeed consider the finite partial products of an infinite Blaschke product.

Proposition 3.2 (Compactness criterion) *Let X, Y be two Banach spaces of analytic functions on $\mathbb{D}$, X having the Fatou property. Let φ be a symbol such that $C_\varphi(f) = f \circ \varphi \in Y$ whenever $f \in X$.*

Then $C_\varphi \colon X \to Y$ is compact if and only if for every bounded sequence $(f_n)_n$ in X, which converges to 0 in $\mathscr{H}(\mathbb{D})$, one has $\|C_\varphi(f_n)\|_Y \to 0$.

Proof: Assume that the above condition is fulfilled. Let $(f_n)_{n\geqslant 1}$ be in the unit ball of X. The assumption on X implies that $(f_n)_n$ is a normal family in $\mathscr{H}(\mathbb{D})$. Montel's Theorem (see Theorem 1.2) allows us to extract a subsequence that we still denote by $(f_n)_n$ to save notation, which converges to some f in $\mathscr{H}(\mathbb{D})$. Since X has the Fatou property, one has $f \in X$. Now, since $(f_n - f)_n$ is a bounded sequence in X, which converges to 0 in $\mathscr{H}(\mathbb{D})$, one has $\|C_\varphi(f_n) - C_\varphi(f)\|_Y = \|C_\varphi(f_n - f)\|_Y \to 0$. Hence C_φ is compact.

Conversely, assume that C_φ is compact and let $(f_n)_n$ be a bounded sequence in X, which converges to 0 in $\mathscr{H}(\mathbb{D})$. By compactness of C_φ, there is at least one convergent subsequence for $\big(C_\varphi(f_n)\big)_n$. Consider any subsequence $\big(C_\varphi(f_{n_k})\big)_k$ converging to some $g \in Y$. A fortiori $(f_{n_k} \circ \varphi)_k$ converges pointwise to g. But g must be 0 since $(f_n)_n$ (and in particular $f_n \circ \varphi$) converges to 0 in $\mathscr{H}(\mathbb{D})$, hence $\|C_\varphi(f_n)\|_Y \to 0$. □

Let us point out that there is another approach to this compactness criterion and its proof. Indeed, most spaces of analytic functions X on $\mathbb{D}$ that we introduced (keep in mind Hardy spaces, Bergman spaces) are dual Banach spaces, and we mentioned that the topology of uniform convergence on compact subsets of $\mathbb{D}$ is nothing but the weak-star topology on the unit ball of X.

With this viewpoint, the criterion reduces to characterizing when C_φ is continuous from the unit ball of X, with the weak-star topology, to Y with the norm topology. Therefore, the proof reduces to the translation of two facts: this map is always continuous when each space is equipped with the weak-star topology; and the unit ball of a dual space equipped with the weak-star topology is compact.

3.2 A Geometric Comparison Principle

We place this remark in a general framework: let X be a Banach space of analytic functions on $\mathbb{D}$ such that every composition operator on X is bounded (of course we have in mind Hardy spaces and Bergman spaces).

Given two symbols φ and ψ, such that φ is univalent and $\psi(\mathbb{D}) \subset \varphi(\mathbb{D})$, C_ψ factorizes through C_φ, these composition operators being viewed acting on X.

More precisely we can define a symbol χ such that $\psi = \varphi \circ \chi$ in the following way: for every $z \in \mathbb{D}$, $\psi(z) \in \varphi(\mathbb{D})$; hence there exists a unique $\chi(z) \in \mathbb{D}$ such that $\psi(z) = \varphi(\chi(z))$. It is easy to check that χ is analytic.

Since $\psi = \varphi \circ \chi$ and we assumed that every composition operator on X is bounded, we have C_χ bounded on X and $C_\psi = C_\chi \circ C_\varphi$.

Therefore, in this situation, we immediately get the following proposition:

Proposition 3.3 *Let φ be a univalent symbol and ψ be a symbol satisfying $\psi(\mathbb{D}) \subset \varphi(\mathbb{D})$. Let $\mathcal{I}$ be an ideal of operators.*
Then $\quad C_\varphi \in \mathcal{I} \implies C_\psi \in \mathcal{I}$.
In particular, if C_φ is compact on X then C_ψ is compact on X.

A natural question then occurs: can we replace *univalent* by *finitely valent*? Using results of Section 4.4, the answer is clearly *no*! Indeed, there exist 4-valent *onto* symbols φ such that C_φ is compact. Actually even with 2-valent this is false: there exist 2-valent symbols φ such that $\varphi(\mathbb{D}) = \mathbb{D} \setminus \{0\}$ and C_φ is compact; and of course there are many symbols ψ, avoiding 0, such that the operator C_ψ is not compact (a suitable affine symbol for example: see Section 4.1).

3.3 $A(\mathbb{D})$ and H^∞

We already pointed out that the boundedness of C_φ on H^∞ and on $A(\mathbb{D})$ (under the additional condition that $\varphi \in A(\mathbb{D})$ in this case) is obvious (see Section 2.2). Moreover we have in either case $\|C_\varphi\| = 1$.

The behavior of composition operators on such spaces is very specific.

Theorem 3.4 *Let $\varphi : \mathbb{D} \to \mathbb{D}$ be analytic.*

(i) *C_φ is compact on H^∞ if and only if* $\|\varphi\|_\infty < 1$.
(ii) *C_φ is compact on $A(\mathbb{D})$ if and only if* $\|\varphi\|_\infty < 1$.

It is a particular case of the following more general result.

Theorem 3.5 *If C_φ is a weakly compact operator on H^∞, then* $\|\varphi\|_\infty < 1$.

Hence we have the following immediate corollary of the previous results.

Corollary 3.6 *Let $C_\varphi : H^\infty \longrightarrow H^\infty$. The following assertions are equivalent:*

(i) $\|\varphi\|_\infty < 1$.
(ii) *C_φ is a compact operator.*
(iii) *C_φ is a nuclear operator.*
(iv) *C_φ is a weakly compact operator.*

The same holds replacing H^∞ by $A(\mathbb{D})$ (once $\varphi \in A(\mathbb{D})$).

Proof of Theorem 3.5: Let us assume that C_φ is weakly compact on H^∞. Assume also that $\|\varphi\|_\infty = 1$. There are $a \in \mathbb{T}$ and $(z_j) \in \mathbb{D}$ such that $\varphi(z_j) \longrightarrow a$.

Consider $f_n(z) = \left(\dfrac{1+\bar{a}z}{2}\right)^n$. We clearly have

$$f_n \in A(\mathbb{D}), \quad \|f_n\|_\infty = 1, \quad f_n(a) = 1, \quad \text{and} \quad f_n(z) \longrightarrow 0 \text{ for all } z \in \mathbb{D}\,.$$

Since C_φ is weakly compact, $f_{n_k} \circ \varphi \overset{w}{\longrightarrow} g \in H^\infty$. In particular, for every $z \in \mathbb{D}$, $f_{n_k}\big(\varphi(z)\big)$ converges both to $g(z)$ and 0, so that $g(z) = 0$.

By the Banach–Mazur theorem, we have for some convex combination

$$\sum_{I_m} c_k f_{n_k} \circ \varphi \overset{\|\cdot\|_\infty}{\longrightarrow} g = 0,$$

where the I_m are finite intervals of integers.

But, for every m,

$$\Big\| \sum_{I_m} c_k f_{n_k} \circ \varphi \Big\|_\infty \geqslant \lim_{j \to +\infty} \Big| \sum_{I_m} c_k f_{n_k}\big(\varphi(z_j)\big)\Big| = \sum_{I_m} c_k = 1.$$

We get the desired contradiction.

Now assume that $\|\varphi\|_\infty < 1$. We shall easily see that C_φ is a nuclear operator. Indeed

$$f \circ \varphi = \sum_{n=0}^{\infty} \hat{f}(n)\varphi^n, \quad \text{where} \quad \sum_{n=0}^{\infty} \|\varphi^n\|_\infty < +\infty \text{ and } f \mapsto \hat{f}(n) \text{ has norm } 1.$$

From that we get immediately the desired consequences.

Indeed, if $\|\varphi\|_\infty < 1$, we just proved that C_φ is a nuclear operator, therefore it is compact, a fortiori weakly compact. Finally, if C_φ is weakly compact, then the first part of the proof shows that we must have $\|\varphi\|_\infty < 1$.

When $\varphi \in A(\mathbb{D})$, C_φ may be viewed as a bounded operator on $A(\mathbb{D})$, and the same proof works (have in mind that the sequence (f_n) lies in $A(\mathbb{D})$). □

We even know the distance from C_φ to compact operators: the essential norm on those spaces was first estimated by Zheng [182] (see also [11] or [93]):

Theorem 3.7 *Let φ be a symbol. Then*

(i) *We have $\|C_\varphi\|_{e,H^\infty} = 1$ if $\|\varphi\|_\infty = 1$, and $\|C_\varphi\|_{e,H^\infty} = 0$ if $\|\varphi\|_\infty < 1$.*
(ii) *When $\varphi \in A(\mathbb{D})$, we have $\|C_\varphi\|_{e,A(\mathbb{D})} = 1$ if $\|\varphi\|_\infty = 1$, and $\|C_\varphi\|_{e,A(\mathbb{D})} = 0$ if $\|\varphi\|_\infty < 1$.*

Proof: We are going to prove it only in the case of H^∞. In the disk algebra, the same proof works. We write $\|\cdot\|_e$ instead of $\|\cdot\|_{e,H^\infty}$ for short.

If $\|\varphi\|_\infty < 1$ then we already know that C_φ is compact so $\|C_\varphi\|_e = 0$.

In the following we assume that $\|\varphi\|_\infty = 1$. Obviously

$$\|C_\varphi\|_e \leqslant \|C_\varphi\| = 1.$$

Let $a \in \mathbb{T}$ satisfy $\varphi(z_j) \longrightarrow a$ for a suitable sequence $(z_j) \in \mathbb{D}$. We use a sequence (g_n) different from the one introduced in the proof of Theorem 3.5:

$$g_n(z) = \frac{n\bar{a}z - (n-1)}{n - (n-1)\bar{a}z} \quad \text{for } n \geqslant 1 \text{ and } z \in \overline{\mathbb{D}}.$$

It satisfies $g_n(a) = 1$ and $g_n(z) \longrightarrow -1$ for every $z \in \overline{\mathbb{D}} \setminus \{a\}$. It lies in the unit ball of H^∞. Actually it even lies in the unit sphere of $A(\mathbb{D})$.

Now consider any compact operator S on H^∞. Up to taking a subsequence: $S(g_{n_k})$ converges to some $g \in H^\infty$. We get

$$\|C_\varphi - S\| \geqslant \overline{\lim}\|g_{n_k} \circ \varphi - S(g_{n_k})\|_\infty = \overline{\lim}\|g_{n_k} \circ \varphi - g\|_\infty.$$

We have two cases: either $\|1 + g\|_\infty \leqslant 1$ or $\|1 + g\|_\infty \geqslant 1$.

Assume first that $\|1 + g\|_\infty \leqslant 1$ and point out that for every k,

$$\begin{aligned}\|g_{n_k} \circ \varphi - g\|_\infty &\geqslant \|g_{n_k} \circ \varphi + 1\|_\infty - \|1 + g\|_\infty \\ &\geqslant \lim_j |g_{n_k}(\varphi(z_j)) + 1| - \|1 + g\|_\infty \\ &\geqslant 2 - 1 = 1.\end{aligned}$$

We conclude that $\|C_\varphi - S\| \geqslant 1$ in this case.

Assume now that $\|1 + g\|_\infty \geqslant 1$ and point out that for every $z \in \mathbb{D}$,

$$\overline{\lim}\,\|g_{n_k} \circ \varphi - g\|_\infty \geqslant \overline{\lim}|g_{n_k}(\varphi(z)) - g(z)| = |1 + g(z)|,$$

since $\varphi(z) \in \mathbb{D}$. Therefore $\|C_\varphi - S\| \geqslant \|1 + g\|_\infty \geqslant 1$ in this case as well.

In either case we get, for every compact operator S,

$$\|C_\varphi - S\| \geqslant 1,$$

hence $\|C_\varphi\|_e \geqslant 1$. □

Let us make some remarks on the choice of the sequence (g_n). The first natural idea is to consider the functions $(nd(z,a) - 1)/(nd(z,a) + 1)$, but the distance $d(z,a)$ from z to a is not an analytic function of z, which explains our choice that appears as a slight modification of this idea.

We could use the sequence in the first part of the proof of Theorem 3.5 up to a slight modification of the last line.

3.4 Prehistory and First Approach on Hardy Spaces

We expose in this section the first (easy) results around compactness on Hardy spaces.

Let us begin with a very simple sufficient criterion. It is not necessary (we shall prove it below) and this makes the situation very different from the H^∞ case.

Proposition 3.8 *Let φ be a symbol such that $\|\varphi\|_\infty < 1$, and $p \geqslant 1$.*
Then C_φ is a nuclear operator (hence a compact operator) on H^p.

Proof: We have $\|\varphi^n\|_p \leqslant \|\varphi^n\|_\infty = \|\varphi\|_\infty^n$, so we can write

$$f \circ \varphi = \sum_{n=0}^{\infty} \hat{f}(n)\varphi^n,$$

where $\sum_{n\geqslant 0} \|\varphi^n\|_p < +\infty$ and the functional $f \in H^p \mapsto \hat{f}(n)$ has norm 1. □

The following proposition gives some more subtle first results.

Theorem 3.9 *Let φ be a symbol and $p \geqslant 1$.*

(i) *The operator $C_\varphi \colon H^p \to H^p$ is compact if and only if for every bounded sequence $\{f_n\}_n$ in H^p converging to 0 in $\mathscr{H}(\mathbb{D})$, we have $f_n \circ \varphi \to 0$ in H^p.*
(ii) *If C_φ is compact on H^p, then $m_\varphi(\mathbb{T}) = 0$. In particular, when φ is inner, C_φ is not compact on H^p, for any value of $p \geqslant 1$.*
(iii) *If C_φ is compact on H^p, then $\displaystyle\lim_{|z|\to 1^-} \frac{1-|z|}{1-|\varphi(z)|} = 0$.*

Several natural remarks and questions immediately occur and we postpone the proof to make first some comments.

The condition $m_\varphi(\mathbb{T}) = 0$ is surely not a characterization: consider the symbol $\varphi(z) = \dfrac{1+z}{2}$ (see Section 4.1): it satisfies $m_\varphi(\mathbb{T}) = 0$ since $(\varphi^*)^{-1}(\mathbb{T}) = \{1\}$. Nevertheless, C_φ cannot be compact thanks to assertion (iii): φ has an angular derivative at point 1.

Indeed, thanks to (iii), it appears that a necessary condition for compactness on H^p is that the symbol φ has no angular derivative at any point of $\mathbb{T}$ (recall Julia–Caratheodory's theorem: Theorem 1.33). We shall see later (see Theorem 3.15 and Proposition 6.7) that this condition is a characterization when φ is finitely valent. Nevertheless, it is not always the case (see Theorem 3.22).

Therefore, this condition turns out to play a key role in the theory of composition operators, and we shall meet it again many times. Let us mention here that this condition is invariant by composition with an automorphism:

Proposition 3.10 *Let φ be a symbol and $a \in \mathbb{D}$.*
Let $\psi = \varphi_a \circ \varphi$ where φ_a is the Möbius map associated to a.
Then

$$\lim_{|z|\to 1^-} \frac{1-|z|}{1-|\varphi(z)|} = 0 \iff \lim_{|z|\to 1^-} \frac{1-|z|}{1-|\psi(z)|} = 0.$$

Proof: Since $\varphi = \varphi_a \circ \psi$, it suffices to prove one implication. Let us assume that φ has no angular derivative at any point of $\mathbb{T}$. We point out that

$$\frac{1-|z|^2}{1-|\psi(z)|^2} = \frac{1-|z|^2}{1-|\varphi(z)|^2} \cdot \frac{|1-\bar{a}\varphi(z)|^2}{1-|a|^2} \leqslant \frac{1+|a|}{1-|a|} \cdot \frac{1-|z|^2}{1-|\varphi(z)|^2},$$

and the conclusion follows. □

Proof of Theorem 3.9: The first assertion is just a direct translation of Proposition 3.2.

(ii) The sequence (z^n) converges to 0 in $\mathscr{H}(\mathbb{D})$, so, from the first assertion, we get

$$\|C_\varphi(z^n)\|_p = \|\varphi^n\|_p \longrightarrow 0,$$

but $\|\varphi^n\|_p^p = \int_{\mathbb{T}} |\varphi^*|^{np}\, dm \longrightarrow m_\varphi(\mathbb{T})$ and we conclude for the first part of (ii).

When φ is inner, we have $m_\varphi(\mathbb{T}) = 1$; hence $C_\varphi : H^p \to H^p$ cannot be compact.

(iii) The functional δ_z has norm $(1-|z|^2)^{-1/p}$ and $C_\varphi^*(\delta_z) = \delta_{\varphi(z)}$. Then for any sequence $z_n \in \mathbb{D}$ such that $|z_n| \to 1^-$, the sequence $\mu_n = (1-|z_n|^2)^{1/p}\delta_{z_n}$ lies in the unit sphere of the dual of H^p. Since C_φ^* is compact on $(H^p)^*$ and μ_n is weak-star convergent to 0, we have

$$\|C_\varphi^*(\mu_n)\|_{(H^p)^*} \longrightarrow 0.$$

But

$$\|C_\varphi^*(\mu_n)\|_{(H^p)^*} = (1-|z_n|^2)^{1/p} \left\|\delta_{\varphi(z_n)}\right\|_{(H^p)^*} = \frac{(1-|z_n|^2)^{1/p}}{\left(1-|\varphi(z_n)|^2\right)^{1/p}}.$$

□

The study of compactness reduces to the Hilbertian case ([166]):

Theorem 3.11 *Let φ be a symbol and $p \geqslant 1$.*
C_φ is compact on H^p if and only if C_φ is compact on H^2.

Proof: We only have to prove that if $p, q \geqslant 1$ and C_φ is compact on H^q then C_φ is compact on H^p.

Consider $(f_n)_n$ a sequence in the unit ball of H^p converging to 0 in $\mathscr{H}(\mathbb{D})$. Write $f_n = B_n . g_n$ with $|B_n^*| = 1$ a.e. and $g_n \in H^p$ without zeros in $\mathbb{D}$. The sequence $G_n = g_n^{p/q}$ is defined and lies in the unit ball of H^q.

Up to subsequences, we may assume that $G_n \to G$ and $B_n \to B$ in $\mathscr{H}(\mathbb{D})$, and that the sequence $(G_n \circ \varphi)$ converges to $G \circ \varphi$ in H^q (since $C_\varphi \colon H^q \to H^q$ is compact). For every $z \in \mathbb{D}$, we get that $|f_n(z)|^p = |G_n(z)|^q |B_n(z)|^p \longrightarrow 0$, hence $|G(z)|^q |B(z)|^p = 0$.

Now, we can compute

$$\int_{\mathbb{T}} |f_n \circ \varphi^*|^p \, dm \lesssim \int_{\mathbb{T}} |(G_n - G) \circ \varphi^*|^q |B_n \circ \varphi^*|^p \, dm + \int_{\mathbb{T}} |G \circ \varphi^*|^q |B_n \circ \varphi^*|^p \, dm,$$

so

$$\int_{\mathbb{T}} |f_n \circ \varphi^*|^p \, dm \lesssim \int_{\mathbb{T}} |(G_n - G) \circ \varphi^*|^q \, dm + \int_{\mathbb{T}} |G \circ \varphi^*|^q |B_n \circ \varphi^*|^p \, dm.$$

Remembering that $m_\varphi(\mathbb{T}) = 0$, we have $\varphi^*(z) \in \mathbb{D}$ for almost every $z \in \mathbb{T}$. Therefore, the dominated convergence theorem gives that

$$\int_{\mathbb{T}} |G \circ \varphi^*|^q |B_n \circ \varphi^*|^p \, dm \longrightarrow \int_{\mathbb{T}} |G \circ \varphi^*|^q |B \circ \varphi^*|^p \, dm,$$

but

$$\int_{\mathbb{T}} |G \circ \varphi^*|^q |B \circ \varphi^*|^p \, dm = \int_{\mathbb{D}} |G(z)|^q |B(z)|^p \, dm_\varphi(z) = 0\,.$$

The compactness of C_φ on H^p is proved. □

Compactness via Hilbert–Schmidt

Recall that an operator $T \colon H \to H$ is *Hilbert–Schmidt* if for one (any) orthonormal basis (b_n) of the Hilbert space H, we have

$$\|T\|_{HS}^2 = \sum_{n \geqslant 0} \|T(b_n)\|_H^2 < +\infty.$$

We already saw that Hilbert–Schmidt operators are compact and one of the main interest here is that the membership in the class of Hilbert–Schmidt operators is very easy to check:

Theorem 3.12 *C_φ is Hilbert–Schmidt on H^2 if and only if*

$$\|C_\varphi\|_{HS}^2 = \int_{\mathbb{T}} \frac{1}{1 - |\varphi^*|^2} \, dm < \infty. \tag{3.1}$$

We point out that when $\|\varphi\|_\infty < 1$, C_φ is Hilbert–Schmidt; hence we recover here that C_φ is compact in this particular case.

Proof: Indeed the sequence $b_n(z) = z^n$ (where $n \in \mathbb{N}$) is an orthonormal basis of H^2. Now, let us compute

$$\|C_\varphi\|_{HS}^2 = \sum_{n=0}^{\infty} \|\varphi^n\|_{H^2}^2 = \sum_{n=0}^{\infty} \int_{\mathbb{T}} |\varphi^*|^{2n}\, dm = \int_{\mathbb{T}} \frac{1}{1 - |\varphi^*|^2}\, dm,$$

and the theorem is proved. □

The expression can also be written as

$$\|C_\varphi\|_{HS}^2 = \int_{\overline{\mathbb{D}}} \frac{1}{1 - |z|^2}\, dm_\varphi.$$

3.4.1 A Necessary Condition for Compactness

Characterizing compactness of C_φ is natural and not always that easy. We first indicate a necessary condition.

Actually there is a general principle that works in a general framework: let H be an analytic Hilbert space with reproducing kernels K_a and normalized reproducing kernels

$$k_a = \frac{K_a}{\|K_a\|}.$$

Let also φ be a symbol such that $C_\varphi : H \to H$.

Theorem 3.13 *Assume that polynomials are dense in H and that the reproducing kernel of H satisfies*

$$\lim_{|a| \to 1^-} \|K_a\| = \infty. \tag{3.2}$$

Let φ be a symbol such that C_φ is bounded on H. If this operator is compact, then

$$\lim_{|a| \to 1^-} \frac{\|K_{\varphi(a)}\|}{\|K_a\|} = 0.$$

Proof: The hypothesis clearly implies that $k_a \overset{w}{\to} 0$ as $|a| \to 1$. Indeed, $\|k_a\| = 1$ and $\langle f, k_a\rangle = \frac{f(a)}{\|K_a\|} \to 0$ when f is a polynomial. Next, since C_φ is compact, so is C_φ^*. Since $k_a \overset{w}{\to} 0$ as $|a| \to 1^-$, it follows that $\|C_\varphi^*(k_a)\| \to 0$. But $C_\varphi^*(k_a) = \frac{K_{\varphi(a)}}{\|K_a\|}$, which gives the result. □

As an immediate corollary, we recover Theorem 3.9 (iii) when $p = 2$:

Corollary 3.14 *If C_φ is compact on H^2, then*

$$\lim_{|a|\to 1^-} \frac{1-|a|}{1-|\varphi(a)|} = 0. \tag{3.3}$$

In particular, φ has an angular derivative at no point of $\mathbb{T}$.

Proof: Recall that the reproducing kernel on H^2 at point α is $K_\alpha(z) = \dfrac{1}{1-\bar{\alpha}z}$, whose norm is

$$\|K_\alpha\|_{H^2} = \frac{1}{\sqrt{1-|\alpha|^2}}\,\cdot$$

Point out too that $1-|\alpha| \leqslant 1-|\alpha|^2 \leqslant 2(1-|\alpha|)$. The conclusion follows.

The final remark comes from the Julia–Carathéodory theorem (see Theorem 1.33). □

Remarks. (1) The assumption $\lim_{|a|\to 1} \|K_a\| = \infty$ is not always satisfied. If $H = \mathscr{K}_\theta$ is the model space associated with an inner function θ, more generally a de Branges space $\mathscr{H}(\theta)$, we know ([55]) that the corresponding reproducing kernel is

$$K_a(z) = \frac{1-\overline{\theta(a)}\theta(z)}{1-\bar{a}z}, \quad \text{hence} \quad \|K_a\|^2 = \frac{1-|\theta(a)|^2}{1-|a|^2}\cdot$$

And condition (3.2) has no reason to be satisfied in general.

(2) As simple as it is, Theorem 3.13 already gives a lot of non-compact composition operators. If H is a Bergman space $\mathscr{B}^2_\alpha$ with $\alpha > -1$ and φ a symbol analytic on $\overline{\mathbb{D}}$, say with positive Taylor coefficients and $\varphi(1) = 1$, for example $\varphi(z) = \frac{1+z}{2}$, C_φ is non-compact. Indeed, as $a \to 1^-$, we see that

$$\frac{\|K_a\|}{\|K_{\varphi(a)}\|} = \frac{(1-\varphi(a))^{\alpha+2}}{(1-a)^{\alpha+2}} \to (\varphi'(1))^{\alpha+2},$$

and the necessary condition for compactness is violated.

3.4.2 Characterization in Hardy Spaces in the Finitely Valent Case

We already saw in Section 3.4.1 the necessary condition (3.3).

Recall that φ is *univalent* means that φ is one-to-one. More generally, denoting $n_\varphi(z)$ the cardinal of the w's in $\mathbb{D}$ such that $\varphi(w) = z$, and given $V \in \mathbb{N}$, we say that φ is *V-valent* when, for each $z \in \mathbb{D}$, $n_\varphi(z) \leqslant V$. Of course, univalent means V-valent with $V = 1$.

We say that φ is *finitely valent* when φ is *V-valent* for some $V \in \mathbb{N}$.

We have a characterization for compact composition operators on Hardy spaces in the finitely valent case.

Theorem 3.15 *Let φ be a finitely valent symbol and $p \geqslant 1$. Then the following assertions are equivalent:*

(i) C_φ *is compact on* H^p.
(ii) C_φ *is compact on* H^2.
(iii) $\displaystyle\lim_{|a|\to 1^-} \frac{1-|a|}{1-|\varphi(a)|} = 0.$

This theorem will be recovered later (see Corollary 6.7). Actually the proof below already gives the spirit of the proof for the general case via Nevanlinna counting functions.

Another important remark is that we do not have an equivalence between (i) (or (ii)) and (iii) in full generality: see (iii) in Theorem 3.22 for a counterexample, so the extra assumption with finite valence is not only technical.

Proof: The fact that (i) $\Leftrightarrow$(ii) $\Rightarrow$(iii) is already known (see Theorem 3.11 and (3.3)).

Let us prove that (iii) is sufficient. The argument relies on the Littlewood–Paley formula (see Theorem 1.79) and the criterion of compactness of Hardy spaces (Theorem 3.9(i)). We assumed that φ is finitely valent, so there exists some $V \in \mathbb{N}$ such that, for every $z \in \mathbb{D}$, $n_\varphi(z) \leqslant V$.

Our hypothesis states that

$$\lim_{|a|\to 1^-} \frac{\log(1/|a|)}{1-|\varphi(a)|} = 0,$$

since $\log(1/|a|) \sim 1-|a|$ when $|a| \to 1^-$.

Let (f_n) be a sequence in the unit ball of H^2, converging to 0 in $\mathscr{H}(\mathbb{D})$. A fortiori, from the classical Weierstrass theorem, the sequence (f_n') converges to 0 in $\mathscr{H}(\mathbb{D})$ too.

We know that

$$\|f_n \circ \varphi\|_2^2 = |f_n(\varphi(0))|^2 + 2\int_{\mathbb{D}} \left|f_n' \circ \varphi(w)\right|^2 |\varphi'(w)|^2 \log\frac{1}{|w|}\, d\sigma(w),$$

and we have $|f_n(\varphi(0))| \longrightarrow 0$.

Fix $\varepsilon > 0$. By hypothesis, there exists some $r \in (0,1)$ such that for every $w \in \mathbb{D}$,

$$|w| > r \implies \log(1/|w|) \leqslant \varepsilon(1-|\varphi(w)|) \leqslant \varepsilon \log(1/|\varphi(w)|).$$

Since φ is V-valent, φ vanishes at most V times: we can assume that $|w| > r$ implies that $\varphi(w) \neq 0$.

On one hand, $K_r = \varphi(\bar{D}(0,r))$ is a compact subset of $\mathbb{D}$ and

$$2\int_{\{|w|\leqslant r\}} \left|f_n' \circ \varphi(w)\right|^2 |\varphi'(w)|^2 \log\frac{1}{|w|}\, d\sigma \leqslant \sup_{z\in K_r} \left|f_n'(z)\right|^2 \left(2\int_{\mathbb{D}} |\varphi'(w)|^2 \log\frac{1}{|w|}\, d\sigma\right).$$

But $2\int_{\mathbb{D}} |\varphi'(w)|^2 \log\frac{1}{|w|}\, d\sigma(w) = \|\varphi\|_2^2 - |\varphi(0)|^2 \leqslant 1$. Therefore,

$$2\int_{\{|w|\leqslant r\}} \left|f_n' \circ \varphi(w)\right|^2 |\varphi'(w)|^2 \log\frac{1}{|w|}\, d\sigma(w) \leqslant \sup_{z\in K_r} \left|f_n'(z)\right|^2,$$

which converges to 0.

On the other hand,

$$\int_{\{|w|>r\}} \left|f_n' \circ \varphi(w)\right|^2 |\varphi'(w)|^2 \log\frac{1}{|w|}\, d\sigma \leqslant \varepsilon \int_{\mathbb{D}} \left|f_n' \circ \varphi(w)\right|^2 |\varphi'(w)|^2 \log(1/|\varphi(w)|)\, d\sigma.$$

The generalized change of variable formula (when φ is univalent it is just a "true" change of variable) leads to

$$\int_{\mathbb{D}} \left|f_n' \circ \varphi(w)\right|^2 |\varphi'(w)|^2 \log(1/|\varphi(w)|)\, d\sigma(w) = \int_{\varphi(\mathbb{D})} \left|f_n'(z)\right|^2 \log(1/|z|)\, n_\varphi(z)\, d\sigma(z),$$

and we get

$$2\int_{\{|w|>r\}} \left|f_n' \circ \varphi(w)\right|^2 |\varphi'(w)|^2 \log\frac{1}{|w|}\, d\sigma(w) \leqslant \varepsilon V \|f_n\|_2^2 \leqslant \varepsilon V.$$

Gathering everything, we proved $\overline{\lim} \|f_n \circ \varphi\|_2^2 \leqslant \varepsilon V$; hence $\lim \|f_n \circ \varphi\|_2 = 0$. □

3.4.3 Weak Compactness Properties on H^1

Since H^p is reflexive for $p > 1$, weak compactness and Dunford–Pettis property of C_φ are not interesting in this case. Indeed, every operator is weakly compact on reflexive spaces, and on such spaces, an operator is Dunford–Pettis if and only if it is compact.

That is why we focus only on the case of $p = 1$ in this section.

Let us recall that an operator T is *Dunford–Pettis* when the sequence $\big(T(x_n)\big)$ converges in norm to 0 for any weakly null sequence (x_n).

Concerning the Dunford–Pettis property, we have the following theorem (see [28]).

Theorem 3.16 *C_φ is Dunford–Pettis on H^1 if and only if $m_\varphi(\mathbb{T}) = 0$.*

In particular, we easily have an example of a composition operator on H^1, which is Dunford–Pettis but not compact: just take the symbol $\varphi(z) = \dfrac{1+z}{2}$.

Proof: Assume first that C_φ is Dunford–Pettis on H^1. Clearly the sequence (z^n) is weakly convergent to 0 thanks to the Riemann–Lebesgue lemma: indeed for every $h \in L^\infty(\mathbb{T})$, we have

$$\langle h, z^n \rangle = \widehat{h}(n) \longrightarrow 0.$$

So, by definition, $\|C_\varphi(z^n)\|_{H^1} \longrightarrow 0$, but

$$\|C_\varphi(z^n)\|_{H^1} = \|\varphi^n\|_{H^1} = \int_{\mathbb{T}} |\varphi|^n \, dm = \int_{\overline{\mathbb{D}}} |z|^n \, dm_\varphi(z) \longrightarrow m_\varphi(\mathbb{T}),$$

and we get $m_\varphi(\mathbb{T}) = 0$.

For the converse, take a sequence (f_n) in H^1 weakly converging to 0. We want to prove that

$$\lim \|C_\varphi(f_n)\|_{H^1} = 0. \tag{3.4}$$

Fix $\varepsilon > 0$. The operator C_φ is continuous from H^1 to $L^1(\mathbb{T}, m)$. Equivalently, the formal identity from H^1 to $L^1(\overline{\mathbb{D}}, m_\varphi)$ is continuous, hence it is $w - w$ continuous (when the spaces are equipped with their weak topology). Therefore, f_n is weakly converging to 0 in $L^1(\overline{\mathbb{D}}, m_\varphi)$, hence it is uniformly integrable (see [180, III.C.12] for instance). There exists $\delta > 0$ such that for every measurable subset $A \subset \overline{\mathbb{D}}$,

$$m_\varphi(A) \leqslant \delta \quad \Longrightarrow \quad \forall n, \quad \int_A |f_n| \, dm_\varphi \leqslant \varepsilon.$$

Since $m_\varphi(\mathbb{T}) = 0$, there exists some $r \in (0,1)$ such that

$$m_\varphi\big(\overline{\mathbb{D}} \setminus K_r\big) \leqslant \delta,$$

where K_r is the closed disk centered at 0 with radius r.

We get, for every n,

$$\|C_\varphi(f_n)\|_{H^1} = \int_{\overline{\mathbb{D}}} |f_n| \, dm_\varphi \leqslant \int_{K_r} |f_n| \, dm_\varphi + \varepsilon.$$

Clearly $\displaystyle\int_{K_r} |f_n| \, dm_\varphi \leqslant \sup_{K_r} |f_n|$. Since (f_n) (which is uniformly bounded in H^1) is weakly converging to 0, it is in particular pointwise converging to 0 on $\mathbb{D}$; hence it converges to 0 uniformly on every compact of $\mathbb{D}$.

We get $\overline{\lim} \|C_\varphi(f_n)\|_{H^1} \leqslant \varepsilon$ and the conclusion follows. □

The case of weakly compact composition operators is settled by the following theorem due to Sarason. We present here a proof using ideas similar to the proof of Theorem 3.16.

Theorem 3.17 *C_φ is weakly compact on H^1 if and only if C_φ is compact on H^1.*

Proof: Only one direction needs to be proved.

Let us assume that C_φ is weakly compact on H^1. We shall use the compactness criterion. But first we claim that

$$m_\varphi(\mathbb{T}) = 0. \tag{3.5}$$

Assume $\delta = m_\varphi(\mathbb{T}) > 0$. Then for every $n \geqslant 1$, since $\mathbb{T}$ is the disjoint union of the arcs $\{w \in \mathbb{T} \mid |\arg(\bar{\xi}_j w)| \leqslant \pi/n\}$ where $\xi_j = \exp(2i\pi j/n)$ (with $0 \leqslant j < n$), there exists (at least) one of these arcs, say I_n, satisfying $m_\varphi(I_n) \geqslant \delta/n$. We denote by $u_n \in \mathbb{T}$ the center of this arc (i.e., the corresponding ξ_j) and consider the sequence of functions

$$\forall z \in \overline{\mathbb{D}}, \quad g_n(z) = \frac{1 - r_n^2}{(1 - r_n \overline{u_n} z)^2} \quad \text{with } r_n = 1 - \frac{1}{n} \cdot$$

Clearly $\|g_n\|_{H^1} = 1$. Point out too that for every $z \in I_n$ we have

$$|1 - r_n \overline{u_n} z| = |u_n - r_n z| \leqslant |u_n - z| + 1 - r_n \leqslant \frac{\pi + 1}{n};$$

so

$$|g_n(z)| \geqslant \frac{n}{(\pi + 1)^2} \cdot$$

We know that C_φ is bounded on H^1 and we have

$$\|C_\varphi\| \geqslant \|C_\varphi(g_n)\|_{H^1} \geqslant \int_{I_n} |g_n| \, dm_\varphi \geqslant \frac{n}{(\pi + 1)^2} \cdot m_\varphi(I_n) \geqslant \frac{\delta}{(\pi + 1)^2} \cdot$$

Therefore,

$$\frac{(\pi + 1)^2 \|C_\varphi\|}{n} \geqslant m_\varphi(I_n),$$

and

$$\int_{I_n} |g_n| \, dm_\varphi \geqslant \frac{\delta}{(\pi + 1)^2} \cdot$$

We have a contradiction. Indeed since C_φ is weakly compact, the formal identity from H^1 to $L^1(m_\varphi)$ is weakly compact as well and the (bounded) sequence g_n should be uniformly integrable in $L^1(m_\varphi)$, but $m_\varphi(I_n) \to 0$ and can be arbitrarily small, whereas $\int_{I_n} |g_n| \, dm_\varphi$ cannot. This proves (3.5).

Let now $(f_n)_{n\in\mathbb{N}}$ be a sequence in the unit ball of H^1 converging to 0 in $\mathscr{H}(\mathbb{D})$. We also want to prove (3.4), and the argument now follows the lines of the proof of Theorem 3.16: let us give the key steps.

Since C_φ is weakly compact, the sequence (f_n) is uniformly integrable in $L^1(m_\varphi)$. Fix $\varepsilon > 0$. There exists $\delta > 0$ such that every measurable subset $A \subset \overline{\mathbb{D}}$ with $m_\varphi(A) \leqslant \delta$ satisfies

$$\forall n \in \mathbb{N}, \qquad \int_A |f_n|\, dm_\varphi \leqslant \varepsilon.$$

Now, since $m_\varphi(\mathbb{T}) = 0$, there exists some $r \in (0,1)$ such that $m_\varphi(\overline{\mathbb{D}} \backslash K_r) \leqslant \delta$ where K_r is the closed disk centered at 0 with radius r. On the other hand, $\sup_{K_r} |f_n| \to 0$ by hypothesis.

We get $\overline{\lim} \| C_\varphi(f_n) \|_{H^1} \leqslant \varepsilon$ and the conclusion follows. □

Let us mention that our proof introduces some ideas linked to the Carleson embedding theory. These aspects will be developed in Chapter 5.

3.5 Compactness on Classical Weighted Bergman Spaces

Compactness via Hilbert–Schmidt

As for Hardy spaces, in the Hilbertian framework, an easy way to produce a sufficient condition for compactness is provided by being an Hilbert–Schmidt operator.

Theorem 3.18 *Let $\alpha > -1$. The composition operator C_φ is Hilbert–Schmidt on $\mathscr{B}^2_\alpha$ if and only if*

$$\|C_\varphi\|_{HS}^2 = \int_{\mathbb{D}} \frac{1}{(1-|\varphi|^2)^{\alpha+2}}\, d\sigma_\alpha < \infty. \tag{3.6}$$

Equivalently

$$\|C_\varphi\|_{HS}^2 = (\alpha+1) \int_{\mathbb{D}} \left(\frac{1-|z|^2}{1-|\varphi|^2} \right)^{\alpha+2} \frac{d\sigma(z)}{(1-|z|^2)^2} \cdot$$

For the classical Hilbertian Bergman space $\mathscr{B}^2$, we have in particular

$$\|C_\varphi\|_{HS}^2 = \int_{\mathbb{D}} \frac{1}{(1-|\varphi|^2)^2}\, d\sigma.$$

Proof: Let n be a non-negative integer and

$$\beta_n := 2(\alpha+1)\int_0^1 r^{2n+1}(1-r^2)^\alpha \, dr = (\alpha+1)B(n+1,\alpha+1) = \frac{n!\;\Gamma(\alpha+2)}{\Gamma(n+\alpha+2)},$$

where Γ is the classical Euler function and B is the beta function.

The functions $b_n(z) = \beta_n^{-\frac{1}{2}} z^n$ form an orthonormal basis of $\mathscr{B}^2_\alpha$. Now, let us compute

$$\|C_\varphi\|^2_{HS} = \sum_{n=0}^{\infty} \|C_\varphi(b_n)\|^2_{\mathscr{B}^2_\alpha} = \sum_{n=0}^{\infty} \beta_n^{-1} \|\varphi^n\|^2_{\mathscr{B}^2_\alpha} = \int_{\mathbb{D}} \sum_{n=0}^{\infty} \beta_n^{-1} |\varphi|^{2n} \, d\sigma_\alpha.$$

But for every $y \in \mathbb{D}$, we have

$$\frac{1}{(1-y)^{\alpha+2}} = \sum_{n=0}^{\infty} \frac{\Gamma(n+\alpha+2)}{n!\;\Gamma(\alpha+2)}\, y^n = \sum_{n=0}^{\infty} \beta_n^{-1}\, y^n.$$

So, for every $z \in \mathbb{D}$, we have

$$\sum_{n=0}^{\infty} \beta_n^{-1} |\varphi(z)|^{2n} = \frac{1}{\left(1-|\varphi(z)|^2\right)^{\alpha+2}},$$

and the theorem is proved. □

Characterization of Compactness on $\mathscr{B}^p_\alpha$

In this section, we shall characterize the compactness of composition operators on classical weighted Bergman spaces.

Theorem 3.19 *Let $\alpha > -1$ and φ be a symbol.*

Then the following assertions are equivalent:

(i) *C_φ is compact on $\mathscr{B}^p_\alpha$ for some $p \geqslant 1$.*
(ii) $\displaystyle\lim_{|a|\to 1^-} \frac{1-|a|}{1-|\varphi(a)|} = 0.$
(iii) *φ has an angular derivative at no point of $\mathbb{T}$.*
(iv) *C_φ is compact on $\mathscr{B}^p_\alpha$ for every $p \geqslant 1$.*

Point out that the condition of compactness (ii) (equivalently (iii)) does not depend on α (as soon as $\alpha > -1$), neither on $p \geqslant 1$. We shall recover this result later using the theory of Carleson measures (which is the classical approach, but requires more elaborate tools). We adopt here an elementary approach.

To deal with the most general situation, we shall use the following lemma.

Lemma 3.20 *Let $p \geqslant 1$, $\alpha > -1$, and $\beta > -(\alpha+1)$. Let φ be a symbol. For every $h \in \mathscr{B}^p_{\alpha+\beta}$, we have*

$$\int_{\mathbb{D}} (1-|\varphi(w)|)^{\beta} |h \circ \varphi(w)|^p \, d\sigma_\alpha(w) \lesssim \int_{\mathbb{D}} |h(z)|^p \, d\sigma_{\alpha+\beta}(z). \tag{3.7}$$

In other words the composition operator C_φ is bounded from $\mathscr{B}^p_{\alpha+\beta}$ to the space $L^p(\mathbb{D}, (1-|\varphi|)^\beta d\sigma_\alpha)$.

Proof: We first claim that for every analytic function $h \in \mathscr{B}^p_\gamma$, where $\gamma > -1$, we have, for $w \in \mathbb{D}$,

$$|h(w)|^p \leqslant (1-|w|^2)^{\gamma+2} \int_{\mathbb{D}} \frac{|h(z)|^p}{|1-\bar{z}w|^{2(\gamma+2)}} \, d\sigma_\gamma(z). \tag{3.8}$$

This is easy to check: it is clear for $w = 0$ and $p = 1$ by the mean average property, and since $\|\cdot\|_1 \leqslant \|\cdot\|_p$, we have the result for $w = 0$ and $p \geqslant 1$. To conclude for any $w \in \mathbb{D}$, just apply this to $h \circ \varphi_w$ and make the obvious change of variable. We leave the details as an exercise.

Now, given any polynomial h, we apply the preceding inequality with some $\gamma > \max(-(\beta+2), \alpha+\beta) > -1$. We get, using Fubini's theorem,

$$\begin{aligned}
&\int_{\mathbb{D}} (1-|\varphi(w)|)^{\beta} |h \circ \varphi(w)|^p \, d\sigma_\alpha(w) \\
&\quad \leqslant \int_{\mathbb{D}} |h(z)|^p \int_{\mathbb{D}} \frac{(1-|\varphi(w)|)^\beta (1-|\varphi(w)|^2)^{\gamma+2}}{|1-\bar{z}\varphi(w)|^{2(\gamma+2)}} \, d\sigma_\alpha(w) \, d\sigma_\gamma(z) \\
&\quad \lesssim \int_{\mathbb{D}} |h(z)|^p \int_{\mathbb{D}} \frac{1}{|1-\bar{z}\varphi(w)|^{\gamma+2-\beta}} \, d\sigma_\alpha(w) \, d\sigma_\gamma(z).
\end{aligned}$$

But we already know that C_φ is bounded on the weighted Bergman space $\mathscr{B}^p_\alpha$ where $p = \gamma + 2 - \beta > 1$. So (with underlying constants depending only on the symbol and the parameters α, β, γ),

$$\begin{aligned}
\int_{\mathbb{D}} \frac{1}{|1-\bar{z}\varphi(w)|^{\gamma+2-\beta}} \, d\sigma_\alpha(w) &\lesssim \int_{\mathbb{D}} \frac{1}{|1-\bar{z}w|^{\gamma+2-\beta}} \, d\sigma_\alpha(w) \\
&\lesssim \frac{1}{(1-|z|^2)^{\gamma-(\alpha+\beta)}}.
\end{aligned}$$

Finally, we proved that, for every polynomial h,

$$\int_{\mathbb{D}} (1-|\varphi(w)|)^{\beta} |h \circ \varphi(w)|^p \, d\sigma_\alpha(w) \lesssim \int_{\mathbb{D}} \frac{|h(z)|^p}{(1-|z|^2)^{\gamma-(\alpha+\beta)}} d\sigma_\gamma(z) \approx \int_{\mathbb{D}} |h|^p \, d\sigma_{\alpha+\beta},$$

and a standard density argument gives the conclusion. □

Proof of Theorem 3.19: The fact that (ii) $\Leftrightarrow$ (iii) follows directly from Julia–Caratheodory's theorem 1.33.

The proof of (i) $\Rightarrow$(ii) follows the lines of the proof in case of Hardy spaces.

We know from Proposition 1.91 that the functional δ_z has norm $(1-|z|^2)^{-\frac{\alpha+2}{p}}$. Since $C_\varphi^*(\delta_z)=\delta_{\varphi(z)}$, the sequence $\mu_n=(1-|z_n|^2)^{\frac{\alpha+2}{p}}\delta_{z_n}$ lies in the unit sphere of the dual of $\mathscr{B}_\alpha^p$ for any sequence $z_n\in\mathbb{D}$ such that $|z_n|\to 1^-$. Moreover, since C_φ^* is compact on $(\mathscr{B}_\alpha^p)^*$ and μ_n is weak-star convergent to 0, we have

$$\|C_\varphi^*(\mu_n)\|_{(\mathscr{B}_\alpha^p)^*}\longrightarrow 0.$$

But

$$\|C_\varphi^*(\mu_n)\|_{(\mathscr{B}_\alpha^p)^*}=\frac{(1-|z_n|^2)^{\frac{\alpha+2}{p}}}{\left(1-|\varphi(z_n)|^2\right)^{\frac{\alpha+2}{p}}},$$

and, since $\dfrac{\alpha+2}{p}>0$, this gives (ii).

Now let us prove that (ii)⇒(iv). The strategy of the proof is similar to the one of Theorem 3.15 and relies on the compactness criterion (see Proposition 3.2) and the Littlewood–Paley type formulas (see (1.62) or (1.75) in the Hilbertian case and (1.76) for $p\geqslant 1$).

We first assume that $\varphi(0)=0$.

Let (f_n) be a sequence in the unit ball of $\mathscr{B}_\alpha^p$, converging to 0 in $\mathscr{H}(\mathbb{D})$. The sequence (f_n') converges to 0 in $\mathscr{H}(\mathbb{D})$ as well.

We know that

$$\|f_n\circ\varphi-f_n(0)\|_{\mathscr{B}_\alpha^p}^p\approx\int_{\mathbb{D}}\left|f_n'\circ\varphi(w)\right|^p|\varphi'(w)|^p\big(1-|w|\big)^{p+\alpha}\,d\sigma(w),$$

and we already know that $|f_n(0)|\longrightarrow 0$. Therefore we focus on the right-hand side term.

Fix $\varepsilon>0$. By hypothesis, there exists some $r\in(0,1)$ such that for every $w\in\mathbb{D}$,

$$|w|>r\implies 1-|w|\leqslant\varepsilon\left(1-|\varphi(w)|\right).$$

On one hand, $K_r=\varphi(\bar{D}(0,r))$ is a compact subset of $\mathbb{D}$ and

$$\begin{aligned}\int_{\{|w|\leqslant r\}}\left|f_n'\circ\varphi(w)\right|^p|\varphi'(w)|^p\big(1-|w|\big)^{p+\alpha}\,d\sigma(w)&\lesssim\sup_{z\in K_r}\left|f_n'(z)\right|^p\|\varphi\|_{\mathscr{B}_\alpha^p}^p\\&\leqslant\sup_{z\in K_r}\left|f_n'(z)\right|^p,\end{aligned}$$

which converges to 0.

On the other hand, we treat the other integral term (over $\{|w|>r\}$), but we split here the proof into two cases. The first proof uses a standard way that treats in particular the case $p=2$, and the second case is a bit more subtle but without change of variable and allows to deal with the case of every $p\geqslant 1$.

Proof: When $p \geqslant 2$.

Here $p + \alpha > 1$ for any $\alpha > -1$. We have

$$\int_{\{|w|>r\}} |f_n' \circ \varphi(w)|^p |\varphi'(w)|^p (1 - |w|)^{p+\alpha} \, d\sigma(w)$$
$$\leqslant \int_{\{|w|>r\}} |f_n' \circ \varphi(w)|^p (1 - |\varphi(w)|^2)^{p-2} |\varphi'(w)|^2 (1 - |w|)^{2+\alpha} \, d\sigma(w),$$

thanks to the Schwarz–Pick inequality ((iii) in Lemma 1.28) and since $p - 2 \geqslant 0$.

Therefore,

$$\int_{\{|w|>r\}} |f_n' \circ \varphi(w)|^p |\varphi'(w)|^p (1 - |w|)^{p+\alpha} \, d\sigma(w)$$
$$\leqslant 2^{p-2} \varepsilon^{\alpha+1} \int_{\{|w|>r\}} |f_n' \circ \varphi(w)|^p |\varphi'(w)|^2 (1 - |\varphi(w)|)^{p+\alpha-1} (1 - |w|) \, d\sigma(w)$$
$$\leqslant 2^{p-2} \varepsilon^{\alpha+1} \int_{\mathbb{D}} |f_n' \circ \varphi(w)|^p |\varphi'(w)|^2 (1 - |\varphi(w)|)^{p+\alpha-1} \log \frac{1}{|w|} \, d\sigma(w).$$

Now the generalized change of variable formula leads to

$$\int_{\mathbb{D}} |f_n' \circ \varphi(w)|^p |\varphi'(w)|^2 \left(1 - |\varphi(w)|\right)^{p+\alpha-1} \log \frac{1}{|w|} \, d\sigma(w)$$
$$= \int_{\varphi(\mathbb{D})} |f_n'(z)|^p (1 - |z|)^{p+\alpha-1} N_\varphi(z) \, d\sigma(z),$$

but the Nevanlinna counting function N_φ is controled thanks to the Littlewood inequality (see (1.37)):

$$\forall z \in \mathbb{D} \setminus \{0\}, \quad N_\varphi(z) \leqslant \log \frac{1}{|z|} \lesssim \frac{1 - |z|}{|z|},$$

and we get

$$\int_{\{|w|>r\}} |f_n' \circ \varphi(w)|^p |\varphi'(w)|^p \left(1 - |w|\right)^{p+\alpha} d\sigma(w)$$
$$\lesssim 2^{p-2} \varepsilon^{\alpha+1} \int_{\mathbb{D}} |f_n'(z)|^p \frac{(1 - |z|)^{p+\alpha}}{|z|} \, d\sigma(z).$$

Now, it is easy to check (but see Proposition 1.92 for a proof) that

$$\int_{\mathbb{D}} |f_n'(z)|^p (1 - |z|)^{p+\alpha} \frac{1}{|z|} \, d\sigma(z) \lesssim \int_{\mathbb{D}} |f_n'(z)|^p (1 - |z|)^{p+\alpha} \, d\sigma(z) \lesssim \|f_n\|_{\mathscr{B}_\alpha^p}^p = 1.$$

Gathering everything, we proved $\overline{\lim} \|f_n \circ \varphi\|_{\mathscr{B}_\alpha^p}^p \lesssim \varepsilon^{\alpha+1}$. Since $\alpha + 1 > 0$, we get $\lim \|f_n \circ \varphi\|_{\mathscr{B}_\alpha^p} = 0$.

Proof: When $p \geqslant 1$ and $\alpha > -1$.

We choose $\delta \in (0, 1 + \alpha)$.

Using similar ideas as in the first case (in particular the Schwarz–Pick inequality), but also Lemma 3.8, we have

$$
\begin{aligned}
\int_{\{|w|>r\}} &|f_n' \circ \varphi(w)|^p |\varphi'(w)|^p (1-|w|)^{p+\alpha}\, d\sigma(w) \\
&\lesssim \varepsilon^{\alpha+1-\delta} \int_{\{|w|>r\}} |f_n' \circ \varphi(w)|^p |\varphi'(w)|^p (1-|\varphi(w)|)^{\alpha+1-\delta} (1-|w|)^{p-1+\delta}\, d\sigma \\
&\lesssim \varepsilon^{\alpha+1-\delta} \int_{\mathbb{D}} |f_n' \circ \varphi(w)|^p (1-|\varphi(w)|)^{p+\alpha+1-\delta}\, d\sigma_{\delta-1}(w) \\
&\lesssim \varepsilon^{\alpha+1-\delta} \int_{\mathbb{D}} |f_n'(z)|^p\, d\sigma_{p+\alpha}(z) \qquad \text{thanks to (3.7)} \\
&\lesssim \varepsilon^{\alpha+1-\delta}\, \|f_n\|^p_{\mathfrak{B}^p_\alpha} \lesssim \varepsilon^{\alpha+1-\delta}\,.
\end{aligned}
$$

Therefore, in every case, we get $\lim \|f_n \circ \varphi\|_{\mathfrak{B}^p_\alpha} = 0$, and C_φ is compact on $\mathfrak{B}^p_\alpha$ by the compactness criterion, when φ fixes the origin.

Now we focus on the case of a general symbol and use the "dévissage" principle: set $a = \varphi(0)$ and consider $\psi = \varphi_a \circ \varphi$, which is a symbol fixing the origin.

Thanks to Proposition 3.10, ψ satisfies condition (ii) as well (in other words ψ has no angular derivative at any point of $\mathbb{T}$); therefore C_ψ is compact thanks to the first case. We conclude that $C_\varphi = C_\psi \circ C_{\varphi_a}$ is compact on $\mathfrak{B}^p_\alpha$. □

Let us mention that the corresponding results relatively to weak compactness properties for Bergman spaces (see Theorem 3.16 and Theorem 3.17) do hold but are obvious. Indeed the Bergman space $\mathfrak{B}^1_\alpha$ is isomorphic to ℓ^1 (recall Theorem 1.90) so it has the Schur property. Therefore, any weakly compact operator from any Banach space X to $\mathfrak{B}^1_\alpha$ is compact, and the situation is not specific to composition operators. In the same way any Dunford–Pettis operator from $\mathfrak{B}^1_\alpha$ to any Banach space is compact. We sum up these remarks in the following statement:

Proposition 3.21 *Let $\alpha > -1$ and φ be a symbol.*
The following assertions are equivalent:

(i) *C_φ is compact on $\mathfrak{B}^1_\alpha$.*
(ii) *C_φ is weakly compact on $\mathfrak{B}^1_\alpha$.*
(iii) *C_φ is Dunford–Pettis on $\mathfrak{B}^1_\alpha$.*

In view of our previous results, we can compare the behavior of C_φ on Hardy spaces and Bergman spaces.

Theorem 3.22 *Let $p, q \geqslant 1$ and $\alpha > -1$.*

(i) *If C_φ is compact on some Hardy space H^p, then it is compact on every weighted Bergman space $\mathscr{B}_\alpha^q$.*
(ii) *When φ is finitely valent, C_φ is compact on H^p if and only if it is compact on $\mathscr{B}_\alpha^q$.*
(iii) *Nevertheless, there exist some (inner) symbol φ such that C_φ is compact on $\mathscr{B}_\alpha^q$ but not on H^p.*

Proof: Assertion (i) is an immediate corollary of Theorem 3.9 and Theorem 3.19.

Assertion (ii) follows from Theorem 3.15 and Theorem 3.19.

For the third assertion, it suffices to consider a slow Blaschke product B. More precisely, take $\rho(t) = \sqrt{t}$ on $(0, 1]$ and apply Theorem 4.11 to follow.

Then C_B is compact on any $\mathscr{B}_\alpha^q$ since B has no angular derivative at any point:

$$\frac{1 - |B(a)|}{1 - |a|} \geqslant \frac{1}{\sqrt{1 - |a|}} \longrightarrow +\infty \quad \text{when } |a| \to 1^-.$$

But C_B cannot be compact on any Hardy space since B is an inner function (see Proposition 2.5). □

Let us mention here another nice geometric consequence of compactness.

Corollary 3.23 *Let φ be a symbol and $p \geqslant 1$. Let us assume that C_φ is compact on H^p or merely compact on some Bergman space $\mathscr{B}_\alpha^p$.*

Then φ must have a unique fixed point in $\mathbb{D}$.

Proof: The symbol φ cannot have (at least) two fixed points; else it would be the identity (see Proposition 1.35(i)), and C_φ would not be compact.

Assume that φ has no fixed point inside $\mathbb{D}$. From Theorem 1.36, we get that φ has an angular derivative at some point of the boundary of the unit disk. This contradicts Theorem 3.19.

Alernatively, since $\displaystyle\lim_{|z| \to 1^-} \frac{1 - |\varphi(z)|}{1 - |z|} = +\infty$, there exists some $r \in (0, 1)$ such that

$$\forall z \in \mathbb{D}, \quad |z| = r \implies |\varphi(z)| < |z|.$$

Therefore, by Rouché's theorem, φ has a fixed point inside $r\mathbb{D}$. □

Point out that in the preceding proof, the second argument only uses $\displaystyle\underline{\lim}_{|z| \to 1^-} \frac{1 - |\varphi(z)|}{1 - |z|} > 1$.

4

Fundamental Examples

In this chapter, we are interested in specific examples of symbols. We already met some. For each of them, we shall sum up here their properties, even if the proofs or results appear later in this book.

In some statements of the present chapter, we shall use the notation $\rho_\varphi(h) = \sup_{\xi \in \mathbb{T}} m_\varphi(W(\xi, h))$, where $W(\xi, h)$ is the Carleson box centered in ξ with opening $h > 0$ (see the dedicated Chapter 5 for details).

4.1 The Affine Symbol

Let $\varphi(z) = \dfrac{z+1}{2}$ for $z \in \mathbb{D}$. It is clearly a conformal map from $\mathbb{D}$ onto a subdisk "smoothly touching" the boundary of $\mathbb{D}$:

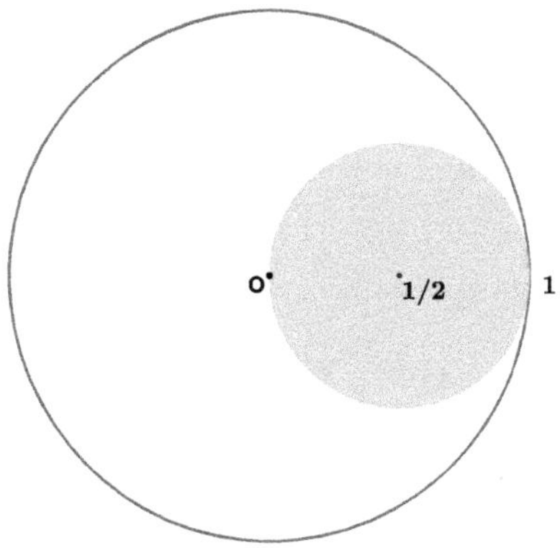

Figure 4.1 The region delimited by an affine symbol is a subdisk.

This particular case provides a natural example of a symbol inducing a non-compact composition operator on the Hardy space, yet touching the boundary in only one point and satisfying $m_\varphi(\mathbb{T}) = 0$; see the remark after Theorem 3.9.

Of course, more generally, we can also consider the symbol defined for $z \in \mathbb{D}$ by

$$\varphi(z) = az + b \qquad \text{with} \qquad |a| + |b| \leqslant 1.$$

When $|a| + |b| = 1$, its geometric image also "smoothly touches" the boundary of $\mathbb{D}$ and it mainly shares the same properties with the previous particular example.

In full generality, there is a dichotomy:

- When $|a| + |b| < 1$, the symbol satisfies $\|\varphi\|_\infty < 1$ and C_φ is compact in a strong way on Hardy and Bergman spaces, since it defines a nuclear operator (see Corollary 3.6 and Theorem 3.8).
- When $|a| + |b| = 1$, then C_φ is not compact and its essential norm is even the "worse" possible on H^2, $\mathscr{B}^2_\alpha$, and on H^∞: $\|C_\varphi\|_e = \|C_\varphi\|$ (see Proposition 2.8, Corollary 6.2, Theorem 6.14(iii), and Theorem 3.7).

  C_φ is not compact on H^1; nevertheless, as soon as φ is not a rotation (i.e., for $a \in \mathbb{D}$) C_φ is a Dunford–Pettis operator on H^1 (see Theorem 3.16).

  In Exercise I.12, it is indicated that

$$\|C_\varphi\|_{B(\mathscr{B}^2_\alpha)} = |a|^{-\frac{\alpha+2}{2}}.$$

4.2 The Lens Map

Given $\theta \in (0, 1]$, we define the lens map as the composition of several maps, translating in $\mathbb{D}$, the power operation $w \in \mathbb{C}_+ \longmapsto \gamma_\theta(w) = w^\theta \in \mathbb{C}_+$ on the right half plane $\mathbb{C}_+$. Concretely, we have

$$\Lambda_\theta(z) = \frac{(1+z)^\theta - (1-z)^\theta}{(1+z)^\theta + (1-z)^\theta}.$$

In other words, with the Cayley transform, that is, the conformal map Φ defined by $z \in \mathbb{D} \mapsto \dfrac{1-z}{1+z} \in \mathbb{C}_+$, we have

$$z \in \mathbb{D} \overset{\Phi}{\longrightarrow} \Phi(z) \in \mathbb{C}_+ \overset{\gamma_\theta}{\longrightarrow} (\Phi(z))^\theta \in \mathbb{C}_+ \overset{\Phi^{-1}}{\longrightarrow} \Lambda_\theta(z) \in \mathbb{D}.$$

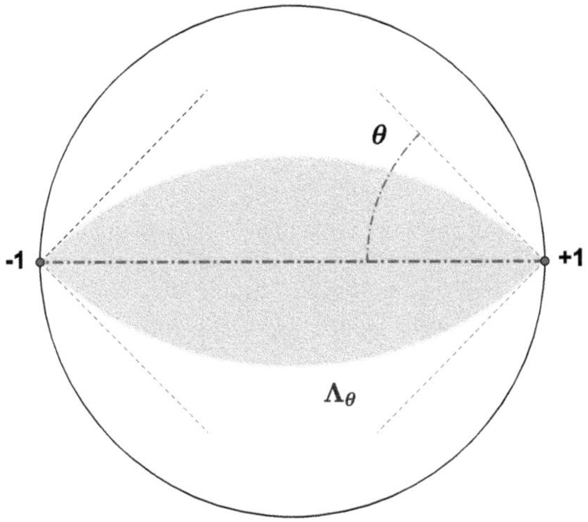

Figure 4.2 The lens map.

The definition of Λ_θ can be extended easily to $\overline{\mathbb{D}}$ and lies in $A(\mathbb{D})$ (even in the Wiener algebra W^+, see Exercise I.15) and has 0 and ± 1 as fixed points inside $\overline{\mathbb{D}}$.

The map Λ_θ is a conformal mapping from $\mathbb{D}$ onto a subdomain like in Figure 4.2.

Proposition 4.1 *Let $\theta \in (0, 1)$. The symbol Λ_θ satisfies:*

(i) $(\Lambda_\theta)_{\theta\in(0,1]}$ *forms a multiplicative and commutative semi-group (conjugate to the semi-group formed by the γ_θ) with unit $\Lambda_1 = I$, namely*

$$\Lambda_{\theta_1\theta_2} = \Lambda_{\theta_1} \circ \Lambda_{\theta_2}.$$

(ii) *For every $z \in \overline{\mathbb{D}}$ with $\operatorname{Re} z \geqslant -\dfrac{1}{2}$, we have*

$$1 - |\Lambda_\theta(z)| \approx |1 - \Lambda_\theta(z)| \approx |1 - z|^\theta . \tag{4.1}$$

(iii) *More precisely, there exists some numerical $\delta > 0$ such that, for every $z \in \overline{\mathbb{D}}$ with $\operatorname{Re}(z) \geqslant -\dfrac{1}{2}$,*

$$1 - |\Lambda_\theta(z)| \geqslant \delta(1 - \theta)\,|1 - z|^\theta. \tag{4.2}$$

(iv) *For every $z, z' \in \overline{\mathbb{D}}$, we have*

$$|\Lambda_\theta(z) - \Lambda_\theta(z')| \lesssim |z - z'|^\theta. \tag{4.3}$$

Actually the last relation in (4.1) holds for every $z \in \overline{\mathbb{D}}$.

Proof: (i) is clear since the properties are inherited from the corresponding semi-group property of $(\gamma_\theta)_{\theta \in (0,1]}$.

(ii) The inequality $|1 - \Lambda_\theta(z)| \geqslant 1 - |\Lambda_\theta(z)|$ is obvious.

We have

$$1 - \Lambda_\theta(z) = \frac{2(1-z)^\theta}{(1+z)^\theta + (1-z)^\theta},$$

and $|(1+z)^\theta + (1-z)^\theta| \leqslant 2^{\theta+1}$ while

$$\begin{aligned}\left|(1+z)^\theta + (1-z)^\theta\right| &\geqslant \mathrm{Re}\left[(1+z)^\theta + (1-z)^\theta\right] \geqslant \left[|1+z|^\theta + |1-z|^\theta\right]\cos\left(\frac{\pi\theta}{2}\right)\\ &\geqslant \left[|1+z| + |1-z|\right]^\theta \cos\left(\frac{\pi\theta}{2}\right) \geqslant 2^\theta \cos\left(\frac{\pi\theta}{2}\right),\end{aligned}$$

which gives $|1 - \Lambda_\theta(z)| \approx |1-z|^\theta$ in (4.1).

The last lower estimate will follow from (iii).

(iii) Clearly $\mathrm{Re}\,[(1+\overline{z})(1-z)] = 1 - |z|^2 > 0$.

We observe that $(1+\overline{z})^\theta\,(1-z)^\theta = [(1+\overline{z})\,(1-z)]^\theta$, so we can write

$$\mathrm{Re}\left((1+\overline{z})^\theta(1-z)^\theta\right) \geqslant \cos\left(\frac{\pi\theta}{2}\right)\left|(1+\overline{z})^\theta(1-z)^\theta\right|.$$

But

$$\cos\left(\frac{\pi\theta}{2}\right) = \sin\left((1-\theta)\frac{\pi}{2}\right) \geqslant 1 - \theta.$$

Therefore, when $\mathrm{Re}\, z \geqslant -\frac{1}{2}$,

$$\begin{aligned}1 - |\Lambda_\theta(z)|^2 &= 4\frac{\mathrm{Re}\left((1+\overline{z})^\theta(1-z)^\theta\right)}{|(1+z)^\theta + (1-z)^\theta|^2} \gtrsim \cos\left(\frac{\pi\theta}{2}\right)\ |(1+\overline{z})(1-z)|^\theta\\ &\geqslant \frac{1-\theta}{2}\,|1-z|^\theta.\end{aligned}$$

To prove (4.3), we can assume $|z - z'| \leqslant 1/2$ (because $|\Lambda_\theta(z) - \Lambda_\theta(z')| \leqslant 2^{1+\theta}|z-z'|^\theta$ for $|z - z'| > 1/2$), and $|z| = |z'| = 1$ by the maximum modulus principle (see Exercise I.13 for details). We can also assume (since Λ_θ is odd) that $\mathrm{Re}\, z \geqslant 0$, implying $\mathrm{Re}\, z' \geqslant -1/2$ since $|z - z'| \leqslant 1/2$. Majorizing $|\Phi'|$, we see that Φ is Lipschitz with constants at most 2 on $\mathbb{C}_+$ and at most 8 on the convex half-plane $\mathbb{C}_{-1/2}$. Let us now set for an instant $w = \Phi(z) =: iy$ and $w' = \Phi(z') =: iy'$. We claim that

$$|w^\theta - w'^\theta| \leqslant 2|w - w'|^\theta. \tag{4.4}$$

Indeed, this is obvious (since $|u^\theta - v^\theta| \leqslant |u-v|^\theta$ for $u, v \geqslant 0$) when y, y' have the same sign. And if, say, $y > 0$ and $y' < 0$, we have

$$|w^\theta - w'^\theta| \leqslant y^\theta + |y'|^\theta \leqslant 2(y + |y'|)^\theta = 2|w - w'|^\theta.$$

We now have $\operatorname{Re} \gamma_\theta[\Phi(z)] \geqslant 0$, $\operatorname{Re} \gamma_\theta[\Phi(z')] \geqslant 0$, so that by (4.4) and the Lipschitz properties of Φ above, we have

$$\begin{aligned} |\Lambda_\theta(z) - \Lambda_\theta(z')| &= |\Phi\gamma_\theta\Phi(z) - \Phi\gamma_\theta\Phi(z')| \leqslant 2|\gamma_\theta\Phi(z) - \gamma_\theta\Phi(z')| \\ &\leqslant 4|\Phi(z) - \Phi(z')|^\theta \leqslant 4.\, 8^\theta|z - z'|^\theta. \end{aligned}$$

□

We sum up here the results obtained for the composition operators where the symbol is the lens map.

Theorem 4.2 *Let $\theta \in (0,1)$ and Λ_θ be the associated lens map. The composition operator with symbol Λ_θ satisfies:*

(i) *The operator C_{Λ_θ} is Hilbert–Schmidt on H^2 with $\|C_{\Lambda_\theta}\|_{HS} \leqslant \dfrac{C}{1-\theta}$ where C is a numerical constant.*
(ii) *On the Hardy space H^p, we have some positive constants a, b, a', b' depending only on θ such that*

$$a' \exp(-b'\sqrt{n}) \leqslant a_n(C_{\Lambda_\theta}) \leqslant a \exp(-b\sqrt{n}).$$

(iii) *C_{Λ_θ} belongs to the Schatten class $\mathscr{S}^r(H^2)$ for every $r > 0$.*
(iv) *C_{Λ_θ} is a nuclear operator on H^p for every $p \geqslant 1$.*
(v) *The operator C_{Λ_θ} is Hilbert–Schmidt on $\mathfrak{B}^2_\alpha$.*
(vi) *C_{Λ_θ} belongs to the Schatten class $\mathscr{S}^r(\mathfrak{B}^2_\alpha)$ for every $r > 0$ and $\alpha > -1$.*
(vii) *C_{Λ_θ} is a nuclear operator on $\mathfrak{B}^p_\alpha$ for every $p \geqslant 1$ and $\alpha > -1$.*

In particular, the composition operator C_{Λ_θ} is compact on Hardy and Bergman spaces.

Proof: (i) By symmetry, we know, from (3.1), that

$$\|C_{\Lambda_\theta}\|^2_{HS} = \frac{2}{\pi}\int_0^{\pi/2} \frac{1}{1 - |\Lambda_\theta(e^{it})|^2} dt.$$

By (4.2), since $|1 - e^{it}| \geqslant (2/\pi)t$ for $0 \leqslant t \leqslant \pi$, this implies

$$\|C_{\Lambda_\theta}\|^2_{HS} \lesssim \frac{1}{1-\theta}\int_0^{\pi/2} t^{-\theta} dt \lesssim \frac{1}{(1-\theta)^2}\cdot$$

(ii) This key (and not obvious at all) result is proved in Theorem 10.25 for H^2, and in Theorem 10.27 and Theorem 10.27 for H^p.

(iii) It follows from (ii) and the definition. Nevertheless, for $r \geqslant 1$, it can also be proved easily from (i) and the semi-group property. Indeed, $C_{\Lambda_\theta} = C^2_{\Lambda_{\sqrt{\theta}}}$ and $C_{\Lambda_{\sqrt{\theta}}}$ is Hilbert–Schmidt, but the composition of two Hilbert–Schmidt operators is nuclear, and hence belongs to $\mathscr{S}^r(H^2)$ for every $r \geqslant 1$. Moreover, this implies that its nuclear norm (on H^2) is $O\big((1-\theta)^{-2}\big)$.

(iv) From (ii) the sequence $\big(a_n(C_{\Lambda_\theta})\big)$ belongs to ℓ^1. Then the result follows from Proposition 1.117.

(v)–(vii) follow the lines of the proof for Hardy spaces (see first (3.6)). □

4.3 The Cusp Map

Given a curve γ with some specific properties (see [53]), we define a conformal mapping χ from $\mathbb{D}$ onto the domain whose boundary is given by the graph of this curve, the graph of $\overline{\gamma}$, and some arc joining the two preceding curves in $\mathbb{D}$. We get Figure 4.3.

The crucial point for the properties of the symbol (and then of the associated composition operator C_χ) is, of course, the behavior at point 1. The philosophy is that the more the curve is flattened, the stronger the operator properties of C_χ will be.

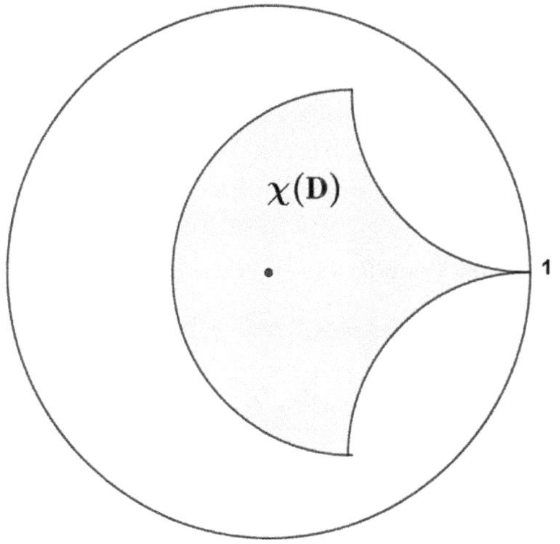

Figure 4.3 A cusp map with a curve $\gamma(x) \approx (1-x)^2$ in the neighborhood of 1.

We shall focus here only on one specific case, the one corresponding to Figure 4.3 and detail the explicit definition of the map and some of its remarkable properties.

We begin by coining

$$\chi_0(z) = \frac{\left(\frac{z-i}{iz-1}\right)^{1/2} - i}{-i\left(\frac{z-i}{iz-1}\right)^{1/2} + 1}. \tag{4.5}$$

That defines a conformal mapping from $\mathbb{D}$ onto the right half-disk $D = \mathbb{D} \cap \mathbb{C}_+$, such that

$\chi_0(1) = 0$, $\chi_0(-1) = 1$, $\chi_0(i) = -i$, $\chi_0(-i) = i$, and $\chi_0(0) = \sqrt{2} - 1$.

Then we set

$$\chi_1(z) = \log \chi_0(z), \quad \chi_2(z) = -\frac{2}{\pi}\, \chi_1(z) + 1, \quad \chi_3(z) = \frac{a}{\chi_2(z)}, \tag{4.6}$$

and finally

$$\chi(z) = 1 - \chi_3(z), \tag{4.7}$$

where

$$a = 1 - \frac{2}{\pi} \log(\sqrt{2} - 1) = 1.56\ldots \in (1, 2) \tag{4.8}$$

is chosen in order that $\chi(0) = 0$.

The image Ω of the (univalent) cusp map χ is formed by the intersection of the interior of the disk $D\big(1 - \frac{a}{2}, \frac{a}{2}\big)$ and the outer part of the two closed disks $\overline{D\big(1 + \frac{ia}{2}, \frac{a}{2}\big)}$ and $\overline{D\big(1 - \frac{ia}{2}, \frac{a}{2}\big)}$.

Since χ is injective, it follows from Zorboska's characterization in [187] (see also [163, section 6.12], or Exercise I.25) that the composition operator C_χ is bounded on the Dirichlet space $\mathcal{D}^2_\alpha$ for $\alpha \geqslant 0$. In particular $\chi \in \mathcal{D}^2_\alpha$ (see Section 10.8 for details).

The behavior near the boundary is emphasized in the following lemma.

Lemma 4.3 *When $r \to 1^-$, it holds:*

$$1 - \chi(r) \approx \frac{1}{\log[1/(1-r)]} \quad \text{and} \quad \chi'(r) \approx \frac{1}{(1-r)\log^2[1/(1-r)]}. \tag{4.9}$$

Proof: For $r \in (0, 1)$, we have the easily checked relation (see also [111, lemma 4.2]):

$$\chi_0(r) = \tan\left[\frac{1}{2} \arctan\left(\frac{1-r}{1+r}\right)\right] = \frac{1-r}{\sqrt{2(1+r^2)} + 1 + r};$$

hence $\chi_0(r) \approx 1 - r$ and

$$\chi_0'(r) = -\frac{1 + [\chi_0(r)]^2}{2(1 + r^2)}\,.$$

Using the definitions (4.6) and (4.7), we get $\chi_2' = -\frac{2}{\pi}\frac{\chi_0'}{\chi_0}$ and $\chi_3' = -\frac{a\,\chi_2'}{\chi_2^2}$. So that, when $r \to 1^-$, $\chi_2(r) \approx \log[1/(1 - r)]$ and

$$\chi_2'(r) = \left(\frac{1}{\pi}\,\frac{1 + [\chi_0(r)]^2}{1 + r^2}\right)\frac{1}{\chi_0(r)} \approx \frac{1}{\chi_0(r)} \approx \frac{1}{1 - r}\,.$$

The result follows. □

Actually the first estimate in Lemma 4.3 can be generalized:

Lemma 4.4 *We have* $1 - |\chi(z)| \gtrsim \dfrac{1}{\log[1 + 1/|1 - z|]}$ *for all* $z \in \mathbb{D}$.

Proof: It suffices to work in a neighborhood V of 1 since on $\mathbb{D} \setminus V$ both functions are continuous and non-vanishing. Setting $u = 1 - z$, an easy computation with Taylor expansions gives that $\frac{z-i}{iz-1} = -1 + iu + o\,(u)$, as $u \to 0$, so $\left(\dfrac{z - i}{iz - 1}\right)^{1/2} = i[1 - iu/2 + o\,(u)] = i + u/2 + o\,(u)$, and (recall (4.5) and (4.6))

$$\chi_0(1 - u) = \frac{u}{4} + o\,(u) \quad \text{as } u \to 0\,;$$

hence $|1 - \chi(z)| = |\chi_3(z)| \approx \dfrac{1}{|\log(\chi_0(z))|} \gtrsim \dfrac{1}{\log(1/|1 - z|)}$ when $z \in \mathbb{D}$ is close to 1.

Finally, since the image of the cusp is contained in an angular sector, there exists $\delta > 0$ such that $1 - |\chi(z)| \geqslant \delta|1 - \chi(z)|$ for every $z \in \mathbb{D}$. □

Let us summarize some of the other properties of the associated composition operator.

Theorem 4.5 *The composition operator C_χ satisfies:*

(i) *On the Hardy space H^2 and on Bergman spaces $\mathfrak{B}_\alpha^2$, the operator C_χ is nuclear. It also belongs to all the Schatten classes.*
(ii) *On the Dirichlet spaces $\mathcal{D}_\alpha^2$, there exists $b, b' > 0$ such that*

$$e^{-b'n/\log n} \lesssim a_n(C_\chi) \lesssim e^{-b\alpha n/\log n} \qquad \text{when } 1 \geqslant \alpha > 0,$$

and there exists $b_0, b_0' > 0$ such that

$$e^{-b_0'\sqrt{n}} \lesssim a_n(C_\chi) \lesssim e^{-b_0\sqrt{n}} \qquad \text{when } \alpha = 0\,.$$

Proof: (i) follows from (ii), which is proved in Theorem 10.36. □

4.4 Surjective Symbols with Remarkable Properties

We present in this section how to produce onto symbols that can be compact on Hardy spaces (subsection below), and even "very compact" (in the second subsection).

An Onto Compact Symbol

We detail here the remarkable example originally due to B. MacCluer and J. Shapiro: they constructed a surjective symbol Φ from $\mathbb{D}$ onto $\mathbb{D}$, which will be actually also finitely valent (4-valent) and induces a compact composition operator on Hardy spaces (hence on Bergman spaces as well).

A first step is to construct another example of symbol φ that is weaker in the sense that $\varphi(\mathbb{D}) = \mathbb{D} \setminus \{0\}$ (the range fails to catch the origin), but is stronger in the sense that φ is 2-valent. Here is the construction.

We first consider the following domain Ω (Figure 4.4) and F any conformal mapping from $\mathbb{D}$ onto this domain.

In the original example of MacCluer–Shapiro, $\gamma(x) = \dfrac{1}{x}$, but the construction works with rather general curves γ.

Now we take the image of Ω by the map $z \longmapsto \check{e}(z) = \exp(-z)$. In Figure 4.5, the longest curve is the image of γ and of $\widetilde{\gamma}$, whereas the shortest one gives an idea of the image of any curve translated (along the imaginary axis) from γ in Ω: we see that the whole disk (except the origin) is reached.

To sum up,

$$\varphi = \check{e} \circ F.$$

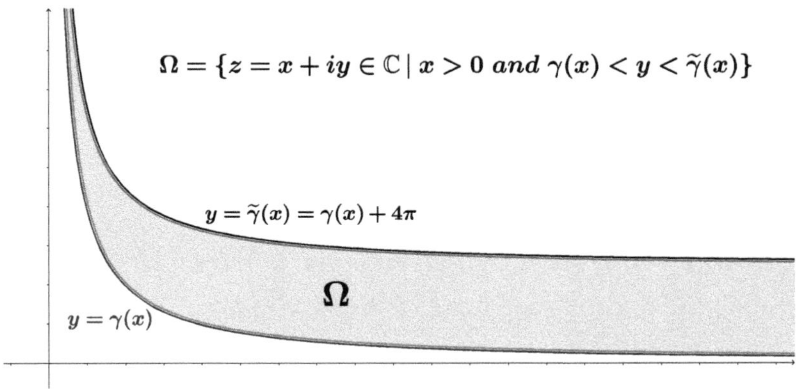

Figure 4.4 The domain Ω.

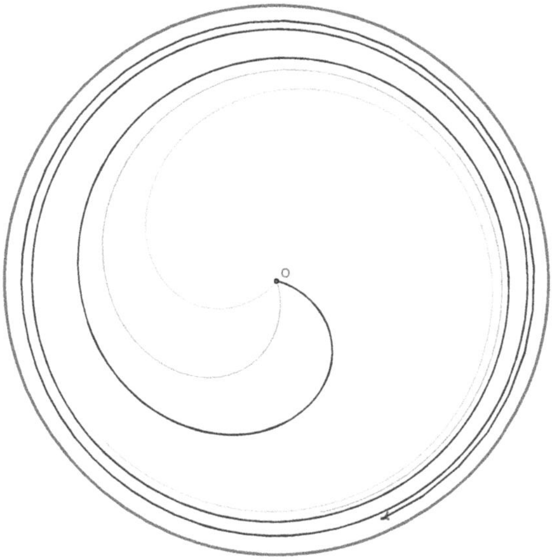

Figure 4.5 Mapping Ω with ě.

It is clear that $\varphi(\mathbb{D}) = \mathbb{D} \setminus \{0\}$ since $\check{e}(\Omega) = \mathbb{D} \setminus \{0\}$. Indeed, on one hand, since Ω is included in the right half plane, $\check{e}(\Omega) \subset \mathbb{D} \setminus \{0\}$. On the other hand, given any $w = \check{e}(x+i\theta) \in \mathbb{D}\setminus\{0\}$ where $x > 0$ and $\theta \in [0, 2\pi)$, it is possible to find exactly two values of y such that $\theta - y \in 2\pi\mathbb{Z}$ and $\gamma(x) < y < \gamma(x)+4\pi$. Hence $\check{e}$ is 2-valent on Ω. Since F is a conformal mapping, φ is 2-valent as well.

Now, in order to reach the whole unit disk, we introduce $\Phi = \varphi_a^2 \circ \varphi$ for some $a \in \mathbb{D} \setminus \{0\}$:

$$\forall z \in \mathbb{D}, \qquad \Phi(z) = \left(\frac{a - \varphi(z)}{1 - \overline{a}\varphi(z)}\right)^2 .$$

Let $w \in \mathbb{D}$. If $w \neq a^2$, then take any squareroot u of w and there is some $z \in \mathbb{D}$ such that $\varphi(z) = \varphi_a(u) \neq 0$: we have $\Phi(z) = w$. If $w = a^2$, there exists some $z \in \mathbb{D}$ such that $\varphi(z) = \varphi_a(-a) = 2a/(1 + |a|^2) \neq 0$. Again $\Phi(z) = (-a)^2 = w$. Therefore Φ is onto.

Since φ_a is univalent, it is clear that φ_a^2 is 2-valent, hence Φ is 4-valent.

Let us prove that C_φ is compact. Once this is proved, we get $C_\Phi = C_\varphi \circ C_{\varphi_a^2}$ is compact as well. Since φ is finitely valent, we know from Corollary 3.15 (iii) and Julia–Caratheodory's theorem (see Theorem 1.33) that it is sufficient to prove that φ has no angular derivative at any point of $\mathbb{T}$. Actually, φ is quite far from having such a derivative. Indeed, let us assume that at some point

$u \in \mathbb{T}$, φ admits some angular limit $v \in \mathbb{T}$. By definition of φ it means in particular that $\angle \lim_{z \to u} \mathrm{Re}\,(F(z)) = 0$. The shape of Ω forces the imaginary part of $F(z)$ to go to $+\infty$ since $\mathrm{Im}\,(F(z)) \geqslant \gamma(\mathrm{Re}\,(F(z))) = 1/\mathrm{Re}\,(F(z))$. Therefore, (by connectedness) $\exp(-i\mathrm{Im}\,(F(z))$ does not have any (angular) limit and this contradicts the fact that $\angle \lim_{z \to u} \varphi(z)$ exists.

Other Onto Compact Symbols

We extend here the previously discussed idea to the construction of same types of examples, but with a prescribed speed of decay for the pull-back measure of an annulus. The following result allows to produce examples with stronger compactness properties.

Theorem 4.6 *Given a continuous increasing function* $\delta : (0,1) \to (0, +\infty)$, *there exist a* 2*-valent symbol* φ *and an* $h_0 > 0$ *such that* $\varphi(\mathbb{D}) = \mathbb{D} \setminus \{0\}$ *and*

$$\forall h \in (0, h_0), \qquad m\left(\{z \in \mathbb{T};\ |\varphi^*(z)| > 1 - h\}\right) \leqslant \delta(h)\,.$$

In other words,

$$\rho_\varphi(h) \leqslant m_\varphi\left(\overline{\mathbb{D}} \setminus (1-h)\overline{\mathbb{D}}\right) \leqslant \delta(h)\,.$$

This is actually a particular case of the following result (see [101, proposition 2.12]).

Theorem 4.7 *Let* $g : (0, +\infty) \to (0, +\infty)$ *be a continuous non-increasing function with* $\lim\limits_{0^+} g = +\infty$ *and* $\lim\limits_{+\infty} g = 0$; *let* $h : (0, +\infty) \to (0, +\infty]$ *be a lower semi-continuous function such that* $\sup\{h(t) : t \geqslant \pi\} < \infty$, *and* $\Omega = \{x + iy : x \in (0, +\infty) \text{ and } g(x) < y < g(x) + h(x)\}$.

Then, given $f : \overline{\mathbb{D}} \to \overline{\Omega} \cup \{\infty\}$, *which is a conformal mapping from* $\mathbb{D}$ *onto* Ω *satisfying* $f(0) = \pi + i(g(\pi) + h(\pi)/2)$, *the symbol* $\varphi = \exp(-f)$ *satisfies, for some constants* $h_0, k_0 > 0$:

(i) $\forall h \in (0, h_0), \qquad m\left(\{z \in \mathbb{T};\ |\varphi^*(z)| > 1 - h\}\right) \leqslant k_0\, e^{-g(2h)}$.
(ii) *Assume that, for some* $r \in (0, +\infty]$ *and integers* $0 \leqslant n < N \leqslant \infty$, *one has* $\{h(t) : t \leqslant r\} \subset (2n\pi, 2N\pi]$; *then, for all* $z \in \mathbb{D}$ *with* $|z| > e^{-r}$, *we have* $n_\varphi(z) \in [n, N]$.

Even though its proof relies on the same ideas, we shall use only the version of Theorem 4.6 in the sequel and that is why we only detail this one. The proof of Theorem 4.6 requires the following lemma, where, given $\alpha \in \mathbb{R}$, we shall use the domain

$$U_\alpha = \{z \in \mathbb{C};\ \mathrm{Re}\,(z) \in (0, \pi)\} \cup \{z \in \mathbb{C};\ \mathrm{Im}(z) < \alpha\}.$$

Lemma 4.8 *There exists some $L > 0$ with the following property: For every $f \in \mathscr{H}(\mathbb{D})$ satisfying, for some $\alpha \in \mathbb{R}$, $\mathrm{Im} f(0) < \alpha$ and $f(\mathbb{D}) \subset U_\alpha$, we have*

$$\forall y \geqslant \alpha\,, \qquad m\left(\{z \in \mathbb{T}\,;\ \mathrm{Im}(f^*(z)) > y\}\right) \leqslant L\, e^{\alpha - y}.$$

Let us mention that an analytic function f satisfying the hypothesis of the lemma admits radial limits. Actually $f \in H^p$ for every $p \in (0, 1)$ because $f(\mathbb{D}) \subset \mathbb{C}_+$: see (5.36).

Proof: We define $F(z) = if(z) - i\dfrac{\pi}{2} + \alpha$ for $z \in \mathbb{D}$. We point out that for every $z \in \mathbb{D}$, we have either $F(z) \in \mathbb{C}_+$ or $\mathrm{Im}\, F(z) \in \left(-\dfrac{\pi}{2}, \dfrac{\pi}{2}\right)$. It implies that $h(z) \in \mathbb{C}_+$ for every $z \in \mathbb{D}$, where h is the function defined by

$$\forall z \in \mathbb{D}\,, \quad h(z) = 1 + \exp\left(-F(z)\right) = 1 + i\mathrm{e}^{-\alpha} \exp\left(-if(z)\right) \cdot$$

Finally we define, for $z \in \mathbb{D}$, $H(z) = h(z) - i\mathrm{Im}\, h(0) \in \mathbb{C}_+$. Point out that $H(0) \in \mathbb{R}$.

Applying Kolmogorov's inequality (see Lemma 5.33), there exists some $k_1 \geqslant 1$ such that

$$\forall t > 0\,, \qquad m\left(\{z \in \mathbb{T}\,;\ |H^*(z)| > t\}\right) \leqslant k_1\, \frac{H(0)}{t} \leqslant \frac{2k_1}{t} \cdot \tag{4.10}$$

Indeed, $\mathrm{Im} f(0) < \alpha$, so $\left|\mathrm{Im}\big(h(0)\big)\right| < 1$ and $H(0) = \mathrm{Re}\,\big(h(0)\big) < 2$.

Now fix $y \geqslant \alpha$. Any $z \in \mathbb{D}$ satisfying $\mathrm{Im}(f(z)) > y \geqslant \alpha$ actually satisfies $\mathrm{Re}\,(f(z)) \in (0, \pi)$ by hypothesis, hence $|h(z)| \geqslant \mathrm{e}^{y-\alpha}$. In other words, we have $\{z \in \mathbb{D}\,;\ \mathrm{Im}(f(z)) > y\} \subset \{z \in \mathbb{D}\,;\ |h(z)| \geqslant \mathrm{e}^{y-\alpha}\}$. A fortiori, using radial limits (since they exist for f a.e., so do they for F, h and H a.e.), we obtain

$$m\left(\{z \in \mathbb{T}\,;\ \mathrm{Im}(f^*(z)) > y\}\right) \leqslant m\left(\{z \in \mathbb{T}\,;\ |h^*(z)| \geqslant \mathrm{e}^{y-\alpha}\}\right).$$

We now split the discussion into two cases: when $\mathrm{e}^{y-\alpha} \geqslant 2$, the condition $|h^*(z)| \geqslant \mathrm{e}^{y-\alpha}$ implies that

$$|H^*(z)| \geqslant |h^*(z)| - \left|\mathrm{Im}\big(h(0)\big)\right| > |h^*(z)| - 1 \geqslant \frac{1}{2}\mathrm{e}^{y-\alpha},$$

and we get, from (4.10),

$$m\left(\{z \in \mathbb{T}\,;\ \mathrm{Im}(f^*(z)) > y\}\right) \leqslant 4k_1\mathrm{e}^{\alpha-y}\,.$$

When $\mathrm{e}^{y-\alpha} < 2$, the preceding inequality is obviously fulfilled since we have $m\left(\{z \in \mathbb{T}\,;\ \mathrm{Im}(f^*(z)) > y\}\right) \leqslant 1 \leqslant 2\mathrm{e}^{\alpha-y}$.

Therefore, in any case, we have $m\left(\{z \in \mathbb{T}\,;\ \mathrm{Im}(f^*(z)) > y\}\right) \leqslant 4k_1\mathrm{e}^{\alpha-y}$.

□

Proof of Theorem 4.6: In the spirit of the example above, we define the same domain

$$\Omega = \{x + iy\,;\ x \in (0, +\infty) \text{ and } \gamma(x) < y < \gamma(x) + 4\pi\,\},$$

where we choose a continuous decreasing function $\gamma : (0, +\infty) \to (0, +\infty)$ satisfying the four conditions

$$\lim_{t\to 0^+} \gamma(t) = +\infty, \qquad \lim_{t\to +\infty} \gamma(t) = 0, \quad \gamma(\pi) = \pi, \quad \text{and} \quad e^{\gamma(t)} \geqslant \frac{e^{5\pi} L}{\delta(t/2)},$$

when t is small enough (say $t \in (0, t_0)$).

As above, we take a conformal mapping $f : \mathbb{D} \to \Omega$ such that $f(0) = \pi + 3i\pi$ and we define the symbol

$$\varphi = e^{-f}.$$

As above φ is 2-valent and $\varphi(\mathbb{D}) = \mathbb{D} \setminus \{0\}$.

Now we check that it works: there exists $\eta_0 > 0$ such that $e^{-2h} < 1 - h$ and $\gamma(2h) > 5\pi$ for every $h \in (0, \eta_0)$. Fix $h \in (0, h_0)$ where $h_0 = \min(\eta_0, t_0/2)$. For any $z \in \mathbb{T}$ such that $|\varphi^*(z)| > 1 - h$, we have Re $(f^*(z)) < 2h$. But since $f^*(z) \in \overline{\Omega}$, we must have $\mathrm{Im}(f^*(z)) > \gamma(2h)$. Writing $y = \gamma(2h) > 5\pi$, we get from Lemma 4.8, applied with $\alpha = 5\pi$,

$$\begin{aligned} m\left(\{z \in \mathbb{T};\ |\varphi^*(z)| > 1 - h\}\right) &\leqslant m\left(\{z \in \mathbb{T};\ \mathrm{Im}(f^*(z)) > y\}\right) \\ &\leqslant L e^{5\pi - y} = L e^{5\pi - \gamma(2h)} \\ &\leqslant \delta(h). \end{aligned}$$

□

As applications, we can construct *surjective* symbols with many strong compactness properties. For instance,

- Given an Orlicz function Ψ, there exists a surjective symbol φ such that $C_\varphi : H^\Psi \to H^\Psi$ is compact.
- There exists an onto symbol φ such that $C_\varphi : H^2 \to H^2$ belongs to every Schatten class.

It suffices to apply Theorem 4.6 and the corresponding criterion (which involves a condition on ρ_φ) to get an almost surjective symbol. To get a "true" surjective symbol, just use the same trick as in the MacCluer–Shapiro's example: take $\Phi = \varphi_a^2 \circ \varphi$ for some $a \in \mathbb{D} \setminus \{0\}$. See details in Proposition 11.27 and Chapter 9.

But let us mention that the choice $\gamma(x) = \dfrac{1}{x}$ in the proof of Theorem 4.6 actually shows that MacCluer–Shapiro's example already satisfies that C_Φ is more than compact on H^2: it belongs to every Schatten class.

4.5 The Shapiro–Taylor Symbols

There are three similar types of such symbols: the original one was introduced by J. Shapiro and P. Taylor in their seminal paper [166], but in [96], some variants are introduced to produce examples for separation of Schatten classes.

We focus below only on the first two, justifying their definition and giving their main properties. The third one will be discussed in Chapter 9.

The First Shapiro–Taylor Map

We shall use the same function as J. Shapiro and P. Taylor in [166, § 4], with a slight modification. This modified function will be easier to analyze.

Let $\theta > 0$. For $\operatorname{Re} z > 0$, $\log z$ will be the principal determination of the logarithm. Let, for $\varepsilon > 0$,

$$V_\varepsilon = \{z \in \mathbb{C};\ \operatorname{Re} z > 0 \text{ and } |z| < \varepsilon\} = \mathbb{C}_+ \cap D(0, \varepsilon), \tag{4.11}$$

and consider, for $\varepsilon > 0$ small enough,

$$f_\theta(z) = z(-\log z)^\theta, \qquad z \in V_\varepsilon. \tag{4.12}$$

Let now $g \colon \mathbb{D} \to V_\varepsilon$ be the conformal map from $\mathbb{D}$ onto V_ε sending $\mathbb{T} = \partial\mathbb{D}$ onto ∂V_ε, and with $g(1) = 0$ and $g'(1) = -\varepsilon/4$. Explicitly,

$$g = \tau_\varepsilon \circ \gamma^{-1} \circ s \circ \gamma \circ q \tag{4.13}$$

is the composition of the following maps:

a) $q \colon z \mapsto -z$ from $\mathbb{D}$ onto itself;

b) $\gamma \colon z \mapsto \dfrac{z + i}{1 + iz}$ from $\mathbb{D}$ onto $P = \{\operatorname{Im} z > 0\}$, with $\operatorname{Im} \gamma(z) = \dfrac{1 - |z|^2}{|1 + iz|^2}$;

c) $s \colon z \mapsto \sqrt{z}$ from P onto $Q = \{\operatorname{Re} z > 0, \operatorname{Im} z > 0\}$;

d) $\gamma^{-1} \colon z \mapsto \dfrac{z - i}{1 - iz}$ from Q onto $V = \big\{|z| < 1, \operatorname{Re} z > 0\big\}$; and

e) $\tau_\varepsilon \colon z \mapsto \varepsilon z$ from V onto V_ε.

We finally set

$$\phi_\theta = \exp(-f_\theta \circ g). \tag{4.14}$$

Setting $h_\theta = f_\theta \circ g$, J. Shapiro and P. Taylor used the symbol $\phi_\theta = 1 - h_\theta$ instead of our $\exp(-h_\theta)$.

We show here, in a precised way, that ϕ_θ maps $\mathbb{D}$ to itself.

Lemma 4.9 *For $\varepsilon > 0$ small enough, one has* $\operatorname{Re} f_\theta(re^{i\alpha}) > 0$ *for* $0 < r < \varepsilon$ *and* $|\alpha| < \pi/2$. *Moreover, one has* $\operatorname{Re} f_\theta^*(z) > 0$ *for all* $z \in \partial V_\varepsilon \setminus \{0\}$ *and Re* $f_\theta'(re^{i\alpha}) > 0$ *for* $0 < r < \varepsilon$, *and* $|\alpha| < \pi/2$.

This lemma will be used again in Chapter 9.

Proof: Actually f_θ can be defined on $\overline{V_\varepsilon} \setminus \{0\}$, and we shall do that.

Let, for $|\alpha| \leqslant \pi/2$ and $0 < r \leqslant \varepsilon$,

$$Z_\alpha = \big(-\log(re^{i\alpha})\big)^\theta = \Big(\log\frac{1}{r} - i\alpha\Big)^\theta.$$

One has $f_\theta(re^{i\alpha}) = re^{i\alpha}|Z_\alpha|e^{i\arg Z_\alpha}$, so that

$$\operatorname{Re} f_\theta(re^{i\alpha}) = r\,|Z_\alpha|\cos(\alpha + \arg Z_\alpha).$$

On the other hand,

$$\arg Z_\alpha = -\theta \arctan\frac{\alpha}{\log 1/r}\cdot$$

Since $\arctan x \geqslant x/2$ for $0 \leqslant x \leqslant 1$, we get, for $0 < r \leqslant \varepsilon \leqslant e^{-\pi/2}$,

$$\begin{aligned} |\alpha + \arg Z_\alpha| &\leqslant |\alpha|\left(1 - \frac{\theta}{\alpha}\arctan\frac{\alpha}{\log 1/r}\right) \\ &\leqslant \frac{\pi}{2}\left(1 - \frac{\theta}{2\log 1/r}\right) =: \Upsilon_r < \pi/2. \end{aligned}$$

Therefore, for $0 < r \leqslant \varepsilon$ and $|\alpha| \leqslant \pi/2$,

$$\operatorname{Re} f_\theta(re^{i\alpha}) \geqslant (\cos\Upsilon_r)\, r\left(\log\frac{1}{r}\right)^\theta > 0.$$

Finally, for $z = re^{i\alpha} \in V_\varepsilon$, we have

$$f_\theta'(z) = (-\log z)^\theta - \theta(-\log z)^{\theta-1},$$

and a similar computation leads to

$$\operatorname{Re} f_\theta'(re^{i\alpha}) \geqslant \delta_r\left(\log\frac{1}{r}\right)^\theta \quad \text{with } \delta_r > 0.$$

□

We summarize some properties of the composition operator associated to ϕ_θ:

(i) ϕ_θ is univalent.

(ii) We have $\rho_{\phi_\theta}(h) \approx \dfrac{h}{(\log 1/h)^\theta}$ when $h \to 0^+$.

(iii) $C_{\phi_\theta} : H^2 \to H^2$ is compact for every $\theta > 0$, and

$$C_{\phi_\theta} \in \mathscr{S}^p(H^2) \qquad \text{if and only if} \qquad p > \frac{4}{\theta}.$$

(iv) $C_{\phi_\theta} : \mathscr{B}^2_\gamma \to \mathscr{B}^2_\gamma$ is compact for every $\theta > 0$, and

$$C_{\phi_\theta} \in \mathscr{S}^p(\mathscr{B}^2_\gamma) \qquad \text{if and only if} \qquad p > \frac{4}{(\gamma+2)\theta}.$$

(v) C_{ϕ_θ} is compact on H^Ψ if and only if, for every $A > 1$, we have, when $u \to +\infty$, $\Psi(Au) = o\Big(\Psi(u)\big[\log \Psi(u)\big]^\theta\Big)$.

Formula (ii) is (9.25), proved in Chapter 9. (i), (iii), and (iv) will be proved in Theorem 9.14. Finally (v) is a clear consequence of (ii) and the characterization (11.28) in Theorem 11.22.

The Second Shapiro–Taylor Map

We introduce now a variant of the preceding Shapiro–Taylor map. It is the modified function $f = f_{\theta,s}$ (in which we drop the dependence in θ and s to ease notations). It is defined as follows:

$$f(z) = z(-\log z)^\theta \big[\log(-\log z)\big]^s, \tag{4.15}$$

where $s > \theta/2 \quad (s = \theta$, for example).

We still denote the corresponding symbol by

$$\phi_\theta = \exp(-f \circ g),$$

where g is the conformal mapping given by (4.13).

We have first to check again that ϕ_θ is indeed a symbol.

Lemma 4.10 *For $\varepsilon > 0$ small enough, one has* $\operatorname{Re} f(z) > 0$ *and* $\operatorname{Re} f'(z) > 0$ *for every $z \in V_\varepsilon$. Moreover $|\phi_\theta^*| < 1$ a.e. on $\mathbb{T}$.*

Proof: Write $z = re^{i\alpha}$ with $0 < r < \varepsilon$ and $|\alpha| \leqslant \pi/2$. One has

$$f(z) = re^{i\alpha}\Big(\log\frac{1}{r} - i\alpha\Big)^\theta \Big(\log\Big(\log\frac{1}{r} - i\alpha\Big)\Big)^s.$$

But

$$\Big(\log\frac{1}{r} - i\alpha\Big)^\theta = \Big(\log\frac{1}{r}\Big)^\theta \Big(1 - \frac{i\alpha\theta}{\log 1/r} + o\Big(\frac{1}{(\log 1/r)}\Big)\Big);$$

and, on the other hand,

$$\log\Big(\log\frac{1}{r}-i\alpha\Big)=\log\sqrt{\Big(\log\frac{1}{r}\Big)^2+\alpha^2}+i\arg\Big(\log\frac{1}{r}-i\alpha\Big)$$
$$=\Big(\log\log\frac{1}{r}\Big)\big(1+o\,(1/\log(1/r))\big)-i\arctan\frac{\alpha}{\log 1/r}$$
$$=\Big(\log\log\frac{1}{r}\Big)\big(1+o\,(1/\log(1/r))\big)-i\frac{\alpha}{\log 1/r}\big(1+o\,(1)\big),$$

and

$$\Big[\log\Big(\log\frac{1}{r}-i\alpha\Big)\Big]^s$$
$$=\Big(\log\log\frac{1}{r}\Big)^s\Big[1+o\Big(\frac{1}{\log 1/r}\Big)-i\frac{1}{(\log 1/r)(\log\log 1/r)}\big(\alpha s+o\,(1)\big)\Big]$$
$$=\Big(\log\log\frac{1}{r}\Big)^s\Big[1+o\Big(\frac{1}{\log 1/r}\Big)\Big]. \tag{4.16}$$

Hence, for $\varepsilon>0$ small enough and $z\in V_\varepsilon$,

$$\mathrm{Re} f(z)=r(\log 1/r)^\theta(\log\log 1/r)^s\Big[\cos\alpha+\frac{\theta\alpha\sin\alpha}{\log 1/r}+o\Big(\frac{1}{\log 1/r}\Big)\Big]$$
$$\geqslant r(\log 1/r)^\theta(\log\log 1/r)^s\Big[\cos\alpha+\frac{\theta(\alpha\sin\alpha-1/4)}{\log 1/r}\Big].$$

That gives the result separating two cases:

If $|\alpha|\leqslant\pi/4$, since $|\theta(\alpha\sin\alpha-1/4)/(\log 1/r)|\leqslant\sqrt{2}/4$ for $\varepsilon>0$ (and hence r) small enough, we get $\cos\alpha+\dfrac{\theta(\alpha\sin\alpha-1/4)}{\log 1/r}\geqslant\sqrt{2}/2-\sqrt{2}/4=\sqrt{2}/4$.

If $\pi/4\leqslant|\alpha|\leqslant\pi/2$, one has

$$\cos\alpha+\frac{\theta(\alpha\sin\alpha-1/4)}{\log 1/r}\geqslant\theta(\alpha\sin\alpha-1/4)\geqslant\theta\big(\pi\sqrt{2}/8-1/4\big)>0.$$

That ends the proof of the lemma, after a similar estimation for $\mathrm{Re} f'(z)$. □

We summarize some properties of the composition operator associated to this symbol:

(i) ϕ_θ is injective.

(ii) We have $\rho_{\phi_\theta}(h)\approx\dfrac{h}{(\log 1/h)^\theta\,(\log(\log 1/h))^s}$ when $h\to 0^+$.

(iii) We have

$$\sigma_{\phi_\theta,\gamma}(h)\approx\frac{h^{\gamma+2}}{(\log 1/h)^{(\gamma+2)\theta}\,(\log(\log 1/h))^{(\gamma+2)s}}\quad\text{when } h\to 0^+.$$

(iv) $C_{\phi_\theta} : H^2 \to H^2$ is compact for every $\theta > 0$ and $s > \dfrac{\theta}{2}$, and

$$C_{\phi_\theta} \in \mathscr{S}^p(H^2) \text{ if and only if } p \geqslant \frac{4}{\theta}.$$

(v) $C_{\phi_\theta} : \mathscr{B}^2_\gamma \to \mathscr{B}^2_\gamma$ is compact for every $\theta > 0$ and $s > \dfrac{\theta}{2}$, and

$$C_{\phi_\theta} \in \mathscr{S}^p(\mathscr{B}^2_\gamma) \text{ if and only if } p \geqslant \frac{4}{(\gamma+2)\theta}.$$

(vi) C_{ϕ_θ} is compact on H^Ψ if and only if, for every $A > 1$, we have $\Psi(Au) = o\left(\Psi(u)\left[\log\Psi(u)\right]^\theta \left[\log\log\Psi(u)\right]^s\right)$ when $u \to +\infty$.

(vii) C_{ϕ_θ} is compact on $\mathscr{B}^\Psi$ if and only if, for every $A > 1$, we have $\Psi(Au) = o\left(\Psi(u)\left[\log\Psi(u)\right]^{2\theta} \left[\log\log\Psi(u)\right]^{2s}\right)$ when $u \to +\infty$.

(ii) and (iii) can be deduced from (9.32) and (9.33).

(iv) and (v) will be proved in Theorem 9.15.

Finally (vi) and (vii) are clear consequences of (ii) and (iii) and the characterizations (11.28) and (11.47)

4.6 The Slow Blaschke Products

The purpose of this section is to show that, for non-boundedly valent symbols φ, the necessary condition (seen in Chapter 2)

$$\lim_{|z|\to 1} \frac{1-|\varphi(z)|}{1-|z|} = +\infty, \tag{4.17}$$

for the compactness of C_φ on H^2 is far from being sufficient.

Our construction of the so-called *slow Blaschke products* can be seen as an elaboration of Shapiro's example in [165, p. 185]. Actually it is used at several places in this book and we detail it here.

Theorem 4.11 *Let* $\Delta : (0,1] \to (0,\infty)$ *be such that* $\lim_{\varepsilon\to 0}\Delta(\varepsilon) = 0$. *Then, there exists a Blaschke product B satisfying*

$$1-|B(z)| \geqslant \Delta(1-|z|) \qquad \textit{for every } z \in \mathbb{D}. \tag{4.18}$$

In particular, there are symbols φ *satisfying* (4.17) *such that* $C_\varphi : H^p \to H^p$ *is not compact for any* $p \geqslant 1$, *whereas* $C_\varphi : \mathscr{B}^q \to \mathscr{B}^q$ *is compact for every* $q \geqslant 1$.

Actually a similar result holds in the framework of Orlicz analytic spaces: see Chapter 11.

Proof: For the application, just take for our symbol the Blaschke product B corresponding to the function $\Delta(\varepsilon) = \sqrt{\varepsilon}$. It satisfies (4.17) and C_B is non-compact on H^p, for any value of $p \geqslant 1$, since B is inner (see Theorem 3.9(ii)). On the other hand, thanks to Theorem 3.19, we know that C_B is compact on every Bergman space $\mathscr{B}^p$, with $p \geqslant 1$.

Now, our slow Blaschke product will be of the form

$$B = \prod_{n=1}^{\infty} B_n,$$

where each B_n has p_n zeros equidistributed on the circumference $|z| = r_n$ with $0 < r_n < 1$, and where the sequences (p_n) and (r_n), $r_n = \mathrm{e}^{-h_n}$, will be specified later, always satisfying the Blaschke condition

$$\sum_{n=1}^{\infty} p_n(1 - r_n) < \infty, \quad \text{or else} \quad \sum_{n=1}^{\infty} p_n\, h_n < \infty. \tag{4.19}$$

We begin with a technical lemma.

Lemma 4.12 *Let $p \in \mathbb{N}$ and $0 < r < 1$. Consider the Blaschke product*

$$G(z) = \prod_{k=1}^{p} \frac{z - r\mathrm{e}^{i\theta_k}}{1 - r\mathrm{e}^{-i\theta_k} z} = \frac{z^p - r^p}{1 - r^p z^p}, \tag{4.20}$$

where $\theta_k = \frac{2k\pi}{p}$ with $1 \leqslant k \leqslant p$.

Then,

(i) *for every z with $|z| \leqslant r$, we have*

$$|G(z)| \leqslant \frac{2r^p}{1 + r^{2p}} \leqslant 1 - \frac{(1 - r^p)^2}{2}. \tag{4.21}$$

(ii) *if $r = \mathrm{e}^{-h}$ with $ph \leqslant 1$, we also have*

$$|G(z)| \leqslant 1 - \frac{(ph)^2}{8}. \tag{4.22}$$

Proof: The second equality in (4.20) is obvious, and is used as follows: for $|z| \leqslant r$, one has

$$|G(z)| = \rho(z^p, r^p) \leqslant \frac{|z|^p + r^p}{1 + r^p |z|^p} \leqslant \frac{2r^p}{1 + r^{2p}} = 1 - \frac{(1 - r^p)^2}{1 + r^{2p}},$$

where we used the property $\rho(a,b) \leqslant \dfrac{|a|+|b|}{1+|a||b|}$ of the pseudo-hyperbolic distance.

We use the elementary inequality $1-\mathrm{e}^{-x} \geqslant \dfrac{x}{1+x}$ for any $x \geqslant 0$ to get

$$1 - r^p \geqslant \frac{ph}{1+ph} \geqslant \frac{ph}{2} \quad \text{when } ph \leqslant 1,$$

and get (4.22) through (4.21). □

Now, replacing Δ by $\tilde{\Delta}(x) = \sup_{0<t\leqslant x} \left[\max(\Delta(t), \sqrt{t})\right]$ if necessary, we can as well assume that Δ is non-decreasing and satisfies $\Delta(t) \geqslant \sqrt{t}$ for each $0 < t \leqslant 1$.

In order to keep things less technical, we will prove (4.18) only for $|z|$ close to 1, which will be sufficient for us. By hypothesis, there exists a decreasing sequence (h_n) of positive numbers such that $\Delta(h_n) \leqslant 2^{-n}$. Let next p_n be the smallest integer such that

$$\frac{(p_n h_n)^2}{8} > 2^{-(n-1)} \quad \text{that is,} \quad p_n h_n > 4.2^{-n/2}.$$

This choice of the integer p_n ensures that (4.19) holds. Indeed, we know that $\sqrt{h_n} \leqslant \Delta(h_n) \leqslant 2^{-n}$, so that $h_n \leqslant 2^{-2n}$ and

$$p_n h_n = (p_n - 1)h_n + h_n = O(2^{-n/2}).$$

Hence, there exists an integer n_0 such that

$$p_n h_n \leqslant 1 \qquad \text{for } n \geqslant n_0. \tag{4.23}$$

Now take $z \in \mathbb{D}$. We can assume that $|z| = \mathrm{e}^{-h}$ with $\mathrm{e}^{-h} \geqslant \mathrm{e}^{-h_{n_0}}$.

We fix $n > n_0$ satisfying $h_n < h \leqslant h_{n-1}$. We obtain, thanks to (4.22) and (4.23), since $|z| \leqslant \mathrm{e}^{-h_n}$,

$$|B(z)| \leqslant |B_n(z)| \leqslant 1 - \frac{(p_n\, h_n)^2}{8} \leqslant 1 - 2^{-(n-1)} \leqslant 1 - \Delta\,(h_{n-1}),$$

so that

$$1 - |B(z)| \geqslant \Delta\,(h_{n-1}) \geqslant \Delta(h) \geqslant \Delta(1-\mathrm{e}^{-h}) = \Delta(1-|z|).$$

This ends the proof. □

5
Carleson Embedding Point of View

In this chapter, we elaborate on the fundamental importance of Carleson measures in complex analysis on the unit disk, concerning here interpolation sequences and embedding theorems. It is convenient to first recall a few classical facts. We will denote by (e^{ia}, e^{ib}) the circular arc $\{e^{it} : a < t < b\}$.

5.1 Reminders on Maximal Functions

We first need a few results on maximal functions and their comparison and begin with the most famous one.

Hardy–Littlewood Maximal Function

Let $g \in L^1(\mathbb{T}, m)$. The Hardy–Littlewood maximal function $Mg : \mathbb{T} \to \mathbb{R}^+$ attached to g is defined as follows:

$$Mg(\xi) = \sup_I \frac{1}{m(I)} \int_I |g|\, dm, \tag{5.1}$$

where I runs over the collection of all arcs (intervals) of $\mathbb{T}$ centered at $\xi \in \mathbb{T}$. If ν is a positive measure on $\mathbb{T}$, we define similarly the maximal function

$$M\nu(\xi) = \sup_I \frac{\nu(I)}{m(I)} \cdot$$

The following basic theorem about this Hardy–Littlewood maximal function will be admitted here [153, pp. 136–137].

Theorem 5.1 (Hardy–Littlewood) *Let $g \in L^1(\mathbb{T}, m)$. Then, $Mg \in L^{1,\infty}(\mathbb{T})$ with $L^{1,\infty}(\mathbb{T})$ the weak L^1-space, namely for some absolute contant $C > 0$ and all $t > 0$, one has*

$$m(\{Mg > t\}) \leqslant \frac{C}{t} \|g\|_1.$$

By the Marcinkiewicz interpolation theorem [58, p. 25] (see also Chapter 11), we infer the following useful result.

Theorem 5.2 *Fix $1 < p < \infty$. Let $g \in L^p(\mathbb{T}, m)$. Then, $Mg \in L^p(\mathbb{T})$ with moreover*

$$\|Mg\|_p \leqslant C_p \|g\|_p,$$

where C_p only depends on p.

Maximal Functions

Other maximal functions, all dominated by the Hardy–Littlewood one, will be used in this chapter. We recall (see Section 1.3.1) that, given $\xi \in \mathbb{T}$ and $a \geqslant 1$, the non-tangential approach region (the radius $[0, \xi]$ if $a = 1$) $\Gamma_a(\xi)$ with vertex ξ is

$$\Gamma_a(\xi) = \left\{ z \in \mathbb{D} : |\xi - z| < a(1 - |z|) \right\}.$$

We now define for any function $f : \mathbb{D} \to \mathbb{C}$:

- The *radial maximal function* $N(f) : \mathbb{T} \to \mathbb{C}$ is defined by

$$N(f)(\xi) = \sup_{0 \leqslant r < 1} |f(r\xi)|.$$

One advantage of the function N is that it obviously satisfies

$$N\left(f^2\right) = (Nf)^2.$$

We will exploit that later in this chapter.

- The *non-tangential maximal function* $N_a f$, $a \geqslant 1$ is defined by

$$N_a f(\xi) = \sup_{z \in \Gamma_a(\xi)} |f(z)|, \text{ where } \xi \in \mathbb{T}. \tag{5.2}$$

Note that $N_1 = N$. It is also useful to note that all those maximal functions N_a on $\mathbb{T}$ are dominated by the Hardy–Littlewood one.

Theorem 5.3 *For any $a \geqslant 1$, there exists a constant $C_a > 0$ such that, for any $f \in H^1$, it holds*

$$N_a f(\xi) \leqslant C_a M f^*(\xi), \qquad \textit{for every } \xi \in \mathbb{T}.$$

This theorem is well known; we yet wish to sketch a proof, based on the following lemma:

Lemma 5.4 *Let ψ be a positive function defined on $\mathbb{T}$ such that the sets $I_c = \{\psi > c\}$ are arcs in $\mathbb{T}$, symmetric around* 1, *for any positive c. Let ν be a positive measure on $\mathbb{T}$. Then, for any $z \in \mathbb{T}$, one has*

$$\int_{\mathbb{T}} \psi(\bar{\xi} z)\, d\nu(\xi) \leqslant \|\psi\|_{L^1(\mathbb{T})} .M\nu(z).$$

Proof: We borrow an elegant proof from Pisier's notes in Texas A&M. We can assume that $\|\psi\|_{L^1(\mathbb{T})} = \int_{\mathbb{T}} \psi\, dm = 1$, and the idea is to write ψ as a "convex combination" of functions of the form $\frac{\mathbb{1}_{I_c}}{m(I_c)}$ · We first note that

$$\forall z \in \mathbb{T}, \qquad \psi(z) = \int_0^{+\infty} \mathbb{1}_{I_c}(z)\, dc = \int_0^{+\infty} \frac{\mathbb{1}_{I_c}(z)}{m(I_c)} m(I_c)\, dc$$

and

$$\int_{\mathbb{T}} \psi\, dm = \int_0^{+\infty} m(I_c)\, dc = 1.$$

This allows us to write, via Fubini,

$$\int_{\mathbb{T}} \psi(\bar{\xi} z)\, d\nu(\xi) = \int_0^{+\infty} m(I_c) \left[\int_{\mathbb{T}} \frac{\mathbb{1}_{I_c}(\bar{\xi} z)}{m(I_c)}\, d\nu(\xi) \right] dc = \int_0^{+\infty} m(I_c) \frac{\nu(J_c(z))}{m(J_c(z))}\, dc,$$

where $J_c(z) = \{z\bar{\xi} : \xi \in I_c\}$ is an arc, symmetric around z, with $m(J_c(z)) = m(I_c)$.

We conclude that $\int_{\mathbb{T}} \psi(\bar{\xi} z)\, d\nu(\xi) \leqslant \int_0^{+\infty} m(I_c)\, M\nu(z)\, dc = M\nu(z)$. □

Proof of Theorem 5.3: We can assume $\xi = 1$ and begin with the radial maximal function Nf. We apply the preceding Lemma 5.4 to the function $\psi = P_r$, the Poisson kernel at $r \in (0, 1)$. Since $\theta \in [-\pi, \pi] \mapsto P_r(\exp(i\theta))$ is even, and also decreasing on $[0, \pi]$, it is clear that the set I_c is an arc for any $c > 0$. Now we can write

$$|f(r)| \leqslant \int_{\mathbb{T}} |f^*(\xi)| P_r(\xi)\, dm(\xi),$$

and, thanks to Lemma 5.4 applied with the measure $\nu = |f^*|\, dm$, we obtain

$$Nf(1) = \sup_{r<1} |f(r)| \leqslant C\, Mf^*(1)\,.$$

The general case follows from this one, if we observe that, setting $r = |z|$, $P_z(\xi) \leqslant (a+2)^2 P_r(\xi)$ for all $z \in \Gamma_a(1)$ and $\xi \in \mathbb{T}$. Indeed, for such z and ξ

$$\begin{aligned} |r - \xi| &\leqslant |r - z| + |z - \xi| \leqslant 1 - r + |1 - z| + |z - \xi| \\ &\leqslant (a+1)(1-r) + |z - \xi| \leqslant (a+2)|z - \xi|. \end{aligned}$$

Therefore, for $z \in \Gamma_a(1)$,

$$|f(z)| \leqslant \int_{\mathbb{T}} |f^*| P_z\, dm \leqslant (a+2)^2 \int_{\mathbb{T}} |f^*| P_r\, dm \leqslant C(a+2)^2 Mf^*(1).$$

□

Here is a reverse comparison result ([153, theorem 11.20, exercise 19]). We recall that h^1 is the set of harmonic functions $u : \mathbb{D} \to \mathbb{R}$ such that

$$\sup_{0<r<1} \int_{\mathbb{T}} |u(r\xi)|\, dm(\xi) < \infty.$$

Any function $u \in h^1$ has a.e. radial limits $u^* \in L^1(\mathbb{T})$. We say that $u \in h^1$ is *regular* if moreover $u = Ref$ for some $f \in H^1$. We have the following lemma:

Lemma 5.5 *Let $u : \mathbb{D} \to \mathbb{R}^+$ be regular. Then we have for any $\xi \in \mathbb{T}$:*

$$Nu(\xi) \leqslant M(u^*)(\xi) \leqslant 11Nu(\xi). \tag{5.3}$$

Proof: The first inequality was proved in Theorem 5.3, even in the more general case of non-tangential maximal functions. Point out that $u \in h^1$ as soon as u is harmonic and non-negative. Next, we can assume $\xi = 1$.

We fix $\delta \in (0,1)$ and let $I_\delta = \left(e^{-i\pi\delta}, e^{i\pi\delta}\right)$ be an arc in $\mathbb{T}$ centered at 1 with $m(I_\delta) = \delta$ and $r = 1 - \delta$. Observe that the Poisson kernel P_r satisfies, for any $t \in \mathbb{R}$,

$$P_r\left(e^{it}\right) = \frac{1 - r^2}{(1-r)^2 + 4r\sin^2 t/2} \geqslant \frac{1-r}{(1-r)^2 + t^2} = \frac{\delta}{\delta^2 + t^2}.$$

Then

$$\begin{aligned} Nu(1) \geqslant u(r) &= \int_{-\pi}^{\pi} P_r\left(e^{it}\right) u^*\left(e^{it}\right) \frac{dt}{2\pi} \\ &\geqslant \int_{-\pi\delta}^{\pi\delta} P_r\left(e^{it}\right) u^*\left(e^{it}\right) \frac{dt}{2\pi} \\ &\geqslant \int_{-\pi\delta}^{\pi\delta} \frac{\delta}{\delta^2 + t^2}\, u^*\left(e^{it}\right) \frac{dt}{2\pi} \\ &\geqslant \frac{1}{11\delta} \int_{-\pi\delta}^{\pi\delta} u^*\left(e^{it}\right) \frac{dt}{2\pi} \\ &= \frac{1}{11} \frac{1}{m(I_\delta)} \int_{I_\delta} u^*(\eta)\, dm(\eta). \end{aligned}$$

Passing to the supremum on δ gives the second inequality of (5.3). □

We can now introduce and discuss the notion of Carleson measure, the main topic of this chapter.

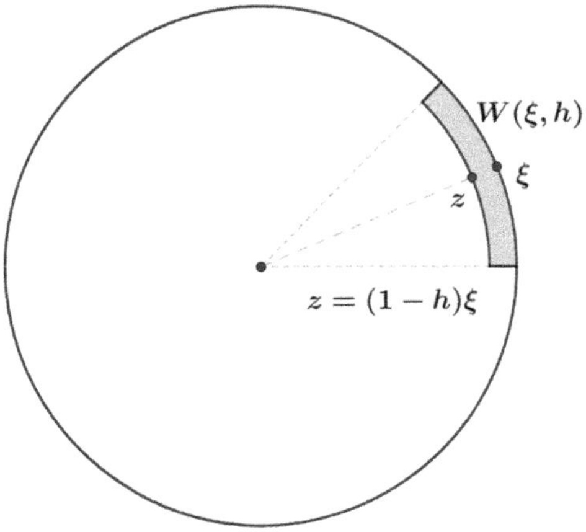

Figure 5.1 $W(\xi, h)$.

5.2 Carleson Measures

We first need the following geometrical tools, which will be repeatedly used in this book:

Definition 5.6 Given $h \in (0,1]$ and $\xi \in \mathbb{T}$, we define the following:

- The Carleson "square" window (or box) with *center* ξ and *opening* h:

$$W(\xi,h) = \left\{z \in \overline{\mathbb{D}} \,:\, 1 - |z| < h \text{ and } |\arg(\bar{\xi} z)| \leqslant \pi h\right\}.$$

  Alternatively, if $(\theta_0 - \pi h, \theta_0 + \pi h)$ is viewed as an arc I in $\mathbb{T}$ with normalized length $m(I) = h$ on $\mathbb{T}$, we have with $\xi = \mathrm{e}^{i\theta_0}$

$$W(I) := W(\xi,h) = \left\{z = r\mathrm{e}^{i\theta} \in \overline{\mathbb{D}} : r > 1 - h \text{ and } |\theta - \theta_0| \leqslant \pi h\right\}. \tag{5.4}$$

- The Carleson "circular" window (or box): $S(\xi,h) = \left\{z \in \overline{\mathbb{D}} : |\xi - z| < h\right\}$, for $h > 0$.

They are – roughly speaking – very similar objects.

In particular, up to uniform constants, they have the same σ-measure and the same side length. More precisely, we clearly have

$$m\big(W(\xi,h) \cap \mathbb{T}\big) = h \qquad \text{and} \qquad \sigma\big(W(\xi,h)\big) = h^2(2 - h),$$

and we observe that $\sigma\big(W(\xi,h)\big) \sim 2h^2$ as $h \to 0^+$.

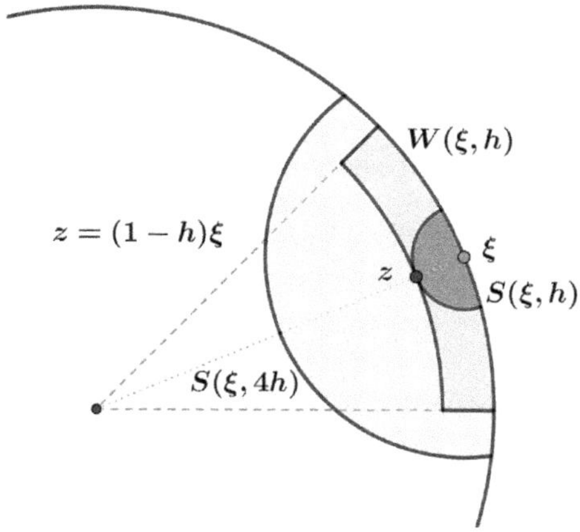

Figure 5.2 $S(\xi,h) \subset W(\xi,h) \subset S(\xi,4h)$.

We can also compute the sizes for the circular windows $S(\xi,h)$, but this requires a bit more attention. We have

$$m\big(S(\xi,h)\cap\mathbb{T}\big) = \frac{2}{\pi}\arcsin(h/2) \tag{5.5}$$

and

$$\sigma\big(S(\xi,h)\big) = 1 - \frac{1}{\pi}\left[h\sqrt{1-\frac{h^2}{4}} + (2-h^2)\arccos\left(\frac{h}{2}\right)\right]. \tag{5.6}$$

We point out that, when $h \to 0^+$, we have

$$m\big(S(\xi,h)\cap\mathbb{T}\big) \sim \frac{h}{\pi} \quad \text{and} \quad \sigma\big(S(\xi,h)\big) \sim \frac{h^2}{2}.$$

We leave the details as an exercise (Exercise I.17).

Let μ be a positive (*a priori* finite or not) measure on $\mathbb{D}$ or $\overline{\mathbb{D}}$, and H be a Hilbert space of analytic functions on $\mathbb{D}$ (i.e., continuously embedded into $\mathscr{H}(\mathbb{D})$); a main concern (in relation, but not only, with composition operators) is to know when the identity map J_μ sends H into $L^2(\mu)$. This leads to the following fundamental notion, which will play a key role in the study of composition operators on the unit disk. It is useful to define two functions ρ_μ and K_μ attached to μ.

Definition 5.7 For $h \in (0,1]$,

- $\rho_\mu(h) = \sup_{\xi\in\mathbb{T}} \mu\big(W(\xi,h)\big)$.

- $K_\mu(h) = \sup\limits_{0<t\leqslant h} \dfrac{\rho_\mu(t)}{t}$.

By definition, we have $\mu\big(W(\xi,t)\big) \leqslant t\, K_\mu(h)$ for every $t \in (0,h]$.

Definition 5.8 The positive measure μ is *a Carleson measure for* H if there is a constant C such that

$$\|f\|_{L^2(\mu)} \leqslant C\|f\|_H \qquad \text{for all } f \in H.$$

The best constant C is equal to $\|J_\mu\|$ with $J_\mu : H \to L^2(\mu)$ the canonical inclusion, usually called a *Carleson embedding*.

This definition also makes sense if we replace H by a Banach space of analytic functions on $\mathbb{D}$, and (say) the L^2-norm by another L^p-norm. This aspect will reappear in the sequel.

Warning. This definition of a Carleson measure is an *analytic* one. Carleson's theorem 5.11 (for Hardy spaces and Theorem 5.15 for his Bergman counterpart) to follow will indicate that, for some spaces H, a *geometric* equivalent definition, more easily tractable on specific examples, can be given.

In Chapter 9, we will also need the notion of *reverse Carleson measure* for H:

$$\|f\|_H \leqslant C\|f\|_{L^2(\mu)} \qquad \text{for all } f \in H.$$

Sometimes, we need to consider measures carried by the closed unit disk $\overline{\mathbb{D}}$, even entirely carried by the circle $\mathbb{T}$, in particular the case of Hardy spaces.

This notion was put into evidence by Carleson right after his work on interpolation sequences (see Chapter 1), notably in his proof of the corona theorem, and his interpolation result there can be rephrased as follows (see [35], pp. 371–374 for a direct proof due to Neville, or Lemma 10.29 in Chapter 10).

Theorem 5.9 (Carleson) *Given a Blaschke sequence* $S = (z_j)_{j\geqslant 1}$ *of distinct points of* $\mathbb{D}$*, the following are equivalent:*

(i) *S is interpolating for H^∞.*
(ii) *S is separated ($j \neq k$ implies $\rho(z_j, z_k) \geqslant \delta > 0$) and, moreover, the discrete measure $\mu = \sum_{j=1}^{\infty}(1-|z_j|^2)\,\delta_{z_j}$ is a Carleson measure for H^2.*

Besides the case of Hardy spaces $H = H^2$, the space H will also often (but not only) be a weighted Bergman space $\mathscr{B}^2_\alpha$, with $\alpha > -1$, corresponding to the probability measure and reproducing kernels

$$d\sigma_\alpha(z) = (\alpha+1)(1-|z|^2)^\alpha d\sigma(z) \quad \text{and} \quad K_{a,\alpha}(z) = (1-\overline{a}z)^{-(\alpha+2)}$$

and the (weighted) norm

$$\Big\| \sum_{n=0}^{\infty} a_n z^n \Big\|^2 = \sum_{n=0}^{\infty} |a_n|^2 \frac{\Gamma(n+1)\Gamma(\alpha+2)}{\Gamma(n+\alpha+2)}. \tag{5.7}$$

The case $\alpha = -1$ appears as the limiting case of the Hardy space H^2: indeed, the RHS of (5.7) tends to $\sum_{n=0}^{\infty} |a_n|^2$ as $\alpha \to -1$. But it will require a separate treatment.

As we already mentioned, the analytic definition above can sometimes, with profit, be replaced by a *geometric one*, in which the "Carleson windows" play a privileged role. To emphasize this role, we begin with a simple lemma.

Lemma 5.10 *Let $r \in (0,1)$ and W be a Carleson window with opening $1-r$ centered at $\xi \in \mathbb{T}$, and consider $f(z) = \frac{1-r}{1-\bar{\xi} r z}$ for $z \in \overline{\mathbb{D}}$. Then*

$$|f| \geqslant \frac{1}{\pi+2} \mathbb{1}_W .$$

Proof: We can assume $\xi = 1$ and note that, for $z = \rho e^{i\theta} \in W$ and hence $\rho \geqslant r$, we have

$$|1 - rz| \leqslant 1 - r\rho + r\rho |1 - e^{i\theta}| \leqslant 1 - r^2 + |\theta| \leqslant (2+\pi)(1-r).$$

This gives the lemma. □

Now, a first answer of the above type was given by Carleson when $H = H^2$ as follows.

5.2.1 The Hardy Case

We must first insist that we use here the versions of Carleson square or circular windows involving $\overline{\mathbb{D}}$, and not only $\mathbb{D}$ (as we shall do when working on Bergman spaces). Indeed, in the framework of Hardy spaces, taking the boundary into account is important. When we focus on compactness of J_μ, remembering $\mu(\mathbb{T}) = 0$ in this case, we shall forget the part in $\mathbb{T}$ of Carleson windows.

Anyway we shall pay attention to this part in the following statements: the general results concerning the boundedness of J_μ on Hardy spaces *involve this boundary*.

We can now state Carleson's theorem for Hardy spaces.

Theorem 5.11 (Carleson 1) *For any positive measure μ on $\overline{\mathbb{D}}$, the following are equivalent:*

(i) *$J_\mu : H^p \to L^p(\mu)$ is bounded for some $p \geqslant 1$ with norm λ_p.*
(ii) *$J_\mu : H^q \to L^q(\mu)$ is bounded for every $q \geqslant 1$ with norm equal to $\lambda_q = \lambda_2^{2/q}$.*
(iii) *There exists a constant C' such that, for every Carleson box W with opening $h \in (0,1)$,*

$$\mu(W) \leqslant C'h. \tag{5.8}$$

In other words, $\rho_\mu(h) = O(h)$, *when* $h \to 0^+$.

Moreover, one has, for every $p \geqslant 1$,

$$\|J_\mu : H^2 \to L^2(\mu)\|^2 = \|J_\mu : H^p \to L^p(\mu)\|^p \approx K_\mu(1). \tag{5.9}$$

Proof: Assume (i). Let $q \geqslant 1$ and $f \in H^q$. By factoring out the Blaschke product B formed with the zeroes of f, we can write, with h analytic on $\mathbb{D}$,

$$f = Bh^{p/q}, \quad |f|^q \leqslant |h|^p, \quad |f^*|^q = |h^*|^p \quad \text{a.e. on } \mathbb{T}$$

so that $h \in H^p$. It ensues that

$$\int_{\overline{\mathbb{D}}} |f|^q \, d\mu \leqslant \int_{\overline{\mathbb{D}}} |h|^p \, d\mu \leqslant \lambda_p^p \int_{\mathbb{T}} |h^*|^p \, dm = \lambda_p^p \int_{\mathbb{T}} |f^*|^q \, dm.$$

This proves that $\lambda_q^q \leqslant \lambda_p^p$ and we have equality by symmetry between p and q.

Simple as it is, this argument will fail for Bergman spaces, and we will have to follow a different route.

Assume (ii). Let $W = W(\xi, h)$ be a Carleson box. We test the assumption with $q = 2$ on the function f of Lemma 5.10, with $r = 1 - h$. This gives, in terms of the reproducing kernel $K_{\overline{\xi}r}$ of H^2,

$$\mu(W) \lesssim \int_{\overline{\mathbb{D}}} |f|^2 \, d\mu \leqslant \lambda_2^2 \|f\|_{H^2}^2 = \lambda_2^2 h^2 \|K_{\overline{\xi}r}\|_{H^2}^2 = \lambda_2^2 \frac{h^2}{1 - r^2} \leqslant \lambda_2^2 h.$$

This also easily gives $K_\mu(1) \lesssim \|J_\mu : H^2 \to L^2(\mu)\|^2$.

The reverse implication (iii) (geometric) implies (i) or (ii) (analytic) lies much deeper. A key ingredient there is another theorem of Carleson. In this theorem, N_a denotes the non-tangential maximal function attached to the region $\Gamma_a(u)$ for $u \in \mathbb{T}$, and M the Hardy–Littlewood maximal function at the boundary. The role of the constant 5 appears in the proof.

Theorem 5.12 (Carleson 2) *Let μ be a positive measure on $\overline{\mathbb{D}}$ such that $K_\mu(1)$ is finite. Then, for any $f \in H^2$ with radial limits f^*, we have*

$$\mu\left(\left\{z \in \overline{\mathbb{D}} : |f(z)| > t\right\}\right) \lesssim K_\mu(1) \, m(\{\xi \in \mathbb{T} : N_5 f(\xi) > t\}). \tag{5.10}$$

$$\lesssim K_\mu(1) \, m(\{\xi \in \mathbb{T} : Mf^*(\xi) > t\}). \tag{5.11}$$

We postpone the proof of this theorem and finish the proof of Theorem 5.11.

For any $f \in H^2$, integration by parts gives

$$\int_{\overline{\mathbb{D}}} |f|^2\, d\mu = \int_0^{+\infty} 2t\, \mu\left(\left\{z \in \overline{\mathbb{D}} : |f(z)| > t\right\}\right) dt$$
$$\lesssim K_\mu(1) \int_0^{+\infty} 2t\, m(\{u \in \mathbb{T} : Mf^*(u) > t\})\, dt,$$

or else, since the Hardy–Littlewood maximal operator maps $L^2(\mathbb{T})$ to itself,

$$\int_{\overline{\mathbb{D}}} |f|^2\, d\mu \lesssim K_\mu(1) \int_{\mathbb{T}} |Mf^*|^2\, dm \lesssim K_\mu(1) \int_{\mathbb{T}} |f^*|^2\, dm = K_\mu(1)\, \|f\|_{H^2}^2.$$

□

Proof of Theorem 5.12: The second inequality (5.11) is easy once we remember that for the H^2-space, the Hardy–Littlewood maximal function dominates all the maximal functions N_a.

To prove (5.10), we argue as follows. Fix $t > 0$. By lower semi-continuity, the set $\{N_5 f > t\}$ is open and can be written as a disjoint union of open arcs I_j. We now claim that

$$\{z \in \mathbb{D} : |f(z)| > t\} \subset \bigcup_{j=1}^{\infty} W(I_j). \tag{5.12}$$

Indeed, if $|f(z)| > t$ with $z = |z|\xi$, where $\xi \in \mathbb{T}$, and $h = 11(1 - |z|)/10$, let A_z be the arc with center ξ and length $2\pi h$, we have $z \in W(A_z)$ by definition. If $\eta \in A_z$, then $z \in \Gamma_5(\eta)$ since

$$|z - \eta| \leqslant |z - \xi| + |\xi - \eta| = h + |\xi - \eta| \leqslant (1 + \pi)h < 5(1 - |z|).$$

So

$$N_5 f(\eta) \geqslant |f(z)| > t \quad \text{and} \quad A_z \subset \{N_5 f > t\}.$$

Now, by connectedness, $A_z \subset I_j$ for some j and finally $z \in W(A_z) \subset W(I_j)$, proving (5.12). Now, (5.11) is easy:

$$\mu(\{z \in \mathbb{D} : |f(z)| > t\}) \leqslant \sum_{j=1}^{\infty} \mu(W(I_j)) \leqslant K_\mu(1) \sum_{j=1}^{\infty} m(I_j) = K_\mu(1)\, m(N_5 f > t).$$

□

Compactness on Hardy Spaces

We now focus on a compact version of Carleson's embedding theorem for Hardy spaces.

We first mention the following lemma, similar to Theorems 3.9 (ii) and 3.11.

Lemma 5.13 *Let μ be a positive finite measure on $\overline{\mathbb{D}}$ and $p,q \geqslant 1$.*

(i) *If $J_\mu : H^p \to L^p(\overline{\mathbb{D}},\mu)$ is compact, then $\mu(\mathbb{T}) = 0$.*
(ii) *$J_\mu : H^p \to L^p(\overline{\mathbb{D}},\mu)$ is compact if and only if $J_\mu : H^q \to L^q(\overline{\mathbb{D}},\mu)$ is compact.*

Proof: (i) The sequence (z^n) is weakly converging to 0 in $L^p(\mathbb{T},m)$, and hence in H^p, thanks to the Riemann–Lebesgue theorem. By compactness of J_μ, we get $\| J_\mu(z^n)\|_{L^p(\mu)} \to 0$. But

$$\| J_\mu(z^n)\|^p_{L^p(\mu)} = \int_{\overline{\mathbb{D}}} |z|^{np}\, d\mu \longrightarrow \mu(\mathbb{T}),$$

and the conclusion follows.

(ii) Assume that J_μ is compact on H^q. Take $(f_n)_n$ a sequence in the unit ball of H^p. Up to an extraction, we can assume that $(f_n)_n$ is converging to some analytic function f in $\mathscr{H}(\mathbb{D})$, and $f \in H^p$ by Fatou; therefore (considering $f_n - f$ instead of f_n), we may assume in the following that $(f_n)_n$ is converging to 0 in $\mathscr{H}(\mathbb{D})$.

Write $f_n = B_n.g_n$ with $|B_n^*| = 1$ a.e. and $g_n \in H^p$ without zeros in $\mathbb{D}$. The sequence $G_n = g_n^{p/q}$ is defined and lies in the unit ball of H^q. Up to subsequences, we may assume that $G_n \to G$ and $B_n \to B$, in $\mathscr{H}(\mathbb{D})$, and that (G_n) converges to G in $L^q(\overline{\mathbb{D}},\mu)$ since J_μ is compact on H^q.

We get $|G(z)|^q |B(z)|^p = \lim |f_n(z)|^p = 0$ for every $z \in \mathbb{D}$.

From (i), we know that $\mu(\mathbb{T}) = 0$ and we can compute

$$\int_{\overline{\mathbb{D}}} |f_n|^p\, d\mu = \int_{\mathbb{D}} |f_n|^p\, d\mu \lesssim \int_{\mathbb{D}} |G_n - G|^q\, d\mu + \int_{\mathbb{D}} |G|^q\, |B_n|^p\, d\mu.$$

We see, thanks to the dominated convergence theorem, that the second term converges to 0. From the compactness of J_μ on H^q, so does the first one. □

For the sequel, we recall that $\rho_\mu(h) = \sup_{\xi\in\mathbb{T}} \mu[W(\xi,h)]$, $K_\mu(h) = \sup_{0<t\leqslant h} \dfrac{\rho_\mu(t)}{t}$ and that μ is Carleson for H^2 if and only if $\rho_\mu(h) = O(h)$, equivalently $K_\mu(1)$ finite.

We will say that μ is a *vanishing Carleson measure for Hardy spaces* if $\rho_\mu(h) = o(h)$ as $h \to 0$, equivalently $K_\mu(h) = o(1)$.

We can then state (see [141], with a different proof).

Theorem 5.14 (Power) *Let $p \geqslant 1$ and μ be a positive finite measure on $\mathbb{D}$. Then the following are equivalent:*

(i) *$J_\mu : H^2 \to L^2(\mu)$ is compact.*
(ii) *μ is a vanishing Carleson measure for Hardy spaces.*
(iii) *$J_\mu : H^p \to L^p(\mu)$ is compact.*

Let us mention that the case of a measure μ on $\overline{\mathbb{D}}$ is also treated by this theorem: if J_μ is compact, we must have $\mu(\mathbb{T}) = 0$, thanks to Lemma 5.13, and we can consider that the measure is carried by $\mathbb{D}$.

We could give a direct proof below, in the spirit of the one given for Bergman spaces (Theorem 5.15), but we chose to present a stronger form in the Hilbertian case $p = 2$ (see (5.14)) and invoke Lemma 5.13 to get other values of p. Let us mention that this relation and the proof below also lead an estimate of the essential norm:

$$\|J_\mu\|_{e,H^2} \approx \overline{\lim_{h\to 0}} \sqrt{\frac{\rho_\mu(h)}{h}}. \tag{5.13}$$

Proof: We already know from Lemma 5.13 that (i) and (iii) are equivalent and we focus on the equivalence between (i) and (ii).

That (i) implies (ii) is easy by testing on reproducing kernels. Alternatively, let W be a Carleson window, which we assume to be centered at 1, without loss of generality, and let

$$f_h(z) = \frac{\sqrt{h}}{1 - rz} \text{ with } r = 1 - h.$$

The functions f_h tend weakly to 0 as $h \to 0$ and are clearly norm-bounded in H^2. Therefore, by compactness of J_μ, we have

$$\lim_{h\to 0} \int_{\mathbb{D}} |f_h|^2 \, d\mu = \lim_{h\to 0} \|J_\mu(f_h)\|^2_{L^2(\mu)} = 0\,.$$

But Lemma 5.10 tells us that $\mathbb{1}_W \lesssim \sqrt{h}\, f_h$, so that

$$\mu(W) \lesssim h \int_{\mathbb{D}} |f_h|^2 \, d\mu = o(h).$$

For the converse, we will prove the following inequality, a *reinforced version* of Theorem 5.14, in which $a_n(J_\mu)$ stands for the nth approximation number of J_μ, that is, the distance of J_μ to operators of rank at most $n - 1$ for the operator norm (those approximation numbers will be studied in full detail in Chapter 10).

$$a_n(J_\mu) \leqslant C \inf_{0<h<1} \left[\mathrm{e}^{-nh} + \sqrt{K_\mu(h)} \right]. \tag{5.14}$$

Indeed, write

$$f = \sum_{k=0}^{n-2} \widehat{f}(k) z^k + z^{n-1} \sum_{k=n-1}^{+\infty} \widehat{f}(k) z^{k-n+1} =: S_{n-2}(f) + z^{n-1} R_n(f)$$

with $\|R_n(f)\|_{H^2} \leqslant \|f\|_{H^2}$.

We see that, assuming $\|f\|_{H^2} \leqslant 1$ and using $1-h \leqslant \mathrm{e}^{-h}$,

$$\begin{aligned}\|z^{n-1}R_n(f)\|^2_{L^2(\mu)} &\leqslant \mathrm{e}^{-2(n-1)h}\int_{(1-h)\mathbb{D}} |R_n(f)|^2\,d\mu + \int_{\mathbb{D}\setminus(1-h)\mathbb{D}} |R_n(f)|^2\,d\mu \\ &\leqslant \mathrm{e}^{-2(n-1)h}\|R_n(f)\|^2_{L^2(\mu)} + \|J_{\mu_h}\big(R_n(f)\big)\|^2_{L^2(\mu_h)},\end{aligned}$$

where μ_h is the restriction of μ to the annulus $1-h \leqslant |z| < 1$.

But $\|R_n(f)\|_{L^2(\mu)} \leqslant \|J_\mu : H^2 \to L^2(\mu)\|$ since $\|R_n(f)\|_{H^2} \leqslant \|f\|_{H^2} \leqslant 1$. Similarly, the second term is dominated by $\|J_{\mu_h} : H^2 \to L^2(\mu_h)\|^2$ and clearly

$$\|J_\mu : H^2 \to L^2(\mu)\|^2 \lesssim K_{\mu_h}(1)\,.$$

But we claim that, for $h \in (0,1)$,

$$K_{\mu_h}(1) \leqslant 2K_\mu(h)\,. \tag{5.15}$$

Indeed, by definition $K_{\mu_h}(1) = \sup\limits_{0<t\leqslant 1} \dfrac{\rho_{\mu_h}(t)}{t}$. but, given $\xi \in \mathbb{T}$,

$$\mu_h(W(\xi,t)) = \mu\big\{z \in \overline{\mathbb{D}} : 1-|z| < \min(t,h) \text{ and } |arg(\bar{\xi}z)| \leqslant \pi t\big\},$$

so that $\mu_h(W(\xi,t)) = \mu(W(\xi,t))$ when $t \leqslant h$.

When $t > h$, there exists some integer $s \geqslant 1$ with $t \in (sh,(s+1)h]$ and the set $\big\{z \in \overline{\mathbb{D}} : 1-|z| < h \text{ and } |\arg(\bar{\xi}z)| \leqslant \pi t\big\}$ can be covered by $(s+1)$ windows with opening h, hence $\rho_{\mu_h}(t) \leqslant (s+1)\rho_\mu(h)$. Therefore, in this case,

$$\frac{\rho_{\mu_h}(t)}{t} \leqslant \frac{(s+1)\rho_\mu(h)}{sh} \leqslant 2K_\mu(h),$$

and the claim (5.15) is proved.

To sum up, we proved that

$$\|J_\mu - S_n\|^2_{L^2(\mu)} \lesssim \mathrm{e}^{-2nh} + K_\mu(h).$$

Since S_{n-2} is an operator of rank at most $n-1$, we have proved (5.14), which in turn proves the compactness of J_μ when μ is vanishing Carleson (the RHS of (5.14) clearly tends to 0 as $n \to \infty$ (taking $h = 1/\sqrt{n}$ for instance), so that J_μ is the uniform limit of finite rank operators). □

5.2.2 The Bergman Case

Theorem 5.11 has the following extension, which is the main result of that subsection. Here again, the Carleson windows will play a crucial role, and in the context of Bergman spaces, these windows must be considered as subsets of $\mathbb{D}$. The best constant C', such that for any Carleson window W of size h one has

$$\mu(W) \leqslant C' h^{2+\alpha},$$

will be called the Carleson constant of μ for $\mathscr{B}^2_\alpha$.
In other words, the Carleson constant is $K_{\mu,\alpha+2}(1)$ where

$$K_{\mu,\beta}(h) = \sup_{0<t\leqslant h} \frac{\rho_\mu(t)}{t^\beta}. \tag{5.16}$$

The following version of Carleson's theorem for Bergman spaces was proved independently by Hastings [70] and Oleinik–Pavlov [132] (for short the HOP theorem in the sequel).

Theorem 5.15 (HOP theorem) *Let μ be a positive measure on $\mathbb{D}$, $\alpha > -1$, and $p \geqslant 1$.*

(i) *The following are equivalent:*
  (a) $J_\mu : \mathscr{B}^p_\alpha \to L^p(\mu)$ *with norm* λ_p.
  (b) *There exists a constant C' such that*

$$\rho_\mu(h) \leqslant C' h^{2+\alpha} \qquad \textit{for any } 0 < h < 1.$$

Moreover $\| J_\mu : \mathscr{B}^p_\alpha \to L^p(\mu) \|^p \approx K_{\mu,\alpha+2}(1)$.

(ii) *The injection* $J_\mu : \mathscr{B}^p_\alpha \to L^p(\mu)$ *is compact if and only if* $\rho_\mu(h) = o(h^{2+\alpha})$.

We mention that a version of (5.14) for Bergman spaces holds: using the inequality $\| R_n(f) \|_{\mathscr{B}^2_\alpha} \lesssim n^{(\alpha+1)/2} \| f \|_{\mathscr{B}^2_\alpha}$, we can easily adapt (we leave it as an exercise: Exercise I.18) the proof of (5.14) to get

$$a_n(J_\mu) \lesssim \inf_{0<h<1} \left[n^{(\alpha+1)/2} \mathrm{e}^{-nh} + \sqrt{K_{\mu,\alpha+2}(h)} \right]. \tag{5.17}$$

From this inequality and the proof below, we easily get an estimate of the essential norm:

$$\| J_\mu \|_{e,\mathscr{B}^2_\alpha} \approx \overline{\lim_{h\to 0}} \sqrt{\frac{\rho_\mu(h)}{h^{\alpha+2}}}. \tag{5.18}$$

Proof: We begin by (ii) assuming we already proved entirely (i) (which will be done below, but with some hard work); in particular we need to know that

$$\| J_\mu : \mathscr{B}^p_\alpha \to L^p(\mu) \|^p \approx K_{\mu,\alpha+2}(1). \tag{5.19}$$

Let us assume that $\rho_\mu(h) = o(h^{2+\alpha})$ when $h \to 0+$. For $h \in (0, 1)$, we introduce the measure μ_h, which is the restriction of μ to the annulus

$1-h<|z|<1$, and ν_h, which is the restriction of μ to the disk $(1-h)\overline{\mathbb{D}}$ so that we can write $J_\mu = \tilde{J}_{\nu_h} + \tilde{J}_{\mu_h}$ where $\tilde{J}_{\nu_h}(f) = J_\mu(f).\mathbb{1}_{(1-h)\overline{\mathbb{D}}}$, and $\tilde{J}_{\mu_h}(f) = J_\mu(f).\mathbb{1}_{|z|>1-h}$.

We claim that $\tilde{J}_{\nu_h}$ is compact for any $h \in (0,1)$. Indeed, take $(f_n)_n$ a sequence in the unit ball of $\mathscr{B}^p_\alpha$. Up to an extraction, we can assume that $(f_n)_n$ is converging to some analytic function f in $\mathscr{H}(\mathbb{D})$, and $f \in \mathscr{B}^p_\alpha$ by Fatou. Clearly

$$\|\tilde{J}_{\nu_h}(f_n) - \tilde{J}_{\nu_h}(f)\|_{L^p(\mathbb{D},\mu)} = \|f_n - f\|_{L^p(\mathbb{D},\nu_h)} \leqslant \mu(\mathbb{D})^{\frac{1}{p}} \sup_{(1-h)\overline{\mathbb{D}}} |f_n - f| \longrightarrow 0,$$

and this proves the claim.

Now we prove that $\|(J_\mu - \tilde{J}_{\nu_h}) : \mathscr{B}^p_\alpha \to L^p(\mu)\|$ converges to 0 when $h \to 0^+$. But

$$\|(J_\mu - \tilde{J}_{\nu_h}) : \mathscr{B}^p_\alpha \to L^p(\mu)\| = \|\tilde{J}_{\mu_h} : \mathscr{B}^p_\alpha \to L^p(\mu)\| = \|J_{\mu_h} : \mathscr{B}^p_\alpha \to L^p(\mu_h)\|$$
$$\approx K_{\mu_h,\alpha+2}(1)$$

since we took (5.19) for granted.

Now we remark that, for $h \in (0,1)$ and $\beta \geqslant 1$, we have

$$K_{\mu_h,\beta}(1) \leqslant 2K_{\mu,\beta}(h). \tag{5.20}$$

We mimick the argument for (5.15): $\mu_h(W(\xi,t)) = \mu(W(\xi,t))$ when $t \leqslant h$, and, when $t > h$, there exists some integer $s \geqslant 1$ with $t \in (sh,(s+1)h]$ and we have $\rho_{\mu_h}(t) \leqslant (s+1)\rho_\mu(h)$. Therefore, in this case, since $\beta \geqslant 1$,

$$\frac{\rho_{\mu_h}(t)}{t^\beta} \leqslant \frac{(s+1)\rho_\mu(h)}{s^\beta h^\beta} \leqslant 2K_{\mu,\beta}(h),$$

and the claim (5.20) is proved.

We conclude that $\|(J_\mu - \tilde{J}_{\nu_h}) : \mathscr{B}^p_\alpha \to L^p(\mu)\| \lesssim K_{\mu,\alpha+2}(h) \longrightarrow 0$ by hypothesis, and J_μ is compact, as a limit of compact operators.

Now let us assume the compactness of J_μ. Fix $\xi \in \mathbb{T}$. We use the functions

$$g_{h,\xi}(z) = \frac{h^{\frac{\alpha+2}{p}}}{\left(1-(1-h)\bar{\xi}z\right)^{\frac{2(\alpha+2)}{p}}} \qquad \text{where } h \in (0,1) \text{ and } z \in \mathbb{D}. \tag{5.21}$$

In other words, we know from (1.70) that $|g_{h,\xi}|^p = h^{\alpha+2}|K_{(1-h)\xi,\alpha}|^2$ where $K_{(1-h)\xi,\alpha}$ is the reproducing kernel in $\mathscr{B}^2_\alpha$ at point $(1-h)\xi$. In particular, $g_{h,\xi}$ lies in the unit ball of $\mathscr{B}^p_\alpha$. It is clear that $g_{h,\xi}$ converges to 0 in $\mathscr{H}(\mathbb{D})$, uniformly in ξ, when $h \to 0^+$.

When $p > 1$ it means that $g_{h,\xi}$ is weakly converging to 0 in $\mathscr{B}^p_\alpha$, uniformly in ξ, when $h \to 0^+$, and by compactness of J_μ, we get that $g_{h,\xi}$ is (norm) converging to 0 in $L^p(\mathbb{D},\mu)$, uniformly in ξ.

For $p = 1$, the weak convergence is not true anymore (if it were true, since $\mathscr{B}^1_\alpha$ is isomorphic to ℓ^1, we would have, by the Schur property, $g_{h,\xi} \to 0$ in $\mathscr{B}^1_\alpha$, which is false). Nevertheless, it is still true that $g_{h,\xi}$ is (norm) converging to 0 in $L^1(\mathbb{D}, \mu)$, uniformly in ξ (our argument works as well for any $p \geqslant 1$ actually). Indeed, J_μ is compact and consider any family of points $(\xi(h))$ and any cluster point g of $J_\mu(g_{h,\xi(h)})$ and fix any $r \in (0,1)$. We have

$$\int_{r\mathbb{D}} |g_{h,\xi(h)} - g| \, d\mu \leqslant \| J_\mu(g_{h,\xi(h)} - g)\|_{L^1(\mathbb{D},\mu)} \longrightarrow 0 \qquad \text{when } h \to 0^+,$$

but $g_{h,\xi(h)}$ is uniformly convergent to 0 on the compact $r\overline{\mathbb{D}}$, so $\int_{r\mathbb{D}} |g| \, d\mu = 0$ and $g = 0$ μ-a.e. on $r\mathbb{D}$ for any $r \in (0,1)$. Hence $g = 0$ in $L^1(\mathbb{D}, \mu)$. Therefore $g_{h,\xi}$ is (norm) converging to 0 in $L^1(\mathbb{D}, \mu)$, uniformly in ξ, in this case too.

Once we know $g_{h,\xi}$ is (norm) converging to 0 in $L^p(\mathbb{D}, \mu)$, uniformly in ξ, $(p \geqslant 1)$, we can estimate, thanks to Lemma 5.10,

$$\sup_{\xi \in \mathbb{T}} \mu\big(W(\xi, h)\big) \lesssim C h^{\alpha+2} \sup_{\xi \in \mathbb{T}} \int_{\mathbb{D}} |g_{h,\xi}|^p \, d\mu\,,$$

and the result follows: $\rho_\mu(h) = o(h^{\alpha+2})$.

Now we focus on (i).

Assume that (a) is fulfilled for some $p \geqslant 1$. Fix $\xi \in \mathbb{T}$, and for $h \in (0,1)$, use again the functions $g_{h,\xi}$ defined in (5.21). With the help of the explanations given above, we compute, using again Lemma 5.10 (and writing $\|J_\mu\|$ instead of $\|J_\mu : \mathscr{B}^p_\alpha \to L^p(\mu)\|$ for short),

$$\begin{aligned}
\mu\big(W(\xi, h)\big) &\lesssim h^{\alpha+2} \int_{\mathbb{D}} |g_{h,\xi}|^p \, d\mu \leqslant h^{\alpha+2} \|J_\mu\|^p \int_{\mathbb{D}} |g_{h,\xi}|^p \, d\sigma_\alpha \\
&= h^{2(\alpha+2)} \|J_\mu\|^p \int_{\mathbb{D}} |K_{(1-h)\xi,\alpha}|^2 \, d\sigma_\alpha \\
&= \|J_\mu\|^p \, h^{\alpha+2} (2-h)^{-(\alpha+2)} \leqslant \|J_\mu\|^p \, h^{\alpha+2},
\end{aligned}$$

and we get (b) with the information that $K_{\mu,\alpha+2}(1) \lesssim \|J_\mu : \mathscr{B}^p_\alpha \to L^p(\mu)\|^p$.

The rest of this section will be devoted to the (more difficult) proof of the converse implications from (b) to (a), and we shall also get the information $\|J_\mu : \mathscr{B}^p_\alpha \to L^p(\mu)\|^p \lesssim K_{\mu,\alpha+2}(1)$.

Remark. We decided to privilege this Carleson window point of view for a specific reason: throughout this book, we will take for μ the pullback measures m_φ or $\sigma_{\varphi,\,\alpha}$ of the measures m and σ_α under a symbol φ. We will often have precise informations at our disposal on the behavior of φ near the boundary, thereby on the geometric properties of m_φ or $\sigma_{\varphi,\,\alpha}$. Then we will be able to use the full (quantitative) strength of Carleson or HOP theorems 5.11 or 5.15.

But, we first prove a theorem in terms of pseudo-hyperbolic disks (*abbreviated hyperbolic* in what follows since we will no longer work with real hyperbolic disks), before going back to the Carleson window point of view. In this presentation (which is inspired by those of Zhu [184, pp. 72–74]) and Duren–Schuster [46, pp. 62–68]), those hyperbolic disks $\Delta(z,r) = \{w : \rho(w,z) < r\}$, where $r \in (0,1)$, play a prominent role for two specific reasons: first they are attached to a metric; and secondly they resist well to the automorphisms φ_z thanks to the obvious relations

$$\Delta(0,r) = D(0,r), \qquad \varphi_z[\Delta(0,r)] = \Delta(z,r). \tag{5.22}$$

We begin by listing some properties of those disks.

The parameter $r \in (0,1)$ will be fixed once and for all. C will denote a constant depending only on r and α, which can vary from a formula to another.

Lemma 5.16 *For any $z \in \mathbb{D}$, it holds*

$$\sigma_\alpha[\Delta(z,r)] \approx (1-|z|^2)^{\alpha+2}.$$

Proof: $\sigma_\alpha[\Delta(z,r)] = \int_{\Delta(z,r)} d\sigma_\alpha(w)$. The change of variable $w = \varphi_z(u)$ gives

$$\sigma_\alpha[\Delta(z,r)] = \int_{\Delta(0,r)} \frac{(1-|z|^2)^{\alpha+2}}{|1-\overline{u}z|^{2(\alpha+2)}}\, d\sigma_\alpha(u).$$

Since $1-r \leqslant |1-\overline{u}z| \leqslant 2$ for $u \in \Delta(0,r) = D(0,r)$, this gives the result. □

Lemma 5.17 *Fix $0 < p < \infty$. Then, for any function f analytic in $\mathbb{D}$ and any $z \in \mathbb{D}$,*

$$|f(z)|^p \leqslant C(1-|z|^2)^{-\alpha-2} \int_{\Delta(z,r)} |f|^p\, d\sigma_\alpha. \tag{5.23}$$

Equivalently

$$|f(z)|^p \leqslant C \frac{1}{\sigma_\alpha[\Delta(z,r)]} \int_{\Delta(z,r)} |f|^p\, d\sigma_\alpha. \tag{5.24}$$

Proof: The equivalence of both inequalities follows from Lemma 5.16. Now, for $z = 0$, the subharmonicity of the function $|f|^p$ gives, after radial integration against the measure $2t(\alpha+1)(1-t^2)^\alpha\, dt$ over $(0,r)$,

$$|f(0)|^p \leqslant \frac{1}{\sigma_\alpha[\Delta(0,r)]} \int_{\Delta(0,r)} |f(w)|^p\, d\sigma_\alpha(w) =: C \int_{\Delta(0,r)} |f(w)|^p\, d\sigma_\alpha(w).$$

Next, change f to $f \circ \varphi_z$ and as in Lemma 5.16 make the (involutive) change of variable $w = \varphi_z(u)$. We get

$$|f(z)|^p \leqslant C \int_{\Delta(z,r)} |f(u)|^p \frac{(1-|z|^2)^{\alpha+2}}{|1-\overline{u}z|^{2(\alpha+2)}} \, d\sigma_\alpha(u).$$

Finally, observe that $|1-\overline{u}z| \geqslant \dfrac{1-|z|^2}{2}$ and this gives (5.23). □

Lemma 5.18 *Fix $0 < p < \infty$ and $a \in \mathbb{D}$. Let $R = r + \dfrac{1-r}{2} = \dfrac{1+r}{2}$.*
Then for any function f analytic in $\mathbb{D}$ and any $z \in \Delta(a,r)$,

$$|f(z)|^p \leqslant C \frac{1}{\sigma_\alpha[\Delta(a,r)]} \int_{\Delta(a,R)} |f|^p \, d\sigma_\alpha.$$

Proof: Set $s = \frac{1-r}{2}$. Observe that if $\rho(z,a) \leqslant r$, we have

$$1-|z| \geqslant s(1-|a|). \tag{5.25}$$

Indeed, we know from (1.14) that $\frac{|z|-|a|}{1-|a||z|} \leqslant \rho(z,a) \leqslant r$, hence $|z| \leqslant \frac{r+|a|}{1+|a|r}$, and (5.25) ensues. By Lemma 5.16, we see that $\sigma_\alpha[\Delta(z,s)] \gtrsim \sigma_\alpha[\Delta(a,r)]$. Finally, use Lemma 5.17 with r changed into s and the inclusion $\Delta(z,s) \subset \Delta(a,R)$ whenever $z \in \Delta(a,r)$. □

The next lemmas connect hyperbolic disks and Carleson windows. The first one is indicated for completeness, but will not be needed here [46, p. 66].

Lemma 5.19 *Let W be a Carleson window of size h. Then, there exist hyperbolic disks $\Delta(z_{j,k},r)_{\substack{k\geqslant 0,\\ 1\leqslant j\leqslant 2^k}}$, with $|z_{j,k}| = 1-2^{-k}h$, such that*

$$W \subset \bigcup_{k\geqslant 0} \bigcup_{1\leqslant j\leqslant 2^k} \Delta(z_{j,k},r).$$

Just indicate a consequence; if μ is a positive Borel measure on $\mathbb{D}$ such that $\mu[\Delta(a,r)] \leqslant M\sigma_\alpha[\Delta(a,r)]$ for all hyperbolic disks of radius r, then, thanks to Lemma 5.16 and since $\alpha+1>0$,

$$\mu(W) \leqslant CM \sum_{k=0}^{\infty} \sum_{1\leqslant j\leqslant 2^k} (2^{-k}h)^{\alpha+2} \leqslant CM \sum_{k\geqslant 0} 2^{-k(\alpha+1)} h^{\alpha+2} \leqslant CMh^{\alpha+2}.$$

Lemma 5.20 *Fix $0 < r < 1$. For any hyperbolic disk $\Delta(a,r)$, with $a \in \mathbb{D}$, there exists a Carleson window W with opening less then $C(1-|a|)$ satisfying*

$$\Delta(a,r) \subset W,$$

where $C = C_r$ is a constant.

Proof: By Proposition 1.24, $\Delta(a,r)$ is the euclidean disk $D(a',r')$ with

$$a' = \frac{(1-r^2)a}{1-r^2|a|^2} \quad \text{and} \quad r' = \frac{r(1-|a|^2)}{1-r^2|a|^2}.$$

We can assume a real with $a > r > 0$ and hence $a' - r' = \dfrac{a-r}{1+ar} > 0$. Let $w = |w|e^{i\theta} \in \Delta(a,r)$, with $|\theta| \leqslant \pi/2$. We know that

$$\frac{a-|w|}{1-a|w|} \leqslant \rho(w,a) \leqslant r,$$

whence $1 - |w| \leqslant 1 - \dfrac{a-r}{1-ar} = \dfrac{(1-a)(1+r)}{1-ar} \leqslant \dfrac{1+r}{1-r}(1-|a|)$.

Moreover, drawing the tangent from 0 to $D(a',r')$, we see that the argument θ of w satisfies

$$\frac{2}{\pi}|\theta| \leqslant |\sin\theta| \leqslant \frac{r'}{a'} = \frac{r(1-a^2)}{(1-r^2)a} \leqslant C(1-|a|). \tag{5.26}$$

This gives the lemma, with here a window centered at 1. □

We will now first present a pseudo-hyperbolic version of Theorem 5.15. It is based on a simple Vitali-type lemma for hyperbolic disks [46, p. 62].

Lemma 5.21 *Let* $0 < r < 1$ *and* $R = \dfrac{1+r}{2}\cdot$ *Then there exists a sequence* $(z_k)_{k\geqslant 1}$ *of points of* $\mathbb{D}$ *and an integer* $N = N_r$ *such that*

$$\mathbb{D} = \bigcup_{k=1}^{\infty} \Delta(z_k, r).$$

Moreover, every point of $\mathbb{D}$ *belongs to no more than* N *disks* $\Delta(z_k, R)$.

Let us add a remark, which confirms the intuition. In this Lemma 5.21, we claim that the sequence satisfies

$$\lim |z_k| = 1. \tag{5.27}$$

Indeed, else we would have a subsequence (z_{k_j}) converging to some $w \in \mathbb{D}$ (for the usual topology), but this implies that $\rho\big(z_{k_j}, w\big) \to 0$ since $|w| < 1$, and for j large enough w would belong to all the disks $\Delta\big(z_{k_j}, R\big)$, but w can belong to at most N of them, and this proves the claim.

With this lemma, we can prove our hyperbolic version of Theorem 5.15.

Theorem 5.22 *Fix* $0 < p < \infty$ *and* $0 < r < 1$. *Then, for any positive Borel measure* μ *on* $\mathbb{D}$ *and* $\alpha > -1$, *the following are equivalent:*

(i) *There exists a constant C' such that, for every hyperbolic disk $\Delta(z,r)$, it holds*

$$\mu[\Delta(z,r)] \leqslant C'\sigma_\alpha[\Delta(z,r)].$$

(ii) $J_\mu : \mathscr{B}^p_\alpha \to L^p(\mu)$ *with norm* $\lambda_p \lesssim C'$.

Proof: (ii) implies (i) thanks to Lemma 5.20 and the already proved implication (a)⇒(b) of Theorem 5.15.

We now prove that (i) implies (ii). Let $f \in \mathscr{H}(\mathbb{D})$. Using successively Lemma 5.18 and the assumption, we get

$$\begin{aligned}\int_{\mathbb{D}} |f(z)|^p \, d\mu(z) &\leqslant \sum_{k=1}^{\infty} \int_{\Delta(z_k,r)} |f(z)|^p \, d\mu(z) \\ &\leqslant C \sum_{k=1}^{\infty} \frac{\mu[\Delta(z_k,r)]}{\sigma_\alpha[\Delta(z_k,r)]} \int_{\Delta(z_k,R)} |f(w)|^p \, d\sigma_\alpha(w).\end{aligned}$$

We now observe that, due to our assumption

$$\frac{\mu[\Delta(z_k,r)]}{\sigma_\alpha[\Delta(z_k,r)]} \leqslant C'$$

and, invoking Lemma 5.21,

$$\begin{aligned}\int_{\mathbb{D}} |f(z)|^p \, d\mu(z) &\leqslant CC' \sum_{k=1}^{\infty} \int_{\Delta(z_k,R)} |f(z)|^p \, d\sigma_\alpha(z) \\ &= CC' \int_{\mathbb{D}} \left(\sum_{k=1}^{\infty} \mathbb{1}_{\Delta(z_k,R)}(z) \right) |f(z)|^p \, d\sigma_\alpha(z) \\ &\leqslant CC'N \int_{\mathbb{D}} |f(z)|^p \, d\sigma_\alpha(z).\end{aligned}$$

Moreover, we got $\|J_\mu : \mathscr{B}^p_\alpha \to L^p(\mu)\|^p \lesssim \sup_{z\in\mathbb{D}} \dfrac{\mu[\Delta(z,r)]}{\sigma_\alpha[\Delta(z,r)]}$. □

End of the Proof of Theorem 5.15.

Assume that $\mu(W) \leqslant Mh^{\alpha+2}$ for every Carleson window $W(a,h)$.

By Lemma 5.20, we get for fixed $0 < r < 1$

$$\mu[\Delta(z,r)] \leqslant CM(1-|z|^2)^{\alpha+2} \leqslant CM\sigma_\alpha[\Delta(z,r)].$$

We are in a position to apply Theorem 5.22, with C' changed into CM, to conclude that J_μ is bounded. □

We can summarize all the results of this section in the following theorem, once we realize that we proved the following inequalities (see successively

Proof of Theorem 5.15, Proof of Theorem 5.22, and the last inequality comes from Lemma 5.20):

$$K_{\mu,\alpha+2}(1) \lesssim \|J_\mu : \mathscr{B}^p_\alpha \to L^p(\mu)\|^p \lesssim \sup_{z\in\mathbb{D}} \frac{\mu[\Delta(z,r)]}{\sigma_\alpha[\Delta(z,r)]} \lesssim \sup_{0<h\leqslant 1} \frac{\rho_\mu(h)}{h^{\alpha+2}} = K_{\mu,\alpha+2}(1).$$

In (ii) below, (c)⇒(a) (or (b)) was actually not proved yet, so we now complete the proof.

Take a sequence (f_n) in the unit ball of $\mathscr{B}^p_\alpha$, converging to 0 in $\mathscr{H}(\mathbb{D})$. Fix $\varepsilon > 0$. Thanks to the remark (5.27) about the sequence (z_k) in Lemma 5.21, the hypothesis provides a k_0 such that $\mu[\Delta(z_k, r)] \leqslant \varepsilon\sigma_\alpha[\Delta(z_k, r)]$ for all $k > k_0$, and we split the sum appearing in the proof of Theorem 5.22. More precisely, we have

$$\begin{aligned}\int_{\mathbb{D}} |f_n(z)|^p \, d\mu(z) &\leqslant \sum_{k=1}^{\infty} \int_{\Delta(z_k,r)} |f(z)|^p \, d\mu(z) \\ &\leqslant \sum_{k=1}^{k_0} \int_{\Delta(z_k,r)} |f_n(z)|^p \, d\mu(z) \\ &\quad + C \sum_{k>k_0} \frac{\mu[\Delta(z_k,r)]}{\sigma_\alpha[\Delta(z_k,r)]} \int_{\Delta(z_k,R)} |f_n(w)|^p \, d\sigma_\alpha(w).\end{aligned}$$

The first sum converges to 0 thanks to the hypothesis $f_n \to 0$ in $\mathscr{H}(\mathbb{D})$; the second sum satisfies

$$\sum_{k>k_0} \frac{\mu[\Delta(z_k,r)]}{\sigma_\alpha[\Delta(z_k,r)]} \int_{\Delta(z_k,R)} |f_n(w)|^p \, d\sigma_\alpha(w) \leqslant \varepsilon N \int_{\mathbb{D}} |f_n(z)|^p \, d\sigma_\alpha(z) \leqslant \varepsilon N$$

using the same computation as in the proof of Theorem 5.22.

We conclude that $J_\mu(f_n) \to 0$ in $\mathscr{B}^p_\alpha$, and $J_\mu : \mathscr{B}^p_\alpha \to L^p(\mu)$ is compact, that is, we get (ii)(a).

All in all, we proved in the present section Theorem 5.23 below

Theorem 5.23 *Let μ be a positive measure on $\mathbb{D}$, $\alpha > -1$, and $p \geqslant 1$.*

(i) *The following are equivalent:*
  - (a) $J_\mu : \mathscr{B}^p_\alpha \to L^p(\mu)$ *is bounded.*
  - (b) $\rho_\mu(h) = O\left(h^{\alpha+2}\right)$ *when* $h \to 0^+$.
  - (c) $\sup_{z\in\mathbb{D}} \dfrac{\mu[\Delta(z,r)]}{\sigma_\alpha[\Delta(z,r)]}$ *is finite.*

Moreover $\|J_\mu : \mathscr{B}^p_\alpha \to L^p(\mu)\|^p \approx \sup_{z\in\mathbb{D}} \dfrac{\mu[\Delta(z,r)]}{\sigma_\alpha[\Delta(z,r)]} \approx \sup_{0<h\leqslant 1} \dfrac{\rho_\mu(h)}{h^{\alpha+2}}.$

(ii) *The following are equivalent:*

(a) $J_\mu : \mathscr{B}_\alpha^p \to L^p(\mu)$ *is compact.*

(b) $\rho_\mu(h) = o(h^{\alpha+2})$ *when* $h \to 0^+$.

(c) $\displaystyle \lim_{|z|\to 1^-} \frac{\mu[\Delta(z,r)]}{\sigma_\alpha[\Delta(z,r)]} = 0.$

5.3 Pullback Measure and Applications

Recall a basic definition, notably familiar to ergodicians.

Definition 5.24 Let X, Y be two measurable spaces, $T : X \to Y$ a measurable map and μ a measure on X. The *image* $\nu = T(\mu) = \mu_T$ of μ, also called the *pullback* of μ under T, is the measure on Y defined by the formula

$$\nu(B) = \mu[T^{-1}(B)] \tag{5.28}$$

for any measurable set $B \subset Y$. An equivalent functional form is

$$\int_Y f d\nu = \int_X (f \circ T)\, d\mu. \tag{5.29}$$

5.3.1 Applications to Composition Operators

Let now φ be a symbol, $C_\varphi : H^p \to H^p$ the attached composition operator, and m_φ the pullback of m under $\varphi^* : \mathbb{T} \to \overline{\mathbb{D}}$. We have by definition

$$\|C_\varphi(f)\|^p = \|f \circ \varphi\|^p = \int_{\mathbb{T}} |f^*(\varphi^*(u))|^p\, dm(u) = \int_{\overline{\mathbb{D}}} |f|^p\, dm_\varphi.$$

From Chapter 2, C_φ is always bounded on H^p; this means that the Carleson embedding $J_{m_\varphi} : H^p \to L^p(m_\varphi)$ is bounded; in other terms m_φ is always a Carleson measure for H^p.

We then write for short ρ_φ instead of ρ_{m_φ}, that is,

$$\rho_\varphi(h) = \sup_{\xi\in\mathbb{T}} m_\varphi[W(\xi,h)]. \tag{5.30}$$

We can do the same with $\sigma_\varphi = \varphi(\sigma)$, with here $\varphi : \mathbb{D} \to \mathbb{D}$ and the parameters

$$\rho_{\varphi,\,2}(h) = \sup_{\xi\in\mathbb{T}} \sigma_\varphi[W(\xi,h)], \tag{5.31}$$

and more generally, with the pullback measure $\sigma_{\varphi,\alpha} = \varphi(\sigma_\alpha) = (\sigma_\alpha)_\varphi$,

$$\rho_{\varphi,\,\alpha+2}(h) = \sup_{\xi\in\mathbb{T}} \sigma_{\varphi,\alpha}[W(\xi,h)], \qquad \text{where } \alpha > -1. \tag{5.32}$$

The Carleson and HOP theorems can now be rephrased as follows:

Theorem 5.25 *Let φ be a symbol, $p \geqslant 1$, and $\alpha > -1$. Then the composition operator C_φ satisfies*

(i) $\|C_\varphi : H^p \to H^p\| \approx \sup_{0<h<1} \left(\frac{\rho_\varphi(h)}{h}\right)^{\frac{1}{p}} \approx \left(K_{m_\varphi}(1)\right)^{\frac{1}{p}}.$

(ii) $\|C_\varphi : \mathscr{B}^p_\alpha \to \mathscr{B}^p_\alpha\| \approx \sup_{0<h<1} \left(\frac{\rho_{\varphi,\,\alpha+2}(h)}{h^{\alpha+2}}\right)^{\frac{1}{p}} \approx \sup_{z\in\mathbb{D}} \left(\frac{\sigma_{\varphi,\alpha}[\Delta(z,r)]}{\sigma_\alpha[\Delta(z,r)]}\right)^{\frac{1}{p}}.$

Using the compact version of the Carleson and HOP theorems proved before, we can also state the following theorem:

Theorem 5.26 *Let φ be a symbol, $p \geqslant 1$, and $\alpha > -1$. Then:*

(i) $C_\varphi : H^p \to H^p$ *is compact if and only* $\rho_\varphi(h) = o(h)$ *as* $h \to 0$.
(ii) $C_\varphi : \mathscr{B}^p_\alpha \to \mathscr{B}^p_\alpha$ *is compact if and only* $\rho_{\varphi,\,\alpha+2}(h) = o\left(h^{\alpha+2}\right)$ *if and only* $\sigma_{\varphi,\alpha+2}[\Delta(z,r)] = o\left(\sigma_\alpha[\Delta(z,r)]\right)$, *when* $|z| \to 1^-$.

5.3.2 Applications to Weighted Composition Operators

Given φ a symbol and w an analytic function on $\mathbb{D}$, we can define the associated weighted composition operator

$$f \in \mathscr{H}(\mathbb{D}) \longmapsto (w.C_\varphi)(f) = w.(f \circ \varphi) \in \mathscr{H}(\mathbb{D}),$$

and we can write, for every analytic function f on $\mathbb{D}$ and for $\alpha > -1$,

$$\int_{\mathbb{D}} |w.(f\circ\varphi)|^p \, d\sigma_\alpha = \int_{\mathbb{D}} |f|^p \, d\mu, \qquad \text{where } \mu = \left(|w|^p \, d\sigma_\alpha\right)_\varphi,$$

and, assuming that w and f have some extension a.e. on $\mathbb{T}$ (for instance they belong to some Hardy space – which is natural in our context),

$$\int_{\mathbb{T}} |w.(f\circ\varphi)|^p \, dm = \int_{\overline{\mathbb{D}}} |f|^p \, d\mu, \qquad \text{where } \mu = \left(|w|^p \, dm\right)_\varphi.$$

Therefore a natural link appears between the boundedness (resp. compactness) of the weighted composition operator on H^p (or on $\mathscr{B}^p_\alpha$) and the boundedness (resp. compactness) of the Carleson embedding (with the corresponding measure as above).

More precisely, we obtain for free, in view of this explanation and the results of the present chapter, the following theorem:

Theorem 5.27 *Let $p \geqslant 1$, φ be a symbol, and $w \in H^p$.*
Then:

- *The following are equivalent:*
  (i) *The weighted composition operator $w.C_\varphi$ is bounded on H^p.*
  (ii) *The measure $\mu = \big(|w|^p\, dm\big)_\varphi$ satisfies $\rho_\mu(h) = O(h)$ when $h \to 0^+$.*
- *The following are equivalent:*
  (i) *The weighted composition operator $w.C_\varphi$ is compact on H^p.*
  (ii) *The measure $\mu = \big(|w|^p\, dm\big)_\varphi$ satisfies $\rho_\mu(h) = o(h)$ when $h \to 0^+$.*

Let us mention that requiring $w \in H^p$ is not really a constraint: $w = (w.C_\varphi)(1)$ must belong to H^p when the operator $w.C_\varphi$ is bounded on H^p. We added the hypothesis $w \in H^p$ to be sure that $w(z)$ makes sense for a.e. $z \in \mathbb{T}$.

Let us mention that $\rho_\mu(h) = \sup_{\xi\in\mathbb{T}} \int_{\varphi^{*-1}(W(\xi,h))} |w(z)|^p\, dm(z)$, where $W(\xi,h)$ is the Carleson box viewed as a subset of $\overline{\mathbb{D}}$. We write especially φ^* to insist on the fact that ${\varphi^*}^{-1}(W(\xi,h)) \subset \mathbb{T}$.

In the framework of Bergman spaces, we have the following theorem:

Theorem 5.28 *Let $p \geqslant 1$, $\alpha > -1$, φ be a symbol, and w be an analytic function on $\mathbb{D}$.*
Then:

- *The following are equivalent:*
  (i) *The weighted composition operator $w.C_\varphi$ is bounded on $\mathscr{B}^p_\alpha$.*
  (ii) *The measure $\mu = \big(|w|^p\, d\sigma_\alpha\big)_\varphi$ satisfies $\rho_\mu(h) = O(h)$ when $h \to 0^+$.*
- *The following are equivalent:*
  (i) *The weighted composition operator $w.C_\varphi$ is compact on $\mathscr{B}^p_\alpha$.*
  (ii) *The measure $\mu = \big(|w|^p\, d\sigma_\alpha\big)_\varphi$ satisfies $\rho_\mu(h) = o(h)$ when $h \to 0^+$.*

Let us mention that here $\rho_\mu(h) = \sup_{\xi\in\mathbb{T}} \int_{\varphi^{-1}(W(\xi,h))} |w(z)|^p\, d\sigma_\alpha(z)$, where $W(\xi,h)$ is the Carleson box viewed as a subset of $\mathbb{D}$.

5.4 Contraction Principle for the Arc-Length

Let φ be a symbol, and m_φ the associated pullback measure. We recall that, for $\xi \in \mathbb{T}$ and $0 < h < 1$, the Carleson circular window $S(\xi,h)$ is defined by

$$S(\xi,h) = \big\{z \in \overline{\mathbb{D}} : |z - \xi| < h\big\}.$$

Previously in this chapter, we saw that m_φ is a Carleson measure for Hardy spaces, equivalently that $m_\varphi(S(\xi,h)) \leqslant Ch$ for all $0 < h < 1$, where C only depends on φ. The aim of this section is to show a remarkable reinforcement of that fact, namely a homogeneity property of this measure m_φ, as expressed by the following theorem:

Theorem 5.29 *There exists $C > 0$ such that, for any symbol φ, one has*

$$m_\varphi\big(S(\xi,\varepsilon h)\big) \leqslant C\varepsilon m_\varphi\big(S(\xi,h)\big) \tag{5.33}$$

for any $h \in (0,1]$, $\varepsilon \in (0,1)$, and any $\xi \in \mathbb{T}$.

Let us mention that Theorem 5.29 is used in a crucial way in Chapter 11 of this book (see also [98, theorem 4.18]) to characterize the symbols φ such that C_φ is compact on Hardy–Orlicz spaces.

We also mention that it is sufficient to prove (5.33) for small h and ε (actually it is even true for any $h \in (0,2]$):

Lemma 5.30 *Let $h_0 \in (0,1]$, $\varepsilon_0 \in (0,1)$, and $\xi \in \mathbb{T}$. We assume that, for some $C_0 > 0$,*

$$\forall h \in (0,h_0],\ \forall \varepsilon \in (0,\varepsilon_0),\quad m_\varphi\big(S(\xi,\varepsilon h)\big) \leqslant C_0\varepsilon m_\varphi\big(S(\xi,h)\big).$$

Then (5.33) is true for every $h \in (0,1]$ and every $\varepsilon \in (0,1)$.

Proof: Take $h \in (0,h_0]$ and ε in $(0,1)$. If $\varepsilon \in (0,\varepsilon_0)$, then it is already known, so we assume that $\varepsilon \geqslant \varepsilon_0$. Obviously,

$$m_\varphi\big(S(\xi,\varepsilon h)\big) \leqslant m_\varphi\big(S(\xi,h)\big) \leqslant \frac{1}{\varepsilon_0}\varepsilon m_\varphi\big(S(\xi,h)\big);$$

hence (5.33) is true for every $h \in (0,h_0]$ and every $\varepsilon \in (0,1)$ with a constant $C = \max(C_0, 1/\varepsilon_0)$.

Now take $h \in (0,1]$ and ε in $(0,h_0) \subset (0,1)$. Point out that $h' = h_0 h \in (0,h_0]$ and, with $\varepsilon' = \dfrac{\varepsilon}{h_0} \in (0,1)$, we have

$$m_\varphi\big(S(\xi,\varepsilon h)\big) = m_\varphi\big(S(\xi,\varepsilon' h')\big) \leqslant C\varepsilon' m_\varphi\big(S(\xi,h')\big) \leqslant \frac{C}{h_0}\varepsilon\, m_\varphi\big(S(\xi,h)\big),$$

since $h' \leqslant h$. It means that (5.33) is true for every $h \in (0,1]$ and $\varepsilon \in (0,h_0)$ with a new constant C.

The final conclusion comes from our first step in the present proof, applied with a new $h_0 = 1$ and a new ε_0. □

For the proof of Theorem 5.29, we need preliminary notations: C will denote as usual a "varying" positive constant, which can change from a formula to another.

For $0 < \theta \leqslant 1$, we will denote by $\mathcal{C}_\theta$ the cone

$$\mathcal{C}_\theta = \left\{ z \in \mathbb{C} : |\arg(z)| < \frac{\pi}{2}\theta \right\},$$

where arg denotes the principal determination of the argument of z. We will also set once and for all, for $z \in \mathbb{C}_+$ and $\eta > 0$,

$$z^\eta = \exp(\eta \log z) = |z|^\eta \exp(i\eta \arg(z)), \tag{5.34}$$

where $\log z$ is the principal determination of the logarithm of z.

Finally, we will denote by $\mathcal{O}_\theta$ the set of $\mathcal{C}_\theta$-valued analytic functions on $\mathbb{D}$:

$$\mathcal{O}_\theta = \{f = u + iv : \mathbb{D} \to \mathcal{C}_\theta,\ f \text{ analytic}\}. \tag{5.35}$$

We recall that

$$f \in \mathcal{O}_\theta \Longrightarrow f \in H^p \text{ for all } p < 1/\theta. \tag{5.36}$$

Indeed, omitting the variable z, if $f = |f|\,\mathrm{e}^{it}$ with $|t| < \frac{\pi}{2}\theta$, we have

$$\mathrm{Re}\left(f^p\right) = |f|^p \cos(pt) \geqslant |f|^p \cos\left(p\frac{\pi}{2}\theta\right) =: \delta |f|^p \quad \text{with } \delta > 0,$$

so that for any $0 < r < 1$ (writing $f_r(z) = f(rz)$), since $\mathrm{Re}\,(f^p)$ is harmonic and positive,

$$\int_{\mathbb{T}} |f_r|^p\, dm \leqslant \delta^{-1} \int_{\mathbb{T}} \mathrm{Re}\big(f_r^p\big)\, dm = \delta^{-1}\mathrm{Re}\,(f^p(0)).$$

Observe that $\mathcal{C}_1 = \mathbb{C}_+$, the right half-plane. We will mainly need the value $\theta = 1/3$ and *will abbreviate $\mathcal{O}_{1/3}$ to $\mathcal{O}$ and $\mathcal{C}_{1/3}$ to $\mathcal{C}$.* The (non-critical) choice of this $1/3$ is motivated by the following: if $f \in \mathcal{O}$, then by (5.36) $f \in H^p$ for $p < 3$, in particular $f \in H^2$: so, if $f = u + iv \in \mathcal{O}$, the harmonic functions $u, v \in h^1$ and have radial limits u^*, v^*. Moreover,

$$|f| \leqslant (2/\sqrt{3})\, u \quad \text{and} \quad |v| \leqslant (1/\sqrt{3})\, u. \tag{5.37}$$

Scheme of Proof

The proof of Theorem 5.29 is long and involved. Hence, we first give a heuristic scheme of this proof, all the more that a different but parallel scheme will occur in the Bergman space case. Things go as follows.

Reformulation.

We can forget the symbol φ and prove instead, for λ positive and large,

$$m\left(\{|f^*| > \lambda\}\right) \lesssim \frac{1}{\lambda^3} m\left(\{|f^*| > 1\}\right) \tag{5.38}$$

for any $f \in \mathcal{O}_{1/3} =: \mathcal{O}$ as in (5.35) (recall that $\mathcal{O} \subset H^2$), with controlled $|f(0)|$. This reformulation is for free. The rest of the proof consists of three steps.

Step 1.
If $f = u + iv \in \mathcal{O}$, we have for λ large

$$m(\{|f^*| > \lambda\}) \leqslant C\left(\frac{u(0)}{\lambda}\right)^3 \leqslant C\left(\frac{|f(0)|}{\lambda}\right)^3. \tag{5.39}$$

Without being difficult, this step (Kolmogorov, Carleson invoked) is non-trivial.

Step 2.
If $f = u + iv \in \mathcal{O}$, we have for λ large and $0 < y \leqslant \lambda\sqrt{3}/2$

$$m(\{|f^*| > \lambda\}) \leqslant C\left(\frac{y}{\lambda}\right)^3 m(\{Mu^* > y\}) \leqslant C\left(\frac{y}{\lambda}\right)^3 m(\{Mf^* > y\}), \tag{5.40}$$

where M designates the Hardy–Littlewood maximal function. This step is delicate and requires a "localization."

Step 3.
We can go back from Mf to f and prove, say for $y = 1$ and λ large,

$$m(\{|f^*| > \lambda\}) \leqslant \frac{C}{\lambda^3} m(\{|f^*| > 1\}), \tag{5.41}$$

which is nothing but the claimed inequality (5.38). This step is not so difficult as Step 2 but nevertheless requires kind of a reverse Tchebycheff inequality, which holds because $(Mf^*)^2 \approx M(f^{*2})$ for $f \in \mathcal{O}$.

Here are the details.

Back to the Main Proof

We will prove the following form (reformulation) of Theorem 5.29:

Theorem 5.31 *There exists δ_0, C and $\lambda_0 > 0$ such that, for any $f \in \mathcal{O}$ satisfying $|f(0)| \leqslant \delta_0$, and any $\lambda > \lambda_0$, it holds*

$$m(\{|f^*| > \lambda\}) \leqslant \frac{C}{\lambda^3} m(\{|f^*| > 1\}). \tag{5.42}$$

Clearly, Theorem 5.31 implies Theorem 5.29. Indeed, the function defined by $h \mapsto m_\varphi(S(\xi, h))$ is non-decreasing, hence (thanks to Lemma 5.30) it is sufficient to prove (5.33) for $0 < h < \delta_0^3(1 - |\varphi(0)|)$ and $0 < \varepsilon < \varepsilon_0$ where ε_0 is small enough. We can assume $\xi = 1$. Then, the function

$$f(z) = \left(\frac{h}{1 - \varphi(z)}\right)^{1/3} \in \mathcal{O}$$

since $\operatorname{Re} \dfrac{1}{1-\varphi(z)} \geqslant 1/2 > 0$. Moreover $|f(0)| \leqslant \delta_0$, hence (5.42) now gives, for $\varepsilon < \lambda_0^{-3}$,

$$m_\varphi\big(S(1,\varepsilon h)\big) = m(\{|f^*| > \varepsilon^{-1/3}\}) \leqslant C\varepsilon m(\{|f^*| > 1\}) = C\varepsilon m_\varphi\big(S(1,h)\big).$$

The proof of Theorem 5.31 will make strong use of the Hardy–Littlewood maximal function M encountered in Section 5.1, but we first need to extend the definitions and properties of Section 5.1 to *harmonic functions*.

The Three Steps

The proof of Theorem 5.31 will now consist of *three* main steps.

Step 1

The first step in the proof of Theorem 5.31 is the following proposition.

Proposition 5.32 *Let $f = u + iv \in \mathcal{O}$. Then*

$$m\left(\{|f^*| > \lambda\}\right) \leqslant C\left(\frac{u(0)}{\lambda}\right)^3 \leqslant C\left(\frac{|f(0)|}{\lambda}\right)^3 \qquad \textit{for all } \lambda > 0.$$

We first need the following lemma:

Lemma 5.33 *Let $g = u + iv \in \mathcal{O}_1$ with $g(0) > 0$. Then*

$$m\left(\{|g^*| > \lambda\}\right) \leqslant \frac{C}{\lambda} g(0) \quad \textit{for all } \lambda > 0.$$

Proof: Since $g(0)$ is real, $v = Hu$ is the conjugate function of u, and by Kolmogorov's theorem ([188, p. 134, Vol. I]), the conjugation operator H is of weak-type $(1,1)$. Hence, using $m(\{|v^*| > \lambda/2\}) \lesssim \dfrac{\|u^*\|_1}{\lambda}$, we obtain

$$m(\{|g^*| > \lambda\}) \leqslant m(\{u^* > \lambda/2\}) + m\left(\{|v^*| > \lambda/2\}\right) \leqslant \frac{C}{\lambda}\|u^*\|_1 = \frac{C}{\lambda} g(0),$$

since u is harmonic and positive, with $u(0) = g(0)$.

Here is an alternative argument: we may assume that $g(0) = 1$ and $\lambda > 2$. The function $\varphi(z) = \frac{g(z)-1}{g(z)+1}$ is a symbol because $\operatorname{Re}(g) \in \mathbb{C}_+$. Moreover, it satisfies, when $|g(z)| > \lambda$,

$$|\varphi(z) - 1| = \frac{2}{|g(z)+1|} \leqslant \frac{2}{\lambda - 1} \leqslant \frac{4}{\lambda} \cdot$$

Hence,

$$m(\{|g^*| > \lambda\}) \leqslant m_\varphi(S(1, 4/\lambda)) \leqslant \frac{C}{\lambda}$$

since we know that m_φ is a Carleson measure. □

Proof of Proposition 5.32: Observe that $|w_1 + w_2| \geqslant |w_1|$ for all $w_1, w_2 \in \mathbb{C}$ since Re $(w_1\overline{w_2}) \geqslant 0$. Now take $w_1 = f(z)$, $w_2 = \overline{f(0)}$ and apply Lemma 5.33 to $g = \left(f + \overline{f(0)}\right)^3 \in \mathcal{O}_1$, which satisfies $g(0) = 8\,u(0)^3 > 0$ to get

$$m(\{|f^*| > \lambda\}) \leqslant m(\{|g^*| > \lambda^3\}) \leqslant C\left(\frac{u(0)}{\lambda}\right)^3 \leqslant C\left(\frac{|f(0)|}{\lambda}\right)^3$$

as claimed. □

Our task will be to replace $|f(0)|^3$ by $m(\{|f^*| > 1\})$. This will be done in two more steps.

Step 2

Here is the next step.

Proposition 5.34 *Let $f = u + iv \in \mathcal{O}$. If $y > 0$ satisfies $m(\{Nu > y\}) < 1/4$, then, for all $\lambda \geqslant \dfrac{2y}{\sqrt{3}}$,*

$$m\left(\{|f^*| > \lambda\}\right) \leqslant C\left(\frac{y}{\lambda}\right)^3 m\left(\{Nu > y\}\right) \tag{5.43}$$

$$\leqslant C\left(\frac{y}{\lambda}\right)^3 m\left(\{Mu^* > y\}\right). \tag{5.44}$$

Once (5.43) is proved, (5.44) (our main interest) follows immediately since $N \leqslant M$. The proposition tells that our wish is met (indeed, the condition $m(\{Nu > y\}) < 1/4$ will be easy to achieve and we will only need the numerical value $y = 2$) when we replace u^* by Mu^*.

We now prove (5.43). For $0 < b - a < \frac{\pi}{2}$, we denote by (e^{ia}, e^{ib}) the circular arc $I = \{e^{it} : a < t < b\}$. We shall first prove a *"localized"* version of (5.43) as follows.

Proposition 5.35 *Let $f = u + iv \in \mathcal{O}$. If $y > 0$ and $I = (e^{ia}, e^{ib})$ is an arc satisfying $b - a < \pi/2$, $Nu(e^{ia}) \leqslant y$ and $Nu(e^{ib}) \leqslant y$, then*

$$m\left(\{|f^*| > \lambda\} \cap I\right) \leqslant C\left(\frac{y}{\lambda}\right)^3 m\,(I) \qquad \textit{for all } \lambda \geqslant \frac{2y}{\sqrt{3}}. \tag{5.45}$$

This localized version implies the global one. Indeed, since Nu is l.s.c. by construction, the set $\{Nu > y\}$ is open and writes as a disjoint union of arcs $I_j = (e^{ia_j}, e^{ib_j})$ satisfying $Nu(e^{ia_j}) \leqslant y$ and $Nu(e^{ib_j}) \leqslant y$.

This implies $m\,(I_j) \leqslant m\left(\{Nu > y\}\right) < 1/4$ and $b_j - a_j < \pi/2$.

Hence by (5.45)

$$m\left(\{|f^*| > \lambda\} \cap I_j\right) \leqslant C\left(\frac{y}{\lambda}\right)^3 m\,(I_j) \qquad \text{for every } j.$$

Adding up those inequalities gives

$$m(\{|f^*| > \lambda\} \cap \{Nu > y\}) \leqslant C\left(\frac{y}{\lambda}\right)^3 m(\{Nu > y\}). \tag{5.46}$$

But, since $f \in \mathcal{O}$, we have $|f^*| \leqslant (2/\sqrt{3})\, u^* \leqslant (2/\sqrt{3})\, Nu$, so that the set $\{|f^*| > \lambda\}$ is contained in the set $\{Nu > y\}$ for $\lambda \geqslant \frac{2y}{\sqrt{3}}$; and inequality (5.46) gives (5.43), as desired:

$$m(\{|f^*| > \lambda\}) \leqslant C\left(\frac{y}{\lambda}\right)^3 m(\{Nu > y\}).$$

Proof of Proposition 5.35: Without loss of generality, we may assume that $a = -\delta$ and $b = \delta$ with $0 < \delta < \pi/4$.

Let I^+ and I^- be the arcs

$$I^+ = (1, \mathrm{e}^{i\delta}), \qquad I^- = (\mathrm{e}^{-i\delta}, 1).$$

To get (5.45), it is enough to prove that, as soon as $Nu(\mathrm{e}^{i\delta}) \leqslant y$, we have

$$m(\{|f^*| > \lambda\} \cap I^+) \leqslant C\left(\frac{y}{\lambda}\right)^3 m(I^+) \quad \text{for all } \lambda \geqslant \frac{2y}{\sqrt{3}} \tag{5.47}$$

with a similar estimate for I^-. To that effect, we consider the right half-disk

$$Q = \{z \in \mathbb{D} \;:\; \operatorname{Re} z > 0\}$$

and $\psi : \mathbb{D} \to Q$ the conformal map

$$\psi = T^{-1} s T,$$

where

$$T(z) = -i\frac{z+i}{z-i} = \frac{1-iz}{z-i}, \quad s(z) = z^{1/2}, \quad T^{-1}(z) = \frac{z-i}{1-iz}.$$

We see that T maps $\mathbb{D}$ onto the upper half-plane, s maps this upper half-plane onto the first quadrant, and T^{-1} this first quadrant onto Q (we happen to have here $T^{-1} = 1/T$). We also see that ψ extends to a homeomorphism $\overline{\mathbb{D}} \to \overline{Q}$ and maps $\partial\mathbb{D} \cap \mathbb{C}_+$ to the circular part of ∂Q. Moreover, we claim that

$$\psi(-1) = 0, \quad \psi(0) = \sqrt{2} - 1, \quad \psi(\mathrm{e}^{i\rho}) = \mathrm{e}^{i\pi/4}, \quad \psi(\mathrm{e}^{-i\rho}) = \mathrm{e}^{-i\pi/4}$$

for an easily computable argument ρ; indeed, denoting c, $c(z) = z^2$, the square function, we have (with $\psi^{-1} = T^{-1} c T$ and hence $\psi(\mathrm{e}^{i\rho}) = \mathrm{e}^{i\pi/4}$)

$$\mathrm{e}^{i\rho} = T^{-1} c T(\mathrm{e}^{i\pi/4}) = T^{-1} c(1 + \sqrt{2}) = T^{-1}(3 + \sqrt{8}) = \frac{1}{3} + i\frac{\sqrt{8}}{3};$$

hence $\rho = \arctan\sqrt{8}$. We also have $\psi(\mathrm{e}^{-i\rho}) = \mathrm{e}^{-i\pi/4}$, since T and ψ commute with the map $S : z \mapsto z^{-1}$.

Let us consider the arc $I_\rho := (\mathrm{e}^{-i\rho}, \mathrm{e}^{i\rho})$, on which ψ is clearly bi-Lipschitz; hence, there are constants $\gamma_1, \gamma_2 > 0$ such that

$$\gamma_1\, m(E) \leqslant m(\psi(E)) \leqslant \gamma_2\, m(E)$$

for every Borel subset E of I_ρ (only the constant γ_2 will be used).

To prove Proposition 5.35, we consider the function $F \in \mathcal{O}$ defined by

$$F(z) = f(\mathrm{e}^{i\delta}\psi^\beta(z)) =: (f \circ \chi)(z), \quad \text{where } \beta = \frac{4\delta}{\pi} = 8m(I^+) < 1.$$

We used (5.34) to define ψ^β and observe that

$$\operatorname{Re} F(0) = u\big((\sqrt{2}-1)^\beta \mathrm{e}^{i\delta}\big) \leqslant Nu\big(\mathrm{e}^{i\delta}\big) \leqslant y.$$

We have by construction the following diagram

$$I_\rho \xrightarrow{\psi} (\mathrm{e}^{-i\pi/4}, \mathrm{e}^{i\pi/4}) \xrightarrow{z^\beta} (\mathrm{e}^{-i\delta}, \mathrm{e}^{i\delta}) \xrightarrow{\mathrm{e}^{i\delta}} (1, \mathrm{e}^{2i\delta}) \supset I^+;$$

in other terms, $I^+ \subset \chi(I_\rho)$. Let us now set

$$A = \{|f^*| > \lambda\} \cap I^+.$$

Then $E = \chi^{-1}(A)$ is a Borel subset of I_ρ satisfying $|F^*(\xi)| = |f^*(\chi(\xi))| > \lambda$ for every $\xi \in E$. This gives (recall that $\beta = 8m(I^+) < 1$)

$$\begin{aligned} m(A) = m\,(\chi(E)) &= m\,\big(\psi^\beta(E)\big) = \beta m\,(\psi(E)) \\ &\leqslant \gamma_2\, \beta\, m(E) \leqslant 8\gamma_2\, m(I^+)\, m(\{F^*| > \lambda\}), \end{aligned}$$

and finally, using Proposition 5.32 for F (step 1) in an essential way,

$$m(\{|f^*| > \lambda\} \cap I^+) \leqslant Cm\,(I^+) \left(\frac{\operatorname{Re} F(0)}{\lambda}\right)^3 \leqslant C \left(\frac{y}{\lambda}\right)^3 m\,(I^+). \qquad \square$$

Step 3

The next proposition allows to go back from Mu^* to u^*.

Proposition 5.36 *Let $f = u + iv \in \mathcal{O}$. Then, for every $y \geqslant 2$, we have*

$$m\,(\{Mu^* > y\}) \leqslant Cm\,(\{u^* > y/2\}). \tag{5.48}$$

It is in the proof of this Proposition 5.36 that the equation $(Nf)^2 = N(f^2)$ and the comparison between the maximal functions M and N will intervene. Indeed, they first imply the following lemma:

Lemma 5.37 *For any function $f = u + iv \in \mathcal{O}$, it holds*

$$M((u^*)^2) \leqslant C\big(M(u^*)\big)^2.$$

Proof: The function $f^2 = (u^2 - v^2) + 2iuv$ is in H^1; in particular the harmonic function $u^2 - v^2$ is regular, we may apply inequalities (5.3). We know by (5.37) that $|v| \leqslant (1/\sqrt{3})\, u$ (hence $u^2 - v^2 \geqslant 2u^2/3 \geqslant 0$) and that $|f| \leqslant (2/\sqrt{3})\, u$. This implies via Lemma 5.5:

$$M((u^*)^2) \leqslant \frac{3}{2} M((u^*)^2 - (v^*)^2) \leqslant 18\, N(u^2 - v^2) \leqslant 18\, N(f^2) = 18\,(Nf)^2$$

$$\leqslant 18 \left(\frac{2}{\sqrt{3}} Nu\right)^2 \leqslant 24\,(Mu^*)^2. \qquad \square$$

End of Proof of Proposition 5.36: We will use a kind of simple reverse Tchebycheff inequality due to Paley and Zygmund ([77, p. 8], [108, p. 23, Vol. I]), which tells the following:

$$\mathbb{P}(X \geqslant \lambda \mathbb{E}(X)) \geqslant (1-\lambda)^2 \frac{\mathbb{E}(X)^2}{\mathbb{E}(X^2)}. \tag{5.49}$$

Here, X is a non-negative, square-integrable, random variable, on a probability space $(\Omega, \mathbb{P})$, $\mathbb{E}$ denotes the expectation, and $0 < \lambda < 1$. This inequality is all the better as $\mathbb{E}(X^2)$ is not too large before $\mathbb{E}(X)^2$. This information is provided here by Lemma 5.37. More precisely, set

$$A = \{Mu^* > y\} \quad \text{and} \quad B = \{u^* > y/2\}.$$

We will show that A and B have comparable sizes, namely that $m(B) \geqslant \delta m(A)$ for some positive constant δ, by using the *Vitali covering lemma.*

To that effect, for each $\xi \in A$, let I_ξ be an open arc centered at ξ such that

$$p := \frac{1}{m(I_\xi)} \int_{I_\xi} u^*\, dm > \frac{Mu^*(\xi)}{2}, \quad \text{and } p > y.$$

Lemma 5.37 implies

$$\frac{1}{m(I_\xi)} \int_{I_\xi} (u^*)^2\, dm \leqslant M((u^*)^2)(\xi) \leqslant C\big(M(u^*)\big)^2(\xi) \leqslant 4Cp^2.$$

Let L be the set $L = \{u^* > p/2\} \cap I_\xi$. We now apply (5.49) to the pair $(I_\xi, \mathbb{P})$ with $\mathbb{P}$ defined for each Borel subset ω of I_ξ by

$$\mathbb{P}(\omega) = \frac{m(\omega \cap I_\xi)}{m(I_\xi)},$$

and we get

$$\mathbb{P}(L) \geqslant \frac{1}{4} \times \frac{(\mathbb{E}u^*)^2}{\mathbb{E}((u^*)^2)} \geqslant \frac{1}{4}\frac{p^2}{4Cp^2} = \frac{1}{16C}.$$

Therefore, $16C\, m(L) \geqslant m(I_\xi)$, and since $B \cap I_\xi \supset L$, we infer

$$16Cm(B \cap I_\xi) \geqslant m(I_\xi).$$

Now, we use Vitali's lemma: there exists a disjoint sequence (I_{ξ_j}) of the arcs I_ξ such that $A \subset \bigcup_j 3I_{\xi_j}$, where $3I_\xi$ denotes as usual the open arc centered at ξ with length three times that of I_ξ. This gives

$$m(A) \leqslant 3 \sum_j m(I_{\xi_j}) \leqslant 3 \times 16C \sum_j m(I_{\xi_j} \cap B) \leqslant 48\, Cm(B)$$

as desired. □

Proof of Theorem 5.31: Proposition 5.34 with $y = 2$ gives us, for $\lambda \geqslant 4/\sqrt{3}$,

$$m\left(\{|f^*| > \lambda\}\right) \leqslant C\lambda^{-3} m\left(\{Mu^* > 2\}\right) \tag{5.50}$$

(provided $m(\{Nu > 2\}) < 1/4$). But Proposition 5.36 shows that

$$\begin{aligned} m\left(\{Nu > 2\}\right) &\leqslant m\left(\{Mu^* > 2\}\right) \leqslant Cm\left(\{u^* > 1\}\right) \leqslant C\|u^*\|_1 = Cu(0) \\ &\leqslant C|f(0)| < \frac{1}{4} \end{aligned}$$

as soon as $|f(0)| \leqslant \dfrac{1}{4C}\cdot$ We may hence use (5.48) and (5.50) to get (for λ large enough)

$$m\left(\{|f^*| > \lambda\}\right) \leqslant C\lambda^{-3} m\left(\{u^* > 1\}\right) \leqslant C\lambda^{-3} m\left(\{|f^*| > 1\}\right),$$

and we get (5.42). □

5.5 Contraction Principle for Weighted Area Measures

Recall that σ_α, $\alpha > -1$, is the probability measure $(\alpha + 1)(1 - |z|^2)^\alpha \, d\sigma(z)$ on $\mathbb{D}$ and that if φ is a symbol, $\sigma_{\varphi,\alpha} = \varphi(\sigma_\alpha)$ is the pullback of the measure σ_α under φ. We know that $\sigma_{\varphi,\,\alpha}$ is a $(\alpha + 2)$-Carleson measure, namely satisfies (i)(b) in Theorem 5.23. In this section, we discuss the following theorem, which is a non-straightforward extension of Theorem 5.29 when one replaces m_φ by $\sigma_{\varphi,\alpha}$. This extension will notably be used in Chapters 9 and 11.

Theorem 5.38 *Let $\alpha > -1$. There exists a constant $C = C(\alpha) > 0$ such that for any symbol φ, any $0 < h \leqslant 1$, and any $0 < \varepsilon \leqslant 1$, we have*

$$\sigma_{\varphi,\alpha}(S(\xi, \varepsilon h)) \leqslant C\varepsilon^{\alpha+2} \sigma_{\varphi,\alpha}(S(\xi, h)) \qquad \textit{for every } \xi \in \mathbb{T}.$$

It is sufficient to prove this theorem when h and ε are "small": this is due to a corresponding version of Lemma 5.30 (that we do not state here), with a very similar proof.

Theorem 5.38 does not follow from, or implies, Theorem 5.29 by putting $\alpha = -1$ since the functions considered live on $\mathbb{T}$ for H^2 (through their radial limits) and on $\mathbb{D}$ for $\mathcal{B}^2_\alpha$. We elaborate below on the details, skipping possibly some (but few) of them. One or two extra (in the end unessential) technical difficulties arise. We also refer to [102] (case of the ordinary Bergman space) or [112] (general case) for additional details.

We first give a heuristics for Theorem 5.38.

Analog Scheme of Proof

The scheme of proof in the weighted Bergman case is quite parallel to the Hardy case, as we emphasize in this "interlude" subsection.

Reformulation.
We can forget the symbol φ and prove instead that there exist two constants $c_0 > 0, \lambda_0 > 1$ such that

$$\sigma_\alpha(\{|g| > \lambda\}) \leqslant \frac{C}{\lambda^{\alpha+2}} \sigma_\alpha(\{|g| > 1\}) \tag{5.51}$$

for any analytic function $g : \mathbb{D} \to \mathbb{C}_+$ with $|g(0)| \leqslant c_0$ and any $\lambda \geqslant \lambda_0$. We will then obtain Theorem 5.38 by taking $g = \dfrac{h}{1-\varphi}$.

The rest of the proof consists of four steps.

Step 1.
Let $T : \mathbb{D} \to \mathbb{C}_+$ be the involutive conformal map $z \mapsto \dfrac{1-z}{1+z}$; we introduce the pullback measure $\tau_\alpha = T(\sigma_\alpha)$. We prove in Step 1 that there are constants $\lambda_1, h_1 > 1$ such that, for $f : \mathbb{C}_+ \to \mathbb{C}_+$ analytic, satisfying $\lambda \geqslant \lambda_1$, we have

$$\tau_\alpha(\{|f| > \lambda\}) \leqslant \frac{k_1}{\lambda^{\alpha+2}} |f(1)|^{\alpha+2}. \tag{5.52}$$

Step 2.
We *first* prove a localization of (5.52), in terms of a dyadic grid of the square $\Omega = (0,2] \times (-1,1] \subset \mathbb{C}_+$. This substep uses a Harnack-type inequality and the invariance properties of an auxiliary measure μ_α. We *next* prove, for any $f : \mathbb{C}_+ \to \mathbb{C}_+$ analytic, with a controlled $|f(1)|$

$$\tau_\alpha(\{|f| > \lambda\} \cap \Omega) \leqslant \frac{C}{\lambda^{\alpha+2}} \tau_\alpha(\{\tilde{M} f > 1\}). \tag{5.53}$$

This second substep, which exploits the above localization, is delicate, uses a *dyadic, maximal function* $\tilde{M} f : \Omega \to \mathbb{C}_+$, attached to the above dyadic *filtration* of the square Ω and then a Calderón–Zygmund decomposition.

Step 3.
We can go back from $\tilde{M}f$ function to f and prove, for any $f : \mathbb{C}_+ \to \mathbb{C}_+$ analytic, with $|f(1)| \leqslant c_1$

$$\forall \lambda \geqslant \lambda_1', \qquad \tau_\alpha(\{|f| > \lambda\} \cap \Omega) \leqslant \frac{C}{\lambda^{\alpha+2}} \tau_\alpha(\{|f| > 1\} \cap \Omega) \tag{5.54}$$

(where $c_1, C_1 > 0$, and $\lambda_1' \geqslant 1$ are constants). This step is not so difficult as Step 2, but again requires a kind of reverse Tchebycheff inequality, which holds using a kind of *reverse subharmonicity* ([112]).

Step 4.
We come back from (5.54) to our aim (5.51).

Back to the Main Proof

We now detail the scheme above, for the full proof of Theorem 5.38.

Step 1

Proposition 5.39 *There exists a constant $C = C(\alpha) > 0$ such that*

$$\sigma_\alpha(\{|g| > \lambda\}) \leqslant \frac{C}{\lambda^{\alpha+2}} |g(0)|^{\alpha+2} \tag{5.55}$$

for any analytic function $g : \mathbb{D} \to \mathbb{C}_+$ and any $\lambda > 0$.

Proof: We may assume that $|g(0)| = 1$. It suffices to consider $\lambda > 2$ (else it is obvious taking $C \geqslant 2^{\alpha+2}$ because $\sigma_\alpha(\{|g| > \lambda\}) \leqslant 1$). Set

$$\varphi(z) = \frac{g(z) - g(0)}{g(z) + \overline{g(0)}}. \tag{5.56}$$

Then $|g(z)| > \lambda$ implies

$$|\varphi(z) - 1| = 2\frac{|\mathrm{Re}\, g(0)|}{|g(z) + g(0)|} < \frac{2}{\lambda - 1} < \frac{4}{\lambda}.$$

But φ is a symbol vanishing at the origin, so that $\|C_\varphi\| = \|C_\varphi\|_{B(\mathscr{B}_\alpha^2)} = 1$, and we know that $\sigma_{\varphi,\alpha}$ is Carleson for the space $\mathscr{B}_\alpha^2$. So

$$\sigma_\alpha\big(\{|g| > \lambda\}\big) \leqslant \sigma_{\varphi,\alpha}\big[S(1, 4/\lambda)\big] \leqslant C_\alpha' \|C_\varphi\|^2 \left(\frac{4}{\lambda}\right)^{\alpha+2} \leqslant \frac{C_\alpha}{\lambda^{\alpha+2}}$$

and we are done. □

We will in the sequel start from $\mathbb{C}_+$ rather than from $\mathbb{D}$ and use the square Ω and the probability measure τ_α (on $\mathbb{C}_+$) defined by

$$\Omega = (0, 2] \times (-1, 1], \qquad \tau_\alpha = T(\sigma_\alpha),$$

with T the conformal map defined by $z \in \mathbb{D} \longmapsto T(z) = \dfrac{1 - z}{1 + z} \in \mathbb{C}_+$.

The restriction of τ_α to Ω (still denoted τ_α) is easily explicited thanks to the change of variable formula, and we have, with A the planar Lebesgue measure,

$$d\tau_\alpha(w) = \frac{4^{\alpha+1}(\alpha+1)}{\pi} \frac{(\mathrm{Re}(w))^\alpha}{|1+w|^{2(\alpha+2)}} \mathbb{1}_\Omega(w)\, dA(w). \tag{5.57}$$

In particular, writing $w = x + iy$, we have (with a clear meaning)

$$d\tau_\alpha(w) \approx x^\alpha\, dxdy = (\mathrm{Re}(w))^\alpha\, dA(w) =: d\mu_\alpha(w) \quad \text{on } \Omega. \tag{5.58}$$

We can reformulate Proposition 5.39 under the following form, just transporting things by T to $g = f \circ T$.

Proposition 5.40 *There exists a constant $C = C_\alpha > 0$ such that*

$$\tau_\alpha(\{|f| > \lambda\} \cap \Omega) \leqslant \tau_\alpha(\{|f| > \lambda\}) \leqslant \frac{C}{\lambda^{\alpha+2}} |f(1)|^{\alpha+2} \tag{5.59}$$

for any analytic function $f : \mathbb{C}_+ \to \mathbb{C}_+$ and any $\lambda > 0$.

As a corollary, we have the following subharmonicity/reverse subharmonicity inequalities:

Lemma 5.41 *There exists a constant $C > 1$ such that, for any analytic function $f : \mathbb{C}_+ \to \mathbb{C}_+$, one has $f \in L^1(\Omega, dA)$ and more precisely*

$$C^{-1}|f(1)| \leqslant \int_\Omega |f(z)|\, dA(z) \leqslant C|f(1)|. \tag{5.60}$$

Moreover, if c is the center of an open square $Q \subset \Omega$, it holds

$$C^{-1}|f(c)| \leqslant \frac{1}{A(Q)} \int_Q |f|\, dA \leqslant C|f(c)|. \tag{5.61}$$

Proof: Indeed, by a change of variable, relation (5.61) is a clear consequence of (5.60). The LHS of (5.60) is just subharmonicity (note that $D(1,1) \subset \Omega$). For the RHS (kind of reverse subharmonicity), we know by (5.58) that τ_0 behaves (up to multiplicative constants) like the area measure A on Ω (i.e., the measure μ_0), and we hence know by Proposition 5.40 that $A(\{|f| > \lambda\} \cap \Omega) \leqslant \frac{C}{\lambda^2}|f(1)|^2$ for all $\lambda > 0$. This implies (note that $A(\Omega) = 4$)

$$\begin{aligned}\int_\Omega |f|\, dA = \int_0^\infty A(\{|f| > \lambda\} \cap \Omega)\, d\lambda \leqslant 4|f(1)| + \int_{|f(1)|}^\infty \frac{C}{\lambda^2}|f(1)|^2 d\lambda \\ = (4 + C)|f(1)|.\end{aligned}$$

□

Step 2

A kind of improvement *(localization)* of Proposition 5.40 is needed. Let us split the square $\Omega = (0,2] \times (-1,1]$ into the dyadic subsquares

$$Q_l = \left(\frac{2j}{2^n}, \frac{2(j+1)}{2^n}\right] \times \left(\frac{2k}{2^n} - 1, \frac{2(k+1)}{2^n} - 1\right]$$

of center

$$c_l = \frac{2j+1}{2^n} + i\left(\frac{2k+1}{2^n} - 1\right)$$

with $n \geqslant 0$, $0 \leqslant j,k \leqslant 2^n - 1$ and where $l = (n,j,k)$. We will prove the following proposition:

Proposition 5.42 *There exists a constant $C_\alpha > 0$ such that for any analytic function $f : \mathbb{C}_+ \to \mathbb{C}_+$, any dyadic subsquare Q_l of Ω, and any $\lambda > 0$,*

$$\tau_\alpha(\{|f| > \lambda\} \cap Q_l) \leqslant \frac{C_\alpha}{\lambda^{\alpha+2}} \tau_\alpha(Q_l)|f(c_l)|^{\alpha+2}.$$

Proof: The proof will depend on the following lemma. Denoting by δ the pseudo-hyperbolic distance in $\mathbb{C}+$,

$$\delta(a,b) = \left|\frac{a-b}{a+\overline{b}}\right|,$$

and by $\Delta(c,s)$, with $0 < s < 1$, the pseudo-hyperbolic disk of $\mathbb{C}+$,

$$\Delta(c,s) = \{z \,:\, \delta(z,c) < s\},$$

we have the symmetric Harnack-type inequality.

Lemma 5.43 *For each $s < 1$, there exists $M_s > 1$ such that, for any analytic $f : \mathbb{C}_+ \to \mathbb{C}_+$ and any $z \in \Delta(c,s)$, we have*

$$\frac{1}{M_s} \leqslant \frac{|f(z)|}{|f(c)|} \leqslant M_s.$$

Proof: The Schwarz–Pick inequality (1.14) reads here, for the half-plane $\mathbb{C}_+$, as

$$\left|\frac{f(z)-f(c)}{f(z)+\overline{f(c)}}\right| = \delta(f(z),f(c)) \leqslant \delta(z,c) = \left|\frac{z-c}{z+\overline{c}}\right|$$

for any $z \in \mathbb{C}_+$. Now, suppose that $z \in \Delta(c,s)$ and set

$$u = \frac{f(z)-f(c)}{f(z)+\overline{f(c)}}, \quad \text{or} \quad f(z) = \frac{f(c)+u\overline{f(c)}}{1-u}.$$

Since $|u| \leqslant s$, we have $|f(z)| \leqslant |f(c)|\dfrac{1+|u|}{1-|u|} \leqslant |f(c)|\dfrac{1+s}{1-s} =: |f(c)|\,M_s$.

□

The derivation of Proposition 5.42 from Lemma 5.43 is done as follows;

(i) if $l = (n, j, k)$ with $j \geqslant 1$, we easily see that $Q_l \subset \Delta(c_l, s)$ with $s = \sqrt{2}/3$. Assume now $\{|f| > \lambda\} \cap Q_l \neq \emptyset$ (else it is trivial) so that for some $z \in Q_l$, $|f(z)|/\lambda > 1$, hence $M_s|f(c_l)|/\lambda > 1$ by Lemma 5.43. Therefore we have $\tau_\alpha(\{|f| > \lambda\} \cap Q_l) \leqslant \tau_\alpha(Q_l)$ and $1 < C_\alpha |f(c_l)|^{\alpha+2}/\lambda^{\alpha+2}$ taking $C_\alpha = M_s^{\alpha+2}$.

(ii) $l = (n, 0, k)$. We first assume that $l = (0, 0, 0)$ or else $Q_l = \Omega$ and just use Proposition 5.40. Next, exploiting the vertical translation invariance and $(\alpha + 2)$-homogeneity properties of μ_α, and again Proposition 5.40, we get the general result (we refer to [112, lemma 3.4] for detailed computations). □

In the next substep, let $f : \mathbb{C}_+ \to \mathbb{C}_+$ be analytic.

We assume that $|f(1)| \leqslant \frac{1}{C}$ where C is the constant in Lemma 5.41 so that

$$\int_\Omega |f(z)|\, dA(z) \leqslant 1.$$

One will replace $|f(c_l)|^{\alpha+2}$ by $\tau_\alpha(\{\tilde{M} f > 1\})$ with a new maximal function $\tilde{M} f : \Omega \to \mathbb{R}^+$, as follows: let $\mathcal{Q}_n$ be the subalgebra of Borel subsets of Ω generated by the squares $Q_{n,j,k}$, with $n \geqslant 0$ and $0 \leqslant j, k \leqslant 2^n - 1$ (note that $\mathcal{Q}_n \subset \mathcal{Q}_{n+1}$) and let E_n be the conditional expectation with respect to $\mathcal{Q}_n$:

$$E_n(g) = \sum_{0 \leqslant j, k \leqslant 2^n - 1} \left(\frac{1}{A(Q_{n,j,k})} \int_{Q_{n,j,k}} g\, dA \right) \mathbb{1}_{Q_{n,j,k}}.$$

Here $g : \Omega \to \mathbb{C}$ denotes a dA-integrable function. Those conditional expectations clearly satisfy

$$\lim_{n\to\infty} E_n(|f|)(z) = |f(z)| \qquad \text{for each } z \in \Omega.$$

We now set

$$\tilde{M} f = \sup_n E_n(|f|) \geqslant |f|$$

and we also have, for any $z \in \Omega$,

$$\tilde{M} f(z) = \sup_{Q_l \ni z} \frac{1}{A(Q_l)} \int_{Q_l} |f|\, dA.$$

This function $\tilde{M} f : \Omega \to \mathbb{R}^+$ is called the *dyadic maximal function* attached to f, and obviously

$$\{|f| > 1\} \cap \Omega \subset \{\tilde{M} f > 1\}.$$

We can now perform a *Calderón–Zygmund decomposition* of the set $\{\tilde{M} f > 1\}$ as follows ([67, pp. 286–288], [112, p. 1384]):

$$\{\tilde{M} f > 1\} = \bigsqcup_{l \in L} Q_l, \tag{5.62}$$

where L is a subset of the indices (n, j, k) for which

$$1 \leqslant \frac{1}{A(Q_l)} \int_{Q_l} |f| \, dA \leqslant 4. \tag{5.63}$$

We can now end the proof of (5.53).

For $\lambda \geqslant 1$, let us set $E^\lambda = \{|f| > \lambda\}$. From (5.63) and subharmonicity, we have, for some $C > 0$ and every $l \in L$,

$$|f(c_l)| \leqslant C \frac{1}{A(Q_l)} \int_{Q_l} |f| \, dA \leqslant 4C,$$

and we get with the help of (5.62) and Proposition 5.42

$$\begin{aligned}
\tau_\alpha(E^\lambda \cap \Omega) \leqslant \tau_\alpha(E^\lambda \cap \{\tilde{M} f > 1\}) &= \sum_{l \in L} \tau_\alpha(E^\lambda \cap Q_l) \\
&\leqslant \frac{C_\alpha}{\lambda^{\alpha+2}} \sum_{l \in L} \tau_\alpha(Q_l) |f(c_l)|^{\alpha+2} \leqslant \frac{C_\alpha (4C)^{\alpha+2}}{\lambda^{\alpha+2}} \sum_{l \in L} \tau_\alpha(Q_l) \\
&= \frac{C_\alpha (4C)^{\alpha+2}}{\lambda^{\alpha+2}} \tau_\alpha(\{\tilde{M} f > 1\}),
\end{aligned}$$

and this proves (5.53), which was *the core* of our Step 2.

Step 3

It remains to replace the set $\{\tilde{M} f > 1\}$ by the set $\{|f| > 1\} \cap \Omega$, and this step will depend on two other simple lemmas.

Lemma 5.44 *There exists a constant $\delta > 0$, depending only on α, such that for every $l = (n, j, k)$ there exists $R_l \subset Q_l$ with $\tau_\alpha(R_l) \geqslant \delta \tau_\alpha(Q_l)$ and, for every analytic map $f : \mathbb{C}_+ \to \mathbb{C}_+$,*

$$|f(z)| \geqslant \delta |f(c_l)| \quad \text{for every } z \in R_l.$$

Proof: This lemma [112, lemma 3.5, p. 1380] can be seen as a substitute to the Paley–Zygmund inequality in the Hardy space case. In view of (5.58), it suffices to prove the lemma with the measure μ_α instead of τ_α. To that effect, we separate two cases.

(i) If $l = (n, j, k)$ with $j \geqslant 1$, we can simply take $R_l = Q_l \subset \Delta(c_l, s)$ where $s = \sqrt{2}/3 < 1$ and use Lemma 5.43.

(ii) If $l = (n, j, k)$ with $j = 0$ (meaning that Q_l touches $\partial\mathbb{C}_+$), we may assume that $Q_l = \Omega$ since either a vertical translation or a dilation with ratio h are isometries for the pseudo-hyperbolic distance and multiply the measure μ_α respectively by 1 or by $h^{\alpha+2}$ (here the advantage of working with μ_α instead of τ_α manifests itself). Now, we can check that

$$\Delta(1, 1/4) \subset \Omega = Q_l.$$

Indeed, set $t = 1/4$ and $u = \dfrac{1-z}{1+z}$, so that $z = \dfrac{1-u}{1+u}\cdot$

If $|u| \leqslant t$, we get $0 < \operatorname{Re} z = \dfrac{1-|u|^2}{|1+u|^2} \leqslant \dfrac{1+t}{1-t} < 2$ and

$$|\operatorname{Im} z| = \frac{2|\operatorname{Im} u|}{|1+u|^2} \leqslant \frac{2t}{(1-t)^2} = \frac{8}{9} < 1.$$

So we can take $R_l = \Delta(1, 1/4)$, using Lemma 5.43, once more. □

We assume again that the analytic function $f : \mathbb{C}_+ \to \mathbb{C}_+$ satisfies $|f(1)| \leqslant \frac{1}{C}$, as in Step 2. Now Lemma 5.41 and (5.63) give us, for $l \in L$,

$$|f(c_l)| \geqslant C^{-1} \frac{1}{A(Q_l)} \int_{Q_l} |f| \, dA \geqslant C^{-1}. \tag{5.64}$$

So, by Lemma 5.44, for $z \in R_l$, we have $|f(z)| \geqslant \delta |f(c_l)| > \dfrac{\delta C^{-1}}{2} =: \delta_1$. This gives (note that the R_l are disjoint since $R_l \subset Q_l$)

$$\begin{aligned}\tau_\alpha(\{|f| > \delta_1\} \cap \Omega) &\geqslant \tau_\alpha\Big(\bigsqcup_{l\in L} R_l\Big) = \sum_{l\in L} \tau_\alpha(R_l) \geqslant \delta \sum_{l\in L} \tau_\alpha(Q_l) \\ &= \delta\, \tau_\alpha\Big(\bigsqcup_{l\in L} Q_l\Big) = \delta\, \tau_\alpha(\{\tilde{M} f > 1\})\end{aligned}$$

proving that $\delta\, \tau_\alpha(\{\tilde{M} f > 1\}) \leqslant \tau_\alpha(\{|f| > \delta_1\} \cap \Omega)$.

Combining this inequality and (5.53), we get, for $\lambda > 1$,

$$\tau_\alpha(\{|f| > \lambda\} \cap \Omega) \leqslant \frac{C_1}{\lambda^{\alpha+2}} \tau_\alpha(\{|f| > \delta_1\} \cap \Omega).$$

Applying this inequality, but replacing f by $\delta_1 f$ and λ by $\delta_1 \lambda$, this leads to (5.54) for $\lambda > 1/\delta_1$, and when f satisfies $|f(1)| \leqslant \frac{1}{\delta_1 C}\cdot$

Step 4

Finally, we wish to finish the proof of Theorem 5.38, so we must explain how (5.54) implies (5.51).

Consider c_1 given by (5.54); we define $c_0 = \min\left\{c_1 \tanh\left(\frac{\pi}{2}\right); \tanh(\pi)\right\}$.

Let $g : \mathbb{D} \to \mathbb{C}_+$ be an analytic function, with $|g(0)| \leqslant c_0$. Applying the Schwarz lemma to the symbol φ defined by (5.56), we point out that

$$|g(z)| \leqslant |g(0)| \frac{1+|z|}{1-|z|} \tag{5.65}$$

for every $z \in \mathbb{D}$, so that

$$|z| \leqslant \mathrm{e}^{-2\pi} \text{ and } |g(0)| \leqslant c_0 \quad \Longrightarrow \quad |g(z)| \leqslant \frac{c_0}{\tanh(\pi)} \leqslant 1. \tag{5.66}$$

On the other hand, denoting U the annulus $U = \{z \in \mathbb{D} : |z| \geqslant \mathrm{e}^{-2\pi}\}$, let $\chi : \mathbb{C}_+ \to \mathbb{D}$ be the exponential function $\chi(z) = \mathrm{e}^{-\pi z}$. Its restriction to Ω defines a bijection from Ω onto U whose inverse $L : U \to \Omega$ is well defined.

We can define the analytic function $f = g \circ \chi : \mathbb{C}_+ \to \mathbb{C}_+$ and we can write $g(z) = f(L(z))$ when $z \in U$. Moreover, we have $|f(1)| = |g(\mathrm{e}^{-\pi})| \leqslant c_1$ thanks to (5.65).

Now we set $\nu_\alpha = L(\sigma_\alpha) = (\sigma_\alpha)_L$ and we observe that for a Borel set $A \subset \Omega$, we have

$$\nu_\alpha(A) \approx \tau_\alpha(A) \tag{5.67}$$

since, by the change of variable $w = L(z)$, we easily see that ν_α is the measure σ with weight $\pi^2(\alpha+1)\mathrm{e}^{-\pi \operatorname{Re}(w)}\left|1-\mathrm{e}^{-2\pi \operatorname{Re}(w)}\right|^\alpha \approx (\operatorname{Re}(w))^\alpha$ for $w \in \Omega$.

Thanks to (5.66), we have $\{|g| > \lambda\} = \{|g| > \lambda\} \cap U$, for any $\lambda \geqslant 1$, and this leads to

$$\sigma_\alpha(\{|g| > \lambda\}) = \sigma_\alpha(\{|g| > \lambda\} \cap U) = \nu_\alpha(\{|f| > \lambda\} \cap \Omega) \approx \tau_\alpha(\{|f| > \lambda\} \cap \Omega).$$

It suffices now to slot this comparison once for $\lambda \geqslant \lambda_1$ and once for $\lambda = 1$ in (5.54) and we get (5.51).

5.6 How to Produce Pullback Measures

A natural question is: which probability measures μ on $\overline{\mathbb{D}}$ can be written as $\mu = m_\varphi$ for some symbol φ? The question seems very difficult in full generality, but a partial answer, yet highly non-trivial, is provided by the following theorem [8, theorem 1.1], in which S_0 denotes the set of symbols vanishing at the origin:

Theorem 5.45 (Bishop) *Let $\varphi \in S_0$. Then, the following are equivalent:*

(i) *The powers φ^n, $n = 0, 1, \ldots$ are orthogonal in H^2.*
(ii) *The measure $\mu = m_\varphi$ is a* radial *probability measure satisfying*

$$\int_{\overline{\mathbb{D}}} \log(1/|z|)\, d\mu(z) < \infty. \tag{5.68}$$

Moreover, any radial *probability measure* μ *on* $\overline{\mathbb{D}}$ *satisfying (5.68) is of the form* $\mu = m_\varphi$ *for some symbol* $\varphi \in S_0$ *with orthogonal powers.*

We do not provide a proof here and we refer to [8].

Symbols $\varphi \in S_0$ with orthogonal powers are called *Rudin functions*, an obvious example being that of inner functions vanishing at the origin. Rudin conjectured that there are no others, but Theorem 5.45 disproves this conjecture. Just take for μ the Lebesgue measure on the union of the two circles $\frac{1}{2}\mathbb{T}$ and $\mathbb{T}$, normalized to give each mass $\frac{1}{2}$. It is clearly radial and satisfies (5.68), hence is of the form m_φ for some $\varphi \in S_0$, but this mysterious φ is certainly not inner. Note in passing that m_φ is very far from determining φ: for example, all inner functions $\varphi \in S_0$ satisfy $m_\varphi = m$.

Rudin functions will reappear in Appendix E in a counterexample about composition operators on the bidisk.

6

Compactness via Nevanlinna Counting Functions and Essential Norms

In this chapter, we restrict our attention to the Hilbertian framework and get some estimates of the essential norms involving the Nevanlinna counting functions. In particular we get (or recover) characterizations of the compactness of composition operators.

We shall often denote by $\|C_\varphi\|_{e,X}$ the essential norm of the operator C_φ acting on the space X, and we mostly write $\overline{\lim}$ instead of $\limsup$ for short.

6.1 The Case of Hardy Spaces

In this section, we investigate a characterization of compact composition operators on Hardy spaces: this will be done here via the Nevanlinna counting function. Moreover, the method gives the exact value of the essential norm. This is a very beautiful result obtained by J. Shapiro in 1987 [164]. Its proof also points out that the normalized reproducing kernel plays a key role and that the estimation of this essential norm, as well as the characterization of the compactness, satisfies what is called *RKT* (the Reproducing Kernel Thesis), also later emphasized by Akeroyd in [1].

Concretely, we have the following statement in the Hilbertian framework:

Theorem 6.1 (Shapiro) *Let φ be a symbol. Then, on the Hardy space H^2, we have*

$$\|C_\varphi\|_{e,H^2} = \overline{\lim}_{|w|\to 1^-} \left(\frac{N_\varphi(w)}{\log(1/|w|)}\right)^{\frac{1}{2}} = \overline{\lim}_{|w|\to 1^-} \|C_\varphi(k_w)\|_{H^2}, \tag{6.1}$$

where $k_w(z) = \dfrac{(1-|w|^2)^{\frac{1}{2}}}{1-\bar{w}z}$ *is the normalized reproducing kernel on* H^2.

Of course, in (6.1), we could use the expression $\dfrac{N_\varphi(w)}{1-|w|}$ instead of $\dfrac{N_\varphi(w)}{\log(1/|w|)}$.

Let us mention the following interesting consequence:

Corollary 6.2 *Let φ be a univalent symbol. Then, on the Hardy space H^2, we have*

$$\|C_\varphi\|_{e,H^2} = \overline{\lim_{|z|\to 1^-}} \left(\frac{1-|z|}{1-|\varphi(z)|}\right)^{\frac{1}{2}}.$$

Proof: First keep in mind (1.16). This corollary is a consequence of (6.1), thanks to the following observation: for every $z \in \mathbb{D}$, since φ is univalent, z is the only $a \in \mathbb{D}$ such that $\varphi(a) = \varphi(z)$; hence

$$N_\varphi(\varphi(z)) = \log(1/|z|) \sim 1 - |z|\,. \qquad \square$$

For the proof of Shapiro's theorem, we need several useful lemmas. The first one is a general result (even though we only use the easier inequality (6.2)).

Lemma 6.3 *Let T be a bounded operator on a Hilbert space H and (R_n) a sequence of normalized self-adjoint operators satisfying $R_n(h) \to 0$ for every $h \in H$ and $I - R_n$ is compact. Then, we have*

$$\|T\|_e = \lim \|TR_n\|\,.$$

In particular, the lemma states that the sequence $(\|TR_n\|)$ converges.

Proof: We shall proceed with an upper and a lower estimation.

The upper estimate is obvious: since $I - R_n$ is compact, $T(I - R_n)$ is compact as well. In other words, T and TR_n have the same essential norm, so

$$\|T\|_e = \|TR_n\|_e \leqslant \|TR_n\|\,.$$

We get

$$\|T\|_e \leqslant \underline{\lim}\, \|TR_n\|\,. \tag{6.2}$$

For the reverse inequality, we first point out that for every compact operator K on H, we have

$$\|T - K\| \geqslant \|TR_n - KR_n\| \geqslant \|TR_n\| - \|KR_n\| = \|TR_n\| - \|R_n K^*\|, \tag{6.3}$$

since $\|R_n\| = 1$ and R_n is self-adjoint.

But $\|R_n K^*\| = \sup_{x\in Q} \|R_n(x)\|$ where $Q = \overline{K^*(B_H)}$ is a compact subset of H since K^* is a compact operator thanks to Schauder's theorem. Since

R_n converges pointwise to 0 and lies in the unit ball of $B(H)$, R_n converges uniformly to 0 on Q. Therefore $\|R_n K^*\| \to 0$ and we get from (6.3)

$$\|T - K\| \geqslant \overline{\lim}\, \|TR_n\| \tag{6.4}$$

for every compact operator K on H, hence the lower estimate.

Finally, from (6.2) and (6.4), we get the conclusion. □

The second lemma provides a sharper estimate of the point evaluation on H^2 when we have information on the valuation of the function.

Lemma 6.4 *Let $d \geqslant 1$ be an integer and $z \in \mathbb{D}$. Let f be a function in H^2 with valuation at least d, that is, f has a zero at 0 with order at least d. Then*

(i) $\left|f(z)\right| \leqslant \dfrac{|z|^d}{\sqrt{1-|z|^2}} \|f\|_2$.

(ii) $\left|f'(z)\right| \leqslant \dfrac{|z|^{d-1}}{(1-|z|^2)^{\frac{3}{2}}} \max\left(\sqrt{2}, d\right) \|f\|_2$.

Proof: (i) can be obtained from a simple Cauchy–Schwarz argument. Alternatively, we can write f as $z^d g$ where g is analytic on $\mathbb{D}$. With the integral definition of the H^2 norm, since $\left|z^d\right| = 1$ on $\mathbb{T}$, we have $\|f\|_2 = \|g\|_2$. Therefore,

$$\left|f(z)\right| = |z|^d \left|g(z)\right| \leqslant |z|^d \frac{\|g\|_2}{\sqrt{1-|z|^2}} = \frac{|z|^d}{\sqrt{1-|z|^2}} \|f\|_2,$$

since we know the norm of the point evaluation (see Proposition 1.76).

For (ii), we could use a similar idea and exploit also the point evaluation for the derivative (see Proposition 1.77). We rather use now a standard Cauchy–Schwarz argument:

$$f'(z) = \sum_{k=d}^{+\infty} k a_k z^{k-1} = \sum_{j=0}^{+\infty} (j+d)\, a_{j+d}\, z^{j+d-1};$$

hence, from Cauchy-Schwarz, we get

$$\left|f'(z)\right| \leqslant \left(\sum_{j=0}^{+\infty} (j+d)^2 |z|^{2(j+d-1)}\right)^{\frac{1}{2}} \|f\|_2$$

$$= |z|^{d-1} \left(\sum_{j=0}^{+\infty} (j+d)^2 |z|^{2j}\right)^{\frac{1}{2}} \|f\|_2 .$$

Let us write $r = |z|^2$ and estimate the sum in the last brackets:

$$\begin{aligned}\sum_{j=0}^{+\infty}(j+d)^2 r^j &= \sum_{j=2}^{+\infty} j(j-1)r^j + (2d+1)\sum_{j=1}^{+\infty} jr^j + d^2\sum_{j=0}^{+\infty} r^j\\ &= \frac{2r^2}{(1-r)^3} + \frac{(2d+1)r}{(1-r)^2} + \frac{d^2}{1-r}\\ &= \frac{(d-1)^2r^2 + (2d+1-2d^2)r + d^2}{(1-r)^3}\\ &\leqslant \frac{(2-d^2)r + d^2}{(1-r)^3} \leqslant \frac{\max(2,d^2)}{(1-r)^3}\,.\end{aligned}$$

Gathering everything, we get the estimate (ii). □

We can now prove the main theorem of this section.

Proof of Theorem 6.1: First let us prove that

$$\|C_\varphi\|_{e,H^2} \leqslant \sqrt{L}, \tag{6.5}$$

where $L = \overline{\lim}_{|w|\to 1^-} \dfrac{N_\varphi(w)}{\log(1/|w|)}$.

In view of Lemma 6.3, we shall actually prove that

$$\lim \|C_\varphi R_n\| \leqslant \overline{\lim}_{|w|\to 1^-} \left(\frac{N_\varphi(w)}{\log(1/|w|)}\right)^{\frac{1}{2}}$$

with suitable operators R_n.

Let us consider the partial sum operator defined by $S_n(f)(z) = \sum_{k=0}^{n} a_k z^k$ where $f(z) = \sum_{k=0}^{+\infty} a_k z^k$ belongs to H^2. This is clearly a bounded operator (with an operator norm equal to 1).

Let us summarize the properties shared by this operator for $n \geqslant 2$, and by $R_n = I - S_n$:

(i) R_n has norm 1, R_n and S_n are self-adjoint, and S_n is compact.
(ii) $S_n(f) \to f$ for every $f \in H^2$.
(iii) For every $z \in \mathbb{D}$, we have $\displaystyle\sup_{\substack{\|f\|\leqslant 1\\ f\in H^2}} \left|R_n(f)(z)\right| \leqslant \frac{|z|^n}{\sqrt{1-|z|^2}}\,.$
(iv) For every $z \in \mathbb{D}$, we have $\displaystyle\sup_{\substack{\|f\|\leqslant 1\\ f\in H^2}} \left|(R_n(f))'(z)\right| \leqslant \frac{n|z|^{n-1}}{(1-|z|^2)^{\frac{3}{2}}}\,.$

Assertion (i) is obvious: the first point comes from Bessel's inequality. The operators R_n and S_n are diagonal and self-adjoint, and S_n has rank $n + 1$. Assertion (ii) comes from Parseval. Assertions (iii) and (iv) are direct consequences of Lemma 6.4 and the fact that R_n has norm less than 1.

Now we can prove (6.5). We fix $\varepsilon > 0$.

Thanks to the Littlewood–Paley formula applied to composition operators (recall (2.8)), we can write, for every $n \geqslant 2$ and every f in the unit ball of H^2,

$$\|R_n(f) \circ \varphi\|_2^2 = \left|R_n(f)(\varphi(0))\right|^2 + 2\int_{\mathbb{D}} |(R_n(f))'(w)|^2 N_\varphi(w)\, d\sigma(w)$$

and $\sup_{\substack{\|f\| \leqslant 1 \\ f \in H^2}} R_n(f)(\varphi(0)) \longrightarrow 0$, when n tends to infinity, thanks to remark (iii) above.

On the other hand there is some $r \in (0,1)$ such that for every $w \in \mathbb{D}$ with $|w| \geqslant r$, we have

$$N_\varphi(w) \leqslant (L + \varepsilon) \log(1/|w|)\,.$$

In particular,

$$\begin{aligned} &\int_{\mathbb{D}\setminus r\mathbb{D}} |(R_n(f))'(w)|^2 N_\varphi(w)\, d\sigma(w) \\ &\qquad \leqslant (L + \varepsilon) \int_{\mathbb{D}\setminus r\mathbb{D}} |(R_n(f))'(w)|^2 \log \frac{1}{|w|}\, d\sigma(w) \\ &\qquad \leqslant (L + \varepsilon) \int_{\mathbb{D}} |(R_n(f))'(w)|^2 \log \frac{1}{|w|}\, d\sigma(w)\,. \end{aligned}$$

Hence

$$\begin{aligned} 2\int_{\mathbb{D}\setminus r\mathbb{D}} |(R_n(f))'(w)|^2 N_\varphi(w)\, d\sigma(w) &\leqslant (L + \varepsilon) \|R_n(f)\|_2^2 \\ &\leqslant (L + \varepsilon) \|R_n\|^2 \\ &\leqslant L + \varepsilon\,. \end{aligned}$$

For the other part of the integral, we simply use assertion (iv) above:

$$\begin{aligned} \int_{r\mathbb{D}} |(R_n(f))'(w)|^2 N_\varphi(w)\, d\sigma(w) &\leqslant \frac{n^2 r^{2(n-1)}}{(1 - r^2)^3} \int_{\mathbb{D}} N_\varphi(w)\, d\sigma(w) \\ &\leqslant \frac{n^2 r^{2(n-1)} \|\varphi\|_2^2}{(1 - r^2)^3} \\ &\leqslant \frac{n^2 r^{2(n-1)}}{(1 - r^2)^3} \longrightarrow 0 \end{aligned}$$

when n tends to infinity.

We get $\lim \|C_\varphi R_n\| \leqslant \sqrt{L+\varepsilon}$ for any arbitrary $\varepsilon > 0$, so that (6.5) is proved.

Now we prove that

$$\|C_\varphi\|_{e,H^2} \geqslant \overline{\lim_{|a|\to 1^-}} \left\| \frac{(1-|a|^2)^{\frac{1}{2}}}{1-\bar{a}\varphi} \right\|_{H^2}. \tag{6.6}$$

This is far easier than (6.5). Indeed, observe that for every compact operator K, we have $K(k_a) \to 0$ since k_a is weakly converging to 0. This latter point comes from the fact that k_a lies in the unit ball of H^2 and is uniformly convergent to 0 on every compact subset of $\mathbb{D}$. Hence

$$\|C_\varphi - K\| \geqslant \overline{\lim_{|a|\to 1^-}} \left\| \frac{(1-|a|^2)^{\frac{1}{2}}}{1-\bar{a}\varphi} \right\|_{H^2} - \left\|K(k_a)\right\|_{H^2} = \overline{\lim_{|a|\to 1^-}} \left\| \frac{(1-|a|^2)^{\frac{1}{2}}}{1-\bar{a}\varphi} \right\|_{H^2},$$

and (6.6) is proved.

Finally we show that

$$\overline{\lim_{|a|\to 1^-}} \left\| \frac{(1-|a|^2)^{\frac{1}{2}}}{1-\bar{a}\varphi} \right\|_{H^2}^2 \geqslant \overline{\lim_{|w|\to 1^-}} \frac{N_\varphi(w)}{1-|w|}. \tag{6.7}$$

For every $a \in \mathbb{D} \setminus \{0\}$, we first point out that (recall (2.8))

$$\left\| \frac{(1-|a|^2)^{\frac{1}{2}}}{1-\bar{a}\varphi} \right\|_{H^2}^2 \geqslant 2 \int_{\mathbb{D}} |k_a'(w)|^2 N_\varphi(w)\, d\sigma(w).$$

But

$$\forall w \in \mathbb{D}, \quad \left|k_a'(w)\right|^2 = \frac{(1-|a|^2)|a|^2}{|1-\bar{a}w|^4} = c(a)\left|\varphi_a'(w)\right|^2$$

where $c(a) = \dfrac{|a|^2}{1-|a|^2}$.

Now we can proceed with a (true) change of variable: $z = \varphi_a(w)$, and we get

$$\left\| \frac{(1-|a|^2)^{\frac{1}{2}}}{1-\bar{a}\varphi} \right\|_{H^2}^2 \geqslant 2c(a) \int_{\mathbb{D}} N_\varphi\big(\varphi_a(z)\big)\, d\sigma(z).$$

Fix any $r \in (0,1)$, for $|a|$ close enough to 1; we will have $|\varphi_a(\varphi(0))| > r$ since $|\varphi_a(u)| = |\varphi_u(a)| \to 1$ for every $u \in \mathbb{D}$. We can write

$$\int_{\mathbb{D}} N_\varphi\big(\varphi_a(z)\big)\, d\sigma(z) \geqslant \int_{r\mathbb{D}} N_{\varphi_a \circ \varphi}(z)\, d\sigma(z)$$

since $\varphi(u) = \varphi_a(z)$ if and only if $\varphi_a \circ \varphi(u) = z$.

Thanks to the sub-mean value property of N_φ (see Theorem 1.38), we get

$$\left\| \frac{(1-|a|^2)^{\frac{1}{2}}}{1-\bar{a}\varphi} \right\|_{H^2}^2 \geqslant 2c(a)r^2 N_{\varphi_a \circ \varphi}(0) = \frac{2|a|^2 r^2}{1+|a|} \cdot \frac{N_\varphi(a)}{1-|a|} .$$

Taking the limsup over a, we get

$$\overline{\lim_{|a|\to 1^-}} \left\| \frac{(1-|a|^2)^{\frac{1}{2}}}{1-\bar{a}\varphi} \right\|_{H^2}^2 \geqslant r^2 \overline{\lim_{|a|\to 1^-}} \frac{N_\varphi(a)}{1-|a|} .$$

Since $r \in (0,1)$ is arbitrary, we proved (6.7), and this clearly ends the proof of Theorem 6.1. □

As an immediate corollary of Theorem 6.1, we get the characterization of the compactness of composition operators on Hardy spaces.

Corollary 6.5 *Let φ be a symbol and $p \geqslant 1$.*
The following assertions are equivalent.

(i) *C_φ is compact on H^p.*
(ii) *C_φ is compact on H^2.*
(iii) *$N_\varphi(w) = o(1-|w|)$ when $|z| \to 1^-$.*
(iv) $\displaystyle \lim_{|a|\to 1^-} \left\| \frac{(1-|a|)^2)^{\frac{1}{2}}}{1-\bar{a}\varphi} \right\|_{H^2} = 0 .$

In the case of a finitely valent symbol (i.e., V-valent for some V), assertion (iii) can be expressed in a simpler way. We recall that we say φ is V-valent for some $V \in \mathbb{N}$ when, for every $w \in \mathbb{D}$,

$$\text{card}\,\{z \in \mathbb{D} \mid \varphi(z) = w\} \leqslant V .$$

In particular, when $V = 1$, we say that φ is univalent (rather than 1-valent).

Proposition 6.6 *Let φ be a symbol. Then:*

(i) *We always have*

$$N_\varphi(z) = o\Big(\log(1/|z|)\Big) \text{ when } |z| \to 1^- \Longrightarrow \lim_{|z|\to 1^-} \frac{1-|z|}{1-|\varphi(z)|} = 0.$$

(ii) *If φ is finitely valent then*

$$\lim_{|z|\to 1^-} \frac{1-|z|}{1-|\varphi(z)|} = 0 \Longleftrightarrow N_\varphi(z) = o\Big(\log(1/|z|)\Big) \quad \text{when } |z| \to 1^- .$$

Point out that $N_\varphi(z) = o\Big(\log(1/|z|)\Big)$ is obviously equivalent to $N_\varphi(z) = o\big(1-|z|\big)$.

Proof: (i) Assume that $N_\varphi(z) = o\big(1 - |z|\big)$. In particular we can write

$$1 - |z| \leqslant N_\varphi(\varphi(z)) = o\big(1 - |\varphi(z)|\big),$$

and we thus conclude easily: when fixing $\varepsilon \in (0,1)$, there exists $\delta \in (0,1)$ such that

$$1 - |\varphi(z)| \leqslant \delta \Longrightarrow 1 - |z| \leqslant \varepsilon\big(1 - |\varphi(z)|\big).$$

Therefore, for every $z \in \mathbb{D}$ satisfying $1-|z| \leqslant \delta\varepsilon$, we have: either $1-|\varphi(z)| \leqslant \delta$ hence $1 - |z| \leqslant \varepsilon\left(1 - |\varphi(z)|\right)$; or $1 - |\varphi(z)| > \delta$ hence

$$\frac{1 - |z|}{1 - |\varphi(z)|} \leqslant \frac{1 - |z|}{\delta} \leqslant \varepsilon,$$

and thus the desired conclusion.

(ii) Assume that $\displaystyle\lim_{|z|\to 1^-} \frac{1 - |z|}{1 - |\varphi(z)|} = 0$. In other words, there exists a function $\varepsilon : (0,1) \to (0, +\infty)$ satisfying

$$\forall u \in \mathbb{D}, \quad \log(1/|u|) = \varepsilon(|u|)(1 - |\varphi(u)|) \quad \text{and} \quad \lim_{t\to 1^-} \varepsilon(t) = 0.$$

Since φ is V-valent for some $V \geqslant 1$, we have

$$N_\varphi(z) = \sum_{\varphi(u)=z} \log(1/|u|) \leqslant V \sup\{\varepsilon(|u|)\,;\ \varphi(u) = z\}.(1 - |z|)\ .$$

But from the definition of the function ε and (1.16), we know that

$$\sup\{\varepsilon(|u|)\,;\ \varphi(u) = z\} \longrightarrow 0 \quad \text{as } |z| \to 1^-,$$

and we get $N_\varphi(z) = o\big(1 - |z|\big)$. □

From this, we easily recover the characterization for compact composition operators on Hardy spaces in the finitely valent case (see Theorem 3.15). We recall the statement for convenience.

Corollary 6.7 *Let φ be a finitely valent symbol and $p \geqslant 1$. Then the following assertions are equivalent:*

(i) *C_φ is compact on H^p.*
(ii) *C_φ is compact on H^2.*
(iii) $\displaystyle\lim_{|a|\to 1^-} \frac{1 - |a|}{1 - |\varphi(a)|} = 0.$

6.2 Characterization via Aleksandrov–Clark's Measures

We come back to the topic of Aleksandrov–Clark's measures $(\mu_\theta)_{\theta\in\mathbb{T}}$ introduced in Section 2.5.1. We mention here a result due to Cima–Matheson [29] giving both a characterization (in terms of Aleksandrov–Clark's measures) of the compactness of C_φ on H^2 (already obtained by Sarason) and the exact value of the distance of C_φ to compact operators, relying on Shapiro's theorem.

Theorem 6.8 (Cima–Matheson, Sarason) *Let φ be a symbol. For every $\theta \in \mathbb{T}$, write $\mu_\theta = h_\theta\, dm + \nu_\theta$ the Lebesgue decomposition so that $\nu_\theta \perp m$ is the singular part of μ_θ.*

Then

$$\|C_\varphi\|_{e,H^2} = \Big(\sup_{\theta\in\mathbb{T}} \nu_\theta(\mathbb{T})\Big)^{\frac{1}{2}}. \tag{6.8}$$

In particular, C_φ is compact if and only if μ_θ is absolutely continuous (with respect to m) for every $\theta \in \mathbb{T}$.

Proof: Below we identifiy φ and φ^*.

Let $\theta \in \mathbb{T}$. We know (see [153] for instance) that, for *a.e.* $\xi \in \mathbb{T}$,

$$h_\theta(\xi) = \lim_{r\to 1^-} \int_{\mathbb{T}} P_{r\xi}(z)\, d\mu_\theta(z),$$

where $P_{r\xi}$ is the Poisson kernel at point $r\xi$.

Therefore, from (2.24), we get for a.e. $\xi \in \mathbb{T}$

$$h_\theta(\xi) = \lim_{r\to 1^-} \operatorname{Re}\left(\frac{\theta+\varphi(r\xi)}{\theta-\varphi(r\xi)}\right) = \frac{1-|\varphi(\xi)|^2}{|1-\overline{\theta}\varphi(\xi)|^2}\cdot$$

Now, we can rewrite (2.26) as

$$\operatorname{Re}\left(\frac{\theta+\varphi(0)}{\theta-\varphi(0)}\right) = \mu_\theta(\mathbb{T}) = \|\mu_\theta\| = \|\nu_\theta\| + \|h_\theta\|_1 = \nu_\theta(\mathbb{T}) + \|h_\theta\|_1 .$$

On the other hand, we know from Shapiro's result (see Theorem 6.1) that

$$\big\|C_\varphi\big\|_{e,H^2} = \overline{\lim_{|w|\to 1^-}} \big\|C_\varphi(k_w)\big\|_{H^2} \geqslant \sup_{\theta\in\mathbb{T}} \overline{\lim_{r\to 1^-}} \big\|C_\varphi(k_{r\theta})\big\|_{H^2}, \tag{6.9}$$

where $k_w(z) = \dfrac{(1-|w|^2)^{\frac{1}{2}}}{1-\bar{w}z}$ is the normalized reproducing kernel on H^2, but

$$\begin{aligned}\|C_\varphi(k_{r\theta})\|_{H^2}^2 &= \int_{\mathbb{T}} \frac{1-r^2}{|1-r\bar\theta\varphi(z)|^2}\,dm(z)\\ &= \int_{\mathbb{T}} \frac{1-|r\varphi(z)|^2}{|1-r\bar\theta\varphi(z)|^2}\,dm(z) - r^2\int_{\mathbb{T}} \frac{1-|\varphi(z)|^2}{|1-r\bar\theta\varphi(z)|^2}\,dm(z)\\ &= \mathrm{Re}\left(\frac{1+r\bar\theta\varphi(0)}{1-r\bar\theta\varphi(0)}\right) - r^2\int_{\mathbb{T}} \frac{1-|\varphi(z)|^2}{|1-r\bar\theta\varphi(z)|^2}\,dm(z).\end{aligned}$$

Clearly, for every $\theta \in \mathbb{T}$, we have

$$\lim_{w\to\theta} \mathrm{Re}\left(\frac{1+\bar w\varphi(0)}{1-\bar w\varphi(0)}\right) = \mathrm{Re}\left(\frac{\theta+\varphi(0)}{\theta-\varphi(0)}\right).$$

It is easy to see that, for every $\lambda \in \overline{\mathbb{D}}\setminus\{1\}$, the map $r \in [0,1] \longmapsto \dfrac{r^2}{|1-r\lambda|^2}$ is increasing, and by the monotone convergence theorem, we get from the preceding estimates and (6.9)

$$\|C_\varphi\|_{e,H^2}^2 \geqslant \sup_{\theta\in\mathbb{T}}\left[\mathrm{Re}\left(\frac{\theta+\varphi(0)}{\theta-\varphi(0)}\right) - \|h_\theta\|_1\right] = \sup_{\theta\in\mathbb{T}} \nu_\theta(\mathbb{T}).$$

In other words, we get immediately the lower estimate in (6.8).

Actually we can see that this is an equality. Indeed, from (6.9), there exists sequences $(r_j)_j$ in $(0,1)$ and $(\theta_j)_j$ in $\mathbb{T}$, with $\lim r_j = 1$ and $\lim \theta_j = \theta' \in \mathbb{T}$, such that

$$\|C_\varphi\|_{e,H^2} = \lim_j \|C_\varphi(k_{r_j\theta_j})\|_{H^2}.$$

The preceding remark on monotony shows that the functions

$$r \in [0,1] \longmapsto g_j(r) = r^2\int_{\mathbb{T}} \frac{1-|\varphi|^2}{|1-r\overline{\theta_j}\varphi|^2}\,dm$$

are continuous and non-decreasing on $[0,1]$, and pointwise converging to the continuous function $r \in [0,1] \mapsto \displaystyle\int_{\mathbb{T}} \frac{1-|\varphi|^2}{|1-r\overline{\theta'}\varphi|^2}\,dm$.

From Dini's (second) theorem, we have uniform convergence and we get

$$\|C_\varphi\|_{e,H^2} = \left[\mathrm{Re}\left(\frac{\theta'+\varphi(0)}{\theta'-\varphi(0)}\right) - \|h_{\theta'}\|_1\right] = \nu_{\theta'}(\mathbb{T}) \leqslant \sup_{\theta\in\mathbb{T}} \nu_\theta(\mathbb{T}).$$

□

The original proof does not invoke Dini's theorem; nevertheless, this is exactly the same idea.

6.3 The Case of (Weighted) Bergman Spaces

In this section, we treat the case of Bergman spaces, in their classical and weighted versions. We already characterized the compactness of composition operators on $\mathscr{B}_\alpha^p$ in Theorem 3.19. Here we shall give some estimates for the essential norm of a composition operator in the framework of Hilbertian Bergman spaces with radial weights. In particular, we get some estimates for the essential norm for classical Bergman spaces $\mathscr{B}^2$ and $\mathscr{B}_\alpha^2$ and recover the characterization of compactness as well.

The methods of this section are inspired from [82].

Before stating the main results of this section, let us recall some facts proved in the first chapter. In the sequel, all our radial weights ω are normalized:

$$\int_{\mathbb{D}} \omega \, d\sigma = 1.$$

The exact Littlewood–Paley formula for $\mathscr{B}_\omega^2$ (see (1.71) and the remarks after) involves another weight function defined by

$$\widetilde{\omega}(z) = 4 \int_{|z|}^{1} x\, \omega(x) \log\left(\frac{x}{|z|}\right) dx.$$

We recall that $\widetilde{\omega}(z) \to 0$ when $|z| \to 1^-$.

In the spirit of this chapter and the work done in the framework of Hardy spaces, we shall use a weighted version of the Nevanlinna counting function: given any function $\Omega : (0,1) \to [0, +\infty)$ and a symbol φ, we can define

$$N_{\varphi,\Omega}(w) = \begin{cases} \sum\limits_{\varphi(z)=w} \Omega(|z|) & \text{if } w \neq \varphi(0),\ w \in \varphi(\mathbb{D}), \\ +\infty & \text{if } w = \varphi(0), \\ 0 & \text{if } w \notin \varphi(\mathbb{D}). \end{cases}$$

Proposition 6.9 *Let φ be a symbol and ω be a radial weight. Then, on the weighted Bergman space $\mathscr{B}_\omega^2$, we have*

$$\|C_\varphi\|_{e,\mathscr{B}_\omega^2} \leqslant \varlimsup_{|z|\to 1^-} \left(\frac{N_{\varphi,\widetilde{\omega}}(z)}{\widetilde{\omega}(z)}\right)^{\frac{1}{2}}. \tag{6.10}$$

We need again a lemma to control the point evaluation relatively to the valuation.

Lemma 6.10 *Let ω be a normalized weight. There exist positive increasing functions K_0 and K_1 on $(0,1)$ with the following properties: for every integer*

$d \geqslant 1$, every $z \in \mathbb{D}$, every function $f \in \mathscr{B}^2_\omega$ such that f has a zero at 0 with order at least d, we have

(i) $|f(z)| \leqslant K_0(|z|)\,|z|^{\frac{d}{2}}\,\|f\|_{\mathscr{B}^2_\omega}$.

(ii) $|f'(z)| \leqslant K_1(|z|)\left(\frac{1+|z|}{2}\right)^{\frac{d}{2}}\|f\|_{\mathscr{B}^2_\omega}$.

Proof: We first recall that $f_r(z) = f(rz)$ for $z \in \mathbb{D}$ and $r \in (0, 1)$.

We are going to prove (i) using the corresponding result for H^2. Indeed, write $z \in \mathbb{D}$ as $\xi|z|$ where $\xi \in \mathbb{T}$. From Lemma 6.4, we know that

$$|f(z)|^2 = \left|f\left(\xi\sqrt{|z|}.\sqrt{|z|}\right)\right|^2 \leqslant \left\|f_{\sqrt{|z|}}\right\|^2_{H^2}\frac{|z|^d}{1-|z|}.$$

But since $r \mapsto \left\|f_r\right\|_{H^2}$ is non-decreasing, we get for every $r \in (\sqrt{|z|}, 1)$

$$|f(z)|^2 \leqslant \|f_r\|^2_{H^2}\frac{|z|^d}{1-|z|},$$

and integrating over $r \in (\sqrt{|z|}, 1)$, we obtain

$$\begin{aligned}|f(z)|^2\int_{\sqrt{|z|}}^1 2r\omega(r)\,dr &\leqslant \frac{|z|^d}{1-|z|}\int_{|u|\geqslant\sqrt{|z|}}|f(u)|^2\omega(|u|)\,d\sigma(u)\\ &\leqslant \frac{|z|^d}{1-|z|}\left\|f\right\|^2_{\mathscr{B}^2_\omega}.\end{aligned}$$

We get

$$|f(z)| \leqslant \frac{|z|^{\frac{d}{2}}}{\sqrt{1-|z|}}\left(\int_{\sqrt{|z|}}^1 2r\omega(r)\,dr\right)^{-\frac{1}{2}}\left\|f\right\|_{\mathscr{B}^2_\omega} \tag{6.11}$$

and, since $t \in [0, 1) \mapsto \int_t^1 2r\omega(r)\,dr$ is non-increasing and positive, (i) is proved with $t \in [0,1) \mapsto K_0(t) = \dfrac{1}{\sqrt{1-t}}\left(\int_{\sqrt{t}}^1 2r\omega(r)\,dr\right)^{-\frac{1}{2}}$.

Now, we could again use the same idea as in (ii) of Lemma 6.4, but we shall obtain the result directly from (i) here via the Cauchy integral formula:

$$\begin{aligned}f'(z) &= \frac{1}{2i\pi}\int_{C(z,(1-|z|)/2)}\frac{f(u)}{(u-z)^2}\,du\\ &= \frac{2}{1-|z|}\int_0^1 f\left(z+\frac{1-|z|}{2}\mathrm{e}^{2i\pi\theta}\right)\mathrm{e}^{-2i\pi\theta}\,d\theta\end{aligned}$$

for every $z \in \mathbb{D}$.

From (i), we get

$$|f'(z)| \leqslant \frac{2}{1-|z|} \int_0^1 K_0\left(\left|z+\frac{1-|z|}{2}\mathrm{e}^{2i\pi\theta}\right|\right) \left|z+\frac{1-|z|}{2}\mathrm{e}^{2i\pi\theta}\right|^{\frac{d}{2}} \|f\|_{\mathscr{B}^2_\omega}\, d\theta.$$

But, since K_0 is non-decreasing, we have

$$K_0\left(\left|z+\frac{1-|z|}{2}\mathrm{e}^{2i\pi\theta}\right|\right) \leqslant K_0\big((1+|z|)/2\big),$$

so

$$|f'(z)| \leqslant K_1(|z|)\left(\frac{1+|z|}{2}\right)^{\frac{d}{2}} \|f\|_{\mathscr{B}^2_\omega}$$

with $K_1(t) = \dfrac{2}{1-t} K_0\big((1+t)/2\big)$. □

We can now prove the upper estimate for the essential norm.

Proof of Proposition 6.9: The strategy is similar to the one used in the proof of Theorem 6.1.

We denote $L = \overline{\lim\limits_{|w|\to 1^-}} \dfrac{N_{\varphi,\widetilde{\omega}}(z)}{\widetilde{\omega}(z)}$.

Thanks to Lemma 6.3, we shall actually prove that

$$\lim \|C_\varphi R_n\| \leqslant \sqrt{L}, \tag{6.12}$$

where $R_n = Id - S_n$ and S_n is the partial sum operator defined by $S_n(f)(z) = \sum_{k=0}^{n} a_k z^k$ where $f(z) = \sum_{k=0}^{+\infty} a_k z^k$ belongs to $\mathscr{B}^2_\omega$. The operator R_n is clearly bounded and satisfies the following properties:

- R_n has norm 1, R_n and S_n are self-adjoint, and S_n is compact.
- $R_n(f) \to 0$, for every $f \in \mathscr{B}^2_\omega$.

Moreover, from Lemma 6.10, we know that

(i) $\forall z \in \mathbb{D}$, $\sup\limits_{\substack{\|f\|\leqslant 1\\ f\in\mathscr{B}^2_\omega}} \big|R_n(f)(z)\big| \longrightarrow 0$, as $n \to +\infty$.

(ii) $\forall r \in (0,1)$, $\sup\limits_{\substack{\|f\|\leqslant 1\\ f\in\mathscr{B}^2_\omega}} \sup\limits_{|z|\leqslant r} \big|(R_n(f))'(z)\big| \longrightarrow 0$, as $n \to +\infty$.

Now we can prove (6.12). We fix $\varepsilon > 0$.

Thanks to the Littlewood–Paley formula (1.71), we can write for every $n \geqslant 2$ and every f in the unit ball of $\mathscr{B}^2_\omega$

$$\|R_n(f)\circ\varphi\|^2_{\mathscr{B}^2_\omega} = \big|R_n(f)(\varphi(0))\big|^2 + \int_{\mathbb{D}} |(R_n(f))'(z)|^2 N_{\varphi,\widetilde{\omega}}(z)\, d\sigma(z).$$

By definition of L, there is some $r \in (0,1)$ such that for every $z \in \mathbb{D}$ with $|z| \geqslant r$, we have

$$N_{\varphi,\widetilde{\omega}}(z) \leqslant (L+\varepsilon)\,\widetilde{\omega}(z)\,.$$

In particular,

$$\begin{aligned}\int_{\mathbb{D}\setminus r\mathbb{D}} |(R_n(f))'(z)|^2 N_{\varphi,\widetilde{\omega}}(z)\,d\sigma(z) &\leqslant (L+\varepsilon)\int_{\mathbb{D}\setminus r\mathbb{D}} |(R_n(f))'(z)|^2\,\widetilde{\omega}(z)\,d\sigma(z)\\ &\leqslant (L+\varepsilon).\end{aligned}$$

From assertion (i) above, we have $\sup\limits_{\substack{\|f\|\leqslant 1\\ f\in\mathscr{B}^2_\omega}} \big|R_n(f)(\varphi(0))\big| \longrightarrow 0$, and from assertion (ii) above, we have

$$\begin{aligned}\int_{r\mathbb{D}} |(R_n(f))'(z)|^2 N_{\varphi,\widetilde{\omega}}(z)\,d\sigma(z) &\leqslant \sup_{\substack{\|f\|\leqslant 1\\ f\in\mathscr{B}^2_\omega}}\ \sup_{|z|\leqslant r}\big|(R_n(f))'(z)\big|\int_{\mathbb{D}} N_{\varphi,\widetilde{\omega}}(z)\,d\sigma(z)\\ &\leqslant \sup_{\substack{\|f\|\leqslant 1\\ f\in\mathscr{B}^2_\omega}}\ \sup_{|z|\leqslant r}\big|(R_n(f))'(z)\big|.\|\varphi\|^2_{\mathscr{B}^2_\omega} \longrightarrow 0\end{aligned}$$

when n tends to infinity.

We get $\lim \|C_\varphi R_n\| \leqslant \sqrt{L+\varepsilon}$ for any $\varepsilon > 0$, so that (6.12) is proved. □

We were not able to prove the equality in (6.10) in full generality, but we shall get it in some particular cases (see Theorem 6.14).

We shall get various consequences of Proposition 6.9. Let us mention that, in general for every $a \in \mathbb{D}$, we have $N_{\varphi,\widetilde{\omega}}(\varphi(a)) \geqslant \widetilde{\omega}(a)$; therefore

$$\overline{\lim_{|z|\to 1^-}}\ \frac{N_{\varphi,\widetilde{\omega}}(z)}{\widetilde{\omega}(z)} \geqslant \overline{\lim_{|a|\to 1^-}}\ \frac{\widetilde{\omega}(a)}{\widetilde{\omega}(\varphi(a))}\cdot \tag{6.13}$$

The previous results emphasize the role of the Nevanlinna counting functions.

The following lemma compares the classical Nevanlinna counting function and its weighted version associated to the weight $\widetilde{\omega}$. We need to introduce, for $z \in \varphi(\mathbb{D})$, the parameter

$$q_z = \frac{1}{1-|z|}\sup\big\{1-|u|\,;\varphi(u)=z\big\}\,.$$

Point out that $0 < q_z(1-|z|) \leqslant 1$.

Lemma 6.11 *Let $z \in \varphi(\mathbb{D}) \setminus \{\varphi(0)\}$. We have*

(i) $N_{\varphi,\widetilde{\omega}}(z) \leqslant \dfrac{\widetilde{\omega}(1-q_z(1-|z|))}{q_z(1-|z|)}\,N_\varphi(z)\,.$

(ii) *If* $q_z \leqslant 1$ *then*

$$\frac{N_{\varphi,\widetilde{\omega}}(z)}{\widetilde{\omega}(z)} \leqslant \frac{N_\varphi(z)}{1-|z|} \,. \tag{6.14}$$

Proof: The key point is that $\widetilde{\omega}$ is convex on $(0,1]$ and vanishes at 1. Therefore, comparing the slopes, we get for every $u, v \in \mathbb{D}$ with $0 < |v| \leqslant |u|$,

$$\widetilde{\omega}(u) \leqslant \frac{1-|u|}{1-|v|}\,\widetilde{\omega}(v)\,.$$

Applying this with $u \in \mathbb{D}$ such that $\varphi(u) = z$ and $v = 1 - q_z(1-|z|)$, we get

$$\begin{aligned} N_{\varphi,\widetilde{\omega}}(z) &\leqslant \frac{\widetilde{\omega}\big(1-q_z(1-|z|)\big)}{q_z(1-|z|)} \sum_{\varphi(u)=z} (1-|u|) \\ &\leqslant \frac{\widetilde{\omega}(1-q_z(1-|z|))}{q_z(1-|z|)} \sum_{\varphi(u)=z} \log\frac{1}{|u|} \\ &\leqslant \frac{\widetilde{\omega}(1-q_z(1-|z|))}{q_z(1-|z|)}\, N_\varphi(z). \end{aligned}$$

Now, if $q_z \leqslant 1$ then

$$\widetilde{\omega}(1-q_z(1-|z|)) = \widetilde{\omega}((1-q_z).1 + q_z.|z|)) \leqslant q_z\,\widetilde{\omega}(z),$$

thanks to the convexity again. We get from (i)

$$\frac{N_{\varphi,\widetilde{\omega}}(z)}{\widetilde{\omega}(z)} \leqslant \frac{N_\varphi(z)}{1-|z|}.$$

□

We can now state the following theorem:

Theorem 6.12 *Let* φ *be a symbol and* ω *be a radial weight. Then:*

(i) *If* $\displaystyle\varlimsup_{|u|\to 1^-} \frac{1-|u|}{1-|\varphi(u)|} < 1$, *then* $\displaystyle\varlimsup_{|z|\to 1^-} \frac{N_{\varphi,\widetilde{\omega}}(z)}{\widetilde{\omega}(z)} \leqslant \varlimsup_{|z|\to 1^-} \frac{N_\varphi(z)}{1-|z|}\,.$
In particular

$$\big\|C_\varphi\big\|_{e,\mathscr{B}^2_\omega} \leqslant \big\|C_\varphi\big\|_{e,H^2}\,. \tag{6.15}$$

(ii) *If* φ *fixes the origin, then for every* $z \in \mathbb{D}$, *we have* $\displaystyle\frac{N_{\varphi,\widetilde{\omega}}(z)}{\widetilde{\omega}(z)} \leqslant \frac{N_\varphi(z)}{1-|z|}\,.$
In particular (6.15) is satisfied.

(iii) *If* C_φ *is compact on* H^2, *then* C_φ *is compact on* $\mathscr{B}^2_\omega$.

Proof: (i) For $|z|$ close enough to 1, we have $z \neq \varphi(0)$. Moreover, either $z \notin \varphi(\mathbb{D})$ and $N_{\varphi,\widetilde{\omega}}(z) = N_\varphi(z) = 0$, or $z \in \varphi(\mathbb{D})$ and $z = \varphi(u)$ for some

$u \in \mathbb{D}$; thus we can assume that $|u|$ is close to 1 so that $1 - |u| < 1 - |\varphi(u)| = 1 - |z|$; hence $q_z \leqslant 1$. We get from (6.14)

$$\frac{N_{\varphi,\widetilde{\omega}}(z)}{\widetilde{\omega}(z)} \leqslant \frac{N_\varphi(z)}{1 - |z|},$$

which implies

$$\varlimsup_{|z|\to 1^-} \frac{N_{\varphi,\widetilde{\omega}}(z)}{\widetilde{\omega}(z)} \leqslant \varlimsup_{|z|\to 1^-} \frac{N_\varphi(z)}{1 - |z|} \cdot$$

But we know from Proposition 6.10 and Theorem 6.1 that

$$\left\| C_\varphi \right\|_{e,\mathscr{B}^2_\omega} \leqslant \varlimsup_{|z|\to 1^-} \left(\frac{N_{\varphi,\widetilde{\omega}}(z)}{\widetilde{\omega}(z)} \right)^{\frac{1}{2}} \quad \text{and} \quad \left\| C_\varphi \right\|_{e,H^2} = \varlimsup_{|w|\to 1^-} \left(\frac{N_\varphi(w)}{1 - |w|} \right)^{\frac{1}{2}}.$$

which proves (i).

Now assume that φ fixes the origin. From the Schwarz lemma, we know that $q_z \leqslant 1$ and we conclude the proof of (ii) in the same way.

Finally assume that C_φ is compact on H^2. Equivalently $\lim\limits_{|w|\to 1^-} \dfrac{N_\varphi(w)}{1 - |w|} = 0$ and this implies that $\lim\limits_{|u|\to 1^-} \dfrac{1 - |u|}{1 - |\varphi(u)|} = 0.$

In other words, we have $\lim\limits_{|z|\to 1^-} q_z = 0$. From (i) (we could also use directly (6.14)), we get $\left\| C_\varphi \right\|_{e,\mathscr{B}^2_\omega} \leqslant \left\| C_\varphi \right\|_{e,H^2} = 0.$ □

From (iii) of Theorem 6.12, a natural question could occur: do we have an equivalence between compactness of C_φ on H^2 and compactness on $\mathscr{B}^2_\omega$? Nevertheless, we already know, having in mind the case of classical weighted Bergman spaces, that it is not the case in general. The following proposition clarifies the link between the key quantities involved in the estimates of the essential norms when φ is finitely valent.

Proposition 6.13 *Let φ be a finitely valent symbol and ω be a radial weight. Then*

$$\lim_{|z|\to 1^-} \frac{1 - |z|}{1 - |\varphi(z)|} = 0 \implies N_{\varphi,\widetilde{\omega}}(z) = o\big(\widetilde{\omega}(z)\big) \text{ when } |z| \to 1^-.$$

In particular, $\lim\limits_{|z|\to 1^-} \dfrac{1 - |z|}{1 - |\varphi(z)|} = 0 \implies C_\varphi$ *is compact on* $\mathscr{B}^2_\omega$.

Proof: Assume that φ is V-valent and that $\lim\limits_{|u|\to 1^-} \dfrac{1 - |u|}{1 - |\varphi(u)|} = 0$. With the notations used for Lemma 6.11, since $\widetilde{\omega}$ is non-increasing and convex, we have

$$N_{\varphi,\widetilde{\omega}}(z) \leqslant V.\widetilde{\omega}\big(1 - q_z(1-|z|)\big) = V.\widetilde{\omega}\big((1-q_z).1 + q_z|z|\big) \leqslant V.q_z\widetilde{\omega}(z),$$

since $q_z \in (0,1)$ for $|z|$ close enough to 1. The conclusion follows.

The final statement comes from Proposition 6.9. □

In the sequel, we focus on standard weights and get sharper estimates.

In the following theorem, $N_{\varphi,\beta}$ stands for $N_{\varphi,\Omega}$ where $\Omega(z) = (1-|z|)^\beta$ and $\beta \geqslant 1$. Be aware that at some other places of the book, we write $N_{\varphi,\beta}$ for $N_{\varphi,\Omega}$ where $\Omega(z) = \big(\log(1/|z|)\big)^\beta$ and $\beta \geqslant 1$ so that $N_{\varphi,1} = N_\varphi$. Anyway these two functions are equivalent when $|z| \to 1^-$, so that the results of Theorem 6.14 are exactly the same with this alternative definition (see Chapter 7 for more comments around this fact). We use below the weight $(1-|z|)^\beta$ since it is the one appearing in the definition of the classical Bergman spaces, and we already computed the associated weight $\widetilde{\omega_\alpha}$.

Theorem 6.14 *Let φ be a symbol and $\alpha > -1$. Then, on the Bergman space $\mathscr{B}^2_\alpha$, we have*

(i) $$\overline{\lim_{|w|\to 1^-}} \left(\frac{1-|w|}{1-|\varphi(w)|}\right)^{\frac{\alpha+2}{2}} \leqslant \|C_\varphi\|_{e,\mathscr{B}^2_\alpha} \leqslant \overline{\lim_{|z|\to 1^-}} \frac{\big(N_{\varphi,\alpha+2}(z)\big)^{\frac{1}{2}}}{(1-|z|)^{\frac{\alpha+2}{2}}}\,.$$

In particular, we have the upper estimate

$$\|C_\varphi\|_{e,\mathscr{B}^2_\alpha} \leqslant \left(\overline{\lim_{|z|\to 1^-}} \frac{N_\varphi(z)}{\log(1/|z|)}\right)^{\frac{1}{2}} \left(\overline{\lim_{|w|\to 1^-}} \frac{1-|w|}{1-|\varphi(w)|}\right)^{\frac{\alpha+1}{2}}.$$

(ii) *C_φ is compact on $\mathscr{B}^2_\alpha$ if and only if* $\displaystyle\lim_{|w|\to 1^-} \frac{1-|w|}{1-|\varphi(w)|} = 0.$

(iii) *When φ is univalent, we have* $\displaystyle\|C_\varphi\|_{e,\mathscr{B}^2_\alpha} = \overline{\lim_{|w|\to 1^-}} \left(\frac{1-|w|}{1-|\varphi(w)|}\right)^{\frac{\alpha+2}{2}}.$

(iv) *We also have the lower estimate*

$$\|C_\varphi\|_{e,\mathscr{B}^2_\alpha} \geqslant \frac{1}{\sqrt{2\alpha+3}} \overline{\lim_{|z|\to 1^-}} \frac{\big(N_{\varphi,\alpha+2}(z)\big)^{\frac{1}{2}}}{(1-|z|)^{\frac{\alpha+2}{2}}}\,.$$

With (ii) we recover the result of Theorem 3.19 in the case $p = 2$.

Keep in mind that, thanks to the Littlewood inequality, $\displaystyle\overline{\lim_{|z|\to 1^-}} \frac{N_\varphi(z)}{\log(1/|z|)}$ is finite. The proof shows that $\displaystyle\lim_{|w|\to 1^-} \frac{1-|w|}{1-|\varphi(w)|} = 0$ if and only if $\displaystyle\lim_{|z|\to 1^-} \frac{N_{\varphi,\alpha+2}(z)}{(1-|z|)^{\frac{\alpha+2}{2}}} = 0$ (see (6.16), also (6.13)). This can also be seen as a consequence of (i) with the lower estimate in (iv).

Point out, too, that the lower estimate in Shapiro's theorem is the limiting case of (iv) when $\alpha \to -1^+$.

Proof: From Proposition 6.9 and the fact that $\widetilde{\omega_\alpha}(z) \sim \frac{1}{\alpha+2}(1-|z|^2)^{\alpha+2}$ (see (1.73)), we already know that $\|C_\varphi\|_{e,\mathscr{B}^2_\alpha} \leqslant \overline{\lim}_{|z|\to 1^-} \dfrac{\big(N_{\varphi,\alpha+2}(z)\big)^{\frac{1}{2}}}{(1-|z|)^{\frac{\alpha+2}{2}}}$.

For the left-hand side of (i), we use ideas of the proof of (2.18) (see also Theorem 2.13): for every compact operator S on $\mathscr{B}^2_\alpha$, we have $\|C_\varphi - S\| = \|C^*_\varphi - S^*\|$ (with the viewpoint of the Hilbertian adjoint) and we "test" this latter norm on the normalized reproducing kernel k_w of $\mathscr{B}^2_\alpha$ at point w and we obtain, for every $w \in \mathbb{D}$,

$$\|C_\varphi - S\|_{B(\mathscr{B}^2_\alpha)} \geqslant \|C^*_\varphi(k_w) - S^*(k_w)\|_{\mathscr{B}^2_\alpha} .$$

But S^* is compact as well and k_w is bounded and converges to 0 in $\mathscr{H}(\mathbb{D})$, hence is weakly convergent to 0, when $|w| \to 1^-$, so that $S^*(k_w)$ is norm-convergent to 0. We get

$$\|C_\varphi - S\|_{B(\mathscr{B}^2_\alpha)} \geqslant \overline{\lim_{|w|\to 1^-}} \frac{\|K_{\varphi(w)}\|_{\mathscr{B}^2_\alpha}}{\|K_w\|_{\mathscr{B}^2_\alpha}},$$

where K_a is the reproducing kernel at point a, and this implies that

$$\|C_\varphi - S\|_{B(\mathscr{B}^2_\alpha)} \geqslant \overline{\lim_{|w|\to 1^-}} \left(\frac{1-|w|^2}{1-|\varphi(w)|^2}\right)^{\frac{\alpha+2}{2}} = \overline{\lim_{|w|\to 1^-}} \left(\frac{1-|w|}{1-|\varphi(w)|}\right)^{\frac{\alpha+2}{2}},$$

thanks to Proposition 1.91. We hence get the left-hand side of (i).

The particular case in (i) follows from the fact that, by definition,

$$N_{\varphi,\alpha+2}(z) \leqslant N_\varphi(z) \big(\sup\{1-|w| :\ \varphi(w) = z\}\big)^{\alpha+1};$$

hence, for every $r \in (0,1)$,

$$\sup_{|z|\geqslant r} \frac{N_{\varphi,\alpha+2}(z)}{(1-|z|)^{\alpha+2}} \leqslant \left(\sup_{|z|\geqslant r} \frac{N_\varphi(z)}{1-|z|}\right)\left(\sup_{|\varphi(w)|\geqslant r} \frac{1-|w|}{1-|\varphi(w)|}\right)^{\alpha+1}, \tag{6.16}$$

and by (1.16), we get the right-hand-side term taking the limit when $r \to 1^-$.

The operator C_φ is compact on $\mathscr{B}^2_\alpha$ if and only if its essential norm is 0; therefore (ii) is clear from (i).

When φ is univalent, $\overline{\lim}_{|z|\to 1^-} \dfrac{N_\varphi(z)}{\log(1/|z|)} = \overline{\lim}_{|w|\to 1^-} \dfrac{1-|w|}{1-|\varphi(w)|}$, so the conclusion for (iii) follows again from (i).

(iv) We use the kernel $K_{z,\alpha}(w) = (1 - \bar{z}w)^{-(\alpha+2)} = K_z^{\alpha+2}$ where K_z is the reproducing kernel for the Hardy space H^2, and we have

$$\|K_{z,\alpha}\|_{\mathscr{B}^2_\alpha} = \frac{1}{\left(1 - |z|^2\right)^{\frac{\alpha+2}{2}}}.$$

Let S be any compact operator on $\mathscr{B}^2_\alpha$ and we introduce the test function $g_a = \left(1 - |a|^2\right)^{\frac{\alpha+2}{2}}.K_{a,\alpha}$, which is clearly converging to 0 in $\mathscr{H}(\mathbb{D})$. In other words, g_a is the normalized reproducing kernel; it lies in the unit ball and is weakly converging to 0 in $\mathscr{B}^2_\alpha$ (it clearly converges to 0 in $\mathscr{H}(\mathbb{D})$), hence $S(g_a) \to 0$ when $|a| \to 1^-$.

We obtain

$$\begin{aligned}\left\|C_\varphi - S\right\|_{B(\mathscr{B}^2_\alpha)} &\geqslant \varlimsup_{|a|\to 1^-} \left\|\left(\frac{(1 - |a|^2)^{\frac{1}{2}}}{1 - \bar{a}\varphi}\right)^{\alpha+2}\right\|_{\mathscr{B}^2_\alpha} - \left\|S(g_a)\right\|_{\mathscr{B}^2_\alpha} \\ &= \varlimsup_{|a|\to 1^-} \left\|\left(\frac{(1 - |a|^2)^{\frac{1}{2}}}{1 - \bar{a}\varphi}\right)^{\alpha+2}\right\|_{\mathscr{B}^2_\alpha}.\end{aligned}$$

Finally we show that

$$\varlimsup_{|a|\to 1^-} \left\|\left(\frac{(1 - |a|^2)^{\frac{1}{2}}}{1 - \bar{a}\varphi}\right)^{\alpha+2}\right\|^2_{\mathscr{B}^2_\alpha} \geqslant \frac{1}{2\alpha + 3} \varlimsup_{|z|\to 1^-} \frac{N_{\varphi,\alpha+2}(z)}{(1 - |z|)^{\alpha+2}}. \tag{6.17}$$

For every $a \in \mathbb{D} \setminus \{0\}$, we first point out that

$$\begin{aligned}&\left\|\left(\frac{(1 - |a|^2)^{\frac{1}{2}}}{1 - \bar{a}\varphi}\right)^{\alpha+2}\right\|^2_{\mathscr{B}^2_\alpha} \\ &\qquad \geqslant (\alpha + 2)^2 \left(1 - |a|^2\right)^{\alpha+2} \int_{\mathbb{D}} |K_a'(w)|^2 |K_a|^{2(\alpha+1)} N_{\varphi,\widetilde{\omega}_\alpha}(w)\, d\sigma(w).\end{aligned}$$

We focus on the integral and observe that $\left(1 - |a|^2\right) K_a' = -\bar{a}\varphi_a'$ so that we can proceed with a (true) change of variable $z = \varphi_a(w)$ to get

$$\begin{aligned}&\left(1 - |a|^2\right)^2 \int_{\mathbb{D}} |K_a'(w)|^2 |K_a|^{2(\alpha+1)} N_{\varphi,\widetilde{\omega}_\alpha}(w)\, d\sigma(w) \\ &\qquad = |a|^2 \int_{\mathbb{D}} |\varphi_a'(w)|^2 |K_a \circ \varphi_a(\varphi_a(w))|^{2(\alpha+1)} N_{\varphi_a\circ\varphi,\widetilde{\omega}_\alpha}(\varphi_a(w))\, d\sigma(w)\end{aligned}$$

$$= |a|^2 \int_{\mathbb{D}} |K_a \circ \varphi_a(z)|^{2(\alpha+1)} N_{\varphi_a \circ \varphi, \widetilde{\omega}_\alpha}(z)\, d\sigma(z)$$
$$= \frac{|a|^2}{(1-|a|^2)^{2\alpha+2}} \int_{\mathbb{D}} |1 - \bar{a}z|^{2(\alpha+1)} N_{\varphi_a \circ \varphi, \widetilde{\omega}_\alpha}(z)\, d\sigma(z),$$

since $K_a \circ \varphi_a(z) = \dfrac{1-\bar{a}z}{1-|a|^2}$.

But, writing $R_a = |\varphi_a(\varphi(0))| = \left|\dfrac{a - \varphi(0)}{1-\bar{a}\varphi(0)}\right|$ (which goes to 1 when $|a| \to 1^-$), we have

$$\int_{\mathbb{D}} |1 - \bar{a}z|^{2(\alpha+1)} N_{\varphi_a \circ \varphi, \widetilde{\omega}_\alpha}(z)\, d\sigma(z)$$
$$\geqslant \int_{\mathbb{D}} (1-|z|)^{2(\alpha+1)} N_{\varphi_a \circ \varphi, \widetilde{\omega}_\alpha}(z)\, d\sigma(z)$$
$$\geqslant \int_0^{R_a} (1-\rho)^{2(\alpha+1)} \int_{\mathbb{T}} N_{\varphi_a \circ \varphi, \widetilde{\omega}_\alpha}(\rho\xi)\, dm(\xi)\, 2\rho d\rho$$
$$\geqslant \int_0^{R_a} (1-\rho)^{2(\alpha+1)} N_{\varphi_a \circ \varphi, \widetilde{\omega}_\alpha}(0)\, 2\rho d\rho$$
$$= N_{\varphi, \widetilde{\omega}_\alpha}(a) \int_0^{R_a} (1-\rho)^{2(\alpha+1)}\, 2\rho d\rho$$

using Proposition 7.9.

When $|a| \to 1^-$, we have $\displaystyle\int_0^{R_a} (1-\rho)^{2(\alpha+1)}\, 2\rho d\rho \sim \frac{1}{(\alpha+2)(2\alpha+3)}$.

We obtain

$$\left\| \left(\frac{(1-|a|^2)^{\frac{1}{2}}}{1-\bar{a}\varphi} \right)^{\alpha+2} \right\|^2_{\mathfrak{B}^2_\alpha} \geqslant \frac{(\alpha+2)\, c(a)}{2\alpha+3} \cdot \frac{N_{\varphi, \widetilde{\omega}_\alpha}(a)}{(1-|a|^2)^{\alpha+2}},$$

where $c(a) \sim 1$ when $|a| \to 1^-$.

Taking into account $\widetilde{\omega_\alpha}(z) \sim \dfrac{1}{\alpha+2}(1-|z|^2)^{\alpha+2} \sim \dfrac{2^{\alpha+2}}{\alpha+2}(1-|z|)^{\alpha+2}$, we have

$$(\alpha+2) \frac{N_{\varphi, \widetilde{\omega}_\alpha}(a)}{(1-|a|^2)^{\alpha+2}} \sim \frac{N_{\varphi, \alpha+2}(a)}{(1-|a|)^{\alpha+2}},$$

and the result follows. □

6.4 The Non-Hilbertian Case

We do not develop the case $p \neq 2$. As far as we know there is no sharp estimate. We just mention the following inequalities and leave the proofs as exercises (Exercise I.19) since this mainly uses the techniques of the present chapter.

In Hardy spaces, we have for $p \geqslant 2$

$$\left(\overline{\lim_{|z|\to 1^-}} \frac{N_\varphi(z)}{1-|z|}\right)^{\frac{1}{p}} \leqslant \|C_\varphi\|_{e,H^p} \leqslant 2^{1-\frac{2}{p}} \left(\overline{\lim_{|z|\to 1^-}} \frac{N_\varphi(z)}{1-|z|}\right)^{\frac{1}{p}}; \tag{6.18}$$

and in Bergman spaces, we have

$$\left(\frac{1}{2\alpha+3}\overline{\lim_{|z|\to 1^-}} \frac{N_{\varphi,\alpha+2}(z)}{(1-|z|)^{\alpha+2}}\right)^{\frac{1}{p}} \leqslant \|C_\varphi\|_{e,\mathfrak{B}^p_\alpha} \leqslant 2^{1-\frac{2}{p}} \left(\overline{\lim_{|z|\to 1^-}} \frac{N_{\varphi,\alpha+2}(z)}{(1-|z|)^{\alpha+2}}\right)^{\frac{1}{p}}. \tag{6.19}$$

Actually the lower estimates in (6.18) and (6.19) are valid for any $p \geqslant 1$.

7

Carleson versus Nevanlinna

In this chapter we focus on two important functions of complex analysis, which play a key role in the theory of composition operators:

- the Nevanlinna counting function N_φ introduced in Section 1.6: it played a key role in Chapter 6.
- the pullback measure associated to the symbol φ, which plays a key role in Chapter 5 to link composition operators and Carleson embeddings.

Actually they appear mainly in the study of composition operators on Hardy spaces, but we shall also investigate below what happens in the framework of (weighted) Bergman spaces, with the corresponding weighted Carleson and Nevanlinna functions.

We shall use several times in this chapter the partial Nevanlinna counting function defined on $\mathbb{D}$, for $r \in (0, 1]$, by 0 outside $\varphi(\mathbb{D})$ and

$$\forall z \in \varphi(\mathbb{D}) \setminus \{\varphi(0)\}, \qquad N_\varphi(z, r) = \sum_{\substack{\varphi(a)=z \\ |a|<r}} \log \frac{r}{|a|} \cdot \tag{7.1}$$

The function $N_\varphi(\cdot, r)$ is nothing but the usual Nevanlinna counting function for the function ϕ_r defined by $z \in \mathbb{D} \mapsto \varphi(rz)$:

$$\forall z \in \mathbb{D}, \qquad N_\varphi(z, r) = N_{\phi_r}(z)\,. \tag{7.2}$$

Point out that $N_\varphi(z, 1) = N_\varphi(z)$.

7.1 N_φ versus m_φ

In this chapter, and as in Chapter 5, the main geometrical tools are the Carleson windows. In the present section, we use the versions involving $\overline{\mathbb{D}}$, and not only $\mathbb{D}$ because m_φ involves φ^*, which is defined on $\overline{\mathbb{D}}$, up to a subset of $\mathbb{T}$

of m-measure 0. It is also linked to the fact that, in the framework of Hardy spaces (where m_φ and N_φ come into play), taking the boundary into account is important.

Therefore, we use again in the sequel of this section the Carleson square box

$$W(\xi,h) = \big\{z \in \overline{\mathbb{D}} \mid 1 - |z| < h \text{ and } |\arg(\bar{\xi} z)| \leqslant \pi h\big\}$$

and the Carleson circular window

$$S(\xi,h) = \big\{z \in \overline{\mathbb{D}};\ |z - \xi| < h\big\}.$$

Once again we recall that these are – roughly speaking – the same objects in the sense that, for every $h \in (0,1)$ and every $\xi \in \mathbb{T}$, they satisfy

$$S(\xi,h) \subset W(\xi,h) \subset S(\xi,4h).$$

Given a symbol $\varphi : \mathbb{D} \to \mathbb{D}$, we already know that m_φ is a Carleson measure for Hardy spaces, that is, satisfies the geometric condition $\sup_{\xi\in\mathbb{T}} m_\varphi\big(W(\xi,h)\big) = O(h)$ as $h \to 0^+$ and, from Littlewood's inequality, that $N_\varphi(w) = O(1 - |w|)$ as $|w| \to 1^-$.

We also know that the compactness of C_φ on H^p can be expressed indifferently by $\sup_{\xi\in\mathbb{T}} m_\varphi\big(W(\xi,h)\big) = o(h)$ as $h \to 0^+$, and by $N_\varphi(w) = o(1 - |w|)$ as $|w| \to 1^-$.

In other words, we have, with $\rho_\varphi(h) = \sup_{\xi\in\mathbb{T}} m_\varphi\big(W(\xi,h)\big)$,

$$\rho_\varphi(h) = O(h) \quad \text{and} \quad \sup_{|w|>1-h} N_\varphi(w) = O(h);$$

moreover,

$$\rho_\varphi(h) = o(h) \iff \sup_{|w|>1-h} N_\varphi(w) = o(h).$$

Therefore, a direct link between these two maximal functions should occur. We shall confirm this intuition and we actually prove a more accurate result: the equivalence concerns not just the maximal Carleson function ρ_φ and maximal Nevanlinna function but the average of those functions on Carleson windows, up to constants and a dilation of the opening.

We shall first prove the following lemma, which shows that a Carleson window centered at a point in the range of φ has positive m_φ-measure.

Lemma 7.1 *Let φ be an analytic self map of $\mathbb{D}$. For every $z \in \mathbb{D}$, one has, if $w = \varphi(z)$, $\xi = w/|w|$ and $h = 1 - |w| \leqslant 1/4$:*

$$m_\varphi\big(W(\xi,6h)\big) \geqslant m_\varphi\big(S(\xi,6h)\big) \geqslant \frac{|w|}{8}(1 - |z|) \geqslant \frac{1}{11}(1 - |z|). \tag{7.3}$$

Proof: Up to a rotation (*i.e.* considering $\bar{\xi}\varphi$), we may assume that w is a positive real number and since $h \in (0, 1/4]$, we have $w = 1 - h \in [3/4, 1)$.

We consider the analytic transformation $T : \mathbb{D} \to \mathbb{D}$ defined by

$$T(u) = \frac{au + 1}{u + a}\,, \text{ where } a = w - \frac{2}{w} < -1\,, \tag{7.4}$$

Point out that $T(w) = \dfrac{w}{2}\cdot$

Let P_z be the Poisson kernel at z, we have

$$\frac{w}{2} = \frac{\varphi(z)}{2} = T\big(\varphi(z)\big) = \int_{\mathbb{T}} (T \circ \varphi)^* P_z \, dm = \int_{\mathbb{T}} \mathrm{Re}[(T \circ \varphi)^*] P_z \, dm.$$

Writing $E = \big\{\mathrm{Re}(T \circ \varphi^*) \geqslant w/4\big\} = \big\{\mathrm{Re}[(T \circ \varphi)^*] \geqslant w/4\big\} \subset \mathbb{T}$, we have

$$\frac{w}{2} \leqslant \int_E P_z \, dm + \frac{w}{4} \int_{E^c} P_z \, dm \leqslant \int_E P_z \, dm + \frac{w}{4} \int_{\mathbb{T}} P_z \, dm = \int_E P_z \, dm + \frac{w}{4}$$

therefore:

$$\frac{w}{4} \leqslant \int_E P_z \, dm \,\cdot$$

But $\|P_z\|_\infty = \dfrac{1 + |z|}{1 - |z|} \leqslant \dfrac{2}{1 - |z|}$ and we get:

$$m(E) \geqslant \frac{w}{8}\,(1 - |z|)\,. \tag{7.5}$$

On the other hand, from (7.4) we have

$$T(u) = v \iff u = T^{-1}(v) = \frac{av - 1}{a - v}$$

hence:

$$|1 - u| = |a + 1| \frac{|1 - v|}{|a - v|} \leqslant \frac{2\,|a + 1|}{|a - v|}\,\cdot$$

When $\mathrm{Re}(v) \geqslant w/4$ we get

$$|a - v| \geqslant \mathrm{Re}(v - a) \geqslant \frac{w}{4} - a = \frac{2}{w} - \frac{3}{4} w \geqslant \frac{5}{4}\,\cdot$$

We also know that $w \geqslant 3/4$ so $|a + 1| = (1 - w)\left(\dfrac{2}{w} + 1\right) \leqslant \dfrac{11}{3}\,(1 - w)$ and we get

$$|1 - u| \leqslant \frac{88}{15}\,(1 - w) \leqslant 6h\,.$$

We conclude here that, when $T(u) = v$ with $\mathrm{Re}(v) \geqslant w/4$, then $u \in S(1, 6h)$. In other words, we have

$$\varphi^*(E) \subset T^{-1}(\{\mathrm{Re}(v) \geqslant w/4\}) \subset S(1, 6h),$$

which means that

$$E \subset (\varphi^*)^{-1}(S(1,6h)).$$

We get

$$m_\varphi\big(W(1,6h)\big) \geqslant m_\varphi\big(S(1,6h)\big) \geqslant m(E) \geqslant \frac{w}{8}\,(1-|z|),$$

thanks to (7.5); and (7.3) follows obviously. □

Majorizing the Nevanlinna Counting Function by the Carleson Function

Here our goal is to prove the following theorem:

Theorem 7.2 *For every symbol φ, every $z \in \mathbb{D}$, and every $\delta \in (0,1/2)$, we have*

$$N_\varphi(z) \leqslant \sup_{w \in W(\xi,h)} N_\varphi(w) \leqslant \frac{3(1+\delta)^4}{\delta^2}\, m_\varphi\big(W(\xi,(1+\delta)h)\big), \tag{7.6}$$

for $0 < h < (1-|\varphi(0)|)/2$, where $z = (1-h)\xi$ with $\xi \in \mathbb{T}$.

In particular, for $0 < h < (1-|\varphi(0)|)/2$,

$$\sup_{|z|>1-h} N_\varphi(z) \leqslant \frac{16}{\delta^2}\, \rho_\varphi\big((1+\delta)h\big). \tag{7.7}$$

This result (up to constants and arbitrary δ) was first proved in [99], but we present here this variant (see [49]).

In order to prove Theorem 7.2, we need the following mix of the Stanton formula (recall Theorem 1.69) and the Green formula (recall (1.59)). Once again, in the following, we consider φ as defined on $\overline{\mathbb{D}}$, via φ^*.

Lemma 7.3 *For every symbol φ and every function $G\colon \mathbb{C} \to \mathbb{R}$, which is $\mathscr{C}^2$, one has*

$$\int_{\overline{\mathbb{D}}} G\, dm_\varphi = G[\varphi(0)] + \frac{1}{2} \int_{\mathbb{D}} \Delta\, G(w)\, N_\varphi(w)\, d\sigma(w),$$

where Δ is the usual Laplacian operator.

Proof: We denote $\phi_r(z) = \varphi(rz)$ for $z \in \overline{\mathbb{D}}$. Applying (1.59) to $G \circ \varphi$ we have

$$\begin{aligned}\int_{\mathbb{T}} G[\phi_r(\xi)]\, dm(\xi) &= G[\varphi(0)] + \frac{1}{2}\int_{D(0,r)} (\Delta\, G)\circ\varphi(z).|\varphi'(z)|^2.\log\left(\frac{r}{|z|}\right) d\sigma(z)\\ &= G[\varphi(0)] + \frac{1}{2}\int_{\mathbb{D}} \Delta\, G(w)\, N_{\phi_r}(w)\, d\sigma(w),\end{aligned}$$

thanks to the change of variable formula applied with φ, using (7.2).

Now letting $r \uparrow 1$, we get

$$\int_{\mathbb{T}} G[\varphi(\xi)]\, dm(\xi) = G[\varphi(0)] + \frac{1}{2}\int_{\mathbb{D}} \Delta\, G(w)\, N_\varphi(w)\, d\sigma(w).$$

On the other hand, $\displaystyle\int_{\mathbb{T}} G \circ \varphi\, dm = \int_{\overline{\mathbb{D}}} G\, dm_\varphi\,.$ □

Proof of Theorem 7.2: Let $u = (1 + qh)\xi \in \mathbb{C}$ and $r = 1 + \delta + q$ where $q = (1 + \sqrt{2})(1 + \delta)$. Point out that $r = q\sqrt{2} = (2 + \sqrt{2})(1 + \delta)$.

For $p \geqslant 2$, we introduce the function

$$G_p(w) = \big(r^2h^2 - |w - u|^2\big)^p\, \mathbb{1}_{D(u,rh)}(w)\,.$$

For $w \in D(u, rh)$, by a straightforward computation (like (1.60)), we have

$$\Delta G_p(w) = 4p\big(p|w - u|^2 - r^2h^2\big)\, G_{p-2}(w). \tag{7.8}$$

When $w \in \overline{\mathbb{D}}$, we have $\sqrt{2}|w - u| \geqslant \sqrt{2}qh = rh$ so that $\Delta G_p(w) \geqslant 0$.

When $w \in D(z, \delta h)$, we have $|w - u| \geqslant |u - z| - |z - w| \geqslant (q + 1 - \delta)h$; hence for every $w \in D(z, \delta h) \subset \overline{\mathbb{D}} \cap D(u, rh)$, we have

$$\Delta G_p(w) \geqslant 16 G_{p-2}(w)(1 - \delta)(3 + 2\sqrt{2})h^2 \geqslant 8c G_{p-2}(w) h^2.$$

where c is a numerical constant (for instance $c = 3 + 2\sqrt{2}$).

When $p > 2$, the function $G_p \in \mathscr{C}^2(\mathbb{C})$ and we apply the preceding lemma with G_p to get, since $G_p(\varphi(0)) \geqslant 0$,

$$\int_{\overline{\mathbb{D}} \cap D(u,rh)} \big(r^2h^2 - |w-u|^2\big)^p\, dm_\varphi(w) \geqslant 8ch^2 \int_{D(z,\delta h)} G_{p-2}(w)\, N_\varphi(w)\, d\sigma(w).$$

Taking the limit when $p \to 2^+$, we have

$$\int_{\overline{\mathbb{D}} \cap D(u,rh)} \big(r^2h^2 - |w - u|^2\big)^2\, dm_\varphi(w) \geqslant 8ch^2 \int_{D(z,\delta h)} N_\varphi(w)\, d\sigma(w).$$

Hence

$$r^4h^4 m_\varphi\big(\overline{\mathbb{D}} \cap D(u, rh)\big) \geqslant 8ch^2 \int_{D(z,\delta h)} N_\varphi(w)\, d\sigma(w).$$

Now, point out that $\varphi(0) \notin D(z, \delta h)$ since $|\varphi(0)| < 1 - 2h$; so

$$r^4h^4 m_\varphi\big(\overline{\mathbb{D}} \cap D(u, rh)\big) \geqslant 8ch^2 \int_{D(z,\delta h)} N_\varphi(w)\, d\sigma(w) \geqslant 8c\delta^2h^4 N_\varphi(z)$$

using the sub-mean property of the Nevanlinna function N_φ (see Theorem 1.38 (ii)). Therefore

$$N_\varphi(z) \leqslant \frac{r^4}{8c\delta^2}\, m_\varphi\big(\overline{\mathbb{D}} \cap D(u, rh)\big)\,.$$

Point out that, with c $= 3 + 2\sqrt{2}$, we have $\dfrac{r^4}{8c} = \dfrac{c}{2}(1+\delta)^4 < 3(1+\delta)^4$.

Now we claim that $\overline{\mathbb{D}} \cap D(u, rh) \subset W\big(\xi, (1+\delta)h\big)$.

Let $w \in \overline{\mathbb{D}} \cap D(u, rh)$: we have $|w| \geqslant (1+qh) - rh = 1 - (1+\delta)h$, so that the modulus constraint is fulfilled. We focus now on the argument of $\bar{\xi} w$. Up to a rotation, it suffices to treat the case $\xi = 1$; equivalently, we wish to estimate the argument of $v \in \overline{\mathbb{D}} \cap D(1+qh, rh)$. We can write $v = |v| \mathrm{e}^{i\theta}$ where $|v| < 1$ and $|\theta| \leqslant \theta_h$, with $\theta_h \in (0, \pi/2)$ satisfying $|\exp(i\theta_h) - (1+qh)| = rh = \sqrt{2} qh$, which leads to

$$\cos(\theta_h) = 1 - \frac{q^2 h^2}{2(1+qh)} \qquad \Longleftrightarrow \qquad \sin(\theta_h/2) = \frac{qh}{2\sqrt{1+qh}} \cdot$$

Therefore, since $\tan(x) \geqslant x$ on $(0, \pi/2)$,

$$\frac{\theta_h}{2}\left(1 - \frac{q^2 h^2}{4(1+qh)}\right)^{\frac{1}{2}} = \frac{\theta_h}{2} \cos(\theta_h/2) \leqslant \frac{qh}{2\sqrt{1+qh}} \leqslant \frac{qh}{2},$$

but $qh < 4$, so

$$\theta_h \leqslant \left(1 + qh - \frac{q^2 h^2}{4}\right)^{-\frac{1}{2}} qh \leqslant qh = (1+\sqrt{2})(1+\delta)h < \pi(1+\delta)h,$$

and (7.6) follows. □

Domination of the Carleson Function by the Nevanlinna Function

Having in mind Theorem 7.2, we could expect a pointwise estimate from above of the m_φ-measure of Carleson windows centered at $\xi = w/|w|$ by $N_\varphi(w)$. Nevertheless it is impossible in full generality.

Indeed, consider a conformal mapping φ from $\mathbb{D}$ onto $\mathbb{D} \setminus [0,1)$, which is a simply connected domain of the complex plane. One has $N_\varphi(w) = 0$ for every $w \in (0,1)$ since $(0,1) \cap \varphi(\mathbb{D}) = \emptyset$. But $m_\varphi\big(W(1,h)\big) > 0$ for every $h > 0$ because $W(1,h) \supset W(\mathrm{e}^{ih/2}, h/2)$ and $m_\varphi\big(W(\mathrm{e}^{ih/2}, h/2)\big) > 0$ by Lemma 7.1.

We can give another more subtle yet concrete and simple example: we consider the affine symbol $\varphi(z) = \dfrac{1+z}{2} \cdot$ We shall estimate in the following the pullback measure of Carleson windows and the Nevanlinna counting function associated to φ at some special points close to 1.

We focus on the point $\xi_0 = \mathrm{e}^{i\sqrt{h}} \in \mathbb{T}$ and $w = \xi_0 \left(1 - \dfrac{h}{2}\right)$ where $h \in (0,1)$, and we shall estimate the growth of the functions when $h \to 0^+$.

For $\xi = \mathrm{e}^{i\theta} \in \mathbb{T}$, we can write $\varphi(\xi) = \varphi^*(\xi) = (\cos\theta/2)\, \mathrm{e}^{i\theta/2}$ where $|\theta| \leqslant \pi$. Hence, writing $\theta_0 = \sqrt{h}$ for short, $\varphi(\xi) \in W(\xi_0, h/2)$ if and only if we have both $\cos(\theta/2) \geqslant 1 - h/2$ and $|(\theta/2) - \theta_0| \leqslant \pi h/2$, that is, $2\theta_0 - \pi h \leqslant$

$\theta \leqslant 2\theta_0 + \pi h$. Now, $1-\cos(\theta/2) \leqslant \theta^2/8$, so the modulus condition is satisfied as soon as $|\theta| \leqslant 2\sqrt{h}$. We actually have just proved that $m_\varphi\big(W(\xi_0, h/2)\big)$ is larger than the length of the interval

$$\left[-2\sqrt{h}, 2\sqrt{h}\right] \cap \left[2\sqrt{h} - \pi h, 2\sqrt{h} + \pi h\right] = \left[2\sqrt{h} - \pi h, 2\sqrt{h}\right],$$

which is πh. Therefore $m_\varphi\big(W(\xi_0, h/2)\big) \geqslant \pi h$, hence $m_\varphi\big(W(\xi_0, h/2)\big) \approx h$ (the reverse inequality comes from the fact that m_φ is a Carleson measure).

Let us now check whether $w = \varphi(z)$ for some $z \in \mathbb{D}$ and estimate its modulus. Since $z = \mathrm{e}^{i\sqrt{h}}(2-h) - 1$, we compute

$$|z|^2 = 1 - \frac{h^2}{6} + o(h^2).$$

This proves that for h small enough $z \in \mathbb{D}$, moreover $1 - |z| \sim \dfrac{h^2}{12}\cdot$

To sum up, we get with our choice of $w = \mathrm{e}^{i\sqrt{h}}\left(1 - \dfrac{h}{2}\right)$ that, when $h \to 0^+$,

$$m_\varphi\left(W\left(\frac{w}{|w|}, 1 - |w|\right)\right) = m_\varphi\big(W(\xi_0, h/2)\big) \approx h, \quad \text{whereas} \quad N_\varphi(w) \sim \frac{h^2}{12}\cdot$$

We conclude that one cannot hence expect to control $m_\varphi\big(W(w/|w|, 1-|w|)\big)$ by $N_\varphi(w)$ in a uniform way.

Yet we focus on a general relation, reverse to (7.6), which, by the previous remarks, cannot be pointwise.

It was proved in [99] (with large constants) and later in [49] (with better constants) that a maximal version of the reverse inequality holds. Nevertheless, it was proved in [99] a deeper result in the sense that $m_\varphi\big(S(\xi, h)\big)$ is controlled by a mean (and not an upper bound) involving the Nevanlinna function. It turns out to be very important in order to obtain a version of these inequalities for weighted area measures.

We now present this strong version (as in [99]) but with better constants.

Theorem 7.4 *Let φ be a symbol. Fix $\xi \in \mathbb{T}$ and $\delta \in \big(0, 2 - \sqrt{2}\big)$.*
Then, for $0 < h < 1 - |\varphi(0)|$, we have

$$m_\varphi\big(S(\xi, (1-\delta)h)\big) \leqslant \frac{1}{\delta^2} \sup_{w \in S(\xi, h) \cap \mathbb{D}} N_\varphi(w). \tag{7.9}$$

Actually

$$m_\varphi\big(S(\xi, (1-\delta)h)\big) \leqslant \frac{2}{\delta^2 h^2} \int_{S(\xi, h)} N_\varphi(z)\, d\sigma(z). \tag{7.10}$$

The original approach due to Lefèvre–Li–Queffélec–Rodríguez Piazza in [99] was inspired by the framework of Chapter 11, but their proof involves "large" constants. The proof presented here is mostly a variant of the version (up to constants) due to El Fallah–Kellay in [49] where they state the "maximal version" (7.9).

We can rewrite (7.10) as

$$m_\varphi\big(S(\xi,(1-\delta)h)\big) \leqslant \frac{C_\delta}{\sigma\big(S(\xi,h)\big)}\int_{S(\xi,h)} N_\varphi(z)\,d\sigma(z),$$

where $C_\delta \approx \dfrac{1}{\delta^2}$ when $\delta \to 0^+$.

Proof: The strategy is quite similar to the proof of Theorems 7.2.

We introduce the function $G(w) = \big(h^2 - |w-\xi|^2\big)^2\,\mathbb{1}_{S(\xi,h)}(w)$ for $w \in \mathbb{C}$ and we point out that $G\big(\varphi(0)\big) = 0$ since $\varphi(0) \notin S(\xi,h)$.

We also know that, for w in the interior of $S(\xi,h)$ (recall (7.8)),

$$\Delta G(w) = 8\big(2|w-\xi|^2 - h^2\big),$$

so that $\Delta G(w) \leqslant 8h^2$ when $w \in S(\xi,h)$, and $\Delta G(w) \leqslant 0$ when $w \in S(\xi,h/\sqrt{2})$.

We denote now $E_h = S(\xi,h)\setminus S(\xi,h/\sqrt{2})$ and point out that $\sigma(E_h) \leqslant \dfrac{h^2}{2}$.

For every $p>1$, $G^p \in \mathscr{C}^2(\mathbb{C})$, so that we can apply Lemma 7.3 to G^p and then, passing to the limit when $p \to 1^+$, we get

$$\begin{aligned}\int_{S(\xi,h)} \big(h^2 - |w-\xi|^2\big)^2\,dm_\varphi(w) &= \frac{1}{2}\int_{\mathbb{D}} \Delta\, G(w)\,N_\varphi(w)\,d\sigma(w)\\ &\leqslant 4h^2\int_{E_h} N_\varphi(w)\,d\sigma(w) \qquad (7.11)\\ &\leqslant 2h^4 \sup_{w\in S(\xi,h)\cap\mathbb{D}} N_\varphi(w)\,. \qquad (7.12)\end{aligned}$$

On the other hand, we obviously have

$$\int_{S(\xi,h)} \big(h^2 - |w-\xi|^2\big)^2\,dm_\varphi(w) \geqslant \int_{S(\xi,(1-\delta)h)} \big(h^2 - |w-\xi|^2\big)^2\,dm_\varphi(w)\,. \qquad (7.13)$$

But for $w \in S(\xi,(1-\delta)h)$, we have

$$\big(h^2 - |w-\xi|^2\big)^2 \geqslant h^4\big(1-(1-\delta)^2\big)^2 = h^4\delta^2(2-\delta)^2\,.$$

Therefore, from (7.12) and (7.13), we obtain

$$m_\varphi\big(S(\xi,(1-\delta)h)\big) \leqslant \frac{2}{\delta^2(2-\delta)^2}\sup_{w\in S(\xi,h)\cap\mathbb{D}} N_\varphi(w) \leqslant \frac{1}{\delta^2}\sup_{w\in S(\xi,h)\cap\mathbb{D}} N_\varphi(w),$$

which is (7.9).

On the other hand, from (7.11), we have

$$m_\varphi\big(S(\xi,(1-\delta)h)\big) \leqslant \frac{4}{\delta^2 h^2 (2-\delta)^2} \int_{E_h} N_\varphi(w)\, d\sigma(w)$$
$$\leqslant \frac{2}{\delta^2 h^2} \int_{S(\xi,h)} N_\varphi(w)\, d\sigma(w),$$

which is (7.10). □

In the following statement, we summarize how one can compare the maximal Carleson function and the maximal Nevanlinna function: this is an immediate consequence of Theorems 7.2 and 7.4. Actually we have a more accurate result.

Corollary 7.5 *Let φ be a symbol.*

For every $\delta \in (0,1/2)$, there exist some positive $c_\delta \approx \delta^2$ and $C_\delta \approx \delta^{-2}$ (when $\delta \to 0^+$), such that, for $h \in \big(0,(1-|\varphi(0)|)/2\big)$,

$$c_\delta \sup_{W(\xi,h/(1+\delta))} N_\varphi(z) \leqslant m_\varphi\big(W(\xi,h)\big) \leqslant \frac{C_\delta}{h^2} \int_{S(\xi,4h/(1-\delta))} N_\varphi(z)\, d\sigma(z).$$

In particular, there exist some constants $c, c' > 0$ such that

$$\rho_\varphi\big(ch\big) \lesssim \sup_{|w|>1-h} N_\varphi(w) \lesssim \rho_\varphi\big(c'h\big) \qquad \text{when } h \to 0^+.$$

7.2 Weighted Versions

We are interested in this section in the versions of Theorems 7.2 and 7.4, or Corollary 7.5, in the framework of weighted Bergman spaces. This will be inherited from the results of Section 7.1. We present the weighted versions in two subsections: first we treat the case of weights associated to the classical Bergman spaces. Then we adapt the computations to get some estimates for general *radial* weights, but the ideas are the same.

Once again we recall that, in the framework of Bergman spaces, functions are defined on $\mathbb{D}$ and not on $\overline{\mathbb{D}}$ in general. That is why, from now on, and yet keeping the same notations, we work with

$$W(\xi,h) = \big\{z \in \mathbb{D} \mid 1-|z| < h \text{ and } |\arg(\bar{\xi} z)| \leqslant \pi h\big\}$$

$$S(\xi,h) = \{z \in \mathbb{D};\ |z-\xi| < h\},$$

and we still have $S(\xi,h) \subset W(\xi,h) \subset S(\xi,4h)$.

Classical Weights

Let $\alpha > -1$ and $\varphi : \mathbb{D} \to \mathbb{D}$ be a symbol. Here we shall compare the σ_α-measure of Carleson's window (or box) with "the" weighted Nevanlinna counting function $N_{\varphi,\alpha+2}$. Actually, in this context, there are two natural versions of $N_{\varphi,\alpha+2}$: it can stand for $N_{\varphi,\Omega}$ where $\Omega(z) = (1 - |z|^2)^{\alpha+2}$ (as the weight in the integral defining Bergman spaces) or for $\Omega(z) = \left(\log \frac{1}{|z|}\right)^{\alpha+2}$ (in this case N_φ appears as the limiting case $N_{\varphi,1}$).

These two functions are equivalent when $|z| \to 1^-$ (it is where their specific behavior mainly comes into play), and we emphasize that for every $\beta \geqslant 1$ we have, when $|z| \to 1^-$,

$$\sum_{\substack{\varphi(a)=z \\ a\in\mathbb{D}}} \left(\log \frac{1}{|a|}\right)^\beta \sim \sum_{\substack{\varphi(a)=z \\ a\in\mathbb{D}}} (1 - |a|)^\beta \sim 2^{-\beta} \sum_{\substack{\varphi(a)=z \\ a\in\mathbb{D}}} \left(1 - |a|^2\right)^\beta . \tag{7.14}$$

Indeed, first point out that $\log \frac{1}{|a|} \sim 1 - |a| \sim \frac{1}{2}(1 - |a|^2)$ when $|a| \to 1^-$ and the conclusion then comes from (1.16).

Anyway, we have a choice to make and we choose here the second option:

$$\forall z \in \varphi(\mathbb{D}) \setminus \{\varphi(0)\}, \qquad N_{\varphi,\alpha+2}(z) = \sum_{\substack{\varphi(a)=z \\ a\in\mathbb{D}}} \left(\log \frac{1}{|a|}\right)^{\alpha+2} .$$

Let us first mention the following lemma making clear the link between the weighted Nevanlinna function and the (non-weighted) partial one.

Lemma 7.6 *Let $\alpha > -1$ and φ be a symbol. We have, for every $z \in \mathbb{D}\setminus\{\varphi(0)\}$,*

$$N_{\varphi,\alpha+2}(z) = (\alpha + 1)(\alpha + 2) \int_0^1 N_\varphi(z,r) \left(\log \frac{1}{r}\right)^\alpha \frac{dr}{r} \tag{7.15}$$

$$= (\alpha + 1)(\alpha + 2) \int_{|u_z|}^1 N_\varphi(z,r) \left(\log \frac{1}{r}\right)^\alpha \frac{dr}{r}, \tag{7.16}$$

where $u_z = \dfrac{\varphi(0) - z}{1 - \overline{\varphi(0)}\, z}\cdot$

Point out that $u_z = \varphi_{\varphi(0)}(z)$. Nevertheless we have to be very careful with the notations here: we avoid for a while the use of the notation φ_s, where $s \in \mathbb{D}$, for the Möbius function since the function $z \mapsto \varphi(rz)$ comes naturally into the story below and the notation φ_r for this function is also very common. That is why we shall rather use the notation $\phi_r(z) = \varphi(rz)$ below.

Proof: By definition, we have, using Fubini,

$$\int_0^1 N_\varphi(z,r)\left(\log\frac{1}{r}\right)^\alpha \frac{dr}{r} = \sum_{\substack{\varphi(a)=z\\ a\in\mathbb{D}}} \int_0^1 \mathbb{1}_{\{|a|<r\}} \log\frac{r}{|a|}\left(\log\frac{1}{r}\right)^\alpha \frac{dr}{r},$$

and let us focus on the integral on the right-hand side.

For $a \in \mathbb{D}$, we have

$$\begin{aligned}\int_0^1 \mathbb{1}_{|a|<r} \log\frac{r}{|a|}\left(\log\frac{1}{r}\right)^\alpha \frac{dr}{r} &= \int_{|a|}^1 \log\frac{r}{|a|}\left(\log\frac{1}{r}\right)^\alpha \frac{dr}{r}\\ &= \log\frac{1}{|a|}\int_{|a|}^1 \left(\log\frac{1}{r}\right)^\alpha \frac{dr}{r} - \int_{|a|}^1 \left(\log\frac{1}{r}\right)^{\alpha+1} \frac{dr}{r}.\end{aligned}$$

But, for $\beta > -1$, we have

$$\int_{|a|}^1 \left(\log\frac{1}{r}\right)^\beta \frac{dr}{r} = \frac{1}{\beta+1}\left(\log\frac{1}{|a|}\right)^{\beta+1},$$

and we get (7.15).

To prove (7.16), we just notice that when $r \leqslant |u_z|$, there is no $a \in \mathbb{D}$ with $|a| < r$ such that $\varphi(a) = z$. Indeed, assume that such an a exists, then we have

$$|u_z| = \left|\frac{\varphi(0)-\varphi(a)}{1-\overline{\varphi(0)}\varphi(a)}\right| = \rho\big(\varphi(0),\varphi(a)\big) \leqslant \rho\big(0,a\big) = |a| < r$$

by the Schwarz–Pick lemma. □

Before stating the main result of this subsection, let us recall that $\sigma_{\varphi,\alpha}$ is the pullback measure associated to φ and the measure σ_α.

Theorem 7.7 *Let φ be a symbol and $\alpha > -1$. Fix $\xi \in \mathbb{T}$ and $\delta \in \big(0, 2-\sqrt{2}\big)$. Then:*

(i) *Let $z = (1-h)\xi$ and $0 < h < \dfrac{1-|\varphi(0)|}{2}$.*
For some constant A_α only depending on α, we have

$$N_{\varphi,\alpha+2}(z) \leqslant \sup_{w\in W(z,h)} N_{\varphi,\alpha+2}(w) \leqslant \frac{A_\alpha}{\delta^2}\,\sigma_{\varphi,\alpha}\big(W(\xi,(1+\delta)h)\big). \tag{7.17}$$

(ii) *Let $0 < h < 1-|\varphi(0)|$. For some $B_\alpha > 0$ only depending on α, we have*

$$\begin{aligned}\sigma_{\varphi,\alpha}\big(S(\xi,(1-\delta)h)\big) &\leqslant \frac{B_\alpha}{\delta^2h^2}\int_{S(\xi,h)} N_{\varphi,\alpha+2}(w)\,d\sigma(w)\\ &\leqslant \frac{B_\alpha}{\delta^2}\sup_{w\in S(\xi,h)} N_{\varphi,\alpha+2}(w).\end{aligned} \tag{7.18}$$

As an immediate consequence, in the very particular case $\alpha = 0$, there exists some constants $c, c' > 0$ such that

$$\rho_{\varphi,2}\big(ch\big) \lesssim \sup_{|w|\geqslant 1-h} N_{\varphi,2}(w) \lesssim \rho_{\varphi,2}\big(c'h\big) \qquad \text{when } h \to 0^+ . \tag{7.19}$$

This is valid whatever the choice of the definition of $N_{\varphi,2}$ (recall (7.14)).

Let us also mention that it is easily seen in the proof below that the constants A_α and B_α are bounded when $\alpha \to -1^+$. The results of Section 7.1 appear one more time as this limiting case: the Hardy space can be seen as a degenerate version of the weighted Bergman when $\alpha \to -1^+$ even though it is required to be careful with this viewpoint.

Proof: Let $w \in W(z,h)$ and write $z = (1-h)\xi$. From (7.16), we have the inequality

$$N_{\varphi,\alpha+2}(w) \leqslant \frac{(\alpha+1)(\alpha+2)}{|u_w|^3} \int_0^1 N_\varphi(w,r) \left(\log\frac{1}{r}\right)^\alpha r^2 dr\,.$$

There exists some $C_\alpha > 0$ such that for every $r \in (0,1)$ we have

$$r(\alpha+2)\left(\log\frac{1}{r}\right)^\alpha \leqslant 2C_\alpha(1-r^2)^\alpha,$$

and we get

$$N_{\varphi,\alpha+2}(w) \leqslant \frac{(\alpha+1)C_\alpha}{|u_w|^3} \int_0^1 N_\varphi(w,r)\, 2r(1-r^2)^\alpha dr\,. \tag{7.20}$$

As already mentioned, $N_\varphi(\cdot,r)$ is the usual Nevanlinna counting function for the function $z \in \mathbb{D} \to \varphi(rz) =: \phi_r(z)$.

Applying Theorem 7.2 and (7.6) to this function, we know that (point out that $\phi_r(0) = \varphi(0)$)

$$\begin{aligned} N_\varphi(w,r) = N_{\phi_r}(w) &\leqslant \frac{3(1+\delta)^4}{\delta^2}\, m_{\phi_r}\big(W(\xi,(1+\delta)h)\big) \\ &\leqslant \frac{12}{\delta^2}\, m_{\phi_r}\big(W(\xi,(1+\delta)h)\big)\,. \end{aligned} \tag{7.21}$$

At first glance, we should be aware of the definition of the Carleson box: it is sligthly different here from the one in Section 7.1 (the difference is between taking into account the part on the circle $\mathbb{T}$ or not), but it turns out to be a "false problem" since for $r < 1$ the symbol ϕ_r cannot reach the boundary of the disk and $m_{\phi_r}(\mathbb{T}) = 0$ for $r < 1$, so that $m_{\phi_r}\big(W(\xi,(1+\delta)h)\big)$ is the same whatever the definition of the Carleson box.

On the other hand, we see that $|u_w| \geqslant \dfrac{1}{3}$ since $1-|w| \leqslant h < \dfrac{1-|\varphi(0)|}{2}$. Indeed $|w-\varphi(0)| \geqslant (1-|\varphi(0)|) - (1-|w|) \geqslant 1-|w|$, so we also have

$$|1-\overline{\varphi(0)}w| \leqslant 1-|w|^2+|w||w-\varphi(0)| \leqslant 2(1-|w|)+|w-\varphi(0)| \leqslant 3|w-\varphi(0)|\,.$$

We obtain from this remark and (7.20)

$$N_{\varphi,\alpha+2}(w) \leqslant \frac{(\alpha+1)A_\alpha}{\delta^2} \int_0^1 m_{\phi_r}\big(W(\xi,(1+\delta)h)\big)\, 2r(1-r^2)^\alpha dr,$$

where $A_\alpha > 0$.

Finally

$$\begin{aligned}
&\int_0^1 m_{\phi_r}\big(W(\xi,(1+\delta)h)\big)\, 2r(1-r^2)^\alpha dr \\
&\qquad = \int_0^1 m\big(\{u \in \mathbb{T} : \varphi(ru) \in W(\xi,(1+\delta)h)\}\big)\, 2r(1-r^2)^\alpha dr \\
&\qquad = \frac{1}{\alpha+1} \int_{\mathbb{D}} \mathbb{1}_{\{s\in\mathbb{D} :\, \varphi(s)\in W(\xi,\,(1+\delta)h)\}}\, d\sigma_\alpha \\
&\qquad = \frac{1}{\alpha+1} \sigma_{\varphi,\alpha}\big(W(\xi,(1+\delta)h)\big),
\end{aligned}$$

and we proved (7.17).

We prove (ii) following a similar strategy, but let us give some details.

We can write

$$\begin{aligned}
\sigma_{\varphi,\alpha}\big(S(\xi,(1-\delta)h)\big) &= \int_{\mathbb{D}} \mathbb{1}_{\{s\in\mathbb{D} :\, \varphi(s)\in S(\xi,\,(1-\delta)h)\}}\, d\sigma_\alpha(s) \\
&= \int_0^1 m_{\phi_r}\big(S(\xi,(1-\delta)h)\big)\, 2(\alpha+1)r(1-r^2)^\alpha dr,
\end{aligned}$$

but we know from (7.10), applied to the function $\phi_r(z) = \varphi(rz)$, that

$$m_{\phi_r}\big(S(\xi,(1-\delta)h)\big) \leqslant \frac{2}{\delta^2 h^2} \int_{S(\xi,h)} N_{\phi_r}(z)\, d\sigma(z) = \frac{2}{\delta^2 h^2} \int_{S(\xi,h)} N_\varphi(z,r)\, d\sigma(z),$$

and the same remark occurs here: $m_{\phi_r}\big(S(\xi,(1-\delta)h)\big)$ has the same value whatever the definition of the circular Carleson window since $m_{\phi_r}(\mathbb{T}) = 0$ for $r < 1$.

Thanks to Fubini, we have

$$\sigma_{\varphi,\alpha}\big(S(\xi,(1-\delta)h)\big) \leqslant \frac{2(\alpha+1)}{\delta^2 h^2} \int_{S(\xi,h)} \left(\int_0^1 N_\varphi(z,r)\, 2r(1-r^2)^\alpha dr \right) d\sigma(z)\,.$$

For some constant $B_\alpha > 0$, we have $4r^2(1-r^2)^\alpha \leqslant B_\alpha\,(\alpha+2)\left(\log\frac{1}{r}\right)^\alpha$ for every $r \in (0,1)$ and we obtain

$$\begin{aligned}
&\sigma_{\varphi,\alpha}\big(S(\xi,(1-\delta)h)\big) \\
&\qquad \leqslant \frac{(\alpha+1)(\alpha+2)B_\alpha}{\delta^2 h^2} \int_{S(\xi,h)} \left(\int_0^1 N_\varphi(z,r) \left(\log\frac{1}{r}\right)^\alpha \frac{dr}{r} \right) d\sigma(z)\,.
\end{aligned}$$

Invoking (7.15), this leads to

$$\sigma_{\varphi,\alpha}\big(S(\xi,(1-\delta)h)\big) \leqslant \frac{B_\alpha}{\delta^2 h^2} \int_{S(\xi,h)} N_{\varphi,\alpha+2}(z)\, d\sigma(z),$$

which is the main inequality in (ii).

The very last inequality is then obvious. □

Let us remark that in the proof of (i) above, we used a trick to bound $\left(\log \frac{1}{r}\right)^{\alpha}$ multiplying by some r to recover the right usual weight associated to the Bergman space. When $\alpha \leqslant 0$, it is unnecessary since we can use directly $\log(1/r) \geqslant 1 - r$. In the proof of (ii), it is unnecessary when $\alpha \geqslant 0$.

General Radial Weights

We now generalize the previous results to the case of a general radial weight. Given a (radial) weight ω, we recall (see Remark 1.94 and (1.72)) that the following associated weight comes naturally in the story through the generalized Littlewood–Paley formula (see Theorem 1.71):

$$\forall s \in (0,1), \qquad \widetilde{\omega}(s) = 4 \int_s^1 r\, \omega(r) \log\left(\frac{r}{s}\right) dr,$$

and is involved in the characterization of compactness of $C\varphi$ via the following weighted Nevanlinna counting function in Proposition 6.13:

$$N_{\varphi,\widetilde{\omega}}(z) = \begin{cases} \sum\limits_{\varphi(a)=z} \widetilde{\omega}(|a|) & \text{if } z \neq \varphi(0),\ z \in \varphi(\mathbb{D}) \\ +\infty & \text{if } z = \varphi(0) \\ 0 & \text{if } z \notin \varphi(\mathbb{D}). \end{cases}$$

Lemma 7.8 *Let φ be a symbol. We have, for every $z \in \mathbb{D} \setminus \{\varphi(0)\}$,*

$$N_{\varphi,\widetilde{\omega}}(z) = 4 \int_0^1 N_\varphi(z,r)\, r\omega(r)\, dr \tag{7.22}$$

$$= 4 \int_{|u_z|}^1 N_\varphi(z,r)\, r\omega(r)\, dr, \tag{7.23}$$

where $u_z = \dfrac{\varphi(0) - z}{1 - \overline{\varphi(0)}\, z}$.

Proof: By definition, we have, using Fubini,

$$4 \int_0^1 N_\varphi(z,r)\, r\omega(r)\, dr = \sum_{\substack{\varphi(a)=z \\ a\in\mathbb{D}}} 4 \int_0^1 \mathbb{1}_{\{|a|<r\}} \log \frac{r}{|a|}\, r\omega(r)\, dr,$$

but, for $a \in \mathbb{D}$, we have

$$4\int_0^1 \mathbb{1}_{|a|<r} \log\frac{r}{|a|}\, r\omega(r)\,dr = 4\int_{|a|}^1 \log\frac{r}{|a|}\, r\omega(r)\,dr = \widetilde{\omega}(|a|),$$

and we get (7.22).

The second relation (7.23) follows from (7.22), with the same argument as for the proof of (7.16). □

We mention here a sub-mean property, which extends Lemma 1.38.

Proposition 7.9 *Let $\varphi : \mathbb{D} \to \mathbb{D}$ be an analytic function, $a \in \mathbb{D}$, and ω be a radial weight.*

Then, for every $r \in (0, |a - \varphi(0)|)$, we have

$$\int_{\mathbb{T}} N_{\varphi,\widetilde{\omega}}(a + r\xi)\, dm(\xi) \geqslant N_{\varphi,\widetilde{\omega}}(a).$$

Proof: We know that, for every $z \in \mathbb{D} \setminus \{\varphi(0)\}$,

$$N_{\varphi,\widetilde{\omega}}(z) = 4\int_0^1 N_\varphi(z,s)\, s\omega(s)\,ds = 4\int_0^1 N_{\phi_s}(z)\, s\omega(s)\,ds,$$

where $\phi_s(z) = \varphi(sz)$.

Therefore, thanks to Lemma 1.38, since $r \in (0, |a-\varphi(0)|) = (0, |a-\phi_s(0)|)$ for every $s \in (0,1)$,

$$\begin{aligned}\int_{\mathbb{T}} N_{\varphi,\widetilde{\omega}}(a + r\xi)\, dm(\xi) &= 4\int_0^1 \int_{\mathbb{T}} N_{\phi_s}(a + r\xi)\, dm(\xi)\, s\omega(s)\,ds \\ &\geqslant 4\int_0^1 N_{\phi_s}(a)\, dm(\xi)\, s\omega(s)\,ds = N_{\varphi,\widetilde{\omega}}(a)\,.\end{aligned}$$

□

Before stating the main result of this subsection, let us define $\sigma_{\varphi,\omega}$ as the pullback measure associated to the symbol φ and the weighted area measure $\sigma_\omega = \omega\, d\sigma$.

Theorem 7.10 *Let φ be a symbol. Fix $\xi \in \mathbb{T}$ and $\delta \in \big(0, 2-\sqrt{2}\big)$.*

Then:

(i) *Let $z = (1-h)\xi$ and $0 < h < \dfrac{1 - |\varphi(0)|}{2}\cdot$ We have*

$$N_{\varphi,\widetilde{\omega}}(z) \leqslant \sup_{w \in W(z,h)} N_{\varphi,\widetilde{\omega}}(w) \leqslant \frac{6(1+\delta)^4}{\delta^2}\, \sigma_{\varphi,\omega}\big(W(\xi, (1+\delta)h)\big)\,. \quad (7.24)$$

(ii) *Let* $0 < h < 1 - |\varphi(0)|$. *We have*

$$\sigma_{\varphi,\omega}\big(S(\xi,(1-\delta)h)\big) \leqslant \frac{1}{\delta^2 h^2}\int_{S(\xi,h)} N_{\varphi,\tilde{\omega}}(z)\,d\sigma(z) \leqslant \frac{1}{\delta^2} \sup_{z\in S(\xi,h)} N_{\varphi,\tilde{\omega}}(z). \tag{7.25}$$

The proof is similar to the one of Theorem 7.7.

Proof: Let $w \in W(z,h)$. We rewrite (7.22), so we have the relation

$$N_{\varphi,\tilde{\omega}}(w) = 4\int_0^1 N_{\phi_r}(w)\,r\omega(r)\,dr,$$

where again ϕ_r is the function $z \mapsto \varphi(rz)$.

Applying Theorem 7.2 to ϕ_r (in other words, using again (7.21)), we have

$$\begin{aligned} N_{\varphi,\tilde{\omega}}(w) &\leqslant \frac{12(1+\delta)^4}{\delta^2}\int_0^1 m_{\phi_r}\big(W(\xi,(1+\delta)h)\big)\,r\omega(r)\,dr \\ &\leqslant \frac{6(1+\delta)^4}{\delta^2}\int_{\mathbb{D}} \mathbb{1}_{\{s\in\mathbb{D}:\,\varphi(s)\in W(\xi,(1+\delta)h)\}}\,d\sigma_\omega \\ &= \frac{6(1+\delta)^4}{\delta^2}\sigma_{\varphi,\omega}\big(W(\xi,(1+\delta)h)\big), \end{aligned}$$

and we proved (7.24).

We prove (ii) mimicking the proof of (ii) in Theorem 7.7. We can write

$$\sigma_{\varphi,\omega}\big(S(\xi,(1-\delta)h)\big) = \int_0^1 m_{\phi_r}\big(S(\xi,(1-\delta)h)\big)\,2r\omega(r)\,dr,$$

but we know from (7.10) that

$$m_{\phi_r}\big(S(\xi,(1-\delta)h)\big) \leqslant \frac{2}{\delta^2 h^2}\int_{S(\xi,h)} N_{\phi_r}(z)\,d\sigma(z),$$

and Fubini's theorem gives

$$\begin{aligned} \sigma_{\varphi,\omega}\big(S(\xi,(1-\delta)h)\big) &\leqslant \frac{4}{\delta^2 h^2}\int_{S(\xi,h)}\left(\int_0^1 N_\varphi(z,r)\,r\omega(r)\,dr\right)d\sigma(z) \\ &= \frac{1}{\delta^2 h^2}\int_{S(\xi,h)} N_{\varphi,\tilde{\omega}}(z)\,d\sigma(z), \end{aligned}$$

thanks to (7.22) and (ii) is proved. □

Clearly we could also deduce here some relations linking some maximal functions associated to weighted Nevanlinna function and area measures.

7.3 Applications

We give here various applications of the preceding results.

Regularity of K_μ-type Ratios

The following results enlighten the (almost) monotony of $t \longmapsto \dfrac{\rho_{\varphi,\beta}(t)}{t^\beta}$. These results will play a crucial role for the characterization of compactness of C_φ in the Orlicz framework.

We begin with the circle case, that is, with

$$K_{m_\varphi}(h) = \sup_{0<t\leqslant h} \frac{\rho_\varphi(t)}{t}\,.$$

Proposition 7.11 *Let φ be a symbol. There exists $C > 0$ such that, for every $h \in (0,1]$,*

$$\frac{\rho_\varphi(h)}{h} \leqslant K_{m_\varphi}(h) \leqslant C\,\frac{\rho_\varphi(h)}{h}\,.$$

Proof: The left-hand-side inequality is obvious.

From (5.33) (actually also valid when $h \in (0,4)$ thanks to the proof of Lemma (5.30)), we know that, for some $c > 0$,

$$\begin{aligned} K_{m_\varphi}(h) &= \sup_{\xi\in\mathbb{T}}\ \sup_{\delta\in(0,1]} \frac{m_\varphi(W(\xi,\delta h))}{\delta h} \leqslant \sup_{\xi\in\mathbb{T}}\ \sup_{\delta\in(0,1]} \frac{m_\varphi(S(\xi,4\delta h))}{\delta h} \\ &\leqslant 4c\sup_{\xi\in\mathbb{T}} \frac{m_\varphi(S(\xi,h))}{h} \leqslant 4c\sup_{\xi\in\mathbb{T}} \frac{m_\varphi(W(\xi,h))}{h} \\ &\leqslant 4c\frac{\rho_\varphi(h)}{h}, \end{aligned}$$

and the constant $C = 4c$ does the job. □

We have the same result for the (weighted) area measure. We already set (see (5.16)), more generally for $h \in (0,1]$ and $\beta \geqslant 1$,

$$K_{\mu,\beta}(h) = \sup_{0<t\leqslant h} \frac{\rho_\mu(t)}{t^\beta}\,. \tag{7.26}$$

It appears that $K_\mu = K_{\mu,1}$.

Proposition 7.12 *Let φ be a symbol.*

There exists $C > 0$ such that, for every $h \in (0,1)$,

$$\frac{\rho_{\varphi,2}(h)}{h^2} \leqslant K_{\sigma_\varphi,2}(h) \leqslant C\,\frac{\rho_{\varphi,2}(h)}{h^2}\,.$$

More generally, for every $\alpha > -1$, there exists $C_\alpha > 0$ such that, for every $h \in (0,1)$,

$$\frac{\rho_{\varphi,\alpha+2}(h)}{h^{\alpha+2}} \leqslant K_{\sigma_{\varphi,\alpha},\alpha+2}(h) \leqslant C_\alpha \, \frac{\rho_{\varphi,\alpha+2}(h)}{h^{\alpha+2}} \,.$$

Proof: The left-hand-side inequality is obvious by definition, and from Theorem 5.38, we know that, for some $c > 0$, we have

$$\begin{aligned} K_{\sigma_{\varphi,\alpha},\alpha+2}(h) &= \sup_{\xi\in\mathbb{T}} \sup_{\delta\in(0,1]} \frac{\sigma_{\varphi,\alpha}(W(\xi,\delta h))}{\delta^{\alpha+2}h^{\alpha+2}} \leqslant \sup_{\xi\in\mathbb{T}} \sup_{\delta\in(0,1]} \frac{\sigma_{\varphi,\alpha}(S(\xi,4\delta h))}{\delta^{\alpha+2}h^{\alpha+2}} \\ &\leqslant c4^{\alpha+2} \sup_{\xi\in\mathbb{T}} \frac{\sigma_{\varphi,\alpha}(S(\xi,h))}{h^{\alpha+2}} \leqslant c4^{\alpha+2} \sup_{\xi\in\mathbb{T}} \frac{\sigma_{\varphi,\alpha}(W(\xi,h))}{h^{\alpha+2}} \\ &\leqslant c4^{\alpha+2} \, \frac{\rho_{\varphi,\alpha+2}(h)}{h^{\alpha+2}} \,. \end{aligned}$$

□

Comparison of Pullback Measures

Since the ℓ^1-norm is greater than the ℓ^2-norm, we clearly have

$$\forall z \in \mathbb{D} \setminus \{\varphi(0)\}, \qquad N_{\varphi,2}(z) \leqslant \big(N_\varphi(z)\big)^2 \,.$$

From (7.18) (with $\alpha = 0$) and (7.6), we obtain the inequality

$$\sigma_\varphi\big(S(\xi,(1-\delta)h)\big) \leqslant \frac{B_0}{\delta^6} \big(m_\varphi\big(W(\xi,(1+\delta)h)\big)\big)^2, \tag{7.27}$$

where $\xi \in \mathbb{T}$, $0 < h < \dfrac{1-|\varphi(0)|}{2}$, and $\delta \in \big(0, 2-\sqrt{2}\big)$.

Of course, using the inclusions between the Carleson box and circular window (up to a change of the opening), we can also compare the m_φ-measure and the σ_φ-measure of these windows: (7.27) implies that, for some numerical $k > 0$, and for every $h \in (0,1)$ small enough, we have

$$\forall \xi \in \mathbb{T}, \qquad \sigma_\varphi\big(W(\xi,h)\big) \leqslant k \left(m_\varphi\big(W(\xi,kh)\big)\right)^2 . \tag{7.28}$$

We can generalize this remarkable inequality to compare the sizes of $W(\xi,h)$ and $W(\xi,kh)$ relatively to different kinds of weighted area measures, and (7.27) appears as a special case of the following theorem.

Theorem 7.13 *Let $\xi \in \mathbb{T}$, $0 < h < \dfrac{1-|\varphi(0)|}{2}$ and $\delta \in \big(0, 2-\sqrt{2}\big)$.*

(i) *Let $\beta > \alpha > -1$. We have*

$$\sigma_{\varphi,\beta}\big(S(\xi,(1-\delta)h)\big) \leqslant K_{\alpha,\beta,\delta} \left(\sigma_{\varphi,\alpha}\big(W(\xi,(1+\delta)h)\big)\right)^{\frac{\beta+2}{\alpha+2}}, \tag{7.29}$$

where the constant $K_{\alpha,\beta,\delta}$ is $O\left(\delta^{-\frac{2(\alpha+\beta+4)}{\alpha+2}}\right)$ when $\delta \to 0^+$.

(ii) *Let $\beta > -1$. We have*

$$\sigma_{\varphi,\beta}\big(S(\xi,(1-\delta)h)\big) \leqslant K_{\beta,\delta}\left(m_\varphi\big(W(\xi,(1+\delta)h)\big)\right)^{\beta+2}, \tag{7.30}$$

where the constant $K_{\beta,\delta}$ is $O\left(\delta^{-2(\beta+3)}\right)$ when $\delta \to 0^+$.

Of course (7.27) appears as (7.30) with $\beta = 0$.

On the other hand, we could state a variant of this result with formulas involving only circular Carleson windows or only Carleson boxes.

Proof: We use Theorem 7.7. From (7.18), we know that

$$\sigma_{\varphi,\beta}\big(S(\xi,(1-\delta)h)\big) \leqslant \frac{B_\beta}{\delta^2} \sup_{w\in S(\xi,h)} N_{\varphi,\beta+2}(w)\,.$$

Since the $\ell^{\alpha+2}$-norm is greater than the $\ell^{\beta+2}$-norm, we also have

$$\forall z \in \mathbb{D}\setminus\{\varphi(0)\}, \qquad N_{\varphi,\beta+2}(z) \leqslant \big(N_{\varphi,\alpha+2}(w)\big)^{\frac{\beta+2}{\alpha+2}}\,.$$

From (7.17), we also know that, with $z = (1-h)\xi$,

$$\sup_{w\in S(\xi,h)} N_{\varphi,\alpha+2}(w) \leqslant \sup_{w\in W(z,h)} N_{\varphi,\alpha+2}(w) \leqslant \frac{A_\alpha}{\delta^2}\,\sigma_{\varphi,\alpha}\big(W(\xi,(1+\delta)h)\big),$$

and we obtain inequality (7.29).

The proof of (7.30) is similar. Actually, it is formally (7.29) (and its proof) with $\alpha = -1$ if we replace σ_{-1} by m_φ. □

Contraction Principle for the Nevanlinna Function

A remarkable contraction principle (5.33) is proved in Theorem 5.29: the Carleson function of an analytic self-map φ turns out to satisfy the following property of homogeneity, improving the fact that m_φ is a Carleson measure:

$$m_\varphi\big(S(\xi,\varepsilon\, h)\big) \leqslant K\,\varepsilon\, m_\varphi\big(S(\xi,h)\big)$$

for h and ε in $(0,1)$, where $\xi \in \mathbb{T}$ and K is a universal constant.

Thanks to the bridge established in the present chapter between Nevanlinna's function and Carleson's function, we can deduce a version of this property for the maximal Nevanlinna function.

Once again, we should clarify which viewpoint we adopt on Carleson's windows. Here they can be either the ones taking into account the boundary part, or not; the statements work anyway. In particular we are safe in the following when taking the values of N_φ at some point z: those z are always in $\mathbb{D}$.

Theorem 7.14 *There exists a universal constant $K > 0$ such that, for every symbol φ, one has, for $\varepsilon \in (0,1)$, $\xi \in \mathbb{T}$, and $t \in \big(0, (1 - |\varphi(0)|)/2\big)$,*

$$\nu_\varphi(\xi, \varepsilon\, t) \leqslant K\, \varepsilon\, \nu_\varphi(\xi, t)\,, \tag{7.31}$$

where $\nu_\varphi(\xi,t) = \sup_{w \in S(\xi,t) \cap \mathbb{D}} N_\varphi(w)$.

In particular, the maximal Nevanlinna function satisfies

$$\sup_{|z| > 1 - \varepsilon\, t} N_\varphi(z) \leqslant K\, \varepsilon \sup_{|z| > 1 - t} N_\varphi(z)\,. \tag{7.32}$$

Proof: Let $w \in S(\xi, \varepsilon t) \subset W(\xi, \varepsilon t)$ and $T = (1 - |\varphi(0)|)/2$.

Assume first that $\varepsilon \in (0, 1/12)$. We know from (7.6) (with $\delta = 1/2$) that

$$N_\varphi(w) \leqslant C\, m_\varphi\big(W(\xi, 3\varepsilon t/2)\big) \leqslant C\, m_\varphi\big(S(\xi, 6\varepsilon t)\big)$$

since $0 < \varepsilon t < t < T$, where $C > 0$ is a (numerical) constant.

Therefore, with (5.33), we have

$$N_\varphi(w) \leqslant 12CK\, \varepsilon\, m_\varphi\big(S(\xi, t/2)\big)\,.$$

Then, thanks to (7.9) (with $\delta = 1/2$ again), we obtain

$$N_\varphi(w) \leqslant 48CK\, \varepsilon \sup_{z \in S(\xi,t) \cap \mathbb{D}} N_\varphi(z) \leqslant A \varepsilon \nu_\varphi(\xi, t),$$

where $A > 0$, and we conclude for these values of ε, taking the supremum over $w \in S(\xi, \varepsilon t)$.

We proved (7.31) in the case where $\varepsilon \in (0, 1/12)$, but this is sufficient (and of course the core of the property). Indeed, when $\varepsilon \in (1/12, 1)$, one has simply to see that

$$\nu_\varphi(\xi, \varepsilon\, t) \leqslant \nu_\varphi(\xi, t) \leqslant 12\varepsilon\, \nu_\varphi(\xi, t),$$

and we get the general case. □

We can also write a version for weighted Nevanlinna functions.

Theorem 7.15 *Let $\beta \geqslant 1$.*

There exists a constant $K_\beta > 0$ such that, for every analytic self-map φ of $\mathbb{D}$, one has, for $\varepsilon \in (0,1)$, $\xi \in \mathbb{T}$, and $t \in \big(0, 1 - |\varphi(0)|)/2\big)$,

$$\nu_{\varphi,\beta}(\xi, \varepsilon\, t) \leqslant K\, \varepsilon^\beta\, \nu_{\varphi,\beta}(\xi, t)\,, \tag{7.33}$$

where $\nu_{\varphi,\beta}(\xi,t) = \sup_{w \in S(\xi,t) \cap \mathbb{D}} N_{\varphi,\beta}(w)$.

In particular, the maximal weighted Nevanlinna function satisfies

$$\sup_{|z| > 1 - \varepsilon\, t} N_{\varphi,\beta}(z) \leqslant C_\beta\, \varepsilon^\beta \sup_{|z| > 1 - t} N_{\varphi,\beta}(z)\,. \tag{7.34}$$

Proof: The proof is similar to the unweighted case and we leave it as an exercise. □

7.4 Mean Property for Nevanlinna's Functions and Applications

Inequalities (7.10) and (7.18) provide an upper estimate of the size of a Carleson box with the help of the mean of the (weighted) Nevanlinna counting function. In Chapter 9, we shall need a version involving not only the normalized L^1-norm of the Nevanlinna counting function but also the normalized L^q-norm of the Nevanlinna counting function for any $q > 0$.

Theorem 7.16 *Let $q > 0$ and $a = (1-h)\xi \in \mathbb{D}$ with $h \in (0, (1-|\varphi(0)|)/10)$ and $\xi \in \mathbb{D}$. Then*

$$m_\varphi(W(\xi,h)) \lesssim \left(\frac{1}{\sigma(W(\xi,10h)} \int_{W(\xi,10h)} \left[N_\varphi(z)\right]^q \, d\sigma(z) \right)^{\frac{1}{q}}.$$

Proof: From (7.10), we have

$$m_\varphi(W(\xi,h)) \leqslant m_\varphi(S(\xi,4h)) \lesssim \sup_{w \in W(\xi,5h)} N_\varphi(w).$$

Hence, there is some $w_0 \in W(\xi,5h)$ such that $m_\varphi(W(\xi,h)) \lesssim N_\varphi(w_0)$.

Take $R = |w_0| + 5h > 1$ and set $\varphi_0(z) = \varphi(z)/R$. One has $N_{\varphi_0}(z) = N_\varphi(Rz)$ for $|z| < 1/R$ and $N_{\varphi_0}(z) = 0$ if $|z| \geqslant 1/R$.

Let now u be the upper subharmonic regularization of N_{φ_0} [123, lemma 1]: u is a subharmonic function on $\mathbb{D} \setminus \{0\}$ such that $u \geqslant N_{\varphi_0}$ and $u = N_{\varphi_0}$ almost everywhere, with respect to $d\sigma$.

Now, a result of C. Fefferman and E. M. Stein, generously attributed by them to Hardy and Littlewood, asserts that for any $q > 0$, there exists a constant $C = C(q)$ such that

$$[u(a)]^q \leqslant \frac{C}{\sigma\big(D(a,r)\big)} \int_{D(a,r)} [u(z)]^q \, d\sigma(z) \tag{7.35}$$

for every non-negative subharmonic function u on a domain G and every disk $D(a,r) \subseteq G$ (see also [123], lemma 3).

Therefore, if Δ is the (euclidean) disk centered at w_0/R and of radius $1 - |w_0|/R$ (which is contained in $\mathbb{D} \setminus \{0\}$ since $R > |w_0|$), one has via (7.35)

$$\begin{aligned}
[N_\varphi(w_0)]^q &= [N_{\varphi_0}(w_0/R)]^q \leqslant [u(w_0/R)]^q \\
&\leqslant \frac{C}{\sigma(\Delta)} \int_\Delta [u(z)]^q \, d\sigma(z) \\
&= \frac{C}{\sigma(\Delta)} \int_\Delta [N_{\varphi_0}(z)]^q \, d\sigma(z) \\
&= \frac{C}{\sigma(\Delta)} \int_{\Delta \cap D(0,1/R)} [N_\varphi(Rz)]^q \, d\sigma(z) \\
&= \frac{C}{\sigma(\tilde{\Delta})} \int_{\tilde{\Delta} \cap \mathbb{D}} [N_\varphi(w)]^q \, d\sigma(w),
\end{aligned}$$

where $\tilde{\Delta} = D(w_0, R - |w_0|) = D(w_0, 5h)$.

Since the center w_0 of $\tilde{\Delta}$ is in $\mathbb{D}$, $\tilde{\Delta} \cap \mathbb{D}$ contains more than a quarter of $\tilde{\Delta}$ (at least for $5h \leqslant 1$), and hence $\sigma(\tilde{\Delta} \cap \mathbb{D}) \geqslant \sigma(\tilde{\Delta})/4 = 25h^2/4\pi$.

Clearly $W(\xi, 10h)$ contains $\tilde{\Delta} \cap \mathbb{D}$ and $\sigma(W(\xi, 10h)) \approx \sigma(\tilde{\Delta})$. We therefore get

$$\big[N_\varphi(w_0)\big]^q \lesssim \frac{1}{\sigma(W(\xi, 10h)} \int_{W(\xi, 10h)} \big[N_\varphi(w)\big]^q \, d\sigma(w).$$

But since $m_\varphi(W(\xi, h)) \lesssim N_\varphi(w_0)$, we get the conclusion. □

8
Spectrum

8.1 Introduction

Whereas it is natural, when one studies a class of operators on a Banach or a Fréchet space, to try and describe the spectrum of elements of this class, we will devote little space to that question in this book (in the case of composition operators), essentially for three reasons:

(i) We will pay much attention to the spectral radius (the sup of the spectrum) with a "spectral radius formula," which involves the Green capacity of the image of the symbol, in Chapter 10.
(ii) The study of the spectrum of composition operators (a huge topic that could occupy an entire volume) has already been done rather extensively, in books or surveys: see in particular the two fundamental treatises [34] and [165].
(iii) In this book, we privileged other aspects, like compactness, singular numbers, membership in Schatten classes, in the nuclear class, and so on. Recall in passing that composition operators are far from being self-adjoint, or even normal. So that the knowledge (which we have) of the pointwise spectrum $\sigma_p(C_\varphi)$ is often of little help to describe the full spectrum $\sigma(C_\varphi)$.

Warning. We use in the sequel the usual notations of spectrum theory. Nevertheless, for instance, please do not confuse σ_p with the area masure σ_α!

Reminder. The iterates of the symbol φ will play an important role here. We recall that the symbol $\varphi^{[n]}$ denotes the nth iterate of φ with itself.

This being said, this chapter will contain few proofs, but rather a condensed (and deliberately incomplete) description of the state of art on the question in two quite different cases:

(i) The case of Hardy spaces H^p, sometimes specifically H^2.
(ii) The case of the Fréchet space $\mathscr{H}(\mathbb{D})$, for which the situation is fairly different.

We shall need below the following general lemma about spectra of isometries:

Lemma 8.1 *Let $T : X \to X$ be an isometry from a Banach space into itself. Then*

$$\sigma(T) \subset \partial\mathbb{D} \quad \textit{or} \quad \sigma(T) = \overline{\mathbb{D}}.$$

Proof: Let R be the complement of $\sigma(T) \cap \mathbb{D}$ in $\mathbb{D}$. This set is closed in $\mathbb{D}$: indeed, given $\lambda_n \in R$ with $\lambda_n \to \lambda \in \mathbb{D}$ and having in mind that T is an isometry, we know that, for every $x \in B_X$, $\|T(x) - \lambda_n x\| \geqslant (1 - |\lambda_n|)\|x\|$. We obtain (with $I = I_X$ the identity on X)

$$\|(T - \lambda_n I)^{-1}\| \leqslant (1 - |\lambda_n|)^{-1} \longrightarrow (1 - |\lambda|)^{-1}$$

and $\|(T - \lambda_n I)^{-1}\| = O(1)$. The sequence $\big((T - \lambda_n I)^{-1}\big)$ is therefore a Cauchy sequence in $B(X)$ since

$$(T - \lambda_p I)^{-1} - (T - \lambda_q I)^{-1} = (T - \lambda_p I)^{-1}(\lambda_p - \lambda_q)(T - \lambda_q I)^{-1}.$$

Hence, it converges in $B(X)$ to some operator $C : X \to X$.

Passing to the limit in

$$(T - \lambda_n I)(T - \lambda_n I)^{-1} = (T - \lambda_n I)^{-1}(T - \lambda_n I) = I$$

gives

$$(T - \lambda I)C = C(T - \lambda I) = I,$$

so that $\lambda \in R$. As a consequence, $\sigma(T) \cap \mathbb{D}$ is open and closed in $\mathbb{D}$. Therefore, either $\sigma(T) \cap \mathbb{D} = \emptyset$, that is, $\sigma(T) \subset \partial\mathbb{D}$, or $\sigma(T) \cap \mathbb{D} = \mathbb{D}$, that is, $\sigma(T) = \overline{\mathbb{D}}$. □

8.2 König–Schröder's Equation

The first step in the study of the spectrum is to determine the point spectrum, that is, the set of the eigenvalues. In this regard, the following result of König, answering a question of Schröder, is very useful.

We recall that for a symbol φ with fixed point $a \in \mathbb{D}$, we have $|\varphi'(a)| \leqslant 1$. Moreover, when φ is not an (elliptic) automorphism, we even have $|\varphi'(a)| < 1$ (see Section 1.5).

Theorem 8.2 (König) *Let φ be a non-constant analytic self-map of $\mathbb{D}$ with a fixed point $a \in \mathbb{D}$. Assume that φ is not an (elliptic) automorphism.*

Consider the König–Schröder equation

$$f \circ \varphi = \lambda . f \qquad \text{with } f \in \mathscr{H}(\mathbb{D}) \text{ and } f \neq 0. \tag{8.1}$$

(i) *Assume that there is a solution. Then*

(a) $\lambda \neq 0$ *and moreover* $\lambda = 1$ *or* $\varphi'(a)^d$ *for some* $d \in \mathbb{N}$.

(b) $\lambda = 1$ *if and only if* f *is constant; and when* $\lambda \neq 1$, $f(a) = 0$.

(ii) *When* $\varphi'(a) = 0$, *(8.1) is satisfied only for* $\lambda = 1$ *and* f *constant.*

(iii) *If* $0 < |\varphi'(a)| < 1$ *then (8.1) is satisfied for* $\lambda = \varphi'(a)^d$ *for some* $d \in \mathbb{N}$.

Conversely:

For $d = 1$, *that is,* $\lambda = \varphi'(a)$, *(8.1) admits a unique solution* $\tau \in \mathscr{H}(\mathbb{D})$, *called the König function of* φ *at* a, *such that* $\tau(a) = 0$ *and* $\tau'(a) = 1$.

For $d \geqslant 1$, *that is,* $\lambda = \varphi'(a)^d$, *the function* τ^d *is a solution of (8.1) and any solution must belong to* $\mathbb{C}.\tau^d$.

Moreover, when φ *is injective, so is* τ.

Remark: We shall use in (i)(a) and (iii) in the proof below that, by conjugation upon φ_a, we can assume that $a = 0$: indeed $\psi = \varphi_a \circ \varphi \circ \varphi_a$ fixes the origin and $\psi'(0) = \varphi'(a)$ belongs to $\mathbb{D}$ by the Schwarz lemma except if ψ is a rotation, which would mean that φ is an (elliptic) automorphism (and it is not).

Proof: (i) If $\lambda = 0$ in (8.1) then $f = 0$ on $\varphi(\mathbb{D})$, which is a non-empty open set so f would be 0, hence $\lambda \neq 0$.

Let f be a solution, write $d \in \mathbb{N}_0$ such that $f(z) = z^d g(z)$ for $z \in V_0$, a neighborhood of 0 with $g(0) \neq 0$ and $g \circ \varphi(z) \neq 0$ on V_0. From (8.1), we have, for every $z \in V_0 \setminus \{0\}$,

$$\left(\frac{\varphi(z)}{z}\right)^d = \lambda \frac{g(z)}{g \circ \varphi(z)},$$

and passing to the limit we get $\lambda = 1$ when $d = 0$ or $\varphi'(0)^d = \lambda$, proving (i)(a).

(i)(b) From (8.1), we see that $f \circ \varphi^{[k]}(z) = \lambda^k f(z)$ for every $z \in \mathbb{D}$, but $\varphi^{[k]}(z) \to a$, thanks to Proposition 1.35 so $f \circ \varphi^{[k]}(z) \longrightarrow f(a)$. Therefore when $\lambda = 1$, we get $f(z) = f(a)$ and f is constant (and it is clear that any constant function f satisfies (8.1) with $\lambda = 1$).

When $\lambda \neq 1$, from (8.1), we know that $f(a) = \lambda f(a)$ hence $f(a) = 0$.

(ii) When $\varphi'(a) = 0$, since λ cannot be 0, we cannot have $\lambda = \varphi'(a)^d$ for some $d \in \mathbb{N}$, hence $d = 0$ and $\lambda = 1$ is the only possibility.

(iii) From now on, we assume that $0 < |\varphi'(a)| < 1$, so that a necessary condition on λ is that this is 1 or $\varphi'(a)^d$ for some $d \geqslant 1$. We may assume $\lambda \neq 1$ since this case is already treated in (i)(b).

We focus first on the existence of a solution for $d = 1$: we claim that there is indeed a solution for $\lambda = \varphi'(a)$.

We may still assume that $a = 0$. With the Picard fixed point theorem in mind, we define inductively

$$f_0(z) = z \qquad \text{and} \qquad f_{n+1}(z) = (1/\lambda) f_n(\varphi(z)).$$

Hence $f_n = \frac{\varphi^{[n]}}{\lambda^n}$ is a renormalization of $\varphi^{[n]}$ with $f_n(0) = 0$ and $f_n'(0) = 1$.

We claim that the series of functions $\sum_{n \geqslant 0}(f_{n+1} - f_n)$ is locally normally convergent on $\mathbb{D}$; and this will show that the sequence (f_n) is locally uniformly convergent on $\mathbb{D}$ to some function $\tau \in \mathscr{H}(\mathbb{D})$ satisfying all our requirements.

Fix $1 < r < |\lambda|^{-1/2}$. There exists $\delta > 0$ such that, by induction,

$$|z| \leqslant \delta \Longrightarrow |\varphi^{[k]}(z)| \leqslant (r|\lambda|)^k |z|, \quad \text{for } k = 1, 2, \ldots. \tag{8.2}$$

Let K be a compact subset of $\mathbb{D}$. There exists M and $p = p_K$ such that

$$z \in \mathbb{D} \Longrightarrow |\varphi(z) - \lambda z| \leqslant M|z|^2 \text{ and } \varphi^{[p]}(K) \subset \overline{D}(0, \delta). \tag{8.3}$$

Indeed, $\varphi^{[n]} \xrightarrow{uc} 0$ since $|\varphi'(0)| < 1$ (Proposition 1.35 again).

This implies via (8.2), for $z \in K$ and $n > p$:

$$|\varphi^{[n]}(z)| \leqslant (r|\lambda|)^{n-p} |\varphi^{[p]}(z)| \leqslant (r|\lambda|)^{n-p} =: C(r|\lambda|)^n,$$

where C only depends on K. For those n and z, we hence get from (8.3):

$$|f_{n+1}(z) - f_n(z)| = \left|\frac{\varphi(\varphi^{[n]}(z)) - \lambda\varphi^{[n]}(z)}{\lambda^{n+1}}\right| \leqslant \frac{M|\varphi^{[n]}(z)|^2}{|\lambda^{n+1}|} \leqslant \frac{MC^2}{|\lambda|}(r^2|\lambda|)^n.$$

This proves our claim.

Now we prove uniqueness (for general $a \in \mathbb{D}$).

Let f be a solution of (8.1) with $f(a) = 0$ and $f'(a) = 1$. Differentiating this equation n times with $n \geqslant 2$, we get

$$\sum_{j=0}^{n-1} \alpha_j(z) f^{(j)}(\varphi(z)) + f^{(n)}(\varphi(z))\varphi'(z)^n = \lambda f^{(n)}(z), \quad \text{for } z \in \mathbb{D},$$

where the functions α_j only depend on φ (this weak form of the Faa di Bruno formula is enough here). Putting $z = a$ gives

$$(\lambda - \lambda^n) f^{(n)}(a) = \sum_{j=0}^{n-1} \alpha_j(a) f^{(j)}(a).$$

Since $\lambda - \lambda^n \neq 0$, this shows that $f^{(n)}(a)$ is inductively determined when $f(a)$ and $f'(a)$ are given. The uniqueness in the case $d = 1$ follows.

If φ is injective, so are the f_n. By Hurwitz's theorem, the limit τ is either injective or constant, and since $\tau'(a) = 1$, it is injective.

We now focus on the existence of a solution for $d \geqslant 2$ assuming again $a = 0$. Clearly τ^d is a solution of (8.1) associated to $\varphi'(0)^d$, and if f is another solution, then the proof in (i) shows that 0 is a root of f with multiplicity d. Therefore $g = f/\tau^d$ is holomorphic in a neighborhood W_0 of 0 and satisfies $g \circ \varphi = g$ on W_0. The iteration argument given in (i) gives that g is constant, that is, $f = c\tau^d$ in W_0, where $c \in \mathbb{C}$. We obtain by connectedness that $f = c\tau^d$ on $\mathbb{D}$. □

Example. Let us focus on what happens for the lens map. Let $0 < \theta < 1$. Consider the lens map $\Lambda_\theta : \mathbb{D} \to \mathbb{D}$ (see Section 4.2) defined by

$$\Lambda_\theta = T^{-1}\gamma_\theta T, \text{ or else } \Lambda_\theta(z) = \frac{(1+z)^\theta - (1-z)^\theta}{(1+z)^\theta + (1-z)^\theta}, \tag{8.4}$$

where $T : \mathbb{D} \to \mathbb{C}_0$ is the conformal mapping $z \mapsto \frac{1+z}{1-z}$ and $\gamma_\theta : \mathbb{C}_0 \to \mathbb{C}_0$ the map $w \mapsto w^\theta$. The Denjoy–Wolff point of Λ_θ is 0 and $\Lambda'_\theta(0) = \theta$. The König function is

$$\tau(z) = \frac{1}{2} \log \frac{1+z}{1-z} \cdot$$

Indeed, $\tau(0) = 0,\ \tau'(0) = 1$ and

$$\tau \circ \Lambda_\theta = \frac{1}{2} \log \circ T \circ T^{-1} \circ \gamma_\theta \circ T = \frac{1}{2} \log \circ \gamma_\theta \circ T = \frac{\theta}{2} \log \circ T = \theta\tau.$$

Observe on this example that $\tau \in \bigcap_{p<\infty} H^p$. Actually, this is always true when C_φ is compact on H^2: see Theorem 8.4. Also observe that $\tau \notin H^\infty$. Actually the situation can be worse: in [13], it is shown that, whenever φ is inner, then $\tau \notin \mathcal{N}$, the Nevanlinna class.

Remark. Theorem 8.2 expresses that, locally, φ is semi-conjugate to the dilation $z \mapsto \varphi'(0)z$ and helps in understanding the behavior of the iterates of φ. This result will also reappear when we study the eigenvalues, spectrum, and singular numbers of (weighted) composition operators. It is interesting to rephrase Theorem 8.2 in terms of operators acting on the Fréchet space $\mathscr{H}(\mathbb{D})$.

Indeed we have the following theorem:

Theorem 8.3 *Let φ be an analytic self-map of $\mathbb{D}$ with a fixed point $a \in \mathbb{D}$ such that $0 < |\varphi'(a)| < 1$; and w analytic in $\mathbb{D}$. Then the eigenvalues of*

$f \in \mathscr{H}(\mathbb{D}) \longmapsto w.C_\varphi(f) \in \mathscr{H}(\mathbb{D})$ are among the numbers $w(a)(\varphi'(a))^n$, where $n \in \mathbb{N}_0$.

Proof: Suppose that $w(f \circ \varphi) = \lambda f$ with $f \in \mathscr{H}(\mathbb{D})$, $f \neq 0$, and $\lambda \in \mathbb{C}$. Let n be the smallest integer such that $f^{(n)}(a) \neq 0$. The same computation as in Theorem 8.2 gives

$$\sum_{j=0}^{n-1} \alpha_j(z) f^{(j)}(\varphi(z)) + f^{(n)}(\varphi(z)) w(z)(\varphi'(z))^n = \lambda f^{(n)}(z), \qquad \text{for } z \in \mathbb{D},$$

where the functions α_j only depend on φ and w.

Putting $z = a$ gives $\lambda = w(a)(\varphi'(a))^n$. □

8.3 Hardy Spaces

This section, mainly based on [80], is itself divided into quite different subcases. We did not try to be exhaustive, but tried to exhibit some typical cases and proofs.

Compact Case

The situation for compact composition operators is rather easy:

Theorem 8.4 *Let $p \geqslant 1$ and φ be a symbol such that $C_\varphi : H^p \to H^p$ is compact. Then the König function τ belongs to $\bigcap_{m<+\infty} H^m$. Moreover*

$$\sigma(C_\varphi) = \{0; 1\} \cup \{\varphi'(a)^n : n \in \mathbb{N}\},$$

where $a \in \mathbb{D}$ is the unique fixed point of φ.

The existence and uniqueness of a are assured by Corollary 3.23. The function τ is given in Theorem 8.2 (φ cannot be an elliptic automorphism).

Proof: We know that $0 \in \sigma(C_\varphi)$ and $\sigma(C_\varphi) \setminus \{0\} = \sigma_p(C_\varphi) \setminus \{0\}$.

When $\varphi'(a) = 0$, then 1 is the only eigenvalue in $\mathscr{H}(\mathbb{D})$ with associated eigenfunctions that are constants (see Theorem 8.2).

We now assume that $\varphi'(a) \neq 0$. Moreover, C_φ is a compact operator on H^2 (since this property does not depend on p).

For $N \in \mathbb{N}_0$, let $\mathscr{P}_N$ be the vector space spanned by monomials $1, \ldots, z^N$. Its orthogonal $\mathscr{P}_N^\perp$ is the vector space spanned by $z^{N+1}, \ldots$

By conjugation upon C_{φ_a}, we can assume that $a = 0$, so that we have $C_\varphi(\mathscr{P}_N^\perp) \subset \mathscr{P}_N^\perp$ for every $N \in \mathbb{N}_0$. Therefore $C_\varphi^*(\mathscr{P}_N) \subset \mathscr{P}_N$. Moreover, for any $N \in \mathbb{N}_0$, the matrix of C_φ^* in the canonical basis of $\mathscr{P}_N$ is upper triangular with the coefficients $\overline{\varphi'(0)}^n$ (where $0 \leqslant n \leqslant N$) on the diagonal, hence those coefficients are eigenvalues of this matrix, a fortiori of $C_\varphi^* : H^2 \to H^2$. As a consequence, the $\varphi'(0)^n$ are eigenvalues of $C_\varphi : H^2 \to H^2$ for every $n \in \mathbb{N}_0$: indeed, it comes from the classical Riesz theory of compact operators (see Theorem 1.101) that $\varphi'(0)^n I - C_\varphi = \varphi'(0)^n (I - \varphi'(0)^{-n} C_\varphi)$ is not one to one, since it is not invertible. We get from Theorem 8.2 that an eigenvector (viewed in $\mathscr{H}(\mathbb{D})$) associated to $\varphi'(0)^n$ must belong to $\mathbb{C}\tau^n$. At the same time, it belongs to H^2 by definition. Therefore $\tau^n \in H^2$ *i.e.* $\tau \in H^{2n}$.

A fortiori, for any $p \geqslant 1$, we have $\tau^n \in H^p$ for every $n \geqslant 1$, so $\varphi'(0)^n$ is an eigenvalue of $C_\varphi : H^p \to H^p$. □

Power-Compact Case

Here we consider *power-compact* symbols φ, that is, satisfying that C_φ^N is compact on H^2 for some positive integer N.

It can very well happen that a non-compact operator C_φ has some compact power, as evidenced by the simple example

$$\varphi(z) = \frac{1 - z}{2}, \qquad \varphi \circ \varphi(z) = \varphi^{[2]}(z) = \frac{1 + z}{4}.$$

Since, for $r \to -1^+$, we have $\frac{1 - |\varphi(r)|}{1 - |r|} = \frac{1}{2}$ (as soon as $r \in (-1, 0)$), so φ has a radial derivative at -1 and hence C_φ is non-compact. But $\|\varphi^{[2]}\|_\infty = 1/2 < 1$, hence $C_{\varphi^{[2]}} = C_\varphi^2$ is compact (even nuclear).

That being said, as is more or less predictable from the general Banach space theory, the situation is essentially that already described in Chapter 1 when C_φ is compact. Recall [84, chapter 1, pp. 19–26] that power-compact operators fall into the category of *Riesz operators* on Banach spaces for which the spectral theory is quite well understood. In particular, if T is Riesz, we have the relation $\sigma(T) \setminus \{0\} = \sigma_p(T) \setminus \{0\}$.

Coming back to our situation, we have more precisely the following theorem:

Theorem 8.5 *Let φ be a symbol such that $C_\varphi^N : H^p \to H^p$ is compact for some integer $N \geqslant 1$. Then*

(i) *φ has a fixed point $z_0 \in \mathbb{D}$.*
(ii) *We have*

$$\sigma(C_\varphi) = \{0, 1\} \cup \left\{ \left(\varphi'(z_0)\right)^n : n \in \mathbb{N} \right\}.$$

Proof: We have $C_{\varphi^{[N]}} = C_\varphi^N$ is compact and we know from Theorem 3.23 that $\varphi^{[N]}$ has a fixed point z_0 inside $\mathbb{D}$. Since

$$\varphi^{[N]}\big(\varphi(z_0)\big) = \varphi\big(\varphi^{[N]}(z_0)\big) = \varphi(z_0),$$

the point $\varphi(z_0)$ is also a fixed point for $\varphi^{[N]}$. The analytic self-map $\varphi^{[N]}$ of $\mathbb{D}$ is surely different from the identity since $C_{\varphi^{[N]}}$ is compact. Then it has at most one fixed point inside $\mathbb{D}$ and we get $\varphi(z_0) = z_0$.

Then $|\varphi'(z_0)^N| = |(\varphi^{[N]})'(z_0)| < 1$, hence $|\varphi'(z_0)| < 1$, so we are back to König's theorem 8.2: all the $(\varphi'(z_0))^n$ are eigenvalues of $C_\varphi : \mathscr{H}(\mathbb{D}) \to \mathscr{H}(\mathbb{D})$. Actually, with τ the König function associated to φ, we have

$$\tau \circ \varphi^{[N]} = \varphi'(z_0)^N \tau = (\varphi^{[N]})'(z_0)\tau.$$

Since $\tau(z_0) = 0$ and $\tau'(z_0) = 1$, Theorem 8.2 (applied to the symbol $\varphi^{[N]}$) states, by uniqueness, that τ must be the König function associated to $\varphi^{[N]}$, but Theorem 8.4 (applied to the symbol $\varphi^{[N]}$) states that this function belongs to all the H^m for $m \geqslant 1$. Therefore $\tau^n \in H^p$ for every $n \geqslant 1$ and we obtain that all the $(\varphi'(z_0))^n$ are eigenvalues of $C_\varphi : H^p \to H^p$.

Moreover $0 \in \sigma(C_\varphi)$ since C_φ^N and hence C_φ is not invertible. □

Here is a variant of this theorem for weighted composition operators on a weighted Hardy space. We consider König's equation when replacing H^p by $H = H^2(\beta) \hookrightarrow \mathscr{H}(\mathbb{D})$, where $H^2(\beta)$ stands for the weighted Hilbert space of analytic functions $f(z) = \sum_{n=0}^{\infty} a_n z^n$ on $\mathbb{D}$ such that

$$\|f\|^2 = \sum_{n=0}^{\infty} |a_n|^2 \beta_n < \infty.$$

We shall come back to this space in Chapter 12 and content ourselves here with the statement, due to Gunatillake [68, theorem 1], below:

Theorem 8.6 *Let φ be an analytic self-map of $\mathbb{D}$ with a Denjoy–Wolff point $a \in \mathbb{D}$; let $w \in \mathscr{H}(\mathbb{D})$ and $T = w.C_\varphi : H \to H$, assumed to be compact.*

Then the spectrum $\sigma(T)$ of $T : H \to H$ is

$$\sigma(T) = \{0\} \cup \big\{w(a).(\varphi'(a))^n : n \in \mathbb{N}\big\}.$$

Inner Case

Let φ be an inner function. Then, $\varphi^{[N]}$ is inner for each positive integer N, and $C_\varphi^N = C_{\varphi^{[N]}}$ is never compact. This case again splits into two subcases:

(i) φ has a fixed point inside $\mathbb{D}$, which is automatically its Denjoy–Wolff point.

(ii) φ has its Denjoy–Wolff point β on $\partial\mathbb{D}$, with automatically $0 < \varphi'(\beta) \leqslant 1$ as we saw in Theorem 1.36.

Here are two precise corresponding theorems.

Theorem 8.7 *Let φ be an inner symbol with a fixed point $\alpha \in \mathbb{D}$, but not a fractional linear map $\frac{az+b}{cz+d}$. Then*

(i) $C_\varphi : H^p \to H^p$ *is similar to an isometry.*
(ii) $\sigma(C_\varphi) = \overline{\mathbb{D}}$.

Proof: Let $\psi = \varphi_\alpha \circ \varphi \circ \varphi_\alpha$, inner as well, with $\psi(0) = 0$. Then

$$C_\varphi = C_{\varphi_\alpha}^{-1} C_\psi C_{\varphi_\alpha}$$

is similar to C_ψ and we know from Proposisition 2.5 that $C_\psi : H^p \to H^p$ is an isometry.

The second item directly follows from Lemma 8.1, since in our case C_φ is not invertible because φ is surely not an automorphism. Therefore $0 \in \sigma(C_\varphi)$ and $\sigma(C_\varphi) = \overline{\mathbb{D}}$. □

Our second theorem goes as follows.

Theorem 8.8 *Let φ be a linear fractional transformation $\varphi(z) = \frac{az+b}{cz+d}$, which is an inner symbol. Then we can assume that φ is of the form*

$$\frac{az+\overline{c}}{cz+\overline{a}} \quad \text{with} \quad \operatorname{Re} a > 0 \text{ and } |a|^2 - |c|^2 = 1.$$

Setting $a = u + iv$, with $u > 0$ and $v \in \mathbb{R}$, the spectrum of $C_\varphi : H^p \to H^p$ has either of the three forms:

(i) *The closure of the (unimodular) sequence $(K^n)_{n\geqslant 1}$ when $|c| < |v|$.*
(ii) *The unit circle when $|c| = |v|$.*
(iii) *The annulus $\{\lambda : L^{-\frac{1}{p}} \leqslant |\lambda| \leqslant L^{\frac{1}{p}}\}$ when $|c| > |v|$,*

where

$$K = \frac{u + i(v^2 - |c|^2)^{1/2}}{u - i(v^2 - |c|^2)^{1/2}} \qquad \text{with } |c| < |v|,\ |K| = 1;$$

$$L = \frac{u + (|c|^2 - v^2)^{1/2}}{u - (|c|^2 - v^2)^{1/2}} \qquad \text{with } |c| > |v|,\ L > 1.$$

Proof: We must have $|b| < |d|$ since $|\varphi(0)| < 1$. Then, it is clear that φ is inner if and only if $|a|^2 + |b|^2 = |c|^2 + |d|^2$ and $a\overline{b} = c\overline{d}$. Since a, b, c, d are determined up to a multiplicative constant, we can assume $b = \overline{c}, d = \overline{a}$ with $|c| < |a|$ and also $|a|^2 - |c|^2 = 1$. We refer to [131] for the detailed proof. □

Mixed Case

Warning. In the sequel of this section, we will only consider symbols φ analytic on $\overline{\mathbb{D}}$, that is, having an analytic extension to some open set of $\mathbb{C}$ containing $\overline{\mathbb{D}}$.

We will content ourselves with the following "mixed" result, in which φ may be inner or not, but is *not* a linear fractional transformation (LFT for short). The proof uses Carleson interpolation sequences already presented in Subsection 1.7.4.

Proposition 8.9 *Let $\varphi : \mathbb{D} \to \mathbb{D}$, not a LFT, with Denjoy–Wolff point $\beta \in \partial\mathbb{D}$. Assume that $\varphi'(\beta) < 1$. The spectrum of $C_\varphi : H^p \to H^p$ satisfies*

$$\overline{D}\left(0, \varphi'(\beta)^{-\frac{1}{p}}\right) \setminus \mathbb{D} \subset \sigma(C_\varphi) \subset \overline{D}\left(0, \varphi'(\beta)^{-\frac{1}{p}}\right).$$

In particular, the spectral radius is $r(C_\varphi) = \varphi'(\beta)^{-\frac{1}{p}}$.

We recall that we know from the Denjoy–Wolff theorem that $0 < \varphi'(\beta) \leqslant 1$.

Let us mention that Kamowitz states in [80] that

$$\sigma(C_\varphi) = \overline{D}\left(0, \varphi'(\beta)^{-\frac{1}{p}}\right).$$

We can give an illustration with the following example:

$$\varphi(z) = \left(\frac{2z+1}{z+2}\right)^2.$$

This is an inner Blaschke product, but not a LFT. It has three fixed points $1, j, j^2$ on $\partial\mathbb{D}$ and the Denjoy–Wolff point is 1, with $\varphi'(1) = \frac{2}{3}\cdot$ We see that

$$r(C_\varphi) = (3/2)^{\frac{1}{p}} < (5/3)^{\frac{1}{p}} = \|C_\varphi\|$$

since we know that, for inner symbols,

$$\|C_\varphi\| = \left(\frac{1 + |\varphi(0)|}{1 - |\varphi(0)|}\right)^{\frac{1}{p}}.$$

The proof below mainly follows the strategy of [80] (with a few extra explanations) and makes use of the following iteration result [81, p.139].

Fact: When $\beta = 1$, there exists a function A, analytic on an open set U_1 containing 1, with $A(1) = 0$, such that (with $c = \varphi'(1)$)

$$\frac{1 - \varphi^{[n]}(z)}{c^n} \to A(z) \quad \text{uniformly on } U_1 \cap \mathbb{D}. \tag{8.5}$$

Moreover, for each $z \in U_1 \cap \mathbb{D}$, we have $\operatorname{Re} A(z) > 0$.

A good test example is the symbol $\varphi(z) = \dfrac{1+z}{2}$, which satisfies $\varphi'(1) = 1/2$, $\varphi^{[n]}(z) = \dfrac{z + 2^n - 1}{2^n}$, and $A(z) = 1 - z$.

Proof of Proposition 8.9: We can assume (up to a similarity that preserves the spectrum) that $\beta = 1$. In the sequel of the proof we write $c = \varphi'(1)$ for short.

The proof splits into two parts.

Step 1. The spectral radius of C_φ satisfies

$$r\left(C_\varphi\right) \leqslant c^{-\frac{1}{p}}. \tag{8.6}$$

Indeed, for any integer $n \geqslant 1$, we know, from the results of Chapter 2 (see (2.14)), that

$$\|C_\varphi^n\|_{B(H^p)}^{1/n} = \|C_{\varphi^{[n]}}\|_{B(H^p)}^{1/n} \leqslant \left(\frac{1 + |\varphi^{[n]}(0)|}{1 - |\varphi^{[n]}(0)|}\right)^{\frac{1}{pn}}.$$

Since $\varphi^{[n]}(0) \to 1$, we have $\varphi^{[n]}(0) \in U_1 \cap \mathbb{D}$ for $n \geqslant n_0$ for some n_0, so that, from (8.5), we have for every $n \geqslant n_0$

$$|\varphi^{[n]}(0)| = |\varphi^{[n-n_0]}(\varphi^{[n_0]}(0))| = 1 - c^{n-n_0}\operatorname{Re} A(\varphi^{[n_0]}(0)) + o\big(c^n\big).$$

In other words, $1 + |\varphi^{[n]}(0)| \to 2$ and

$$1 - |\varphi^{[n]}(0)| \sim c^n A_0 \tag{8.7}$$

for some positive number A_0, implying

$$\lim_n \left(\frac{1 + |\varphi^{[n]}(0)|}{1 - |\varphi^{[n]}(0)|}\right)^{1/pn} = \lim_n \left(\frac{2}{c^n A_0}\right)^{1/pn} = c^{-\frac{1}{p}}.$$

The spectral radius formula now gives (8.6).

Step 2. We claim the following:

$$1 < |\lambda_0| < c^{-\frac{1}{p}} \Rightarrow \lambda_0 \in \sigma(C_\varphi). \tag{8.8}$$

Suppose, to the contrary, that $\left(\lambda_0 I - C_\varphi\right)^{-1}$ exists for such a λ_0. Write $\lambda_0 = c^\alpha$ with $|\lambda_0| = c^{\operatorname{Re}\alpha} < c^{-\frac{1}{p}}$ so that

$$0 > \operatorname{Re}\alpha > -\frac{1}{p}\cdot$$

Let U be an open set containing λ_0 such that the operator-valued function

$$\lambda \in U \longmapsto \left(\lambda I - C_\varphi\right)^{-1}$$

is defined and analytic.

From (8.7), we have

$$\overline{\lim_k} \frac{1 - |\varphi^{[k+1]}(0)|}{1 - |\varphi^{[k]}(0)|} = c < 1.$$

This means that the sequence of points $\big(\varphi^{[k]}(0)\big)$ is a Newman sequence, a fortiori a Carleson interpolation sequence for H^∞.

Now, for any $\lambda \in U$ with $|\lambda| > |\lambda_0|$, consider the sequence

$$\gamma_k = \frac{|c^\alpha|^k}{\big(1 - \varphi^{[k]}(0)\big)^\alpha} \times \frac{\lambda^k}{|\lambda|^k}.$$

Since, by (8.5), $\dfrac{1 - \varphi^{[k]}(0)}{c^k} \to A(0) =: A \neq 0$ and

$$|\gamma_k| = \frac{|c^\alpha|^k}{\big|(1 - \varphi^{[k]}(0))^\alpha\big|} \to |A|^{-\mathrm{Re}(\alpha)},$$

we see that (γ_k) is bounded, and there exists a bounded analytic function, say $h \in H^\infty$, such that, for every $k \in \mathbb{N}$,

$$h\left(\varphi^{[k]}(0)\right) = \gamma_k.$$

Let us now define $g(z) = (1 - z)^\alpha h(z)$. We have $g \in H^p$ since $\mathrm{Re}\,\alpha > -\frac{1}{p}$. This function satisfies

$$g\left(\varphi^{[k]}(0)\right) = |c^\alpha|^k \times \frac{\lambda^k}{|\lambda|^k}. \tag{8.9}$$

The rest of this step 2 is itself divided into several substeps.

Substep 1.
We claim that

$$\sum_{k\geqslant 1} \lambda^{-k} g \circ \varphi^{[k]} \text{ converges in } H^p, \tag{8.10}$$

and in particular $\lambda^{-k} g \circ \varphi^{[k]}$ converges to 0 in H^p.

Indeed, choose $\delta > 0$ small enough to ensure that $|\lambda| > |\lambda_0|^{1+\delta}$ and then choose $q > p$ such that $\mathrm{Re}\,\alpha > -\frac{1}{q} > (1 + \delta)\mathrm{Re}\,\alpha$. We have $g \in H^q$ and

$$\left\| g \circ \varphi^{[k]} \right\|_{H^p} \leqslant \left\| g \circ \varphi^{[k]} \right\|_{H^q} \leqslant \|g\|_{H^q} \left(\frac{1 + |\varphi^{[k]}(0)|}{1 - |\varphi^{[k]}(0)|} \right)^{\frac{1}{q}}$$

using (2.14). But for k large enough, thanks to (8.7), we have

$$\left(\frac{1 + |\varphi^{[k]}(0)|}{1 - |\varphi^{[k]}(0)|} \right)^{\frac{1}{q}} \leqslant \frac{2^{\frac{1}{q}}}{(1 - |\varphi^{[k]}(0)|)^{\frac{1}{q}}} \lesssim c^{-\frac{k}{q}} \leqslant |\lambda_0|^{k(1+\delta)},$$

and this proves the claim (8.10).

Substep 2. We claim that, with convergence in H^p,

$$(\lambda I - C_\varphi)^{-1} g = \lambda^{-1} \sum_{k \geqslant 0} \lambda^{-k} g \circ \varphi^{[k]}. \tag{8.11}$$

We see that (by telescoping), for all $n \geqslant 1$,

$$\begin{aligned}(\lambda I - C_\varphi)\left(\lambda^{-1} \sum_{k=0}^{n} \lambda^{-k} g \circ \varphi^{[k]}\right) &= \sum_{k=0}^{n} \lambda^{-k} g \circ \varphi^{[k]} - \sum_{k=1}^{n+1} \lambda^{-k} g \circ \varphi^{[k]} \\ &= g(z) - \lambda^{-(n+1)} g \circ \varphi^{[n+1]},\end{aligned}$$

so that

$$\begin{aligned}\lim_{n\to\infty} \sum_{k=0}^{n} \lambda^{-k} g \circ \varphi^{[k]} &= (\lambda I - C_\varphi)^{-1} g - \lim_{n\to\infty} (\lambda I - C_\varphi)^{-1} \lambda^{-(n+1)} g \circ \varphi^{[n+1]} \\ &= (\lambda I - C_\varphi)^{-1} g,\end{aligned}$$

and the claim is proved.

Substep 3. We test (8.11) (for $\lambda \in U$ with $|\lambda| > |\lambda_0|$ and $\lambda \to \lambda_0$) at point 0 and get the relation

$$\left[(\lambda I - C_\varphi)^{-1} g\right](0) = \lambda^{-1} \sum_{k \geqslant 0} g\left(\varphi^{[k]}(0)\right)\lambda^{-k}. \tag{8.12}$$

Since $g \in H^p$ and the point evaluation at 0 is continuous on H^p, the LHS of (8.12) tends to $\left[(\lambda_0 - C_\varphi)^{-1} g\right](0)$. But, thanks to (8.9), the value of the RHS is the sum of the geometric series

$$\lambda^{-1} \sum_{k \geqslant 0} |c^\alpha|^k \times \frac{1}{|\lambda|^k} = \lambda^{-1} \sum_{k \geqslant 0} \left(\frac{|\lambda_0|}{|\lambda|}\right)^k,$$

whose modulus is $\dfrac{1}{|\lambda| - |\lambda_0|}$, a quantity that tends to $+\infty$ as $|\lambda| \to |\lambda_0|^+$.

This contradiction proves (8.8) and step 2, namely we have the inclusion $\overline{D}\left(0, \varphi'(\beta)^{-\frac{1}{p}}\right) \setminus \mathbb{D} \subset \sigma(C_\varphi)$.

Together with step 1, this ends the proof of Proposition 8.9. □

Non-inner Case

We will content ourselves with quoting the following theorem [80, theorem 3.8]:

Theorem 8.10 *Let $\varphi : \mathbb{D} \to \mathbb{D}$ be a non-inner symbol with no power of $C_\varphi : H^2 \to H^2$ compact, and a fixed point $z_0 \in \mathbb{D}$. Then:*

(i) *There exists a positive integer N such that the set*

$$S_N = \left\{\varphi^{[N]}(w) : w \in \mathbb{T} \text{ and } |\varphi^{[N]}(w)| = 1\right\}$$

is finite, non-empty, and consists only of fixed points of $\varphi^{[N]}$.

(ii) *If* $c := \min_{z \in S_N} |\varphi'(z)|$, *then* $c > 1$ *and*

$$\sigma(C_\varphi) = \overline{D}\left(0, c^{-1/2N}\right) \cup \left\{\left(\varphi'(z_0)^n\right) : n \in \mathbb{N}\right\} \cup \left\{0, 1\right\}.$$

A simple example of such a φ is the non-inner symbol (with $0 < a < 1$) defined by $\varphi(z) = z\,(az + 1 - a)$, which satisfies $\varphi(0) = 0$, $\varphi'(0) = 1 - a$, $\varphi(1) = 1$, and $\varphi'(1) = 1 + a$. Every iterate $\varphi^{[M]}$ of φ is a polynomial with non-negative coefficients, satisfying $\varphi^{[M]}(1) = 1$, hence $\varphi^{[M]}$ has a radial derivative at 1 and C_φ^M is non-compact. Here, $N = 1$, $c = 1 + a$, and $S_1 = \{1\}$.

8.4 The $\mathscr{H}(\mathbb{D})$-Case

In this section, the symbol φ is *no longer assumed to be analytic on* $\overline{\mathbb{D}}$. We will denote $e_n(z) = z^n$ for all $z \in \mathbb{D}$ and $n \in \mathbb{N}_0$ (so that e_1 is the identity of $\mathscr{H}(\mathbb{D})$); and I the identity operator of $\mathscr{H}(\mathbb{D})$, equipped with its Fréchet topology of uniform convergence on compacta of $\mathbb{D}$ (for which it is a Fréchet algebra without divisors of 0). We will only consider the linear maps $T : \mathscr{H}(\mathbb{D}) \to \mathscr{H}(\mathbb{D})$ that are continuous for that topology.

The notions of spectrum $\sigma(T)$, pointwise spectrum $\sigma_p(T)$, are the same as in the Banach space case. Such an operator T is said to be compact if, for some (equivalently any) neighborhood V of 0 in $\mathscr{H}(\mathbb{D})$, $T(V)$ is relatively compact in $\mathscr{H}(\mathbb{D})$.

Among those continuous operators are the multiplication operators M_w,

$$M_w(f) = wf, \qquad \text{where } w \in \mathscr{H}(\mathbb{D}),$$

and the composition operators $C_\varphi : C_\varphi(f) = f \circ \varphi$, where φ is a *non-constant* analytic self-map of $\mathbb{D}$, called a symbol, so that C_φ is always injective.

Those composition operators T are characterized by their injectivity and the identity $T(fg) = T(f)T(g)$: in that case, $T = C_\varphi$ where $\varphi = T(e_1)$. Indeed, we first observe that φ is a symbol:

$$T\left(\sum_{n=0}^{N} e_n\right)(z) = \sum_{n=0}^{N} (T(e_1))^n(z) = \sum_{n=0}^{N} \varphi^n(z)$$

converges for each $z \in \mathbb{D}$, because $\sum_{n\geqslant 0} e_n$ converges in $\mathcal{H}(\mathbb{D})$, so that $|\varphi(z)| < 1$. Moreover, this φ is non-constant since T is assumed to be injective.

Combining those two types gives the so-called weighted composition operators $T = M_w C_\varphi$, namely

$$T(f) = w.(f \circ \varphi).$$

The following simple proposition and lemma will be constantly used in the sequel.

Proposition 8.11 *The following holds.*

(i) M_w *is invertible if and only if* w *never vanishes and* $M_w^{-1} = M_{\frac{1}{w}}$.
(ii) C_φ *is invertible if and only if* φ *is an automorphism of* $\mathbb{D}$ *and* $(C_\varphi)^{-1} = C_{\varphi^{-1}}$.
(iii) $T = M_w C_\varphi$ *is invertible if and only if* w *does not vanish and* φ *is an automorphism. Moreover*

$$T^{-1} = C_{\varphi^{-1}} M_{w^{-1}} = M_{w_1} C_{\varphi^{-1}} \qquad \textit{with } w_1 = \frac{1}{w \circ \varphi^{-1}}.$$

Proof: The first item is obvious. We just detail the second one, even if it works as in the Hardy case. If C_φ is invertible, let $S = C_\varphi^{-1}$, injective, and $f, g \in \mathcal{H}(\mathbb{D})$ with $f = C_\varphi(u), g = C_\varphi(v)$, where $u, v \in \mathcal{H}(\mathbb{D})$. We see that

$$S(fg) = S\left(C_\varphi(u)C_\varphi(v)\right) = SC_\varphi(uv) = uv = S(f)S(g),$$

so that $S = C_\psi$ for some symbol ψ, and we see that $C_{\psi\circ\varphi} = C_\varphi C_\psi = C_\varphi S = I$ giving $\psi \circ \varphi = e_1$ and similarly $\varphi \circ \psi = e_1$, so that $\varphi \in \text{Aut}(\mathbb{D})$ and $\psi = \varphi^{-1}$. The third item is an obvious consequence of the first two. □

Lemma 8.12 *Let* $T = M_w C_\varphi$.
If $\lambda \in \sigma_p(T)$ *and* $\mu \in \sigma_p(C_\varphi)$, *then* $\lambda\mu \in \sigma_p(T)$.

Proof: Let $f, g \in \mathcal{H}(\mathbb{D})$ with $f \neq 0$, $g \neq 0$ and $w.(f \circ \varphi) = \lambda f$, $g \circ \varphi = \mu g$. Then $h := fg \neq 0$ and $T(h) = \lambda\mu h$. □

Here is another proposition [5, proposition 2.4].

Proposition 8.13 *Let* $T = M_w C_\varphi$.

The following assertions are equivalent:

(i) T *is compact on* $\mathcal{H}(\mathbb{D})$;
(ii) $w = 0$ *or* $\|\varphi\|_\infty < 1$.

Proof: Assume that T is compact. Let V be a neighborhood of 0: there exist $r > 0$ and $\lambda > 1$ such that the sequence $(r\lambda^n z^n)$ belongs to V. By compactness, for some subsequence, $w.\lambda^{n_j}\varphi^{n_j}$ converges in $\mathscr{H}(\mathbb{D})$: for every $z \in \mathbb{D}$, we have $|\lambda\varphi(z)| \leqslant 1$ or $w(z) = 0$. Therefore $w = 0$ or $\|\varphi\|_\infty \leqslant 1/\lambda < 1$. The converse is clear by Montel's theorem. □

Here is now an important tool for going further. It is an extension of the König theorem, namely a weighted König-type theorem, which we will revisit later. We give here an *alternative proof*, borrowed from [5, proposition 3.2].

Lemma 8.14 *Let $\alpha \in \mathbb{D}, w \in \mathscr{H}(\mathbb{D})$ with $w(\alpha) \neq 0$, and let φ be a symbol with $\varphi(\alpha) = \alpha$ and $|\varphi'(\alpha)| < 1$.*

Then, the sequence

$$\eta_n(z) := \frac{w(z)\, w(\varphi(z)) \cdots w\big(\varphi^{[n-1]}(z)\big)}{w(\alpha)^n} \tag{8.13}$$

converges in $\mathscr{H}(\mathbb{D})$ to a function $\eta \in \mathscr{H}(\mathbb{D})$, called the weighted König function associated with w and φ, which is the unique solution in $\mathscr{H}(\mathbb{D})$ of

$$w.(\eta \circ \varphi) = w(\alpha)\, \eta \quad \text{and} \quad \eta(\alpha) = 1.$$

Proof: We sketch a proof. We can assume $\alpha = 0$. Since $|\varphi'(0)| < 1$, using (8.2) gives that, for each $0 < r < 1$, there exists $\delta = \delta(r) \in (0, 1)$ such that

$$|z| \leqslant r \Rightarrow |\varphi^{[n]}(z)| \leqslant \delta^n |z|.$$

This implies the convergence in $\mathscr{H}(\mathbb{D})$ of the infinite product

$$\prod_{n\geqslant 0} \frac{w\big(\varphi^{[n]}(z)\big)}{w(0)}$$

and that of η_n. Then, the inductive relation $w.\,(\eta_n \circ \varphi) = w(0)\,\eta_{n+1}$ gives the equation $w.(\eta \circ \varphi) = w(0)\eta$ by passing to the limit (note that $\eta_n(0) = 1$ for each $n \in \mathbb{N}$). □

Here is first an application of Lemma 8.14: this is an extension of Gunatillake's theorem (see Theorem 8.6) with the description of the associated eigenspace, borrowed again from [5, theorem 3.4]. A more complete form can be found there.

Theorem 8.15 *Let $T = M_w C_\varphi$. Let $\alpha \in \mathbb{D}, w \in \mathscr{H}(\mathbb{D})$ with $w(\alpha) \neq 0$, and let φ be a symbol with $\varphi(\alpha) = \alpha$ and $|\varphi'(\alpha)| < 1$. Let τ and η be the associated König and weighted König functions.*

Then

$$\sigma_p(T) = \left\{w(\alpha).\varphi'(\alpha)^N : N \in \mathbb{N}_0\right\}.$$

Moreover,

$$Ker\left(T - w(\alpha).\varphi'(\alpha)^N I\right) = \mathbb{C}\eta\tau^N.$$

Proof: Let first $N \in \mathbb{N}_0$ and $u = \eta\tau^N$. We have $w(\alpha) \in \sigma_p(T)$ by Lemma 8.14, and $\varphi'(\alpha)^N \in \sigma_p(C_\varphi)$ by Theorem 8.2. Therefore, by Lemma 8.12, we get $w(\alpha)\,\varphi'(\alpha)^N \in \sigma_p(T)$.

Assume conversely that $Tf = \lambda f$ with $f \neq 0, f \in \mathscr{H}(\mathbb{D})$. We may assume $\alpha = 0$ and we write $f(z) = z^N g(z)$ with $g \in \mathscr{H}(\mathbb{D})$ and $g(0) \neq 0$.

Therefore, for $|z| \neq 0$ small enough, we can write

$$\lambda = w(z)\left(\frac{\varphi(z)}{z}\right)^N \frac{g(\varphi(z))}{g(z)}.$$

Passing to the limit as $z \to 0$ gives $\lambda = w(0)\,(\varphi'(0))^N$. In other words,

$$\sigma_p(T) \subset \{w(0)\,\varphi'(0)^N : N \in \mathbb{N}_0\}.$$

Moreover, our function f satisfies $T^n f = w(0)^n \left(\varphi'(0)\right)^{Nn} f$ for every $n \in \mathbb{N}$, or else, for all $z \in \mathbb{D}$ and $n \in \mathbb{N}$,

$$w(0)^n \left(\varphi'(0)^{Nn}\right) z^N g(z) = w(z) \cdots w(\varphi^{[n-1]}(z))\,\varphi^{[n]}(z)^N g\left(\varphi^{[n]}(z)\right). \tag{8.14}$$

This gives, when $\varphi'(0) \neq 0$,

$$f(z) = z^N g(z) = \eta_n(z)\left(\frac{\varphi^{[n]}(z)}{\varphi'(0)^n}\right)^N g\left(\varphi^{[n]}(z)\right),$$

where η_n is as in Lemma 8.14. Taking the limit as $n \to \infty$ gives

$$f = g(0)\,\eta\,\tau^N$$

using the way the function τ is built in the proof of Theorem 8.2.

Since λ cannot be 0, we have $N = 0$ when $\varphi'(0) = 0$. In that case, (8.14) leads also to $f = g(0)\,\eta$. □

Let φ be a symbol, with Denjoy–Wolff point ω.

If $\omega \in \partial\mathbb{D}$, we know that $0 < \varphi'(\omega) \leqslant 1$.

If $\omega \in \mathbb{D}$, we know that $0 \leqslant |\varphi'(\omega)| < 1$ unless φ is an automorphism.

We present below some examples of symbols with various kinds of behavior.

(i) *Situation where $\omega \in \mathbb{D}$ with φ not an automorphism.*
Easy examples are the symbols $\varphi(z) = \lambda z^d$ with $d \geqslant 2$ and $|\lambda| \leqslant 1$.

(ii) *Situation where* $\omega \in \partial\mathbb{D}$ *with* $0 < \varphi'(\omega) < 1$.

An example is the automorphism $\varphi(z) = \dfrac{z+r}{1+rz}$, with $0 < r < 1$, for which $\omega = 1$. We have $\varphi'(1) = \dfrac{1-r}{1+r}\cdot$

Another example is the non-automorphic symbol $\varphi(z) = \dfrac{1+z}{2}$, which corresponds to $\omega = 1$ and $\varphi'(1) = 1/2$.

(iii) *Situation where* $\omega \in \partial\mathbb{D}$ *with* $\varphi'(\omega) = 1$.

An example is the automorphism $\varphi(z) = \dfrac{(2-i)z+i}{-iz+i+2}$, with $\omega = 1$.

A non-automorphic one is given by $\varphi(z) = \dfrac{1}{2-z}$, with $\omega = 1$.

Another class of examples is that of Blaschke products B with associated sequence of zeros denoted (z_n) and Denjoy–Wolff point $\omega = 1$. Then [31, theorem 7.1], $B'(1) = \displaystyle\sum_{n=1}^{\infty} \frac{1-|z_n|^2}{|1-z_n|^2}$ so that

$$B'(1) = 1 \quad \Longleftrightarrow \quad \sum_{n=1}^{\infty} \frac{1-|z_n|^2}{|1-z_n|^2} = 1.$$

We mention the following results corresponding to the situation (i).

Theorem 8.16 *Let* $w \in \mathscr{H}(\mathbb{D})$ *and* φ *be a symbol (not an automorphism) with Denjoy–Wolff point* $\alpha \in \mathbb{D}$.

(i) $\sigma(T) = \sigma_p(T) \cup \{0\}$.

In other words, for $\lambda \notin \sigma_p(T)) \cup \{0\}$ *and every* $g \in \mathscr{H}(\mathbb{D})$, *there exists some solution in* $\mathscr{H}(\mathbb{D})$ *to the equation* $w.f \circ \varphi - \lambda f = g$.

(ii) *Assume that* $w(\alpha) \neq 0$ *and* $\varphi'(\alpha) \neq 0$. *Then*

$$\sigma(T) = \left\{w(\alpha)\,\varphi'(\alpha)^n : n \in \mathbb{N}\right\} \cup \{0\}.$$

The proofs will not be given here (see [5, theorems 4.1 and 4.4]).

We switch to the situation (ii), which relies on the following important theorem due to G. Valiron [177], where $\mathbb{C}_+ = \{z \in \mathbb{C} : \operatorname{Re}(z) > 0\}$.

Theorem 8.17 (Valiron) *Let* φ *be a symbol with Denjoy–Wolff point* $\omega \in \partial\mathbb{D}$ *satisfying* $0 < \varphi'(\omega) < 1$.

Then there exists a holomorphic map $\Psi : \mathbb{D} \to \mathbb{C}_+$ *such that*

$$\Psi \circ \varphi = \frac{1}{\varphi'(\omega)}\Psi.$$

The main point for the applications is that the semi-conjugation map Ψ takes its values in the right half-plane, not only in the complex plane. Therefore Ψ belongs to H^p for every $p \in (0,1)$.

Here is now a spectacular application of Theorem 8.17 which shows to what extent the spectrum of composition operators can be different when one switches from the Banach case (for which it is a compact subset of $\mathbb{C}$) to the Fréchet case, where it can be the whole complex plane.

Theorem 8.18 *Let φ be a symbol with Denjoy–Wolff point $\omega \in \partial\mathbb{D}$ satisfying $0 < \varphi'(\omega) < 1$. Then*

$$\sigma_p(C_\varphi) = \mathbb{C} \setminus \{0\}.$$

Moreover, if φ is not an automorphism, we have $\sigma(C_\varphi) = \mathbb{C}$.

Proof: Let Ψ be as in Theorem 8.17 and let $\beta = 1/\varphi'(\omega) > 1$. For any complex number s, the function $\Psi^s := \exp(s \log \Psi)$ (where log stands for the principal determination of the logarithm) belongs to $\mathscr{H}(\mathbb{D})$ and satisfies

$$C_\varphi(\Psi^s) = \Psi^s \circ \varphi = \beta^s \, \Psi^s,$$

so that $\beta^s \in \sigma_p(C_\varphi)$. But clearly β^s can be any non-zero complex number, so that $\sigma_p(C_\varphi) = \mathbb{C} \setminus \{0\}$ since we know that C_φ is injective, that is, $0 \notin \sigma_p(C_\varphi)$. Observe in passing that, for any $\lambda \in \mathbb{C} \setminus \{0\}$, there are infinitely many s, say $s \in E$, such that $\beta^s = \lambda$ and that the corresponding functions Ψ^s, $s \in E$ are linearly independent (by a standard argument using Vandermonde determinants) and satisfy $C_\varphi(\Psi^s) = \lambda\Psi^s$. This means that $\dim \operatorname{Ker}(C_\varphi - \lambda I) = \infty$.

If moreover φ is not an automorphism, we know from Proposition 8.11 that $0 \in \sigma(C_\varphi)$, which ends the proof. □

Similar conjugation theorems (due to Baker and Pommerenke) exist in the situation (iii), depending on whether those maps have respectively positive, or zero, hyperbolic step [5, theorems 5.11 and 5.12].

Recall first that, when φ is a symbol with a Denjoy–Wolff point $\omega \in \partial\mathbb{D}$ satisfying $\varphi'(\omega) = 1$, we say that

(i) the symbol φ has *positive hyperbolic step* if, for all (equivalently one) $z \in \mathbb{D}$, we have

$$\lim_n \rho\,(\varphi^{[n+1]}(z), \varphi^{[n]}(z)) > 0\,.$$

(ii) the symbol φ has *zero hyperbolic step* if, for all (equivalently one) $z \in \mathbb{D}$, we have

$$\lim_n \rho\,(\varphi^{[n+1]}(z), \varphi^{[n]}(z)) = 0\,.$$

The limit exists by the Schwarz–Pick lemma, which tells that the sequence $\rho(\varphi^{[n+1]}(z), \varphi^{[n]}(z))$ is non-increasing.

A class of examples for the first case is given by the parabolic automorphisms with Denjoy–Wolff point ω, namely:

$$\varphi(z) = \mathrm{e}^{i\theta} \frac{p - z}{1 - \overline{p}z},$$

where $|\theta| \in (0, \pi)$ and $|p| = \cos\theta/2 \in (0, 1)$. Moreover, in this case, $\omega = \dfrac{1 + \mathrm{e}^{i\theta}}{2\overline{p}}\cdot$

In particular, let φ be the symbol

$$\varphi(z) = i\left(\frac{p - z}{1 - pz}\right),$$

with $p = \dfrac{1}{\sqrt{2}}\cdot$ We have $\varphi(\omega) = \omega$ with $\omega = \dfrac{1 + i}{\sqrt{2}}\cdot$ Indeed, for $z \in \mathbb{D}$, since $\varphi^{[n]}$ is an isometry for the pseudo-hyperbolic metric, we see that

$$\rho\left(\varphi^{[n+1]}(z), \varphi^{[n]}(z)\right) = \rho(\varphi(z), z) > 0,$$

since φ has no fixed point inside $\mathbb{D}$. The converse holds true for LFT symbols [31, proposition 6.1]:

Proposition 8.19 *Let φ be a LFT symbol with a Denjoy–Wolff point $\omega \in \partial\mathbb{D}$ satisfying $\varphi'(\omega) = 1$.*

Then, the following are equivalent:

(i) *φ has positive hyperbolic step.*
(ii) *φ is an automorphism of $\mathbb{D}$.*

The situation remains quite mysterious for non-automorphic symbols. A wealth of other examples and related questions can be found in the nice paper [31], including a specific connection between the step and the interpolation properties of the sequence $\left(\varphi^{[n]}(z_0)\right)$ of iterates of a given point.

An explicit example in the second case (zero step) is given by (with $\omega = 1$)

$$\varphi(z) = \frac{1}{2 - z},$$

which satisfies

$$\varphi^{[n]}(z) = \frac{n - (n-1)z}{n + 1 - nz} \quad \text{and} \quad \rho\left(\varphi^{[n+1]}(0), \varphi^{[n]}(0)\right) = \frac{1}{2(n+1)}\cdot$$

It is interesting to observe that this function φ generates a continuous semi-group of symbols

$$\varphi^{[t]}(z) = \frac{t-(t-1)z}{t+1-tz} \qquad \text{for } t \geqslant 0.$$

Now, the theorems of Baker and Pommerenke show that φ is semi-conjugate to the translation by i, namely that there exists $\Lambda \in \mathscr{H}(\mathbb{D})$ such that

$$\Lambda \circ \varphi = \Lambda + i$$

with Λ taking its values respectively in $\mathbb{C}_+$ (positive step) and $\mathbb{C}$ (zero step).

Still for symbols φ with a Denjoy–Wolff point $\omega \in \partial\mathbb{D}$ satisfying $\varphi'(\omega) = 1$, we have the same spectacular phenomenon [5, theorem 7.1], whether φ has positive hyperbolic step or not.

Theorem 8.20 *Let φ be a symbol with a Denjoy–Wolff point $\omega \in \partial\mathbb{D}$ satisfying $\varphi'(\omega) = 1$.*

Then

$$\sigma_p(T) = \mathbb{C} \setminus \{0\} \qquad \text{with dim Ker } (C_\varphi - \lambda I) = \infty \qquad \text{for all } \lambda \neq 0.$$

Moreover, $\sigma(C_\varphi) = \mathbb{C}$ if φ is not an automorphism and $\sigma(C_\varphi) = \sigma_p(C_\varphi)$ otherwise.

Proof: By the Baker and Pommerenke theorem mentioned above, and changing Λ to $-i\Lambda$, there exists a holomorphic map $\Lambda : \mathbb{D} \to \mathbb{C}$ such that

$$\Lambda \circ \varphi = \Lambda + 1.$$

Let $a = |a|\,e^{i\theta} \in \mathbb{C} \setminus \{0\}$. Set $f(z) = |a|^{\Lambda(z)}\,e^{i\theta\Lambda(z)}$. Then $f \in \mathscr{H}(\mathbb{D})$ and

$$f \circ \varphi = |a|^{\Lambda(z)+1} e^{i\theta(\Lambda(z)+1)} = |a| e^{i\theta} f = af.$$

The rest of the proof goes as in the hyperbolic case. □

When we add a weight, the situation can change dramatically [5, theorem 7.2].

Theorem 8.21 *Suppose that φ is symbol with positive hyperbolic step.*

Then there exists a Blaschke product B such that, if $w \in B\mathscr{H}(\mathbb{D})$, the operator $T = M_w C_\varphi$ has no eigenvalue : $\sigma_p(T) = \emptyset$.

The Annulus Case

Here is an interesting complement to the preceding, once more illustrating the difference between the spectrum and point spectrum in the Fréchet case, borrowed from [22]. Here, we fix a number $R > 1$ and consider the symmetric annulus

$$\mathbb{A} = \left\{z \in \mathbb{C} : R^{-1} < |z| < R\right\}$$

and the associated Fréchet space $\mathscr{H}(\mathbb{A})$, equipped with the sequence of seminorms

$$\|f\|_n = \sup_{z \in K_n} |f(z)| \quad \text{with } K_n = \left\{z \in \mathbb{C} : R^{\frac{1}{n}-1} \leqslant |z| \leqslant R^{1-\frac{1}{n}}\right\}.$$

Two natural, simple, weighted composition operators on $\mathscr{H}(\mathbb{A})$ emerge, namely

$$T_{\alpha,\beta} f(z) = \alpha f(\beta z) \quad \text{and} \quad S_{\alpha,\beta} f(z) = \alpha f(\beta/z) \qquad \text{where } \alpha, \beta \in \partial\mathbb{D}. \tag{8.15}$$

The relevance of those operators is evidenced by [22, theorem 4.1].

Theorem 8.22 *Let $T : \mathscr{H}(\mathbb{A}) \to \mathscr{H}(\mathbb{A})$ be a linear and continuous operator, isometric in the sense that*

$$\forall f \in \mathscr{H}(\mathbb{A}), \forall n \in \mathbb{N}, \qquad \|T(f)\|_n = \|f\|_n.$$

Then there exist $\alpha, \beta \in \partial\mathbb{D}$ such that T is either $T_{\alpha,\beta}$ or $S_{\alpha,\beta}$.

The pointwise spectra σ_p and spectra σ of $S_{\alpha,\beta}$ and $T_{\alpha,\beta}$ are given by the following theorems, respectively [22, theorem 6.1] and [22, theorem 6.2].

Theorem 8.23 *Let $\alpha, \beta \in \partial\mathbb{D}$.*

(i) *One always has*

$$\sigma_p(S_{\alpha,\beta}) = \sigma(S_{\alpha,\beta}) = \{\alpha, -\alpha\}.$$

(ii) *Two cases occur*

(a) $\beta^n = 1$ *for some $n \in \mathbb{N}$. Then*

$$\sigma_p(T_{\alpha,\beta}) = \sigma(T_{\alpha,\beta}) = \left\{\alpha\beta^k, 0 \leqslant k \leqslant n-1\right\}.$$

(b) $\beta^n \neq 1$ *for all $n \in \mathbb{N}$. Then*

$$\sigma_p(T_{\alpha,\beta}) = \left\{\alpha\beta^k : k \in \mathbb{Z}\right\} \subset \sigma(T_{\alpha,\beta}) \subset \partial\mathbb{D}.$$

Moreover, the inclusion $\sigma_p(T_{\alpha,\beta}) \subset \sigma(T_{\alpha,\beta})$ can be strict, depending on the diophantine nature of the unimodular number β.

An example of strict inclusion is given by the following situation: take $\alpha = 1$, $\beta = e^{i\pi\theta}$, with θ irrational and "highly Liouville." Then, it is proved (see [22, example 6.4]) that

$$-1 \in \sigma(T_{\alpha,\beta}) \setminus \sigma_p(T_{\alpha,\beta}).$$

9

Schatten Classes for Composition Operators

9.1 Introduction

The purpose of this chapter is to study in some detail the membership of composition operators C_φ in the Schatten class $\mathscr{S}^p(H)$, where H is a weighted Hilbert space (equivalently a rotation-invariant Hilbert space) of analytic functions on $\mathbb{D}$. This is the next step after J. Shapiro's characterization of compactness, and is hence both important and non-trivial. We will also, for a given C_φ, compare this membership when *two* weighted Hilbert spaces come into play.

Some results fall into the general scope of the comparison results of Chapter 12 and allow, for example, simplified and unified proofs of results of Zorboska [187]. Some others do not, and one needs specific facts on those composition operators. Indeed, there are special necessary and sufficient conditions, which will allow us to prove special Schattenness results for composition operators acting mainly on $H = H^2$ but not only. Those criteria are respectively due to Luecking (in terms of Carleson windows or half-windows), and to Luecking–Zhu (in terms of weighted Nevanlinna counting functions). We will reprove them in detail and will also dwell on quite a few of their applications. We will show in particular that those composition operators "separate" Schatten classes: if $q < p$, there exists some C_φ belonging to $\mathscr{S}^p(H)$ but not to $\mathscr{S}^q(H)$.

The context will be that of spaces H_α defined as follows: those are the spaces of functions (analytic on the disk) $f(z) = \sum_{k\geqslant 0} a_k z^k$ such that

$$\sum_{k\geqslant 0} (k+1)^\alpha |a_k|^2 < \infty, \tag{9.1}$$

where $\alpha \in \mathbb{R}$. When $\alpha < 1$, H_α is the weighted Hilbert space $H^2(\beta)$ (see Chapter 12) equipped with the norm

$$\left(\sum_{k=0}^{+\infty} \beta_k |a_k|^2\right)^{\frac{1}{2}},$$

where (see (1.69))

$$\beta_k = \frac{k!\,\Gamma(1-\alpha)}{\Gamma(k+1-\alpha)} \approx (k+1)^{\alpha}.$$

The associated reproducing kernel is $K_\alpha(z,w) = K_{w,\alpha}(z) = (1-\overline{w}z)^{\alpha-1}$.

In particular (with $\gamma = -\alpha - 1$),

$$\text{when } \alpha < 0, \qquad f(z) = \langle f, K_{z,\alpha}\rangle = \int_{\mathbb{D}} f(w)(1-\overline{w}z)^{\alpha-1}\, d\sigma_\gamma(w);$$

$$\text{when } \alpha = 0, \qquad f(z) = \langle f, K_{z,0}\rangle = \int_{\mathbb{T}} f(w)(1-\overline{w}z)^{-1}\, dm(w).$$

For $\alpha < 1$, and $f(z) = \sum_{k\geqslant 0} a_k z^k \in H_\alpha$, the norm of f is "equivalent to" (9.1) (with constants depending only on α)

$$\|f\|_{H_\alpha}^2 \approx \sum_{k\geqslant 0} (k+1)^{\alpha} |a_k|^2 < \infty\,.$$

Actually we shall almost exclusively use the case $\alpha \leqslant 0$.

Warning 1. The space H_α has the advantage to gather in one viewpoint two different kind of spaces: Hardy and Bergman; nevertheless we must from time to time split the discussion or explicit the underlying objects.

For instance, the letters α, γ will always be related as

$$\gamma = -\alpha - 1.$$

Indeed we prefer to keep both in order to facilitate, for the reader, the comparison with Luecking's original paper, which mainly concerns Toeplitz operators T_μ on spaces H_α, whereas the point of view of this book is mainly that of composition operators C_φ on those spaces. We will see a relation of the type

$$C_\varphi^* C_\varphi = T_\mu,$$

which perfectly relates both viewpoints.

Concretely, we are motivated by the fact that H_α is none other than:

- for $\alpha < 0$, the weighted Bergman space $\mathscr{B}_\gamma^2$. We recall that, for $\gamma > -1$,

$$\|f\|_{\mathscr{B}_\gamma^2}^2 = \int_{\mathbb{D}} |f(z)|^2\, d\sigma_\gamma(z) < \infty$$

  with $d\sigma_\gamma(z) := (\gamma+1)(1-|z|^2)^\gamma\, d\sigma(z)$, a probability measure on $\mathbb{D}$.

- For $\alpha = 0$, $H_\alpha = H^2$, the usual Hardy space, and we recall that

$$\|f\|_{H^2}^2 = \int_{\mathbb{T}} |f(z)|^2 \, dm(z) < \infty .$$

Warning 2. In many of the forthcoming theorems, the membership of some Toeplitz operator T_μ in a $\mathscr{S}^p$-class is expressed in terms of an ℓ^p-summability of the sequence $\mu(R_i)\, l(R_i)^{\alpha-1}$ where $l(R_i)$ denotes the side-length of a half Carleson window R_i. It is good to keep in mind the relation

$$\mu(R_i)\, l(R_i)^{\alpha-1} \approx \frac{\mu(R_i)}{\sigma_\gamma(R_i)} \qquad \text{when } \alpha < 0, \quad \text{that is, } \gamma > -1,$$

which expresses this membership in terms of the discrete derivative $\dfrac{\mu(R_i)}{\sigma_\gamma(R_i)}$ of μ with respect to σ_γ.

Some Reminders

Throughout this chapter, we only consider measures μ carried by the *open* disk $\mathbb{D}$. Actually it is mandatory when working on Bergman spaces. When working with Carleson embeddings on Hardy spaces, we usually consider measures carried by $\overline{\mathbb{D}}$ (see Chapter 5). Nevertheless, we restrict ourselves here to the case where $J_\mu : H^2 \to L^2(\mu)$ is compact, which infers that $\mu(\mathbb{T}) = 0$. That is why the Carleson windows and boxes are defined below as subsets of the open disk $\mathbb{D}$ (and not as subsets of $\overline{\mathbb{D}}$).

In order to properly understand the Luecking criterion to follow, it is useful to first recall some definitions and notations.

The Carleson window of size h, $0 < h < 1$, centered at $\xi \in \mathbb{T}$ is (with $-\pi < \arg \leqslant \pi$)

$$W(\xi,h) = \{z \in \mathbb{D} : |z| > 1-h \text{ and } |\arg(\bar{\xi} z)| \leqslant \pi h\}. \tag{9.2}$$

We also use the Carleson circular boxes (with again $\xi \in \mathbb{T}$)

$$S(\xi,h) = \{z \in \mathbb{D} : |\xi - z| < h\}, \tag{9.3}$$

which satisfy $S(\xi,h) \subseteq W(\xi,h) \subseteq S(\xi,4h)$; hence we will often confuse windows and boxes.

We recall that;

- For $\alpha < 0$ (i.e., $\gamma > -1$), a positive measure μ on $\mathbb{D}$ is a *Carleson measure for $\mathscr{B}^2_\gamma$* if

$$\rho_\mu(h) = \sup_{\xi \in \mathbb{T}} \mu\big[W(\xi,h)\big] = \mathrm{O}\,(h^{2+\gamma}).$$

It is vanishing Carleson for $\mathscr{B}^2_\gamma$ if $\rho_\mu(h) = \mathrm{o}\,(h^{2+\gamma})$ as $h \to 0$. By the generalized Carleson embedding theorem (see Chapter 5), μ is a Carleson measure for $\mathscr{B}^2_\gamma$ if and only if the Carleson embedding $J_\mu : \mathscr{B}^2_\gamma \to L^2(\mu)$ is bounded, and it is vanishing Carleson if and only if this Carleson embedding is compact. Since composition operators C_φ are automatically bounded on $\mathscr{B}^2_\gamma$, the maximal function $\rho_{\varphi,\gamma+2}$ of the pullback measure $\mu = \sigma_{\varphi,\gamma} = \varphi(\sigma_\gamma)$, namely

$$\rho_{\varphi,\gamma+2}(h) = \sup_{\xi\in\mathbb{T}} \sigma_{\varphi,\gamma}\big(W(\xi,h)\big), \tag{9.4}$$

satisfies this Carleson condition: $\rho_{\varphi,\gamma+2}(h) = \mathrm{O}\,(h^{2+\gamma})$; and C_φ is compact on $\mathscr{B}^2_\gamma$ if and only if $\rho_{\varphi,\gamma+2}(h) = \mathrm{o}\,(h^{2+\gamma})$.

- For $\alpha = 0$, a measure μ on $\mathbb{D}$ is a *Carleson measure for H^2* if

$$\rho_\mu(h) = \mathrm{O}(h).$$

It is vanishing Carleson for H^2 if $\rho_\mu(h) = \mathrm{o}(h)$ as $h \to 0$. By the Carleson embedding theorem, μ is a Carleson measure for H^2 if and only if the Carleson embedding $J_\mu : H^2 \to L^2(\mu)$ is bounded, and it is vanishing Carleson if and only if this Carleson embedding is compact. Since composition operators C_φ are automatically bounded on H^2, the maximal function ρ_φ defined in (5.30) satisfies this condition $\rho_\varphi(h) = \mathrm{O}(h)$; and C_φ is compact on H^2 if and only if $\rho_\varphi(h) = \mathrm{o}(h)$.

We know that if a composition operator C_φ is compact on the Hardy space H^2, then it is compact on the Bergman space $\mathscr{B}^2$ (see Theorem 3.22 and Chapter 12). Theorems 9.12 and 9.18 show that we can expect a bit better in terms of Schatten classes, but not much better.

9.2 The Luecking Criterion

There are two main criteria related to a symbol φ for membership of the associated composition operator C_φ in a Schatten class $\mathscr{S}^p(H_\alpha)$; a first one by Luecking in terms of Carleson measures, a second one three years later by Luecking and Zhu in terms of Nevanlinna functions. Given the utility of those criteria, we will prove them at the end of this chapter. Indeed, the first one will be proved in three ways. And we will recover directly the second one by the comparison between Nevanlinna counting functions and Carleson measures established in Chapter 7. For now, we state both criteria and show how to exploit them. Some new definitions are needed.

We set $W_{0,0} = \mathbb{D}$ and, for every integer $n \geqslant 1$ and for $j = 0, \ldots, 2^n - 1$,

$$W_{n,j} = W\big(\mathrm{e}^{(2j+1)i\pi/2^n}, 2^{-n}\big). \tag{9.5}$$

The half Carleson windows are defined as $R_{0,0} = \overline{D}(0, 1/2)$ and for all integers $n \geqslant 1$ and j with $0 \leqslant j \leqslant 2^n - 1$ (with here $0 \leqslant \arg z < 2\pi$) as

$$R_{n,j} = \Big\{z \in \mathbb{D} : 1 - \frac{1}{2^n} < |z| \leqslant 1 - \frac{1}{2^{n+1}} \text{ and } \frac{2j\pi}{2^n} \leqslant \arg z < \frac{2(j+1)\pi}{2^n}\Big\}. \tag{9.6}$$

Those boxes are disjoint, contrary to the $W_{n,j}$, and we always have

$$R_{n,j} \subset W_{n,j}\,.$$

We will index them by the set of pairs

$$I = \big\{i = (n, j) : n \geqslant 0,\ 0 \leqslant j \leqslant 2^n - 1\big\} \tag{9.7}$$

and will denote by a_i the center of R_i, that is, if $i = (n, j)$,

$$|a_i| = 1 - (3/4)2^{-n}, \quad \arg a_i = \frac{(2j+1)\pi}{2^n}, \quad l(R_i) = 2^{-n}.$$

The following slow oscillation result is useful.

Lemma 9.1 *The points a_i, $i \in I$, are separated for the metric ρ.*

Moreover, there is a constant $C > 1$ such that, for any $z \in R_i$ and any $a \in \mathbb{D}$, we have

$$C^{-1}|1 - \overline{a}z| \leqslant |1 - \overline{a}a_i| \leqslant C|1 - \overline{a}z|. \tag{9.8}$$

Actually, there exist three constants $0 < r_1 < r_2 < r_3 < 1$ such that

$$\Delta(a_i, r_1) \subset R_i \subset \Delta(a_i, r_2) \subset \overline{\Delta}(a_i, r_2) \subset \Delta(a_i, r_3) =: R_i', \tag{9.9}$$

and there exists an integer $M \geqslant 1$ such that each point of $\mathbb{D}$ belongs at most to M of the R_i' (finite multiplicity property).

Proof: If $i = (n, j)$ and $i' = (n', j')$ are in I with, for example, $n' \geqslant n$, distinguishing the cases $n' > n$ and $n' = n$, $j' \neq j$, we easily check that

$$|a_i - a_{i'}| \geqslant c \max(1 - |a_i|, 1 - |a_{i'}|)$$

for some $c > 0$ and get $\rho(a_i, a_{i'}) \geqslant c(c^2 + 4)^{-1/2}$ by Lemma 1.25.

Next, if $z \in R_i$, we have $|z - a_i| \lesssim 2^{-n}$ and $\min(1 - |z|, 1 - |a_i|) \gtrsim 2^{-n}$. This implies, by the same lemma, that $\rho(z, a_i) \leqslant r < 1$ where r does not depend on $i \in I$. Hence, we can write

$$z = \frac{a_i + w}{1 + \overline{a_i}w} \quad \text{with } |w| \leqslant r, \quad \text{and } 1 - \overline{a}z = \frac{(1 - \overline{a}a_i) + (\overline{a_i} - \overline{a})w}{1 + \overline{a_i}w}.$$

So that

$$\frac{|1-\overline{a}z|}{|1-\overline{a}a_i|}=\left|1+\frac{\overline{a_i}-\overline{a}}{1-\overline{a}a_i}w\right|\times|1+\overline{a_i}w|^{-1}.$$

The RHS is less than $\dfrac{1+|w|}{1-|w|}\leqslant\dfrac{1+r}{1-r}$ and is greater than $\dfrac{1-|w|}{1+|w|}\geqslant\dfrac{1-r}{1+r}\cdot$ This gives (9.8) with $C=\dfrac{1+r}{1-r}\cdot$

For (9.9), only the first inclusion deserves a proof: we only have to choose $r_1>0$ small enough to have $D(a_i,Cr_1(1-|a_i|))\subset R_i$. But it is clear, arguing as in the proof of (5.26), using that $|a_i|\geqslant\dfrac{1}{4}\cdot$ Then from (9.8), we know that $\Delta(a_i,r_1)\subset D(a_i,Cr_1(1-|a_i|))$. □

Remark. The a_i do not form a Blaschke sequence. Indeed,

$$\sum_{i\in I}(1-|a_i|)\approx\sum_{n\geqslant 1}2^n\times 2^{-n}=\infty.$$

Nevertheless, if $s>1$, one has

$$\sum_{i\in I}(1-|a_i|)^s\lesssim\sum_{n\geqslant 1}2^n\times 2^{-ns}<\infty.$$

This means that the measure $\sum_{i\in I}(1-|a_i|)^s\delta_{a_i}$ is finite.

In passing from compactness to the more delicate membership in Schatten classes, we rely on a criterion of Luecking, again in terms of measures.

We need the following family of (non-classical) Toeplitz operators defined on H_α, for a complex or positive measure μ on $\mathbb{D}$, by

$$T_\mu f(z)=\int_{\mathbb{D}}f(w)\overline{K_z(w)}\,d\mu(w)=\int_{\mathbb{D}}f(w)(1-\overline{w}z)^{\alpha-1}\,d\mu(w).$$

This criterion [122, p. 347] is as follows:

Theorem 9.2 (Luecking) *Let μ be a positive, finite measure on $\mathbb{D}$. Let $p>0$ and $\alpha\leqslant 0$.*

The following are equivalent;

(i) $T_\mu\in\mathscr{S}^p(H_\alpha)$.
(ii) *The following sum is finite:*

$$\sum_i\left[\mu(R_i)\times l(R_i)^{\alpha-1}\right]^p<\infty. \tag{9.10}$$

Moreover

$$\mathfrak{s}_p(T_\mu) \approx \left(\sum_i \left[\mu(R_i) \times l(R_i)^{\alpha-1} \right]^p \right)^{1/p}. \tag{9.11}$$

In particular, when $\alpha < 0$,

$$\mathfrak{s}_p(T_\mu) \approx \left(\sum_i \left[\frac{\mu(R_i)}{\sigma_\gamma(R_i)} \right]^p \right)^{1/p}.$$

This statement is the original one in Luecking's paper. Let us rephrase it in terms of Carleson embedding $J_\mu : H_\alpha \to L^2(\mu)$. Indeed, we notice that for every $f \in H_\alpha$ (with $\alpha \leqslant 0$), we have, for every $z \in \mathbb{D}$,

$$\begin{aligned} J_\mu^* J_\mu(f)(z) &= \langle J_\mu^* J_\mu(f), K_z \rangle_{H_\alpha} = \langle J_\mu(f), J_\mu(K_z) \rangle_{L^2(\mu)} \\ &= \int_{\mathbb{D}} f(w) \overline{K_z(w)} \, d\mu(w). \end{aligned}$$

In other words,

$$J_\mu^* J_\mu = T_\mu \qquad \text{and} \qquad a_n(J_\mu) = \sqrt{a_n(T_\mu)} \quad \text{for every } n \in \mathbb{N}. \tag{9.12}$$

Hence we have the following theorem:

Theorem 9.3 *Let* μ *be a positive, finite measure on* $\mathbb{D}$. *Let* $p > 0$ *and* $\alpha \leqslant 0$. *The following are equivalent:*

(i) $J_\mu \in \mathscr{S}^p(H_\alpha)$.

(ii) *The following sum is finite:* $\displaystyle\sum_i \left[\mu(R_i) \times l(R_i)^{\alpha-1} \right]^{\frac{p}{2}} < \infty$.

Moreover, $\displaystyle \mathfrak{s}_p(J_\mu) = \sqrt{\mathfrak{s}_{\frac{p}{2}}(T_\mu)} \approx \left(\sum_i \left[\mu(R_i) \times l(R_i)^{\alpha-1} \right]^{\frac{p}{2}} \right)^{\frac{1}{p}}$.

The preceding version of Luecking's criterion applies in particular to weighted composition operators. We shall not specify how it reads in this case (it is easy following the ideas of Section 5.3.2), but focus below on the versions for composition operators: if we consider a composition operator C_φ on H^2 or on $\mathscr{B}_\gamma^2$, we can apply directly Theorem 9.3, but we also mention the link with Toeplitz operators in this case.

$$\text{For } \gamma > -1,\ C_\varphi^* C_\varphi = T_{\sigma_{\varphi,\gamma}} \qquad \text{and} \qquad C_\varphi^* C_\varphi = T_{m_\varphi} \ \text{ when } \alpha = 0.$$

Hence $C_\varphi \in \mathscr{S}^p(\mathscr{B}_\gamma^2)$ if and only if $T_{\sigma_{\varphi,\gamma}} \in \mathscr{S}^{p/2}(\mathscr{B}_\gamma^2)$ for $\gamma > -1$; similarly $C_\varphi \in \mathscr{S}^p(H^2)$ if and only if $T_{m_\varphi} \in \mathscr{S}^{p/2}(H^2)$.

Therefore, in terms of composition operators, Luecking's criterion becomes Theorem 9.4, since one can use $W_{n,j}$ or $R_{n,j}$ indifferently, as shown by Proposition 9.6 to follow:

Theorem 9.4 (Luecking) *Let φ be a symbol, $p > 0$, and $\gamma > -1$. The following are equivalent:*

(i) $C_\varphi \in \mathscr{S}^p(\mathscr{B}^2_\gamma)$.
(ii) *The following double sum is finite:*

$$\sum_{n=0}^{\infty}\left(\sum_{j=0}^{2^n-1} 2^{n(\gamma+2)p/2}\left[\sigma_{\varphi,\gamma}(R_{n,j})\right]^{p/2}\right) < \infty.$$

Moreover

$$\mathfrak{s}_p(C_\varphi) \approx \left(\sum_{n=0}^{\infty}\sum_{j=0}^{2^n-1} 2^{n(\gamma+2)p/2}\left[\sigma_{\varphi,\gamma}(R_{n,j})\right]^{p/2}\right)^{1/p}$$
$$\approx \left(\sum_{n=0}^{\infty}\sum_{j=0}^{2^n-1} 2^{n(\gamma+2)p/2}\left[\sigma_{\varphi,\gamma}(W_{n,j})\right]^{p/2}\right)^{1/p}.$$

It is worth expliciting the limiting case $H^2 = H_0$: the Carleson measure m_φ is this time carried by the closed disk in general. Nevertheless it is natural to restrict our study to compact composition operators, and, in this case, we know that the necessary condition $m_\varphi(\mathbb{T}) = 0$ holds (see Theorem 3.9 (ii)).

Theorem 9.5 (Luecking) *Let φ be a symbol such that $m_\varphi(\mathbb{T}) = 0$ and $p > 0$.*
The following are equivalent:

(i) $C_\varphi \in \mathscr{S}^p(H^2)$.
(ii) *The following double sum is finite:*

$$\sum_{n=0}^{\infty}\left(\sum_{j=0}^{2^n-1} 2^{np/2}[m_\varphi(R_{n,j})]^{p/2}\right) < \infty. \tag{9.13}$$

Moreover

$$\mathfrak{s}_p(C_\varphi) \approx \left(\sum_{n=0}^{\infty}\sum_{j=0}^{2^n-1} 2^{np/2}\left[m_\varphi(R_{n,j})\right]^{p/2}\right)^{1/p}$$
$$\approx \left(\sum_{n=0}^{\infty}\sum_{j=0}^{2^n-1} 2^{np/2}\left[m_\varphi(W_{n,j})\right]^{p/2}\right)^{1/p}. \tag{9.14}$$

The following proposition proved in [96] justifies that the criterion can be stated indifferently with $R_{n,j}$'s and $W_{n,j}$'s.

Proposition 9.6 *Let μ be a finite positive measure on $\mathbb{D}$, β be positive, and $q > \max\left\{1, 2^{1-\frac{1}{\beta}}\right\}$. The following are equivalent:*

(i) $\mathscr{R} = \sum_{n\geqslant 1} \sum_{0\leqslant j\leqslant 2^n-1} q^{n\beta}[\mu(R_{n,j})]^\beta < \infty.$

(ii) $\mathscr{W} = \sum_{n\geqslant 1} \sum_{0\leqslant j\leqslant 2^n-1} q^{n\beta}[\mu(W_{n,j})]^\beta < \infty.$

Moreover $\mathscr{R} \approx \mathscr{W}$ with underlying constants depending only on β and q.

In particular, we can apply this statement with any $q \geqslant 2$.

The case $q = 2$ justifies the equivalence above in the Luecking criterion in the framework of Hardy spaces, whereas the case $q = 2^{\gamma+2}$ justifies the equivalence in the framework of Bergman spaces.

Proof: (ii) implies (i) since $R_{n,j} \subset W_{n,j}$. For the converse, we define, for positive integers l, n, j with $l \geqslant n$ and $0 \leqslant j \leqslant 2^n - 1$, the set of integers

$$H_{l,n,j} = \left\{k \in \{0, 1, \ldots, 2^l - 1\};\ \frac{j}{2^n} \leqslant \frac{k}{2^l} < \frac{j+1}{2^n}\right\}.$$

It is clear that, for every n and j,

$$W_{n,j} = \bigcup_{l\geqslant n} \bigcup_{k\in H_{l,n,j}} R_{l,k}$$

so that

$$\mu(W_{n,j}) = \sum_{l\geqslant n} \sum_{k\in H_{l,n,j}} \mu(R_{l,k}).$$

We now separate two cases.

- $\beta \leqslant 1$. Here we simply need $q > 1$. Using $\left(\sum_{j=1}^N x_j\right)^\beta \leqslant \sum_{j=1}^N x_j^\beta$ for any $x_1, \ldots, x_N \geqslant 0$ gives

$$\begin{aligned}
\sum_{n\geqslant 1} \sum_{0\leqslant j\leqslant 2^n-1} q^{n\beta}[\mu(W_{n,j})]^\beta &= \sum_{n\geqslant 1} \sum_{0\leqslant j\leqslant 2^n-1} q^{n\beta} \left(\sum_{l\geqslant n} \sum_{k\in H_{l,n,j}} \mu(R_{l,k})\right)^\beta \\
&\leqslant \sum_{n\geqslant 1} \sum_{0\leqslant j\leqslant 2^n-1} q^{n\beta} \sum_{l\geqslant n} \sum_{k\in H_{l,n,j}} \mu(R_{l,k})^\beta \\
&= \sum_{l\geqslant 1} \sum_{0\leqslant k\leqslant 2^l-1} \mu(R_{l,k})^\beta \sum_{\substack{n\leqslant l \\ \{j:k\in H_{l,n,j}\}}} q^{n\beta}.
\end{aligned}$$

But, for any $n \leqslant l$, there is only one j such that $k \in H_{l,n,j}$. Hence

$$\sum_{\substack{n \leqslant l \\ \{j : k \in H_{l,n,j}\}}} q^{n\beta} \leqslant \sum_{n=1}^{l} q^{n\beta} \leqslant C_{\beta,q} q^{l\beta},$$

and the result follows.

- If $\beta > 1$, we can use Hölder's inequality. Let β' be the conjugate exponent of β. Choose a such that $q > a > 2^{\frac{1}{\beta'}}$. Then

$$\begin{aligned} \mu(W_{n,j}) &= \sum_{\substack{l \geqslant n \\ k \in H_{l,n,j}}} \mu(R_{l,k}) \\ &\leqslant \left(\sum_{\substack{l \geqslant n \\ k \in H_{l,n,j}}} a^{-l\beta'} \right)^{1/\beta'} \left(\sum_{\substack{l \geqslant n \\ k \in H_{l,n,j}}} a^{l\beta} \big(\mu(R_{l,k})\big)^{\beta} \right)^{1/\beta}. \end{aligned}$$

Using $|H_{l,n,j}| = 2^{l-n}$, we get

$$\left[\sum_{\substack{l \geqslant n \\ k \in H_{l,n,j}}} a^{-l\beta'} \right]^{1/\beta'} = \left[\sum_{l \geqslant n} 2^{l-n} a^{-l\beta'} \right]^{1/\beta'} = \left(2^{-n} \frac{(2a^{-\beta'})^n}{1 - 2a^{-\beta'}} \right)^{1/\beta'} = C'_{\beta} a^{-n}.$$

Therefore we have

$$\begin{aligned} \sum_{n=1}^{\infty} \sum_{j=0}^{2^n-1} q^{n\beta} \big(\mu(W_{n,j})\big)^{\beta} &\leqslant \sum_{n=1}^{\infty} \sum_{j=0}^{2^n-1} q^{n\beta} C''_{\beta} a^{-n\beta} \sum_{\substack{l \geqslant n \\ k \in H_{l,n,j}}} a^{l\beta} \big(\mu(R_{l,k})\big)^{\beta} \\ &= C''_{\beta} \sum_{l=1}^{\infty} \sum_{k=0}^{2^l-1} \big(\mu(R_{l,k})\big)^{\beta} a^{l\beta} \sum_{\substack{n \leqslant l \\ \{j : k \in H_{l,n,j}\}}} \left(\frac{q}{a}\right)^{n\beta} \\ &\lesssim \sum_{l=1}^{\infty} \sum_{k=0}^{2^l-1} \big(\mu(R_{l,k})\big)^{\beta} a^{l\beta} \left(\frac{q}{a}\right)^{l\beta} \\ &= \sum_{l=1}^{\infty} \sum_{k=0}^{2^l-1} q^{l\beta} \big(\mu(R_{l,k})\big)^{\beta}. \end{aligned}$$

We have proved that (i) implies (ii) for $\beta > 1$ and Proposition 9.6 follows. □

Proposition 9.6 could be expressed in terms of membership in ℓ^{β} (with comparable "norm") for *finite* β. Then a natural question occurs: is it true for $\beta = +\infty$? In other words, is it equivalent to be a Carleson measure relative to the dyadic windows and relative to the dyadic half-windows ? Actually it depends on the framework: it is false (in general) for Hardy spaces, but true for Bergman spaces (as the formulation with hyperbolic disks in HOP theorem may suggest). We provide below a simple argument avoiding hyperbolic disks.

Before that we insist on the fact that it is clearly equivalent to be a Carleson measure relative to the dyadic windows and all Carleson windows (we are still dealing with measures on $\mathbb{D}$ here): any Carleson window is comparable with at most two (well-chosen) consecutive dyadic windows with comparable opening, and this idea will be exploited several times in the proofs of the present chapter.

Proposition 9.7 *Let μ be a finite positive measure on $\mathbb{D}$ and $\gamma > -1$.*

(i) *If* $\sup_{0\leqslant j<2^n} \mu(R_{n,j}) = O\big(2^{-n(\gamma+2)}\big)$ *then* $J_\mu : \mathscr{B}^2_\gamma \to L^2(\mu)$ *is bounded.*

(ii) *If* $\sup_{0\leqslant j<2^n} \mu(R_{n,j}) = o\big(2^{-n(\gamma+2)}\big)$ *then* $J_\mu : \mathscr{B}^2_\gamma \to L^2(\mu)$ *is compact.*

(iii) *There exist some finite measures μ on $\mathbb{D}$ with* $\sup_{0\leqslant j<2^n} \mu(R_{n,j}) = o\left(2^{-n}\right)$, *but the Carleson embedding $J_\mu : H^2 \to L^2(\mu)$ is not even-bounded (hence far from compact).*

In the framework of Hardy spaces, the general question of boundedness of the Carleson embedding $J_\mu : H^2 \to L^2(\mu)$ involves positive (finite) measures on $\overline{\mathbb{D}}$. We could of course consider a quick yet a bit artificial (but giving a clear intuition) example of a measure carried by $\mathbb{T}$. Nevertheless, we recall that in the spirit of the present chapter, we only work here with measures on $\mathbb{D}$.

Proof: (i) For fixed (n, j), with $n \geqslant 1$ and $j < 2^n$, we have

$$W_{n,j} = \bigcup_{\substack{m\geqslant n \\ j2^{m-n}\leqslant k<(j+1)2^{m-n}}} R_{m,k};$$

hence, by hypothesis, for some constant $C > 0$,

$$\mu(W_{n,j}) \leqslant \sum_{m\geqslant n} 2^{m-n} \sup_k \mu(R_{m,k}) \leqslant C \sum_{m\geqslant n} 2^{m-n} 2^{-m(\gamma+2)},$$

but $\sum_{m\geqslant n} 2^{-m(\gamma+1)} = O\big(2^{-n(\gamma+1)}\big)$ since $\gamma + 1 > 0$, and we get

$$\sup_{j<2^n} \mu(W_{n,j}) = O\big(2^{-n(\gamma+2)}\big).$$

Now, as mentioned, we can easily recover the condition for arbitrary Carleson square window: let $h < 1/4$, choose $n \geqslant 1$ such that $2^{-(n+1)} < h \leqslant 2^{-n}$:

for any $\xi \in \mathbb{T}$, $W(\xi,h)$ can be covered by $W_{n-1,j} \cup W_{n-1,j+1}$ for a well-chosen j (viewed modulo 2^{n-1}). Therefore

$$\sup_{\xi\in\mathbb{T}} \mu(W(\xi,h)) = O\left(h^{\gamma+2}\right),$$

and μ is a Carleson measure for $\mathscr{B}^2_\gamma$.

(ii) It suffices to adapt (easily) the previous proof.

(iii) We denote $a_{n,j}$ the center of the dyadic half-window $R_{n,j}$ for $n \in \mathbb{N}$ and $j < 2^n$. We define the (discrete) positive measure

$$\mu = \sum_{n\geqslant 1} \frac{1}{2^n\sqrt{n}} \sum_{0\leqslant j<2^{n/2}} \delta_{a_{n,j}},$$

where δ_a is the Dirac measure at point a. This is a finite positive measure since $\mu(\mathbb{D}) = \sum \dfrac{2^{-n}.2^{n/2}}{\sqrt{n}} < \infty$. This also satisfies $\mu(R_{n,j}) \leqslant \dfrac{1}{2^n\sqrt{n}}$ for every (n,j).

On the other hand,

$$\mu(W_{n,0}) = \sum_{m\geqslant n}\sum_{k<2^{m-n}} \mu(R_{m,k}) = \sum_{m\geqslant n} \frac{1}{2^m\sqrt{m}} \sum_{k<2^{m-n}} \mathbb{1}_{[0,2^{m/2}[}(k).$$

But, when $m < 2n$, we have $2^{m-n} < 2^{m/2}$; therefore

$$\mu(W_{n,0}) \geqslant \sum_{n\leqslant m<2n} \frac{2^{-m}.2^{m-n}}{\sqrt{2n}} = 2^{-n}\frac{\sqrt{n}}{2},$$

and μ is not a Carleson measure for H^2. □

A *sufficient*, easier to test condition emerges from Theorem 9.5.

Corollary 9.8 *Let φ be a symbol, with $m_\varphi(\mathbb{T}) = 0$, and assume that m_φ is a β-Carleson measure, with $\beta > 1$, namely*

$$\rho_\varphi(h) = O(h^\beta) \qquad \text{as } h \to 0.$$

Then $C_\varphi \in \mathscr{S}^p(H^2)$ for every $p > \dfrac{2}{\beta - 1}$.

Proof: Since $\rho_\varphi(h) \lesssim h^\beta$, one gets

$$\sum_{n\geqslant 0} 2^{np/2}\left(\sum_{j=0}^{2^n-1}\left[m_\varphi(R_{n,j})\right]^{p/2}\right) \lesssim \sum_{n\geqslant 0} 2^{np/2}\,2^n\,(2^{-n\beta})^{p/2} = \sum_{n\geqslant 0} 2^{n[1-(\beta-1)p/2]},$$

which is finite since $1 - (\beta - 1)\dfrac{p}{2} < 0$. □

Remark. We could also deduce Corollary 9.8 from the following result, which we will then apply to $\mu = m_\varphi$.

Proposition 9.9 *If μ is a β-Carleson probability measure on $\mathbb{D}$, with $\beta > 2$, then the Poisson integral $\mathcal{P}\colon L^2(\mathbb{T}) \to L^2(\mu)$ belongs to the Schatten class $\mathscr{S}^p$ for any $p > 2/(\beta - 1)$.*

Proof: We may assume that $p \leqslant 2$ since $\mathscr{S}^{p_1} \subseteq \mathscr{S}^{p_2}$ when $p_1 \leqslant p_2$. To get $\mathcal{P} \in \mathscr{S}^p$, it then suffices to have $\sum_{n\in\mathbb{Z}} \|\mathcal{P}(e_n)\|^p_{L^2(\mu)} < +\infty$, where $e_n(\xi) = \xi^n$ for $\xi \in \mathbb{T}$. We have $(\mathcal{P}e_n)(z) = z^n$ for $n \geqslant 0$ and $(\mathcal{P}e_n)(z) = \bar{z}^{|n|}$ for $n \leqslant -1$. Hence

$$\sum_{n\in\mathbb{Z}} \|\mathcal{P}(e_n)\|^p_{L^2(\mu)} \leqslant 2\sum_{n=0}^{\infty} \left(\int_{\mathbb{D}} |z|^{2n}\, d\mu\right)^{p/2}.$$

Now

$$\begin{aligned}\int_{\mathbb{D}} |z|^{2n}\, d\mu &= \int_0^1 2nt^{2n-1}\mu\{|z| \geqslant t\}\, dt \\ &= 2n\int_0^1 (1-x)^{2n-1}\mu\{|z| \geqslant 1-x\}\, dx \\ &\lesssim n\int_0^1 (1-x)^{2n-1}\frac{1}{x}x^\beta\, dx \approx n.n^{-\beta},\end{aligned}$$

since $\{|z| \geqslant 1-x\}$ can be split into $O(1/x)$ windows $W(a,x)$.

Hence, since $(\beta - 1)p/2 > 1$,

$$\sum_{n\in\mathbb{Z}} \|\mathcal{P}(e_n)\|^p_{L^2(\mu)} \lesssim \sum_{n\geqslant 1} \frac{1}{n^{(\beta-1)p/2}} < \infty. \qquad \square$$

Here is now a sometimes useful corollary of Theorem 9.4: a *necessary* condition for membership in $\mathscr{S}^p(\mathscr{B}^2)$. This is, when $C_\varphi \in \mathscr{S}^p(\mathscr{B}^2_\gamma)$, a (slight) reinforcement of the fact that the compactness of $C_\varphi\colon \mathscr{B}^2_\gamma \to \mathscr{B}^2_\gamma$ implies $\rho_{\varphi,\gamma}(h) = o(h^{2+\gamma})$.

Proposition 9.10 (i) *If the composition operator $C_\varphi\colon \mathscr{B}^2_\gamma \to \mathscr{B}^2_\gamma$ is in the Schatten class $\mathscr{S}^p(\mathscr{B}^2_\gamma)$ for some $p \in (0,\infty)$, then*

$$\rho_{\varphi,\gamma+2}(h) = \mathrm{o}\left(h^{2+\gamma}\left(\log\frac{1}{h}\right)^{-2/p}\right). \tag{9.15}$$

(ii) *If $C_\varphi \in \mathscr{S}^p(H^2)$, it holds*

$$\rho_\varphi(h) = \mathrm{o}\left(h\left(\log\frac{1}{h}\right)^{-2/p}\right). \tag{9.16}$$

Proof: (i) We follow the proof of [96, proposition 3.4]. By Theorem 9.4 and Proposition 9.6, we have since $C_\varphi \in \mathcal{S}^p(\mathcal{B}^2_\gamma)$ (see Luecking's condition (9.10))

$$\sum_{n=1}^{\infty}\left(\sum_{j=0}^{2^n-1} 2^{n(\gamma+2)p/2}[\sigma_{\varphi,\gamma}(W_{n,j})]^{p/2}\right) < \infty.$$

Observing that, for $h = 2^{-n}$, each window $W(\xi, h)$ is contained in the union of at most three of the $W_{n,j}$'s, we get

$$\left[\rho_{\varphi,\gamma+2}(2^{-n})\right]^{p/2} \leqslant \left(3 \max_{0\leqslant j\leqslant 2^n-1} \sigma_{\varphi,\gamma}(W_{n,j})\right)^{p/2} \leqslant 3^{p/2}\sum_{j=0}^{2^n-1}\left[\sigma_{\varphi,\gamma}(W_{n,j})\right]^{p/2},$$

and Luecking's condition yields

$$\sum_{n=1}^{\infty}\left[\rho_{\varphi,\gamma+2}(2^{-n})\right]^{p/2} 2^{n(\gamma+2)p/2} < +\infty\,.$$

By the contraction principle established in Theorem 5.38 for the Bergman spaces $\mathcal{B}^2_\gamma$ for any $\gamma > -1$, we have a constant $C_0 > 0$ depending only on γ such that

$$\rho_{\varphi,\gamma+2}(\varepsilon h) \leqslant C_0\, \varepsilon^{2+\gamma}\, \rho_{\varphi,\gamma+2}(h)$$

for $0 < \varepsilon \leqslant 1$ and $0 < h < 1$. Hence, if we set

$$u_n = \left(\frac{\rho_{\varphi,\gamma+2}(2^{-n})}{2^{-n(\gamma+2)}}\right)^{p/2},$$

we have for $n \geqslant k$, $\quad u_n \leqslant C_0^{p/2}\, u_k\,.$

We recall here the following elementary fact: let $\sum u_n$ be a convergent series of positive numbers such that $u_n \leqslant C\, u_k$ for $n \geqslant k$, for some positive constant C. Then $u_n = \mathrm{o}\,(1/n)$.

Hence we get $n\left(\dfrac{\rho_{\varphi,\gamma+2}(2^{-n})}{2^{-n(\gamma+2)}}\right)^{p/2} \underset{n\to\infty}{\longrightarrow} 0$, ending the proof of (i).

The proof of (ii) follows the same ideas and arguments: this is formally the preceding one with $\gamma = -1$, but we use the corresponding contraction principle Theorem 5.29 proved for measures m_φ. □

9.3 The Luecking–Zhu Criterion

The Criterion

Whereas Luecking's criterion was expressed in terms of half Carleson windows, the present criterion is given in terms of Nevanlinna counting functions. It must be noticed that, whereas Luecking's criterion applies to general Carleson embeddings, the Luecking–Zhu criterion concerns composition operators.

Denote by $d\nu(w) = \dfrac{d\sigma(w)}{(1-|w|^2)^2}$ the hyperbolic (Möbius invariant) measure on $\mathbb{D}$. This means that

$$\int_{\mathbb{D}} (f \circ \varphi)\, d\nu = \int_{\mathbb{D}} f\, d\nu$$

for every automorphism φ of $\mathbb{D}$ and every test function f, and this is easily checked (with φ^{-1} instead of φ) using the change of variable formula and the identity

$$\frac{|\varphi'(z)|}{1-|\varphi(z)|^2} = \frac{1}{1-|z|^2}.$$

In the sequel, we denote by $N_{\varphi,\beta}$ and $M_{\varphi,\beta}$, with $\beta \geqslant 1$, the *weighted and normalized weighted Nevanlinna counting functions* that follow

$$N_{\varphi,\beta}(w) = \sum_{\varphi(z)=w} \left(\log\frac{1}{|z|}\right)^{\beta} \tag{9.17}$$

if $w \in \varphi(\mathbb{D}) \setminus \varphi(0)$, and $N_{\varphi,\beta}(w) = 0$ otherwise.

$$M_{\varphi,\beta}(w) = \sqrt{\frac{N_{\varphi,\beta}(w)}{(\log 1/|w|)^{\beta}}}. \tag{9.18}$$

Observe that $N_{\varphi,1}$ is the classical Littlewood function N_φ encountered several times in this book. In [123, theorems 1 and 3], the following is proved.

Theorem 9.11 (Luecking–Zhu) *Let $\gamma \geqslant -1$, $p > 0$, and let φ be a symbol. The following are equivalent;*

(i) $C_\varphi \in \mathscr{S}^p(H_{-\gamma-1})$.
(ii) *The normalized counting function $M_{\varphi,\gamma+2}$ belongs to $L^p(\nu)$.*

We recall that, for $\gamma = -1$, we have $H_0 = H^2$ and, for $\gamma > -1$, we have $H_{-\gamma-1} = \mathscr{B}^2_\gamma$.

In the special case $\gamma = -1$, the proof below is taken from [100] (exploiting the results of Chapter 7).

Proof: Assume first that condition (i) is satisfied.

We begin with the case $\gamma = -1$: $C_\varphi \in \mathcal{S}^p(H^2)$, so we know that, thanks to Theorem 9.5 and Proposition 9.6,

$$\sum_{n=0}^{\infty} \left(\sum_{j=0}^{2^n-1} 2^{np/2} [m_\varphi(R_{n,j})]^{p/2} \right) < \infty .$$

By Theorem 7.2 and (7.6), for n large enough (say $n \geqslant n_0 \geqslant 1$), there is a constant $C > 0$ such that $N_\varphi(z) \leqslant C\, m_\varphi(\widetilde{W}_{n,j})$, for every $z \in R_{n,j}$, where $\widetilde{W}_{n,j}$ is the Carleson box with same center as $W_{n,j}$, but with opening 2.2^{-n}. Though the boxes $W_{n-1,j}$, with $j = 0, 1, \ldots, 2^{n-1} - 1$, have the same size as $\widetilde{W}_{n,j}$, they have a different center; nevertheless, each $\widetilde{W}_{n,j}$ can be covered by the boxes $W_{n-1,l}$ and $W_{n-1,l+1}$, for some $l = l_j \in \{0, \ldots, 2^{n-1} - 1\}$ (where $l + 1$ is understood as 0 if $l = 2^{n-1} - 1$), and we must add that each value $l \in \{0, \ldots, 2^{n-1} - 1\}$ is taken at most three times (for a fixed n), that is, there are at most three j such that $l = l_j$.

We get, with $\Omega = \mathbb{D} \setminus D\big(0, 1 - 2^{-n_0}\big)$,

$$\begin{aligned}
\int_\Omega \left(\frac{N_\varphi(z)}{\log(1/|z|)} \right)^{p/2} dv(z) &\lesssim \int_\Omega \frac{(N_\varphi(z))^{p/2}}{(1 - |z|)^{\frac{p}{2}+2}}\, d\sigma(z) \\
&\lesssim \sum_{\substack{n \geqslant n_0 \\ j < 2^n}} \int_{R_{n,j}} (2^n)^{\frac{p}{2}+2} \big(N_\varphi(z)\big)^{p/2}\, d\sigma(z) \\
&\lesssim \sum_{n \geqslant 1} \sum_{j < 2^n} \int_{R_{n,j}} (2^n)^{\frac{p}{2}+2} \big(m_\varphi(\widetilde{W}_{n,j})\big)^{p/2}\, d\sigma(z) \\
&\lesssim \sum_{n \geqslant 1} \sum_{j < 2^n} 2^{\frac{np}{2}} \big(m_\varphi(\widetilde{W}_{n,j})\big)^{p/2} \\
&\lesssim \sum_{n \geqslant 1} \sum_{l < 2^{n-1}} 2^{\frac{np}{2}} \big(m_\varphi(W_{n-1,l})\big)^{p/2} < \infty .
\end{aligned}$$

The integral over $D\big(0, 1 - 2^{-n_0}\big)$ is clearly finite and (ii) holds for $\gamma = -1$.

Now, for $\gamma > -1$, the strategy is the same: we invoke Theorem 7.7 and (7.17) to get, for $n \geqslant n_0$ and every $z \in R_{n,j}$,

$$N_{\varphi,\gamma+2}(z) \leqslant C\, \sigma_{\varphi,\gamma}(\widetilde{W}_{n,j}) .$$

Then, we mimic the previous computations (skipping details):

$$\begin{aligned}
&\int_\Omega \left(\frac{N_{\varphi,\gamma+2}(z)}{\big(\log(1/|z|)\big)^{\gamma+2}} \right)^{p/2} dv(z) \\
&\quad \lesssim \sum_{\substack{n\geqslant n_0 \\ j}} \int_{R_{n,j}} (2^n)^{\frac{p(\gamma+2)}{2}+2} \big(N_{\varphi,\gamma+2}(z)\big)^{p/2}\, d\sigma(z) \\
&\quad \lesssim \sum_{n,j} (2^n)^{p(\gamma+2)/2} \big(\sigma_{\varphi,\gamma}(\widetilde{W}_{n,j})\big)^{p/2} \\
&\quad \lesssim \sum_{n,l} (2^n)^{p(\gamma+2)/2} \big(\sigma_{\varphi,\gamma}(W_{n,l})\big)^{p/2} < \infty,
\end{aligned}$$

which is finite thanks to Theorem 9.4 and Proposition 9.6; hence (ii) holds for $\gamma > -1$.

Conversely, assume that (ii) is satisfied.

Once again we begin with the case $\gamma = -1$. We shall use the following inequality proved in Section 7.4 (take $q = p/2$; Theorem 7.16):

$$[m_\varphi(W_{n,j})]^{p/2} \lesssim \frac{1}{\sigma(\widetilde{W}_{n,j})} \int_{\widetilde{W}_{n,j}} [N_\varphi(z)]^{p/2}\, d\sigma(z), \tag{9.19}$$

where $\widetilde{W}_{n,j}$ has same center as $W_{n,j}$ but an opening $2^L.2^{-n}$ for some fixed integer $L \geqslant 1$. We obtain

$$\begin{aligned}
\sum_{n,j} [2^n\, m_\varphi(W_{n,j})]^{p/2} &\lesssim \sum_{n,j} 2^{np/2}\, 2^{2n} \int_{\widetilde{W}_{n,j}} [N_\varphi(z)]^{p/2}\, d\sigma(z) \\
&= \int_{\mathbb{D}} \left(\sum_n 2^{n(2+\frac{p}{2})} \Big[\sum_j \mathbb{1}_{\widetilde{W}_{n,j}}(z) \Big] \right) [N_\varphi(z)]^{p/2}\, d\sigma(z).
\end{aligned}$$

But, for $k \in \mathbb{N}_0$, when z belongs to $\Gamma_k = \{z \in \mathbb{C} : 2^{-k} > 1 - |z| \geqslant 2^{-(k+1)}\}$, one has $z \in \widetilde{W}_{n,j}$ only if $n \leqslant k + L$, and then, for each such n, z belongs at most to 2^L boxes $\widetilde{W}_{n,j}$. It follows that for $z \in \Gamma_k$:

$$\sum_n 2^{n(2+\frac{p}{2})} \sum_j \mathbb{1}_{\widetilde{W}_{n,j}}(z) \lesssim 2^{k(2+\frac{p}{2})} \lesssim \frac{1}{(1-|z|)^{2+\frac{p}{2}}};$$

hence

$$\sum_{n,j} \big[2^n\, m_\varphi(W_{n,j})\big]^{p/2} \lesssim \int_{\mathbb{D}} \frac{[N_\varphi(z)]^{p/2}}{(1-|z|)^{\frac{p}{2}+2}}\, d\sigma(z) < \infty,$$

and (i) holds.

Finally, the case $\gamma > -1$ can be proven following the sames lines when $p \geqslant 2$, starting from the weighted version of (9.19) (obvious consequence from the case $p = 2$, which is (7.18)).

When $p < 2$, we refer to [123]. □

Applications

In this subsection, we first show that this Luecking–Zhu criterion leads to self-improvements of the behavior of C_φ when we switch from H^2 to $\mathscr{B}^2$, more generally from $\mathscr{B}^2_{\gamma_1}$ to $\mathscr{B}^2_{\gamma_2}$ with $\gamma_2 > \gamma_1$, which seems out of reach with the general comparison results of Chapter 12.

Theorem 9.12 ([105]) *Let γ_1 and γ_2, with $\gamma_2 > \gamma_1 \geqslant -1$. Then, for every $p > 0$ and any symbol φ, we have:*

(i) *$C_\varphi \in \mathscr{S}^p(H_{-\gamma_1-1})$ implies $C_\varphi \in \mathscr{S}^{\tilde{p}}(H_{-\gamma_2-1})$, with $\tilde{p} = \frac{\gamma_1+2}{\gamma_2+2}p < p$;*
(ii) *when φ is finitely valent, in particular injective, the converse holds;*
(iii) *in particular, $C_\varphi \in \mathscr{S}^p(H^2)$ implies $C_\varphi \in \mathscr{S}^{p/2}(\mathscr{B}^2)$;*
(iv) *when $\gamma_1 > -1$, $C_\varphi \in \mathscr{S}^p(\mathscr{B}^2_{\gamma_1})$ implies $C_\varphi \in \mathscr{S}^{\tilde{p}}(\mathscr{B}^2_{\gamma_2})$, with $\tilde{p} = \frac{\gamma_1+2}{\gamma_2+2}p < p$.*

Proof: The proof goes as follows:

(i) We use Theorem 9.11. Set $\beta_j = \gamma_j + 2$, with $1 \leqslant \beta_1 < \beta_2$. The ℓ_{β_2} norm is smaller than the ℓ_{β_1} norm; so that

$$[N_{\varphi,\,\beta_2}]^{1/\beta_2} \leqslant [N_{\varphi,\,\beta_1}]^{1/\beta_1} \quad \text{and} \quad [M_{\varphi,\,\beta_2}]^{1/\beta_2} \leqslant [M_{\varphi,\,\beta_1}]^{1/\beta_1}. \tag{9.20}$$

It follows that, for $-1 \leqslant \gamma_1 < \gamma_2$,

$$\begin{aligned}\int_{\mathbb{D}} \big(M_{\varphi,\,\beta_2}(w)\big)^{\tilde{p}} d\nu(w) &= \int_{\mathbb{D}} \big(M_{\varphi,\,\beta_2}(w)\big)^{p\beta_1/\beta_2} d\nu(w)\\ &\leqslant \int_{\mathbb{D}} \big(M_{\varphi,\,\beta_1}(w)\big)^{p} d\nu(w),\end{aligned}$$

and Theorem 9.11 shows that $C_\varphi \in \mathscr{S}^{\tilde{p}}(H_{-\gamma_2-1})$ whenever $C_\varphi \in \mathscr{S}^p(H_{-\gamma_1-1})$.

(ii) Assume that φ is s-valent; it means that, for every $w \in \mathbb{D}$,

$$n_\varphi(w) := \sum_{\varphi(z)=w} 1 = \text{card}\,\{z \in \mathbb{D}; \varphi(z) = w\} \leqslant s.$$

Using Hölder's inequality, we get, since $1 \leqslant \beta_1 < \beta_2$,

$$\begin{aligned}N_{\varphi,\,\beta_1}(w) &\leqslant \big[n_\varphi(w)\big]^{(\beta_2-\beta_1)/\beta_2}\big[N_{\varphi,\,\beta_2}(w)\big]^{\beta_1/\beta_2}\\ &\leqslant s^{(\beta_2-\beta_1)/\beta_2}\big[N_{\varphi,\,\beta_1}(w)\big]^{\beta_1/\beta_2}.\end{aligned}$$

Therefore

$$\int_{\mathbb{D}} \big(M_{\varphi,\,\beta_1}(w)\big)^p dv(w) \leqslant s^{p(\beta_2-\beta_1)/(2\beta_2)} \int_{\mathbb{D}} \big(M_{\varphi,\,\beta_2}(w)\big)^{\tilde{p}} dv(w),$$

and $C_\varphi \in \mathscr{S}^{\tilde{p}}(H_{-\gamma_2-1})$ implies that $C_\varphi \in \mathscr{S}^p(H_{-\gamma_1-1})$.

(iii) Just take $\gamma_1 = -1$ and $\gamma_2 = 0$. And (iv) is obvious from (i). □

Remark. We can also give a proof based on Luecking's criterion: we only detail it for the pair $(H^2, \mathscr{B}^2)$ to simplify; that is, we only justify (iii) in Theorem 9.12, but this is easy to adapt to get the weighted version (i).

We start from the following inequality (this is the special case (7.28) of Theorem 7.13 in Chapter 7):

There exists some positive integer N such that, for all $\xi \in \mathbb{T}$ and $0 < h < 1$ small enough,

$$\sigma_\varphi[W(\xi, h)] \leqslant C\,\big(m_\varphi[W(\xi, 2^N h)]\big)^2. \tag{9.21}$$

Hence, for $n \geqslant N$ (where $N \geqslant 1$ is large enough), introducing the box $W'_{n,j}$ that has same center $W_{n,j}$ but with opening $2^N 2^{-n}$, it holds that

$$\sum_{j=0}^{2^n-1} \big(4^n \sigma_\varphi(W_{n,j})\big)^{p/4} \leqslant C \sum_{j=0}^{2^n-1} \big(2^n m_\varphi(W'_{n,j})\big)^{p/2},$$

but each box $W'_{n,j}$ is contained in the union of at most two consecutive dyadic Carleson boxes W_{n-N,l_j} and W_{n-N,l_j+1} of size 2^{N-n} (conversely, the value l_j is taken at most $2^{N+1} - 1$ times, when n is fixed) and we get

$$\begin{aligned}\sum_{j=0}^{2^n-1} \big(4^n \sigma_\varphi(W_{n,j})\big)^{p/4} &\lesssim \sum_{l=0}^{2^{n-N}-1} \big(2^n m_\varphi(W_{n-N,j})\big)^{p/2}\\ &\lesssim \sum_{l=0}^{2^{n-N}-1} \big(2^{n-N} m_\varphi(W_{n-N,j})\big)^{p/2}.\end{aligned}$$

Summing over $n \geqslant N$, and thanks to Luecking's Theorem 9.5, we conclude again that $C_\varphi \in \mathscr{S}^p(H^2)$ implies $C_\varphi \in \mathscr{S}^{p/2}(\mathscr{B}^2)$.

We also mention another application of the Luecking–Zhu Criterion due to Zhu [183] for Bergman spaces:

Theorem 9.13 *Let $\gamma > -1$, $p \geqslant 2$, and let φ be a finitely valent symbol. The following are equivalent:*

(i) $C_\varphi \in \mathscr{S}^p(\mathscr{B}^2_\gamma)$.

(ii) $\displaystyle\int_{\mathbb{D}} \left(\frac{1-|z|^2}{1-|\varphi(z)|^2}\right)^{\frac{p(\gamma+2)}{2}} dv(z) < \infty.$

Actually, when $p = 2$ (i.e., when we focus on the Hilbert–Schmidt character), the result is clearly true without any valence condition.

9.4 The Shapiro–Taylor Maps

We refer to Chapter 4 where special symbols were introduced; here, we focus in particular on three variants of Shapiro–Taylor maps: they are designed to exploit further the Luecking criterion and emphasize separation phenomena in Schatten classes.

The First Shapiro–Taylor Map

We use here the map ϕ_θ, where $\theta > 0$, introduced in Chapter 4 (see Section 4.5) and its basic properties.

We use below the notations of Section 4.5 and prove the following theorem, in which ε is implicit in the definition of ϕ_θ (assertion (ii) was proved by J. Shapiro and P. Taylor in [166] and assertion (iii) appears in [96]).

Theorem 9.14 *Let $\gamma > -1$. For ε small enough, we have:*

(i) *The map ϕ_θ is injective.*
(ii) *$C_{\phi_\theta} : H^2 \to H^2$ is compact for every $\theta > 0$.*
(iii) *C_{ϕ_θ} belongs to the Schatten class $\mathscr{S}^p(H^2)$ if and only if $p > \dfrac{4}{\theta}\cdot$*
(iv) *C_{ϕ_θ} belongs to the Schatten class $\mathscr{S}^p(\mathscr{B}^2_\gamma)$ if and only if $p > \dfrac{4}{(\gamma+2)\theta}\cdot$*

As a consequence, for each $p_0 > 0$, there exists a symbol φ such that we have $C_\varphi \in \mathscr{S}^p(H^2)$ for all $p > p_0$ but $C_\varphi \notin \mathscr{S}^{p_0}(H^2)$.
Similarly there exists a symbol φ such that we have $C_\varphi \in \mathscr{S}^p(\mathscr{B}^2_\gamma)$ for all $p > p_0$ but $C_\varphi \notin \mathscr{S}^{p_0}(\mathscr{B}^2_\gamma)$.

Proof: The consequences are obvious: for instance, for H^2, just take $\theta = 4/p_0$.

Lemma 4.9 shows that f_θ is injective on V_ε, as well as $f_\theta \circ g$ on $\mathbb{D}$. Since clearly $|f_\theta \circ g| < \pi$ on $\mathbb{D}$ for ε small enough, the injectivity of ϕ_θ follows. Since our symbol is $\phi_\theta = \exp(-h_\theta)$, and $\exp\big(h_\theta(\mathrm{e}^{it})\big) \to 1$ as $t \to 0$, it follows that the derivatives have the same behaviour when $t \to 0$. However, we are going to recover this compactness result by another method. For convenience, we shall write the boundary values $f_\theta^*, g^*, \ldots$ of the different analytic functions $f_\theta, g, \ldots$ in the same way as the analytic functions, without the exponent *.

We detail here (ii).

Note that, by Lemma 4.9, $|\phi_\theta|$ is far from 1 when $g(z)$ belongs to the half-circle $\{\varepsilon e^{i\alpha}\,;\ |\alpha| \leqslant \pi/2\}$. Moreover, since $f_\theta(it) = it\left(\log\frac{1}{|t|} - i\,\mathrm{sgn}\,(t)\frac{\pi}{2}\right)^\theta$, one has

$$|f_\theta(it)| \approx |t|\left(\log\frac{1}{|t|}\right)^\theta,$$

so that $|\phi_\theta\big(g^{-1}(it)\big)|$ is far from 1 when t is away from 0. Therefore, it suffices to study what happens when t is in a neighbourhood of 0.

Also note that g is bi-Lipschitz in a neighborhood of 1 (so $g(e^{it}) \approx it$ when $|t| \leqslant \pi/2$), so we may ignore it and only consider the measure of those t's for which $f_\theta(it)$ belongs to the suitable sets. Moreover, for convenience, we only write the proof for $t > 0$.

Since

$$\begin{aligned} f_\theta(it) &= it\left(\log\frac{1}{t}\right)^\theta\left[\left(1 - \frac{i\pi/2}{\log 1/t}\right)^\theta\right] \\ &= it\left(\log\frac{1}{t}\right)^\theta\left[1 - \frac{i\pi\theta/2}{\log 1/t} + O\left(\frac{1}{\log 1/t}\right)^2\right], \end{aligned} \tag{9.22}$$

one has, when $t \to 0^+$,

$$\mathrm{Re}\, f_\theta(it) \sim \frac{\pi\theta\, t}{2}\left(\log\frac{1}{t}\right)^{\theta-1}$$

$$A(t) := \mathrm{Im}\, f_\theta(it) = t\left(\log\frac{1}{t}\right)^\theta + O\left(t\left(\log\frac{1}{t}\right)^{\theta-2}\right) \sim t\left(\log\frac{1}{t}\right)^\theta$$

$$A'(t) \sim \left(\log\frac{1}{t}\right)^\theta.$$

Let us mention that for every $\delta \in \mathbb{R}$, the function $t \in (0,1) \mapsto t\big(\log(1/t)\big)^\delta$ is increasing in a neighborhood of 0^+.

Now, we have $\exp\big(-f_\theta(it)\big) \in W(e^{-i\alpha}, h)$ $(0 \leqslant \alpha < 2\pi)$ if and only if

$$\mathrm{Re}\, f_\theta(it) \leqslant h\big(1 + o(1)\big) \tag{9.23}$$

$$|\mathrm{Im}\, f_\theta(it) - \alpha| \leqslant \pi h. \tag{9.24}$$

We will argue as follows. First take $\alpha = 0$. Then, (9.24) (and therefore (9.23)) holds as soon as $t \leqslant \dfrac{h}{(\log(1/h)^\theta}$ (and t small enough), so that

$$\rho_{\phi_\theta}(h) \gtrsim \frac{h}{(\log(1/h))^\theta}.$$

Now, take α arbitrary with $\alpha > 0$ (without loss of generality, by symmetry), and assume that conditions (9.23), (9.24) are satisfied. It follows that $\dfrac{\pi\theta\, t}{2}(\log 1/t)^{\theta-1} \leqslant h\,\big(1 + o(1)\big)$; hence

$$t\left(\log\frac{1}{t}\right)^{\theta}\leqslant\frac{2h}{\pi\theta}\left(\log\frac{1}{h}\right)\big(1+o(1)\big).$$

So that the implied argument α must be small:

$$\alpha\leqslant h+\frac{2h}{\pi\theta}\left(\log\frac{1}{h}\right)\big(1+o(1)\big)\approx h\log\frac{1}{h}\cdot$$

Then condition (9.24), namely $A(t)\in[\alpha-h,\alpha+h]$, gives us

$$t\in I(h):=[A^{-1}(\alpha-h),A^{-1}(\alpha+h)],$$

an interval whose length $l(h)$ equals, for some $c\in[\alpha-h,\alpha+h]$,

$$l(h)=2h(A^{-1})'(c)=\frac{2h}{A'(A^{-1}(c))}\leqslant\frac{2h}{A'(A^{-1}(\alpha+h))}\lesssim\frac{h}{(\log 1/h)^{\theta}}$$

given that, near 0, A^{-1} increases, A' decreases, with (for $u>0$ and small):

$$A^{-1}(u)\sim\frac{u}{(\log 1/u)^{\theta}}\quad\text{and}\quad A'(u)\sim\big(\log(1/u)\big)^{\theta}\cdot$$

This implies that $m_{\varphi}(W(e^{-i\alpha},h))\leqslant l(h)\lesssim\dfrac{h}{(\log 1/h)^{\theta}}\cdot$

All in all, we have proved that

$$\rho_{\phi_\theta}(h)\approx\frac{h}{(\log 1/h)^{\theta}}\cdot\tag{9.25}$$

Since $\rho_{\phi_\theta}(h)=o(h)$, the compactness criterion (Theorem 5.14) gives the compactness of C_{ϕ_θ}. Since the symbol is univalent, we could also justify the compactness of C_{ϕ_θ} proving that $\lim\limits_{|z|\to 1^-}\dfrac{1-|z|}{1-|\phi_\theta(z)|}=0$. We shall use this idea for the third Shapiro–Taylor map below. Anyway, we got more information for our symbol with (9.25).

Now we shall examine the membership of C_{ϕ_θ} in the Schatten class $\mathscr{S}^p$ and shall use Luecking's theorem (Theorem 9.5) to that effect. We have to analyze the behavior of ϕ_θ with respect to the Luecking sets $R_{n,j}$; actually, for convenience, we shall work with

$$R'_{n,j}=R_{n,\,2^n-1-j}\,.$$

We must consider the modulus and argument constraints and the ideas developed above, but in a more detailed and accurate way.

Modulus constraint. The condition $\dfrac{h}{2}\leqslant 1-\exp(-\mathrm{Re}\,f_\theta(it))<h$ writes

$$\frac{h}{2}\leqslant\frac{\pi\theta\,t}{2}\left(\log\frac{1}{t}\right)^{\theta-1}\big(1+o(1)\big)<h,$$

and reads (observing first that this imposes $\log(1/t) \sim \log(1/h)$)

$$\frac{h/2}{(\log 1/h)^{\theta-1}}\big(1+o(1)\big) \leqslant \frac{\pi\theta\, t}{2} < \frac{h}{(\log 1/h)^{\theta-1}}\big(1+o(1)\big),$$

or, when $h = h_n = 2^{-n}$,

$$\frac{2^{-n}}{\pi\theta n^{\theta-1}}\big(1+o(1)\big) \leqslant t < \frac{2^{-(n-1)}}{\pi\theta n^{\theta-1}}\big(1+o(1)\big).$$

In other words, t must belong to $I_n = [a_n, b_n]$, with

$$a_n \sim \frac{2^{-n}}{\pi\theta n^{\theta-1}}, \qquad b_n \sim \frac{2^{-(n-1)}}{\pi\theta n^{\theta-1}}, \tag{9.26}$$

and we have $|I_n| \approx \dfrac{2^{-n}}{n^{\theta-1}}$.

Argument constraint. We are looking for the set J_n of those indices j, with $j = 0, 1, \ldots, 2^n - 1$ for which $\exp\big(-f_\theta(it)\big) \in R'_{n,j}$ for some t. This imposes both the modulus constraint $t \in I_n$ and the constraint

$$2\pi j h_n < A(t) \leqslant 2\pi(j+1)h_n, \tag{9.27}$$

which can be rephrased as $j = j(t)$, the largest integer strictly smaller than $2^n A(t)/2\pi$. This $j(t)$ is $o(2^n)$ (see below) and then

$$J_n = \{j(t) : t \in I_n\} = \{j(a_n), \ldots, j(b_n)\} \subset \{0, \ldots, 2^n - 1\}.$$

We have the informations

$$j(a_n) \sim 2^n A(a_n)/2\pi \sim \kappa n \quad \text{and} \quad j(b_n) \sim 2^n A(b_n)/2\pi \sim 2\kappa n,$$

where $\kappa = (\log(2))^\theta/2\theta\pi^2$; hence

$$\min J_n \approx n \quad \text{and} \quad \max J_n \approx n \quad \text{so} \quad |J_n| \approx n. \tag{9.28}$$

For such $j \in J_n$, we must estimate $m_{\phi_\theta}(R'_{n,j})$: it amounts to evaluate the length of the set of those $t \in I_n$ (thanks to the modulus constraint (9.26)) such that the argument constraint (9.27) is fulfilled, that is,

$$A^{-1}\big(2\pi j h_n\big) < t \leqslant A^{-1}\big(2\pi(j+1)h_n\big).$$

The (normalized) length of the corresponding interval is

$$A^{-1}\big(2\pi(j+1)h_n\big) - A^{-1}\big(2\pi j h_n\big) = h_n\,(A^{-1})'(u_{n,j})$$

for some $u_{n,j} \in]2\pi j h_n, 2\pi(j+1)h_n)[$ so that $2\pi j(a_n)2^{-n} < u_{n,j} < 2\pi j(b_n)2^{-n}$; hence we have $u_{n,j} \approx n2^{-n}$ (with underlying constants uniform in j).

Since $(A^{-1})'(u) \sim \big(\log(1/u)\big)^{-\theta}$, when $u \to 0^+$, it follows that

$$m_{\phi_\theta}(R'_{n,j}) \approx 2^{-n} n^{-\theta},$$

and we get, using (9.28),

$$\sum_{n=0}^{\infty} 2^{np/2} \sum_{j=0}^{2^n-1} \Big[m_{\phi_\theta}(R'_{n,j})\Big]^{p/2} = \sum_{n=0}^{\infty} 2^{np/2} \sum_{j\in J_n} \Big[m_{\phi_\theta}(R'_{n,j})\Big]^{p/2}$$
$$\approx \sum_{n=1}^{\infty} 2^{np/2} n \left(\frac{2^{-n}}{n^\theta}\right)^{p/2} = \sum_{n=1}^{\infty} \frac{1}{n^{\theta\frac{p}{2}-1}},$$

which is finite if and only if $\dfrac{\theta p}{2} - 1 > 1$, that is, $p > 4/\theta$. □

Remark. Theorem 9.14 shows that, for these symbols ϕ_θ, the necessary condition (9.16) of Proposition 9.10 is not sharp; in fact, we have proved in (9.25) that $\rho_{\phi_\theta}(h) \approx h/(\log 1/h)^\theta$; hence (9.16) gives $\theta > 2/p$ when $C_{\phi_\theta} \in \mathscr{S}^p$, even though we must have $\theta > 4/p$.

The Second Shapiro–Taylor Map

In Theorem 9.14, we got, for any $p_0 > 0$, a composition operator that is in $\mathscr{S}^p$ if and only if $p > p_0$. We can slightly modify this operator so that it belongs to $\mathscr{S}^p$ if and only if $p \geqslant p_0$.

Theorem 9.15 *For every $p_0 > 0$, there exists an* injective *symbol φ such that the composition operator $C_\varphi\colon H^2 \to H^2$ is in $\mathscr{S}^{p_0}(H^2)$, but not in $\mathscr{S}^p(H^2)$, for $p < p_0$.*

Similarly, for $\gamma > -1$, there exists an injective *symbol φ such that the composition operator $C_\varphi\colon \mathscr{B}^2_\gamma \to \mathscr{B}^2_\gamma$ is in the Schatten class $\mathscr{S}^{p_0}(\mathscr{B}^2_\gamma)$, but not in $\mathscr{S}^p(\mathscr{B}^2_\gamma)$, for $p < p_0$.*

Let us mention in passing that, in the case of Bergman spaces, we curiously did not find any reference for this result in the literature.

Proof: We explain the proof first for H^2 and indicate the necessary adjustments for $\mathscr{B}^2_\gamma$. We shall use the same method as in Theorem 9.14, but will replace the map used for Theorem 9.14 by the *second Shapiro–Taylor map* ϕ_θ introduced in Section 4.5: see definition (4.15).

Recall that

$$f(z) = z(-\log z)^\theta\big[\log(-\log z)\big]^s, \tag{9.29}$$

where $s > \theta/2$ ($s = \theta$, for example) and that the corresponding symbol ϕ_θ is defined by

$$\phi_\theta = \exp(-f \circ g),$$

where g is given by (4.13).

We already know from the proof of Lemma 4.10 that, for $t > 0$ going to zero,

$$\operatorname{Re}\left[\log\left(\log\frac{1}{t} - i\frac{\pi}{2}\right)\right]^s \sim \left(\log\log\frac{1}{t}\right)^s$$
$$\operatorname{Im}\left[\log\left(\log\frac{1}{t} - i\frac{\pi}{2}\right)\right]^s \sim -\frac{\pi s}{2}\frac{(\log\log 1/t)^{s-1}}{\log 1/t},$$

and, using (9.22) (or (4.16), with $\alpha = \pi/2$), we get

$$\operatorname{Re} f(it) \approx \frac{\pi\theta t}{2}\left(\log\frac{1}{t}\right)^\theta \frac{(\log\log 1/t)^s}{\log 1/t} = \frac{\pi\theta t}{2}\left(\log\frac{1}{t}\right)^{\theta-1}\left(\log\log\frac{1}{t}\right)^s$$
$$\operatorname{Im} f(it) \sim t\left(\log\frac{1}{t}\right)^\theta\left(\log\log\frac{1}{t}\right)^s.$$

Hence the *modulus constraint* gives, with $h = h_n = 2^{-n}$,

$$t \approx \frac{h}{\big(\log 1/h\big)^{\theta-1}\big(\log\log 1/h\big)^s}, \tag{9.30}$$

and the *argument constraint*

$$2\pi jh < t\left(\log\frac{1}{t}\right)^\theta\left(\log\log\frac{1}{t}\right)^s\big(1 + o(1)\big) \leqslant 2\pi(j+1)h.$$

As for the first Shapiro–Taylor map, the set J_n of those $j \in \{0, \dots, 2^n - 1\}$ such that $\exp\big(-f_\theta(it)\big) \in R'_{n,j}$ for some t, satisfies

$$\min J_n \approx n \quad \text{and} \quad \max J_n \approx n \quad \text{so} \quad |J_n| \approx n, \tag{9.31}$$

and we get, for $j \in J_n$,

$$m_{\phi_\theta}(R'_{n,j}) \approx \frac{2^{-n}}{n^\theta(\log n)^s}. \tag{9.32}$$

It follows that Luecking's criterion for H^2 becomes

$$\sum_{n=1}^{\infty}|J_n|2^{np/2}\frac{2^{-np/2}}{n^{\theta p/2}(\log n)^{sp/2}} \approx \sum_{n=1}^{\infty}\frac{1}{n^{\theta p/2-1}(\log n)^{sp/2}} < \infty.$$

This series converges if and only if $p \geqslant 4/\theta$ since, in the particular case where $p = 4/\theta$, we have $sp/2 = s/(\theta/2) > 1$.

Hence $C_{\phi_\theta} \in \mathscr{S}^p(H^2)$ if and only if $p \geqslant 4/\theta$, and that proves Theorem 9.15 for H^2 by taking $\theta = 4/p_0$.

Now switch to $\mathscr{B}^2_\gamma$. A quite similar computation shows that, with again $h = 2^{-n}$,

$$\sigma_{\phi_\theta,\gamma}(R'_{n,j}) \approx \frac{h^{2+\gamma}}{\big(\log 1/h\big)^{(\gamma+2)\theta}\big(\log\log 1/h\big)^{(\gamma+2)s}} \approx \frac{2^{-n(\gamma+2)}}{n^{(\gamma+2)\theta}(\log n)^{(\gamma+2)s}}. \tag{9.33}$$

Again, the windows $R_{n,j}$, with $j \in J_n$, are involved in Luecking's criterion, which reads as

$$\sum_{n=1}^{\infty}\left(\sum_{j\in J_n} 2^{(\gamma+2)np/2}\left[\sigma_{\phi_\theta,\gamma}(R_{n,j})\right]^{p/2}\right) = \sum_{n=1}^{\infty}\left(\sum_{j=0}^{2^n-1} 2^{(\gamma+2)np/2}\left[\sigma_{\phi_\theta,\gamma}(R_{n,j})\right]^{p/2}\right) < \infty$$

for membership in $\mathscr{S}^p(\mathscr{B}^2_\gamma)$ reads

$$\sum_{n=1}^{\infty} n.2^{(\gamma+2)np/2}\frac{2^{-(\gamma+2)np/2}}{n^{(\gamma+2)\theta p/2}(\log n)^{(\gamma+2)sp/2}} < \infty.$$

Take this time $\theta = \dfrac{4}{(\gamma+2)p_0}$ and $s > \dfrac{2}{(\gamma+2)p_0}$ to get the result. □

A Third Shapiro–Taylor Map

In this subsection, we will prove the following theorem (see also [96, theorem 5.6]):

Theorem 9.16 *There exists a* univalent *symbol φ for which the composition operator C_φ is compact on the Hardy space H^2, with moreover*

$$\forall \gamma > -1, \quad \rho_{\varphi,\gamma+2}(h) \geqslant \sigma_{\varphi,\gamma}\big(W(1,h)\big) \gtrsim \frac{h^{\gamma+2}}{\big(\log\log(1/h)\big)^{\gamma+2}}.$$

As a consequence, C_φ is compact on H^2 and all $\mathscr{B}^2_\gamma$ (for $\gamma > -1$), but in no Schatten class $\mathscr{S}^p(\mathscr{B}^2_\gamma)$, a fortiori in no Schatten class $\mathscr{S}^p(H^2)$.

The consequence is no scoop. As an answer to a question of Sarason, Carroll and Cowen [18] showed the existence of a symbol φ such that $C_\varphi : H^2 \to H^2$ is compact but in no Schatten class.

Given an arbitrary sequence (ε_n) of positive numbers converging to 0, we will even construct in Chapter 10 an example for which $C_\varphi : H^2 \to H^2$ is compact, but with singular numbers satisfying

$$a_n(C_\varphi) \gtrsim \varepsilon_n \qquad \text{for all } n \geqslant 1.$$

Hence, the interest of Theorem 9.16 is mainly to give an explicit and simple example, again in terms of a Shapiro–Taylor map, and to make use of the Luecking criterion.

Proof: Let once more V_ε be the convex open set defined by

$$V_\varepsilon = \{z \in \mathbb{C} : \operatorname{Re} z > 0 \text{ and } |z| < \varepsilon\}. \tag{9.34}$$

We set this time

$$f(z) = z \log(-\log z)\,, \tag{9.35}$$

where log is the principal determination of the logarithm.

For $\varepsilon > 0$ small enough, we have $\operatorname{Re} f(z) > 0$ for $z \in V_\varepsilon$. This is detailed below. Moreover, if $z = re^{i\alpha} \in V_\varepsilon$ with $|\alpha| < \pi/2$, we have $f'(z) = \log(-\log z) + \dfrac{1}{\log z}$ and

$$\operatorname{Re} f'(z) = \log\left(\sqrt{\log^2(1/r) + \alpha^2}\right) + o_r(1) \quad \text{when } r \to 0^+,$$

so that $\operatorname{Re} f'(z) > 0$ on V_ε for ε small enough, and that f is injective on V_ε.

Let again, as for the two previous Shapiro–Taylor maps (see Section 4.5), $g \colon \mathbb{D} \to V_\varepsilon$ be the conformal map from $\mathbb{D}$ onto V_ε sending $\mathbb{T} = \partial\mathbb{D}$ onto ∂V_ε, and with $g(1) = 0$ and $g'(1) = -\varepsilon/4$.

Actually, we mention here that a quick look at the definition of g shows that g^{-1} is $\mathcal{C}^2$ at 0 and in particular

$$\forall w \in V_\varepsilon\,, \qquad g^{-1}(w) = 1 - \frac{4}{\varepsilon} w + O\big(w^2\big). \tag{9.36}$$

We take again

$$\varphi = \exp(-f \circ g)\,. \tag{9.37}$$

This analytic function φ maps $\mathbb{D}$ into itself (is a symbol).

For $z = re^{i\alpha} \in V_\varepsilon$, an easy computation gives

$$\operatorname{Re} f(z) = r\left(\log\log\frac{1}{r}\right)\left(\cos\alpha + \frac{\alpha\sin\alpha}{\log(1/r)\log\log(1/r)}\right) + \mathrm{o}\left(\frac{1}{\log(1/r)}\right)$$

and

$$\operatorname{Im} f(z) = r\left(\log\log\frac{1}{r}\right)\left(\sin\alpha - \frac{\alpha\cos\alpha}{\log(1/r)\log\log(1/r)}\right) + \mathrm{o}\left(\frac{1}{\log(1/r)}\right).$$

It follows that

$$0 < \operatorname{Re} f(z) \lesssim r\log\log\frac{1}{r} \qquad \text{and} \qquad |\operatorname{Im} f(z)| \lesssim r\log\log\frac{1}{r}\,\cdot$$

We already know that f is injective on V_ε. This shows that $f\circ g$ and the symbol $\varphi=\exp(-f\circ g)$ are injective on $\mathbb{D}$, as claimed.

Now assume that $r\leqslant r(h):=h/\log\log(1/h)$. Then $r\log\log(1/r)\lesssim h$. Since $g'(1)\neq 0$, g is bi-Lipschitz in a neighborhood of 1; hence $|\varphi(u)|\sim 1-\operatorname{Re} f[g(u)]$ and $\arg\varphi(u)=-\operatorname{Im} f[g(u)]$. Therefore, for some constant $\delta>0$,

$$z\in W(1,\delta r(h))\Longrightarrow \varphi(z)\in W(1,h).$$

This implies

$$\sigma_{\varphi,\gamma}\big(W(1,h)\big)\gtrsim\sigma_\gamma\big(W(1,\delta r(h))\big)\approx r(h)^{\gamma+2}=\big(h/\log\log(1/h)\big)^{\gamma+2}.$$

As a consequence

$$\rho_{\varphi,\gamma+2}(h)\geqslant\sigma_{\varphi,\gamma}\big(W(1,h)\big)\gtrsim\frac{h^{\gamma+2}}{\big(\log\log(1/h)\big)^{\gamma+2}}\cdot \tag{9.38}$$

Clearly condition (9.38) is incompatible with the necessary condition (9.15) of membership in some $\mathscr{S}^p(\mathscr{B}^2_\gamma)$, a fortiori of membership in some $\mathscr{S}^p(H^2)$.

We now prove that C_φ is compact on H^2: since φ is univalent, it suffices to use Theorem 3.15; equivalently it amounts to prove

$$\lim_{\substack{w\to 0\\ w\in V_\varepsilon}}\frac{1-|g^{-1}(w)|}{\operatorname{Re} f(w)}=0\cdot \tag{9.39}$$

Let us write $w=re^{i\alpha}\in V_\varepsilon$, where $r\to 0^+$ and $|\alpha|\leqslant\pi/2$.

We know from (9.36) that $g^{-1}(w)=1-\dfrac{4}{\varepsilon}w+O\big(r^2\big)$; hence

$$1-|g^{-1}(w)|\leqslant 1-|g^{-1}(w)|^2=\frac{8}{\varepsilon}r\cos\alpha+O\big(r^2\big).$$

On the other hand, we know that

$$\operatorname{Re} f(w)\sim r\left(\log\log\frac{1}{r}\right)\left(\cos\alpha+\frac{\alpha\sin\alpha}{\log(1/r)\log\log(1/r)}\right);$$

so

$$\frac{1-|g^{-1}(w)|}{\operatorname{Re} f(w)}\lesssim\frac{\big(\cos(\alpha)+r\big)}{\log\log\frac{1}{r}}\left(\cos\alpha+\frac{\alpha\sin\alpha}{\log(1/r)\log\log(1/r)}\right)^{-1}.$$

When $\cos(\alpha)\geqslant r/2$, we have $\cos(\alpha)+r\leqslant 3\cos(\alpha)$; hence

$$\frac{1-|g^{-1}(w)|}{\operatorname{Re} f(w)}\lesssim\frac{3}{\log\log\frac{1}{r}},$$

and when $\cos(\alpha) \leqslant r/2$, we have $|\alpha| \geqslant \pi/4$ (since $r \in (0,1)$):

$$\frac{1 - |g^{-1}(w)|}{\operatorname{Re} f(w)} \lesssim \frac{3r \log(1/r)}{2\alpha \sin\alpha} \lesssim r \log(1/r)\,.$$

Gathering everything, we have

$$\frac{1 - |g^{-1}(w)|}{\operatorname{Re} f(w)} \lesssim \max\left(\frac{1}{\log\log\frac{1}{r}}; r\log(1/r)\right) \longrightarrow 0$$

when $r \to 0^+$; and (9.39) is proved; hence C_φ is compact on H^2 and on every weighted Bergman space. □

9.5 Separation Results for Schatten Classes

We begin with the following result, enunciated in terms of the spaces H_α.

Theorem 9.17 *Let $\alpha \leqslant 0$. The composition operators C_φ with* injective symbol *separate Schatten classes of spaces H_α. Namely if $0 < q < p$, there exists an injective φ such that $C_\varphi \in \mathscr{S}^p(H_\alpha)$ and $C_\varphi \notin \mathscr{S}^q(H_\alpha)$. We can even have $C_\varphi \notin \bigcup_{q<p} \mathscr{S}^q(H_\alpha)$.*

Proof: This is the content of Theorem 9.15, which we find worth rephrasing under this form. □

The second one is enunciated in terms of the Hardy and Bergman spaces, with again $\gamma_2 > \gamma_1$ and $\tilde{p} = p\frac{\gamma_1+2}{\gamma_2+2}$, as another consequence of Theorem 9.15, and shows the optimality of Theorem 9.12.

Theorem 9.18 *Let $\gamma_2 > \gamma_1 > -1$. For every p with $0 < p < \infty$, there exists an* injective symbol *φ for which the composition operator C_φ is in the Schatten class $\mathscr{S}^p(\mathscr{B}^2_{\gamma_1})$, but in no Schatten class $\mathscr{S}^q(\mathscr{B}^2_{\gamma_2})$ with $q < \tilde{p}$.*

Similarly, we can have $C_\varphi \in \mathscr{S}^p(H^2)$, but $C_\varphi \notin \mathscr{S}^q(\mathscr{B}^2)$ for $q < p/2$.

Proof: Let φ be an injective symbol with $C_\varphi \in \mathscr{S}^{\tilde{p}}(\mathscr{B}^2_{\gamma_2}) \setminus \bigcup_{q<\tilde{p}} \mathscr{S}^q(\mathscr{B}^2_{\gamma_2})$. Such a symbol exists by Theorem 9.15. By Theorem 9.12, $C_\varphi \in \mathscr{S}^p(\mathscr{B}^2_{\gamma_1})$ since φ is injective. And by construction, $C_\varphi \notin \bigcup_{q<\tilde{p}} \mathscr{S}^q(\mathscr{B}^2_{\gamma_2})$. This ends the proof. □

9.6 Proof of Luecking's Theorem

Here we discuss Luecking's theorem 9.2. This theorem is definitely difficult and, when we try to estimate $\mathfrak{s}_p(T_\mu)$ as in (9.11), four issues more or less mandatorily show up:

1. Upper bound (sufficiency) for $p \geqslant 1$.
2. Lower bound (necessity) for $p \geqslant 1$.
3. Upper bound (sufficiency) for $0 < p < 1$.
4. Lower bound (necessity) for $0 < p < 1$.

We will present *three* fairly different proofs of this theorem. They are all long and involved, and none of them, except perhaps the first one for H^2, will be entirely self-contained.

(i) The first one privileges composition operators (with the help of Carleson embeddings or else forward Carleson measures).
(ii) The second one privileges Toeplitz operators (with the help of forward *and reverse* Carleson measures).
(iii) The third one also privileges Toeplitz operators (but is more focused on eigenvalues and reproducing kernels).

The first proof (difficult to find in the literature) [63] is due to González–Doña (when he was a master student of Rodríguez–Piazza and Lacruz–Martín, whom we thank for letting us include it here) and has the merit of ignoring the difference between $p \geqslant 1$ and $p < 1$; only two cases (upper and lower bound) have to be considered. This proof is essentially self-contained (even if it is written only for H^2). The prize to pay is that it is very long, and we will have to (slightly) abbreviate some details.

The second proof (also very long) is due to Luecking and follows more or less closely his initial paper of 1987; it is elegant, even if sometimes sketchy and referring to other papers. We added quite a few details for the convenience of the reader.

The third proof, due to El-Fallah and El Ibbaoui [48], uses a result of Lin and Rochberg and that of Rotfel'd–Thompson discussed in Appendix G.

9.6.1 Carleson Embeddings Approach

The presentation here is due to González–Doña [63]. We recall that $\mathfrak{s}_p(T)$ stands for the Schatten norm (or quasi-norm) of an operator T between two Hilbert spaces H_1 and H_2.

Theorem 9.19 *Let μ be a probability measure on $\mathbb{D}$ and J_μ the Carleson embedding: $H^2 \to L^2(\mu)$. Let also $0 < p < \infty$. Then the following are equivalent:*

(i) $J_\mu \in \mathscr{S}^p(H^2, L^2(\mu))$.

(ii) $\displaystyle\sum_{n=0}^{\infty} \sum_{j=0}^{2^n-1} \left(2^n \mu(R_{n,j})\right)^{p/2} < \infty$.

Moreover, it holds

$$\mathfrak{s}_p(J_\mu) \approx \left(\sum_{n=0}^{\infty} \sum_{j=0}^{2^n-1} \left(2^n \mu(R_{n,j})\right)^{p/2} \right)^{1/p} \approx \left(\sum_{n=0}^{\infty} \sum_{j=0}^{2^n-1} \left(2^n \mu(W_{n,j})\right)^{p/2} \right)^{1/p}.$$

Thanks to (9.12), this is equivalent to Luecking's theorem.
The proof consists of two main steps.

(i) Purely Hilbertian considerations.
(ii) Specialization to H^2 and $L^2(\mu)$.

In the present section, we shall often work with sequences of Hilbert spaces (H_n): please do not confuse H_n with H_α! We recall that in Section 9.6.1 we aim to treat the case of the Hardy space H^2 only.

Hilbertian Results

The first (easy) result we need is the following proposition:

Proposition 9.20 *Let $(H_n)_{n\geqslant 0}$ and $(K_n)_{n\geqslant 0}$ be two sequences of Hilbert spaces, with direct Hilbertian sums*

$$H = \bigoplus_{n\geqslant 0} H_n \qquad \text{and} \qquad K = \bigoplus_{n\geqslant 0} K_n.$$

Let $T_n \in B(H_n, K_n)$ with $\sup_n \|T_n\| < \infty$; define the direct sum $T = \bigoplus_{n\geqslant 0} T_n$ by

$$T((x_n)) = (T_n(x_n)).$$

Then $T : H \to K$ is bounded and we have, for every $0 < p < \infty$,

$$\mathfrak{s}_p(T) = \left(\sum_{n=0}^{\infty} \mathfrak{s}_p(T_n)^p \right)^{1/p}.$$

Proof: The boundedness of T is obvious. Next, for each $n \geqslant 0$, consider a Schmidt decomposition of T_n:

$$T_n = \sum_{m=1}^{\infty} a_m(T_n) f_m^{(n)} \otimes e_m^{(n)}.$$

Then set

$$e_{m,n} = (0, \ldots, e_m^{(n)}, 0, \ldots) \in H,$$

where $e_m^{(n)}$ is in nth position. Define $f_{m,n}$ similarly. Both systems $(e_{m,n})_{m,n}$ and $(f_{m,n})_{m,n}$ are orthonormal in H and K, respectively, and T has the Schmidt decomposition

$$T(x) = \sum_{\substack{m\geqslant 1\\ n\geqslant 0}} a_m(T_n)\langle x, e_{m,n}\rangle f_{m,n}.$$

The result follows. □

Now, the main issue (in Luecking's approach as well) in this Schatten business is the case $0 < p < 1$ due to the lack of a classical triangle inequality. To remediate this, here is a second proposition:

Proposition 9.21 *Let* $0 < p < \infty$ *and* $(T_n)_{n\geqslant 0}$ *be a sequence in* $\mathscr{S}^p(H_1, H_2)$. *Assume that*

$$\sum_{n=0}^{\infty} 2^{n/p} \mathfrak{s}_p(T_n) < \infty.$$

Then the operator $T = \sum_{n=0}^{\infty} T_n$ *is well defined, belongs to* $\mathscr{S}^p(H_1, H_2)$, *and moreover*

$$\mathfrak{s}_p(T) \leqslant \sum_{n=0}^{\infty} 2^{(n+1)/p} \mathfrak{s}_p(T_n).$$

Proof: This is obvious (triangle inequality) for $p \geqslant 1$, without the factor $2^{(n+1)/p}$, so that $T \in \mathscr{S}^{\max(p,1)}(H_1, H_2) \subset B(H_1, H_2)$.

For $0 < p < 1$, we could use the following substitute to the triangle inequality for the quasi-norm $\mathfrak{s}_p$:

$$\mathfrak{s}_p(S+T) \leqslant 4^{1/p}\big(\mathfrak{s}_p(S) + \mathfrak{s}_p(T)\big). \tag{9.40}$$

Indeed, since $a_{2k}(S+T) \leqslant a_{2k-1}(S+T) \leqslant a_k(S) + a_k(T)$, we get

$$\begin{aligned}\mathfrak{s}_p(S+T)^p &\leqslant 2\sum_{k=1}^{\infty}\big(a_k(S)+a_k(T)\big)^p \leqslant 2\sum_{k=1}^{\infty}\big(a_k(S)^p + a_k(T)^p\big)\\ &= 2\big[\mathfrak{s}_p(S)^p + \mathfrak{s}_p(T)^p\big] \leqslant 4\big[\mathfrak{s}_p(S)+\mathfrak{s}_p(T)\big]^p\end{aligned}$$

But we *can do better* thanks to the following relations (G.30) proved in Theorem G.24:

$$\mathfrak{s}_p(S+T)^p \leqslant \mathfrak{s}_p(S)^p + \mathfrak{s}_p(T)^p; \qquad \text{hence} \qquad \mathfrak{s}_p(S+T) \leqslant 2^{1/p}\big(\mathfrak{s}_p(S)+\mathfrak{s}_p(T)\big).$$

Then by induction we get

$$\mathfrak{s}_p\left(\sum_{n=0}^{N} T_n\right) \leqslant \sum_{n=0}^{N} 2^{(n+1)/p} \mathfrak{s}_p(T_n).$$

For the induction step, just write

$$\mathfrak{s}_p\left(\sum_{n=0}^{N+1} T_n\right) \leqslant 2^{1/p}\left[\mathfrak{s}_p(T_0) + \mathfrak{s}_p(\sum_{n=0}^{N} T_{n+1})\right]$$
$$\leqslant 2^{1/p}\left[\mathfrak{s}_p(T_0) + \sum_{n=0}^{N} 2^{(n+1)/p}\mathfrak{s}_p(T_{n+1})\right]$$
$$= \sum_{n=0}^{N+1} 2^{(n+1)/p}\mathfrak{s}_p(T_n).$$

□

Warning. The proof of the nice inequality (G.30) of Rotfel'd theorem G.24 will be given in Appendix G. This interesting proof is fairly long and elaborate. But, as far as only the membership in Schatten classes is considered, the reader who does not wish to rely on (G.30) can harmlessly proceed, replacing everywhere $2^{n/p}$ by $4^{n/p}$ (starting from (9.40)).

A key result for the proof of Theorem 9.19 is as follows:

Theorem 9.22 *Let $p > 0$ and $(H_n)_{n\geqslant 0}$ and $(K_j)_{j\geqslant 0}$ be two sequences of Hilbert spaces, with direct Hilbertian sums*

$$H = \bigoplus_{n\geqslant 0} H_n \qquad \text{and} \qquad K = \bigoplus_{j\geqslant 0} K_j.$$

Let $T \in B(H, K)$ and define

$$T_{n,j} : H_n \to K_j$$

by $T_{n,j} = \pi_{K_j} \circ T \circ i_{H_n}$, where $\pi_B : K \to B$ is the orthogonal projection and i_{H_n} the canonical injection.

Assume that, for some constant C and some $a \in (0,1)$, we have for all integers n, $j \geqslant 0$,

$$\mathfrak{s}_p(T_{n,j}) \leqslant C a^{|n-j|}\mathfrak{s}_p(T_{j,j}). \tag{9.41}$$

Then

$$\mathfrak{s}_p(T) \approx \left(\sum_{n=0}^{\infty} \mathfrak{s}_p(T_{n,n})^p\right)^{1/p},$$

where the underlying constants only depend on p, C, a.

Proof: We first prove the upper bound

$$\mathfrak{s}_p(T) \lesssim \left(\sum_{n=0}^{\infty} \mathfrak{s}_p(T_{n,n})^p\right)^{1/p}.$$

For that, we define $T_j : H \to K$ by

$$T_j((x_n)_n) = \big(T_{n+j,\,n}(x_{n+j})\big)_n \qquad \text{when } j \geqslant 0;$$

and

$$T_j((x_n)_n) = \sum_{n=0}^{\infty}(i_{K_{n-j}} \circ T_{n,\,n-j})(x_n) \qquad \text{when } j < 0.$$

We observe that T_0 is the diagonal of the matrix of operators $(T_{n,\,j})$, whereas the T_j's represent the upper diagonals of this matrix for $j > 0$ and the lower diagonals for $j < 0$. So that

$$T = \sum_{j=-\infty}^{\infty} T_j$$

and that, thanks to Proposition 9.20 and our assumption,

$$\mathfrak{s}_p(T_j) = \bigg(\sum_{n=0}^{\infty} \mathfrak{s}_p(T_{n+j,\,n})^p\bigg)^{1/p} \leqslant C a^{|j|} \bigg(\sum_{n=0}^{\infty} \mathfrak{s}_p(T_{n,n})^p\bigg)^{1/p}$$

for $j \geqslant 0$, with a similar inequality for $j < 0$. We would like to sum up and apply the triangle inequality, but cannot do that directly for $0 < p < 1$. Hence, we first select an index $j_0 \geqslant 1$ such that $2^{1/p} a^{j_0} \leqslant 1/2$; and we set, for $k \geqslant 0$,

$$U_k = \sum_{|j|=j_0 k}^{j_0(k+1)-1} T_j, \quad \text{so that} \quad T = \sum_{k=0}^{\infty} U_k.$$

We now use Proposition 9.21 twice to get first

$$\mathfrak{s}_p(U_k) \leqslant C 2^{(j_0+1)/p}(2j_0 + 1))\, a^{j_0 k} \bigg(\sum_{n=0}^{\infty} \mathfrak{s}_p(T_{n,n})^p\bigg)^{1/p}$$

and then

$$\mathfrak{s}_p(T) \leqslant C 2^{(j_0+2)/p}(2j_0 + 1) \bigg(\sum_{n=0}^{\infty} \mathfrak{s}_p(T_{n,n})^p\bigg)^{1/p}.$$

We now prove the lower bound

$$\mathfrak{s}_p(T) \gtrsim \bigg(\sum_{n=0}^{\infty} \mathfrak{s}_p(T_{n,n})^p\bigg)^{1/p}. \tag{9.42}$$

For this, we fix $\delta > 0$ with $2^{1/p}\delta \leqslant 1/2$ and then a large integer M such that

$$(4C2^{2/p})\, a^M \leqslant \delta \quad \text{and} \quad 2^{1/p} a^M \leqslant 1/2.$$

We now decompose H and K as follows:

$$H = \bigoplus_{j=0}^{M-1} A_j \quad \text{and} \quad K = \bigoplus_{j=0}^{M-1} B_j,$$

where

$$A_j = \bigoplus_{l=0}^{\infty} H_{Ml+j} \quad \text{and} \quad B_j = \bigoplus_{l=0}^{\infty} K_{Ml+j}.$$

By the ideal property of the quasi-norms $\mathfrak{s}_p$ for $0 < p < \infty$, we have, setting

$$S^{(j)} := \pi_{B_j} \circ T \circ i_{A_j}, \tag{9.43}$$

$$\mathfrak{s}_p(T) \geqslant \mathfrak{s}_p(\pi_{B_j} \circ T \circ i_{A_j}) = \mathfrak{s}_p(S^{(j)}).$$

The matricial entries of $S^{(j)} = (S^{(j)}_{n,k})_{n,k}$ are

$$S^{(j)}_{n,k} = T_{j+(n-1)M,\ j+(k-1)M}.$$

Hence, by our hypothesis,

$$\mathfrak{s}_p(S^{(j)}_{n,k}) = \mathfrak{s}_p(T_{j+(n-1)M,\ j+(k-1)M}) \leqslant C a^{M\,|n-k|} \mathfrak{s}_p(S^{(j)}_{k,\ k}).$$

Temporarily fix $j, 0 \leqslant j \leqslant M-1$, and set $S^{(j)} = S$, dropping the dependence in j. We write $S = \sum_{-\infty}^{\infty} S_r$ where $S_r = S^{(j)}_r$ are the analogues for S of the T_r for T. By Proposition 9.20, we know that, since $S_0 = \bigoplus_{n \geqslant 0} S_{n,n}$,

$$\mathfrak{s}_p(S_0) = \left(\sum_{n=0}^{\infty} \mathfrak{s}_p(S_{n,n})^p \right)^{1/p}.$$

By our choice of δ and M, Proposition 9.21, and the assumption, we have

$$\begin{aligned}
\mathfrak{s}_p(S - S_0) = \mathfrak{s}_p\left(\sum_{r \neq 0} S_r \right) &\leqslant 2.2^{1/p} \sum_{r \geqslant 1} 2^{r/p} \mathfrak{s}_p(S_r) \\
&\leqslant 2C2^{1/p} \times \sum_{r=1}^{\infty} (2^{1/p} a^M)^r \mathfrak{s}_p(S_0) \\
&= 2C2^{1/p} \times \frac{2^{1/p} a^M}{1 - 2^{1/p} a^M} \mathfrak{s}_p(S_0) \\
&\leqslant (4C2^{2/p})\, a^M \mathfrak{s}_p(S_0) \leqslant \delta \mathfrak{s}_p(S_0).
\end{aligned}$$

We now pursue by "extrapolation" (domination of an unknown quantity by itself under a damped form):

$$\begin{aligned} \mathfrak{s}_p(S_0) &\leqslant 2^{1/p}\big(\mathfrak{s}_p(S-S_0)+\mathfrak{s}_p(S)\big) \\ &\leqslant (2^{1/p}\delta)\mathfrak{s}_p(S_0)+2^{1/p}\mathfrak{s}_p(S) \\ &\leqslant (1/2)\mathfrak{s}_p(S_0)+2^{1/p}\mathfrak{s}_p(S). \end{aligned}$$

So that

$$\mathfrak{s}_p(S_0) \leqslant 2^{(1/p)+1}\mathfrak{s}_p(S)\cdot$$

Coming back to our initial notations, we have proved that, for $0 \leqslant j \leqslant M-1$,

$$\begin{aligned} 2^{(1/p)+1}\,\mathfrak{s}_p(S^{(j)}) \geqslant \mathfrak{s}_p(S_0^{(j)}) &= \left(\sum_{n=0}^{\infty}\mathfrak{s}_p(S_{n,n}^{(j)})^p\right)^{1/p} \\ &= \left(\sum_{l=0}^{\infty}\mathfrak{s}_p(T_{lM+j,\,lM+j})^p\right)^{1/p}. \end{aligned}$$

All in all, for some other constant C' depending only on p, C, a,

$$\begin{aligned} \left(\sum_{n=0}^{\infty}\mathfrak{s}_p(T_{n,n})^p\right)^{1/p} &= \left(\sum_{j=0}^{M-1}\sum_{l=0}^{\infty}\mathfrak{s}_p(T_{lM+j,\,lM+j})^p\right)^{1/p} \\ &= \left(\sum_{j=0}^{M-1}\mathfrak{s}_p(S_0^{(j)})^p\right)^{1/p} \\ &\leqslant C'\left(\sum_{j=0}^{M-1}\mathfrak{s}_p(S^{(j)})^p\right)^{1/p} \\ &\leqslant C'\left(\sum_{j=0}^{M-1}\mathfrak{s}_p(T)^p\right)^{1/p} \\ &\leqslant C'M^{1/p}\mathfrak{s}_p(T). \end{aligned}$$

This proves (9.42), and thereby Theorem 9.22. □

Specialization of Hilbertian Results

We now apply Theorem 9.22 as follows: we denote Γ_n, $n \geqslant 0$, the dyadic annulus $\Gamma_0 = \overline{D}(0,1/2)$ and, for $n \geqslant 1$, $\Gamma_n = \{z : 1-2^{-n} < |z| \leqslant 1-2^{-n-1}\} = \bigsqcup_{j=0}^{2^n-1} R_{n,j}$

so that

$$\mathbb{D} = \bigsqcup_{n \geqslant 0} \Gamma_n.$$

Then we set, for $N = 1, 2, \ldots$,

$$H = H^2, \qquad K = L^2(\mu), \qquad H_2^N = \operatorname{span}\{z^k \,: 0 \leqslant k \leqslant N-1\},$$

$$H_n = z^{2^n - 1} H_2^{2^n} = \operatorname{span}\{z^k : 2^n - 1 \leqslant k < 2^{n+1} - 1\}, \quad K_n = L^2(\mu_{|\Gamma_n})$$

(again, please do not confuse H_n with H_α!) and finally

$$G_1 = \mathbb{D} \qquad \text{and, for } N \geqslant 2, \quad G_N = \{z \in \mathbb{D} : 1 - N^{-1} < |z| < 1\}.$$

So that, in the Hilbertian sense,

$$H = \bigoplus_{n \geqslant 0} H_n, \qquad K = \bigoplus_{n \geqslant 0} K_n.$$

Then, we will check the following, in writing $j_{n,m}$ for $\pi_m \circ J_\mu \circ i_n$ (where $\pi_m = \pi_{K_m}$ and $i_n = i_{H_n}$ for short):

$$\mathfrak{s}_p(j_{n,n}) \approx \left(\sum_{j=0}^{2^n - 1} (2^n \mu(R_{n,j}))^{p/2} \right)^{1/p} \,; \tag{9.44}$$

and, for some $a \in (0, 1)$,

$$\mathfrak{s}_p(j_{n,m}) \lesssim C a^{|m-n|} \mathfrak{s}_p(j_{n,n}) \tag{9.45}$$

so that all the assumptions of Theorem 9.22 will be met, and we will obtain Theorem 9.19 under its quantitative form. But checking (9.44) and (9.45) is long and technical, and will occupy us for several pages. We begin by a key lemma, which is Theorem 9.19 for special measures μ.

Lemma 9.23 *Let $p > 0$ and let N be an integer $\geqslant 2$, $z_0 \in G_N$, as well as $\omega = \exp(2i\pi/N)$, and $\delta > 0$, independent of N, so small that $8^{1+1/p} e\, \delta \leqslant 1$ and that the euclidean disks $B_j := D(\omega^j z_0, \delta/N)$ are disjoint.*

Let μ be a positive and finite measure supported by

$$\bigcup_{j=0}^{N-1} B_j \cap G_N,$$

and consider the Carleson embedding $T : H_2^N \to L^2(\mu)$.

Then

$$\mathfrak{s}_p(T) \approx \left(\sum_{j=0}^{N-1} (N \mu(B_j))^{p/2} \right)^{1/p}.$$

Proof: Note that δ exists since $j \neq k \Longrightarrow |\omega^j z_0 - \omega^k z_0| \geqslant \dfrac{c}{N}$ with $c > 0$ numerical.

We first prove the upper bound

$$\mathfrak{s}_p(T) \lesssim \left(\sum_{j=0}^{N-1} (N\mu(B_j))^{p/2} \right)^{1/p}. \tag{9.46}$$

To that effect, we can write, in $L^2(\mu)$, the operator T as

$$Tf = \sum_{j=0}^{N-1} f.\mathbb{1}_{B_j}, \qquad \text{for } f \in H_2^N.$$

By Taylor's formula for polynomials, we can further write

$$T = \sum_{l=0}^{N-1} T_l, \qquad T_l = S_l \circ P \circ A_l$$

with the diagram

$$H_2^N \xrightarrow{A_l} H_2^N \xrightarrow{P} \ell_2^N \xrightarrow{S_l} L^2(\mu),$$

where A_l, P, S_l are defined by (with $f^{(l)}$ the lth derivative of f):

$$A_l(f) = f^{(l)}.$$

$$P(g) = \left(\frac{g(\omega^j z_0)}{\sqrt{N}} \right)_{0 \leqslant j \leqslant N-1}.$$

$$S_l(a)(z) = \sum_{j=0}^{N-1} \sqrt{N} a_j \frac{(z - \omega^j z_0)^l}{l!} \mathbb{1}_{B_j}, \qquad \text{for } a = (a_j)_{0 \leqslant j \leqslant N-1}.$$

Clearly $\|A_l\| \leqslant N^l$ (Parseval's relation), $\|P\| \leqslant 1$, and more precisely

$$\|P(g)\|^2_{\ell_N^2} = \sum_{k=0}^{N-1} |\widehat{g}(k)|^2 |z_0|^{2k} \tag{9.47}$$

(Parseval's relation for the finite group $\{\omega^j, 0 \leqslant j \leqslant N-1\}$). Finally, S_l has the Schmidt-type decomposition

$$S_l = \sum_{j=0}^{N-1} \sqrt{N} \|u_j\|_{L^2(\mu)}\, e_j \otimes \frac{u_j}{\|u_j\|_{L^2(\mu)}},$$

where (e_j) denotes the canonical basis of ℓ_N^2 and

$$u_j(z) = \frac{(z - z_0\omega^j)^l}{l!} \mathbb{1}_{B_j}(z)$$

(note that u_j are orthogonal in $L^2(\mu)$ since the B_j are disjoint). So that

$$\mathfrak{s}_p(S_l) = \left(\sum_{j=0}^{N-1} N^{p/2} \|u_j\|_{L^2(\mu)}^p\right)^{1/p} \leqslant \frac{\delta^l}{N^l\, l!}\left(\sum_{j=0}^{N-1} (N\mu(B_j))^{p/2}\right)^{1/p},$$

where we used

$$\|u_j\|_{L^2(\mu)} \leqslant \frac{\delta^l}{N^l\, l!}\sqrt{\mu(B_j)}$$

since μ is supported in $\bigcup_{j=0}^{N-1} B_j$.

The ideal property for $\mathfrak{s}_p$ now gives

$$\mathfrak{s}_p(T_l) \leqslant \frac{\delta^l}{l!}\left(\sum_{j=0}^{N-1} (N\mu(B_j))^{p/2}\right)^{1/p}.$$

Adding up those inequalities and using Proposition 9.21 finally gives, with the help of the relation $2^{1/p}\delta \leqslant 1$,

$$\mathfrak{s}_p(T) \leqslant \sum_{l=0}^{N-1} 2^{\frac{l+1}{p}} \frac{\delta^l}{l!}\left(\sum_{j=0}^{N-1} (N\mu(B_j))^{p/2}\right)^{1/p} \leqslant 2^{1/p}\mathrm{e}\left(\sum_{j=0}^{N-1} (N\mu(B_j))^{p/2}\right)^{1/p},$$

and this proves (9.46).

It remains to prove the reverse inequality (lower bound). For this, we observe that (9.47) implies, since $|z_0| \geqslant 1 - 1/N$ and $(1 - 1/x)^x \geqslant 1/4$ when $x \geqslant 2$, for $g \in H_2^N$,

$$\|P(g)\|^2 = \sum_{k=0}^{N-1} |\widehat{g}(k)|^2 |z_0|^{2k} \geqslant \sum_{k=0}^{N-1} |\widehat{g}(k)|^2 (1 - 1/N)^{2N} \geqslant \frac{1}{16}\|g\|^2.$$

So that P is invertible with $\|P^{-1}\| \leqslant 4$. Hence, since $S_0 = T_0 \circ P^{-1}$, we get

$$\mathfrak{s}_p(T_0) \geqslant \frac{1}{4}\mathfrak{s}_p(S_0) = \frac{1}{4}\left(\sum_{j=0}^{N-1} (N\mu(B_j))^{p/2}\right)^{1/p} =: \rho.$$

Now we write $T_0 = T - \sum_{l=1}^{N-1} T_l$, and we can majorize $\mathfrak{s}_p(\sum_{l=1}^{N-1} T_l)$ by $\frac{\rho}{2.2^{1/p}}$, thanks to the previous estimations and the hypothesis on δ; now we dominate ρ by $\mathfrak{s}_p(T)$, with the same extrapolation trick as in the proof of Theorem 9.22. This gives

$$\rho \leqslant \mathfrak{s}_p(T_0) \leqslant 2^{1/p}\left(\mathfrak{s}_p(T) + \mathfrak{s}_p\left(\sum_{l=1}^{N-1} T_l\right)\right) \leqslant 2^{1/p}\mathfrak{s}_p(T) + \frac{\rho}{2}$$

and $\rho \leqslant 2.2^{1/p}\mathfrak{s}_p(T)$ as desired. □

In the following important proposition, we get rid of some restrictions on μ.

Proposition 9.24 *Let $p > 0$, $N = 2^n$ with $n \geqslant 0$, and assume that the probability measure μ is supported by G_N. Let again $T : H_2^N \to L^2(\mu)$ be the Carleson embedding. Fix $0 < p < \infty$. Then*

$$\mathfrak{s}_p(T) \approx \left(\sum_{j=0}^{N-1} (N\mu(W_{n,j}))^{p/2}\right)^{1/p}.$$

Proof: The case $N = 1$ is clear. If $N \geqslant 2$, we note that

$$G_N = \bigcup_{j=0}^{N-1} W_{n,j}.$$

We fix a numerical $\delta \in (0,1)$ so that the assumptions of Lemma 9.23 are satisfied. Simple geometric considerations show that there exist an integer M (depending on δ, but not on N) and points $z_1, \ldots, z_M \in W_{n,0}$ such that

$$W_{n,0} \subset \bigcup_{l=1}^{M} B(z_l, \delta/N) =: \bigcup_{l=1}^{M} B^l,$$

and hence

$$W_{n,j} \subset \bigcup_{l=1}^{M} B(z_l\omega^j, \delta/N) = \bigcup_{l=1}^{M} \omega^j B^l =: \bigcup_{l=1}^{M} B_j^l.$$

Let $A_l' = \bigcup_{j=0}^{N-1} B_j^l$, so that $G_N \subset \bigcup_{l=1}^{M} A_l'$. We can write

$$\mu = \mu_1 + \cdots + \mu_M$$

with positive measures μ_l supported by A_l' and, for each 1, we define the Carleson embedding

$$E_l : H_2^N \to L^2(\mu_l), \text{ so that } T = \sum_{l=1}^{M} E_l$$

(with the obvious abuse of notations) with clearly $\|Tf\| \geqslant \|E_l f\|$ for each $f \in H_2^N$, implying $\mathfrak{s}_p(T) \geqslant \mathfrak{s}_p(E_l)$ for each l. Select an index l_0 that maximizes

$$\sum_{j=0}^{N-1} \left(N\mu(B_j^l)\right)^{p/2}, \quad \text{with } 1 \leqslant l \leqslant M.$$

Since $W_{n,j} \subset \bigcup_{l=1}^{M} B_j^l$, we have $\mu(W_{n,j}) \leqslant \sum_{l=1}^{M} \mu(B_j^l)$.

It easily follows that, for each $0 < p < \infty$, there is a constant $C_p > 0$ such that

$$\sum_{j=0}^{N-1} \left(N\mu(W_{n,\,j})\right)^{p/2} \leqslant C_p \sum_{j=0}^{N-1} \left(N\mu(B_j^{l_0})\right)^{p/2}$$

so that, with the help of Lemma 9.23 applied to μ_{l_0},

$$\mathfrak{s}_p(T) \geqslant \mathfrak{s}_p(E_{l_0}) \geqslant C_p^{-1/p} \left(\sum_{j=0}^{N-1} (N\mu(W_{n,j}))^{p/2}\right)^{1/p}.$$

The reverse inequality follows from Lemma 9.23 and the triangle inequality

$$\mathfrak{s}_p(T) \leqslant \sum_{l=1}^{M} 2^{l/p} \mathfrak{s}_p(E_l),$$

and we only need to compare $\mathfrak{s}_p(E_l)$ and $\left(\sum_{j=0}^{N-1} (N\mu(W_{n,j}))^{p/2}\right)^{1/p}$, but this is a consequence of the following fact:

Fact: There exists an integer $D \geqslant 1$, independent of N, such that, for every $0 \leqslant j \leqslant N-1$ and every $1 \leqslant l \leqslant M$, one has

$$G_N \cap B\left(z_l \omega^j, \frac{\delta}{N}\right) \subset \bigcup_{|u| \leqslant D} W_{n,\,j+u} \tag{9.48}$$

(agreeing that $W_{n,k} = W_{n,2^n+k}$ when $-2^n \leqslant k \leqslant -1$).

Indeed, by rotation, we can assume that $j = 0$. We can also assume that $N \geqslant N_0$ where $N_0 = N_0(\delta) \geqslant 2$ is a large integer, to be adjusted. Write

$$z_l = r_l e^{i\theta_l} \quad \text{with } 0 \leqslant \theta_l \leqslant \frac{2\pi}{N} \qquad \text{and} \qquad r_l > 1 - \frac{1}{N}.$$

Let $z \in G_N \cap B\left(z_l, \frac{\delta}{N}\right)$. We see that, for $N \geqslant N_0(\delta) \geqslant 3$,

$$\text{Re}\,(z) \geqslant \text{Re}\,(z_l) - \frac{\delta}{N} \geqslant \left(1 - \frac{1}{N}\right) \cos\frac{2\pi}{N} - \frac{\delta}{N} \geqslant 1 - \frac{1+2\delta}{N} \geqslant 0.$$

We can hence write $z = re^{i\theta}$ with $r \geqslant 1 - \dfrac{1}{N}$ and $|\theta| \leqslant \dfrac{\pi}{2}\cdot$

But then, for $N \geqslant 4$,

$$|\theta - \theta_l| \leqslant \frac{\pi}{2} + \frac{2\pi}{N} \leqslant \pi \qquad \text{and} \qquad \left|\frac{\theta - \theta_l}{2}\right| \leqslant \frac{\pi}{2}\cdot \tag{9.49}$$

We also have, for $N \geqslant N_0(\delta)$ large,

$$\left|\frac{\theta + \theta_l}{2}\right| \leqslant \frac{\pi}{6}\cdot \tag{9.50}$$

Indeed, observe that $\cos\theta \geqslant r\cos\theta \geqslant 1 - \dfrac{1+2\delta}{N} =: \cos\theta_N$ with $0 \leqslant \theta_N = o(1)$ so that

$$\left|\frac{\theta + \theta_l}{2}\right| \leqslant \frac{\theta_N}{2} + \frac{\pi}{N} \leqslant \frac{\pi}{6}$$

when N is large enough, as claimed. Moreover, it holds

$$|e^{i\theta} - e^{i\theta_l}| \leqslant \frac{3}{N}\cdot \tag{9.51}$$

Just write

$$\left|e^{i\theta} - e^{i\theta_l}\right| \leqslant \left|e^{i\theta} - re^{i\theta}\right| + \left|z - z_l\right| + \left|r_l e^{\theta_l} - e^{i\theta_l}\right| \leqslant \frac{2+\delta}{N} \leqslant \frac{3}{N},$$

which gives in view of (9.49) and (9.50)

$$|\sin\theta - \sin\theta_l| = \left|2\sin\frac{\theta_l - \theta}{2}\cos\frac{\theta_l + \theta}{2}\right| \geqslant \sqrt{3}\left|\sin\frac{\theta_l - \theta}{2}\right|\cdot$$

So that, in view of (9.49) and (9.51),

$$|\theta - \theta_l| \leqslant \pi\left|\sin\frac{(\theta_l - \theta)}{2}\right| \leqslant \frac{\pi}{\sqrt{3}}\left|\sin\theta - \sin\theta_l\right| \leqslant \frac{\sqrt{3}\pi}{N} < \frac{2\pi}{N}$$

and so $D = 1$ is suitable in (9.48) when N is large.

This ends the proof of (9.48) and we justified the fact.

Now the proof of the upper bound clearly follows. □

As a corollary, we obtain the following (remember Proposition 9.6):

Proposition 9.25 *Let $n \leqslant m$ and $j_{n,m} : H_n \to L^2(\mu_{|\Gamma_m})$ where we have set $j_{n,m} = \pi_m \circ J_\mu \circ i_n$. We have*

$$\mathfrak{s}_p(j_{n,m}) \approx \left(\sum_{j=0}^{2^n-1} \left(2^n \mu_{|\Gamma_m}(W_{n,j})\right)^{p/2}\right)^{1/p}$$

and in particular

$$\mathfrak{s}_p(j_{n,n}) \approx \left(\sum_{j=0}^{2^n-1} \left(2^n \mu(R_{n,j})\right)^{p/2} \right)^{1/p} .$$

Proof: We first consider the case $n = 0$. We have $H_2^1 = H_0$ and the measure $\mu_{|\Gamma_m}$ is supported by G_1. By Proposition 9.24, we have

$$\mathfrak{s}_p(j_{0,m}) = \left(\sum_{j=0}^{2^n-1} 2^n \mu_{|\Gamma_m}(W_{n,j})^{p/2} \right)^{1/p} .$$

If $n \geqslant 1$, we again observe that $\mu_{|\Gamma_m}$ is supported by $\Gamma_m \subset G_{2^m}$ and Proposition 9.24 again tells that the injection $T : H_2^N \to L^2(\mu_{|\Gamma_m})$ satisfies, with $N = 2^n$,

$$\mathfrak{s}_p(T) \approx \left(\sum_{j=0}^{2^n-1} 2^n \mu_{|\Gamma_m}(W_{n,j})^{p/2} \right)^{1/p} .$$

We define two operators $V : H_n \to H_2^N$ and $W : L^2(\mu_{|\Gamma_m}) \to L^2(\mu_{|\Gamma_m})$ by the formulas

$$Vf = z^{-2^n+1} f, \quad Wf = z^{2^n-1} f \quad \text{with} \quad j_{m,n} = W \circ T \circ V.$$

V is an isometry and W a contraction. So that

$$\mathfrak{s}_p(j_{m,n}) \leqslant \|W\| \, \mathfrak{s}_p(T) \, \|V\| \leqslant \mathfrak{s}_p(T).$$

For reverse inequality, we coin $\widehat{V} : H_2^N \to H_n$ and $\widehat{W} : L^2(\mu_{|\Gamma_m}) \to L^2(\mu_{|\Gamma_m})$, $\widehat{V} f = z^{2^n-1} f$, $\widehat{W} f = z^{-2^n+1} f$, the respective inverses of V and W, and we observe that $\widehat{V}$ is an isometry while

$$\|\widehat{W} f\|^2_{L^2(\mu_{|\Gamma_m})} \leqslant \int_{\Gamma_m} (|z|^{-2^n})^2 |f(z)|^2 d\mu \leqslant 16 \int_{\Gamma_m} |f(z)|^2 d\mu = 16 \|f\|^2_{L^2(\mu_{|\Gamma_m})}$$

since $m \geqslant n$.

Since $T = \widehat{V} \circ j_{m,n} \circ \widehat{W}$, this gives $\mathfrak{s}_p(T) \lesssim \mathfrak{s}_p(j_{n,m})$. Hence

$$\mathfrak{s}_p(j_{n,m}) \approx \mathfrak{s}_p(T),$$

which proves the first assertion. The second one follows from the relation

$$W_{n,j} \cap \Gamma_n = R_{n,j} \quad \text{for} \quad j = 0, \ldots, 2^n - 1. \qquad \square$$

As a consequence, we can prove the following key proposition, which will allow us to exploit the previous Hilbertian generalities:

Proposition 9.26 *Let $p > 0$. There exists $C > 0$ and $a \in (0,1)$ satisfying*

$$\mathfrak{s}_p(j_{n,m}) \leqslant C a^{|m-n|} \mathfrak{s}_p(j_{m,m}) \quad \text{for all } m, n \in \mathbb{N}_0.$$

Proof: We can assume $m \neq n$, and we separate two cases:

(i) For $n < m$. We know by Proposition 9.24 that

$$\mathfrak{s}_p(j_{n,m})^p \approx \sum_{l=0}^{2^n-1} \big(2^n \, \mu_{|\Gamma_m}(W_{n,l})\big)^{p/2}.$$

Similarly $\mathfrak{s}_p(j_{m,m})^p \approx \sum_{l=0}^{2^m-1} \big(2^m \, \mu(R_{m,l})\big)^{p/2}$. Moreover, for $l = 0, \dots, 2^n - 1$, we have

$$W_{n,l} \cap \Gamma_m = \bigsqcup_{k \in \mathcal{E}_l} R_{m,k},$$

where

$$\mathcal{E}_l = \big\{2^{m-n} \times l, \dots, 2^{m-n}(l+1) - 1\big\}.$$

This implies

$$\mu_{|\Gamma_m}(W_{n,l}) = \sum_{k \in \mathcal{E}_l} \mu(R_{m,k})$$

and

$$\sum_{l=0}^{2^n-1} \sum_{k \in \mathcal{E}_l} \mu(R_{m,k}) = \sum_{l=0}^{2^m-1} \mu(R_{m,l}).$$

If $0 < p < 2$, we get by subadditivity of the map $x \mapsto x^{p/2}$:

$$\begin{aligned}
\mathfrak{s}_p(j_{n,m})^p \approx \sum_{l=0}^{2^n-1} \big(2^n \mu_{|\Gamma_m}(W_{n,l})\big)^{p/2} &= \sum_{l=0}^{2^n-1} \left(2^n \sum_{k \in \mathcal{E}_l} \mu(R_{m,k})\right)^{p/2} \\
&\leqslant \sum_{l=0}^{2^n-1} 2^{np/2} \sum_{k \in \mathcal{E}_l} \mu(R_{m,k})^{p/2} \\
&\leqslant 2^{(n-m)p/2} \sum_{l=0}^{2^m-1} \big(2^m \mu(R_{m,l})\big)^{p/2} \\
&\leqslant (1/\sqrt{2})^{(m-n)p} \mathfrak{s}_p(j_{m,m})^p.
\end{aligned}$$

If $p \geqslant 2$, we have

$$\left(\frac{1}{2^{m-n}}\sum_{k\in\mathcal{E}_l}\mu(R_{n,\,k})\right)^{p/2} \leqslant \frac{1}{2^{m-n}}\sum_{k\in\mathcal{E}_l}\mu(R_{n,k})^{p/2}. \tag{9.52}$$

It follows that

$$\begin{aligned}\mathfrak{s}_p(j_{n,m})^p \approx \sum_{l=0}^{2^n-1}\left(2^n\mu_{|\Gamma_m}(W_{n,l})\right)^{p/2} &\leqslant \frac{1}{2^{m-n}}\sum_{k=0}^{2^m-1}\left(2^m\mu(R_{m,k})\right)^{p/2}\\ &\lesssim \frac{1}{2^{m-n}}\mathfrak{s}_p(j_{m,m})^p;\end{aligned}$$

that is

$$\mathfrak{s}_p(j_{n,m}) \lesssim 2^{-\frac{m-n}{p}}\mathfrak{s}_p(j_{m,m}). \tag{9.53}$$

(ii) For $n > m$. The idea is to write H_n as a direct sum of spaces with the same dimension as H_m:

$$H_n = \bigoplus_{k=0}^{2^{n-m}-1} z^{l_k}H_m =: \bigoplus_{k=0}^{2^{n-m}-1} Y_k,$$

where $l_k = 2^n + (k-1)2^m$, with $0 \leqslant k \leqslant 2^{n-m}-1$, for example, $2^m - 1 + l_0 = 2^n - 1$. With those notations, it holds

$$j_{n,m} = \sum_{k=0}^{2^{n-m}-1}(j_{n,m})_{|Y_k}.$$

We can write $(j_{n,m})_{|Y_k} = B \circ j_{m,m} \circ A$ where

$$A : Y_k \to H_m \text{ and } B : L^2(\mu_{|\Gamma_m}) \to L^2(\mu_{|\Gamma_m})$$

are defined by $Af = z^{-l_k}f$ and $Bf = z^{l_k}f$. The map A is an isometry. We now recall that $\Gamma_m = \{z \in \mathbb{D} : 1 - 2^{-m} < |z| \leqslant 1 - 2^{-m-1}\}$ and use the simple inequality

$$z \in \Gamma_m \Longrightarrow |z|^{l_k} \leqslant (1 - 2^{-m-1})^{l_k} \leqslant (1/\mathrm{e})^{\frac{l_k}{2^{m+1}}}$$

to get

$$\|B\| \leqslant (1/\mathrm{e})^{\frac{l_k}{2^{m+1}}}$$

and hence

$$\mathfrak{s}_p((j_{n,m})_{|Y_k}) \leqslant \|A\|\,\mathfrak{s}_p(j_{m,m})\,\|B\| \leqslant (1/\mathrm{e})^{\frac{l_k}{2^{m+1}}}\mathfrak{s}_p(j_{m,m}).$$

In other terms

$$\mathfrak{s}_p((j_{n,m})_{|Y_k}) \leqslant \left(1/\sqrt{\mathrm{e}}\right)^{2^{n-m}+k-1} \mathfrak{s}_p(j_{m,m}).$$

We now consider two subcases.

(a) If $1 \leqslant p < \infty$, summing up over k with $0 \leqslant k \leqslant 2^{n-m} - 1$ and using $2^{n-m} \geqslant n+1$ gives

$$\mathfrak{s}_p(j_{n,m}) \leqslant \frac{\sqrt{\mathrm{e}}}{\sqrt{\mathrm{e}}-1}\left(\frac{1}{\sqrt{\mathrm{e}}}\right)^{n-m} \mathfrak{s}_p(j_{m,m}). \tag{9.54}$$

(b) Suppose now that $0 < p < 1$. Fix an integer N (only depending on p) such that $2^{1/p}\mathrm{e}^{-N/2} \leqslant 1/2$ and set

$$\alpha = \left[\frac{2^{n-m}-1}{N}\right]$$

where [.] denotes the integer part. We write our sum $j_{n,m}$ by blocks:

$$j_{n,m} = \sum_{r=0}^{\alpha-1} \sum_{k=rN}^{(r+1)N-1} (j_{n,m})_{|Y_k} + \sum_{k=\alpha N}^{2^{n-m}-1} (j_{n,m})_{|Y_k}.$$

By the Rotfel'd theorem G.24, we have

$$\begin{aligned}\mathfrak{s}_p(j_{n,m}) \leqslant & \sum_{r=0}^{\alpha-1} 2^{(r+1)/p} \mathfrak{s}_p\left(\sum_{k=rN}^{(r+1)N-1} (j_{n,m})_{|Y_k}\right) \\ & + 2^{(\alpha+1)/p} \mathfrak{s}_p\left(\sum_{k=\alpha N}^{2^{n-m}-1} (j_{n,m})_{|Y_k}\right).\end{aligned}$$

By applying again Theorem G.24 to inner sums, we get, for the values $r = 0, \ldots, \alpha - 1$:

$$\begin{aligned}&\mathfrak{s}_p\left(\sum_{k=rN}^{(r+1)N-1} (j_{n,m})_{|Y_k}\right) \\ &\leqslant 2^{1/p} \sum_{k=rN}^{(r+1)N-1} (2^{1/p})^{k-rN} \mathfrak{s}_p\big(j_{n,m}\big)_{|Y_k} \\ &\leqslant 2^{1/p} \sum_{k=rN}^{(r+1)N-1} (2^{1/p})^{k-rN} \left(\frac{1}{\sqrt{\mathrm{e}}}\right)^{2^{n-m+k-1}} \mathfrak{s}_p(j_{m,m}) \\ &\leqslant 2^{N/p} N\sqrt{\mathrm{e}}\left(\frac{1}{\sqrt{\mathrm{e}}}\right)^{2^{n-m+rN}} \mathfrak{s}_p(j_{m,m}).\end{aligned}$$

We obtain in a similar fashion

$$\mathfrak{s}_p\left(\sum_{k=\alpha N}^{2^{n-m-1}} (j_{n,m})_{|Y_k}\right) \leqslant 2^{N/p} N\sqrt{\mathrm{e}}\left(\frac{1}{\sqrt{\mathrm{e}}}\right)^{2^{n-m+\alpha N}} \mathfrak{s}_p(j_{m,m}).$$

The preceding gives, after some simple computations,

$$\begin{aligned}\mathfrak{s}_p\big(j_{n,m}\big) &\leqslant 2^{(N+1)/p} N\sqrt{\mathrm{e}}\left(\frac{1}{\sqrt{\mathrm{e}}}\right)^{n-m}\sum_{r=0}^{\infty}\left(\frac{2^{1/p}}{\mathrm{e}^{N/2}}\right)^{r} \mathfrak{s}_p\big(j_{m,m}\big)\\ &\leqslant 2.2^{(N+1)/p} N\sqrt{\mathrm{e}}\left(\frac{1}{\sqrt{\mathrm{e}}}\right)^{n-m} \mathfrak{s}_p(j_{m,m}) \qquad (9.55)\end{aligned}$$

because $2^{1/p}\mathrm{e}^{-N/2} \leqslant \frac{1}{2}$.

To summarize (observe in passing that m and n play symmetric roles): we obtained for any $p \in (0, +\infty)$, for some $C > 0$ and $a = \max\big(1/\sqrt{\mathrm{e}}, (1\sqrt{2})^p\big) < 1$,

$$\mathfrak{s}_p(j_{n,m}) \leqslant Ca^{|n-m|}\mathfrak{s}_p(j_{m,m}).$$

Finally, Propositions 9.25 and 9.26 allow us to apply the purely Hilbertian Theorem 9.22.

If we now remember that (via the relation $W_{n,j} \cap \Gamma_n = R_{n,j}$)

$$\mathfrak{s}_p(j_{n,n}) \approx \left(\sum_{j=0}^{2^n-1} \big(2^n\mu(R_{n,j})\big)^{p/2}\right)^{1/p},$$

we are done. □

All in all, gathering everything, we have completed the proof of Theorem 9.19.

9.6.2 Toeplitz Operators Approach

The results of that subsection are due to Luecking. In what follows, let μ be a *complex measure on* $\mathbb{D}$. We assume throughout this subsection that

$$\alpha \leqslant 0; \quad \text{we set again} \quad \gamma = -\alpha - 1 \geqslant -1. \qquad (9.56)$$

Recall that $H_0 = H^2$ and that $H_\alpha = \mathcal{B}^2_\gamma$ when $\alpha < 0$, that is, $\gamma > -1$.

We are concerned with the (non-classical) Toeplitz operators T_μ defined as

$$T_\mu f(z) = \int_{\mathbb{D}} f(w)(1 - z\overline{w})^{\alpha-1}\, d\mu(w), \qquad \text{for } f \in H_\alpha.$$

By Fubini's theorem, those operators satisfy the key identity (with $\langle . , . \rangle$ the scalar product in H_α)

$$\langle T_\mu f, g\rangle = \int_{\mathbb{D}} f(z)\overline{g(z)}\, d\mu(z), \qquad \text{for } f,\, g \in H_\alpha. \tag{9.57}$$

Indeed, let us explain it when $\alpha < 0$ (this is similar for $\alpha = 0$ replacing $(\mathbb{D}, \sigma_\gamma)$ by $(\mathbb{T}, m)$); we have, since the reproducing kernel of H_α is $(1 - \overline{w}z)^{\alpha-1}$,

$$\begin{aligned}
\langle T_\mu f, g\rangle &= \int_{\mathbb{D}} T_\mu f(z)\overline{g(z)}\, d\sigma_\gamma(z) \\
&= \int_{\mathbb{D}} \left[\int_{\mathbb{D}} f(w)(1 - z\overline{w})^{\alpha-1}\overline{g(z)}\, d\mu(w)\right] d\sigma_\gamma(z) \\
&= \int_{\mathbb{D}} f(w)\left[\int_{\mathbb{D}} \overline{g(z)}(1 - z\overline{w})^{\alpha-1}\, d\sigma_\gamma(z)\right] d\mu(w) \\
&= \int_{\mathbb{D}} f(w)\overline{g(w)}\, d\mu(w).
\end{aligned}$$

We shall exploit Carleson's theorem and HOP theorem from Chapter 5, also keeping in mind how we can use (or not!) half-windows for boundedness issues (recall Proposition 9.7).

We will also need *reverse Carleson inequalities* of the form

$$\|f\|_{H_\alpha}^2 \leqslant C' \int_{\mathbb{D}} |f|^2 d\mu, \qquad \text{for all } f \in H_\alpha. \tag{9.58}$$

We will return to that later on.

We first mention a Riesz–Thorin interpolation theorem for Schatten classes ([184, p. 40] or [62, theorem 13.1]). For the commutative case of functions, we refer to [109, theorem IV.1.2].

Theorem 9.27 *Let $\mathscr{S}^r$, $1 \leqslant r \leqslant \infty$, be the Schatten r-class of a separable Hilbert space H, and $1 \leqslant p_0,\ p_1 \leqslant \infty$, $\theta \in (0,1)$. Then, in the complex interpolation sense, we have*

$$\left[\mathscr{S}^{p_0}, \mathscr{S}^{p_1}\right]_\theta = \mathscr{S}^p \qquad \text{where } \frac{1}{p} = \frac{1-\theta}{p_0} + \frac{\theta}{p_1}.$$

But we will rather use the form that follows:

Theorem 9.28 *Let Ω be the strip $\Omega = \{z \in \mathbb{C} : 0 < \operatorname{Re} z < 1\}$ and $\big(T(z)\big)_{z\in\overline{\Omega}}$, be a bounded family of bounded operators on H, depending on z analytically on Ω and continuously on $\overline{\Omega}$ (for the weak operator topology). Assume that $1 \leqslant p_0, p_1 \leqslant \infty$, and that*

$$\mathfrak{s}_{p_0}(T(z)) \leqslant M_0 \quad \textit{for } \operatorname{Re} z = 0 \quad \textit{and} \quad \mathfrak{s}_{p_1}(T(z)) \leqslant M_1 \quad \textit{for } \operatorname{Re} z = 1.$$

Then, if $0 < \theta < 1$ *and* $\dfrac{1}{p} = \dfrac{1-\theta}{p_0} + \dfrac{\theta}{p_1}$*, we have* $T(z) \in \mathscr{S}^p$*, for every* z *with* $\mathrm{Re}\,(z) = \theta$ *and*

$$\sup_{\mathrm{Re}(z)=\theta} \mathfrak{s}_p(T(z)) \leqslant M_0^{1-\theta} M_1^{\theta}.$$

In the hypothesis, it is implicit that $T(z) \in \mathscr{S}^{p_j}$ when $\mathrm{Re}\, z = j$, for $j \in \{0; 1\}$.

Proof: We follow the presentation of [66, theorem 2.3], a bit more elementary than the proof in [62, theorem 13.1] though the crucial ideas are the same: the main idea is to complexify the interpolation parameter θ and then use Hadamard's three-lines theorem [153, chapter 11]. We give the proof for a finite p_0, but it is easy to adapt the proof when $p_0 = +\infty$.

Fix $N \in \mathbb{N}$, $a = (a_n)_{n\in\mathbb{N}}$ in the unit ball of $\ell^{p'}$ and two arbitrary orthonormal sequences $(h_n)_{n\in\mathbb{N}}$ and $(k_n)_{n\in\mathbb{N}}$ of H. Write $a_n = u_n|a_n|$ with $|u_n| = 1$, and, for $j \in \{0; 1\}$, denote by $S_j = \sup_{\Re e\,(z)=\alpha_j} \mathfrak{s}_{p_j}\big(T(z)\big)$. Let us now introduce the function

$$z \in \overline{\Omega} \longmapsto \Phi(z) = \sum_{n=1}^{N} u_n |a_n|^{p'\rho(z)+i\beta} \langle T(z)(h_n), k_n\rangle,$$

where $\beta \in \mathbb{R}$ and $\rho(z) = 1 - \dfrac{z}{p_1} - \dfrac{1-z}{p_0} = \dfrac{z}{p_1'} + \dfrac{1-z}{p_0'}$.

The function Φ is holomorphic on Ω, and moreover it is also clearly continuous and bounded on $\overline{\Omega}$ (with a bound maybe depending on N). Actually, on the boundary of Ω, we have some estimates not depending on N. Indeed, from Hölder's inequality and the fact that $\|a\|_{\ell^{p'}} \leqslant 1$, we have, for $j \in \{0; 1\}$ and every $\sigma \in \mathbb{R}$,

$$\begin{aligned}
|\Phi(j+i\sigma)| &\leqslant \sum_{n=1}^{N} |a_n|^{\frac{p'}{p_j'}} |\langle T(j+i\sigma)(h_n), k_n\rangle| \\
&\leqslant \left(\sum_{n=1}^{N} |\langle T(j+i\sigma)(h_n), k_n\rangle|^{p_j}\right)^{\frac{1}{p_j}} \\
&\leqslant \mathfrak{s}_{p_j}(T(j+i\sigma)) \leqslant S_j
\end{aligned}$$

by Theorem 1.111 (iii).

From the three-lines theorem, we get

$$\forall \theta \in [0,1], \forall \sigma \in \mathbb{R}, \qquad |\Phi(\theta+i\sigma)| \leqslant S_0^{1-\theta} S_1^{\theta}\,.$$

For any $\sigma \in \mathbb{R}$, we choose in the sequel $\beta = p'\sigma\left(\frac{1}{p_1} - \frac{1}{p_0}\right)$ so that we check that $p'\rho(\theta + i\sigma) + i\beta = 1$. Thus, for every $\sigma \in \mathbb{R}$, we get

$$\left|\sum_{n=1}^{N} a_n \langle T(\theta + i\sigma)(h_n), k_n\rangle\right| \leqslant S_0^{1-\theta} S_1^{\theta} .$$

Taking the supremum over a in the unit ball of $\ell^{p'}$ and then on $N \in \mathbb{N}$, we get

$$\left(\sum_{n=1}^{+\infty} |\langle T(\theta + i\sigma)(h_n), k_n\rangle|^p\right)^{\frac{1}{p}} \leqslant S_0^{1-\theta} S_1^{\theta} .$$

Finally, taking the supremum over all the orthonormal sequences and using again Theorem 1.111(iii), we obtain

$$\mathfrak{s}_p\big(T(\theta + i\sigma)\big) \leqslant S_0^{1-\theta} S_1^{\theta} .$$

□

Upper Bound when $p \geqslant 1$

We have to prove the following, in which once again i runs over the set I defined in (9.7);

Proposition 9.29 *Assume that* $1 \leqslant p < \infty$. *Then, given a* complex measure μ *on* $\mathbb{D}$, *we have*

$$\mathfrak{s}_p(T_\mu) \leqslant C\left(\sum_i \left[|\mu|(R_i) \times l(R_i)^{\alpha-1}\right]^p\right)^{1/p} .$$

Let us mention that when the RHS is finite, then, thanks to Proposition 9.6, $\sum_i \left[|\mu|(W_i) \times l(R_i)^{\alpha-1}\right]^p$ is finite as well (where W_i is the square Carleson window corresponding to R_i) and in particular $|\mu|$ is a Carleson measure for H_α. Indeed

$$\sup_I \frac{|\mu|(W_i)}{l(W_i)^{1-\alpha}} < \infty,$$

but since any Carleson square box is comparable to a union of a (boundedly) finite number of dyadic Carleson boxes W_i, we have the same inequality replacing W_i by an arbitrary Carleson window. Then the conclusion comes from Carleson theorem, for $\alpha = 0$, and HOP theorem, for $\alpha < 0$, in Chapter 5; and $|\mu|$ is Carleson for H_α, that is, $J_\mu : H_\alpha \to L^2(\mu)$ is bounded, hence T_μ too.

Proof: Without loss of generality, we assume that the RHS is finite:

$$\sum_n \sum_{j<2^n} \left[|\mu|(R_{n,j}) \times l(R_{n,j})^{\alpha-1}\right]^p = \sum_i \left[|\mu|(R_i) \times l(R_i)^{\alpha-1}\right]^p < \infty. \tag{9.59}$$

We first assume that $\alpha < 0$. In other words, we first prove it for Bergman spaces.

We will rely on Theorem 9.28 and use its notations. To that effect, we coin the family of complex measures

$$\mu_z = \sum_{i \in I} \left[|\mu|(R_i) l(R_i)^{\alpha-1}\right]^{pz-1} \mathbb{1}_{R_i}\, \mu, \qquad \text{for } z \in \overline{\Omega},$$

where the coefficient of $\mathbb{1}_{R_i}$ is taken to be zero if $|\mu|(R_i) = 0$, and the family of Toeplitz operators

$$T(z) = T_{\mu_z}.$$

We first claim that, for any $z \in \overline{\Omega}$, the measure $|\mu_z|$ is a Carleson measure for H_α (so that $T(z)$ is bounded on H_α): indeed

$$\begin{aligned}\sup_{i \in I} |\mu_z|(R_i) | l(R_i)^{\alpha-1} &= \left[|\mu|(R_i) l(R_i)^{\alpha-1}\right]^{p\mathrm{Re}(z)} \\ &\leqslant \max\left\{1, \left[|\mu|(R_i) l(R_i)^{\alpha-1}\right]^p\right\},\end{aligned}$$

which is bounded thanks to (9.59), hence thanks to Proposition 9.7, and since $\alpha < 0$ (i.e., the corresponding $\gamma > -1$), this justifies our claim.

We now check that the assumptions of Theorem 9.28 are met.

By definition, given $i \in I$, it holds that

$$|\mu_{z_0}|(R_i) = [|\mu|(R_i)\, l(R_i)^{\alpha-1}]^{-1} |\mu|(R_i) = l(R_i)^{1-\alpha}.$$

This means that $|\mu_{z_0}|$ is a Carleson measure for H_α with a norm $\|J_{|\mu_{z_0}|}\| \leqslant M_0$, where M_0 is a numerical constant. Now, given $f, g \in H_\alpha$, the identity (9.57) provides us with the inequalities (for arbitrary $f, g \in H_\alpha$)

$$\begin{aligned}\left|\langle T(z_0) f, g\rangle\right| = \left|\int_D f(z)\overline{g(z)} d\mu_{z_0}(z)\right| &\leqslant \|f\|_{L^2(|\mu_{z_0}|)} \|g\|_{L^2(|\mu_{z_0}|)} \\ &\leqslant M_0^2 \|f\|_{H_\alpha} \|g\|_{H_\alpha}.\end{aligned}$$

This means that $\|T(z_0)\| \leqslant M_0^2$.

Therefore, we showed that, for z_0 pure imaginary,

$$\mathfrak{s}_\infty(T(z_0)) = \|T(z_0)\| \leqslant M_0^2.$$

We next claim that, for $z_1 = 1 + w$, with w pure imaginary,

$$\mathfrak{s}_1(T(z_1)) \lesssim \sum_{i\in I} \left(|\mu|(R_i)|l(R_i)^{\alpha-1}\right)^p .$$

We have $|\mu_{z_1}|(R_i) = \left(|\mu|(R_i)\right)^p \left(l(R_i)\right)^{(\alpha-1)(p-1)}$. Now take two orthonormal bases $(f_n), (g_n)$ in H_α. The relation (9.57) and a double application of Cauchy–Schwarz give us

$$\begin{aligned}\sum_n |\langle T(z_1)f_n, g_n\rangle| &\leqslant \left(\int_{\mathbb{D}} (\sum_n |f_n|^2)\, d|\mu_{z_1}|\right)^{1/2} \left(\int_{\mathbb{D}} (\sum_n |g_n|^2)\, d|\mu_{z_1}|\right)^{1/2}\\ &= \int_{\mathbb{D}} (1-|z|^2)^{\alpha-1} d|\mu_{z_1}|(z),\end{aligned}$$

where in the last step we used

$$\sum_n |f_n(z)|^2 = \sum_n |g_n(z)|^2 = K_\alpha(z,z) = (1-|z|^2)^{\alpha-1}.$$

Now, the R_i partition $\mathbb{D}$ and if $z \in R_i$ with $i = (n, j) \in I$, we have $1 - |z|^2 \approx 2^{-n} = l(R_i)$, so that

$$\begin{aligned}\int_{\mathbb{D}} (1-|z|^2)^{\alpha-1} d|\mu_{z_1}|(z) &\approx \sum_i l(R_i)^{\alpha-1} |\mu_{z_1}|(R_i)\\ &= \sum_i |\mu|(R_i)^p\, l(R_i)^{(\alpha-1)(p-1+1)},\end{aligned}$$

or else

$$\sum_n |\langle T(z_1)f_n, g_n|\rangle \leqslant C \sum_i \left[|\mu|(R_i)\, l(R_i)^{\alpha-1}\right]^p.$$

Taking the supremum over $(f_n), (g_n)$ gives (having in mind Theorem 1.111)

$$\mathfrak{s}_1(T(z_1)) \leqslant C \sum_i \left[|\mu|(R_i) l(R_i)^{\alpha-1}\right]^p,$$

which was our claim.

We now apply Theorem 9.28 with $p_0 = \infty$, $p_1 = 1$, $\theta = 1/p$, observing that $T_\mu = T(\theta)$. This gives

$$\mathfrak{s}_p(T_\mu) = \mathfrak{s}_p(T(\theta)) \leqslant M_0^{2(1-1/p)} \left(\sum_i \left[|\mu|(R_i) l(R_i)^{\alpha-1}\right]^p\right)^{1/p},$$

and the proof is finished in this case.

We now assume $\alpha = 0$. The case $p = 1$ is actually already settled: the proof made in case $\alpha < 0$ works as well for $\alpha = 0$ (simply take $p = z = 1$ there: $\mu_1 = \mu$).

We now assume that $p > 1$, and the proof requires a special treatment, in view of the counterexample (iii) in Proposition 9.7. Actually the first part of the proof below looks like what we did for $\alpha < 0$.

We give an argument, borrowed from Luecking, with some additional details: fix *once and for all* an auxiliary positive number ε with $0 < \varepsilon < \frac{1}{p-1}$. For ζ in the strip $\overline{\Omega}$, we set $\alpha(\zeta) = 2\varepsilon(p \operatorname{Re} \zeta - 1) < 1$, and we write for short $\alpha_0 = \alpha(0) = -2\varepsilon$, and $\alpha_1 = \alpha(1) = 2\varepsilon(p-1)$.

We also define an operator of "fractional" derivation by the formula

$$D_\zeta \left(\sum_{n=0}^{\infty} a_n z^n \right) = \sum_{n=0}^{\infty} (n+1)^{\varepsilon(1-p\zeta)} a_n z^n.$$

By definition, $D_\zeta = D(\zeta)$ is equivalent to a unitary operator: $H_0 = H^2 \to H_{\alpha(\zeta)}$ (keep in mind that the norm is given by (9.1) for general α), and $D_{1/p}$ is the identity on H_0.

We correspondingly define complex measures

$$\mu_\zeta = \sum_{i \in I} \left[|\mu|(R_i) l(R_i)^{-1} \right]^{p\zeta - 1} \mathbb{1}_{R_i} \, \mu. \tag{9.60}$$

Next, we define an analytic operator valued $S(\zeta)$ on H_0 by the formula

$$S(\zeta) f(w) = \int_{\mathbb{D}} (D_\zeta f)(z) \overline{(D_{\overline{\zeta}} k_w)(z)} . (1 - |z|)^{2\varepsilon(1-p\zeta)} \, d\mu_\zeta(z). \tag{9.61}$$

We note that $S(1/p) = T_\mu$, and by complex interpolation it will be enough to show two things:

(i) $\operatorname{Re} \zeta = 0 \Longrightarrow \mathfrak{s}_\infty(S(\zeta)) = \|S(\zeta)\| \leqslant M_0$;

(ii) $\operatorname{Re} \zeta = 1 \Longrightarrow \mathfrak{s}_1(S(\zeta)) \leqslant M_1 \sum_{i \in I} \left(|\mu|(R_i) |l(R_i)^{-1} \right)^p$,

where M_0 is an absolute constant and M_1 only depends on p and ε.

Indeed, once (i) and (ii) are proved, Theorem 9.28 will give the conclusion.

To prove (i), we observe the identity (a consequence of Fubini's theorem) in which the scalar product is that of H_0:

$$\langle S(\zeta) f, g \rangle = \int_{\mathbb{D}} (D_\zeta f)(z) \overline{(D_{\overline{\zeta}}(g)(z)} . (1 - |z|)^{2\varepsilon(1-p\zeta)} \, d\mu_\zeta(z). \tag{9.62}$$

Since $\operatorname{Re} \zeta = 0$, this implies

$$|\langle S(\zeta) f, g \rangle| \leqslant \int_{\mathbb{D}} |D_\zeta f| . |D_{\overline{\zeta}} g| . (1 - |z|)^{2\varepsilon} \, d|\mu_\zeta|(z).$$

Let now $\nu = (1 - |z|)^{2\varepsilon} d|\mu_\zeta|$. This is a positive measure satisfying, for each $i \in I$,

$$\nu(R_i) \leqslant C[l(R_i)]^{1+2\varepsilon}$$

with C depending only on ε, and we know that this is sufficient, thanks to Theorem 9.7, to imply that this is a Carleson measure for the space $H_{-2\varepsilon}$. So that Cauchy–Schwarz now gives

$$|\langle S(\zeta)f, g\rangle| \leqslant \left[\int_{\mathbb{D}} |D_\zeta f|^2 \, d\nu\right]^{1/2} \left[\int_{\mathbb{D}} |D_\zeta g|^2 d\nu\right]^{1/2} \leqslant C\|D_\zeta f\|_{H_{-2\varepsilon}} \|D_\zeta g\|_{H_{-2\varepsilon}},$$

or else (using the norm expressed in (9.1))

$$|\langle S(\zeta)f, g\rangle| \leqslant M_0 \|f\|_{H_0} \|g\|_{H_0}.$$

This gives our first item.

To prove the second item, let $\zeta \in \overline{\Omega}$ with Re $\zeta = 1$. We first observe that

$$\frac{|\mu_\zeta|(R_i)}{l(R_i)} = \left(\frac{|\mu|(R_i)}{l(R_i)}\right)^p.$$

This easily implies that

$$\int_{\mathbb{D}} \frac{d|\mu_\zeta|(z)}{1 - |z|^2} \leqslant 2\sum_i \left(\frac{|\mu|(R_i)}{l(R_i)}\right)^p. \tag{9.63}$$

Now, choose bounded invertible operators A and B on H_0 such that

$$V = D_\zeta A \text{ and } W = D_{\overline{\zeta}} B : H_0 \to H_{\alpha_1}$$

are unitary operators. We set

$$T = B^* S(\zeta) A : H_0 \to H_0.$$

If (f_n) and (g_n) are orthonormal bases in H_0, the sequences (Vf_n) and (Wg_n) are orthonormal bases in H_{α_1}, and by Cauchy–Schwarz, we see that, since $\alpha_1 < 1$,

$$\sum_n (|Vf_n|.|Wg_n|)(z) \leqslant \|K_{z,\alpha_1}\|^2_{H_{\alpha_1}} = (1 - |z|^2)^{\alpha_1 - 1} \tag{9.64}$$

so that, with scalar product in H_0,

$$\begin{aligned}\sum_n |\langle Tf_n, g_n\rangle| &= \sum_n |\langle S(\zeta)Af_n, Bg_n\rangle| \\ &= \sum_n \left| \int_{\mathbb{D}} (D_\zeta Af_n)(z)\overline{(D_{\overline{\zeta}}(Bg_n)(z)}.(1-|z|)^{2\varepsilon(1-p\zeta)}\, d\mu_\zeta(z) \right| \\ &\leqslant \sum_n \int_{\mathbb{D}} |D_\zeta Af_n)(z)|.|(D_{\overline{\zeta}}(Bg_n)(z)|(1-|z|)^{2\varepsilon(1-p)}\, d|\mu_\zeta|(z) \\ &\leqslant \int_{\mathbb{D}} \sum_n (|Vf_n|.|Wg_n|)(z)\,(1-|z|^2)^{-\alpha_1}\, d|\mu_\zeta|(z),\end{aligned}$$

thanks to (9.62); or else, via (9.64),

$$\sum_n |\langle Tf_n, g_n\rangle| \leqslant \int_{\mathbb{D}} (1-|z|^2)^{\alpha_1-1}(1-|z|^2)^{-\alpha_1}\, d|\mu_\zeta|(z) \leqslant \int_{\mathbb{D}} \frac{d|\mu_\zeta|(z)}{(1-|z|^2)}.$$

By Theorem 1.111 and (9.63), this gives, since the orthonormal sequences (f_n), (g_n) are arbitrary,

$$\mathfrak{s}_1(T) \leqslant 2\sum_i \left(\frac{|\mu|(R_i)}{l(R_i)}\right)^p. \tag{9.65}$$

And we conclude by the ideal property of the one-Schatten norm:

$$\mathfrak{s}_1(S(\zeta)) \leqslant \|(B^*)^{-1}\| \times \sum_i \left(\frac{|\mu|(R_i)}{l(R_i)}\right)^p \times \|A^{-1}\| = M_1,$$

which proves (ii). □

Lower Bound when $p \geqslant 1$

We now prove a converse of Proposition 9.29.

Proposition 9.30 *Let $p \geqslant 1$ and $\alpha \leqslant 0$. Then, given a positive and finite measure μ on $\mathbb{D}$, we have*

$$\left(\sum_{i\in I} [\mu(R_i) \times l(R_i)^{\alpha-1}]^p\right)^{1/p} \leqslant C^2 \mathfrak{s}_p(T_\mu).$$

Proof: We first consider the case $\alpha < 0$.

Assume that $T_\mu \in \mathscr{S}^p(H_\alpha)$. Let $(e_i)_{i\in I}$ be an orthonormal basis of H_α, which we can index by the set I of pairs appearing in (9.7). We know from Theorem 1.111 that

$$\sum_{i\in I} |\langle T_\mu(e_i, e_i\rangle|^p \leqslant \left(\mathfrak{s}_p(T_\mu)\right)^p.$$

Let (with a_i the center of R_i)

$$g_i(z) = (1 - |a_i|)^{\frac{1-\alpha}{2}} (1 - \overline{a_i} z)^{\alpha - 1} = (1 - |a_i|)^{\frac{1-\alpha}{2}} K(z, a_i), \tag{9.66}$$

where $K(z, w)$ denotes the reproducing kernel of H_α. We claim that there exists a bounded operator $A : H_\alpha \to H_\alpha$ such that $A(e_i) = g_i$ for all $i \in I$. Indeed, the (so far formal) adjoint A^* of A is given by

$$\langle A^* f, e_i \rangle = \langle f, g_i \rangle = (1 - |a_i|)^{\frac{1-\alpha}{2}} f(a_i),$$

and we know by Lemma 10.29 of Chapter 10 (applicable since a_i are separated according to Lemma 9.1) that

$$\|A^* f\|^2 = \sum_i |\langle A^* f, e_i \rangle|^2 = \sum_i (1 - |a_i|)^{1-\alpha} |f(a_i)|^2 \leqslant C' \|f\|^2_{H_\alpha}.$$

Hence, A^* is bounded and so is A. Moreover, the ideal property of Schatten classes gives us

$$\sum_i |\langle A^* T_\mu A e_i, e_i \rangle|^p \leqslant \|A\|^{2p} \big(\mathfrak{s}_p(T_\mu)\big)^p,$$

or else

$$\sum_i |\langle T_\mu g_i, g_i \rangle|^p \leqslant \|A\|^{2p} \big(\mathfrak{s}_p(T_\mu)\big)^p =: C \big(\mathfrak{s}_p(T_\mu)\big)^p.$$

Now, we focus on a lower estimate for $|\langle T_\mu g_i, g_i \rangle|$. Using (9.57) gives (since μ is positive)

$$|\langle T_\mu g_i, g_i \rangle| = \int_{\mathbb{D}} |g_i|^2 \, d\mu \geqslant \int_{R_i} |g_i|^2 \, d\mu \gtrsim \int_{R_i} (1 - |a_i|)^{1-\alpha+2\alpha-2} \, d\mu,$$

that is,

$$|\langle T_\mu g_i, g_i \rangle| \geqslant C^{-p} \mu(R_i) (l(R_i))^{\alpha-1}$$

where we used, according to Lemma 9.1, $|1 - \overline{a_i} z| \approx 1 - |a_i| \approx l(R_i)$ when $z \in R_i$. We finally get

$$\sum_i \big[\mu(R_i) \times l(R_i)^{\alpha-1}\big]^p \leqslant C^p \sum_i |\langle T_\mu g_i, g_i \rangle|^p \leqslant C^{2p} \big(\mathfrak{s}_p(T_\mu)\big)^p,$$

which ends the proof when $\alpha < 0$.

We now consider the case $\alpha = 0$.

We just have to modify the test functions g_i by taking

$$g_i(z) = (1 - |a_i|)^{\frac{3}{2}} z (1 - \overline{a_i} z)^{-2} = (1 - |a_i|)^{\frac{3}{2}} \partial_{\overline{a}} [K(z, a)]_{a = a_i}$$

and the operator A with $A(e_i) = g_i$ for all $i \in I$. In view of the identity

$$\langle f, \partial_{\overline{a}}[K(z,a)]_{a=a_i}\rangle = \int_{\mathbb{T}} \frac{f(w)\,\overline{w}}{(1-a_i\overline{w})^2}\, dm(w) = f'(a_i),$$

we can indeed see that

$$\langle A^* f, e_i\rangle = \langle f, g_i\rangle = (1-|a_i|)^{\frac{3}{2}} f'(a_i).$$

Lemma 10.29 in Chapter 10 gives this time

$$\|A^* f\|^2 = \sum_i |\langle A^* f, e_i\rangle|^2 = \sum_i (1-|a_i|)^3 |f'(a_i)|^2 \leqslant C\|f\|_{H^2}^2,$$

and A is bounded on H^2. A computation similar to that of the case $\alpha < 0$ gives

$$|\langle T_\mu g_i, g_i\rangle| \gtrsim \int_{R_i} |g_i|^2\, d\mu \gtrsim \int_{R_i} (1-|a_i|)^3 (1-|a_i|)^{-4}\, d\mu \gtrsim \mu(R_i) l(R_i)^{-1},$$

and we conclude as in the case $\alpha < 0$. □

Upper Bound when $0 < p < 1$

In this section, we prove the following:

Theorem 9.31 *Assume that* $0 < p < 1$ *and* $\alpha \leqslant 0$. *Then, given a* complex measure μ *on* $\mathbb{D}$, *we have for some constant* C *depending only on* α *and* p

$$\mathfrak{s}_p(T_\mu) \leqslant C\left(\sum_i \left[|\mu|(R_i) \times l(R_i)^{\alpha-1}\right]^p\right)^{1/p}.$$

The proof is based on the central theorem below, whose elaborate proof requires several propositions and lemmas (and is postponed: see the subsection "The Missing Proof" at the end of the present section).

Theorem 9.32 *Fix* $\alpha < 0$. *Then there exists a separated sequence* (b_k) *of distinct points of* $\mathbb{D}$ *such that (constants* C *and* M *below depending only on* α*):*

(i) *The measure*

$$\tau = \sum_k (1-|b_k|^2)^{1-\alpha}\delta_{b_k} = \sum_k (1-|b_k|^2)^{\gamma+2}\delta_{b_k}$$

is reverse Carleson for $H_\alpha = \mathscr{B}^2_\gamma$, *that is, for all* $f \in H_\alpha$,

$$\|f\|^2_{\mathscr{B}^2_\gamma} = \|f\|^2_{H_\alpha} \leqslant C\int_{\mathbb{D}} |f|^2\, d\tau = C\sum_k (1-|b_k|^2)^{\gamma+2}|f(b_k)|^2.$$

(ii) *Each dyadic half-windows R_i contains at most M (and at least one) of the points b_k.*

In order to proceed, we still need two auxiliary lemmas.

In the next one, (i) contains a classical integral estimate. The point is that the exponents in the RHS *no longer depend* on $r > 1$.

Lemma 9.33 *Let $r > 1$ and $s > 0$. It holds for any $a \in \mathbb{D}$:*

(i) *With some underlying constants only depending on (r,s), we have*

$$\int_{\mathbb{D}} \frac{(1-|z|^2)^{r-2}}{|1-\bar{a}z|^{r+s}}\, d\sigma(z) \approx \frac{1}{(1-|a|)^s}. \tag{9.67}$$

(ii) *If (b_k) is a separated sequence in $\mathbb{D}$, there is a constant $C = C(r,s) > 0$ such that*

$$\sum_k (1-|b_k|)^r\, |1-\bar{a}b_k|^{-(r+s)} \leqslant C(1-|a|)^{-s}. \tag{9.68}$$

Proof: We mention that when $s \geqslant r$, the upper estimate can be directly deduced from Proposition 1.91.

By rotation invariance, we can assume that $a > 0$ and even that $1/2 \leqslant a < 1$ (the estimates are obvious when $a \leqslant 1/2$). Similarly, we can estimate only the quantity

$$E = \int_{1/2 \leqslant |z| < 1} (1-|z|)^{r-2} |1-\bar{a}z|^{-(r+s)}\, d\sigma(z).$$

Passing to polar coordinates gives

$$E \lesssim \int_{\frac{1}{2}}^{1} \int_{|\theta| \leqslant \pi} (1-t)^{r-2} |1-a\,t e^{i\theta}|^{-(r+s)}\, d\theta dt.$$

But, in the domain of integration,

$$|1-a\,t e^{i\theta}|^2 = (1-at)^2 + 4at\sin^2(\theta/2) \gtrsim (1-at)^2 + \theta^2$$

so that, changing t into $1-t$, we get for $0 \leqslant t \leqslant 1/2$ and $|\theta| \leqslant \pi$

$$|1-a\,(1-t)e^{i\theta}|^2 \gtrsim (1-a+at)^2 + \theta^2 \gtrsim (1-a)^2 + t^2 + \theta^2$$

and

$$E \lesssim \int_0^{\frac{1}{2}} \int_{|\theta| \leqslant \pi} t^{r-2} \left((1-a)^2 + t^2 + \theta^2\right)^{-(r+s)/2} d\theta dt.$$

Passing again to polar coordinates, we obtain

$$E \lesssim \int_0^\infty \int_{|u|\leqslant \pi/2} (\rho \cos u)^{r-2}\big[(1-a)^2+\rho^2\big]^{-(r+s)/2} \rho\, d\rho du$$
$$\leqslant C \int_0^\infty \rho^{r-2}\big[(1-a)^2+\rho^2\big]^{-(r+s)/2} \rho\, d\rho,$$

where we used $\int_{|u|\leqslant \pi/2} (\cos u)^{r-2}\, du < \infty$ since $r > 1$. The ultimate change of variable $\rho = (1-a)\,x$ finally gives

$$E \lesssim (1-a)^{r-2-r-s+2} \int_0^\infty x^{r-2}(1+x^2)^{-(r+s)/2} x\, dx \lesssim (1-a)^{-s}.$$

The lower estimate is easier:

$$\int_{\mathbb{D}} \frac{(1-|z|^2)^{r-2}}{|1-\overline{a}z|^{r+s}}\, d\sigma(z) \geqslant \int_{D(a,(1-|a|)/2} \frac{(1-|z|^2)^{r-2}}{|1-\overline{a}z|^{r+s}}\, d\sigma(z)$$
$$\approx \frac{1}{(1-|a|^2)^{s+2}} \sigma\big(D(a,(1-|a|)/2\big)$$
$$\approx \frac{1}{(1-|a|^2)^{s}}.$$

Now, we derive (9.68) from (9.67) by applying Lemma 10.29(i) with $\gamma = r-2 > -1$ and the holomorphic function $f(z) = (1-\overline{a}z)^{-(r+s)/2}$. Indeed, the LHS appears as $\|f\|^2_{\mathcal{B}^2_{r-2}}$ up to the factor $(r-1)$. □

In [3, proposition 2.2.], in the study of interpolation sequences in Bergman-type spaces $\mathcal{B}^p$, Amar introduced the following device: starting from $p < 1$, multiply by a large integer N so that $Np =: q > 1$, use what you already know for $\mathcal{B}^q$, and in the end divide licitly by N to fall again in $\mathcal{B}^p$ with the desired conclusion. The lemma below uses a similar procedure.

Lemma 9.34 *Let $\alpha \leqslant 0$. Let (b_k) be a separated sequence in $\mathbb{D}$, and let $N > 1$ be a real number. Set, for $z \in \mathbb{D}$,*

$$h_k(z) = (1-|b_k|)^{\frac{1-\alpha}{2}+N}.(1-\overline{b_k}z)^{\alpha-1-N}. \tag{9.69}$$

Let (e_k) be an orthonormal basis of ℓ^2. Then:

- (i) *There is a bounded operator $A : \ell^2 \to H_\alpha$ such that $A(e_k) = h_k$ for all k.*
- (ii) *If (b_k) is as in Theorem 9.32 with γ replaced by $2N+\gamma = 2N-\alpha-1$, the operator A is moreover onto.*

Proof: We first assume $\alpha < 0$.

Let $D_N : H_\alpha \to H_{\alpha-2N}$ be defined by

$$D_N f(z) = \int_{\mathbb{D}} f(w)\,(1 - \overline{w}z)^{\alpha-1-N} d\sigma_\gamma(w).$$

In view of the equivalent definitions of H_α, or $H_{\alpha-2N}$ in terms of Taylor coefficients, we can see that, for $f(z) = \sum_k f_k z^k$, we have $D_N f(z) = \sum_k g_k z^k$ with $|g_k| \approx (k+1)^N |f_k|$, and this makes the boundedness and invertibility of D_N obvious, since

$$\sum_k |g_k|^2 (k+1)^{\alpha-2N} \approx \sum_k |f_k|^2 (k+1)^\alpha.$$

As a consequence, Lemma 10.29 with $\alpha' = \alpha - 2N$, that is, $\gamma' = -\alpha' - 1 = 2N - \alpha - 1$, gives

$$\sum_k (1 - |b_k|)^{1-\alpha+2N} |D_N f(b_k)|^2 \leqslant C\|D_N f\|^2_{H_{\alpha-2N}} \leqslant C'\|f\|^2_{H_\alpha}. \quad (9.70)$$

Now, the (so far formal) adjoint $A^* : H_\alpha \to \ell^2$ satisfies

$$\begin{aligned}\langle A^* f, e_k\rangle = \langle f, h_k\rangle &= \int_{\mathbb{D}} f(w)\,(1 - |b_k|)^{\frac{1-\alpha}{2}+N}.(1 - \overline{w}b_k)^{\alpha-1-N}\, d\sigma_\gamma(w) \\ &= (1 - |b_k|)^{\frac{1-\alpha}{2}+N}.D_N f(b_k).\end{aligned}$$

So that, thanks to (9.70),

$$\sum_k |\langle A^* f, e_k\rangle|^2 = \sum_k (1 - |b_k|)^{1-\alpha+2N} |D_N f(b_k)|^2 \leqslant C'\|f\|^2_{H_\alpha}.$$

This shows that A^* and hence A are indeed defined and continuous.

If (b_k) moreover satisfies the reverse Carleson inequality of Theorem 9.32, we get this time, for all $f \in H_\alpha$,

$$\begin{aligned}\|A^* f\|^2_{H_\alpha} &= \sum_k |\langle A^* f, e_k\rangle|^2 \\ &= \sum_k (1 - |b_k|)^{1-\alpha+2N} |D_N f(b_k)|^2 \\ &\geqslant \delta\|D_N f\|^2_{H_{\alpha-2N}} \geqslant \delta'\|f\|^2_{H_\alpha}.\end{aligned}$$

This means that $A^* : H_\alpha \to \ell^2$ is bounded below. We now recall that an operator $A : X \to Y$ between Banach spaces is onto if and only if its adjoint $A^* = Y^* \to X^*$ is bounded below [154, theorem 4.13].

In the case $\alpha = 0$, we use the same kind of trick as in Proposition 9.30: we apply the result to $H_{-2} = \mathscr{B}^2_1$ (keep in mind the Littlewood–Paley formula for H^2) with N replaced by $N - 1$. Modulo constants, the derivative $D : H_\alpha \to H_{\alpha-2}$ is a bounded and invertible map. This ends the proof of Lemma 9.34. □

Proof of Theorem 9.31: We make use of a necessary condition for membership in $\mathscr{S}^p$ (see Theorem 1.111), namely: let (e_n) denote the canonical orthonormal basis of ℓ^2 and let T be a bounded operator on ℓ^2. Then

$$0 < p \leqslant 2 \Rightarrow (\mathfrak{s}_p(T))^p \leqslant \sum_{n,k} |\langle Te_n, e_k\rangle|^p. \tag{9.71}$$

Indeed, we proved in the abovementioned references that

$$(\mathfrak{s}_p(T))^p \leqslant \sum_n \|Te_n\|^p,$$

and since $p \leqslant 2$, $\|\cdot\|_{\ell^2} \leqslant \|\cdot\|_{\ell^p}$ and we have for each n

$$\|Te_n\|^p = \left(\sum_k |\langle Te_n, e_k\rangle|^2\right)^{p/2} \leqslant \sum_k |\langle Te_n, e_k\rangle|^p.$$

Now, we set $T = T_\mu$; we take a sequence (b_k) as in Lemma 9.32, so that the operator A^* of Lemma 9.34 is bounded below and hence has a left inverse, say L. This allows us to write

$$T = L\big(A^*TA\big)L^*,$$

and by the ideal property of Schatten classes it is enough to show that A^*TA is in $\mathscr{S}^p$ with controlled norm. We have (with h_n as in (9.69))

$$\langle A^*TAe_n, e_k\rangle = \langle TAe_n, Ae_k\rangle = \langle Th_n, h_k\rangle = \int_{\mathbb{D}} h_n\overline{h_k}\, d\mu = \sum_i \int_{R_i} h_n\overline{h_k}\, d\mu$$

so that, since $p < 1$, $\quad |\langle A^*TAe_n, e_k\rangle|^p \leqslant \sum_i \left(\int_{R_i} |h_n|\,|h_k|\, d|\mu|\right)^p.$

Set $r = \left(\dfrac{1-\alpha}{2} + N\right)p$, and $s = \left(\dfrac{1-\alpha}{2}\right)p$, where we choose N large enough to ensure $r > 1$.

We have $s > 0$ and $r + s = \big(1 - \alpha + N\big)p$.

We know that $|h_n(z)| = (1 - |b_n|)^{r/p}\,|1 - \overline{b_n}z|^{-(r+s)/p}$ behaves uniformly as $(1 - |b_n|)^{r/p}|1 - b_n\overline{a_i}|^{-(r+s)/p}$ on R_i. So that

$$\left(\int_{R_i} |h_n|\,|h_k|\, d|\mu|\right)^p$$
$$\lesssim (1-|b_n|)^r\,|1 - b_n\overline{a_i}|^{-(r+s)}\,(1-|b_k|)^r\,|1 - b_k\overline{a_i}|^{-(r+s)}\,(|\mu|(R_i))^p.$$

The indices n and k in the RHS split, which gives

$$\sum_{n,k} |\langle A^*TAe_n, e_k\rangle|^p \lesssim \sum_i (|\mu|(R_i))^p \left(\sum_n (1-|b_n|)^r\,|1 - b_n\overline{a_i}|^{-(r+s)}\right)^2.$$

Now, by Lemma 9.33(ii), applicable since $r > 1$, and in which s no longer depends on N, we know that

$$\sum_n (1 - |b_n|)^r |1 - b_n \overline{a_i}|^{-(r+s)} \lesssim (1 - |a_i|)^{-s} \approx (l(R_i))^{-s}$$

so that

$$\sum_{n,k} |\langle A^* T A e_n, e_k \rangle|^p \lesssim \sum_i \left(|\mu|(R_i)\right)^p (l(R_i))^{-2s} = \sum_i \left[|\mu|(R_i) l(R_i)^{\alpha-1}\right]^p,$$

and this ends the proof of Theorem 9.31. □

Lower Bound when $0 < p < 1$

It remains to prove the following:

Proposition 9.35 *Let $p \in (0, 1)$ and $\alpha \leqslant 0$. Then, given a* positive *and* finite *measure μ on $\mathbb{D}$, we have for some constant C depending only on α and p*

$$\left(\sum_i \left[\mu(R_i) l(R_i)^{\alpha-1}\right]^p\right)^{1/p} \leqslant C \mathfrak{s}_p(T_\mu).$$

The proof of this proposition is postponed: we first need a small auxiliary lemma. Recall that I denotes the set of pairs (n, k) with $n \in \mathbb{N}_0$ and $0 \leqslant k < 2^n$.

Lemma 9.36 *Let $\delta > 0$. Then, there exists a positive integer $M = M(\delta)$ for which the set I can be divided into M disjoint subsets I_j such that the sequences $S_j = \{a_i : i \in I_j\}$ are $(1 - \delta)$-separated for each $1 \leqslant j \leqslant M$:*

$$i, i' \in I_j \text{ and } i \neq i' \implies \rho(a_i, a_{i'}) \geqslant 1 - \delta.$$

Proof: Let $a, b \in \mathbb{D}$. Observe that, if $|a| \leqslant |b|$,

$$1 - \rho(a, b) \leqslant 1 - \rho^2(a, b) = \frac{(1 - |a|^2)(1 - |b|^2)}{|1 - \overline{a}b|^2} \leqslant 4 \frac{1 - |b|}{1 - |a|} \tag{9.72}$$

and also recall that (see Lemma 1.25) if $|a| = |b|$ and $|a - b| \geqslant P(1 - |a|)$, then

$$\rho(a, b) \geqslant \frac{P}{\sqrt{P^2 + 4}}. \tag{9.73}$$

Let now P be a large positive integer such that $4.2^{-P} \leqslant \delta$ and $\dfrac{P}{\sqrt{P^2 + 4}} \geqslant 1 - \delta$. Select n_0 satisfying $2^{n_0} \geqslant P$. Let E_n be the set of centers a_i lying on the circle $C(0, 1 - (3/4)2^{-n})$. We now write $E := X \sqcup Y$ where $X = \cup_{n \leqslant n_0} E_n$ is finite and $Y = \cup_{n > n_0} E_n$.

If we take E_n with $n > n_0$, from P to P both radially and angularly, that is, consider first the sets E_{Pn+r} with $r = 0 \ldots, P-1$ and for each such E_{Pn+r} we select angularly points a_i from P to P and get P sets $A_{n,r,k}$ $(0 \leqslant r, k < P)$ and, if we define $F_{r,k}$ as the union of such sets $A_{n,r,k}$ for $n > n_0$, we get the desired decomposition.

Indeed, take $a \neq b$ in $F_{r,k}$. If, for example, $a \in E_{Pn+r}$ and $b \in E_{Pm+r}$ with $m > n$, (9.72) gives

$$1 - \rho(a,b) \leqslant 4\frac{2^{-Pm-r}}{2^{-Pn-r}} \leqslant 4.2^{-P} \leqslant \delta.$$

And, if a, b belong to the same E_{Pn+r} then they are angularly separated by at least $2\pi\, P\, 2^{-n}$; we see that $|a-b| \geqslant P2^{-n}$ while $1-|a| = 1-|b| \leqslant 2^{-n}$; so (9.73) gives

$$\rho(a,b) \geqslant \frac{P}{\sqrt{P^2+4}} \geqslant 1-\delta,$$

and this proves that the set $F_{r,k}$ meets the constraints of S_j in the statement.

Connecting each remaining point of X to a far-away E_n with fixed $n > n_0$, we get a possible decomposition (with $M \approx P^2$). □

Proof of Proposition 9.35: We can assume that $\mathfrak{s}_p(T_\mu)$ is finite. In particular T_μ is (bounded and) compact.

It will be convenient to set, for $i \in I$,

$$d_i = \mu(R_i)l(R_i)^{\alpha-1}. \tag{9.74}$$

We also set $\mu_j = \sum_{i \in I_j} \mathbb{1}_{R_i}\, \mu$, where the sets I_j come from the preceding lemma, so that $\mu = \sum_{j=1}^{M} \mu_j$ and

$$T_\mu = \sum_{j=1}^{M} T_{\mu_j}.$$

Let us note that $\sum_i d_i^p = \sum_{j=1}^{M} \sum_{i \in I_j} d_i^p.$

Therefore, it is sufficient to control each of the inner sums $\sum_{i \in I_j} d_i^p$.

We obviously have $0 \leqslant T_{\mu_j} \leqslant T_\mu$ for $j = 1, \ldots, M$. Indeed, for each $f \in H_\alpha$, it holds

$$\langle T_{\mu_j} f, f\rangle = \int_{\mathbb{D}} |f|^2 d\mu_j \leqslant \int_{\mathbb{D}} |f|^2 d\mu = \langle T_\mu f, f\rangle.$$

But we recall that, when $0 \leqslant A \leqslant B$ are two positive operators A, B on a Hilbert space, we have $s_n(A) \leqslant s_n(B)$ for all $n \in \mathbb{N}$ by the minimax formula (G.9); hence, $\mathfrak{s}_p(T_{\mu_j}) \leqslant \mathfrak{s}_p(T_\mu)$ and it will suffice to prove that

$$\sum_{i\in I_j} d_i^p \leqslant C\big(\mathfrak{s}_p(T_{\mu_j})\big)^p \qquad \text{for each } 1 \leqslant j \leqslant M. \tag{9.75}$$

Let us now fix j, with $1 \leqslant j \leqslant M$, and we focus on μ_j.

Actually, we shall work with the measure

$$\nu_j = \sum_{i\in F_j} \mathbb{1}_{R_i}\, \mu\,,$$

where F_j is a finite subset of I_j. We obviously have $\nu_j \leqslant \mu_j$ and $T_{\nu_j} \leqslant T_{\mu_j} \leqslant T_\mu$.

Denote a_i the center of R_i for $i \in I$. The finite sequence $(a_i)_{i\in F_j}$ is $(1-\delta)$-separated. The proof of (9.75) makes strong use of that extreme separation. We once more write

$$r = \Big(\frac{1-\alpha}{2} + N\Big)\, p \qquad \text{and} \qquad s = \Big(\frac{1-\alpha}{2}\Big)\, p, \tag{9.76}$$

with $N \geqslant 1$ large enough to have $r > 2$, and not only $r > 1$.

We have $r + s = \big(1 - \alpha + N\big)p$ and we recall that, for $k \in F_j$,

$$h_k(z) = (1-|a_k|)^{r/p}(1-\overline{a_k}z)^{-(r+s)/p} = (1-|a_k|)^{\frac{1-\alpha}{2}}(1-\overline{a_k}z)^{1-\alpha+N}.$$

Let now A be as in Lemma 9.34(i) and $\varepsilon > 0$ associated to $(a_i)_{i\in I_j}$, which is a bounded operator from ℓ^2 to H_α with norm, say, K. We still denote A the restriction of A to $\ell^2(F_j)$, so that

$$A(e_i) = h_i \qquad \text{for every } i \in F_j\,.$$

Moreover, $T_{\nu_j} \in \mathcal{S}^p$; we have $A^* T_{\nu_j} A \in \mathcal{S}^p$ with

$$\mathfrak{s}_p(A^* T_{\nu_j} A) \leqslant K^2 \mathfrak{s}_p(T_{\nu_j}) \leqslant K^2 \mathfrak{s}_p(T_{\mu_j}) \leqslant K^2 \mathfrak{s}_p(T_\mu).$$

The matrix of this operator on the orthonormal basis (e_n) of $\ell^2(F_j)$ is

$$\big(\langle T_{\nu_j} h_k, h_n\rangle\big)_{n,k\in F_j} =: D + E,$$

where $D = (\langle T_{\nu_j} h_k, h_k\rangle)$ is the diagonal part and where $E = (\langle T_{\nu_j} h_k, h_n\rangle)$ is the off-diagonal matrix.

We write as well $\quad A^* T_{\nu_j} A = D + E.$

The operators $A^* T_{\nu_j} A$, D, and E are operators on $\ell^2(F_j)$.

Now we use an *extrapolation trick* as in the proof for H^2 in Section 9.6.1, namely we will show below that, with c and C positive constants depending only on p and α,

$$\mathfrak{s}_p(D)^p \gtrsim c \sum_{i \in F_j} d_i^p. \tag{9.77}$$

$$\mathfrak{s}_p(E)^p \leqslant C\varepsilon \sum_{i \in F_j} d_i^p. \tag{9.78}$$

Assuming this is proved, since, by Rotfel'd's inequality (see Theorem G.24), we have

$$\mathfrak{s}_p(U+V)^p \leqslant \mathfrak{s}_p(U)^p + \mathfrak{s}_p(V)^p$$

for all U, V, and this implies

$$\begin{aligned}\sum_{i \in F_j} d_i^p \leqslant c^{-1}\mathfrak{s}_p(D)^p \leqslant c^{-1}\big[\mathfrak{s}_p(A^*T_{\nu_j}A)^p + \mathfrak{s}_p(E)^p\big]& \\ \leqslant c^{-1}\mathfrak{s}_p(A^*T_{\nu_j}A)^p + c^{-1}C\varepsilon \sum_{i \in F_j} d_i^p& \\ \leqslant c^{-1}\mathfrak{s}_p(A^*T_{\nu_j}A)^p + \frac{1}{2}\sum_{i \in F_j} d_i^p&\end{aligned}$$

if we adjust ε small enough. So that

$$\sum_{i \in F_j} d_i^p \leqslant 2c^{-1}\mathfrak{s}_p(A^*T_{\nu_j}A)^p \leqslant C'\mathfrak{s}_p(T_\mu)^p.$$

Since $\sum_{i \in F_j} d_i^p \leqslant C'\mathfrak{s}_p(T_\mu)^p$ for arbitrary finite subsets F_j of I_j, we have

$$\sum_{i \in I_j} d_i^p \leqslant C'\mathfrak{s}_p(T_\mu)^p,$$

and after summing up with respect to $j \in \{1, \ldots, M\}$, we are done.

Nevertheless, it remains to establish (9.77) and (9.78). We first see that, since D is a positive operator,

$$\mathfrak{s}_p(D)^p = \sum_{k \in F_j} |\langle T_{\nu_j} h_k, h_k\rangle|^p = \sum_{k \in F_j}\left(\int_{\mathbb{D}} |h_k|^2\, d\nu_j\right)^p \gtrsim \sum_{k \in F_j}\left(\int_{R_k} |h_k|^2\, d\nu_j\right)^p.$$

We saw that

$$|h_k|^2 \gtrsim (1-|a_k|)^{2r/p}(1-|a_k|)^{-2(r+s)/p} = (1-|a_k|)^{-2s/p} \approx l(R_k)^{\alpha-1}$$

on R_k; this implies

$$\mathfrak{s}_p(D)^p \geqslant c \sum_{k \in F_j} \left[\nu_j(R_k) l(R_k)^{\alpha-1} \right]^p = c \sum_{i \in F_j} d_i^p.$$

We switch to (9.78), which is much more elaborate since the indices k and n in the double sum $\sum_{k \neq n}$ no longer split. We begin, using (9.71), as in the proof of Theorem 9.31 to get, using (9.8),

$$\begin{aligned}
\mathfrak{s}_p(E)^p &\leqslant \sum_{k \neq n} |\langle T_{\nu_j} h_k, h_n \rangle|^p \\
&\leqslant \sum_{i \in F_j} \sum_{k \neq n} \left(\int_{R_i} |h_k|\,|h_n|\, d\nu_j \right)^p \\
&\leqslant C \sum_{i \in F_j} (\nu_j(R_i))^p \sum_{k \neq n} \left[(1-|a_k|)\,(1-|a_n|) \right]^r . \left[|1-\overline{a_i} a_k|\,|1-\overline{a_i} a_n| \right]^{-(r+s)} \\
&\lesssim \sum_{i \in F_j} d_i^p (1-|a_i|)^{2s}\, S_i
\end{aligned}$$

with

$$S_i = \sum_{k \neq n} \left[(1-|a_k|)\,(1-|a_n|) \right]^r \left[|1-\overline{a_i} a_k|\,|1-\overline{a_i} a_n| \right]^{-(r+s)}, \tag{9.79}$$

where r, s are as in (9.76). The issue now is to estimate the double sum S_i of (9.79), which follows the pattern of Lemma 9.1 but in a more elaborate way. We first use the bi-subharmonicity of the function

$$\varphi_i(u,v) = \left[|1-\overline{a_i} u| . |1-\overline{a_i} v| \right]^{-(r+s)}$$

to get

$$\begin{aligned}
\varphi_i(a_k, a_n) &\leqslant \frac{1}{\sigma(R_k)\sigma(R_n)} \int_{R_k \times R_n} \varphi_i(u,v)\, d\sigma(u)\, d\sigma(v) \\
&\leqslant \left[(1-|a_k|)\,(1-|a_n|) \right]^{-2} \int_{R_k \times R_n} \varphi_i(u,v)\, d\sigma(u)\, d\sigma(v).
\end{aligned}$$

Moreover, $1-|u|^2$ behaves like $1-|a_k|$ on $R_k \times R_n$ (see Lemma 9.1), and by summation over $(k,n) \in F_j \times F_j$ with $k \neq n$, we get for $i \in F_j$

$$S_i \lesssim C \iint_U \left[(1-|u|^2)\,(1-|v|^2) \right]^{r-2} \left[|1-\overline{a_i} u| . |1-\overline{a_i} v| \right]^{-(r+s)} d\sigma(u)\, d\sigma(v),$$

where

$$U = \bigcup_{k \neq n} R_k \times R_n \subset G := \left\{ (u,v) \in \mathbb{D}^2 \ :\ \rho(u,v) \geqslant 1-\delta \right\}.$$

So that

$$S_i \leqslant C \iint_G \left[(1-|u|^2)(1-|v|^2)\right]^{r-2}\left[|1-\overline{a_i}u|.|1-\overline{a_i}v|\right]^{-(r+s)} d\sigma(u)\, d\sigma(v).$$

We perform the change of variables

$$(u,v) = \left(\frac{z+a_i}{1+\overline{a_i}z}, \frac{w+a_i}{1+\overline{a_i}w}\right)$$

whose jacobian is

$$\left(\frac{1-|a_i|^2}{|1+\overline{a_i}z|^2}\cdot\frac{1-|a_i|^2}{|1+\overline{a_i}w|^2}\right)^2$$

with

$$1-|u|^2 = \frac{(1-|a_i|^2)(1-|z|^2)}{|1+\overline{a_i}z|^2} \quad \text{and} \quad 1-|v|^2 = \frac{(1-|a_i|^2)(1-|w|^2)}{|1+\overline{a_i}w|^2}$$

and

$$|1-\overline{a_i}u| = \frac{1-|a_i|^2}{|1+\overline{a_i}z|}, \quad |1-\overline{a_i}v| = \frac{1-|a_i|^2}{|1+\overline{a_i}w|}.$$

The domain of integration does not change since $\rho(u,v) = \rho(z,w)$ (automorphisms are ρ-isometries). By the above, during this change of variable, the power of $1-|a_i|^2$ that emerges is $2(r-2)-2(r+s)+4 = -2s$. Similarly, the powers of $|1+\overline{a_i}z|$ or $|1+\overline{a_i}w|$ that emerge are $s-r = -Np$. So that

$$\begin{aligned} S_i &\leqslant \frac{C}{(1-|a_i|)^{2s}} \int_G \left[(1-|z|^2)(1-|w|^2)\right]^{r-2}\left[|1+\overline{a_i}z|\,|1+\overline{a_i}w|\right]^{-Np} d\sigma(z) d\sigma(w) \\ &=: \frac{C}{(1-|a_i|)^{2s}}\psi(a_i) \leqslant \frac{C}{(1-|a_i|)^{2s}}\psi(1). \end{aligned}$$

Indeed, the function ψ defined above as a function of a_i is subharmonic on $\mathbb{D}$, continuous on $\overline{\mathbb{D}}$, and rotationally invariant. Hence, it takes its maximum at 1. We have

$$\begin{aligned} \psi(1) &= \int_G \left[(1-|z|^2)((1-|w|^2)\right]^{r-2}\left[|1+z|\,|1+w|\right]^{-Np} d\sigma(z)\, d\sigma(w) \\ &=: \int_G h(z,w)\, d\sigma(z)\, d\sigma(w). \end{aligned}$$

Here, the fact that $r > 2$ intervenes in a decisive way, to show that h is integrable with respect to the measure $\sigma \otimes \sigma$. Because, using $1-|z|^2 \leqslant 2(1-|z|)$ for $z \in \mathbb{D}$, we get

$$h(z,w) \lesssim \left[|1+z|\,|1+w|\right]^{r-2-Np} =: f(z)\, f(w),$$

and $f \in L^1(\sigma)$ since $r - Np = \frac{1-\alpha}{2} > 0$, so that

$$\int_{\mathbb{D}} f d\sigma \leqslant \int_{|Z|\leqslant 2} |Z|^{r-2-Np} \, d\sigma(Z) = 2 \int_0^2 x^{r-1-Np} \, dx < \infty.$$

But obviously $(\sigma \otimes \sigma)(G) \to 0$ with δ; therefore, we can adjust this $\delta = \delta(\varepsilon)$ so as to achieve $\psi(1) \leqslant \varepsilon$, giving us

$$S_i \leqslant C\varepsilon(1 - |a_i|)^{-2s},$$

where C only depends on α and p. Finally, we get

$$(\mathfrak{s}_p(E))^p \leqslant C\varepsilon \sum_{i \in F_j} d_i^p.$$

We get (9.78) and this ends the proof of Proposition 9.35. □

The Missing Reverse Carleson Proof

It remains to prove the previously admitted Theorem 9.32. This theorem played a *key role* in the previous proof of Theorem 9.31.

To that effect, we will need three results on reverse Carleson measures, not mandatorily with the form of τ.

Some Preliminary Notations

We take $0 < h < 1/4$ to be adjusted later and denote $\Delta(a,h)$ the (open) pseudo-hyperbolic disk of center a and radius h. The "varying" positive constants c, C to follow will only depend on $\gamma > -1$. Finally, let χ_h be the function defined by

$$\chi_h(w,z) = \begin{cases} 1 & \text{if } \rho(z,w) < h \\ 0 & \text{if } \rho(z,w) \geqslant h. \end{cases}$$

To ease reading, we will abbreviate $\chi_{1/2}$ to χ, $\Delta(a, 1/2)$ to $\Delta(a)$.

Now, the first needed result is as follows:

Proposition 9.37 *Let $\gamma > -1$ and G be a Borel subset of $\mathbb{D}$ such that, for some $\delta > 0$,*

$$\sigma_\gamma(G \cap W) \geqslant \delta\sigma_\gamma(W) \tag{9.80}$$

for all Carleson windows W. Then there exists $c = c(\delta) > 0$ such that, for all $f \in \mathscr{B}_\gamma^2$,

$$\int_G |f|^2 \, d\sigma_\gamma \geqslant c \int_{\mathbb{D}} |f|^2 \, d\sigma_\gamma.$$

Let us agree to say that sets G satisfying (9.80) are *δ-thick in $\mathbb{D}$*. Proposition 9.37 is indeed a special case (with the measure $\mu = \mathbb{1}_G \, d\sigma_\gamma$) of Proposition 9.38 to follow, and *will be used in its proof.*

But to keep this chapter a reasonable length, we will not detail the proof of Proposition 9.37 here, and refer to [121], in which the proof is given with W replaced by $S(u,r)$ where $|u| = 1$. But as we saw in Chapter 5, Carleson square windows and Carleson circular windows are "equivalent."

The second needed result is a key step for dominating an error term.

Proposition 9.38 *Let $\gamma > -1$, $B_1 > 0$, and μ be a positive, finite measure on $\mathbb{D}$ satisfying*

$$\mu[\Delta(a)] \leqslant B_1 \sigma_\gamma[\Delta(a)] \qquad \text{for all } a \in \mathbb{D}.$$

Then, for all functions $f \in \mathscr{B}^2_\gamma$ and all $0 < h \leqslant 1/4$, it holds with $C = C(\mu) > 0$

$$\iint_{\mathbb{D}^2} \frac{\chi_h(w,z)}{\sigma_\gamma[\Delta(z,h)]} |f(z) - f(w)|^2 d\mu(z)\, d\sigma_\gamma(w) \leqslant C B_1 h^2 \int_{\mathbb{D}} |f|^2\, d\sigma_\gamma.$$

For the proof, we first need the auxiliary lemma as follows:

Lemma 9.39 *For any $f \in \mathscr{B}^2_\gamma$ and any $w, a \in \mathbb{D}$, it holds, for $|h| \leqslant 1/4$,*

$$\chi_h(w,a)\, |f(w) - f(a)|^2 \leqslant \chi_h(w,a)\, C h^2 \int_{\Delta(a)} \frac{|f(\zeta)|^2}{(1-|\zeta|^2)^{\gamma+2}}\, d\sigma_\gamma(\zeta). \tag{9.81}$$

Proof: For this lemma, we might use as Luecking a compactness (normal family) argument, but we prefer to give an explicit proof in this Hilbertian case. We first prove, for $g(z) = \sum_{n\geqslant 0} a_n z^n \in \mathscr{B}^2_\gamma$ and $|z| \leqslant h$,

$$|g(z) - g(0)|^2 \leqslant C h^2 \int_{\Delta(0)} |g(u)|^2 d\sigma(u). \tag{9.82}$$

Indeed, let L, R be respectively the LHS and the integral appearing in the RHS of (9.82). Then

$$R = \sum_{n\geqslant 0} |a_n|^2 \int_0^{1/4} t^n dt = \frac{1}{4} \sum_{n\geqslant 0} \frac{|a_n|^2 4^{-n}}{n+1}.$$

While by Cauchy–Schwarz

$$\begin{aligned} L &\leqslant \left(\sum_{n\geqslant 1} |a_n|\, h^n\right)^2 = \left(\sum_{n\geqslant 1} |a_n| \frac{2^{-n}}{\sqrt{n+1}} \cdot \sqrt{n+1}(2h)^n\right)^2 \\ &\leqslant CR \left(\sum_{n\geqslant 1} (n+1)(2h)^{2n}\right), \end{aligned}$$

implying (9.82) as claimed (since $2h \leqslant 1/2$).

Let now $w \in \Delta(a,h)$ and let $z = \varphi_a(w)$ so that $|z| = \rho(w,a) \leqslant h$, as well as $g = f \circ \varphi_a$, that is, $g \circ \varphi_a = f$. The inequality (9.82) reads, by the change of variable $u = \varphi_a(\zeta)$, and remembering that φ_a is a ρ-isometry,

$$|f(w) - f(a)|^2 \leqslant Ch^2 \int_{\Delta(a)} |f(\zeta)|^2 \left(\frac{1-|a|^2}{|1-\overline{a}\zeta|^2}\right)^2 d\sigma(\zeta).$$

So that, using $\dfrac{1-|a|^2}{|1-\overline{a}\zeta|^2} \leqslant \dfrac{1}{1-|\zeta|^2}$, we get

$$\begin{aligned}|f(w) - f(a)|^2 &\leqslant Ch^2 \int_{\Delta(a)} \frac{|f(\zeta)|^2}{(1-|\zeta|^2)^2}\, d\sigma(\zeta)\\ &= \frac{Ch^2}{\gamma+1}\int_{\Delta(a)} \frac{|f(\zeta)|^2}{(1-|\zeta|^2)^{\gamma+2}}\, d\sigma_\gamma(\zeta).\end{aligned}$$

Multiplying by $\chi_h(w,a)$, this proves Lemma 9.39. □

Proof of Proposition 9.38: This lemma is used as follows: change a to z, divide both members of (9.81) by $\sigma_\gamma[\Delta(z,h)]$, and integrate with respect to σ_γ in the variable w to get, using that $\int_{\mathbb{D}} \chi_h(w,z)\, d\sigma_\gamma(w) = \sigma_\gamma[\Delta(z,h)]$,

$$\int_{\mathbb{D}} \frac{\chi_h(w,z)}{\sigma_\gamma[\Delta(z,h)|} |f(w) - f(z)|^2 d\sigma_\gamma(w) \leqslant Ch^2 \int_{\mathbb{D}} \chi(z,\zeta) \frac{|f(\zeta)|^2}{(1-|\zeta|^2)^{\gamma+2}} d\sigma_\gamma(\zeta).$$

Next, integrate with respect to μ in the variable z and use Fubini's theorem in the RHS; this gives

$$\begin{aligned}&\iint_{\mathbb{D}^2} \frac{\chi_h(w,z)}{\sigma_\gamma[\Delta(z,h)|}\, |f(w) - f(z)|^2\, d\sigma_\gamma(w)\, d\mu(z)\\ &\qquad\leqslant Ch^2 \int_{\mathbb{D}} |f(\zeta)|^2 \mu[\Delta(\zeta)](1-|\zeta|^2)^{-\gamma-2}\, d\sigma_\gamma(\zeta).\end{aligned}$$

Now, we use our assumption $\mu[\Delta(\zeta)] \leqslant B_1 \sigma_\gamma[\Delta(\zeta)]$. And we notably have [184, (4.9)]

$$\sigma_\gamma[\Delta(\zeta)] \leqslant C(1-|\zeta|^2)^{\gamma+2}.$$

In fact, the change of variables $u = \varphi_a(\eta)$ gives, for any $a \in \mathbb{D}$, the useful relation, uniform as long as $|h| \leqslant 1/4$,

$$\begin{aligned}\sigma_\gamma[\Delta(a,h)] &= (\gamma+1)(1-|a|^2)^{\gamma+2} \int_{|\eta|\leqslant h} \frac{(1-|\eta|^2)^\gamma}{|1-\overline{a}\eta|^{2\gamma+4}}\, d\sigma(\eta)\\ &\approx (1-|a|^2)^{\gamma+2} \int_{|\eta|\leqslant h} d\sigma(\eta),\end{aligned}$$

that is (with constant depending on γ only),

$$\sigma_\gamma[\Delta(a,h)] \approx (1-|a|^2)^{\gamma+2}\, h^2. \tag{9.83}$$

Hence, we end up with

$$\iint_{\mathbb{D}^2} \frac{\chi_h(w,z)}{\sigma_\gamma[\Delta(z,h)|} |f(w)-f(z)|^2\, d\sigma_\gamma(w)\, d\mu(z) \leqslant C B_1\, h^2 \int_{\mathbb{D}} |f(\zeta)|^2\, d\sigma_\gamma(\zeta),$$

proving Proposition 9.38. □

A first consequence of Proposition 9.38 is the following general theorem on reverse Carleson measures, our third needed result, which will be later on specialized to (finite) discrete measures μ of the form

$$\mu = \tau := \sum_k (1-|b_k|^2)^{\gamma+2} \delta_{b_k}.$$

We recall that a Borel set G of $\mathbb{D}$ is δ-*thick* if

$$\sigma_\gamma(W \cap G) \geqslant \delta \sigma_\gamma(W)$$

for all Carleson windows W.

For any positive finite measure μ on $\mathbb{D}$, $z \in \mathbb{D}$, and $h > 0$, we set

$$k_h(z) = \frac{\mu[\Delta(z,h)]}{\sigma_\gamma[\Delta(z,h)]} \qquad \text{and} \qquad \|\mu\|_* := \sup_{z\in\mathbb{D}} \frac{\mu[\Delta(z)]}{\sigma_\gamma[\Delta(z)]} = \sup_{z\in\mathbb{D}} k_{\frac{1}{2}}(z)\,.$$

We can then state the following:

Theorem 9.40 *Let $\gamma > -1$ and $\delta > 0$.*

Then there exists $h > 0$ with the following property:

For any positive finite measure μ on $\mathbb{D}$ satisfying both $\|\mu\|_* > 0$ and that $G := \{z : k_h(z) \geqslant \|\mu\|_*/3\}$ is δ-thick, μ is reverse Carleson *for $\mathscr{B}^2_\gamma$.*

In other words, there exists $C = C_\delta > 0$ such that, for all functions $f \in \mathscr{B}^2_\gamma$, it holds

$$\int_{\mathbb{D}} |f(w)|^2\, d\sigma_\gamma(w) \leqslant \frac{C}{\|\mu\|_*} \int_{\mathbb{D}} |f(z)|^2\, d\mu(z). \tag{9.84}$$

Proof: We will transform both sides of (9.84) into double integrals. We start with the obvious inequality

$$\begin{aligned}\iint_{\mathbb{D}^2} &\frac{\chi_h(w,z)}{\sigma_\gamma[\Delta(z,h)]} |f(w)|^2\, d\sigma_\gamma(w)\, d\mu(z)\\ &\leqslant 2 \iint_{\mathbb{D}^2} \frac{\chi_h(w,z)}{\sigma_\gamma[\Delta(z,h)]} |f(w)-f(z)|^2\, d\sigma_\gamma(w)\, d\mu(z)\\ &\quad + 2 \iint_{\mathbb{D}^2} \frac{\chi_h(w,z)}{\sigma_\gamma[\Delta(z,h)]} |f(z)|^2\, d\sigma_\gamma(w)\, d\mu(z),\end{aligned}$$

say

$$L \leqslant 2M_1 + 2M_2.$$

Starting from this, the idea lurking behind the proof of Theorem 9.40 is the following: we will minorize L through Theorem 9.37, majorize M_1 through Proposition 9.38 and M_2 trivially to infer an inequality of the form

$$c\|\mu\|_* \int_{\mathbb{D}} |f|^2 \, d\sigma_\gamma \leqslant C\|\mu\|_* h^2 \int_{\mathbb{D}} |f|^2 \, d\sigma_\gamma + 2\int_{\mathbb{D}} |f|^2 \, d\mu.$$

Then we will adjust h small and conclude with the usual extrapolation trick. Here are the details.

Using $\sigma_\gamma[\Delta(z,h)] \approx \sigma_\gamma[\Delta(w,h)]$ for $z \in \Delta(w,h)$, we get

$$\begin{aligned}\int_{\Delta(w,h)} \frac{1}{\sigma_\gamma[\Delta(z,h)]} \, d\mu(z) &\geqslant c \int_{\Delta(w,h)} \frac{1}{\sigma_\gamma[\Delta(w,h)]} \, d\mu(z) \\ &= c k_h(w) \\ &\geqslant \frac{c}{3} \|\mu\|_* \mathbb{1}_G(w).\end{aligned}$$

So that, using Fubini and Theorem 9.37,

$$L \geqslant \frac{c}{3} \|\mu\|_* \int_G |f(w)|^2 \, d\sigma_\gamma(w) \geqslant c' \|\mu\|_* \int_{\mathbb{D}} |f(w)|^2 \, d\sigma_\gamma(w).$$

Next, the error term M_1 satisfies

$$M_1 \leqslant C\|\mu\|_* \, h^2 \int_{\mathbb{D}} |f|^2 d\sigma_\gamma$$

by Proposition 9.38 (with $B_1 = \|\mu\|_*$).

Finally, by Fubini, we have

$$M_2 = \int_{\mathbb{D}} |f(z)|^2 \frac{\sigma_\gamma[\Delta(z,h)]}{\sigma_\gamma[\Delta(z,h)]} \, d\mu(z) = \int_{\mathbb{D}} |f(z)|^2 \, d\mu(z).$$

So that the inequality $L \leqslant 2M_1 + 2M_2$ reads

$$c' \|\mu\|_* \int_{\mathbb{D}} |f|^2 \, d\sigma_\gamma \leqslant C\|\mu\|_* \, h^2 \int_{\mathbb{D}} |f|^2 \, d\sigma_\gamma + 2\int_{\mathbb{D}} |f|^2 \, d\mu.$$

Now, adjust h small enough to satisfy $Ch^2 \leqslant c'/2$ and get

$$\int_{\mathbb{D}} |f|^2 \, d\sigma_\gamma \leqslant \frac{4}{c' \|\mu\|_*} \int_{\mathbb{D}} |f|^2 \, d\mu.$$

□

Let us switch to Theorem 9.32. As a preliminary, we will need the following Vitali-type lemma of purely metric type (communicated to us by Rodríguez–Piazza).

Lemma 9.41 *Let (X,d) be a metric space, and (B_j) a sequence of balls of (X,d) with* bounded radii. *Then there exists a subsequence (B'_k) of (B_j) such that*

(i) *The B'_k are disjoint;*
(ii) *We have the inclusion* $\bigcup_j B_j \subset \bigcup_k 5B'_k$,

where in general $5B$ denotes the ball with the same center as B and five times its radius.

Proof: We denote $r(B)$ the radius of a ball B and begin by a simple observation: let B and B' be two intersecting balls. Then

$$r(B) \leqslant 2r(B') \Rightarrow B \subset 5B'. \tag{9.85}$$

Indeed, let $x \in B$ and $y \in B \cap B'$. We see that (denoting c, c' the centers of B, B')

$$d(x,c') \leqslant d(x,y) + d(y,c') \leqslant 2r(B) + r(B') \leqslant 5r(B').$$

Now, let $r = \sup_j r(B_j) < \infty$ and let $\mathcal{F}_1$ be the (non-empty) family of balls B_j such that $r/2 < r(B_j) \leqslant r$, inside which we choose a maximal disjoint subfamily $\mathcal{G}_1$. By (9.85), any $B \in \mathcal{F}_1$ is included in $5B'$ for some $B' \in \mathcal{G}_1$. Next, let $\mathcal{F}_2$ be the family of those B_j (if any) such that $r/4 < r(B_j) \leqslant r/2$, $\mathcal{F}'_2$ be the subfamily of $\mathcal{F}_2$ of those balls that do not intersect the balls of $\mathcal{G}_1$, and $\mathcal{G}_2$ be a maximal disjoint subfamily of $\mathcal{F}'_2$. We claim that

$$B \in \mathcal{F}_2 \Rightarrow B \subset 5B' \text{ for some } B' \in \mathcal{G}_1 \cup \mathcal{G}_2.$$

Indeed, if $B \notin \mathcal{F}'_2$, B intersects some $B' \in \mathcal{G}_1$ and we have $r(B) \leqslant r/2 \leqslant r(B')$ so that $B \subset 5B'$. If $B \in \mathcal{F}'_2$, B intersects some $B' \in \mathcal{G}_2$ satisfying $r(B) \leqslant r/2 = 2(r/4) \leqslant 2r(B')$ and again, by (9.85), $B \subset 5B'$, proving our claim. We continue in a clear inductive fashion and the disjoint union

$$\mathcal{G}_1 \sqcup \mathcal{G}_1 \cdots \sqcup \mathcal{G}_p \sqcup \cdots$$

constitutes our subsequence (B'_k). □

Proof of Theorem 9.32: We will choose here the points b_k. We refine each R_i, with $i = (n, j)$, into M subboxes denoted by $S_k := R_{i,l}$ with $1 \leqslant l \leqslant M$, where the integer M is large but will be fixed, and where

$$k = (i,l) = (n,j,l)\,, \quad n \geqslant 1\,, \quad 0 \leqslant j \leqslant 2^n - 1\,, \quad \text{and } 1 \leqslant l \leqslant M.$$

Denote α_k the center of the subbox S_k. It is clear that (α_k) is separated, that each R_i contains at most M of the points α_k, and that we can find $h \in (0, 1/4)$ and $\delta > 0$ such that the set

$$G = \bigcup_k \Delta(\alpha_k, h)$$

is δ-thick. By Lemma 9.41, we can find a subsequence (b_j) of (α_k) such that:

(i) The disks $\Delta(b_j, h)$ are disjoint.
(ii) $G \subset \bigcup_j \Delta(b_j, 5h)$.

This implies that the set $G' = \bigcup_j \Delta(b_j, h)$ is δ_0-thick with $\delta_0 = \delta/25$.

Now set

$$\mu = \sum_j \sigma_\gamma\big[\Delta(b_j, h)\big]\,\delta_{b_j}.$$

We see that $\|\mu\|_* \geqslant c > 0$ since, for one j_0 (and because $h < 1/2$),

$$\|\mu\|_* \geqslant \frac{\mu\big[\Delta(b_{j_0})\big]}{\sigma_\gamma\big[\Delta(b_{j_0})\big]} \geqslant \frac{\mu\big[\Delta(b_{j_0}, h)\big]}{\sigma_\gamma\big[\Delta(b_{j_0})\big]} = \frac{\sigma_\gamma\big[\Delta(b_{j_0}, h)\big]}{\sigma_\gamma\big[\Delta(b_{j_0}, 1/2)\big]},$$

which is a constant since h, j_0, and γ are fixed.

On the other hand, G' is none other than the set of points

$$\big\{z : k_h(z) > \|\mu\|_*/3\big\}.$$

Hence, Theorem 9.40 gives us, for all $f \in H_\alpha$,

$$\begin{aligned}
\int_{\mathbb{D}} |f|^2 d\sigma_\gamma \leqslant C \int_{\mathbb{D}} |f|^2 d\mu &= C \sum_j |f(b_j)|^2 \mu\big[\Delta(b_j, h)\big] \\
&= C \sum_j |f(b_j)|^2 \sigma_\gamma\big[\Delta(b_j, h)\big] \\
&\leqslant C \sum_k |f(b_j)|^2 (1 - |b_j|^2)^{\gamma+2} \\
&= C \int_{\mathbb{D}} |f|^2\, d\tau,
\end{aligned}$$

which is our Theorem 9.32. We used in passing (an already mentioned fact) for given $h > 0$: $\sigma_\gamma[\Delta(b_j, h)] \approx \big(1 - |b_j|^2\big)^{\gamma+2}$ (with underlying constants depending only on γ). □

9.6.3 Reproducing Kernels Approach

Preliminaries

This approach is due to El-Fallah and El Ibbaoui. It is quite elegant, but once more we will not be able to give all the details and have to admit some results.

The center of R_i is again denoted by a_i, R_i' is as in Lemma 9.1, and σ_γ is the Bergman measure $d\sigma_\gamma(z) = (\gamma + 1)(1 - |z|^2)^\gamma d\sigma(z)$. And H denotes the weighted Bergman space $H = H_\alpha = \mathscr{B}^2_\gamma$ (the case $H = H^2$ is not considered here).

We recall that

$$\alpha + \gamma = -1, \qquad \gamma > -1 \text{ (i.e., } \alpha < 0),$$

associated to the measure σ_γ, with the reproducing kernel

$$K(z, w) = (1 - \overline{w}z)^{-\gamma-2}.$$

If $(f_n)_{n\geqslant 1}$ is an orthonormal basis of H, the Parseval identity implies the useful identity, of which we will make again strong use:

$$\sum_{n\geqslant 1} |f_n(z)|^2 = \|K_z\|^2 \qquad \text{for each } z \in \mathbb{D}. \tag{9.86}$$

The dyadic half-windows R_i are indexed by the set I of pairs

$$i = (n, j), \quad \text{with } n \geqslant 0 \text{ and } 0 \leqslant j \leqslant 2^n - 1.$$

We know from Proposition 1.91 that

$$z \in R_i' \Rightarrow \sigma_\gamma(R_i') \times \|K_z\|^2 \approx 1. \tag{9.87}$$

Indeed, for $z \in R_i'$, we have $\|K_z\|^2 = (1 - |a_i|^2)^{-(\gamma+2)} \approx \left(\sigma_\gamma(R_i')\right)^{-1}$. Same for $R_i \subset R_i'$ instead of R_i', since $\sigma_\gamma(R_i') \approx \sigma_\gamma(R_i)$.

The eigenvalues of a *positive and compact* operator T on H, rearranged in non-increasing order, will be denoted $\lambda_n(T)$ with $\lambda_1(T) = \|T\|$.

The letter h will always denote a non-negative, increasing function h defined on $\mathbb{R}^+$ with $h(0) = h(0^+) = 0$. Since we are mainly interested in Schatten classes, that is, in the function $h(x) = x^p$, $0 < p < \infty$, in order to simplify the statements, we will additionally assume that h is *slowly oscillating*, meaning that $\dfrac{h(y)}{h(x)} \leqslant \rho\left(\dfrac{y}{x}\right)$ with $\rho : (0, \infty) \to (0, \infty)$ locally bounded above. We shall also assume that, for some $q \in]0, 1]$, the function h satisfies the subhomogeneity property: $x > 0 \mapsto h(x)/x^q$ is non-decreasing; equivalently, for every $t \in [0, 1]$ and $x > 0$, $h(tx) \leqslant t^q h(x)$. This condition is then obviously satisfied (with $q = 1$) when h is convex, whereas it really is a constraint when h is concave (the concave function $x \to \ln(x + 1)$ does not satisfy this constraint).

The general proof is the same, up to the replacement of $h(d_i)$ respectively by $h(c_1 d_i)$ and $h(c_2 d_i)$ in the LHS and RHS in the double inequality of Theorem 9.42 to follow.

We note that, for T positive and compact with Schmidt decomposition denoted $T = \sum_k s_k(u_k \otimes u_k)$ and, by functional calculus, $h(T) := \sum_k h(s_k)(u_k \otimes u_k)$, it holds

$$\operatorname{Tr} h(T) = \sum_{n \geqslant 1} h(\lambda_n(T)),$$

and that, if S, T are such operators,

$$S \leqslant T \Rightarrow \operatorname{Tr} h(S) \leqslant \operatorname{Tr} h(T).$$

Indeed, for each $n \geqslant 1$, one has $\lambda_n(S) \leqslant \lambda_n(T)$ by the minimax formula, so that $h\big(\lambda_n(S)\big) \leqslant h\big(\lambda_n(T)\big)$ since h increases; and the result follows by summation.

If μ is a positive measure on $\mathbb{D}$, the Toeplitz operator $T_\mu : H \to H$ is again defined, for $f \in H$, by

$$T_\mu f(z) = \int_{\mathbb{D}} K(z, w) f(w)\, d\mu(w)$$

and satisfies, for all $f, g \in H$,

$$\langle T_\mu f, g\rangle = \int_{\mathbb{D}} f(z)\overline{g(z)}\, d\mu(z).$$

In particular $T_\mu \geqslant 0$. The main result of El-Fallah and El Ibbaoui now goes as follows, where, as before, h is an increasing (slowly oscillating) function $\mathbb{R}^+ \to \mathbb{R}^+$ with $h(0) = h(0^+) = 0$, with the subhomogeneity property.

Theorem 9.42 *Let μ be a positive finite measure on $\mathbb{D}$. Assume that the function h is either* convex or concave. *Set here*

$$d_i = \frac{\mu(R_i)}{\sigma_\gamma(R_i)}, \quad \textit{for } i \in I.$$

Then

$$A \sum_i h(d_i) \leqslant \operatorname{Tr} h(T_\mu) \leqslant B \sum_i h(d_i),$$

where A, B are two positive constants.

We take the liberty of replacing in (9.74) $\mu(R_i) l(R_i)^{\alpha-1}$ by $\dfrac{\mu(R_i)}{\sigma_\gamma(R_i)}$ since $\alpha = -\gamma - 1 < 0,\ l(R_i)^{1-\alpha} \approx \sigma_\gamma(R_i)$, and h is slowly oscillating.

Observe that specializing with $h(x) = x^p$ for $0 < p < \infty$, one recovers the results of Luecking on Schattenness of Toeplitz operators on H_α, where $\alpha < 0$. The already mentioned equation

$$C_\varphi^* C_\varphi = T_\mu, \quad \mu = \varphi(\sigma_\gamma),$$

also recovers the previous results for composition operators C_φ: just change p to $p/2$ in the Toeplitz characterizations.

Now the proof will face the same difficulties as Luecking's one. We must give separate proofs, according to whether h is convex or concave. And each case is divided in two subcases: lower bound, upper bound.

For the upper bound, we can assume $\sum_i h(d_i) < \infty$, hence $h(d_i) \to 0$, a fortiori $d_i \to 0$.

We start with the convex case.

The Convex Case

Let first

$$\mu = \sum_i \mu_i, \text{ with } \mu_i = \mu_{|R_i} \quad \text{and} \quad \tau = \sum \tau_i, \text{ with } \tau_i = d_i\,(\sigma_\gamma)_{|R'_i}.$$

We begin by showing that $T_\tau = J_\tau^* J_\tau$ is compact. Indeed

$$\tau(R_i) = \sup_{R_i \cap R'_j \neq \emptyset} d_j\,(\sigma_\gamma)_{|R'_j}(R_i) \leqslant M\sigma_\gamma(R_i)\sup\{d_j;\ R_i \cap R'_j \neq \emptyset\}.$$

But, since $d_i \to 0$, the center $a_{n,k}$ of R_i must satisfy $n \to \infty$. Therefore

$$\sup\{d_j;\ R_i \cap R'_j \neq \emptyset\} \longrightarrow 0,$$

and we get $\tau(R_i) = o\left(l(R_i)^{1-\alpha}\right)$.

We obtain that τ is a vanishing Carleson measure for H_α thanks to Proposition 9.7, hence $J_\tau : H_\alpha \to L^2(\tau)$ is compact.

Now we claim that, for each $i \in I$,

$$T_{\mu_i} \leqslant C T_{\tau_i}. \tag{9.88}$$

Indeed, for $f \in H$ and $i \in I$, we see that (recall Lemma 5.18)

$$\langle T_{\mu_i} f, f\rangle = \int_{R_i} |f(z)|^2\,d\mu(z) \leqslant C\frac{1}{\sigma_\gamma(R_i)}\int_{R'_i \times R_i} |f(w)|^2\,d\sigma_\gamma(w)\,d\mu(z)$$

where we used the subharmonicity of $|f|^2$ via the inclusions of Lemma 9.1, $R_i \subset \Delta_{r_2}(a_i) \subset \Delta_{r_3}(a_i) = R'_i$, and the fact that $\sigma_\gamma(R'_i) \approx \sigma_\gamma(R_i)$. Or again

$$\langle T_{\mu_i} f, f\rangle \leqslant C\frac{\mu(R_i)}{\sigma_\gamma(R_i)}\int_{R'_i} |f(w)|^2 d\sigma_\gamma(w),$$

which gives

$$\langle T_{\mu_i} f, f\rangle \leqslant C d_i \int_{R'_i} |f(w)|^2 d\sigma_\gamma(w) = C\langle T_{\tau_i} f, f\rangle,$$

proving (9.88). By summation, we get $T_\mu \leqslant C T_\tau$ and

$$\operatorname{Tr} h(T_\mu) \leqslant C \operatorname{Tr} h(T_\tau).$$

We now provide an upper bound for $\operatorname{Tr} h(T_\tau)$.

To that effect, we see that (with (f_n) an orthonormal basis of H formed by eigenvectors of T_τ)

$$\operatorname{Tr} h(T_\tau) = \sum_n h\left(\int |f_n|^2\, d\tau\right) = \sum_n h\left(\sum_i d_i \int_{R_i'} |f_n(z)|^2\, d\sigma_\gamma(z)\right).$$

But for each fixed n, we have, thanks to the finite multiplicity of R_i' (recall that each point of $\mathbb{D}$ belongs to at most M of the R_i'),

$$\sum_i \int_{R_i'} |f_n|^2 d\sigma_\gamma \leqslant M \int_{\mathbb{D}} |f_n|^2 d\sigma_\gamma = M$$

so that the convexity and slow oscillation of h give you, for some constant $C = C(M)$,

$$h\left(\sum_i d_i \int_{R_i'} |f_n|^2\, d\sigma_\gamma(z)\right) \leqslant C \sum_i h(d_i) \int_{R_i'} |f_n|^2\, d\sigma_\gamma(z).$$

Now, we sum over n, permute the sums, and use the relations (9.86) and (9.87), which give (for some other constant C')

$$\int_{R_i'} \sum_n |f_n|^2\, d\sigma_\gamma(z) = \int_{R_i'} \|K_z\|^2\, d\sigma_\gamma(z) \leqslant C'.$$

This gives

$$\operatorname{Tr}\ h(T_\tau) \leqslant B \sum_i h(d_i)$$

(with $B = CC'$) and the RHS in Theorem 9.42 is proved.

The LHS is slightly more involved. We set again

$$\mu_i = \mu_{|R_i} \quad \text{with} \quad \mu = \sum_i \mu_i \quad \text{and} \quad T_\mu = \sum_i T_{\mu_i}.$$

The Thompson inequality (G.27) gives us

$$\operatorname{Tr} h(T_\mu) \geqslant \sum_i \operatorname{Tr} h(T_{\mu_i}). \tag{9.89}$$

But for each i, we have

$$\operatorname{Tr} h(T_{\mu_i}) \geqslant h\left(\lambda_1(T_{\mu_i})\right) = h\left(\|T_{\mu_i}\|\right).$$

Set $h_i = \dfrac{K_{a_i}}{\|K_{a_i}\|}$ so that $\|h_i\|_H = 1$ and $h_i(a_i) = \|K_{a_i}\|$. We see that

$$\|T_{\mu_i}\| \geqslant \langle T_{\mu_i}(h_i), h_i\rangle = \int_{\mathbb{D}} |h_i|^2\, d\mu_i = \int_{R_i} |h_i|^2\, d\mu.$$

But we know from (9.87) that $|h_i|^2 \approx \|K_{a_i}\|^2 \approx \dfrac{1}{\sigma_\gamma(R_i)}$ on R_i, so that it holds

$$\int_{R_i} |h_i|^2 d\mu \approx \frac{\mu(R_i)}{\sigma_\gamma(R_i)} = d_i.$$

Hence

$$\|T_{\mu_i}\| \geqslant c\, d_i \quad \text{and} \quad h\left(\|T_{\mu_i}\|\right) \geqslant h\,(cd_i).$$

This, the slow oscillation of h and (9.89) now give the desired lower bound.

The Concave Case

We will need two important lemmas.

Lemma 9.43 *Let ω be the auxiliary measure*

$$\omega = \sum_i \frac{1}{d_i}\, \mu_{|R_i} = \sum_i \frac{\sigma_\gamma(R_i)}{\mu(R_i)}\, \mu_{|R_i}.$$

Then T_ω is bounded on H_α.

Proof: Since $\omega(R_i) \approx l(R_i)^{1-\alpha}$, we know from Proposition 9.7 ($\alpha < 0$ here) that the Carleson embedding $J_\omega : H_\alpha \to L^2(\omega)$ is bounded. Hence $T_\omega = J_\omega^* J_\omega$ is bounded as well. □

Observe that it is $1/d_i$ (instead of d_i in the convex case) that appears here. This can be seen as a discrete hint to the reverse Carleson measures in Luecking's approach when $p < 1$.

The second lemma goes as follows.

Lemma 9.44 *Let $\nu_i = (\sigma_\gamma)_{|R_i}$. Then*

$$\lambda_k(T_{\nu_i}) \leqslant B\, e^{-\delta k} \qquad \textit{for all } k,$$

where $B, \delta > 0$ do not depend on $i \in I$.

Proof: Fix i. Write λ_k for $\lambda_k(T_{\nu_i})$. Following [48, lemma 4.3], we use (the adaptation of) a result of Lin and Rochberg, which tells that there exist a constant $C > 0$ such that

$$\mathfrak{s}_p[T_{\nu_i}]^p \leqslant \frac{C}{p} \qquad \text{for } p \in (0, 1/2).$$

In particular, for each positive integer k,

$$k\, \lambda_k^p \leqslant \sum_{n=1}^{\infty} \lambda_n^p = \mathfrak{s}_p[T_{\nu_i}]^p \leqslant \frac{C}{p}.$$

We optimize in p for given k large enough by choosing $p = \dfrac{\mathrm{e}\,C}{k} < 1/2$. This gives the result for some B and $\delta = \dfrac{1}{\mathrm{e}\,C}\cdot$ □

We can now prove the LHS in Theorem 9.42. Let $(f_n)_{n\geqslant 1}$ be an orthonormal basis of H formed by eigenvectors of T_μ:

$$T_\mu(f_n) = \lambda_n f_n \quad \text{with} \quad \lambda_n = \lambda_n(T_\mu).$$

We have

$$\operatorname{Tr} h(T_\mu) = \sum_{n\geqslant 1} h\left(\lambda_n\right) = \sum_{n\geqslant 1} h\left(\langle T_\mu f_n, f_n\rangle\right) = \sum_{n\geqslant 1} h\left(\int_{\mathbb{D}} |f_n(z)|^2\, d\mu(z)\right).$$

Or again (with ω as in Lemma 9.43)

$$\operatorname{Tr} h(T_\mu) = \sum_{n\geqslant 1} h\left(\sum_i d_i \int_{R_i} |f_n(z)|^2\, d\omega(z)\right).$$

But for each n

$$\sum_i \int_{R_i} |f_n(z)|^2\, d\omega(z) = \langle T_\omega f_n, f_n\rangle \leqslant C.$$

Hence, the concavity and slow oscillation of h gives

$$\operatorname{Tr} h(T_\mu) \geqslant c \sum_n \sum_i h(d_i) \int_{R_i} |f_n(z)|^2\, d\omega(z) \geqslant c \sum_i h(d_i) \int_{R_i} \|K_z\|^2\, d\omega(z).$$

Now, thanks to (9.87), we have

$$\int_{R_i} \|K_z\|^2\, d\omega(z) \approx \frac{1}{\sigma_\gamma(R_i)}\omega(R_i) \approx \frac{1}{\sigma_\gamma(R_i)} \frac{\sigma_\gamma(R_i)}{\mu(R_i)}\mu(R_i) = 1,$$

and we get our lower bound.

For the upper bound (RHS), we argue as follows: we know by (9.88) that $T_\mu = \sum_i T_{\mu_i}$ with $T_{\mu_i} \leqslant C T_{\tau_i}$ for each $i \in I$.

Using the previous notations, we can write

$$\tau_i = d_i \sum_{j\in E_i} (\nu_j)_{|R_i'} \leqslant d_i \sum_{j\in E_i} \nu_j,$$

where $E_i = \{j;\ R_j \cap R_i' \neq \emptyset\}$ has cardinality bounded by an absolute constant D. Therefore

$$T_{\tau_i} \leqslant d_i \sum_{j\in E_i} T_{\nu_j}; \quad \text{hence} \quad h(T_{\tau_i}) \leqslant h\left(d_i \sum_{j\in E_i} T_{\nu_j}\right)$$

and we get, by Theorem G.24,

$$\operatorname{Tr} h(T_{\tau_i}) \leqslant \sum_{j\in E_i} \operatorname{Tr} h(d_i T_{\nu_j}) \leqslant D \sup_{j\in E_i} \sum_{k\geqslant 1} h\left(\lambda_k(d_i T_{\nu_j})\right)$$
$$\leqslant D \sum_{k\geqslant 1} h\left(d_i B \mathrm{e}^{-\delta k}\right)$$

by Lemma 9.44.

This gives, by the subhomogeneity property and then by the slow oscillation of the increasing function h,

$$\operatorname{Tr} h(T_{\tau_i}) \leqslant D \sum_{k\geqslant 1} h(Bd_i)\mathrm{e}^{-q\delta k} \leqslant DC_{\delta,q}\, h(d_i).$$

Next, (9.88) gives us

$$h\left(T_\mu\right) \leqslant h\left(C\sum_i T_{\tau_i}\right).$$

We finally use again the Rotfel'd inequality (Theorem G.24) and the above to get (through slow oscillation again)

$$\operatorname{Tr} h(T_\mu) \leqslant \sum_i \operatorname{Tr} h(CT_{\tau_i}) \leqslant C' \sum_i \operatorname{Tr} h(T_{\tau_i}) \lesssim \sum_i h(d_i).$$

This ends the proof of Theorem 9.42.

10
Approximation Numbers on Hardy and Bergman Spaces

Given a Banach space X of analytic functions on $\mathbb{D}$ and a symbol φ, and, say, $w \in H^\infty, w \neq 0$, we denote by C_φ and $M_w C_\varphi$ the composition and weighted composition operators

$$C_\varphi(f) = f \circ \varphi, \quad M_w C_\varphi(f) = w(f \circ \varphi).$$

We know from the preceding chapters that $M_w C_\varphi : X \to X$ is bounded when X is a Hardy space or a Bergman space. We already explored the connections between the symbol φ and the operator $C_\varphi : X \to X$, in terms of compactness, membership in a Schatten class (when X is Hilbert), and so on, for this operator.

Here we investigate the connections between φ and the decay rate of the approximation numbers of $M_w C_\varphi : X \to X$.

To avoid trivialities, we shall again assume that φ is "non-constant" in the sequel.

10.1 Hilbertian Framework: Generalities

We focus first on the Hilbertian framework, and some results and proofs available for the Hardy space H^2 work also for Bergman spaces. Therefore, we shall say in the sequel that a Hilbert space H is a *standard Hilbert space of functions* if H contains polynomials (with $\|\mathbb{1}\|_H = 1$) and

- $H \subset \mathscr{H}(\mathbb{D})$ continuously (see Section 1.11.1).
- The space $\mathscr{M}(H)$ of multipliers of H is H^∞ isometrically.
- Every composition operator C_φ maps H to H (and is automatically bounded).

This includes Hardy spaces H^2, Bergman spaces $\mathscr{B}^2_\alpha$ where $\alpha > -1$, and more generally some of the weighted Hardy spaces (which always satisfy the inclusion $\mathscr{M}(H) \subset H^\infty$). This does not include the usual Dirichlet space.

We will come later to other non-Hilbertian, Banach spaces, like the Bergman or Hardy spaces $\mathscr{B}^p_\alpha$ or H^p, with $p \neq 2$.

We begin with a general observation.

Theorem 10.1 *For all weights $w \neq 0, w \in H^\infty$, and symbol φ, the singular numbers $a_n(M_w C_\varphi)$ satisfy, for some positive numbers δ, r, the inequality*

$$a_n(M_w C_\varphi) \geqslant \delta r^n, \quad \text{where } n \in \mathbb{N}. \tag{10.1}$$

We can assume that $T := M_w C_\varphi$ is compact, since otherwise we have $a_n(T) \geqslant \operatorname{dist}(T, \mathscr{K}(H)) > 0$. We give two proofs.

First proof.: Let $a \in \mathbb{D}$ be such that $w(a)\,\varphi'(a) \neq 0$ ($\mathscr{H}(\mathbb{D})$ is a division ring and $\varphi' \not\equiv 0$, $w \not\equiv 0$). Let $b = \varphi(a)$ and $\tau \in \operatorname{Aut}\mathbb{D}$ with $\tau(b) = a$. We set

$$\psi = \tau \circ \varphi \quad \text{and} \quad S = M_w C_\psi = T C_\tau.$$

This operator S is compact because T is.

Since $\psi(a) = a$ and $\psi'(a) = \tau'(b)\varphi'(a) \neq 0$, Theorem 8.6 says that the non-zero eigenvalues of S, arranged in non-increasing order, are the numbers $\lambda_n = w(a)\,[\psi'(a)]^{n-1}$, $n \geqslant 1$. We now recall that, by Weyl's inequality, it holds

$$a_1(S)\,a_n(S) \geqslant \Big[\prod_{1\leqslant j\leqslant 2n} a_j(S)\Big]^{1/n} \geqslant \Big[\prod_{1\leqslant j\leqslant 2n} |\lambda_j|\Big]^{1/n} \geqslant |\lambda_{2n}|^2 \geqslant \delta\,\rho^n\,,$$

with $\delta = \dfrac{|w(a)|^2}{|\psi'(a)|^2} > 0$ and $\rho = |\psi'(a)|^4 > 0\,.$

To finally prove (10.1), just observe that $a_n(S) \leqslant a_n(T)\,\|C_\tau\|$ by the ideal property of approximation numbers.

This approach will be extended to several variables in Appendix E. □

Our second proof relies on the following lemma.

Lemma 10.2 *Let H be a standard Hilbert space of functions. Let $(u_j)_{1\leqslant j\leqslant n}$ and $(v_j) = (\varphi(u_j))_{1\leqslant j\leqslant n}$ be two sequences of length n of points of $\mathbb{D}$ with v_j's (and hence u_j's) distinct. Let I_v be the interpolation constant of (v_j). Then if $T = M_w\,C_\varphi : H \to H$, it holds*

$$a_n(T) \gtrsim \Big(\inf_{1\leqslant j\leqslant n} |w(u_j)|\Big) \times \inf_{1\leqslant j\leqslant n} \frac{\|K_{v_j}\|}{\|K_{u_j}\|} \times I_v^{-2} =: \eta \times \delta \times I_v^{-2}. \tag{10.2}$$

In particular, when $H = H^2$, we have

$$a_n(T) \gtrsim \Big(\inf_{1\leqslant j\leqslant n} |w(u_j)|\Big) \times \left(\inf_{1\leqslant j\leqslant n} \sqrt{\frac{1-|u_j|^2}{1-|v_j|^2}}\right) \times I_v^{-2} =: \eta \times \delta \times I_v^{-2},$$

and, when $H = \mathscr{B}^2_\alpha$,

$$a_n(T) \gtrsim \Big(\inf_{1\leqslant j\leqslant n} |w(u_j)|\Big) \times \left(\inf_{1\leqslant j\leqslant n} \left(\frac{1-|u_j|^2}{1-|v_j|^2}\right)^{\frac{1}{\alpha+2}}\right) \times I_v^{-2} =: \eta \times \delta \times I_v^{-2}.$$

We also recall the following classical lemma. Let $z = (z_j)_{1\leqslant j\leqslant n}$ be a finite sequence of points of $\mathbb{D}$ with Carleson interpolation constant I_z. Then the following lemma arises:

Lemma 10.3 *Let H be a standard Hilbert space of functions. For any sequence $(\lambda_j)_{1\leqslant j\leqslant n}$, we have*

$$I_z^{-2} \sum_{j=1}^n |\lambda_j|^2 \|K_{z_j}\|^2 \leqslant \Big\|\sum_{j=1}^n \lambda_j K_{z_j}\Big\|^2 \leqslant I_z^2 \sum_{j=1}^n |\lambda_j|^2 \|K_{z_j}\|^2,$$

where K_a is the reproducing kernel (in H) at point a.

Proof: Indeed, let $(\varepsilon_j)_{1\leqslant j\leqslant n}$ be a sequence of ± 1 signs, and let $\psi \in H^\infty$ with $\psi(z_j) = \varepsilon_j$ and $\|\psi\|_\infty \leqslant I_z$. Let $M_\psi : H \to H$ with $\|M_\psi\| = \|\psi\|_\infty \leqslant I_z$ be the operator of multiplication by ψ. Then

$$\Big\|\sum_{j=1}^n \lambda_j \varepsilon_j K_{z_j}\Big\| = \Big\|M_\psi^*\Big(\sum_{j=1}^n \lambda_j K_{z_j}\Big)\Big\| \leqslant I_z \Big\|\sum_{j=1}^n \lambda_j K_{z_j}\Big\|.$$

Squaring and integrating with respect to independent Rademacher variables ε_j gives

$$\sum_{j=1}^n |\lambda_j|^2 \|K_{z_j}\|^2 = \mathbb{E}\left(\Big\|\sum_{j=1}^n \lambda_j \varepsilon_j K_{z_j}\Big\|^2\right) \leqslant I_z^2 \Big\|\sum_{j=1}^n \lambda_j K_{z_j}\Big\|^2,$$

which implies the LHS of Lemma 10.3. Change λ_j to $\varepsilon_j \lambda_j$ to obtain the RHS. □

Proof of Lemma 10.2: Consider the "model space" E generated by the reproducing kernels $K_{u_1}, \ldots, K_{u_n}$ of H; we have the "mapping equation"

$$T^*(K_a) = \overline{w(a)} K_{\varphi(a)}, \tag{10.3}$$

a readily verified fact since for all $g \in H$, $\langle g, T^*(K_a)\rangle = \langle g, \overline{w(a)} K_{\varphi(a)}\rangle$.

We shall use the definition of $a_n(T)$ as a Bernstein number. Let us now consider $f = \sum_{j=1}^n x_j K_{u_j} \in S_E$, the unit sphere of E. Now (10.3) and the previous lemma give us

$$\begin{aligned}\|T^*(f)\|^2 = \left\| \sum_{j=1}^n x_j \overline{w(u_j)} K_{v_j} \right\|^2 &\geqslant I_v^{-2} \sum_{j=1}^n |x_j|^2 |w(u_j)|^2 \|K_{v_j}\|^2 \\ &\geqslant I_v^{-2} \eta^2 \delta^2 \sum_{j=1}^n |x_j|^2 \|K_{u_j}\|^2 \\ &\geqslant I_v^{-2} I_u^{-2} \eta^2 \delta^2 \|f\|^2 \geqslant I_v^{-4} \eta^2 \delta^2.\end{aligned}$$

Indeed, $I_u \leqslant I_v$ because if $f \in H^\infty$ interpolates the values $w_j, 1 \leqslant j \leqslant n$ at points v_j, the function $f \circ \varphi$ interpolates the same values at points u_j with $\|f \circ \varphi\|_\infty \leqslant \|f\|_\infty$.

We finally get, proving Lemma 10.2,

$$a_n(T) = b_n(T) = b_n(T^*) \geqslant I_v^{-2} \eta\, \delta.$$

The particular cases follow obviously simply expliciting the norm of the reproducing kernels for Hardy and Bergman spaces. □

Second proof of Theorem 10.1: We now argue as follows: let $\Delta \subset \mathbb{D}$ be a closed disk on which w does not vanish, and let $\rho = \inf_\Delta |w| > 0$. The image $\varphi(\Delta)$ contains a disk $\overline{D}(a,r)$. Set

$$\omega = \mathrm{e}^{2i\pi/n}, \quad v_j = a + r\omega^j = \varphi(u_j) \quad \text{with } u_j \in \Delta, \text{ where } 1 \leqslant j \leqslant n.$$

We know (see (1.42)) that $I_v \leqslant C^n$ where C only depends on a and r. Moreover, with the notations of Lemma 10.2, we have $\eta \geqslant \rho$ and

$$\delta \geqslant \inf_{1 \leqslant j \leqslant n} \frac{\|K_{v_j}\|}{\|K_{u_j}\|} \geqslant \frac{1}{\sup_{u \in \Delta} \|K_u\|} > 0,$$

thanks to (1.54) since $\operatorname{dist}(\Delta, \partial\mathbb{D}) > 0$.

So this lemma gives us our second proof of Theorem 10.1. □

10.2 The Case $\|\varphi\|_\infty < 1$

Let H be a standard Hilbert space of functions, which is also $H^2(\beta)$ for some sequence β of positive real numbers.

Theorem 10.4 *Suppose* $\|\varphi\|_\infty = r < 1$. *Then* $T = M_w C_\varphi : H \to H$, *and for any* $r < s < 1$, *there is a constant* $C = C_s$ *such that*

$$a_n(T) \leqslant C \|w\|_\infty \, s^n.$$

Proof: Let S_{n-2} be the partial sum operator of index $n-2$ and $R = M_w C_\varphi S_{n-2}$ an operator of rank $\leqslant n-1$. We have $a_n(T) \leqslant \|T - R\|$. For f in the unit ball of H, the Cauchy–Schwarz inequality first gives, since $\beta_j \geqslant \delta_s \big(r/s\big)^j$ with $\delta_s > 0$,

$$\sum_{j \geqslant n-1} |\widehat{f}(j)| r^j \leqslant \Big(\sum_{j \geqslant n-1} r^{2j}/\beta_j\Big)^{1/2} \leqslant C'_s \Big(\sum_{j \geqslant n-1} s^{2j}\Big)^{1/2} \leqslant C_s \, s^n.$$

But then, since $\|g\|_H \lesssim \|g\|_\infty$ for any $g \in H$ (it holds $H^\infty \subset H$ via $g = M_g(\mathbb{1})$),

$$\begin{aligned}\|T(f) - R(f)\|_H &\leqslant \|w\|_\infty \sum_{j \geqslant n-1} |\widehat{f}(j)| \, \|\varphi^j\|_H \lesssim \|w\|_\infty \sum_{j \geqslant n-1} |\widehat{f}(j)| r^j \\ &\leqslant \|w\|_\infty \, C_s \, s^n.\end{aligned}$$

□

Theorems 10.1 and 10.4 induce us to coin, for an arbitrary operator T (here we consider $T = M_w C_\varphi$), the parameters

$$\beta^-(T) = \varliminf_{n \to \infty} \big(a_n(T)\big)^{1/n}, \quad \beta^+(T) = \varlimsup_{n \to \infty} \big(a_n(T)\big)^{1/n}, \tag{10.4}$$

since they respectively state that $\beta^-(T) > 0$, and $\beta^+(T) < 1$ when $\|\varphi\|_\infty < 1$. Those parameters clearly satisfy an "ideal property"

$$\beta^\pm(UTV) \leqslant \beta^\pm(T).$$

We will show that $\beta^-(T) = \beta^+(T)$ in the setting of composition operators, and will denote this common value $\beta(T)$ in this case.

This is of course wrong for an arbitrary operator: take a diagonal operator T on an orthonormal basis of H with decreasing diagonal elements $\varepsilon_n \in (0,1)$, satisfying

$$\varliminf_{n \to \infty} (\varepsilon_n)^{1/n} = 0, \qquad \varlimsup_{n \to \infty} (\varepsilon_n)^{1/n} = 1$$

and note that $a_n(T) = \varepsilon_n$.

Moreover, we will give a closed formula for that common value. This forces us to a detour through the Green capacity.

The Green Capacity

The *Green capacity* $\mathcal{C}(K)$ of a Borel subset K of $\mathbb{D}$ is defined as the maximum charge that K can receive while keeping the Green potential less than 1 everywhere on $\mathbb{D}$. In symbols: $\mathcal{C}(K) \in [0, +\infty]$ is defined by

$$\mathcal{C}(K) = \sup \{\mu(K) : G_\mu(z) \leqslant 1 \text{ for all } z \in \mathbb{D}\},$$

where μ is a positive measure supported by K, and its Green potential G_μ on $\mathbb{D}$ is defined by

$$G_\mu(z) = \int_K \log \left|\frac{1 - \overline{w}z}{z - w}\right| d\mu(w) = \int_K \log \frac{1}{\rho(z, w)}\, d\mu(w)$$
$$=: \int_K G(z, w)\, d\mu(w). \tag{10.5}$$

We will call μ "admissible" when $G_\mu \leqslant 1$ throughout $\mathbb{D}$.

The function $G(z, w)$ is none other than the Green function of $\mathbb{D}$ at z. This definition (potential approach) is not the one that extends best to several variables, and we have alternatively [90], [157]

$$\mathcal{C}(K) = \sup_{\substack{0 \leqslant u \leqslant 1 \\ u \in sh(\mathbb{D})}} \int_K \Delta u(z)\, d\sigma(z), \tag{10.6}$$

where the laplacian Δ is taken in the sense of distributions (indeed, since $\Delta u \geqslant 0$ for u subharmonic, Δu is a non-negative measure [83]). Here is a list of some basic properties of this capacity to be used [90, chapter 1].

(i) $K \subset L \Longrightarrow \mathcal{C}(K) \leqslant \mathcal{C}(L)$.
(ii) $K_j \uparrow K \Longrightarrow \mathcal{C}(K_j) \uparrow \mathcal{C}(K)$.
(iii) $\mathcal{C}(K) = \mathcal{C}(\partial K) = \mathcal{C}(\overline{K})$ when K is *connected* and $\overline{K} \subset \mathbb{D}$.
(iv) $\mathcal{C}[M(K)] \leqslant \mathcal{C}(K)$ when $M(z) = |z|$ with $z \in \mathbb{D}$.
(v) $\mathcal{C}(K)$ is a conformal invariant: $\mathcal{C}(K) = \mathcal{C}[T(K)]$ for all $T \in \text{Aut}\,(\mathbb{D})$.
(vi) If $K = \overline{\Delta}(a, r)$ is a closed hyperbolic disk, $\mathcal{C}(K) = \dfrac{1}{\log(1/r)}\cdot$
(vii) If $K_h = [0, h]$ with $0 < h < 1$, $\mathcal{C}(K_h) \approx \log \dfrac{1}{1 - h}$ as $h \to 1^-$.

(iii) is non-trivial and not so well known: we refer to [113, theorem 2.3, p. 6]. (iv) is proved as follows: let τ be a positive measure carried by $M(K)$. We can write $\tau = M(\mu)$ where μ is a positive measure carried by K. And for $z \in \mathbb{D}$

$$G_\tau(z) = \int_{M(K)} \log \frac{1}{\rho(z, w)}\, d\tau(w) = \int_K \log \frac{1}{\rho(M(z), M(w))}\, d\mu(w)$$
$$\geqslant \int_K \log \frac{1}{\rho(z, w)}\, d\mu(w) = G_\mu(z),$$

where we used the fact that M contracts the pseudo-hyperbolic distance ρ. This means that there are more admissible measures for K than for $M(K)$ and gives the result. (v) is obvious since automorphisms are ρ-isometries.

For (vi), we can assume $a = 0$ by (v), so that $\partial K = \mathcal{C}(0, r)$. Let m_r be the uniform distribution on ∂K, namely

$$\int_K f\, dm_r = \int_{\mathbb{T}} f(rw)\, dm(w).$$

We have by Jensen's formula

$$\begin{aligned} G_{m_r}(z) &= \int_{\mathbb{T}} \log|1 - r\overline{w}z|\, dm(w) - \int_{\mathbb{T}} \log|z - rw|\, dm(w) \\ &= -\log\max\{|z|, r\} \leqslant \log\frac{1}{r}; \end{aligned}$$

hence the measure $\mu = \dfrac{1}{\log 1/r} m_r$ is admissible, and $\mu(K) = \dfrac{1}{\log 1/r}\cdot$ Since μ is "clearly" optimal, we get the result.
(vii) This is more elaborate [90]. We have precisely

$$\operatorname{diam}_\rho(K_h) = h \text{ and } \mathcal{C}(K_h) = \frac{1}{\pi}\frac{I'}{I},$$

where I and I' are the elliptic integrals

$$I = \int_0^1 \frac{1}{\sqrt{(1-t^2)(1-k^2t^2)}}\, dt \quad \text{and} \quad I' = \int_0^1 \frac{1}{\sqrt{(1-t^2)(1-k'^2t^2)}}\, dt$$

with $k = \dfrac{1-h}{1+h}$ and $k'^2 = 1 - k^2$. Now, as $h \to 1^-$, we have $k \to 0,\ k'^2 \to 1$, and

$$I \to \int_0^1 \frac{1}{\sqrt{1-t^2}}\, dt = \pi/2, \qquad I' \approx \log\frac{1}{1-k'^2} \approx \log\frac{1}{1-h}\cdot$$

An important corollary of properties (v), (vi), (vii) is the following.

Proposition 10.5 *Let (K_j) be a sequence of compact and connected subsets of $\mathbb{D}$ such that $\operatorname{diam}_\rho(K_j) \to 1$. Then*

$$\mathcal{C}(K_j) \to \infty.$$

Proof: By conformal invariance, we may assume that $0 \in K_j$. Let $L_j = M(K_j)$ with again $M(z) = |z|$. Since K_j is connected, $L_j = [0, h_j]$ for some $0 \leqslant h_j < 1$. Now, $h_j \geqslant \operatorname{diam}_\rho(K_j)$ so that $h_j \to 1$. But then

$$\mathcal{C}(K_j) \geqslant \mathcal{C}(L_j) \approx \log 1/(1-h_j)$$

and we are done. □

It will be convenient to use later the notation

$$\Gamma(K) = \exp\left(-\frac{1}{\mathcal{C}(K)}\right). \tag{10.7}$$

As a transition, we observe the following, in the framework of this section:

Theorem 10.6 *Fix* $|\lambda| < 1$ *and let* $\varphi(z) = \lambda z$. *Let* $K = \varphi(\mathbb{D})$. *Then*

$$\beta(C_\varphi) = |\lambda| = \Gamma(K).$$

More generally, if $\varphi(z) = \dfrac{az+b}{cz+d}$ *is a fractional linear map from* $\mathbb{D}$ *to itself, that is,*

$$|a|^2 + |b|^2 + 2|\overline{a}b - \overline{c}d| \leqslant |c|^2 + |d|^2 \quad \text{and} \quad |c| \leqslant |d|,$$

then again $\beta(C_\varphi) = \Gamma\,[\varphi(\mathbb{D})]$.

Proof: Indeed, the matrix of C_φ on the canonical basis of H is the diagonal matrix with diagonal entries λ^n, $n \geqslant 0$. So that $a_n(C_\varphi) = |\lambda|^{n-1}$, $n \geqslant 1$, and $\beta(C_\varphi) = |\lambda| = \Gamma(K)$ where we used item (vi).

For the general case, observe that $\varphi(\mathbb{D})$ is a hyperbolic disk $\Delta(a,r)$, so that $\varphi(\mathbb{D}) = (\varphi_a \circ h)(\mathbb{D})$ where $h(z) = rz$. Hence, $\varphi = \varphi_a \circ h \circ \psi$ for some $\psi \in$ Aut $(\mathbb{D})$. This gives $C_\varphi = C_\psi C_h C_{\varphi_a}$ and, since $\beta(C_{\psi\circ\varphi}) = \beta(C_{\varphi\circ\psi}) = \beta(C_\varphi)$, we get

$$\beta(C_\varphi) = \beta(C_h) = \Gamma\,[h(\mathbb{D})] = \Gamma\,[\varphi(\mathbb{D})]$$

by conformal invariance of the Green capacity. □

10.3 The Spectral Radius Formula

The Case $\|\varphi\|_\infty < 1$

We show here that Theorem 10.6 can be fully generalized, at least when we take $H = H^2$. We prefer to split things into two theorems.

Theorem 10.7 *Let* $T = M_w C_\varphi$ *where* $w \in H^\infty$ *and* $\|\varphi\|_\infty < 1$. *Set* $\Omega = \varphi(\mathbb{D})$. *Then, we have*

$$\beta^+(T) \leqslant \Gamma\left(\overline{\Omega}\right).$$

We will use the following important lemma (see [179]).

Lemma 10.8 (Widom) *Let* $K \subset \mathbb{D}$ *be compact. Then, given* $\varepsilon > 0$, *there exists a cycle* γ *(a finite union of disjoint Jordan curves contained in* $\mathbb{D}$*) whose* *interior U contains K, such that for $n \in \mathbb{N}$ large enough, there exists a rational function R with degree $< n$, having no zero on γ and all poles on $\partial\mathbb{D}$, satisfying*

$$|R(z)| \geqslant e^{-\varepsilon n} \text{ for } z \notin U \quad \text{and} \quad |R(z)| \leqslant e^{\varepsilon n}[\Gamma(K)]^n \text{ for } z \in K.$$

This lemma has itself an important corollary.

Theorem 10.9 *Let $K \subset \mathbb{D}$ be a compact set, and $\varepsilon > 0$. There exist a constant $C_\varepsilon > 0$ and, for any integer n large enough, a rational function R of degree $< n$ with all poles on $\partial\mathbb{D}$, and finitely many points $\zeta_i \in \mathbb{D} \setminus K$ such that, for any $g \in H^\infty$, it holds*

$$\|g - h\|_{C(K)} \leqslant C_\varepsilon e^{\varepsilon n}\big[\Gamma(K)\big]^n \|g\|_\infty,$$

where $h(w) = R(w) \sum_{\substack{i,k \\ k \leqslant m_i}} c_{i,k}(g)(w - \zeta_i)^{-k}$ *with* $\sum_i m_i < n$ *and* $h \in \mathscr{H}(\mathbb{D})$, *the maps $g \mapsto c_{i,k}(g)$ being continuous linear forms on H.*

Proof: Consider (Hermite interpolation formula), for $w \in K$,

$$G(w) = R(w)\left[\frac{1}{2i\pi}\int_\gamma \frac{g(\zeta)}{R(\zeta)(\zeta - w)}\, d\zeta\right], \tag{10.8}$$

where γ and R are given by Lemma 10.8.

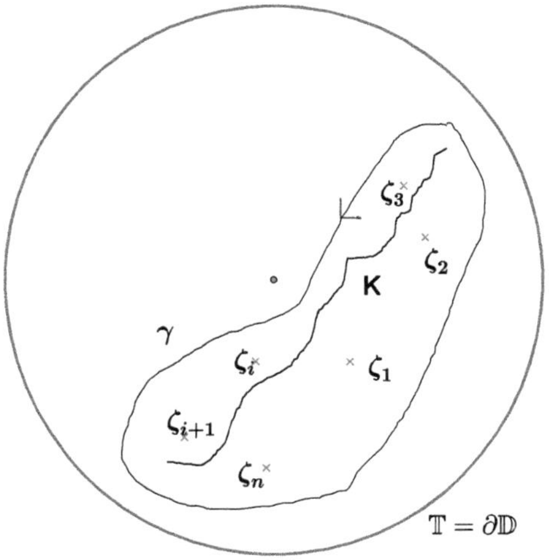

Figure 10.1 The curve γ surrounding K and the ζ_i's.

If (ζ_i) are the zeros of R inside γ with multiplicities m_i, the residue theorem (with poles at w and the ζ_i) gives, since $\operatorname{Res}(G,w) = R(w)\dfrac{g(w)}{R(w)} = g(w)$,

$$G(w) =: g(w) - R(w)\left(\sum_{\substack{i,k\\ k\leqslant m_i}} c_{i,k}(g)(w-\zeta_i)^{-k}\right) =: g(w) - h(w),$$

where

$$c_{i,k}(g) = \sum_{j\leqslant m_i-k} \lambda_{i,j,k}\, g^{(j)}(\zeta_i).$$

Since clearly, for each point $a \in \mathbb{D}$ and each integer $j \geqslant 0$, the linear form $f \mapsto f^{(j)}(a)$ is continuous on H, $c_{i,k}$'s are continuous on H. Next, denoting $|\gamma|$ the length of γ and δ the distance of γ to K, the estimates of Lemma 10.8 and (10.8) give

$$\|G\|_{C(K)} \leqslant \mathrm{e}^{2\varepsilon n}[\Gamma(K)]^n\, |\gamma|\, \frac{\|g\|_\infty}{\delta}.$$

Finally, $h \in C(K)$ and even $h \in \mathscr{H}(\mathbb{D})$ since the zeros of R compensate the poles $(w-\zeta_i)^{-k}$ with the right multiplicity. This proves Theorem 10.9, changing ε into $\varepsilon/2$. □

Proof of Theorem 10.7: Using $a_n(T) \leqslant \|M_w\|\, a_n(C_\varphi)$ and consequently $\beta^+(T) \leqslant \beta^+(C_\varphi)$, we only have to show that $\beta^+(C_\varphi) \leqslant \Gamma\left(\overline{\Omega}\right)$. We rely for that on a second important lemma of Widom, which belongs to the approximation theory by rational functions. *It is here that the Green capacity shows up.* Recall that the degree of a rational function $R = P/Q$ in reduced terms is $\max(\deg P, \deg Q)$.

Set $K = \overline{\Omega} \subset \mathbb{D}$. Let F be the $(n-1)$-dimensional subspace formed by rational functions of the form h as in Theorem 10.9, with $F \subset \mathscr{H}(\mathbb{D})$ as we noted.

Let $\widetilde{F} = \{v = h\circ\varphi : h \in F\}$. We see that $\widetilde{F} \subset H^\infty \subset H$, since $\|\varphi\|_\infty < 1$, and that $\dim \tilde{F} \leqslant \dim F < n$. Now, let $l = l(n)$ be an integer to be adjusted, $r = \|\varphi\|_\infty$, $r < s < 1$; and for $f \in B_H$, let

$$g(z) = \sum_{0\leqslant\alpha<l} \widehat{f}(\alpha) z^\alpha.$$

As we already saw, by Cauchy–Schwarz,

$$\|g\|_{C(K)} \leqslant \|g\|_\infty \leqslant \sqrt{l} \quad \text{and} \quad \|f\circ\varphi - g\circ\varphi\|_\infty \leqslant C_s s^l. \tag{10.9}$$

By Theorem 10.9, there exists $h \in F$ such that

$$\|g-h\|_{C(K)} \leqslant C_\varepsilon \mathrm{e}^{\varepsilon n}[\Gamma(K)]^n \sqrt{l}.$$

Setting $v = h \circ \varphi \in \widetilde{F}$, this gives

$$\|g \circ \varphi - v\|_H \leqslant \|g \circ \varphi - v\|_\infty \leqslant \|g - h\|_{C(K)} \leqslant C_\varepsilon e^{\varepsilon n} [\Gamma(K)]^n \sqrt{l}. \quad (10.10)$$

Next, (10.9) and (10.10) imply

$$\|f \circ \varphi - v\|_H \leqslant C_s s^l + C_\varepsilon e^{\varepsilon n} [\Gamma(K)]^n \sqrt{l}.$$

In terms of Kolmogorov numbers, $d_n(C_\varphi) \leqslant C_s s^l + C_\varepsilon e^{\varepsilon n} [\Gamma(K)]^n \sqrt{l}$. Since $a_n(C_\varphi) = d_n(C_\varphi)$, taking nth roots gives

$$\big(a_n(C_\varphi)\big)^{1/n} \leqslant (C_s)^{1/n} s^{l/n} + (C_\varepsilon)^{1/n} e^\varepsilon [\Gamma(K)]\, l^{1/2n}.$$

Now, take for l the integer part of $n \log n$ and let $n \to \infty$ to get

$$\beta^+(C_\varphi) \leqslant e^\varepsilon [\Gamma(K)].$$

Letting ε tend to 0 gives the result. □

We now turn to the lower bound. We shall make use of a third result of Widom (see [115]):

Theorem 10.10 *Let K be a compact subset of $\mathbb{D}$. If E is a subspace of H^2 with* $\dim E < n$, *there exists f in the unit ball of H^∞, such that (a being an absolute positive constant)*

$$\|f - h\|_{C(K)} \geqslant a\,[\Gamma(K)]^n \qquad \textit{for all } h \in E. \qquad (10.11)$$

Now we can state the following:

Theorem 10.11 *Let $T = M_w C_\varphi$ where $w \in H^\infty$ and $\|\varphi\|_\infty < 1$. Set $\Omega = \varphi(\mathbb{D})$. Then, we have*

$$\beta^-(T) \geqslant \Gamma(\Omega).$$

Proof: Since w is not identically 0, we can find an increasing sequence (r_j) with limit 1, such that $w(z) \neq 0$ for $|z| = r_j$, implying

$$\delta_j := \inf_{|z| = r_j} |w(z)| > 0.$$

Set $K_j = \varphi(r_j \mathbb{T})$ and let $E \subset H^2$ with $\dim E < n$. By (10.11), we can find $f \in B_{H^\infty} \subset B_{H^2}$ satisfying, for all $h \in E$,

$$\|f - h\|_{C(K_j)} \geqslant a\big[\Gamma(K_j)\big]^n.$$

It ensues that

$$\begin{aligned}\|w(f \circ \varphi - h \circ \varphi)\|_{C(r_j \mathbb{T})} &\geqslant (\inf_{r_j \mathbb{T}} |w|)(\sup_{r_j \mathbb{T}} |f \circ \varphi - h \circ \varphi|) \\ &\geqslant \delta_j \|f - h\|_{C(K_j)} \geqslant a \delta_j \big[\Gamma(K_j)\big]^n.\end{aligned}$$

Moreover, we know that, for some positive constant L_j depending only on j, the inequality $\|g\|_{C(r_j\mathbb{T})} \leqslant L_j\|g\|_2$ holds for all $g \in H^2$. This implies

$$a\delta_j\big[\Gamma(K_j)\big]^n \leqslant \|w(f\circ\varphi - h\circ\varphi)\|_{C(r_j\mathbb{T})} \leqslant L_j\|w(f\circ\varphi - h\circ\varphi)\|_2 = L_j\|Tf - Th\|_2$$

so that, for some positive constant δ'_j depending only on j,

$$d(Tf, TE) \geqslant \delta'_j\,\big[\Gamma(K_j)\big]^n.$$

Since this holds for every subspace E of H^2 with $\dim E < n$, we derive from the above that $a_n(T) = d'_n(T) \geqslant \delta'_j\,\big[\Gamma(K_j)\big]^n$ (see Theorem G.7). Taking nth roots and passing to $\underline{\lim}$ as $n \to \infty$, we get

$$\beta^-(T) \geqslant \Gamma(K_j). \tag{10.12}$$

To finish, we set $\omega_j = r_j\,\mathbb{D}$, $K'_j = \varphi\big(\overline{\omega_j}\big) = \overline{\varphi(\omega_j)} \supset K_j$.

We obviously have $\mathcal{C}(K'_j) \geqslant \mathcal{C}(K_j)$. However, since $\partial\varphi(\omega_j) \subset \varphi(\partial\omega_j) = K_j$, we get, using the reminded result (iii) on capacity,

$$\mathcal{C}(K'_j) = \mathcal{C}[\varphi(\omega_j)] = \mathcal{C}[\partial\varphi(\omega_j)] \leqslant \mathcal{C}(K_j).$$

So that $\mathcal{C}(K'_j) = \mathcal{C}(K_j)$ and that, using (10.12),

$$\beta^-(T) \geqslant \Gamma(K'_j). \tag{10.13}$$

Letting $j \to \infty$ and using (ii) of the reminders, we obtain $\beta^-(T) \geqslant \Gamma(\Omega)$. □

Theorems 10.7 and 10.11 combine to finally give the following:

Theorem 10.12 *Let $T = M_w C_\varphi$ where $w \in H^\infty$ and $\|\varphi\|_\infty < 1$. Set $\Omega = \varphi(\mathbb{D})$. Then, the limit $\beta(T)$ exists, and its value is*

$$\beta(T) = \Gamma\left(\overline{\Omega}\right) = \Gamma(\Omega) = \exp\left(-\frac{1}{\mathcal{C}\big(\varphi(\mathbb{D})\big)}\right).$$

Proof: It follows from Theorems 10.7 and 10.11, and from the basic property (iii) of the Green capacity, that (note that $\overline{\Omega} \subset \mathbb{D}$)

$$\beta^+(T) \leqslant \Gamma\left(\overline{\Omega}\right) = \Gamma(\Omega) \leqslant \beta^-(T).$$

□

The Case $\|\varphi\|_\infty = 1$

Theorem 10.1 proved that one never has superexponential decay for the approximation numbers $a_n(M_w C_\varphi)$. The following result indicates that, when $\|\varphi\|_\infty = 1$, whatever the non-zero weight w, the numbers $a_n(M_w C_\varphi)$ indeed have at most subexponential decay.

Theorem 10.13 *Suppose* $\|\varphi\|_\infty = 1$. *Let* $T = M_w C_\varphi$ *where* $w \in H^\infty$, *assumed to be compact. Then*

$$\beta(T) = 1,$$

that is, there exists a sequence (ε_n) *of positive numbers with limit* 0 *such that*

$$a_n(T) \geqslant e^{-n\,\varepsilon_n}.$$

Proof: Let r_j be a sequence of positive numbers with $r_j \to 1^-$, and let us set $h_j(z) = r_j z$ and $\varphi_j = \varphi \circ h_j$, with $\|\varphi_j\|_\infty < 1$, as well as $K_j = \overline{\varphi_j(\mathbb{D})}$ and

$$T_j = C_{h_j} T = M_{w\circ h_j} C_{\varphi_j}.$$

By the ideal property, we have $\beta^-(T_j) \leqslant \beta^-(T)$ and we know, by Theorem 10.11, $\Gamma(K_j) \leqslant \beta^-(T)$. But $\mathrm{diam}_\rho(K_j) \to 1$ since $\|\varphi\|_\infty = 1$. The K_j being connected, this implies $\mathcal{C}(K_j) \to \infty$ by Proposition 10.5; hence $\Gamma(K_j) \to 1$ and we get $\beta^-(T) \geqslant 1$, ending the proof. □

We shall give examples showing the optimality of the result: for any sequence (ε_n) with limit 0, there are composition operators C_φ with $\|\varphi\|_\infty = 1$ such that

$$a_n(C_\varphi) \leqslant C\mathrm{e}^{-n\,\varepsilon_n}.$$

10.4 The Case of Entropy Numbers

We know from Theorem G.13 that the following inequality comparing approximation and entropy numbers (where α, β are positive numerical contants) holds:

$$\alpha \sup_{k\geqslant 1}\left[2^{-n/2k}\left(\prod_{j=1}^{k} a_j(T)\right)^{1/k}\right] \leqslant \mathfrak{e}_n(T) \leqslant \beta \sup_{k\geqslant 1}\left[2^{-n/2k}\left(\prod_{j=1}^{k} a_j(T)\right)^{1/k}\right],$$

This inequality shows that another parameter than $\beta(T)$ emerges for entropy numbers. Set (when the limit exists)

$$\gamma(T) = \lim_{n\to\infty} \left[\mathfrak{e}_n(T)\right]^{1/\sqrt{n}} \tag{10.14}$$

and also consider in general the $\gamma^+(T), \gamma^-(T)$ versions.

We now have the following:

Theorem 10.14 *Let* $\varphi\colon \mathbb{D} \to \mathbb{D}$ *be a symbol,* $\Omega = \varphi(\mathbb{D})$, $w \in H^\infty, w \neq 0$, *and* $T = M_w C_\varphi$. *Then:*

(1) $\gamma^-(T) = \gamma^+(T) = \gamma(T)$ *and*

$$\gamma(T) = \exp\Big[-\sqrt{\log 2/\mathcal{C}(\Omega)}\Big] \in (0,1]. \tag{10.15}$$

(2) *In particular, one has the equivalence:*

$$\gamma(T) = 1 \iff \|\varphi\|_\infty = 1. \tag{10.16}$$

Proof: We first assume that $\|\varphi\|_\infty < 1$, so that $0 < \mathcal{C}(\Omega) < 1$, and we set $\eta = 1/\mathcal{C}(\Omega)$ for simplicity of notations. Let $0 < \varepsilon < \eta$, and C_ε be a positive constant that depends only on ε and can vary from one formula to another. Theorem 10.12 implies $a_k \leqslant C_\varepsilon\, \mathrm{e}^{\varepsilon k}\mathrm{e}^{-k\eta}$, whence

$$(a_1 \cdots a_k)^{1/k} \leqslant C_\varepsilon\, \mathrm{e}^{\varepsilon k/2}\mathrm{e}^{-\eta k/2}\,.$$

Theorem G.13 now gives

$$\begin{aligned} \mathfrak{e}_n(T) &\leqslant C_\varepsilon \sup_{k\geqslant 1} \exp\left[-\frac{1}{2}\left(\frac{n\log 2}{k} + (\eta-\varepsilon)k\right)\right] \\ &\leqslant C_\varepsilon \sup_{t>0} \exp\left[-\frac{1}{2}\left(\frac{n\log 2}{t} + (\eta-\varepsilon)t\right)\right]. \end{aligned}$$

This supremum is attained for $t = \sqrt{n\log 2/(\eta-\varepsilon)}$ and this gives

$$\mathfrak{e}_n(T) \leqslant C_\varepsilon \exp\Big[-(\eta-\varepsilon)\sqrt{n\log 2/(\eta-\varepsilon)}\Big].$$

This implies $\gamma^+(T) \leqslant \mathrm{e}^{-\sqrt{(\eta-\varepsilon)\log 2}}$ and finally $\gamma_1^+(T) \leqslant \mathrm{e}^{-\sqrt{\eta\log 2}}$.

The lower bound $\gamma^-(T) \geqslant \mathrm{e}^{-\sqrt{\eta\log 2}}$ is proved similarly.

When $\|\varphi\|_\infty = 1$, we argue as in the proof of Theorem 10.13, approximating Ω by $\varphi(r_j\mathbb{D})$ with $r_j \to 1^-$. In that case, the formula is still valid, with $\mathcal{C}(\Omega) = \infty$ and $\gamma(T) = 1$. □

10.5 Examples with Slow Decay

In this subsection, we shall see that the convergence to 0 of the approximation numbers of a compact composition operator C_φ can be as slow as one wishes, as soon as $\|\varphi\|_\infty = 1$. This answers in the positive to a question first asked by C. Le Merdy.

Theorem 10.15 *Let $(\varepsilon_n)_{n\geqslant 1}$ be a sequence of positive numbers with limit* 0. *Then:*

(i) *There exists a univalent symbol φ such that $\varphi(0) = 0$, $\|\varphi\|_\infty = 1$, and $C_\varphi\colon H^2 \to H^2$ is compact, that is, $a_n(C_\varphi) \to 0$, but*

$$\underline{\lim}_{n\to\infty} \frac{a_n(C_\varphi)}{\varepsilon_n} > 0. \tag{10.17}$$

Equivalently, we have for some positive number $\delta > 0$, independent of n,

$$a_n(C_\varphi) \geqslant \delta\, \varepsilon_n \qquad \textit{for all } n \geqslant 1.$$

(ii) *There exists a univalent symbol φ such that $\varphi(0) = 0$, $\|\varphi\|_\infty = 1$, and $C_\varphi\colon \mathscr{B}^2_\alpha \to \mathscr{B}^2_\alpha$ is compact and (10.17) holds.*

In order to ease the presentation, we only prove it for $H = H^2$.

As in the case of Hankel operators considered by Peller (see [134]), an immediate consequence of Theorem 10.15 is the following:

Corollary 10.16 *There exists a composition operator $C_\varphi\colon H^2 \to H^2$ that is compact, but in no Schatten class.*

This corollary, which Theorem 10.15 reinforces and precises, was an answer to a question of Sarason, and has been first proved in [18]. Other proofs appeared in [1], [96], [114], [184] (for a positive result on Schattenness, we refer to [123]).

In the present book, we also proved this result in Chapter 9 with the help of Shapiro–Taylor maps (see Theorem 9.16).

The construction of the symbol φ in Theorem 10.15 follows the principle given in Carroll and Cowen's paper. But the final argument is different, and is instead based on Lemma 10.2.

We will need in the proof the following simple lemma (where d is the hyperbolic distance on $\mathbb{D}$):

Lemma 10.17 *For $0 < a < b < 1$, we have*

$$e^{-2d(a,b)} \leqslant \frac{1-b}{1-a} \leqslant 2e^{-2d(a,b)}.$$

Proof: By definition of the hyperbolic and pseudo-hyperbolic distances d and ρ, we have

$$e^{-2d(a,b)} = \frac{1-\rho(a,b)}{1+\rho(a,b)} = \frac{(1-b)(1+a)}{(1-a)(1+b)},$$

and the result follows. □

Now, we shall use as Carroll and Cowen a *comparison principle* due to Hayman [71], concerning the hyperbolic distance in a simply connected domain U. Recall that the hyperbolic distance $d(w_1, w_2; U)$ of two points $w_1, w_2 \in U$ is defined by

$$d(w_1, w_2; U) = d(z_1, z_2; \mathbb{D}),$$

where φ is a Riemann map from $\mathbb{D}$ to U with $z_j \in \mathbb{D}$ and $\varphi(z_j) = w_j$ (the result does not depend on the choice of φ) and where $d(z_1, z_2; \mathbb{D}) =: d(z_1, z_2)$.

This comparison principle is two-fold:

Proposition 10.18 *Suppose that the simply connected domain U contains the rectangle*

$$R = \{z \in \mathbb{C} \ : \ a_1 - b < \operatorname{Re} z < a_2 + b, \text{ and } |\operatorname{Im} z| < b\},$$

where $a_1 < a_2$ and $b > 0$. Then, we have the upper estimate

$$d(a_1, a_2; U) \leqslant \frac{\pi}{4b}(a_2 - a_1) + \frac{\pi}{2}. \tag{10.18}$$

Proposition 10.19 *Suppose the simply connected domain U contains the rectangle*

$$R = \{z \in \mathbb{C} : a_1 - c < \operatorname{Re} z < a_2 + c, \text{ and } |\operatorname{Im} z| < c\},$$

where $a_1 < a_2$ and $c > 0$, but that the horizontal sides

$$\{z \in \mathbb{C} : a_1 - c \leqslant \operatorname{Re} z \leqslant a_2 + c, |\operatorname{Im} z| = c\}$$

of that rectangle are disjoint from U. Then, we have the lower estimate

$$d(a_1, a_2; U) \geqslant \frac{\pi}{4c}(a_2 - a_1) - \frac{\pi}{2}. \tag{10.19}$$

We turn to the construction of our symbol φ.

We first define a continuous map $\psi \colon \mathbb{R} \to \mathbb{R}$ as follows. Let $(A_n)_{n\geqslant 1}$ be an increasing sequence of positive numbers tending to ∞, which is moreover concave. Let $A \colon [0, \infty) \to [0, \infty)$ be the increasing piecewise linear continuous function on the intervals $(0, 1)$ and $(\mathrm{e}^{n-1}, \mathrm{e}^n)$ such that

$$A(0) := A_0 = 0, \quad A(\mathrm{e}^{n-1}) := A_n \quad \text{for } n \geqslant 1, \quad \text{and set} \quad 2K = 1/A(1).$$

The sequence of slopes $\dfrac{A_n - A_{n-1}}{\mathrm{e}^n - \mathrm{e}^{n-1}}$ is decreasing, since

$$A_{n+1} - A_n \leqslant A_n - A_{n-1} \leqslant \mathrm{e}\,(A_n - A_{n-1}).$$

The function A is hence increasing and concave on $(0, \infty)$ and vanishes at 0. This implies that $A(t)/t$ is decreasing on $(0, \infty)$.

We set

$$\psi(t) = \begin{cases} K(1 + |t|) & \text{if } |t| \leqslant 1 \\ |t|/A(|t|) & \text{if } |t| > 1, \end{cases}$$

By the previous discussion, ψ is continuous and increasing on $(0, \infty)$.

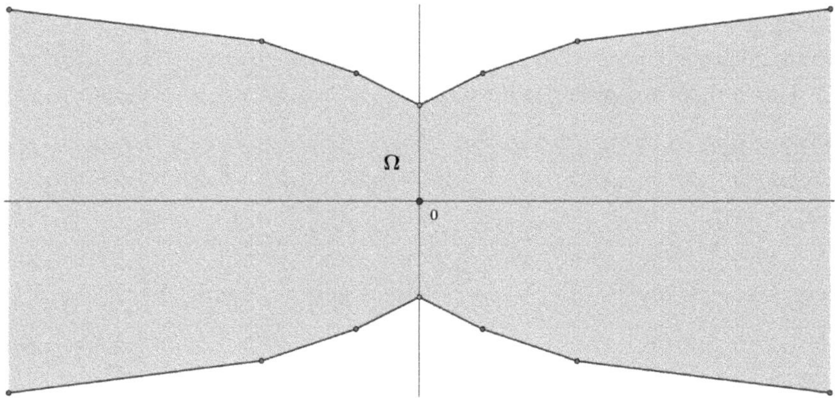

Figure 10.2 The region Ω and its boundary.

We then define a domain Ω of the complex plane by (see Figure 10.2)

$$\Omega = \{w \in \mathbb{C} \,:\, |\mathrm{Im}\, w| < \psi(|\mathrm{Re}\, w|)\}. \tag{10.20}$$

Let $S : \mathbb{D} \to \Omega$ be the unique Riemann map such that $S(0) = 0$ and $S'(0) > 0$.

This map exists in view of the following simple fact.

Lemma 10.20 *The domain Ω defined by (10.20) is star-shaped with respect to the origin, and the map $S\colon (-1, 1) \to \mathbb{R}$ is an increasing bijection such that $S(-1) = -\infty$ and $S(1) = +\infty$.*

Proof: The star-shaped character of Ω will follow from the implication

$$|\mathrm{Im}\, w| < \psi(|\mathrm{Re}\, w|) \quad \text{and} \quad 0 < \lambda < 1 \quad \Longrightarrow \quad |\mathrm{Im}\,(\lambda w)| < \psi(|\mathrm{Re}\,(\lambda w)|).$$

We may assume that both $\mathrm{Re}\, w, \mathrm{Im}\, w$ are positive, and it is enough to prove

$$\forall \lambda \in [0, 1],\ \forall x > 0, \qquad \lambda\psi(x) \leqslant \psi(\lambda x). \tag{10.21}$$

This is easy to check separating three cases:

(1) $x \leqslant 1$; then $\lambda\psi(x) = \lambda K(1 + x) \leqslant K(1 + \lambda x) = \psi(\lambda x)$;

(2) $\lambda x \leqslant 1 < x$; then, since $A(x) > A(1) = 1/(2K)$,

$$\lambda\psi(x) = \lambda\frac{x}{A(x)} < 2K\lambda x \leqslant K(1 + \lambda x) = \psi(\lambda x);$$

(3) $\lambda x > 1$; we then have, since A increases,

$$\lambda\psi(x) = \lambda\frac{x}{A(x)} \leqslant \frac{\lambda x}{A(\lambda x)} = \psi(\lambda x).$$

and this ends the proof of (10.21). Now, since S is determined by the value of $S(0)$ and the sign of $S'(0)$, we have $S(\overline{z}) = \overline{S(z)}$ for all $z \in \mathbb{D}$, so that $S[(-1,1)] \subset \mathbb{R}$. And since the derivative of an injective analytic function does not vanish and $S'(0) > 0$, we find that S is increasing on $(-1,1)$. Finally, if $w \in \mathbb{R}$, there exists $z \in \mathbb{D}$ with $w = S(z)$, we have $\overline{w} = w$, so that $S(\overline{z}) = S(z)$ and $\overline{z} = z$, which proves the surjectivity of $S\colon (-1,1) \to \mathbb{R}$. □

We now begin our construction, and we can assume that (ε_n) is non-increasing and log-convex (see Lemma F.7 of Appendix F), and also that $\varepsilon_n < 1$ for all n.

We now choose A_n as follows, η denoting a positive numerical constant to be specified later.

$$A_n = \eta \log \frac{1}{\varepsilon_n}, \quad \text{for } n \geqslant 1. \tag{10.22}$$

Observe that this is a well-defined, increasing, concave sequence tending to ∞ since we assumed $0 < \varepsilon_n < 1$ and since (ε_n) is log-convex and decreasing to 0.

Finally, we define our symbol φ and an increasing sequence (r_n) with $0 < r_n < 1$ under the form of the following crucial lemma, in which the increasing character of ψ is important.

Lemma 10.21 *Let φ be defined, for $z \in \mathbb{D}$, by*

$$\varphi(z) = S^{-1}\big(e^{-1}\, S(z)\big),$$

and let $r_n = S^{-1}(e^n)$. Then we have:

(i) *φ is univalent and maps $\mathbb{D}$ to $\mathbb{D}$, (r_n) increases, and $\varphi(0) = 0$.*
(ii) *$\varphi(r_{n+1}) = r_n$.*
(iii) *$\dfrac{1 - r_{n+1}}{1 - r_n} \to 0$ and therefore (r_n) is an interpolation sequence.*
(iv) *$C_\varphi\colon H^2 \to H^2$ is compact.*
(v) *$\dfrac{1 - r_{n+1}}{1 - r_n} \geqslant \exp(-2C_0 A_n)$ with C_0 a positive absolute constant.*

Proof: (i) Since Ω is star-shaped, $\mathrm{e}^{-1}S(z) \in \Omega$ when $z \in \mathbb{D}$, so φ is well defined and maps $\mathbb{D}$ to itself in a univalent way. Moreover, $\varphi(0) = S^{-1}(0) = 0$, and (r_n) increases since S^{-1} increases on $\mathbb{R}$.
(ii) We have $\varphi(r_{n+1}) = S^{-1}\big(\frac{1}{\mathrm{e}} S(r_{n+1})\big) = S^{-1}\big(\frac{1}{\mathrm{e}}\, \mathrm{e}^{n+1}\big) = S^{-1}(\mathrm{e}^n) = r_n$. Note the backward shift character of this relation.
(iii) This assertion is more delicate and relies on Proposition 10.19 as follows.

Set $d_n = \psi(\mathrm{e}^n)$. We have clearly $\mathrm{e}^{n+1} + d_{n+2} < \mathrm{e}^{n+2}$ for large n (recall that $\psi(t) = o(t)$ as $t \to \infty$), so that $\psi(\mathrm{e}^{n+1} + d_{n+2}) < \psi(\mathrm{e}^{n+2}) = d_{n+2}$ since ψ is increasing. By the intermediate value theorem for the function $\psi(\mathrm{e}^{n+1}+x)-x$, we can find a positive number $c_n < d_{n+2}$ such that $\psi(\mathrm{e}^{n+1}+c_n) = c_n$.

Now, consider the open sets

$$R_n = \left\{z \in \mathbb{C} : \mathrm{e}^n - c_n < \operatorname{Re} z < \mathrm{e}^{n+1} + c_n \text{ and } |\operatorname{Im} z| < c_n\right\}$$

as well as $U_n = R_n \cup \Omega$.

Those sets U_n satisfy the assumptions of Proposition 10.19 in view of (10.20). Indeed, if z belongs to the horizontal sides of R_n, we have $z \notin U_n$ since

$$\mathrm{e}^n - c_n \leqslant \operatorname{Re} z \leqslant \mathrm{e}^{n+1} + c_n \quad \Longrightarrow \quad \psi(\operatorname{Re} z) \leqslant \psi(\mathrm{e}^{n+1} + c_n) = c_n = |\operatorname{Im} z|.$$

This proposition then gives, since $\Omega \subset U_n$ and $c_n < d_{n+2}$, and since the hyperbolic metric is conformally invariant,

$$\begin{aligned} d(r_n, r_{n+1}) &= d(\mathrm{e}^n, \mathrm{e}^{n+1}; \Omega) \geqslant d(\mathrm{e}^n, \mathrm{e}^{n+1}; U_n) \\ &\geqslant \frac{\pi}{4c_n}(\mathrm{e}^{n+1} - \mathrm{e}^n) - \frac{\pi}{2} \\ &\geqslant \kappa \, \frac{\mathrm{e}^{n+2}}{\psi(\mathrm{e}^{n+2})} = cA(\mathrm{e}^{n+2}) \geqslant \kappa A_n, \end{aligned}$$

where κ is a positive constant. Now, we use Lemma 10.17 to obtain

$$\frac{1 - r_{n+1}}{1 - r_n} \leqslant 2\,\mathrm{e}^{-2d(r_n, r_{n+1})} \leqslant 2\,\mathrm{e}^{-2\kappa A_n},$$

which proves $\frac{1-r_{n+1}}{1-r_n} \to 0$ and implies that (r_n) is an *interpolation sequence.*
(iv) Since φ is univalent, the compactness of $C_\varphi \colon H^2 \to H^2$ amounts to proving that $\lim_{|z|\to 1} \frac{1-|\varphi(z)|}{1-|z|} = \infty$. This in turn is equivalent to proving that for any u, v on the unit circle, the quotient $\frac{\varphi(z)-v}{z-u}$ has no finite limit as z tends to u radially. Indeed suppose $\underline{\lim}_{|z|\to 1} \frac{1-|\varphi(z)|}{1-|z|} < \infty$. By compactness, we can find a sequence (z_n) of $\mathbb{D}$ and a point $u \in \partial\mathbb{D}$ with $\underline{\lim}_{n\to\infty} \frac{1-|\varphi(z_n)|}{1-|z_n|} < \infty$ and $z_n \to u$. So that, with the notations of the Julia–Carathéodory Theorem in Chapter 1, we have $d(u) < \infty$ and there exists a point $v \in \partial\mathbb{D}$ such that the angular limit, a fortiori the radial limit, of $\frac{\varphi(z)-v}{z-u}$ exists in $\mathbb{C}$, which is a contradiction. Now, the latter fact requires some precise justification.

First we notice that S extends continuously to an injective map of the open upper half of the unit circle onto the upper part of the boundary of $\partial\Omega$ (and similarly for lower parts). This follows from the Carathéodory extension theorem [153, p. 290] applied to the restriction of S^{-1} to the Jordan region limited by $\partial\Omega$ and two vertical lines $\operatorname{Re} w = \pm R$ where $R > 0$ is arbitrarily large. Next, let $u \in \partial\mathbb{D}$ with $u \neq \pm 1$. Then $S(ru) \to w \in \partial\Omega$ as $r \to 1^-$, so that $e^{-1}S(ru) \to e^{-1}w = w' \in \Omega$ and $\varphi(u) \to S^{-1}(w') \in \mathbb{D}$. Therefore, the assumptions of the Julia–Carathéodory theorem are met if $u \neq \pm 1$. By symmetry, it remains to test the point $u = 1$ for which we have

$$\varlimsup_{r\to 1^-} \frac{1-\varphi(r)}{1-r} \geqslant \varlimsup_{n\to\infty} \frac{1-\varphi(r_{n+1})}{1-r_{n+1}} = \varlimsup_{n\to\infty} \frac{1-r_n}{1-r_{n+1}} = +\infty$$

by the preceding point (iii). Since $|v-\varphi(r)| \geqslant 1-|\varphi(r)| = 1-\varphi(r)$, we have our justification.

(v) We now want a good *lower* bound for the quotient $\frac{1-r_{n+1}}{1-r_n}$. To that effect, we apply Proposition 10.19 with

$$U = \Omega, \quad a_1 = e^n, \quad a_2 = e^{n+1}, \qquad \text{and} \quad b_n := \psi(e^{n-1}),$$

as well as

$$R_n' = \left\{ z \in \mathbb{C} : e^n - b_n < \operatorname{Re} z < e^{n+1} + b_n \quad \text{and} \quad |\operatorname{Im} z| < b_n \right\}.$$

We have $e^n - b_n > e^{n-1}$ for large n since this amounts to

$$e^n - e^{n-1} > b_n = \frac{e^{n-1}}{A(e^n)}, \quad \text{that is,} \quad e - 1 > \frac{1}{A(e^n)}$$

and $A(t)$ tends to ∞ with t. We then observe that $R_n' \subset \Omega$. Indeed $z \in R_n'$ implies $\operatorname{Re} z > e^n - b_n > e^{n-1}$ and, since ψ is increasing, we have

$$\psi(\operatorname{Re} z) > \psi(e^{n-1}) = b_n > |\operatorname{Im} z|.$$

Therefore, we can apply (10.18) and get, for all $n \geqslant 1$,

$$d(e^n, e^{n+1}; \Omega) \leqslant \frac{\pi}{4\psi(e^{n-1})}(e^{n+1} - e^n) + \frac{\pi}{2} \leqslant C_0 A(e^{n-1}) = C_0 A_n$$

with C_0 a numerical constant.

By conformal invariance, we have as well $d(r_n, r_{n+1}) \leqslant C_0 A_n$. It then follows from Lemma 10.17 that

$$\frac{1-r_{n+1}}{1-r_n} \geqslant \exp\big(-2d(r_n, r_{n+1})\big) \geqslant \exp(-2C_0 A_n).$$

This ends the proof of Lemma 10.21. □

We can now easily finish the proof of Theorem 10.15 with the use of Lemma 10.2, in which we take

$$u_j = r_{j+1}, \quad \text{so that} \quad v_j = \varphi(u_j) = r_j, \quad \text{for } 1 \leqslant j \leqslant n.$$

Here there is no weight and the interpolation constant I_v is uniformly bounded above since the whole sequence (r_n) is interpolating. Moreover, adjusting now $\eta = 1/C_0$ where C_0 is as in (v) of Lemma 10.21, we get, for $1 \leqslant j \leqslant n$,

$$\sqrt{\frac{1-u_j^2}{1-v_j^2}} \geqslant \frac{1}{\sqrt{2}}\sqrt{\frac{1-u_j}{1-v_j}} = \frac{1}{\sqrt{2}}\sqrt{\frac{1-r_{j+1}}{1-r_j}} \geqslant \frac{1}{\sqrt{2}}\exp(-C_0 A_n) = \frac{\varepsilon_n}{\sqrt{2}}.$$

So Lemma 10.2 immediately ends the proof of Theorem 10.15 for H^2.

The proof for classical Bergman spaces is similar. □

Remark. The same results as Theorem 10.15 or Corollary 10.16 hold for the Dirichlet space, as was proved in [114, theorem 1.2], Corollary 10.16 being an immediate consequence of the theorem.

10.6 A Class of Sharp Examples

This subsection presents, in a fairly sketchy way, two results [146], which are closely related with the issues of this chapter.

A word about the proofs: even though they use the same tools (Blaschke products and model spaces for upper bounds; Bernstein numbers and reproducing kernels for lower bounds) as in the rest of this chapter, they are technically delicate, long, and also need sharp estimates on the hyperbolic length $l_P(\Gamma)$ of a curve $\Gamma \subset \mathbb{D}$ in the Poincaré metric, given by the formula

$$l_P(\Gamma) = 2\int_\Gamma \frac{|dz|}{1-|z|^2}.$$

In order to keep this chapter to a reasonable length, we will not detail those proofs in this book.

Here is the context: we consider symbols of the form $\varphi = \exp(-u - i\tilde{u})$ where the real-valued function u is in $C(\mathbb{T})$, with conjugate function $\tilde{u}$, moreover satisfies $u(z) = u(\overline{z})$, and is smooth except possibly at $z = 1$. We set

$$U(t) = u(\mathrm{e}^{it}), \quad \text{for } -\pi < t \leqslant \pi,$$

and we assume that $U(0) = 0$, that U increases on $[0, \pi]$, and that

$$h_U(t) := \int_t^\pi \frac{U(x)}{x^2}dx, \quad 0 < t \leqslant \pi, \text{ is unbounded as } t \to 0^+. \tag{10.23}$$

Let $\mathcal{U}$ be the set of such functions U, and denote by φ_U the symbol attached to some $U \in \mathcal{U}$ as above. Now (10.23) amounts to saying that the operator C_{φ_U} is compact on H^2.

To state our first theorem, we need to coin two more functions η_U and ω_U, the latter being defined implicitely:

(i) $U(t) = \exp\big(-\eta_U(|\log t|)\big)$, where $0 < t \leqslant 1$.

(ii) $\eta_U\left(\dfrac{x}{\omega_U(x)}\right) = \omega_U(x)$ for those $x \geqslant 0$ with $\eta_U(x) \geqslant 1$.

We obtain the most precise results if either h_U grows slowly to ∞, or U tends slowly to 0, at 0. The two extreme cases are covered by the following theorem.

Theorem 10.22 *Suppose that $U \in \mathcal{U}$. Then:*

(i) *If $tU'(t)/U(t) \leqslant 1 + c/|\log t|$ and $U(t)/(th_U(t)) \leqslant C(|\log t| \log |\log t|)^{-1}$ for some $c > 1, C > 0$, and $t > 0$ sufficiently small, then*

$$a_n(C_{\varphi_U}) = \frac{\exp\big(O(1)\big)}{\sqrt{h_U(e^{-\sqrt{n}})}} \qquad \text{as } n \to \infty.$$

(ii) *If $\eta'_U(x)/\eta_U(x) = o(1/x)$ as $x \to +\infty$, then*

$$a_n(C_{\varphi_U}) = \exp\left(-\frac{\big(\pi^2/2 + o(1)\big)n}{\omega_U(n)}\right) \qquad \text{as } n \to \infty.$$

Theorem 10.22 may seem a little daunting (the proof is not!), due to the technicity and generality of its assumptions and conclusions. But it has, among other things, the following interesting corollary, much more explicit, which shows that certain sequences $(g(n))$ can nearly be claimed to be the sequence of approximation numbers of some composition operator. For example, we can take for g below any negative power of the logarithm or slower. And this provides yet another proof of the Carroll–Cowen theorem already mentioned in this chapter: there are compact composition operators on H^2 that belong to no Schatten class $\mathcal{S}^p, p < \infty$. This may also be seen as an alternative approach to Theorem 10.15.

Corollary 10.23 *Let g be a function on $\mathbb{R}^+$ such that $g(x)$ decreases to 0 as $x \to +\infty$ and $\dfrac{g(x^2)}{g(x)}$ is bounded below. Then there exists a compact composition operator on H^2 with the two-sided estimate*

$$a_n(C_\varphi) = e^{O(1)} g(n) \qquad \text{as } n \to \infty.$$

This corollary says that we may prescribe any slow rate of decay and find a function $U \in \mathcal{U}$ such that the approximation numbers $a_n(\varphi_U)$ descend accordingly.

Proof: Just set $h_U(t) = \big[g(\log^2 t)\big]^{-2}$ for small $t > 0$ in (i) of Theorem 10.22. This defines a function $U \in \mathcal{U}$ and the associated symbol φ_U does the job. □

Another more explicit variant of Theorem 10.22, close to the lens map example of the next subsection, is the following;

Theorem 10.24 *Let $0 < \alpha < 1$. Take $\varphi(z) = \big(1 + (1 - z)^\alpha\big)^{-1}$. Then, φ is a symbol and we have the two-sided estimate*

$$\exp\big(-b'_\alpha \sqrt{n}\big) \lesssim a_n(C_\varphi) \lesssim \exp\big(-b_\alpha \sqrt{n}\big)$$

with $b'_\alpha = \pi(1-\alpha)\sqrt{2/\alpha}$ and $b_\alpha = \pi(1-\alpha)\sqrt{1/2\alpha}$.

Proof: That φ is a symbol is clear since $\varphi = \varphi_U$ with

$$U(t) = \operatorname{Re}\log\big(1 + (1 - \mathrm{e}^{it})^\alpha\big), \text{ since } \operatorname{Re}\,(1 - \mathrm{e}^{it})^\alpha \geqslant 0$$

associated with the analytic function

$$u(z) + i\tilde{u}(z) = \log\big(1 + (1 - z)^\alpha\big).$$

The corresponding auxiliary functions h_U and ω_U satisfy $U \in \mathcal{U}$,

$$h_U(t) \sim \int_t^\pi \frac{x^\alpha}{x^2} dx \approx t^{\alpha-1}, \text{ when } t \to 0^+,$$

which suggests, having in mind (i) of Theorem 10.22, why the factor $1 - \alpha$ appears in the constants b_α, b'_α above.

On the other hand, we have $\eta_U(x) \approx \alpha x$, so that

$$\omega_U(x) = \big(1 + o(1)\big)\sqrt{\alpha x},$$

since $\eta_U\left(\frac{x}{\omega_U(x)}\right) = \omega_U(x)$ reads as $\frac{\alpha x}{\omega_U(x)} \sim \omega_U(x)$. This suggests, having in mind (ii) of Theorem 10.22, why the factor $1/\sqrt{\alpha}$ appears in b_α, b'_α. Of course, we cannot apply directly Theorem 10.22: for example, the second assumption is violated since $\eta'_U(x)/\eta_U(x) \approx 1/x$. A full proof of Theorem 10.24 has to be given. We refer for that to the original paper.

Observe that $b'_\alpha \neq b_\alpha$; we were not able to prove that $\dfrac{\log a_n(C_\varphi)}{\sqrt{n}}$ has a limit. Just point out that b_α degenerates correctly when $\alpha \to 0$: heuristically, we have $b_\alpha \to \infty$ and C_φ "becomes" C_{Ψ_0} with $\Psi_0(z) = 1/2$ and $a_n(C_{\Psi_0}) = 0$ for $n \geqslant 2$. In the same spirit, when $\alpha \to 1$, we have $b'_\alpha \to 0$ and C_φ "becomes" C_{Ψ_1} with $\Psi_1(z) = 1/(2 - z)$ and $a_n(C_{\Psi_1}) \geqslant c > 0$ (see Example I.14). □

10.7 The Lens Map on Hardy Spaces

Let $0 < \theta < 1$. Recall that the lens map Λ_θ, presented in detail in Section 4.2 of Chapter 4, is the symbol defined by

$$\Lambda_\theta(z) = \frac{(1+z)^\theta - (1-z)^\theta}{(1+z)^\theta + (1-z)^\theta} = (\Phi^{-1} \circ \gamma_\theta \circ \Phi)(z),$$

where $\Phi : \mathbb{D} \to \mathbb{C}_+$ is the conformal map $z \mapsto \dfrac{1-z}{1+z}$ and $\gamma_\theta : \mathbb{C}_+ \to \mathbb{C}_+$ is defined by $\gamma_\theta(w) = w^\theta$.

We recall the estimate (4.2) enlighted in Chapter 4:

$$1 - |\Lambda_\theta(z)|^2 \geqslant \delta(1-\theta)\, |1-z|^\theta \qquad \text{for } z \in \mathbb{D},\ \mathrm{Re}(z) \geqslant 0. \tag{10.24}$$

The main result of this section is the following:

Theorem 10.25 *Let $p \geqslant 1$. The approximation numbers of $C_{\Lambda_\theta} : H^p \to H^p$ satisfy*

$$a' \exp(-b' \sqrt{n}) \leqslant a_n(C_{\Lambda_\theta}) \leqslant a \exp(-b \sqrt{n})$$

for positive constants a, b, a', b' depending only on θ and p.

We first consider the (easier) *Hilbertian case* and show that the semi-group property leads to a slightly less precise result, namely

$$a_n(C_{\Lambda_\theta}) \leqslant \alpha \exp(-\beta n^{1/3}). \tag{10.25}$$

This is already enough to recapture a result of Shapiro and Taylor ([166]):

$$C_{\Lambda_\theta} \in \bigcap_{r>0} \mathscr{S}^r.$$

To prove (10.25), we rely on the following lemma:

Lemma 10.26 *The Hilbert–Schmidt norm of C_{Λ_θ} satisfies*

$$\|C_{\Lambda_\theta}\|_{HS} \leqslant \frac{C}{1-\theta},$$

where C is a numerical constant.

Proof: By symmetry

$$\|C_{\Lambda_\theta}\|_{HS}^2 = \frac{2}{\pi} \int_0^{\pi/2} \frac{1}{1 - |\Lambda_\theta(\mathrm{e}^{it})|^2} dt.$$

By (4.2), since $|1 - \mathrm{e}^{it}| \geqslant (2/\pi)\, t$ for $0 \leqslant t \leqslant \pi$, this implies

$$\|C_{\Lambda_\theta}\|_{HS}^2 \lesssim \frac{1}{1-\theta} \int_0^{\pi/2} t^{-\theta} dt \lesssim \frac{1}{(1-\theta)^2},$$

proving the lemma. □

Now let k be a positive integer and set

$$\alpha_k = \theta^{1/k}.$$

Proposition G.9 and the semi-group property of lens maps give us, via Lemma 10.26 for any pair (k, m) of integers,

$$a_{km}(C_{\Lambda_\theta}) \leqslant \left[a_m(C_{\alpha_k})\right]^k \leqslant \left[\frac{C}{(1-\alpha_k)\sqrt{m}}\right]^k.$$

Now

$$1-\alpha_k \geqslant \frac{1-\alpha_k^k}{k} = \frac{1-\theta}{k}.$$

We then get, with $c = c_\theta$,

$$a_{km}(C_{\Lambda_\theta}) \leqslant \left(\frac{k}{c\sqrt{m}}\right)^k.$$

Set $d = c/\mathrm{e}$ and take $k = d\sqrt{m}$, ignoring the questions of integer part. This gives

$$a_{dm^{3/2}}(C_{\Lambda_\theta}) \leqslant \mathrm{e}^{-k} = \mathrm{e}^{-d\sqrt{m}}.$$

Finally adjust $m = (nd^{-1})^{2/3}$ so $dm^{3/2} = n$, and we get (10.25).

The full result of Theorem 10.25 will be a consequence of the results of the forthcoming section (H^p-case).

The Case of Hardy Spaces H^p

We now switch to the (Hilbertian or not) case of H^p-spaces. The proofs are slightly more complicated, partly because various singular numbers (Gelfand, approximation, Bernstein, etc.) are possibly no longer equal, and partly because we must appeal to the notion of type and cotype for Banach spaces, be it only in the simple context of subspaces of L^p-spaces.

We begin with a lower bound.

Proposition 10.27 *Let $\theta \in (0,1)$ and $p \in [1,\infty)$. Then, with $C_{\Lambda_\theta} : H^p \to H^p$,*

$$a_n(C_{\Lambda_\theta}) \geqslant a\exp(-b\sqrt{n}) \tag{10.26}$$

for all integers $n \geqslant 1$ (with $a, b > 0$ depending only on θ and p).

Proof: We shall use Bernstein numbers and the obvious inequalities, for an operator $U : X \to X$ from a Banach space to itself and its adjoint (transposed) $U^* : X^* \to X^*$,

$$a_n(U) \geqslant a_n(U^*) \geqslant b_n(U^*). \tag{10.27}$$

We already know that, for any $p \in [1, \infty)$ and any symbol φ,

$$\|\delta_a\|_{(H^p)^*} = (1 - |a|^2)^{-1/p} \quad \text{and} \quad C_\varphi^*(\delta_a) = \delta_{\varphi(a)}. \tag{10.28}$$

We need some unconditionality of the δ_a given by the

Lemma 10.28 *Let $Z = (z_1, \ldots, z_n)$ be a finite sequence of distinct points in $\mathbb{D}$ with Carleson interpolation constant I_Z. Then, for any scalars $\lambda_1, \ldots, \lambda_n$ and any unimodular scalars $\varepsilon_1, \ldots, \varepsilon_n$, it holds*

$$I_Z^{-1} \left\| \sum_{j=1}^{n} \lambda_j \delta_{z_j} \right\|_{(H^p)^*} \leqslant \left\| \sum_{j=1}^{n} \varepsilon_j \lambda_j \delta_{z_j} \right\|_{(H^p)^*} \leqslant I_Z \left\| \sum_{j=1}^{n} \lambda_j \delta_{z_j} \right\|_{(H^p)^*}. \tag{10.29}$$

Proof: Set $L = \sum_{j=1}^{n} \lambda_j \delta_{z_j}$ and $L_\varepsilon = \sum_{j=1}^{n} \varepsilon_j \lambda_j \delta_{z_j}$. Take $g \in H^\infty$ such that $g(z_j) = \varepsilon_j$, $1 \leqslant j \leqslant n$, and $\|g\|_\infty \leqslant I_Z$. Now observe that, for any $f \in H^p$:

$$L_\varepsilon(f) = L(fg).$$

Hence, since H^∞ is isometrically the space of mulipliers of H^p:

$$|L_\varepsilon(f)| \leqslant \|L\|_{(H^p)^*}.\|fg\|_{H^p} \leqslant \|L\|_{(H^p)^*}.\|g\|_\infty \|f\|_{H^p} \leqslant I_Z \|L\|_{(H^p)^*}.\|f\|_{H^p}.$$

This gives the RHS of (10.29) and the LHS follows by changing λ_j into $\lambda_j \varepsilon_j$. □

If $L = \sum_j \lambda_j \delta_{z_j}$, we will write

$$L(f) = \sum_{j=1}^{n} \lambda_j f(z_j) = \sum_{j=1}^{n} \widetilde{\lambda}_j F_j$$

with $\widetilde{\lambda}_j = \lambda_j \|\delta_{z_j}\|_{(H^p)^*}$ and $F_j = \|\delta_{z_j}\|_{(H^p)^*}^{-1} f(z_j)$.

We will set

$$V = \left(\widetilde{\lambda}_1, \ldots, \widetilde{\lambda}_n\right), \quad \text{and} \quad F = (F_1, \ldots, F_n).$$

So that,

$$|L(f)| \leqslant \|V\|_{\ell^{p'}} \|F\|_{\ell^p} \tag{10.30}$$

To exploit (10.30), we will need a good lower bound on $\|V\|_{\ell^{p'}}$ and hence a good upper bound on $\|F\|_{\ell^p}$. This upper bound is provided by a second lemma.

Lemma 10.29 *Let $Z = (z_j)$ be a δ-separated sequence (finite or not) of distinct points in $\mathbb{D}$. Then*

(i) *any* $f \in \mathscr{B}^2_\gamma$, *with* $\gamma > -1$, *satisfies*

$$\sum_j (1 - |z_j|^2)^{\gamma+2} \, |f(z_j)|^2 \leqslant C_\gamma \, \delta^{-2} \, \|f\|^2_{\mathscr{B}^2_\gamma}; \tag{10.31}$$

(ii) *any function* $f \in H^2$ *satisfies*

$$\sum_j (1 - |z_j|^2)^3 \, |f'(z_j)|^2 \leqslant C\delta^{-2} \, \|f\|^2_{H^2}; \tag{10.32}$$

(iii) *In the* H^p*-setting: if now* $\delta(Z) > 0$ *denotes the* Carleson constant *of* Z, *then, for any function* $f \in H^p$, *it holds*

$$\sum_j (1 - |z_j|^2)|f(z_j)|^p \leqslant C\delta(Z)^{-4} \, \|f\|^p_{H^p} \, . \tag{10.33}$$

Proof: We prove (10.31) by the Neville trick alluded to in Chapter 5. This relation (10.31) is due to Luecking [122] (see also [3]), is used in Chapter 9, and is not especially needed now. The sequence Z has not to be interpolating, or even Blaschke, and is only *separated*: $\rho(z_j, z_k) \geqslant \delta$ for $j \neq k$. We do not aim to get an optimal dependence in δ, but in counterpart the proof is very simple, and this statement suffices here for our purposes.

We set $d_j = 1 - |z_j|^2$ and observe that the euclidean disks $\Delta_j := D(z_j, (\delta/4)d_j)$ are disjoint since, for $j \neq k$,

$$|z_j - z_k| \geqslant \delta|1 - z_j\overline{z_k}| \geqslant \delta(1 - |z_j|\,|z_k|) \geqslant (\delta/2)(d_j + d_k).$$

We have $\Delta_j \subset \mathbb{D}$ since $|z_j| + \frac{\delta}{2}(1 - |z_j|^2) < |z_j| + 1 - |z_j| = 1$. Note that

$$d_j = 1 - |z_j|^2 \leqslant 2(1 - |z|^2) \tag{10.34}$$

for any $z \in \Delta_j$ since

$$d_j = (1 - |z|^2) + (|z|^2 - |z_j|^2) \leqslant (1 - |z|^2) + 2|z - z_j| \leqslant (1 - |z|^2) + \frac{d_j}{2}.$$

Conversely $1 - |z|^2 \leqslant (1 + \delta/2)d_j \leqslant 2d_j$.

Now, subharmonicity gives, for given j.

$$\begin{aligned} d_j^{\gamma+2}|f(z_j)|^2 &\lesssim d_j^{\gamma+2}(\delta d_j)^{-2} \int_{\Delta_j} |f(z)|^2 \, d\sigma(z) \\ &\lesssim \delta^{-2} \int_{\Delta_j} (1 - |z|^2)^\gamma \, |f(z)|^2 \, d\sigma(z). \end{aligned}$$

Now, summing up over j, using the disjointness of the Δ_j, we obtain (10.31).

When $f \in H^2$, we have $f' \in \mathscr{B}^2_1$ thanks to the Littlewood–Paley formula. Therefore (10.32) follows from (10.31) applied to f' with $\gamma = 1$.

Finally, in case when Z is $\delta(Z)$-Carleson (hence interpolating), we are allowed to take $\delta = \delta(Z)$, and Neville's idea is to apply (10.32) to $g = Bf$, where B is the Blaschke product formed with (z_j), which as we saw in Chapter 1 satisfies $|B'(z_j)| \geqslant \frac{\delta}{d_j}$. This erases the derivative of f since $g'(z_j) = B'(z_j) f(z_j)$. We hence get, after simplification by δ^2,

$$\delta^2 \sum d_j \, |f(z_j)|^2 \leqslant C \sum d_j^3 \, |g'(z_j)|^2 \leqslant C\delta^{-2} \|g\|_{H^2}^2 = C\delta^{-2} \|f\|_{H^2}^2,$$

and we are done.

The case of a general p follows by writing $f = B_0 G^{2/p}$ where B_0 is the (convergent) Blaschke product formed with the zeroes of f and G is a zero-free analytic function satisfying $\|G\|_{H^2}^2 = \|f\|_{H^p}^p$. This gives, with $\delta = \delta(Z)$,

$$\begin{aligned} \sum_j (1 - |z_j|^2) \, |f(z_j)|^p &= \sum_j (1 - |z_j|^2) \, |B_0(z_j)|^p \, |F(z_j)|^2 \\ &\leqslant \sum_j (1 - |z_j|^2) \, |F(z_j)|^2 \\ &\leqslant C\delta(Z)^{-4} \, \|G\|_{H^2}^2 = C\delta(Z)^{-4} \, \|f\|_{H^p}^p . \qquad \square \end{aligned}$$

Remark. With our previous notation $F = (F_1, \ldots, F_n)$, inequality (10.33) clearly implies (taking the sequence $Z = u$)

$$\|F\|_{\ell^p} \leqslant C\delta(u)^{-4/p}. \tag{10.35}$$

The last to final step in view of Proposition 10.27 is the following:

Lemma 10.30 *Let φ be a symbol and $C_\varphi : H^p \to H^p$, with $1 \leqslant p < \infty$. Let $u_1, \ldots, u_n \in \mathbb{D}$ be such that the points $v_j = \varphi(u_j)$, where $1 \leqslant j \leqslant n$, are distinct and let*

$$\mu_n = \inf_{1 \leqslant j \leqslant n} \frac{\|\delta_{v_j}\|_{(H^p)^*}}{\|\delta_{u_j}\|_{(H^p)^*}} = \inf_{1 \leqslant j \leqslant n} \left(\frac{1 - |u_j|^2}{1 - |v_j|^2} \right)^{1/p}.$$

Then, for some $\alpha = \alpha(p) > 0$,

$$b_n(C_\varphi^*) \geqslant c I_v^{-\alpha} \mu_n. \tag{10.36}$$

Proof: Let $E \subset (H^p)^*$ be the n-dimensional span of $\delta_{u_1}, \ldots, \delta_{u_n}$ and let $L = \sum_{j=1}^n \lambda_j \delta_{u_j}$ be in the unit sphere of E. For $f \in H^p$ with $\|f\|_{H^p} = 1$, we know by (10.30) (for $z_j = u_j$) and by (10.35) in the remark following the proof of Lemma 10.29 that

$$|L(f)| \leqslant \|V\|_{\ell^{p'}} \|F\|_{\ell^p} \leqslant C\delta(u)^{-4/p} \|V\|_{\ell^{p'}}. \tag{10.37}$$

Taking the supremum for $\|f\|_{H^p} = 1$, we get

$$\|V\|_{\ell^{p'}} \geqslant c\delta(u)^{4/p}. \tag{10.38}$$

Now we know that $\|C_\varphi^*(L)\|_{(H^p)^*} = \|\sum_{j=1}^n \lambda_j \delta_{v_j}\|_{(H^p)^*}$ and we split the proof into three cases.

Case 1. $1 < p \leqslant 2$.
Then H^p is of type p and its dual space of cotype p' ([108, vol. 1, chapter V, section IV]). So that averaging the minoration

$$\Big\|\sum_{j=1}^n \lambda_j\, \delta_{v_j}\Big\|_{(H^p)^*} \geqslant I_v^{-1}\Big\|\sum_{j=1}^n \varepsilon_j \lambda_j\, \delta_{v_j}\Big\|_{(H^p)^*}$$

of Lemma 10.28 with respect to Rademacher variables ε_j, we get via (10.38)

$$\begin{aligned}
\|C_\varphi^*(L)\|_{(H^p)^*} &\geqslant c I_v^{-1}\left(\sum_{j=1}^n |\lambda_j|^{p'} \|\delta_{v_j}\|_{(H^p)^*}^{p'}\right)^{1/p'} \\
&\geqslant c I_v^{-1}\left(\sum_{j=1}^n |\lambda_j|^{p'} \mu_n^{p'} \|\delta_{u_j}\|_{(H^p)^*}^{p'}\right)^{1/p'} \\
&= c I_v^{-1}\mu_n \|V\|_{\ell^{p'}} \geqslant c\mu_n \delta(u)^{4/p} I_v^{-1} \\
&\geqslant c\mu_n I_v^{-\alpha}
\end{aligned}$$

with $\alpha = 1 + \dfrac{4}{p} > 0$, where we used (see Theorem 1.43) $\delta(u)^{-1} \leqslant I_u \leqslant I_v$. Indeed, if $\sup|w_j| = 1$ and $f \in H^\infty$ satisfies $f(v_j) = w_j$ for every f, then $g = f \circ \varphi$ satisfies $g(u_j) = w_j$ for every j and $\|g\|_\infty \leqslant \|f\|_\infty$, so that $I_u \leqslant I_v$.

Case 2. $2 < p \leqslant \infty$.
Then H^p is of type 2 and its dual space of cotype 2 ([108]); we get exactly in the same way

$$\|C_\varphi^*(L)\|_{(H^p)^*} \geqslant c I_v^{-1}\mu_n \|V\|_{\ell^2}.$$

But we have this time, by (10.37) for $p = 2$, for any $f \in H^p$ with $\|f\|_{H^p} = 1$,

$$|L(f)| = \Big|\sum_{j=1}^n \widetilde{\lambda_j} F_j\Big| \leqslant \|V\|_{\ell^2}\, \|F\|_{\ell^2} \leqslant \|V\|_{\ell^2}\, C\delta(u)^{-2}\, \|f\|_{H^2} \leqslant \|V\|_{\ell^2}\, C\delta(u)^{-2}$$

because $\|f\|_{H^2} \leqslant \|f\|_{H^p}$.

Recall that $\widetilde{\lambda_j} := \lambda_j \|\delta_{z_j}\|$ and take the supremum on f

$$1 \leqslant \|V\|_{\ell^2}\, C\delta(u)^{-2} \leqslant C\|V\|_{\ell^2} I_v^2$$

This gives again (10.36) with $\alpha = 1$, reasoning as in the first case.

Case 3. $p = 1$.
Then $(H^1)^*$, isomorphic to the space $BMOA$ of analytic functions with bounded mean oscillation [184, p. 272], has no finite cotype. Nevertheless, the unconditionality expressed by Lemma 10.28 clearly implies

$$\sup_k |\lambda_k| . \|\delta_{v_k}\|_{(H^p)^*} \leqslant I_v \Big\| \sum_{j=1}^{n} \lambda_j \delta_{v_j} \Big\|_{(H^p)^*} = I_v \, \|C_\varphi^*(L)\|_{(H^p)^*}.$$

Hence (with here $V = (\|\delta_{u_j}\|_{(H^p)^*} \lambda_j)_{1\leqslant j\leqslant n}$)

$$\|C_\varphi^*(L)\|_{(H^p)^*} \geqslant I_v^{-1} \mu_n \, \|V\|_{\ell^\infty}.$$

Finally, using again (10.37) for $p = 1$ (i.e., $p' = \infty$), we get as in the first case (10.36) with $\alpha = 5$. □

We can now end the proof of (10.26) in Proposition 10.27 by a proper choice of points u_j for the lens map $\varphi = \Lambda_\theta$.

Fix $t \in (0, 1)$. By the intermediate value theorem (recall that $\Lambda_\theta(0) = 0$ and $\Lambda_\theta(1) = 1$), we can inductively find an increasing sequence $u_j \in [0, 1)$ with $u_0 = 0$ and

$$1 - \varphi(u_{j+1}) = t\,(1 - \varphi(u_j)), \text{ or } 1 - v_{j+1} = t(1 - v_j) \text{ and } 1 - v_j = t^j.$$

By (1.41), we know that, with $v = (v_j)$,

$$\delta(v) \geqslant \prod_{j=1}^{\infty} \left(\frac{1 - t^j}{1 + t^j} \right)^2.$$

This implies

$$\delta(v) \geqslant \exp\left(-\frac{a}{1 - t} \right) \quad \text{with} \quad a = \pi^2/2. \tag{10.39}$$

Just use the relation

$$2 \sum_{j=1}^{\infty} \log\left(\frac{1 + t^j}{1 - t^j} \right) = 4 \sum_{\substack{j\geqslant 1 \\ k\geqslant 0}} \frac{t^{j(2k+1)}}{(2k+1)} = 4 \sum_{k\geqslant 0} \frac{t^{2k+1}}{(2k+1)(1 - t^{2k+1})}$$

$$\leqslant \sum_{k\geqslant 0} \frac{4}{(2k+1)^2(1-t)} = \frac{\pi^2/2}{1 - t}$$

since $1 - t^{2k+1} \geqslant (1 - t)(2k + 1)t^{2k+1}$.

On the other hand, we have by definition

$$1 - v_n = \frac{2(1 - u_n)^\theta}{(1 + u_n)^\theta + (1 - u_n)^\theta} \sim 2^{1-\theta}(1 - u_n)^\theta \quad \text{as } n \to \infty.$$

Equivalently $1-u_n \sim 2^{-\rho}(1-v_n)^{1/\theta}$ with $\rho := (1/\theta)-1 > 0$, which implies

$$\frac{1-u_n}{1-v_n} \sim 2^{-\rho}(1-v_n)^{\rho} = 2^{-\rho}t^{n\rho} \quad \text{and} \quad \mu_n \geqslant c t^{\frac{n\rho}{p}}. \tag{10.40}$$

Next, set $t = \mathrm{e}^{-\varepsilon}$. We know that $I_v \leqslant C\delta(v)^{-2}$; therefore (changing the constants), from (10.39),

$$I_v^{-2} \geqslant c\delta(v)^4 \geqslant c\exp\left(-\frac{4a}{1-t}\right) \geqslant c\exp\left(-\frac{a'}{\varepsilon}\right).$$

Using (10.36), we obtain via (10.40)

$$b_n(C_\varphi^*) \gtrsim \exp\left[-b\left(n\varepsilon + \frac{1}{\varepsilon}\right)\right].$$

Optimizing with the choice $\varepsilon = 1/\sqrt{n}$ finally gives

$$b_n(C_\varphi^*) \gtrsim \exp(-2b\sqrt{n}),$$

ending the proof if we remember (10.27). □

We switch to the upper bound.

Proposition 10.31 *Let $\theta \in (0,1)$ and $p \in [1,\infty)$. Then with $C_{\Lambda_\theta} : H^p \to H^p$,*

$$a_n(C_{\Lambda_\theta}) \leqslant \alpha \exp(-\beta\sqrt{n}) \tag{10.41}$$

for all integers $n \geqslant 1$ (with $\alpha = \alpha(\theta, p) > 0$ and $\beta = \beta(\theta, p) > 0$).

Proof: The proof is not that easy: if $p \neq 2$, we are no longer in the Hilbertian case, and do not even know if $a_n(C_{\Lambda_\theta}) \leqslant a\exp(-bn^{1/3})$, since the previous Hilbert–Schmidt trick is no longer available. The general results of Chapter 5 can yet give something as $a_n(C_{\Lambda_\theta}) \leqslant n^{-\beta}$ for some $\beta = \beta(p,\theta) > 0$ (based on the property $\rho_\varphi(h) \leqslant Ch^{1/\theta}$), but reveal insufficient here, mainly because using only the partial sums operator composed with Λ_θ as an approximation to C_{Λ_θ} is too primitive. Some properties of the lens map are needed before proceeding. We refer to Chapter 4: see Section 4.2 and Proposition 4.1. We notably recall that the following estimates hold:

We have, for every $z \in \overline{\mathbb{D}}$ with $\operatorname{Re} z \geqslant -1/2$,

$$|1-\Lambda_\theta(z)| \approx 1-|\Lambda_\theta(z)| \approx C|1-z|^{\theta} \tag{10.42}$$

and, for every $z, z' \in \overline{\mathbb{D}}$,

$$|\Lambda_\theta(z) - \Lambda_\theta(z')| \leqslant C|z-z'|^{\theta}. \tag{10.43}$$

Next, in order to majorize $a_n(C_{\Lambda_\theta})$, we will rely on the following general lemma.

Lemma 10.32 *Let B be a Blaschke product with less than N zeroes. Then, for any symbol φ, one has, for some constant $K > 0$,*

$$\left[c_N(C_\varphi)\right]^p \leqslant K \sup_{\substack{0<h<1 \\ \xi\in\mathbb{T}}} \frac{1}{h}\int_{S(\xi,h)} |B|^p \, dm_\varphi, \tag{10.44}$$

where $c_n(C_\varphi)$ denotes the nth Gelfand number of the operator $C_\varphi : H^p \to H^p$.

Proof: The subspace BH^p is of codimension $\leqslant N-1$ in H^p, so that, by definition, $c_N(C_\varphi) \leqslant \|C_{\varphi|BH^p}\|$. Now, since $\|Bf\|_{H^p} = \|f\|_{H^p}$ for any $f \in H^p$, we have

$$\begin{aligned} \|C_{\varphi|BH^p}\|^p &= \sup_{\|f\|_{H^p}\leqslant 1} \int_{\mathbb{T}} |B\circ\varphi|^p |f\circ\varphi|^p dm \\ &= \sup_{\|f\|_{H^p}\leqslant 1} \int_{\mathbb{T}} |B|^p |f|^p dm_\varphi = \|J_\mu\|^p, \end{aligned}$$

where $\mu = |B|^p dm_\varphi$ and where $J_\mu : H^p \to L^p(\mu)$ is the Carleson embedding. Of course, μ is a Carleson measure since $\mu \leqslant m_\varphi$. Now, Carleson's embedding theorem (see Theorem 5.11) tells us that

$$\|J_\mu\|^p \leqslant K \sup_{\substack{0<h<1 \\ \xi\in\mathbb{T}}} \frac{\mu[S(\xi,h)]}{h},$$

which ends the proof of Lemma 10.32. □

We now give a rough idea for the proof of Proposition 10.31; we will construct a Blaschke product B_0 of length $4N$, with zeroes regularly distributed on the curve $t \mapsto \Lambda_\theta(e^{it})$ and satisfying $|B_0| \leqslant \chi < 1$ on "most of " this curve, with χ as in Lemma 1.25. To accentuate this χ-effect, we raise B_0 at the power N and get a Blaschke product $B = B_0^N$ of length $4N^2$ satisfying $|B| \leqslant \chi^N =: \exp(-bN)$ on "most of" the curve γ, where we set to abbreviate notations

$$\gamma(t) = \Lambda_\theta(e^{it}), \quad \text{for } -\pi \leqslant t \leqslant \pi.$$

We will infer from that the estimate $c_{4N^2+1}(C_{\Lambda_\theta}) \leqslant \chi^N$, which is nearly the claimed result (we will still have to replace c_n by a_n) by a change in the indices.

Now to the details: set once and for all

$$t_k = \pi 2^{-k} \quad \text{and} \quad q_k = \gamma(t_k), \text{ for } 1 \leqslant k \leqslant N.$$

Define the sequence $\bar{q} = (\overline{q_k})_{1\leqslant k\leqslant N}$ and the Blaschke product

$$B_q(z) = \prod_{1\leqslant k\leqslant N} \frac{z - q_k}{1 - \overline{q_k} z}.$$

Define simililarly the Blaschke products $B_{\bar{q}}$, B_{-q}, $B_{-\bar{q}}$ and set

$$B_0 = B_q\, B_{\bar{q}}\, B_{-q}\, B_{-\bar{q}}.$$

Note the double symmetry

$$B_0(\overline{z}) = \overline{B_0(z)}, \quad \text{and} \quad B_0(-z) = B_0(z). \tag{10.45}$$

Now, our final Blaschke product is

$$B = B_0^N.$$

and the smallness of $|B|$ on most of the curve γ is expressed by the following lemma:

Lemma 10.33 *There exists some* $\lambda \in (0,1)$ *such that the following estimate holds:*

$$t_N \leqslant |t| \leqslant \pi - t_N \Longrightarrow |B(\gamma(t))| \leqslant \lambda^N. \tag{10.46}$$

Proof: We can assume $t_N \leqslant t \leqslant \pi/2 = t_1$. For if t belongs respectively to the intervals $[\pi/2, \pi - t_N]$, $-[\pi - t_N, -\pi/2]$, $[-\pi/2, -t_N]$, we will resort respectively to the zeroes of $B_{-\bar{q}}, B_{-q}, B_{\bar{q}}$.

Let k satisfy $t_{k+1} \leqslant t \leqslant t_k$. The assumptions of Lemma 1.25 are satisfied by $a = \gamma(t)$ and $b = \gamma(t_k)$ since, by (10.42) and (10.43),

$$|a - b| \leqslant M|\mathrm{e}^{it} - \mathrm{e}^{it_k}|^\theta \leqslant M|t - t_k|^\theta \leqslant M2^{-k\theta}$$

while, for example,

$$1 - |a| \geqslant c|1 - \mathrm{e}^{it}|^\theta \geqslant 2^\theta c|\sin t/2|^\theta \geqslant c'|t|^\theta \geqslant c''2^{-k\theta}.$$

So that, as claimed, with $\lambda = \dfrac{M}{\sqrt{M^2+1}} < 1$,

$$|B_0(\gamma(t))| \leqslant |B_q(\gamma(t))| \leqslant \rho(a,b) \leqslant \lambda \quad \text{and} \quad |B(\gamma(t))| \leqslant \lambda^N. \qquad \square$$

Next, following Lemma 10.32, we have to estimate

$$\sup_{\substack{0<h<1\\ \xi\in\mathbb{T}}} \frac{1}{h}\int_{S(\xi,h)} |B|^p\, dm_\varphi = \sup_{\substack{0<h<1\\ \xi\in\mathbb{T}}} \frac{1}{h}\int_{|\gamma(t)-\xi|\leqslant h} \big|B[\gamma(t)]\big|^p \frac{dt}{2\pi}.$$

By symmetry, we may assume that $\operatorname{Re}\xi \geqslant 0$ (if $\operatorname{Re}\xi < 0$, we will resort to B_{-q} or $B_{-\bar{q}}$). We may also assume that $h = 2^{-n\theta} \leqslant 1/2$ and that $\mu\big[S(\xi,h)\big] > 0$ (recall that $\mu = |B|^p dm_\varphi$).

If $|\gamma(t) - \xi| \leqslant h$, we have $\operatorname{Re}\gamma(t) \geqslant -1/2$, and we can use the estimates (10.42) and (10.43). The implied ξ satisfy $|\xi - 1| \leqslant (C+1)h$, where the constant C is such that $|1-\gamma(t)| \leqslant C\big(1-|\gamma(t)|\big)$ whenever $\operatorname{Re}\gamma(t) \geqslant -1/2$. Indeed, suppose that $|\gamma(t)-\xi| \leqslant h$. Then

$$|1-\gamma(t)| \leqslant C(1-|\gamma(t)|) \leqslant C|\gamma(t)-\xi| \leqslant Ch$$

$$\text{and } |\xi-1| \leqslant |\xi-\gamma(t)| + |\gamma(t)-1| \leqslant h + Ch = (C+1)h.$$

We henceforth have

$$\int_{|\gamma(t)-\xi|\leqslant h} |B[\gamma(t)]|^p \frac{dt}{2\pi} \leqslant \int_{|\gamma(t)-1|\leqslant (C+2)h} |B(\gamma(t))|^p \frac{dt}{2\pi}$$
$$\leqslant \int_{|t|\leqslant Dt_n} |B(\gamma(t))|^p \frac{dt}{2\pi} =: I_h$$

with some constant $D \geqslant 1$. Indeed, since $|\gamma(t)-1| \geqslant c|t|^\theta$ whenever $\operatorname{Re}\gamma(t) \geqslant -1/2$, the relation $|\gamma(t)-1| \leqslant (C+2)h$ implies $|t| \leqslant Ch^{1/\theta} = \pi^{-1}t_n$. To finish the discussion, we separate two cases.

Case 1. $n \geqslant N$.
We simply majorize $|B|$ by 1 and get, keeping in mind that $2^{\theta-1} < 1$,

$$\frac{1}{h} I_h \leqslant C\frac{1}{h}\int_{|t|\leqslant Dt_n} |B(\gamma(t))|^p \frac{dt}{2\pi} \leqslant C\frac{t_n}{h} \leqslant C2^{(\theta-1)n} \leqslant C2^{(\theta-1)N}.$$

Case 2. $n < N$.
We write

$$\frac{1}{h} I_h = \frac{2}{h}\int_0^{Dt_N} |B(\gamma(t))|^p \frac{dt}{2\pi} + \frac{2}{h}\int_{Dt_N}^{Dt_n} |B(\gamma(t))|^p \frac{dt}{2\pi} =: J_N + K_N.$$

The term J_N is estimated as in case (1), $J_N \leqslant C2^{(\theta-1)N}$. The term K_N is estimated through Lemma 10.33, which gives

$$K_N \leqslant Ct_n 2^{n\theta}\lambda^{pN} \leqslant C\lambda^{pN} \quad \text{since} \quad t_n 2^{n\theta} \leqslant \pi.$$

In view of Lemma 10.32, we obtain (for some $\beta > 0$)

$$\big[c_{4N^2+1}(C_{\Lambda_\theta})\big]^p \leqslant C\left(\max\left\{2^{(\theta-1)N}, \lambda^p\right\}\right)^N =: C\exp(-\beta N). \qquad (10.47)$$

We finally recall the general inequality $a_n(T) \leqslant 2\sqrt{n}\,c_n(T)$ relating approximation and Gelfand numbers of an operator T between Banach spaces (see Theorem G.5). With $n = 4N^2+1$, we get in view of (10.47)

$$\big[a_{4N^2+1}(C_{\Lambda_\theta})\big]^p \leqslant CN\exp(-\beta N),$$

and the factor N is absorbed by the exponential (just change β), which ends the proof of Proposition 10.31 by a change in the indices, and the monotony of the approximation numbers. □

Comparison with the Weighted Hardy

We now give a simple result showing the efficiency of Theorem G.23 in comparing approximation numbers of (weighted) composition operators between weighted (Hilbertian) Hardy spaces. We enunciate here the result of a comparison with the Bergman space, but the same proof gives the case of weighted Hilbert spaces $H^2(\beta)$ with a non-increasing sequence (β_k). The statement makes sense in view of the fact that here $d_n = \sqrt{\frac{n+1}{1}}$ is increasing, so that we will be allowed to use Theorem G.23. Moreover, good (if not optimal) estimates on a and b below are accessible as we will see.

Theorem 10.34 *Let $a, b > 0$ and $\theta \in (0,1)$. Then*

$$a_n(C_{\Lambda_\theta} : H^2 \to H^2) \leqslant a \exp\big(-b\sqrt{n}\big)$$
$$\Longrightarrow a_n(C_{\Lambda_\theta} : \mathscr{B}^2 \to \mathscr{B}^2) \leqslant a \exp\left(-\frac{2b}{3}\sqrt{n}\right).$$

Similarly, for the cusp map χ, we have

$$a_n(C_\chi : H^2 \to H^2) \leqslant a \exp(-b\frac{n}{\log n}) \Longrightarrow a_n(C_\chi : \mathscr{B}^2 \to \mathscr{B}^2)$$
$$\leqslant a(1+\varepsilon_n)\exp\left[-\frac{b}{2}(1-\varepsilon_n)\frac{n}{\log n}\right]$$

where $\varepsilon_n \to 0^+$.

Proof: For short, denote respectively by s_n and t_n the approximation numbers of the lens map Λ_θ on the Hardy and Bergman spaces H^2 and $\mathscr{B}^2$.

We simply write with help of Theorem G.23

$$t_n^n \leqslant t_1 \cdots t_n \leqslant s_1 \cdots s_n \leqslant a^n \exp\big(-b\sum_{j=1}^n \sqrt{j}\big) \leqslant a^n \exp\left(-\frac{2b}{3}n^{3/2}\right)$$

so that

$$t_n \leqslant a \exp\left(-\frac{2b}{3}\sqrt{n}\right),$$

and we are done. We proceed similarly for the cusp map, using

$$\sum_{j=2}^n \frac{j}{\log j} = \frac{n^2}{2\log n} + O\left(\frac{n^2}{\log^2 n}\right).$$

□

10.8 The Cusp Map on Dirichlet Spaces

We focus in this section on an example (see also [104]), which must be compared to the lens map, studied in the previous section. But let us first precise our framework here. We focus here on (Hilbertian) Dirichlet spaces:

Definition 10.35 Let $\alpha > -1$.

We say that $f \in \mathscr{H}(\mathbb{D})$ belongs to the (weighted) Dirichlet spaces $\mathcal{D}^2_\alpha$ if f' belongs to the Bergman space $\mathscr{B}^2_\alpha$. Moreover

$$\|f\|_{\mathcal{D}^2_\alpha} = \left(|f(0)|^2 + \|f'\|^2_{\mathscr{B}^2_\alpha}\right)^{\frac{1}{2}}.$$

For $\alpha = 0$, we simply write $f \in \mathcal{D}$, the classical Dirichlet space.

Equivalently, $f \in \mathscr{H}(\mathbb{D})$ belongs to $\mathcal{D}^2_\alpha$ if and only if

$$\int_{\mathbb{D}} |f'(z)|^2 \, d\sigma_\alpha(z) < \infty.$$

We will also denote

$$\mathcal{D}^2_{\alpha,0} = \{f \in \mathcal{D}^2_\alpha : f(0) = 0\}$$

so that $f \mapsto f' : \mathcal{D}^2_{\alpha,0} \to \mathscr{B}^2_\alpha$ is a bijective isometry, denoted D.

We did not really pay attention to these spaces before, but they are naturally linked to Bergman spaces: for $\alpha > 1$, $\mathcal{D}^2_\alpha = \mathscr{B}^2_{\alpha-2}$. It can be seen in many ways: for instance, with the corresponding Littlewood–Paley formula in Bergman spaces. It can also be seen with the viewpoint of Chapter 12 expliciting the norm in $\mathcal{D}^2_\alpha$ as the $H^2(\beta)$ norm with $\beta_n \approx (n+1)^{1-\alpha}$.

Composition operators on Dirichlet spaces may be unbounded in full generality. Actually, given a symbol φ, the operator $C_\varphi : \mathcal{D}^2_\alpha \to \mathcal{D}^2_\alpha$ is bounded if and only if the weighted composition operator $\varphi' C_\varphi : \mathscr{B}^2_\alpha \to \mathscr{B}^2_\alpha$ is bounded, so that it is easy to characterize concretely boundedness with help of the results of Chapter 5 (we leave it as an exercise, see also Exercise I.25).

We shall pay attention to a special symbol, already introduced in Section 4.3 (see Chapter 4), the cusp map χ. We recall here useful properties: $\chi(0) = 0$ and, when $r \to 1^-$,

$$\chi'(r) \approx \frac{1}{(1-r)\log^2[1/(1-r)]}. \tag{10.48}$$

Since χ is injective, it follows that the composition operator C_χ is bounded on $\mathcal{D}^2_\alpha$ for $\alpha \geqslant 0$. In particular $\chi \in \mathcal{D}^2_\alpha$.

We will assume in the sequel that $0 \leqslant \alpha \leqslant 1$.

Theorem 10.36 *Let χ be the cusp map acting on the Dirichlet space $\mathcal{D}^2_\alpha$. Then, for some positive constants $b' > b > 0$, we have, for all $n \geqslant 1$,*

$$e^{-b'n/\log n} \lesssim a_n(C_\chi) \lesssim e^{-b\alpha n/\log n} \qquad \text{for } 1 \geqslant \alpha > 0,$$

where the underlying constants depend on α only,

and, for some positive constants $b'_0 > b_0 > 0$,

$$e^{-b'_0\sqrt{n}} \lesssim a_n(C_\chi) \lesssim e^{-b_0\sqrt{n}} \qquad \text{for } \alpha = 0. \tag{10.49}$$

We mention here that we have a similar theorem for lens maps Λ_θ with fixed $0 < \theta < 1$. Since both proofs are quite similar, we only detail the first one (cusp case) and refer to [105] for the second one.

Theorem 10.37 *Let Λ_θ be a lens map acting on the Dirichlet space $\mathcal{D}^2_\alpha$. Then, for some positive constants $c' > c > 0$, we have, for all $n \geqslant 1$,*

$$e^{-c'\sqrt{n}} \lesssim a_n(C_{\Lambda_\theta}) \lesssim e^{-c\sqrt{\alpha n}} \qquad \textit{for } 1 \geqslant \alpha > 0,$$

and on $\mathcal{D}^2_0 = \mathcal{D}$, C_{Λ_θ} is bounded, but no longer compact.

Those theorems say the following: the singular numbers of C_χ, roughly speaking, decay like $\exp(-b_\alpha n/\log n)$, with b_α decreasing linearly (in α) to 0 as α decreases from 1 to 0^+. But a drop occurs when $\alpha = 0$ (the ordinary Dirichlet space): those singular numbers now behave as $\exp(-b\sqrt{n})$, like the singular numbers of the lens map Λ_θ on H^2. The associated lens map C_{Λ_θ} is no longer compact on $\mathcal{D}$, while Theorem 10.37 says in a similar way that in turn $a_n(C_{\Lambda_\theta})$ behaves like $\exp(-c\sqrt{n})$ on $\mathcal{D}^2_\alpha$ for $0 < \alpha \leqslant 1$!!

Does this stuff mean something? This motivates a general study of the comparison of the singular numbers of composition operators between two different spaces: see Chapter 12.

The proof of Theorem 10.36 is long and involved, and naturally splits into two parts: the lower bound, then the (much more delicate) upper bound. The proof of this lower bound is based on the study of weighted composition operators acting on $\mathscr{B}^2_\alpha$ (and this works for all $\alpha \geqslant 0$). Since $\chi \in \mathcal{D}^2_\alpha$, we have $\chi' \in \mathscr{B}^2_\alpha$, and this χ' will be our weight on $\mathscr{B}^2_\alpha$.

Proof of Theorem 10.36: We begin with the lower bound.

We choose $0 < u_j < 1$ so as to get via (10.48), with $v_j = \chi(u_j)$ and $\varepsilon > 0$ to be adjusted later,

$$1 - v_j = \mathrm{e}^{-j\varepsilon}. \tag{10.50}$$

Since $\chi(0) = 0$ and $\chi(1) = 1$, this is possible by the intermediate value theorem. Hence, from (4.9) (see Lemma 4.3),

$$\log\left(\frac{1}{1-u_j}\right) \approx \mathrm{e}^{j\varepsilon}. \tag{10.51}$$

The interpolation constant I_v of the sequence $v = (v_j)_j$ satisfies $I_v \lesssim 1/\delta_v^2$, where δ_v is its Carleson constant. Since $\frac{1-v_{j+1}}{1-v_j} = \mathrm{e}^{-\varepsilon}$, we know that, for some positive constants c_1, c_2,

$$\delta_v \geqslant \prod_{j=1}^{\infty}\left(\frac{1-\mathrm{e}^{-j\varepsilon}}{1+\mathrm{e}^{-j\varepsilon}}\right)^2 \geqslant \exp\left(-\frac{c_1}{1-\mathrm{e}^{-\varepsilon}}\right) \geqslant \mathrm{e}^{-c_2/\varepsilon};$$

hence, with $c_0 = 2\,c_2$,

$$I_v \lesssim \mathrm{e}^{c_0/\varepsilon}.$$

The reproducing kernel K_a of the Bergman space $\mathscr{B}^2_\alpha$ satisfies

$$\|K_a\|_{\mathscr{B}^2_\alpha} = \frac{1}{(1-|a|^2)^{1+\alpha/2}} \approx \frac{1}{(1-|a|)^{1+\alpha/2}}\,.$$

Using (10.48) and (10.50), we get, for $1 \leqslant j \leqslant n$ (and with some $\delta_0 > 0$),

$$\begin{aligned}
|\chi'(u_j)|\,\frac{\|K_{v_j}\|_{\mathscr{B}^2_\alpha}}{\|K_{u_j}\|_{\mathscr{B}^2_\alpha}} &\gtrsim \frac{1}{(1-u_j)\,\log^2[1/(1-u_j)]}\,\frac{(1-u_j)^{1+\alpha/2}}{(1-v_j)^{1+\alpha/2}}\\
&= \frac{1}{\log^2[1/(1-u_j)]}\,\frac{(1-u_j)^{\alpha/2}}{(1-v_j)^{1+\alpha/2}}\\
&\gtrsim \frac{1}{\mathrm{e}^{2j\varepsilon}}\,\frac{\exp(-\frac{\alpha\delta_0}{2}\,\mathrm{e}^{j\varepsilon})}{\mathrm{e}^{-j\varepsilon(1+\alpha/2)}} = \exp\Big(-\Big[\frac{\alpha\delta_0}{2}\,\mathrm{e}^{j\varepsilon} + j\varepsilon\Big(1-\frac{\alpha}{2}\Big)\Big]\Big)\\
&\gtrsim \exp\Big(-\Big[\frac{\alpha\delta_0}{2}\,\mathrm{e}^{n\varepsilon} + n\varepsilon\Big(1-\frac{\alpha}{2}\Big)\Big]\Big),
\end{aligned}$$

since the function $t \mapsto \frac{\alpha\delta_0}{2}\,\mathrm{e}^t + t\left(1-\frac{\alpha}{2}\right)$ is increasing for large t (which is fulfilled in the end with our choices of ε).

Lemma 10.2 with $w = \chi'$ gives, by definition of the implied norms, since $0 < \alpha \leqslant 1$ and $\chi(0) = 0$,

$$\begin{aligned}
a_n(C_\chi : \mathcal{D}^2_{\alpha,0} \to \mathcal{D}^2_{\alpha,0}) &\geqslant a_n(M_{\chi'}C_\chi : \mathscr{B}^2_\alpha \to \mathscr{B}^2_\alpha)\\
&\gtrsim \exp\Big(-\Big[\frac{\alpha\delta_0}{2}\,\mathrm{e}^{n\varepsilon} + n\varepsilon + \frac{2\,c_0}{\varepsilon}\Big]\Big). \qquad (10.52)
\end{aligned}$$

Indeed, if $R : \mathcal{D}^2_{\alpha,0} \to \mathcal{D}^2_{\alpha,0}$, with rank $< n$, approaches C_χ on $\mathcal{D}^2_{\alpha,0}$, the operator $S = DRD^{-1} : \mathscr{B}^2_\alpha \to \mathscr{B}^2_\alpha$, with the same rank, approaches $M_{\chi'}C_\chi$ with the same precision. We now distinguish two cases:

Case $\alpha = 0$. In this case, we have

$$a_n(M_{\chi'}C_\chi : \mathscr{B}^2_\alpha \to \mathscr{B}^2_\alpha) \gtrsim \exp\Big(-\Big[n\varepsilon + \frac{2\,c_0}{\varepsilon}\Big]\Big).$$

Taking $\varepsilon = 1/\sqrt{n}$, we get $a_n(C_\chi : \mathcal{D}^2_{\alpha,0} \to \mathcal{D}^2_{\alpha,0}) \gtrsim \mathrm{e}^{-c\sqrt{n}}$, for some positive absolute constant c.

Case $\alpha > 0$. We take $\varepsilon = \frac{1}{n}\log\left(\frac{n}{\log n}\right)$ and we get

$$\begin{aligned}
&a_n(M_{\chi'}C_\chi : \mathscr{B}^2_\alpha \to \mathscr{B}^2_\alpha)\\
&\quad\gtrsim \exp\Big(-\Big[\frac{\alpha\delta_0}{2}\,\frac{n}{\log n} + \log\Big(\frac{n}{\log n}\Big) + 2\,c_0\frac{n}{\log n - \log(\log n)}\Big]\Big).
\end{aligned}$$

Since $\log\left(\frac{n}{\log n}\right) + 2\,c_0 \frac{n}{\log n - \log(\log n)} = O\left(\frac{n}{\log n}\right)$, this gives, since $\alpha \leqslant 1$,

$$a_n(C_\chi : \mathcal{D}^2_{\alpha,0} \to \mathcal{D}^2_{\alpha,0}) \gtrsim \exp(-b'n/\log n),$$

with b' some positive absolute constant.

Proof of the rather more delicate upper bound.

First we make some heuristic comments, also valid for the case $\alpha = 0$, provided that we define the counting function $N_{\chi,0} =: n_\chi$ by the formula $n_\chi(w) = \sum_{\varphi(z)=w} 1$, where the preimages z of w are counted with their multiplicity. The strategy for obtaining a sharp upper bound is the following:

(i) One approximates in operator norm C_χ by $C_\chi S_N$ where N is a large integer and S_N the Nth partial sum operator. This infers an upper bound on $a_n(C_\chi)$ through the obvious relation (special case of $a_{p+q-1}(S+T) \leqslant a_p(S) + a_q(T)$)

$$a_n(C_\chi) \leqslant \|C_\chi - C_\chi S_N\| + a_n(C_\chi S_N).$$

Now, this approximation is made possible by good estimates on the norms $\|\chi^n\|_{\mathcal{D}^2_\alpha}$, themselves possible thanks to estimates of the weighted counting function $N_{\chi,\alpha}$.

(ii) One shows that $C_\chi S_N$ is kind of a contraction of the Carleson embedding $J : H^2 \to L^2(\mu)$, with $\mu = N_{\chi,\alpha} d\sigma$. More precisely, $C_\chi S_N = T_N J \Delta_N$ with $T_N : L^2(\mu) \to \mathcal{D}^2_{\alpha,0}$ a contraction, and where $\Delta_N = DS_N : \mathcal{D}^2_{\alpha,0} \to H^2$. This will imply

$$a_n(C_\chi S_N) \leqslant N^{\frac{\alpha+1}{2}} a_n(J).$$

(iii) One bounds $a_n(J)$ through Carleson's embedding theorem for H^2 (this is the most delicate part).

(iv) One finally adjusts N to get the result.

We begin with the case $\alpha = 0$. We need some lemmas.

Lemma 10.38 *We have for any $n \in \mathbb{N}$*

$$\|\chi^n\|_{\mathcal{D}} = O\left(\frac{\log(n)^{\frac{3}{2}}}{\sqrt{n}}\right). \tag{10.53}$$

Proof: Since χ is univalent and $\chi(0) = 0$, we have for every $0 < h < 1$, by a change of variable,

$$\begin{aligned}\|\chi^n\|^2_{\mathcal{D}} &= \int_{\mathbb{D}} n^2|w|^{2n-2} n_\chi(w)\,d\sigma(w)\\ &\leqslant n^2(1-h)^{2n-2} + n^2 \int_{|w|\geqslant 1-h} n_\chi(w)\,d\sigma(w).\end{aligned}$$

But the latter integral is the area of $\chi(\mathbb{D}) \cap \{|w| \geqslant 1-h\}$, which is $\approx h^3$ since $\chi(\mathbb{D})$ is delimited at the cuspidal point 1 by two circular arcs. Hence

$$\|\chi^n\|_{\mathcal{D}} \lesssim n[\mathrm{e}^{-nh} + h^{3/2}],$$

and the choice $h = \dfrac{2\log n}{n}$ gives the result. □

An immediate corollary is the following:

Lemma 10.39 *We have, for any $\delta \in (0, 1/2)$,*

$$\|C_\chi - C_\chi S_N\|_{B(\mathcal{D})} = O\left(\frac{\log(N)^{\frac{3}{2}}}{\sqrt{N}}\right) = o\left(N^{-\delta}\right). \tag{10.54}$$

Proof: Just use the Hilbert–Schmidt norm in $\mathcal{D}$ to get

$$\|C_\chi - C_\chi S_N\|^2_{B(\mathcal{D})} \leqslant \|C_\chi - C_\chi S_N\|^2_{HS} = \sum_{n>N} \frac{\|\chi^n\|^2_{\mathcal{D}}}{n} \leqslant \sum_{n>N} \frac{\log(n)^3}{n^2}$$

using the previous lemma, and the conclusion follows. □

The end of the proof of Theorem 10.36 now exactly follows (with fewer technical difficulties) the pattern of the forthcoming case $\alpha > 0$, except that here we use the measure $\mu = n_\chi d\sigma$ instead of the forthcoming measure $\mu = N_{\chi,\alpha} d\sigma$. Hence we just enunciate the final necessary statement in the case $\alpha = 0$: for some numerical constant $\varepsilon > 0$, we have

$$a_n(C_\chi S_N) \lesssim \sqrt{N} a_n(J) \quad \text{with} \quad a_n(J) = c_n(J) \lesssim \|J_{|E}\| \leqslant \mathrm{e}^{-\varepsilon\sqrt{n}}$$

where $E = BH^2$ and B is some Blaschke product of length $< n$.

Therefore, from Lemma 10.39, we get

$$a_n(C_\chi) \leqslant \|C_\chi - C_\chi S_N\| + a_n(C_\chi S_N) \lesssim N^{-\frac{1}{4}} + \sqrt{N}\mathrm{e}^{-\varepsilon\sqrt{n}}.$$

By adjusting $N = [\mathrm{e}^{\varepsilon\sqrt{n}}]$, we get the desired upper bound: $a_n(C_\chi) \lesssim \mathrm{e}^{-\frac{\varepsilon}{4}\sqrt{n}}$.

For $\alpha > 0$, we will follow the same strategy, but with the weighted Nevanlinna counting function $N_{\chi,\alpha}$ instead of the counting one n_χ. Here, the weighted counting function $N_{\varphi,\alpha}$ of the analytic function $\varphi\colon \mathbb{D} \to \mathbb{D}$ is chosen as

$$N_{\varphi,\alpha}(w) = \sum_{\varphi(z)=w} \left(1 - |z|^2\right)^\alpha.$$

Step 1. Since the cusp map χ is univalent, we have

$$N_{\chi,\alpha}(w) = \begin{cases} (1 - |\chi^{-1}(w)|^2)^\alpha & \text{for } w \in \chi(\mathbb{D}), \\ 0 & \text{otherwise.} \end{cases} \tag{10.55}$$

The Schwarz lemma already gives $N_{\chi,\alpha}(w) \leqslant (1-|w|^2)^\alpha$, but thanks to Lemma 4.4 , we have a better estimate:

Since $1-|\chi(z)| \gtrsim \dfrac{1}{\log[1+1/|1-z|]}$ for all $z \in \mathbb{D}$, we have, for some positive constant c_0,

$$1-|\chi^{-1}(w)| \leqslant |1-\chi^{-1}(w)| \lesssim \exp\left(-\frac{c_0}{1-|w|}\right) \quad \text{for all } w \in \chi(\mathbb{D}).$$

As a consequence,

$$N_{\chi,\alpha}(w) \lesssim \exp\left(-\frac{c_0\alpha}{1-|w|}\right), \quad \text{for } w \in \chi(\mathbb{D}), \tag{10.56}$$

and this allows us to get the following intermediate estimate.

Lemma 10.40 *We have, for some numerical constant* c_0,

$$\|\chi^n\|^2_{\mathcal{D}^2_\alpha} \lesssim e^{-\sqrt{\alpha c_0 n}}.$$

Proof: Using the change of variable formula, (10.56), and $\chi(0)=0$, we have

$$\begin{aligned}
\|\chi^n\|^2_{\mathcal{D}^2_\alpha} &= \int_{\mathbb{D}} |n\chi^{n-1}(z)\,\chi'(z)|^2\,d\sigma_\alpha(z)\\
&= \int_{\chi(\mathbb{D})} n^2|w|^{2n-2}N_{\chi,\alpha}(w)\,d\sigma(w)\\
&\leqslant n^2(1-h)^{2n-2} + n^2\int_{\chi(\mathbb{D})\cap\{|w|\geqslant 1-h\}} \exp(-c_0\alpha/(1-|w|))\,d\sigma(w)\\
&\lesssim n^2 \mathrm{e}^{-2nh} + n^2\mathrm{e}^{-c_0\alpha/h}.
\end{aligned}$$

Choosing $h=\sqrt{c_0\alpha/2n}$ gives $\|\chi^n\|^2_{\mathcal{D}^2_\alpha} \lesssim \mathrm{e}^{-\sqrt{\alpha c_0 n}}$. □

As a consequence of Lemma 10.40, we have the following majorization.

Lemma 10.41 *We have*

$$\|C_\chi - C_\chi S_N\|_{B(\mathcal{D}^2_\alpha)} \lesssim e^{-\sqrt{\alpha c_0 N}}. \tag{10.57}$$

Proof: It suffices to use again the Hilbert–Schmidt norm and Lemma 10.40, remembering that $\frac{z^n}{(n+1)^{(1-\alpha)/2}}$ is (up to constants) an orthonormal basis of $\mathcal{D}^2_\alpha$; we get, provided that we change c_0,

$$\begin{aligned}
\|C_\chi - C_\chi S_N\|^2_{B(\mathcal{D}^2_\alpha)} &\leqslant \|C_\chi - C_\chi S_N\|^2_{HS} \approx \sum_{n>N} \frac{\|\chi^n\|^2_{\mathcal{D}^2_\alpha}}{n^{1-\alpha}} \lesssim \sum_{n>N}\frac{1}{n^{1-\alpha}}\mathrm{e}^{-\sqrt{\alpha c_0 n}}\\
&\lesssim \mathrm{e}^{-\sqrt{\alpha c_0 N}}.
\end{aligned}$$

□

Step 2. We now compare $C_\chi S_N \in B(\mathcal{D}^2_\alpha)$ and the Carleson embedding J.

Lemma 10.42 *Let J be the Carleson embedding $J\colon H^2 \to L^2(\mu)$ with $d\mu = N_{\chi,\alpha}\, d\sigma$. Then for all integers $n \geqslant 1$,*

$$a_n(C_\chi S_N) \lesssim N^{\frac{1+\alpha}{2}} a_n(J).$$

Proof: Let $f \in \mathcal{D}^2_{\alpha,0}$ and write $f(z) = \sum_{j=1}^{\infty} c_j z^j$; we have

$$\|C_\chi S_N f\|^2_{\mathcal{D}^2_\alpha} = \int_{\mathbb{D}} \left| \sum_{j=1}^{N} j c_j w^{j-1} \right|^2 N_{\chi,\alpha}(w)\, d\sigma(w) = \int_{\mathbb{D}} |(\Delta_N f)(w)|^2 d\mu(w),$$

where $\Delta_N\colon \mathcal{D}^2_{\alpha,0} \to H^2$ is defined by

$$(\Delta_N f)(w) = \sum_{j=1}^{N} j c_j w^{j-1}.$$

We hence have $\|C_\chi S_N f\|_{\mathcal{D}^2_\alpha} = \|J \Delta_N f\|_{L^2(\mu)}$ for all $f \in \mathcal{D}^2_{\alpha,0}$. From Douglas lemma (see Lemma G.6), there exists a contraction $T_N\colon L^2(\mu) \to \mathcal{D}^2_{\alpha,0}$ such that

$$C_\chi S_N = T_N J \Delta_N.$$

Now

$$\|\Delta_N f\|^2_{H^2} = \sum_{j=1}^{N} j^{1+\alpha} j^{1-\alpha} |c_j|^2 \leqslant N^{1+\alpha} \sum_{j=1}^{N} j^{1-\alpha} |c_j|^2 \lesssim N^{1+\alpha} \|f\|^2_{\mathcal{D}^2_\alpha}$$

hence, by the ideal property of approximation numbers,

$$a_n(C_\chi S_N) \approx a_n\big((C_\chi S_N)_{|\mathcal{D}^2_{\alpha,0}}\big) \leqslant \|T_N\|\, \|\Delta_N\|\, a_n(J) \lesssim N^{\frac{\alpha+1}{2}} a_n(J). \qquad \square$$

Step 3. We now exploit Carleson's embedding theorem.

Proposition 10.43 *Let J be the Carleson embedding $J : H^2 \to L^2(\mu)$ with $d\mu = N_{\chi,\alpha}\, d\sigma$, where $\alpha > 0$. Then, for some absolute positive constant c and all integers $n \geqslant 2$,*

$$a_n(J) \lesssim \exp\left(-c\alpha \frac{n}{\log n}\right). \tag{10.58}$$

Proof: Let r be the greatest integer such that $2^r < n$ and let B_0 be the Blaschke product with simple zeros at the points

$$z_j = 1 - 2^{-j}, \qquad \text{where } 1 \leqslant j \leqslant r.$$

We consider the Blaschke product $B = B_0^n$.

Let $E = BH^2$, a subspace of H^2 of codimension nr.

The Carleson embedding theorem for H^2 gives us, since B is inner,

$$\| J_{|E} \|^2 \lesssim \sup_{\substack{0<h<1 \\ \xi \in \mathbb{T}}} \frac{1}{h} \int_{S(\xi,h)\cap\Omega} |B|^2 N_{\chi,\alpha}\, d\sigma\,, \tag{10.59}$$

where $S(\xi,h) = \{z \in \mathbb{D};\ |z-\xi| < h\}$ and $\Omega = \chi(\mathbb{D})$.

Recall (cf. proof of Lemma 10.38) that $\sigma[S(\xi,h)\cap\Omega] \lesssim h^3$.

Now we majorize the RHS of (10.59). For that, we first note that, since Ω is contained in an angular sector, there is an absolute positive constant $\delta_0 \in (0,1)$ such that $1-|w| \geqslant \delta_0|1-w|$ for all $w \in \Omega$. Hence if $w \in S(\xi,h)\cap\Omega$, we have

$$\delta_0\,|1-w| \leqslant 1-|w| \leqslant |\xi - w| < h\,,$$

and $w \in S(1, h/\delta_0)$. This gives $S(\xi,h)\cap\Omega \subseteq S(1,h/\delta_0)\cap\Omega$.

Moreover, we may assume that $h = \delta_0 2^{-l}$ and we separate two cases.

• $l \geqslant r$.

We simply majorize $|B|$ by 1, and (10.56) leads to

$$\begin{aligned}\frac{1}{h}\int_{S(\xi,h)\cap\Omega} |B|^2 N_{\chi,\alpha}\, d\sigma &\lesssim \frac{1}{h}\int_{S(\xi,h)\cap\Omega} \exp(-\alpha\, c_0/h)\, d\sigma \\ &\lesssim \frac{1}{h}\,\mathrm{e}^{-\alpha c_0/h}\,\sigma[S(\xi,h)\cap\Omega] \\ &\approx h^2\,\mathrm{e}^{-\alpha c_0/h} \\ &\leqslant \mathrm{e}^{-\alpha\,(c_0/\delta_0)\,2^l} \\ &\leqslant \mathrm{e}^{-\alpha\,(c_0/\delta_0)\,2^r} \\ &\leqslant \mathrm{e}^{-\alpha\,(c_0/2\delta_0)n}\,.\end{aligned}$$

• $l < r$.

We write

$$\begin{aligned}\int_{S(\xi,h)\cap\Omega} |B|^2 N_{\chi,\alpha}\, d\sigma &\leqslant \int_{S(1,h/\delta_0)\cap\Omega} |B|^2 N_{\chi,\alpha}\, d\sigma \\ &= \int_{S(1,2^{-r})\cap\Omega} |B|^2 N_{\chi,\alpha}\, d\sigma \\ &\quad + \int_{\{w\in\Omega;\ 2^{-r}\leqslant|w-1|<2^{-l}\}} |B|^2 N_{\chi,\alpha}\, d\sigma\,.\end{aligned}$$

By the previous case, the first integral in the RHS, divided by h, is $O\left(\mathrm{e}^{-\alpha\,(c_0/2\delta_0)n}\right)$.

Now, if $2^{-r} \leqslant |w-1| < 2^{-l}$, there is some $j \in \{l+1, \dots, r\}$ such that $2^{-j} \leqslant |w-1| < 2^{-j+1}$. Then

$$|w - z_j| \leqslant |w-1| + |1 - z_j| \leqslant 2^{-j+1} + 2^{-j} = 3\,.\,2^{-j}\,;$$

hence, since $2^{-j} \leqslant (1/\delta_0)\,(1-|w|)$ and $1-|z_j| = 1 - z_j = 2^{-j}$, we have, with $M = 3/\delta_0$: $|w - z_j| \leqslant M \min(1-|w|, 1-|z_j|)$. Therefore, Lemma 1.25 shows that

$$\rho(w, z_j) \leqslant \kappa := \frac{M}{\sqrt{M^2+1}} < 1,$$

and the modulus of the jth factor of $B_0(w)$ is less than or equal to κ. It follows that $|B(w)| = |B_0(w)|^n \leqslant \kappa^n$. Using $N_{\chi,\alpha}(w) \leqslant 1$, we obtain

$$\frac{1}{h}\int_{\{w\in\Omega\,;\;2^{-r}\leqslant|w-1|<2^{-l}\}} |B|^2 N_{\chi,\alpha}\,d\sigma \lesssim \frac{1}{h}\,\sigma[S(\xi,h)\cap\Omega]\,\kappa^{2n} \lesssim \kappa^{2n} = \mathrm{e}^{-\varepsilon_0 n}$$

for some absolute constant $\varepsilon_0 > 0$.

We get

$$\frac{1}{h}\int_{S(\xi,h)\cap\Omega} |B|^2 N_{\chi,\alpha}\,d\sigma \lesssim \mathrm{e}^{-\alpha\,(c_0/2\delta_0)n} + \mathrm{e}^{-\varepsilon_0 n} \lesssim \mathrm{e}^{-c_1\alpha\,n}\,, \tag{10.60}$$

where $c_1 > 0$ does not depend on α.

In either case, we obtain

$$\|J_{|E}\| \lesssim \mathrm{e}^{-c_2\alpha n}\,.$$

That means that the Gelfand number $c_{n\,r}(J) = O\left(\mathrm{e}^{-c_2\alpha\,n}\right)$. Since Gelfand numbers coincide with approximation numbers on Hilbert spaces, we get

$$a_{nr}(J) \lesssim \mathrm{e}^{-c_2\alpha n},$$

or, since $2^r < n \leqslant 2^{r+1}$, making a change of variables in the indices

$$a_n(J) \lesssim \exp\left(-c_3\alpha\frac{n}{\log n}\right),$$

as claimed in Proposition 10.43. □

Step 4. End of the proof of Theorem 10.36. We shall choose N. We already saw that

$$a_n(C_\chi) \leqslant \|C_\chi - C_\chi S_N\| + a_n(C_\chi S_N)\,.$$

Using Lemma 10.41, Lemma 10.42, and Proposition 10.43, we hence obtain

$$a_n(C_\chi) \lesssim \mathrm{e}^{-\sqrt{\alpha c_0 N}} + N^{\frac{\alpha+1}{2}} \exp\left(-c_3\,\alpha \frac{n}{\log n}\right) \lesssim N^{-1} + N \exp\left(-c_3\,\alpha \frac{n}{\log n}\right)$$

with underlying constants depending only on α.

Choosing N as the integral part of $\exp\left(\frac{c_3\,\alpha}{2} \cdot \frac{n}{\log n}\right)$, we get

$$a_n(C_\chi) \lesssim \exp\left(-\frac{c_3\,\alpha}{2} \cdot \frac{n}{\log n}\right).$$

This ends the proof of Theorem 10.36. □

11
Composition Operators on Hardy–Orlicz and Bergman–Orlicz Spaces

In this chapter, we extend to the Orlicz spaces framework the results presented in the preceding chapters about composition operators on Hardy spaces H^p (where $p \in [1, +\infty)$) and on Bergman spaces $\mathfrak{B}^p$ for $p \geqslant 1$. The results mainly come from [98, 100, 102], where the framework is a bit more general than the viewpoint we adopt in the present chapter: for instance there are some results for $C_\varphi : H^{\psi_1} \to H^{\psi_2}$, but we choose to focus here on $\psi_1 = \psi_2$ (one may consult these papers, as well as [24, 25, 27] for extensions in higher dimension). We refer the reader to *Appendix H* for the reminders about classical Orlicz spaces.

An important original motivation comes from the following remark about compactness of composition operators:

Given a symbol φ, we have, for $r > p > 1$,

$$H^\infty \subset H^r \subset H^p \subset H^1$$

with strict inclusion, but the characterization of the compactness of C_φ is the same on H^p for every finite p (and in particular, touching the boundary is allowed), whereas this characterization becomes "suddenly" very restrictive for H^∞. Therefore, one can ask whether there is an intermediate scale of characterizations corresponding to an intermediate scale of Hardy-type spaces.

In particular, for a sufficiently fast-growing Orlicz function Ψ, we have (see Section 11.1 for definitions)

$$H^\infty \subset H^\Psi \subset H^p \subset H^1,$$

and one can expect a new zoology of behaviors according to the growth of Ψ. Do we get something looking "like H^p" or "like H^∞"?

Of course the same questions occur with the viewpoint of Bergman spaces.

In the whole chapter, Ψ is an Orlicz function and we shall always require that Ψ *satisfies the* ∇_2*-condition* (as all the most common examples of Orlicz functions do), so that L^Ψ is the dual of $M^\Phi = L^\Phi$, where Φ is the conjugate function of Ψ, and L^Φ is itself the dual space of M^ψ. Actually, if we had a general Orlicz function, we could consider the function $x \mapsto \Psi(x^2)$ that satisfies ∇_2, and it sometimes allows to reduce to the case ∇_2: see [98, p. 45] for details.

We recall (see Appendix H) that $\Psi : [0, +\infty) \to [0, +\infty)$ is convex, increasing, satisfies $\Psi(0) = 0$ (hence continuous) and that $\Psi^{-1}: \mathbb{R}^+ \to [0,+\infty)$ is well defined. Moreover, to avoid pathologies and because, in the spirit of the book, we are interested in a better understanding of the difference between the behavior of composition operators in an "L^p framework" and the one on the space H^∞, we assume that we work with an Orlicz function Ψ having the following additional properties: *strictly convex (hence increasing) and such that*

$$\frac{\Psi(x)}{x} \underset{x\to+\infty}{\longrightarrow} +\infty.$$

Even though it would be straightforward to extend some (many) of the results of this chapter to more general Orlicz functions, it will be sufficient for our purpose and applications to many interesting non-trivial situations.

Let us first clarify our new framework and we present the construction of Orlicz spaces of analytic functions. This presentation is mainly twofold: first the Hardy–Orlicz spaces and secondly the Bergman–Orlicz spaces. But we also mention some results to compare them in the end.

11.1 Orlicz Spaces of Analytic Functions

Hardy–Orlicz Spaces

In Chapter 1, we introduced H^p spaces enlightening two viewpoints: the unit disk, introducing H^p as the space of analytic functions $f : \mathbb{D} \to \mathbb{C}$ for which the dilated functions $f_r(z) = f(rz)$ satisfy $\|f\|_{H^p} = \sup_{r\in(0,1)} \|f_r\|_p < \infty$. Then the Fatou–Riesz Theorem asserts that the boundary limits exist almost everywhere. On the other hand, the circle point of view allows to see H^p as the space of functions $g \in L^p(\mathbb{T})$ whose Fourier coefficients $\widehat{g}(n)$ vanish when $n < 0$, and then extended to the unit disk thanks to the Poisson transform. See Section 1.11.4 for details.

Hardy–Orlicz spaces H^Ψ are defined in a similar way. Nevertheless, the lack of homogeneity of the function Ψ and the presence of the two distinct (in general) spaces M^Ψ and L^Ψ invite caution, and we shall give some details. We recall that $M^\Psi(\mathbb{T}) = \overline{C(\mathbb{T})}^{\|.\|_\Psi}$.

Definition 11.1 • The *Hardy–Orlicz* space H^Ψ associated to Ψ is the space of analytic functions $f : \mathbb{D} \to \mathbb{C}$ such that one of the equivalent following conditions is fulfilled:

(i) $\sup_{r\in(0,1)} \|f_r\|_{L^\Psi(\mathbb{T})} < +\infty$, where $f_r(z) = f(rz)$ for $z \in \overline{\mathbb{D}}$ and $r \in (0,1)$.

(ii) There exists $f^* \in L^\Psi(\mathbb{T})$ such that $\widehat{f^*}(n) = 0$ for $n < 0$ and for which $f(z) = \sum_{n\geqslant 0} \widehat{f^*}(n)\, z^n$, where $z \in \mathbb{D}$.

When these conditions are satisfied, one has $\|f^*\|_{L^\Psi(\mathbb{T})} = \sup_{r\in(0,1)} \|f_r\|_{L^\Psi(\mathbb{T})}$.

In other words, we have

$$H^\Psi = \left\{ f \in H^1 :\ f^* \in L^\Psi(\mathbb{T}) \right\}. \tag{11.1}$$

The norm of f is then defined by $\|f\|_{H^\Psi} = \left\|f^*\right\|_\Psi = \sup_{r\in(0,1)} \|f_r\|_{L^\Psi(\mathbb{T})}$.

• The *Hardy–Morse–Transue space* HM^Ψ is the subspace of H^Ψ defined by

$$HM^\Psi = \left\{ f \in H^\Psi :\ f^* \in M^\Psi(\mathbb{T}) \right\}.$$

We mention that the definition of H^Ψ is not exactly the same as the one given in [148, s 9.1].

We need to justify that conditions (i) and (ii) are indeed equivalent.

Proof of the equivalence: Assume that (i) is satisfied. Since $\|f_r\|_1 \leqslant C\|f_r\|_{L^\Psi(\mathbb{T})}$ for some $C > 0$, one has $f \in H^1$, and hence, by Fatou–Riesz theorem, f has almost everywhere a boundary limit $f^* \in L^1(\mathbb{T})$.

By definition of the Orlicz norm, we have

$$\forall r \in (0,1), \qquad \int_{\mathbb{T}} \Psi\left(\frac{|f_r|}{C}\right) dm \leqslant 1,$$

where $C = \sup_{r\in(0,1)} \|f_r\|_{L^\Psi(\mathbb{T})}$. Hence, by Fatou's lemma, we get

$$\int_{\mathbb{T}} \Psi\left(\frac{|f^*|}{C}\right) dm \leqslant 1.$$

In other words, $f^* \in L^\Psi$ and $\|f^*\|_{L^\Psi(\mathbb{T})} \leqslant C$.

Assume now that (ii) is satisfied. In particular $f^* \in L^1(m)$; hence $f \in H^1$ and $f^* = \lim_{r\to 1} f_r$ a.e.

One has $f_r = f^* * P_r$ (recall P_a is the Poisson kernel at point $a \in \mathbb{D}$).

Thanks to the Jensen convex inequality applied with the probability measure $P_{r\xi}(z)\,dm(z)$, we have, with $N = \|f^*\|_\Psi$,

$$\begin{aligned}\int_{\mathbb{T}} \Psi\Big(\frac{|(f^* * P_r)(\xi)|}{N}\Big)\,dm(\xi) &\leqslant \int_{\mathbb{T}} \Psi\Big(\int_{\mathbb{T}} \frac{|f^*(z)|}{N} P_{r\xi}(z)\,dm(z)\Big)\,dm(\xi)\\ &\leqslant \int_{\mathbb{T}}\Big(\int_{\mathbb{T}} \Psi\Big(\frac{|f^*(z)|}{N}\Big)\,P_{r\xi}(z)\,dm(z)\Big)\,dm(\xi)\\ &= \int_{\mathbb{T}}\Big(\int_{\mathbb{T}} P_{rz}(\xi)\,dm(\xi)\Big)\,\Psi\Big(\frac{|f^*(z)|}{N}\Big)\,dm(z)\\ &= \int_{\mathbb{T}} \Psi\Big(\frac{|f^*(z)|}{N}\Big)\,dm(z) \leqslant 1.\end{aligned}$$

using Fubini so that, for every $r \in (0,1)$,

$$\|f_r\|_{L^\Psi(\mathbb{T})} \leqslant N = \|f^*\|_{L^\Psi(\mathbb{T})}\,. \tag{11.2}$$

Therefore we have (i), and $\|f\|_{H^\Psi} \leqslant \|f^*\|_{L^\Psi(\mathbb{T})}$.

Gathering everything, we actually proved that $\|f\|_{H^\Psi} = \|f^*\|_{L^\Psi(\mathbb{T})}$. □

In the sequel, we shall make no distinction between f and f^*, unless there may be some ambiguity, and most of the time we shall write f instead of f^* for the boundary limit.

In particular, H^Ψ can be viewed as a subspace of $L^\Psi(\mathbb{T})$ and HM^Ψ can be written as $H^\Psi \cap M^\Psi(\mathbb{T})$. Therefore, these two spaces are closed subspaces of $L^\Psi(\mathbb{T})$ and we get that H^Ψ and HM^Ψ are Banach spaces.

We already mentioned that $L^\Psi(\mathbb{T})$ is a dual space; moreover

$$L^\Psi(\mathbb{T}) = \big(L^\Phi(\mathbb{T})\big)^* = \big(M^\Psi(\mathbb{T})\big)^{**}.$$

Therefore, L^Ψ is equipped with the weak-star topology in a natural way.

Corollary 11.2 *H^Ψ is weak star closed in $L^\Psi = (M^\Phi)^*$.*

When Ψ satisfies ∇_2, it is isometrically isomorphic to the bidual of HM^Ψ.

Proof: The weak-star closed character of H^Ψ is clear using (ii) in Definition 11.1 since $\bar{z}^n \in M^\Psi(\mathbb{T})$ for every $n > 1$.

Since $\Phi \in \Delta_2$, we have $M^\Phi(\mathbb{T}) = L^\Phi(\mathbb{T})$. It is clear that $(HM^\Psi)^\perp$ (viewed in the predual $L^\Phi(\mathbb{T})$ of $L^\Psi(\mathbb{T})$) is the closed subspace of L^Φ generated by $\bar{z}^n$ with $n \geqslant 1$. As $H^\Psi \subset L^\Psi = (M^\Phi)^* = (L^\Phi)^*$ is the orthogonal of this latter subspace, we have $(HM^\Psi)^{\perp\perp} = H^\Psi$. □

We recall our notation $f_r(z) = f(rz)$ for $r \in (0,1)$ and $z \in \overline{\mathbb{D}}$ where f is analytic on $\mathbb{D}$.

Proposition 11.3 *For every $f \in HM^\Psi$, one has $\|f_r - f\|_{H^\Psi} \underset{r\to 1}{\longrightarrow} 0$.*

In particular, analytic polynomials are dense in HM^Ψ.

In other words, $HM^\Psi = \overline{A(\mathbb{D})}^{L^\Psi}$.

Proof: Here $\|\cdot\|_\Psi$ means $\|\cdot\|_{L^\Psi(\mathbb{T})}$. Let $f \in HM^\Psi$ and $\varepsilon > 0$. Since $M^\Psi = \overline{C(\mathbb{T})}^{L^\Psi}$, there exists a continuous function h on $\mathbb{T}$ such that $\|f - h\|_\Psi \leqslant \varepsilon$. Therefore, for every $r \in (0,1)$, we have

$$\|P_r * f - f\|_\Psi \leqslant \|P_r * (f-h)\|_\Psi + \|P_r * h - h\|_\Psi + \|h - f\|_\Psi \leqslant 2\varepsilon + \|P_r * h - h\|_\Psi$$

because $\|P_r * g\|_\Psi \leqslant \|g\|_\Psi$, for every $r \in (0,1)$ and every $g \in L^\Psi(\mathbb{T})$.

But now, $P_r * h \underset{r\to 1}{\longrightarrow} h$ uniformly. The conclusion follows from the fact that $P_r * f$ is nothing but f_r. □

The following simple result (see 3.1) will yet be very useful to apply later Proposition 3.2 on the compactness of composition operators.

Proposition 11.4 *Both spaces H^Ψ and HM^Ψ satisfy the Fatou Property.*

Proof: Let (f_n) be a bounded sequence in H^Ψ such that $f_n \to f$ a.e. We write $C = \underline{\lim}\|f_n\|_{H^\Psi}$ and, taking an appropriate subsequence, we may write $C = \lim\|f_{n_k}\|_{H^\Psi}$. Using Fatou's lemma, we have, for every $r \in (0,1)$ and $\varepsilon > 0$,

$$\begin{aligned}
\int_{\mathbb{T}} \Psi\left(\frac{|f(r\xi)|}{C+\varepsilon}\right) dm(\xi) &= \int_{\mathbb{T}} \underline{\lim}\, \Psi\left(\frac{|f_{n_k}(r\xi)|}{C+\varepsilon}\right) dm(\xi) \\
&\leqslant \underline{\lim} \int_{\mathbb{T}} \Psi\left(\frac{|f_{n_k}(r\xi)|}{C+\varepsilon}\right) dm(\xi) \\
&\leqslant \underline{\lim} \int_{\mathbb{T}} \Psi\left(\frac{|f_{n_k}(r\xi)|}{\|f_{n_k}\|_{H^\Psi}}\right) dm(\xi) \leqslant 1
\end{aligned}$$

using also that Ψ is non-increasing and the fact that $\|f_{n_k}\|_{H^\Psi} < C + \varepsilon$ for k large enough. We get $\|f_r\|_{L^\Psi(\mathbb{T})} \leqslant C + \varepsilon$ and the conclusion follows from the definition of the norm.

The preceding argument works as well for HM^Ψ. □

We use the preceding result in the following property, which extends Proposition 1.78 to our Orlicz framework.

Proposition 11.5 *On the unit ball of H^Ψ, the weak-star topology is the topology of uniform convergence on every compact subset of $\mathbb{D}$.*

In other words, the weak-star topology is the Fréchet topology of $\mathscr{H}(\mathbb{D})$.

Proof: First we notice that those topologies are metrizable since M^Φ is separable, so that the weak-star topology is metrizable on the unit ball of its dual space L^Ψ and *a fortiori* on that of H^Ψ.

Let $(f_k)_{k\geqslant 1}$ be in the unit ball of H^Ψ and weak-star convergent to $f \in H^\Psi$. Let us fix a compact subset K of $\mathbb{D}$ and we may assume that K is the closed ball of center 0 and radius $r < 1$. First, testing the weak-star convergence on characters (a.e. both with z^n and $\bar{z}^n$ where $n \in \mathbb{N}_0$), we have $\widehat{f_k}(n) \underset{k\to\infty}{\longrightarrow} \hat{f}(n)$ for every $n \in \mathbb{Z}$. Then

$$\sup_{|z|\leqslant r} |f_k(z) - f(z)| \leqslant \sum_{n\geqslant 0} r^n |\widehat{f_k}(n) - \hat{f}(n)| \longrightarrow 0$$

when k tends to infinity. Hence the result follows.

Conversely, let $(f_k)_{k\geqslant 1}$ be in the unit ball of H^Ψ and converging to some analytic function f in $\mathscr{H}(\mathbb{D})$. We first notice that f actually lies in the unit ball of H^Ψ by Fatou's lemma (see Proposition 11.4). Fix $h \in M^\Phi$ and $\varepsilon > 0$. There exists some $r < 1$ such that $\|P_r * h - h\|_\Phi \leqslant \varepsilon/8$, where P_r is the Poisson kernel at r. Then, using (H.4),

$$\begin{aligned} |\langle f_k - f, h\rangle| &\leqslant |\langle f_k - f, P_r * h - h\rangle| + |\langle f_k - f, P_r * h\rangle| \\ &\leqslant 2\|f_k - f\|_\Psi \,\|P_r * h - h\|_\Phi + |\langle P_r * (f_k - f), h\rangle| \\ &\leqslant \frac{\varepsilon}{2} + 2\,\|(f_k)_r - f_r\|_\Psi \,\|h\|_\Phi \leqslant \frac{\varepsilon}{2} + 2\,\|(f_k)_r - f_r\|_\Psi \,\|h\|_\Phi \\ &= \frac{\varepsilon}{2} + 2\alpha\,\|h\|_\Phi \sup_{|z|=r} |f_k(z) - f(z)|, \end{aligned}$$

where $\alpha = 1/\Psi^{-1}(1)$ (recall (H.2)).
By uniform convergence of (f_k) to f, on $r\mathbb{T}$, one has $\alpha\|h\|_\Phi \sup_{r\mathbb{T}} |f_k - f| \leqslant \varepsilon/4$ for k large enough. It follows that $(f_k)_k$ weak-star converges to f. □

The following family of analytic functions plays a key role as test functions, as we often need, when estimating norms of operators, for instance, or searching necessary conditions.

Given $a \in \mathbb{T}$ and $r \in (0,1)$, we define the functions

$$u_{a,r}(z) = \left(\frac{1-r}{1-r\bar{a}z}\right)^2 \qquad \text{for } z \in \mathbb{C} \setminus \{a/r\}. \tag{11.3}$$

They are analytic on $\overline{\mathbb{D}}$, and $u_{a,r}$ is nothing but the square of test functions used in the framework of classical Hardy spaces, or, up to a constant depending on r only, the derivative of φ_{ra}. The following lemma gathers several basic properties of these functions.

Lemma 11.6 *With the previous notations, we have:*

(i) $u_{a,r}(ra) = \dfrac{1}{(1+r)^2}$, $\|u_{a,r}\|_\infty = 1$ *and* $\|u_{a,r}\|_1 = \dfrac{1-r}{1+r}\cdot$

(ii) *When* $z \in \mathbb{D} \cap \overline{D}(a, 1-r)$, *we have* $|u_{a,r}(z)| \geqslant \dfrac{1}{(1+r)^2} \geqslant \dfrac{1}{4}\cdot$

(iii) *We have*

$$\|u_{a,r}\|_{H^\Psi} \leqslant \frac{1}{\Psi^{-1}\left(\frac{1+r}{1-r}\right)}\cdot$$

In particular, $u_{a,r}(ra) \approx \|u_{a,r}\|_\infty$ and $\|u_{a,r}\|_{H^\Psi} \leqslant \dfrac{1}{\Psi^{-1}\left(\frac{1}{1-r}\right)}\cdot$

Proof: (i) The value of the sup-norm is clear and

$$\|u_{a,r}\|_1 = (1-r)^2 \left\| \frac{1}{1-r\bar{z}} \right\|_2^2 = \frac{(1-r)^2}{1-r^2} = \frac{1-r}{1+r}\cdot$$

(ii) When $z \in D(a, 1-r)$, we can write $z = a + u$ with $|u| < 1-r$ so that $1 - r\bar{z} = 1 - r + \bar{a}ru$. But $|\bar{a}ru| < r(1-r)$. Hence $|1-r\bar{z}| < (1-r)(1+r) < 2(1-r)$. We get the result.

(iii) From (H.2), thanks to (i), we immediately have

$$\|u_{a,r}\|_{H^\Psi} \leqslant \frac{1}{\Psi^{-1}\left(\frac{1+r}{1-r}\right)}\cdot \qquad \square$$

We can now obtain an estimate for the growth of the norm of the point evaluation on Hardy–Orlicz spaces when the point is close to the boundary. Recall our notation $\delta_z(f) = f(z)$ for $z \in \mathbb{D}$.

Proposition 11.7 (Norm of point evaluation in Hardy–Orlicz spaces) *Let* $z \in \mathbb{D}$. *The point evaluation* δ_z *is bounded on* H^Ψ *and we have*

$$\|\delta_z\|_{(H^\Psi)^*} = \|\delta_z\|_{(HM^\Psi)^*} \approx \Psi^{-1}\left(\frac{1}{1-|z|}\right)\cdot \tag{11.4}$$

Actually

$$\frac{1}{4}\Psi^{-1}\left(\frac{1+|z|}{1-|z|}\right) \leqslant \|\delta_z\|_{(H^\Psi)^*} \leqslant \Psi^{-1}\left(\frac{1+|z|}{1-|z|}\right)\cdot \tag{11.5}$$

Proof: For every $f \in H^\Psi$ and every $r \in (0,1)$, we have $f_r \in H^\infty \subset HM^\Psi$ and we can write

$$|f(rz)| \leqslant \|\delta_z\|_{(HM^\Psi)^*} \|f_r\|_{H^\Psi} \leqslant \|\delta_z\|_{(HM^\Psi)^*} \|f\|_{H^\Psi}\,.$$

Letting r go to 1, we get $|f(z)| \leqslant \|\delta_z\|_{(HM^\Psi)^*} \|f\|_{H^\Psi}$ so that $\|\delta_z\|_{(H^\Psi)^*} \leqslant \|\delta_z\|_{(HM^\Psi)^*}$. Since the reverse inequality is obvious, we proved the equality.

Given $f \in H^\Psi$ with norm 1, we can write $f(z) = \displaystyle\int_{\mathbb{T}} P_z(\xi) f(\xi)\, dm(\xi)$ (where P_z is the Poisson kernel at z), so we know that

$$\Psi(|f(z)|) \leqslant \Psi\left(\int_{\mathbb{T}} |f(\xi)| P_z(\xi)\, dm(\xi)\right)$$

since Ψ is increasing. The Jensen inequality then implies that

$$\Psi(|f(z)|) \leqslant \int_{\mathbb{T}} \Psi(|f|)\, P_z\, dm \leqslant \int_{\mathbb{T}} \Psi(|f|) \frac{1+|z|}{1-|z|}\, dm = \frac{1+|z|}{1-|z|}.$$

Rearranging, and since f is arbitrary in the unit sphere, we get

$$\|\delta_z\|_{(H^\Psi)^*} \leqslant \Psi^{-1}\left(\frac{1+|z|}{1-|z|}\right).$$

We write $z = ar$ with $a \in \mathbb{T}$ and $r \in (0,1)$ and we use the function $u_{a,r}$. From Lemma 11.6 (i) and (iii), we have

$$\|\delta_z\|_{(H^\Psi)^*} \geqslant \frac{|u_{a,r}(z)|}{\|u_{a,r}\|_\Psi} \geqslant \frac{1/(1+r)^2}{1/\Psi^{-1}\left(\frac{1+r}{1-r}\right)} \geqslant \frac{1}{4}\Psi^{-1}\left(\frac{1+r}{1-r}\right). \qquad \square$$

As an immediate corollary, we also have

$$\frac{1}{(1+r)^2} \frac{1}{\Psi^{-1}\left(\frac{1+r}{1-r}\right)} \leqslant \|u_{a,r}\|_{H^\Psi} \leqslant \frac{1}{\Psi^{-1}\left(\frac{1+r}{1-r}\right)}. \tag{11.6}$$

Indeed, this is a direct consequence of the upper estimate of $\|\delta_z\|_{(H^\Psi)^*}$.

Bergman-Orlicz Spaces

Definition 11.8 The *Bergman–Orlicz space* $\mathscr{B}^\Psi$ denotes the space of analytic functions $f : \mathbb{D} \to \mathbb{C}$, which are in the Orlicz space $L^\Psi(\mathbb{D}, \sigma)$:

$$\mathscr{B}^\Psi = \mathscr{H}(\mathbb{D}) \cap L^\Psi(\mathbb{D}, \sigma).$$

The *Bergman–Morse–Transue space* is the subspace

$$\mathscr{B}M^\Psi = \mathscr{B}^\Psi \cap M^\Psi(\mathbb{D}, \sigma) = \mathscr{H}(\mathbb{D}) \cap M^\Psi(\mathbb{D}, \sigma).$$

Let us point out that we have, with contractive inclusion,

$$H^\Psi \subset \mathscr{B}^\Psi \qquad \text{and} \qquad HM^\Psi \subset \mathscr{B}M^\Psi.$$

Indeed, when $f \in H^\Psi$, say with norm 1, we have $\|f_r\|_{H^\Psi} \leqslant \|f\|_{H^\Psi} = 1$; hence

$$\int_{\mathbb{D}} \Psi(|f(w)|)\, d\sigma(w) = \int_0^1 \int_{\mathbb{T}} \Psi(|f(r\xi)|)\, dm(\xi)\, 2r dr \leqslant \int_0^1 2r dr = 1.$$

The second inclusion (between Morse–Transue spaces) is now clear. The space $\mathscr{B}^\Psi$, equipped with the induced norm of $L^\Psi(\mathbb{D},\sigma)$, is a Banach space as an obvious consequence of the following lemma, analogous to Proposition 11.7.

Proposition 11.9 (Norm of the point evaluation in Bergman–Orlicz spaces) *Let $z \in \mathbb{D}$. The point evaluation δ_z is bounded on $\mathscr{B}^\Psi$ and we have*

$$\|\delta_z\|_{(\mathscr{B}^\Psi)^*} = \|\delta_z\|_{(\mathscr{B}M^\Psi)^*} \approx \Psi^{-1}\left(\frac{1}{(1-|z|)^2}\right). \tag{11.7}$$

Actually, we have

$$\frac{1}{(1+|z|)^4}\Psi^{-1}\left(\left(\frac{1+|z|}{1-|z|}\right)^2\right) \leqslant \|\delta_z\|_{(\mathscr{B}^\Psi)^*} \leqslant \Psi^{-1}\left(\left(\frac{1+|z|}{1-|z|}\right)^2\right). \tag{11.8}$$

Proof: Assume that we already proved that the point evaluation δ_z is bounded on $\mathscr{B}M^\Psi$; then the functional δ_z is bounded on $\mathscr{B}^\Psi$ as well, following the same argument as in the proof of Proposition 11.7.

Take f in the unit ball of $\mathscr{B}M^\Psi$. For every analytic function $g\colon \mathbb{D} \to \mathbb{C}$, the mean value property gives $|g(0)| \leqslant \int_{\mathbb{D}} |g(w)|\, d\sigma(w)$; applying this to the function $g = f \circ \varphi_z = f \circ \varphi_z^{-1}$, we get (with the change of variable formula)

$$|f(z)| \leqslant \int_{\mathbb{D}} |f(w)|.|\varphi_z'(w)|^2\, d\sigma(w).$$

Thanks to the Jensen inequality (recall that $|\varphi_z'(w)|^2\, d\sigma(w)$ is a probability measure), we obtain

$$\begin{aligned}\Psi(|f(z)|) &\leqslant \int_{\mathbb{D}} \Psi(|f(w)|).|\varphi_z'(w)|^2\, d\sigma(w) \\ &\leqslant \left(\frac{1+|z|}{1-|z|}\right)^2 \int_{\mathbb{D}} \Psi(|f(w)|)\, d\sigma(w) \\ &\leqslant \left(\frac{1+|z|}{1-|z|}\right)^2.\end{aligned} \tag{11.9}$$

Rearranging, this leads to the upper bound in (11.8).

For the lower bound, we test the functional δ_z on the function $h_z = \varphi_z'^2$, which belongs to $H^\infty \subset \mathscr{B}M^\Psi$. We have

$$\|h_z\|_{\mathscr{B}^1} = 1, \quad h_z(z) = \frac{1}{(1-|z|^2)^2}, \quad \text{and} \quad \|h_z\|_\infty = \left(\frac{1+|z|}{1-|z|}\right)^2. \tag{11.10}$$

Using (H.2), this leads to

$$\|h_z\|_{\mathscr{B}M^\Psi} \leqslant \frac{\left(\dfrac{1+|z|}{1-|z|}\right)^2}{\Psi^{-1}\left(\left(\dfrac{1+|z|}{1-|z|}\right)^2\right)};$$

hence $\|\delta_z\|_{(\mathscr{B}M^\Psi)^*} \geqslant \dfrac{1}{(1+|z|)^4}\Psi^{-1}\left(\left(\dfrac{1+|z|}{1-|z|}\right)^2\right)$ and (11.8) is proved.

From this, we obtain (11.7) easily since Ψ is increasing and convex. Indeed observe that, for any $x \geqslant 0$, $\Psi^{-1}(4x) \leqslant 4\Psi^{-1}(x)$. □

We mention (and leave it as an exercise) that we could also use (H.5) instead of Jensen's inequality: it adds a factor 2 in (11.8), but it does not affect (11.7).

Let us emphasize that in this proof the *Berezin kernel* $|\varphi_z'|^2 = P_z^2 = |h_z|$ plays for $\mathscr{B}^\Psi$ the role the Poisson kernel P_z plays for H^Ψ. Up to a constant depending on $|z|$ only, it is also the modulus of the square of the reproducing kernel K_z for $\mathscr{B}^2$. We shall meet it again.

We set once again $f_r(z) = f(rz)$ for $z \in \mathbb{D}$ and $r \in (0,1)$ and we show below that, in the space $\mathscr{B}M^\Psi$, we have convergence to f.

Proposition 11.10 *Let $f \in \mathscr{B}M^\Psi$. We have* $\lim_{r\to 1^-} f_r = f$ *in $\mathscr{B}M^\Psi$.*

Proof: Let us fix $\varepsilon > 0$. We know (for instance applying Theorem H.3) that there is some $R \in (0,1)$ such that

$$\int_{\mathbb{D}\setminus R\mathbb{D}} \Psi\left(\frac{|f|}{\varepsilon}\right) d\sigma \leqslant 1.$$

For $r \geqslant \dfrac{2R}{R+1} \in (R,1)$, we have, using the change of variable formula,

$$\int_{\mathbb{D}\setminus\frac{R+1}{2}\mathbb{D}} \Psi\left(\frac{|f_r|}{\varepsilon}\right) d\sigma = \frac{1}{r^2}\int_{\mathbb{D}\setminus\frac{r(R+1)}{2}\mathbb{D}} \Psi\left(\frac{|f|}{\varepsilon}\right) d\sigma \leqslant \frac{1}{r^2},$$

since $\dfrac{r(R+1)}{2} > R$. We get using the convexity of Ψ

$$\begin{aligned}\int_{\mathbb{D}} \Psi\left(\frac{|f_r - f|}{3\varepsilon}\right) d\sigma &\leqslant \int_{\frac{R+1}{2}\mathbb{D}} \Psi\left(\frac{|f_r - f|}{3\varepsilon}\right) d\sigma \\ &\quad + \frac{1}{3}\int_{\mathbb{D}\setminus\frac{R+1}{2}\mathbb{D}} \Psi\left(\frac{|f|}{\varepsilon}\right) + \Psi\left(\frac{|f_r|}{\varepsilon}\right) d\sigma \\ &\leqslant \int_{\frac{R+1}{2}\mathbb{D}} \Psi\left(\frac{|f_r - f|}{3\varepsilon}\right) d\sigma + \frac{1}{3} + \frac{1}{3r^2}.\end{aligned}$$

Since $f_r - f$ tends to 0 uniformly on the compact set $\frac{1+R}{2}\overline{\mathbb{D}} \subset \mathbb{D}$, we obtain for r close enough to 1

$$\int_{\mathbb{D}} \Psi\left(\frac{|f_r - f|}{3\varepsilon}\right) d\sigma \leqslant 1.$$

In other words, $\|f_r - f\|_{\mathscr{B}^\Psi} \leqslant 3\varepsilon$. □

Let us mention another result about the dilation of f:

Lemma 11.11 *Let $f \in \mathscr{B}^\Psi$ and $r \in (0,1)$. We have $\|f_r\|_{\mathscr{B}^\Psi} \leqslant \|f\|_{\mathscr{B}^\Psi}$.*

Proof: We assume that $f \neq 0$ and we write $C = \|f\|_{\mathscr{B}^\Psi} > 0$. From the properties of the Poisson kernel and Jensen's inequality (recall the proof of (11.2) for instance), we have

$$\int_{\mathbb{T}} \Psi\left(\frac{|f(r\rho\xi)|}{C}\right) dm(\xi) \leqslant \int_{\mathbb{T}} \Psi\left(\frac{|f(\rho\xi)|}{C}\right) dm(\xi)$$

for every $\rho \in (0,1)$; hence

$$\begin{aligned}\int_{\mathbb{D}} \Psi\left(\frac{|f(rz)|}{C}\right) d\sigma(z) &= \int_0^1 \int_{\mathbb{T}} \Psi\left(\frac{|f(r\rho\xi)|}{C}\right) dm(\xi)\, 2\rho d\rho \\ &\leqslant \int_0^1 \int_{\mathbb{T}} \Psi\left(\frac{|f(\rho\xi)|}{C}\right) dm(\xi)\, 2\rho d\rho \\ &= \int_{\mathbb{D}} \Psi\left(\frac{|f(z)|}{C}\right) d\sigma(z) \leqslant 1,\end{aligned}$$

and this implies the claim. □

Proposition 11.12 *We have the following properties:*

(i) *$\mathscr{B}M^\Psi$ is the closure of $H^\infty(\mathbb{D})$ in $L^\Psi(\mathbb{D},\sigma)$ and actually the analytic polynomials are dense in $\mathscr{B}M^\Psi$.*
(ii) *On the unit ball of $\mathscr{B}^\Psi$, the weak-star topology coincides with the topology of uniform convergence on compact subsets of $\mathbb{D}$.*
(iii) *$\mathscr{B}^\Psi$ is closed in $L^\Psi(\mathbb{D},\sigma)$ for the weak-star topology.*
(iv) *$\mathscr{B}^\Psi$ is (isometric to) the bidual of $\mathscr{B}M^\Psi$.*

In other words, from (i), we have $\overline{A(\mathbb{D})}^{\|\cdot\|_\Psi} = \mathscr{B}M^\Psi$.

Proof: (i) Obviously $H^\infty \subset \mathscr{B}M^\Psi$ and, by Proposition 11.10, each function $f \in \mathscr{B}M^\Psi$ can be approximated by some f_r, which belongs to $A(\mathbb{D}) \subset H^\infty$.

Since f is analytic, f_r can also be uniformly approximated on $\overline{\mathbb{D}}$ by the partial sums of its Taylor series.

(ii) It suffices to use a sequential argument since the topologies are metrizable on balls (recall that the space $M^\Phi(\mathbb{D},\sigma)$ is separable).

Assume that f belongs to the unit ball of $\mathscr{B}^\Psi \subset L^\Psi(\mathbb{D},\sigma)$ and is the weak-star limit of a sequence of analytic functions f_n in the unit ball of $\mathscr{B}^\Psi$. We test this weak-star convergence with the function $h_k(z) = (k+1)z^k \in L^\Phi(\mathbb{D})$, where $k \in \mathbb{N}_0$ (recall Section H.2):

$$a_k(n) = \int_{\mathbb{D}} f_n(z) h_k(\bar{z})\, d\sigma(z) \longrightarrow \int_{\mathbb{D}} f(z) h_k(\bar{z})\, d\sigma(z) = a_k \text{ when } n \to +\infty,$$

and we obtain that the kth Taylor coefficient $a_k(n)$ of f_n converges to the kth Taylor coefficient a_k of f. Fix a compact $K \subset \mathbb{D}$; there exists an $r \in (0,1)$ such that $K \subset r\mathbb{D}$. We obtain

$$\sup_{z\in K} |f_n(z) - f(z)| \leqslant \sum_{k\geqslant 0} |a_k(n) - a_k| r^k \longrightarrow 0$$

by the dominated convergence theorem (observe that $|a_k - a_k(n)| \leqslant 2(k+1)$ for every $k \geqslant 0$ and every $n \geqslant 0$).

For the converse, suppose now that f_n, in the unit ball of $\mathscr{B}^\Psi$, converges in $\mathscr{H}(\mathbb{D})$ to $f \in \mathscr{B}^\Psi$ (with $\|f\|_{\mathscr{B}^\Psi} \leqslant 1$). Fixing $g \in M^\Phi(\mathbb{D},\sigma) \subset L^1(\mathbb{D},\sigma)$ and $\varepsilon > 0$, there exists an $r \in (0,1)$ such that $\|g \mathbb{1}_{\mathbb{D}\setminus r\mathbb{D}}\|_\Phi \leqslant \varepsilon$. We have

$$\begin{aligned}\left|\int_{\mathbb{D}} (f_n - f)(z)\, g(z)\, d\sigma\right| &\leqslant \left|\int_{r\mathbb{D}} (f_n - f)(z)\, g(z)\, d\sigma(z)\right| + 2\varepsilon \\ &\leqslant \sup_{z\in r\mathbb{D}} \big|f_n(z) - f(z)\big| . \|g\|_1 + 2\varepsilon,\end{aligned}$$

but by hypothesis $\sup_{z\in r\mathbb{D}} \big|f_n(z) - f(z)\big| \longrightarrow 0$. The conclusion follows.

(iii) It is sufficient to prove that the balls are weak-star closed thanks to the theorem of Banach–Dieudonné. On the other hand, $M^\Phi(\mathbb{D},\sigma)$ is separable so the weak-star topology is metrizable on balls. From (ii) it is sufficient to prove that the unit ball of $\mathscr{B}^\Psi$ is compact for the topology of uniform convergence on compact subsets of $\mathbb{D}$. Let (f_n) be a sequence in the unit ball of $\mathscr{B}^\Psi$. This is a normal family thanks to Proposition 11.9, and we can apply Montel's theorem 1.2: a subsequence converges to an analytic function f on compact subsets of $\mathbb{D}$. Fatou's property implies that f actually lies in the unit ball of $\mathscr{B}^\Psi$.

(iv) In our framework, Ψ satisfies ∇_2; hence $\big(M^\Psi(\mathbb{D},\sigma)\big)^{**} = L^\Psi(\mathbb{D},\sigma)$ and $(\mathscr{B}M^\Psi)^{**} = \overline{\mathscr{B}M^\Psi}^{w^*}$, viewed in the space $L^\Psi(\mathbb{D},\sigma)$. Therefore, we concretely aim to show $\overline{\mathscr{B}M^\Psi}^{w^*} = \mathscr{B}^\Psi$ and we already know that $\mathscr{B}^\Psi$ is weak-star closed. For any $f \in \mathscr{B}^\Psi$, the functions f_r belong to $H^\infty \subset \mathscr{B}M^\Psi$ for every $r \in (0,1)$ with $\|f_r\|_{\mathscr{B}^\Psi} \leqslant \|f\|_{\mathscr{B}^\Psi}$, thanks to Lemma 11.11. Moreover f_r converges to f in $\mathscr{H}(\mathbb{D})$, when $r \to 1^-$. The conclusion follows. □

11.2 Composition Operators on Hardy–Orlicz Spaces

In this section, we focus on composition operators on Hardy–Orlicz spaces introduced earlier. We shall see that they are always bounded, sharing this remarkable property with classical Hardy spaces. Among other things, we shall also give some necessary and/or sufficient conditions for compactness in full generality. But for many families of Orlicz functions, such as the important examples in Appendix H, these are characterizations.

11.2.1 Boundedness

We prove here that every composition operator is bounded on H^Ψ. We could prove first the result for automorphisms and invoke the Dévissage principle 2.3.2; nevertheless, we give a direct proof below.

Theorem 11.13 *Let $\varphi\colon \mathbb{D} \to \mathbb{D}$ be a symbol.*

(i) *The composition operator C_φ is bounded on H^Ψ and*

$$\frac{1}{4\Psi^{-1}(1)}\Psi^{-1}\left(\frac{1+|\varphi(0)|}{1-|\varphi(0)|}\right) \leqslant \|C_\varphi\|_{B(H^\Psi)} \leqslant \frac{1+|\varphi(0)|}{1-|\varphi(0)|}.$$

(ii) *Assume that φ fixes the origin. Then the composition operator C_φ is a contraction on H^Ψ.*

(iii) *C_φ also maps HM^Ψ into HM^Ψ, and $C_\varphi\colon H^\Psi \to H^\Psi$ is the bi-adjoint of the composition operator $C_\varphi\colon HM^\Psi \to HM^\Psi$.*
Moreover $\|C_\varphi\|_{B(H^\Psi)} = \|C_\varphi\|_{B(HM^\Psi)}$.

(iv) *When $\Psi \in \nabla_1$, that is, $\Psi(x)\Psi(y) \leqslant \Psi(bxy)$ for all $x, y \geqslant 0$, and for some $b > 0$, we have*

$$\|C_\varphi\|_{B(H^\Psi)} \approx \Psi^{-1}\left(\frac{1+|\varphi(0)|}{1-|\varphi(0)|}\right).$$

Note that $\Psi(x) = x^p$ for $p \geqslant 1$ and the functions $\Psi = \Psi_r$, with $r > 0$ (see Appendix H), satisfy $\Psi \in \nabla_1$. In particular, from (iv), we have

$$\|C_\varphi\|_{B(H^{\Psi_r})} \approx \left[\log\left(\frac{2}{1-|\varphi(0)|}\right)\right]^{\frac{1}{r}}.$$

Let us also emphasize that the proof below actually gives the lower estimate:

$$\|C_\varphi\|_{B(H^\Psi)} \geqslant \sup_{z\in\mathbb{D}} \frac{\Psi^{-1}\left(\dfrac{1+|\varphi(z)|}{1-|\varphi(z)|}\right)}{4\Psi^{-1}\left(\dfrac{1+|z|}{1-|z|}\right)}. \tag{11.11}$$

Proof: (i) Let $f \in H^\Psi$. Since $f \in H^1$, f^* is well defined and $(f \circ \varphi)^* = f^* \circ \varphi^*$ from Corollary 1.21. Having this in mind, we write f instead of f^* in the sequel. For every $r \in (0,1)$, inspired by Sarason's approach 2.3.4, we write, like (2.16),

$$f[\varphi(r\xi)] = \int_{\mathbb{T}} f(w)\, P_{\varphi(r\xi)}(w)\, dm(w)$$

for every $\xi \in \mathbb{T}$.

We use the Jensen inequality to get

$$\Psi\left(|f \circ \varphi(r\xi)|\right) \leqslant \int_{\mathbb{T}} \Psi(|f(w)|)\, P_{\varphi(r\xi)}(w)\, dm(w);$$

hence, via Fubini's theorem and thanks to (2.11),

$$\begin{aligned}\int_{\mathbb{T}} \Psi(|f \circ \varphi(r\xi)|)\, dm(\xi) &\leqslant \int_{\mathbb{T}} \Psi(|f(w)|) \int_{\mathbb{T}} P_{\varphi(r\xi)}(w)\, dm(\xi)\, dm(w) \\ &= \int_{\mathbb{T}} \Psi(|f(w)|)\, \mathrm{Re}\frac{1+\varphi(0)\overline{w}}{1-\varphi(0)\overline{w}}\, dm(w) \\ &\leqslant \frac{1+|\varphi(0)|}{1-|\varphi(0)|} \int_{\mathbb{T}} \Psi(|f(w)|)\, dm(w). \qquad (11.12)\end{aligned}$$

From now on assume that f belongs to the unit ball of H^Ψ; we obtain

$$\int_{\mathbb{T}} \Psi(|f \circ \varphi(r\xi)|)\, dm(\xi) \leqslant \frac{1+|\varphi(0)|}{1-|\varphi(0)|}. \qquad (11.13)$$

This implies that $\|f \circ \varphi\|_{H^\Psi} \leqslant \dfrac{1+|\varphi(0)|}{1-|\varphi(0)|}$ (see remark after Definition H.1) and it proves the upper estimate.

Following an idea already used for classical Hardy spaces in (2.17) and emphasized in (2.28) in Proposition 2.15, we know that

$$\|\delta_{\varphi(z)}\|_{(H^\Psi)^*} \leqslant \|C_\varphi\|_{B(H^\Psi)} \|\delta_z\|_{(H^\Psi)^*}. \qquad (11.14)$$

We also know from (11.5) that, for every $a \in \mathbb{D}$,

$$\frac{1}{4}\Psi^{-1}\left(\frac{1+|a|}{1-|a|}\right) \leqslant \|\delta_a\|_{(H^\Psi)^*} \leqslant \Psi^{-1}\left(\frac{1+|a|}{1-|a|}\right). \qquad (11.15)$$

From relations (11.14) and (11.15), we obtain

$$\frac{1}{4}\Psi^{-1}\left(\frac{1+|\varphi(z)|}{1-|\varphi(z)|}\right) \leqslant \|C_\varphi\|_{B(H^\Psi)}\, \Psi^{-1}\left(\frac{1+|z|}{1-|z|}\right);$$

hence, we get (11.11). The specific case $z = 0$ gives the lower estimate.

(ii) This is an obvious consequence of (i).

We mention that, without (i), we could also use that the function $g = \Psi(|f|)$ is subharmonic on $\mathbb{D}$ (thanks to the Jensen inequality) and apply (2.15).

(iii) Let $f \in HM^\Psi$. For every $C > 0$, $Cf \in H^\Psi$, and the computation (11.12) is still valid replacing f by Cf, so that we can write

$$\int_{\mathbb{T}} \Psi(C|f \circ \varphi(r\xi)|)\, dm(\xi) \leqslant \frac{1 + |\varphi(0)|}{1 - |\varphi(0)|} \int_{\mathbb{T}} \Psi(C|f(w)|)\, dm(w) < \infty$$

since f belongs to HM^Ψ and we conclude that $f \circ \varphi \in HM^\Psi$ as well.

The result on the bi-adjoint is straightforward.

(iv) Let f be in the unit ball of H^Ψ. When $\Psi \in \nabla_1$, we go back to (11.13) and get

$$\begin{aligned}\Psi(C) \int_{\mathbb{T}} \Psi\left(\frac{|f \circ \varphi(r\xi)|}{bC}\right) dm(\xi) &\leqslant \int_{\mathbb{T}} \Psi(|f \circ \varphi(r\xi)|)\, dm(\xi) \\ &\leqslant \frac{1 + |\varphi(0)|}{1 - |\varphi(0)|} = \Psi(C),\end{aligned}$$

where $C = \Psi^{-1}\left(\dfrac{1 + |\varphi(0)|}{1 - |\varphi(0)|}\right) > 0$.

This means that $\|(f \circ \varphi)_r\|_{H^\Psi} \leqslant bC$ for every $r \in (0, 1)$ and, with the help of the lower estimate in (i), we proved (iv). □

Let us specify the particular case of inner symbols.

Proposition 11.14 *Let φ be an inner symbol and Ψ an Orlicz function. For every f in H^Ψ, we have*

$$\|f \circ \varphi\|_{H^\Psi} \geqslant \frac{1 - |\varphi(0)|}{1 + |\varphi(0)|}\, \|f\|_{H^\Psi}. \tag{11.16}$$

When φ is an inner symbol fixing the origin, C_φ is an isometry on H^Ψ.

This obviously implies that C_φ cannot be compact on H^Ψ when φ is inner.

Proof: This is mainly a consequence of Lemma 2.4 stating that m_φ is $P_a\, m$ with $a = \varphi(0)$. For every $f \in H^\Psi$ and every $c > 0$, we have

$$\int_{\mathbb{T}} \Psi(c|f \circ \varphi|)\, dm = \int_{\mathbb{T}} \Psi(c|f|)\, dm_\varphi = \int_{\mathbb{T}} \Psi(c|f(z)|) \frac{1 - |a|^2}{|1 - \bar{a}z|^2}\, dm(z);$$

and when $a = \varphi(0) = 0$, the conclusion is clear: $\|f \circ \varphi\|_{H^\Psi} = \|f\|_{H^\Psi}$.

In general, we have

$$\int_{\mathbb{T}} \Psi(c|f \circ \varphi|)\, dm \geqslant \frac{1 - |a|}{1 + |a|} \int_{\mathbb{T}} \Psi(c|f|)\, dm \geqslant \int_{\mathbb{T}} \Psi(k|f|)\, dm,$$

where $k = c\,\dfrac{1 - |a|}{1 + |a|}$ and we get (11.16). □

11.2.2 Carleson Measures on H^Ψ

We wish to provide a full characterization of the compactness of composition operators on H^Ψ, but we require the powerful tool of Carleson measures. In the present section, we focus not only on the pullback measure m_φ but more generally on a (positive finite) measure μ on $\overline{\mathbb{D}}$ as in Chapter 5.

In the Hardy space framework (and because f can be defined on the whole closed unit disk), a Carleson embedding on H^p is a map

$$f \in H^p \longmapsto f^* = f \in L^p\left(\overline{\mathbb{D}}, \mu\right).$$

In the same spirit, we shall work here with the "embedding" map

$$J_\mu : \quad \begin{array}{ccc} H^\Psi & \longrightarrow & L^\Psi\left(\overline{\mathbb{D}}, \mu\right) \\ f & \longmapsto & f^* = f. \end{array} \tag{11.17}$$

We should write $J_{\mu,\Psi}$ since this map also depends on Ψ, but we shall lighten the notation this way.

Let us mention here the following lemma:

Lemma 11.15 *Assume that* $J_\mu : H^\Psi \longrightarrow L^\Psi\left(\overline{\mathbb{D}}, \mu\right)$ *is bounded. Then it is weak-star continuous.*

Proof: Indeed since here the weak-star topologies, both on H^Ψ and $L^\Psi\left(\overline{\mathbb{D}}, \mu\right)$ (separable duals), are metrizable, it suffices to use a sequential argument. Let (f_n) be a sequence in H^Ψ, uniformly convergent to 0 on every compact subset of $\mathbb{D}$, and fix $g \in M^\Phi\left(\overline{\mathbb{D}}, \mu\right) = L^\Phi\left(\overline{\mathbb{D}}, \mu\right)$ (we may assume that g actually belongs to the unit ball) and $\varepsilon > 0$. We recall that $\Phi \in \Delta_2$ (in our framework, $\Psi \in \nabla_2$). We can also assume that (f_n) (which is bounded) lies in the unit ball of H^Ψ. Since $g \in M^\Phi\left(\overline{\mathbb{D}}, \mu\right)$, there exists a compact $K \subset \mathbb{D}$ such that

$$\int_{\overline{\mathbb{D}}\setminus K} \Phi(|g|/\varepsilon)\, d\mu \leqslant 1.$$

In other words, $\|g\mathbb{1}_{\overline{\mathbb{D}}\setminus K}\|_\Phi \leqslant \varepsilon$. We then have, from (H.4),

$$\left|\langle f_n, g\rangle\right| \leqslant 2\int_K |g|.|f_n|\, d\mu + 2\|g\mathbb{1}_{\overline{\mathbb{D}}\setminus K}\|_\Phi.\|f_n\|_\Psi \leqslant 2\sup_K |f_n|.\|g\|_1 + 2\varepsilon,$$

and we conclude easily thanks to the uniform convergence of f_n to 0, on K. □

The main geometrical tools are still the Carleson windows, introduced in Chapter 5, but we recall their definitions, all the more as we must emphasize that we use now the versions involving $\overline{\mathbb{D}}$, and not only $\mathbb{D}$ (see the comments below):

The Carleson "square" window (or box)

$$W(\xi,h) = \big\{z \in \overline{\mathbb{D}} : 1 - |z| < h \text{ and } |\arg(\bar{\xi} z)| \leqslant \pi h\big\},$$

and the "circular" Carleson windows

$$S(\xi,h) = \big\{z \in \overline{\mathbb{D}} \;:\; |z - \xi| < h\big\}.$$

We also recall that, for every $h \in (0,1)$ and every $\xi \in \mathbb{T}$,

$$W(\xi,h/4) \subset S(\xi,h) \subset W(\xi,h).$$

In the framework of Hardy spaces, taking into account the boundary is important. Actually it is especially important when we focus on boundedness. When we focus on compactness for a composition operator C_φ, remembering $m_\varphi(\mathbb{T}) = 0$ in this case, we can forget the part in $\mathbb{T}$ of the Carleson windows (for a compact embedding J_μ, we shall also see that $\mu(\mathbb{T}) = 0$).

Anyway we shall pay attention to this part in the following statements, and the general results concerning the boundedness of J_μ involve this boundary.

We recall the definitions of Carleson's maximal functions; they were already introduced in Chapter 5. For $h \in (0,1]$,

$$\rho_\mu(h) = \sup_{\xi\in\mathbb{T}} \mu\big(W(\xi,h)\big)$$

and

$$K_\mu(h) = \sup_{0<t\leqslant h} \frac{\rho_\mu(t)}{t}.$$

By definition, we have $\mu\big(W(\xi,t)\big) \leqslant K_\mu(h)\,t$ for every $t \in (0,h]$.

The measure μ is a Carleson measure (for classical Hardy spaces) if and only if, for some constant $C > 0$, we have $\mu\big(S(\xi,h)\big) \leqslant Ch$ for every $\xi \in \mathbb{T}$. In other words, μ is a Carleson measure if and only if $K_\mu(h)$ is bounded on $(0,1]$, and this happens as soon as $K_\mu(h_0)$ is finite for some $h_0 \in (0,1)$.

In the same way, the results of Chapter 5 on compactness on Hardy spaces express that μ is a vanishing Carleson measure (for classical Hardy spaces) if and only if we have $\rho_\mu(h) = o(h)$ when $h \to 0$. In other words μ is vanishing Carleson if and only if $K_\mu(h) \longrightarrow 0$ when $h \to 0$.

Before stating the main theorem of this section, let us mention the rather easy following result, which will be useful for the sequel. It can be viewed as a kind of test result: it suffices to test the restriction operator defined by

$$I_r(f) = f\,\mathbb{1}_{\overline{\mathbb{D}}\setminus r\mathbb{D}}, \qquad \text{where } r \in (0,1).$$

Proposition 11.16 *Let μ be a positive finite Borel measure on $\overline{\mathbb{D}}$. The following assertions are equivalent:*

(i) $J_\mu : H^\Psi \to L^\Psi(\mu)$ *is (bounded and) compact.*
(ii) *Every sequence in the unit ball of* H^Ψ*, which is convergent to* 0 *in* $\mathscr{H}(\mathbb{D})$*, is norm convergent to* 0 *in* $L^\Psi(\mu)$.
(iii) J_μ *is bounded and* $\lim_{r\to 1^-} \|I_r\|_{B(H^\Psi, L^\Psi(\mu))} = 0$.

We wrote for short $L^\Psi(\mu)$ instead of $L^\Psi(\overline{\mathbb{D}}, \mu)$.

Proof: Let us assume (i) and let $(f_n)_{n\geqslant 1}$ be a sequence in the unit ball of H^Ψ, convergent to 0 in $\mathscr{H}(\mathbb{D})$. It means that it is weak-star convergent to 0 by Proposition 11.5. Thanks to Lemma 11.15, J_μ is continuous relatively to the (w^*, w^*) topologies, and it is also compact relatively to the norm topologies by assumption; hence $\big(J_\mu(f_n)\big)$ is norm convergent to 0 in $L^\Psi(\mu)$ and (ii) is proved.

Now assume (ii). Suppose that (iii) is not true, then there exists a sequence $(f_n)_{n\geqslant 1}$ in the unit ball of H^Ψ, a sequence (r_n) in $(0,1)$ converging to 1 and $\delta > 0$ with $\|f_n \mathbb{1}_{\overline{\mathbb{D}}\setminus r_n\mathbb{D}}\|_{L^\Psi(\mu)} > \delta$, for every $n \geqslant 1$. Up to an extraction, we can assume that, for each integer $n \geqslant 1$, we have $r_n \geqslant 1/\sqrt[n]{2}$ and we introduce the analytic functions $g_n = z^n f_n$. The sequence $(g_n)_{n\geqslant 1}$ lies in the unit ball of H^Ψ and is convergent to 0 in $\mathscr{H}(\mathbb{D})$ since the point evaluation (hence f_n) is uniformly bounded on any compact of $\mathbb{D}$. But

$$\|g_n\|_{L^\Psi(\mu)} \geqslant \|z^n f_n \mathbb{1}_{\overline{\mathbb{D}}\setminus r_n\mathbb{D}}\|_{L^\Psi(\mu)} \geqslant r_n^n \|f_n \mathbb{1}_{\overline{\mathbb{D}}\setminus r_n\mathbb{D}}\|_{L^\Psi(\mu)} \geqslant \delta/2,$$

and this contradicts (ii).

(iii) clearly implies (ii).

(ii) implies (i) thanks to Propositions 3.2 and 11.4. □

Now we can state the main result of this section.

Theorem 11.17 *Let* μ *be a finite Borel measure on* $\overline{\mathbb{D}}$*. Then:*

(i) (a) *If* $J_\mu : H^\Psi \to L^\Psi(\overline{\mathbb{D}}, \mu)$ *is (defined and) bounded, then there exists some* $A > 0$ *such that*

$$\rho_\mu(h) \leqslant \frac{1}{\Psi\big(A\Psi^{-1}(1/h)\big)} \qquad \textit{for every } h \in (0,1]. \tag{11.18}$$

(b) *If there exists some* $A > 0$ *such that*

$$K_\mu(h) \leqslant \frac{1/h}{\Psi\big(A\Psi^{-1}(1/h)\big)} \qquad \textit{for every } h \in (0,1], \tag{11.19}$$

then $J_\mu : H^\Psi \to L^\Psi(\overline{\mathbb{D}}, \mu)$ *is bounded.*

(ii) (a) *If $J_\mu : H^\Psi \to L^\Psi\left(\overline{\mathbb{D}}, \mu\right)$ is compact, then*

$$\rho_\mu(h) = o\left(\frac{1}{\Psi\left(A\Psi^{-1}(1/h)\right)}\right) \qquad \text{as } h \to 0, \text{ for every } A > 0. \tag{11.20}$$

(b) *If, for every $A > 0$,*

$$K_\mu(h) = o\left(\frac{1/h}{\Psi\left(A\Psi^{-1}(1/h)\right)}\right) \qquad \text{as } h \to 0, \tag{11.21}$$

then $J_\mu : H^\Psi \to L^\Psi\left(\overline{\mathbb{D}}, \mu\right)$ is compact.

(c) *Condition (11.20) is equivalent to*

$$\lim_{h\to 0} \frac{\Psi^{-1}\left(\dfrac{1}{h}\right)}{\Psi^{-1}\left(\dfrac{1}{\rho_\mu(h)}\right)} = 0. \tag{11.22}$$

(iii) *Assume that $\Psi \in \nabla_0$. The compactness of $J_\mu : H^\Psi \to L^\Psi\left(\overline{\mathbb{D}}, \mu\right)$ is equivalent to the condition (11.20) (hence to (11.22)) and to (11.21).*

Let us mention immediate consequences or remarks:

- In (ii)(c), if $\rho_\mu(h) = 0$ for h small enough (think of the case of $\mu = m_\varphi$ with $\|\varphi\|_\infty < 1$), then the whole quotient expression must be understood as 0.
- When $x \mapsto \log\left(\Psi\left(\mathrm{e}^x\right)\right)$ is convex (see Proposition H.13), conditions (11.20), (11.21), and (11.22) are equivalent and characterize compactness.
- For $\Psi(x) = x^p$ where $p \geqslant 1$, we recover the results of Chapter 5.
- In conditions (11.20) and (11.19), it is equivalent to consider a uniform big-oh condition. For instance, (11.20) is equivalent to the fact that $\rho_\mu(h) \leqslant 1/\Psi\left(A\Psi^{-1}(1/h)\right)$ in a neighborhood of 0 for any $A > 0$. This is clear using the fact that it is true for *every* $A > 0$ and the convexity of Ψ. Actually, it is the meaning of (11.22). In the same spirit, we could also give a formula similar to (11.22), using $K_\mu(h)$, equivalent to (11.19).
- If μ is a Carleson measure, then conditions (11.18) and (11.19) are fulfilled. In other words, if $J_\mu : H^p \to L^p\left(\overline{\mathbb{D}}, \mu\right)$ is bounded for some $p \geqslant 1$ then $J_\mu : H^\Psi \to L^\Psi\left(\overline{\mathbb{D}}, \mu\right)$ is bounded.
- Conditions (11.18) and (11.19) are not equivalent in full generality. Neither are (11.20) and (11.21). See [98] for counter-examples.

To prove this theorem, we need auxiliary results, and the following first result is a localized version of Theorem 5.12. Here again, $u \in \mathbb{T} \mapsto N_a f(u)$ denotes the non-tangential maximal function attached to the region $\Gamma_a(u)$ where $u \in \mathbb{T}$.

Theorem 11.18 (Carleson's Theorem) *For every $f \in H^1$ and every finite positive measure μ on $\overline{\mathbb{D}}$, one has, for every $h \in (0,1]$ and every $t > 0$,*

$$\mu\left(\left\{z \in \overline{\mathbb{D}} : |f(z)| > t \text{ and } |z| > 1-h\right\}\right) \leqslant 2K_\mu(h)\, m\left(\{N_3 f > t\}\right).$$

Proof: We shall handle some families of Carleson's windows so, for convenience, we shall denote, when I is a subarc of $\mathbb{T}$,

$$W(I) = \left\{z \in \overline{\mathbb{D}} : |z| > 1 - m(I) \text{ and } \frac{z}{|z|} \in I\right\}. \tag{11.23}$$

In other words, we have $W(I) = W(\xi, m(I))$, where ξ is the center of I.

Since it is clear that $z \in \Gamma_3(u)$ when $z = |z|\, u$ with $u \in \mathbb{T}$, the condition $|f(z)| > t$ implies that $N_3 f\left(\frac{z}{|z|}\right) > t$ when $z \neq 0$. Hence it suffices to majorize the measure of the set

$$\mathcal{M} = \left\{z \in \overline{\mathbb{D}} : |z| > 1-h \text{ and } N_3 f(z/|z|) > t\right\}.$$

The open set $\{N_3 f > t\}$ is the disjoint union of a countable family of open arcs $I_j \subset \mathbb{T}$.

Now, when $m(I_j) > h$, we may cover I_j in such a way that we can write $I_j \subset A_{j,1} \cup \cdots \cup A_{j,n_j}$ with the arcs $A_{j,1}, \ldots, A_{j,n_j}$ satisfying $m(A_{j,k}) = h$ for every $k = 1, \ldots, n_j$ and $2m(I_j) \geqslant m(I_j) + h \geqslant \sum_{k=1}^{n_j} m(A_{j,k})$.

When $m(I_j) \leqslant h$, we simply define $n_j = 1$ and $A_{1,n_j} = I_j$.

Therefore, for every j, we have $m(A_{j,k}) \in (0,h]$ and, by definition of $K_\mu(h)$, we get

$$\mu\big(W(A_{j,k})\big) \leqslant K_\mu(h)\, m\big(A_{j,k}\big)\cdot \tag{11.24}$$

Given $E \subset \overline{\mathbb{D}}$, let us denote by E_h the set of points $z \in E$ such that $|z| > 1-h$. Since, for every j, we have $\big(W(I_j)\big)_h \subset \bigcup_{1 \leqslant k \leqslant n_j} W(A_{j,k})$, we get, using (11.24),

$$\mu\left(W(I_j)_h\right) \leqslant \sum_{k=1}^{n_j} \mu\left(W(A_{j,k})\right) \leqslant \sum_{k=1}^{n_j} K_\mu(h)\, m\big(A_{j,k}\big) \leqslant 2K_\mu(h)\, m(I_j).$$

On the other hand, $\mathcal{M} \subset \bigcup_j \big(W(I_j)\big)_h$; hence

$$\mu(\mathcal{M}) \leqslant \sum_j \mu\big(W(I_j)_h\big) \leqslant 2K_\mu(h) \sum_j m(I_j) = 2K_\mu(h)\, m(\{N_3 f > t\}),$$

since $\{N_3 f > t\}$ is the disjoint union of the I_j's.

A fortiori, $\mu\big(\{z \in \overline{\mathbb{D}} : |z| > 1-h \text{ and } |f(z)| > t\}\big) \leqslant 2K_\mu(h)\, m(\{N_3 f > t\})$. □

We shall also use the following lemma.

Lemma 11.19 *Let μ be a positive finite Borel measure on $\overline{\mathbb{D}}$. Assume that there exist $A > 0$ and $h_A \in (0,1)$ such that*

$$K_\mu(h) \le \frac{1/h}{\Psi\big(A\Psi^{-1}(1/h)\big)} \qquad \text{for every } h \in (0, h_A).$$

Then, for every f in the unit ball of H^Ψ and every Borel $E \subset \overline{\mathbb{D}}$, we have

$$\int_E \Psi\left(\frac{A}{4}|f|\right) d\mu \leqslant \mu(E)\Psi(x_A) + \int_{\mathbb{T}} \Psi(N_3 f)\, dm$$

where $x_A = \dfrac{A}{2}\Psi^{-1}(1/h_A)$.

Proof: For every $s > 0$ and every $z \in \mathbb{D}$, the inequality $|f(z)| > s$ implies that the norm of the point evaluation at z is greater than s. By Proposition 11.7, we must have

$$|z| > \frac{\Psi(s) - 1}{\Psi(s) + 1} = 1 - \frac{2}{\Psi(s) + 1} > 1 - \frac{1}{\Psi(s/2)},$$

and this inequality still obviously holds when $z \in \mathbb{T}$.

When $s > 2\Psi^{-1}(1)$, we know from Theorem 11.18 that

$$\mu\big(\{|f(z)| > s\}\big) \leqslant 2K_\mu\left(\frac{1}{\Psi(s/2)}\right) m(\{N_3 f > s\}).$$

But our hypothesis means that, as soon as $s > 2\Psi^{-1}(1/h_A)$,

$$K_\mu\left(\frac{1}{\Psi(s/2)}\right) \leqslant \frac{\Psi(s/2)}{\Psi(As/2)}.$$

We then have $\mu\big(\{|f(z)| > 4t/A\}\big) \leqslant 2\dfrac{\Psi(2t/A)}{\Psi(2t)} m\big(\{N_3 f > 4t/A\}\big)$ and we can compute

$$\begin{aligned}
\int_E \Psi\left(\frac{A}{4}|f|\right) d\mu &= \int_0^\infty \Psi'(t)\, \mu(\{|f| > 4t/A\} \cap E)\, dt \\
&\leqslant \int_0^{x_A} \Psi'(t)\, \mu(E)\, dt \\
&\quad + 2\int_{x_A}^{+\infty} \Psi'(t) \frac{\Psi(2t/A)}{\Psi(2t)} m(\{N_3 f > 4t/A\})\, dt \\
&\leqslant \Psi(x_A)\, \mu(E) + 2\int_{x_A}^{+\infty} \frac{\Psi'(t)}{\Psi(2t)} \Psi(2t/A)\, m(\{N_3 f > 4t/A\})\, dt\,.
\end{aligned}$$

But $\Psi(x) \leqslant x\Psi'(x) \leqslant \Psi(2x)$ for any Orlicz function Ψ, so the second integral is controlled by

$$\begin{aligned}
&\int_{x_A}^{\infty} \frac{\Psi'(t)}{\Psi(2t)}\,\Psi(2t/A)\,m(\{N_3 f > 4t/A\})\,dt \\
&\qquad\qquad \leqslant \int_0^{\infty} \frac{\Psi(2t/A)}{t}\,m(\{N_3 f > 4t/A\})\,dt \\
&\qquad\qquad \leqslant \frac{2}{A}\int_0^{\infty} \Psi'(2t/A)\,m(\{N_3 f > 4t/A\})\,dt \\
&\qquad\qquad = \int_0^{\infty} \Psi'(x)\,m(\{N_3 f > 2x\})\,dx = \int_{\mathbb{T}} \Psi\left(\frac{1}{2}N_3 f\right)\,dm \\
&\qquad\qquad \leqslant \frac{1}{2}\int_{\mathbb{T}} \Psi(N_3 f)\,dm.
\end{aligned}$$

We get the conclusion. □

Let us also mention the following remark.

Lemma 11.20 *Let μ be a positive finite Borel measure on $\overline{\mathbb{D}}$ such that the embedding $J_\mu : H^\Psi \to L^\Psi\left(\overline{\mathbb{D}}, \mu\right)$ is compact.*

Then $\mu(\mathbb{T}) = 0$.

Proof: We test the compactness on the sequence (z^n) that lies in the unit ball of H^∞, hence is bounded in $HM^\Psi \subset H^\Psi$. This sequence is weakly converging to 0 by the Riemann–Lebesgue theorem. Therefore $\|z^n\|_{L^\Psi(\overline{\mathbb{D}},\mu)} \to 0$ by compactness of J_μ.

For every $\varepsilon \in (0,1)$, we have

$$\int_{\overline{\mathbb{D}}} \Psi\left(\frac{|z^n|}{\varepsilon}\right)\,d\mu(z) \geqslant \int_{\mathbb{T}} \Psi\left(\frac{1}{\varepsilon}\right)\,d\mu(z) = \mu(\mathbb{T})\,\Psi\left(\frac{1}{\varepsilon}\right).$$

But the LHS becomes lower than 1, since $\|z^n\|_{L^\Psi(\overline{\mathbb{D}},\mu)} \leqslant \varepsilon$ as soon as n is large enough. Recall that $\Psi(x) \to +\infty$ when $x \to \infty$, so we get the conclusion. □

Proof of Theorem 11.17: (i) We first focus on boundedness.

(a) We assume that J_μ is bounded with norm $C > 0$. We fix $\xi \in \mathbb{T}$ and $h \in (0,1]$, and we wish to test the boundedness on the function $f = \Psi^{-1}(1/h)\,u_{\xi,1-h}$ (see (11.3)). Write $a = (1-h)\xi$. From Lemma 11.6 (iii), we know that f lies in the unit ball of HM^Ψ.

For every $z \in W(\xi,h)$, we have $|1-\bar{a}z| \leqslant 6h$. Indeed

$$\begin{aligned}
|1-\bar{a}z| &\leqslant |1-\bar{a}\xi| + |\bar{a}(\xi - z)| \leqslant h + (1-h)\left[\left|\xi - \frac{z}{|z|}\right| + \left|\frac{z}{|z|} - z\right|\right] \\
&\leqslant h + (1-h)[\pi h + (1-|z|)] \leqslant 6h.
\end{aligned}$$

Therefore $|u_{\xi,1-h}(z)| \geqslant \dfrac{1}{36}$ and

$$|f(z)| \geqslant \frac{1}{36}\Psi^{-1}(1/h) \qquad \text{on } W(\xi,h). \tag{11.25}$$

From the boundedness of J_μ, we must have

$$1 \geqslant \int_{\overline{\mathbb{D}}} \Psi\left(\frac{|f|}{C}\right) d\mu \geqslant \Psi\left(\frac{1}{36C}\Psi^{-1}(1/h)\right) \mu\big(W(\xi,h)\big),$$

which is (11.18).

(b) Thanks to the Marcinkiewicz theorem (see Theorem H.15), the maximal (non-tangential) map $f \mapsto N_3 f$ is bounded on $L^\Psi(\mathbb{T})$: there exists a constant $C > 1$ such that $\|N_3 f\|_\Psi \leqslant C\|f\|_\Psi$ for every $f \in L^\Psi(\mathbb{T})$.

We fix f in the unit ball of H^Ψ and $\|f/C\|_\Psi$ remains lower than 1. From Lemma 11.19, with $E = \overline{\mathbb{D}}$ and f replaced by f/C (in our situation $h_A = 1$), denoting $\widetilde{C} = 1 + \mu\big(\overline{\mathbb{D}}\big)\Psi(x_A) > 1$, we have

$$\begin{aligned}
\int_{\overline{\mathbb{D}}} \Psi\left(\frac{A}{4C\widetilde{C}}|f|\right) d\mu &\leqslant \frac{1}{\widetilde{C}} \int_{\overline{\mathbb{D}}} \Psi\left(\frac{A}{4C}|f|\right) d\mu \\
&\leqslant \frac{1}{\widetilde{C}}\left(\mu\big(\overline{\mathbb{D}}\big)\Psi(x_A) + \int_{\mathbb{T}} \Psi\Big(\frac{1}{C}N_3 f\Big)\, dm\right) \\
&\leqslant \frac{1}{\widetilde{C}}\big(\mu\big(\overline{\mathbb{D}}\big)\Psi(x_A) + 1\big) = 1.
\end{aligned}$$

In other words $\|f\|_{L^\Psi(\mu)} \leqslant \dfrac{4C\widetilde{C}}{A}$.

(ii) (a) Suppose that J_μ is compact, but that condition (11.20) is not satisfied. In other words, there exist $\varepsilon_0 \in (0,1)$, $A > 0$, a sequence of positive numbers $(h_n)_n$ decreasing to 0, and a sequence of $\xi_n \in \mathbb{T}$, such that

$$\mu\big(W(\xi_n,h_n)\big) \geqslant \frac{\varepsilon_0}{\Psi\big(A\Psi^{-1}(1/h_n)\big)}.$$

We now use the test functions (11.3), write $a_n = (1-h_n)\xi_n$, and define

$$f_n(z) = \Psi^{-1}\left(\frac{1}{h_n}\right) u_{\xi_n,1-h_n}(z) = \Psi^{-1}\left(\frac{1}{h_n}\right) \frac{h_n^2}{(1-\bar{a}_n z)^2}.$$

We know from Lemma 11.6 that f_n lies in the unit ball of H^Ψ. On the other hand, $(f_n)_n$ clearly converges to 0 in $\mathscr{H}(\mathbb{D})$.

By the compactness criterion (Proposition 3.2), $(f_n)_n$ is norm-converging to 0 in $L^\Psi(\overline{\mathbb{D}},\mu)$.

But, using the estimates already made in the proof of (i) (see (11.25)), we have, for every $n \geqslant 1$,

$$|f_n(z)| \geqslant \frac{1}{36}\Psi^{-1}\left(\frac{1}{h_n}\right)$$

when $z \in W(\xi_n, h_n)$. Hence

$$\begin{aligned}\int_{\overline{\mathbb{D}}} \Psi\left(\frac{36A}{\varepsilon_0}|f_n|\right) d\mu &\geqslant \Psi\left(\frac{A}{\varepsilon_0}\Psi^{-1}(1/h_n)\right) \mu\big(W(\xi_n, h_n)\big) \\ &\geqslant \Psi\left(\frac{A}{\varepsilon_0}\Psi^{-1}(1/h_n)\right) \frac{\varepsilon_0}{\Psi\left(A\Psi^{-1}(1/h_n)\right)} \geqslant 1\end{aligned}$$

by the convexity of Ψ, and this implies that $\|f_n\|_{L^\Psi(\mu)} \geqslant \varepsilon_0/36A$ and gives a contradiction.

(b) We already know that J_μ is bounded thanks to (i)(b) since (11.19) is fulfilled (as a particular case of (11.21)).

Let us also observe that $\mu(\mathbb{T}) = 0$. Indeed, for every $n \geqslant 1$, we can cover $\mathbb{T}$ by n (closed) arcs with arc length $2\pi/n$, so that there exists $\xi_1, \ldots, \xi_n \in \mathbb{T}$ such that

$$\mathbb{T} \subset \bigcup_{j=1}^{n} W(\xi_j, 1/n).$$

A fortiori,

$$\mu(\mathbb{T}) \leqslant \sum_{j=1}^{n} \mu\left(W(\xi_j, 1/n)\right) \leqslant \sum_{j=1}^{n} \frac{1}{n} K_\mu(1/n) = K_\mu(1/n) \longrightarrow 0$$

using (11.21) with $A = 1$, hence $\mu(\mathbb{T}) = 0$, as claimed.

We now use Proposition 11.16: given $\varepsilon > 0$, we shall prove that, for $r \in (0,1)$ small enough, the norm of the injection $I_r : H^\Psi \longrightarrow L^\Psi\left(\overline{\mathbb{D}} \setminus r\mathbb{D}, \mu\right)$ is smaller than ε.

Again (as in (i)(b)) $f \mapsto N_3(f)$ is bounded on $L^\Psi(\mathbb{T})$: there exists $C \geqslant 1$ such that $\|N_3 f\|_\Psi \leqslant C\|f\|_\Psi$ for every $f \in L^\Psi(\mathbb{T})$.

We know from our assumption (11.21), applied with $A = \dfrac{8C}{\varepsilon}$, that there exists $h_0 > 0$ such that

$$\forall h \in (0, h_0) \qquad K_\mu(h) \leqslant \frac{1}{2h} \cdot \frac{1}{\Psi\left(A\Psi^{-1}(1/h)\right)}.$$

Let f be in the unit ball of H^Ψ and $r \in (0,1)$. By Lemma 11.19

$$\begin{aligned}\int_{\overline{\mathbb{D}}\setminus r\mathbb{D}} \Psi\left(\frac{|f|}{\varepsilon}\right) d\mu &= \int_{\overline{\mathbb{D}}\setminus r\mathbb{D}} \Psi\left(\frac{A}{8C}|f|\right) d\mu \leqslant \frac{1}{2}\int_{\overline{\mathbb{D}}\setminus r\mathbb{D}} \Psi\left(\frac{A}{4C}|f|\right) d\mu \\ &\leqslant \frac{1}{2}\mu\big(\overline{\mathbb{D}}\setminus r\mathbb{D}\big)\,\Psi(x_A) + \frac{1}{2}\int_{\mathbb{T}} \Psi\left(\frac{M_f}{C}\right) dm \\ &\leqslant \frac{1}{2}\mu\big(\overline{\mathbb{D}}\setminus r\mathbb{D}\big)\,\Psi(x_A) + \frac{1}{2}.\end{aligned}$$

As $\mu(\mathbb{T}) = 0$, there exists some $r_0 \in (0,1)$ such that $\mu\big(\overline{\mathbb{D}}\setminus r\mathbb{D}\big)\,\Psi(x_A) \leqslant 1$, for every $r \in (r_0, 1)$ and this implies that $\|I_r(f)\|_{L^\Psi(\overline{\mathbb{D}},\mu)} \leqslant \varepsilon$ for every f in the unit ball of H^Ψ and every $r \in (r_0, 1)$.

(c) Proving this is just giving the details of one of the remarks after the statement of Theorem 11.17.

Assume that (11.20) holds and fix any $A > 0$. For h small enough, we have

$$\rho_\mu(h) \leqslant 1/\Psi\big(A\Psi^{-1}(1/h)\big).$$

If $\rho_\mu(h) \neq 0$ (see the remark after the statement of Theorem 11.17), we get

$$\frac{\Psi^{-1}\left(\dfrac{1}{h}\right)}{\Psi^{-1}\left(\dfrac{1}{\rho_\mu(h)}\right)} \leqslant \frac{1}{A},$$

and (11.22) is satisfied.

Now assume that (11.22) is satisfied and fix $\varepsilon \in (0,1)$ and $A > 0$. For h small enough, we have

$$\frac{\Psi^{-1}\left(\dfrac{1}{h}\right)}{\Psi^{-1}\left(\dfrac{1}{\rho_\mu(h)}\right)} \leqslant \frac{\varepsilon}{A};$$

equivalently

$$\rho_\mu(h) \leqslant \frac{1}{\Psi\left(\frac{A}{\varepsilon}\Psi^{-1}(1/h)\right)},$$

but Ψ is convex so $\Psi\left(\dfrac{A}{\varepsilon}\Psi^{-1}(1/h)\right) \geqslant \dfrac{1}{\varepsilon}\Psi\left(A\Psi^{-1}(1/h)\right)$ and we have (11.20).

(iii) Assume that Ψ satisfies condition ∇_0 and that condition (11.20) is fulfilled: for every $\alpha > 0$ and every $h \in (0, h_\alpha)$ (h_α small enough), we have

$$\rho_\mu(h) \leqslant \frac{1}{\Psi\big[\alpha \Psi^{-1}(1/h)\big]} \cdot$$

This implies that

$$K_\mu(h) = \sup_{0<s\leqslant h} \frac{\rho_\mu(s)}{s} \leqslant \sup_{0<s\leqslant h} \frac{1/s}{\Psi\big[\alpha\Psi^{-1}(1/s)\big]} = \sup_{x\geqslant \Psi^{-1}(1/h)} \frac{\Psi(x)}{\Psi(\alpha\, x)} \cdot$$

Fix an arbitrary $A > 1$. From (H.6), we can easily find some $C_A > 1$ such that, as soon as $x \geqslant u$ are large enough,

$$\frac{\Psi(Au)}{\Psi(u)} \leqslant \frac{\Psi(AC_A x)}{\Psi(x)} \cdot \tag{11.26}$$

Now choose $\alpha = AC_A > 1$. Thanks to (11.26), we have, for h small enough and $x \geqslant \Psi^{-1}(1/h)$,

$$\frac{\Psi\big[A\Psi^{-1}(1/h)\big]}{\Psi\big[\Psi^{-1}(1/h)\big]} \leqslant \frac{\Psi(AC_A\, x)}{\Psi(x)} = \frac{\Psi(\alpha x)}{\Psi(x)} \cdot$$

We get, for h small enough,

$$K_\mu(h) \leqslant \frac{\Psi\big[\Psi^{-1}(1/h)\big]}{\Psi\big[A\Psi^{-1}(1/h)\big]} = \frac{1/h}{\Psi\big[A\Psi^{-1}(1/h)\big]},$$

and condition (11.21) is fulfilled.

A fortiori, the compactness of J_μ, (11.20), and (11.21) are all equivalent. □

11.2.3 Compactness of Composition Operators on H^Ψ

We shall focus on the compactness issues for composition operators. Concretely, we shall specialize the results of Section 11.2.2 to the case $\mu = m_\varphi$ where φ is a symbol. Indeed, we recall that, for every $f \in H^\Psi$ and every symbol,

$$\int_{\mathbb{T}} \Psi\left(|f \circ \varphi|\right) dm \leqslant 1 \quad \Longleftrightarrow \quad \int_{\overline{\mathbb{D}}} \Psi\left(|f|\right) dm_\varphi \leqslant 1.$$

In other words, for $f \in H^\Psi$, $\|f \circ \varphi\|_{H^\Psi} = \|f\|_{L^\Psi(\overline{\mathbb{D}}, m_\varphi)}$; therefore $C_\varphi : H^\Psi \to H^\Psi$ is compact if and only if $J_{m_\varphi} : H^\Psi \to L^\Psi(\overline{\mathbb{D}}, m_\varphi)$ is compact and we could apply this general principle as well with many other operator ideals.

We first mention an easy necessary condition for compactness:

Proposition 11.21 *Let φ be a symbol and Ψ be an Orlicz function. Assume that $C_\varphi : H^\Psi \to H^\Psi$ is compact. Then:*

(i) *We have $m_\varphi(\mathbb{T}) = 0$.*
(ii) *We have*

$$\lim_{|z|\to 1^-} \frac{\Psi^{-1}\left(\dfrac{1}{1-|\varphi(z)|}\right)}{\Psi^{-1}\left(\dfrac{1}{1-|z|}\right)} = 0\,. \tag{11.27}$$

In particular, we recover with (i) that, when φ is inner, C_φ cannot be compact on H^Ψ for any Orlicz function Ψ.

Proof: (i) This is just the result of Lemma 11.20 with $\mu = m_\varphi$ (we could also write a variant, invoking Proposition 3.2).

(ii) We quickly reproduce the argument given in the framework of classical Hardy spaces: C_φ^* is also compact and thanks to Proposition 11.7, we know how the norm of the functional δ_z behaves. We also know that $C_\varphi^*(\delta_z) = \delta_{\varphi(z)}$. Then for any sequence $z_n \in \mathbb{D}$ such that $|z_n| \to 1^-$, the sequence

$$\mu_n = \left(\Psi^{-1}\left(\frac{1}{1-|z_n|}\right)\right)^{-1} \delta_{z_n}$$

is bounded in the dual of H^Ψ and is weak-star convergent to 0; hence we must have

$$\|C_\varphi^*(\mu_n)\|_{(H^\Psi)^*} \longrightarrow 0.$$

Using again Proposition 11.7, we get the conclusion. □

We recall the notation (see (5.30)): $\rho_\varphi(h) = \sup_{\xi\in\mathbb{T}} m_\varphi\Big(W(\xi,h)\Big)$.

We shall now apply the results of the preceding section to characterize compactness of $C_\varphi : H^\Psi \to H^\Psi$ and we can state the following theorem:

Theorem 11.22 *Let φ be a symbol and Ψ be an Orlicz function.*

Then the following assertions are equivalent:

(i) *$C_\varphi : H^\Psi \to H^\Psi$ is compact.*
(ii) *For every $A > 0$,*

$$\rho_\varphi(h) = o\left(\frac{1}{\Psi\big(A\Psi^{-1}(1/h)\big)}\right) \qquad \text{as } h \to 0. \tag{11.28}$$

(iii) *We have*

$$\lim_{h\to 0} \frac{\Psi^{-1}\left(\dfrac{1}{h}\right)}{\Psi^{-1}\left(\dfrac{1}{\rho_\varphi(h)}\right)} = 0. \tag{11.29}$$

(iv) *We have*

$$\lim_{|z|\to 1^-} \frac{\Psi^{-1}\left(\dfrac{1}{1-|z|}\right)}{\Psi^{-1}\left(\dfrac{1}{N_\varphi(z)}\right)} = 0. \tag{11.30}$$

(v) *When $h \to 0$, we have*

$$\sup_{\xi\in\mathbb{T}} \|C_\varphi(u_{\xi,1-h})\|_{H^\Psi} = o\left(\frac{1}{\Psi^{-1}(1/h)}\right). \tag{11.31}$$

As mentioned earlier, after Theorem 11.17, when $\rho_\varphi(h) = 0$ the whole quotient in (11.29) has to be understood as 0. Similarly, when $N_\varphi(z) = 0$, the quotient in (11.30) is understood as 0. This is the case when $z \notin \varphi(\mathbb{D})$. In particular, when $\|\varphi\|_\infty < 1$, this whole quotient is equal to 0 on an annulus near the boundary.

Proof: The equivalence between (i)–(ii)–(iii) is a direct consequence of Theorem 11.17, applied to the measure $\mu = m_\varphi$, as soon as we have proved that (11.28) implies that, for every $A > 0$, we have

$$K_{m_\varphi}(h) = o\left(\frac{1/h}{\Psi\big(A\Psi^{-1}(1/h)\big)}\right) \qquad \text{as } h \to 0,$$

but this latter fact is nothing but a direct consequence of Proposition 7.11.

Now assume that $C_\varphi : H^\Psi \to H^\Psi$ is compact or, equivalently, that (11.29) is satisfied.

From Theorem 7.2, we know that, for some constants $C, k > 1$, we have $N_\varphi(z) \leqslant C\rho_\varphi(k(1-|z|))$ when $|z|$ is close enough to 1.

Therefore, with $h = k(1-|z|)$ small enough, we have

$$\frac{\Psi^{-1}\left(\dfrac{1}{1-|z|}\right)}{\Psi^{-1}\left(\dfrac{1}{N_\varphi(z)}\right)} \leqslant \frac{\Psi^{-1}\left(\dfrac{k}{h}\right)}{\Psi^{-1}\left(\dfrac{1}{C\rho_\varphi(h)}\right)} \leqslant Ck\frac{\Psi^{-1}\left(\dfrac{1}{h}\right)}{\Psi^{-1}\left(\dfrac{1}{\rho_\varphi(h)}\right)}$$

since $\Psi^{-1}(Tx) \leqslant T\Psi^{-1}(x)$ when $T > 1$ because Ψ is convex, and we get (11.30).

Now, assume that (11.30) is satisfied. From Theorem 7.4, we know that, for some constants $C, k > 1$, we have

$$\rho_\varphi(h) \leqslant C \sup_{|w| \geqslant 1-kh} N_\varphi(w)$$

as soon as h is small enough.

Therefore, for h small enough we have

$$\frac{\Psi^{-1}\left(\frac{1}{h}\right)}{\Psi^{-1}\left(\frac{1}{\rho_\varphi(h)}\right)} \leqslant \sup_{|w| \geqslant 1-kh} \frac{\Psi^{-1}\left(\frac{k}{1-|w|}\right)}{\Psi^{-1}\left(\frac{1}{CN_\varphi(w)}\right)} \leqslant kC \sup_{|w| \geqslant 1-kh} \frac{\Psi^{-1}\left(\frac{1}{1-|w|}\right)}{\Psi^{-1}\left(\frac{1}{N_\varphi(w)}\right)} \longrightarrow 0$$

when $h \to 0$ by hypothesis and (iv) is proved.

Assuming that $C_\varphi : H^\Psi \to H^\Psi$ is compact, point out that the family of functions $\Psi^{-1}(1/h)\, u_{\xi, 1-h}$ is converging to 0 in $\mathscr{H}(\mathbb{D})$, uniformly in ξ. More precisely, whatever the sequences (ξ_n) in $\mathbb{T}$ and (h_n) converging to 0, the sequence of analytic functions $g_n = \Psi^{-1}(1/h_n)\, u_{\xi_n, 1-h_n}$ converges to 0 in $\mathscr{H}(\mathbb{D})$. This is clear by the definition of the test functions (recall (11.3)) and the fact that $\Psi^{-1}(1/h_n)h_n^2 \to 0$. Thanks to the compactness criterion (Proposition 3.2), $(C_\varphi(g_n))_n$ is norm-converging to 0 in H^Ψ, and we get (11.31).

Conversely, assume that condition (11.31) is satisfied. It means that we have for every $A > 0$, as soon as h is small enough,

$$1 \geqslant \int_{\overline{\mathbb{D}}} \Psi\left(4A\Psi^{-1}(1/h)\, u_{\xi, 1-h}\right) dm_\varphi$$

for every $\xi \in \mathbb{T}$. But, thanks to Lemma 11.6 (see (ii)), this implies that

$$1 \geqslant \int_{S(\xi,h)} \Psi\left(A\Psi^{-1}(1/h)\right) dm_\varphi = m_\varphi(S(\xi,h))\, \Psi\left(A\Psi^{-1}(1/h)\right).$$

Recall that $W(\xi,h) \subset S(\xi,4h)$, so we have

$$\rho_\varphi(h) \leqslant \frac{1}{\Psi\big(A\Psi^{-1}(1/4h)\big)} \quad \text{when } h \to 0,$$

but since $A > 0$ is arbitrary and Ψ is convex, this implies (11.28). □

Let us now mention an immediate corollary. Actually the condition expressed in its statement exactly means that C_φ is order-bounded into $M^\Psi(\mathbb{T}, m)$. We chose to not develop here (see [98] to know more about this notion especially on H^Ψ spaces), but point out that for $\psi(x) = x^p$, this exactly means that $C_\varphi : H^2 \to H^2$ is Hilbert–Schmidt, whereas for very fast-growing function ($\Psi \in \Delta^2$), it is equivalent to compactness (see Proposition 11.26).

Corollary 11.23 *Let Ψ be an Orlicz function.*

Assume that $\Psi^{-1}\left(\dfrac{1}{1-|\varphi^*|}\right) \in M^\Psi(\mathbb{T}, m)$. *Then* $C_\varphi : H^\Psi \to H^\Psi$ *is compact.*

Proof: For every $A > 0$, the integral $\displaystyle\int_{\mathbb{T}} \Psi\left(A\Psi^{-1}\left(\frac{1}{1-|\varphi^*|}\right)\right) dm$ is finite.

Therefore, we have for h small enough (denoting Ω_h the annulus $\overline{\mathbb{D}} \setminus (1-h)\mathbb{D}$)

$$1 \geqslant \int_{\Omega_h} \Psi\left(A\Psi^{-1}\left(\frac{1}{1-|z|}\right)\right) dm_\varphi \geqslant m_\varphi(\Omega_h)\Psi\left(A\Psi^{-1}\left(\frac{1}{h}\right)\right),$$

but $m_\varphi(\Omega_h) \geqslant \rho_\varphi(h)$ and we get (11.28) and $C_\varphi : H^\Psi \to H^\Psi$ is compact. □

Let us mention another interesting consequence of Theorem 11.22 for finitely valent symbols. In that case, the characterization can be written in a simpler way.

Theorem 11.24 *Let Ψ be an Orlicz function.*

(i) *Let φ be a finitely valent symbol.*
Then $C_\varphi : H^\Psi \to H^\Psi$ is compact if and only if (11.27) is satisfied, that is,

$$\lim_{|z|\to 1^-} \frac{\Psi^{-1}\left(\dfrac{1}{1-|\varphi(z)|}\right)}{\Psi^{-1}\left(\dfrac{1}{1-|z|}\right)} = 0\,.$$

(ii) *This is not an equivalence in general: there are some inner functions φ (hence C_φ cannot be compact) satisfying (11.27).*

Proof: (i) From Proposition 11.21 (see (ii)), we already know that (11.27) is a necessary condition.

Now assume that (11.27) is satisfied and that φ is V-valent for some $V \geqslant 1$. In other words, we have, for $z \in \mathbb{D}$ with $|z| \geqslant 1/2$,

$$N_\varphi(z) \leqslant 2V \sup\{1-|u| : \varphi(u) = z\},$$

and there exists a function $\varepsilon : (0,1) \to (0,+\infty)$, with $\lim\limits_{t\to 1^-} \varepsilon(t) = 0$, satisfying

$$\forall u \in \mathbb{D}, \quad \frac{1}{1-|u|} = \Psi\left(\frac{1}{\varepsilon(|u|)}\Psi^{-1}\left(\frac{1}{1-|\varphi(u)|}\right)\right).$$

A fortiori

$$\begin{aligned}\Psi^{-1}\left(\frac{1}{N_\varphi(z)}\right) &\geqslant \inf_{\varphi(u)=z} \Psi^{-1}\left(\frac{1}{2V(1-|u|)}\right)\\ &\geqslant \inf_{\varphi(u)=z} \Psi^{-1}\left(\frac{1}{2V}\Psi\left(\frac{1}{\varepsilon(|u|)}\Psi^{-1}\left(\frac{1}{1-|\varphi(u)|}\right)\right)\right)\\ &\geqslant \inf_{\varphi(u)=z} \frac{1}{2V\varepsilon(|u|)}\Psi^{-1}\left(\frac{1}{1-|z|}\right)\end{aligned}$$

because Ψ is convex, and we obtain

$$\frac{\Psi^{-1}\left(\frac{1}{1-|z|}\right)}{\Psi^{-1}\left(\frac{1}{N_\varphi(z)}\right)} \leqslant 2V \sup_{\varphi(u)=z} \varepsilon(|u|)\,.$$

But keeping (1.16) in mind, that is, $\lim_{r\to 1^-} \inf\{|a| : a \in \mathbb{D} \text{ and } |\varphi(a)| \geqslant r\} = 1$, we have

$$\lim_{|z|\to 1^-} \sup_{\varphi(u)=z} \varepsilon(|u|) = 0;$$

hence (11.30) is satisfied and C_φ is compact.

(ii) Theorem 4.11 provides us with a slow Blaschke product B such that

$$1 - |B(z)| \geqslant H(1-|z|),$$

where

$$\forall x \in (0,1)\,, \qquad H(x) = \frac{1}{\Psi\left[\left(\Psi^{-1}(1/x)\right)^{1/2}\right]}\,. \tag{11.32}$$

C_B cannot be compact thanks to Proposition 11.21 (see (i)).

On the other hand (11.27) is satisfied. □

We recall that an important motivation of this chapter is to try to understand the discontinuity in the charaterization of compactness between H^∞ and classical H^p spaces.

In that spirit, let us first mention an interesting consequence of Theorem 11.22 concerning the specific case of the very fast-growing Orlicz functions Ψ_r (see Appendix H). The proof of this theorem is straightforward using the former results (use also (9.10)) or Proposition 11.26 below and we do not give the details.

Proposition 11.25 *Let $r > 0$ and let φ be a symbol.*
Then the following assertions are equivalent:

(i) $C_\varphi : H^{\Psi_r} \to H^{\Psi_r}$ *is compact.*
(ii) *For every* $\gamma \geqslant 1$, *we have* $\rho_\varphi(h) = o\left(h^\gamma\right)$ *when* $h \to 0$.
(iii) *For every* $\gamma \geqslant 1$, *we have* $\displaystyle\int_{\mathbb{T}} \frac{1}{(1-|\varphi^*|)^\gamma}\, dm < \infty$.

In that case, $C_\varphi : H^2 \to H^2$ *belongs to all the Schatten classes.*

Point out that the compactness of $C_\varphi : H^{\Psi_r} \to H^{\Psi_r}$ does not depend on r: this is a similar phenomenon as for the scale of classical Hardy spaces H^r (even though the concrete characterization is very different).

More generally, for very fast-growing functions, we have the following proposition:

Proposition 11.26 *Let $\Psi \in \Delta^2$ and φ be a symbol.*
Then the following assertions are equivalent:

(i) $C_\varphi : H^\Psi \to H^\Psi$ *is compact.*
(ii) $C_\varphi : H^\Psi \to H^\Psi$ *is weakly compact.*
(iii) $\Psi^{-1}\left(\dfrac{1}{1-|\varphi^*|}\right) \in M^\Psi(\mathbb{T}, m)$.

In that case, in particular, $\|\varphi^n\|_{H^p} = o\left(n^{-s}\right)$ *for every* $p, s \geqslant 1$ *when* $n \to +\infty$. *A fortiori* $C_\varphi : H^p \to H^p$ *is nuclear for any* $p \geqslant 1$.

A variant of this result could also be stated for the embedding J_μ: (iii) then becomes $\Psi^{-1}\left(\dfrac{1}{1-|z|}\right) \in M^\Psi(\mathbb{T}, \mu)$ and the proof would be quite the same.

Proof: $\Psi \in \Delta^2$ so $\Psi(\beta x) \geqslant \left(\Psi(x)\right)^3$ for some $\beta \geqslant 1$ and x large enough.

Assertions (i) and (ii) are equivalent as a special case of Theorem 11.29 since condition Δ^2 implies condition Δ^0.

Assume (i) and fix $C \geqslant 1$. We already know that $|\varphi^*| < 1$ a.e. For $h \in (0,1)$, point out that the annulus $\mathbb{D} \setminus (1-h)\mathbb{D}$ can be covered by N Carleson windows $W(\xi, h)$ with $N \approx \dfrac{1}{h}\cdot$ Therefore, using (11.28) with $A = C\beta$, we have

$$m\{1-|\varphi^*| < h\} \leqslant N\rho_\varphi(h) \lesssim \frac{\rho_\varphi(h)}{h} \leqslant \frac{1}{h\Psi\left(C\beta\Psi^{-1}(1/h)\right)}$$

for h small enough. Then

$$\int_{\mathbb{T}} \Psi\left(C\Psi^{-1}\left(\frac{1}{1-|\varphi^*|}\right)\right) dm < +\infty$$

if and only if

$$\int_1^{+\infty} \Psi'(Ct)\, m\big\{1 - |\varphi^*| < 1/\Psi(t)\big\}\, dt < +\infty. \tag{11.33}$$

But, when t is large enough, we have, since $\Psi(C\beta t) \geqslant \big(\Psi(Ct)\big)^3$,

$$\Psi'(Ct)\, m\big\{1 - |\varphi^*| < 1/\Psi(t)\big\} \leqslant \frac{\Psi'(Ct)\Psi(t)}{\Psi(C\beta t)} \leqslant \frac{\Psi'(Ct)}{\Psi^2(Ct)};$$

hence the integral (11.33) is clearly convergent and (iii) is satisfied.

The converse is a direct consequence of Corollary 11.23.

Concerning the other assertions: the first one comes from Lemma H.9 for $p = 1$. The case $p \geqslant 1$ obviously follows since $\|\varphi\|_\infty \leqslant 1$. Once we know that the sequence $\big(\|\varphi^n\|_{H^p}\big) \in \ell^1$ it is clear: C_φ is the absolutely convergent series of rank 1 operators given by $f \mapsto a_n(f)\, \varphi^n$ where $a_n(f)$ is the nth Taylor coefficient of f. □

The next result focuses on what happens in the framework of Hardy–Orlicz spaces concerning the key difference between the behavior of composition operators on H^∞ and H^p spaces, namely the possibility, or not, to have a symbol "touching the boundary," that is, satisfying $\|\varphi\|_\infty = 1$. It shows that H^∞ is special.

Theorem 11.27 (i) *Given an Orlicz function Ψ, there exists a symbol φ such that φ is* onto *(a fortiori $\|\varphi\|_\infty = 1$) and $C_\varphi : H^\Psi \to H^\Psi$ is compact. Moreover φ is* 4*-valent.*

(ii) *Given a symbol φ such that $\|\varphi\|_\infty = 1$, there exists an Orlicz function Ψ such that $C_\varphi : H^\Psi \to H^\Psi$ is not compact.*

Let us mention that this proposition is proved in [97], but with a weaker statement than (i) (φ is not onto there, but still $\|\varphi\|_\infty = 1$). Actually the argument there is based on Corollary 11.23.

Proof: (i) Take the function $\delta(h) = H(h)$ like in (11.32) and apply Theorem 4.6 to produce an almost surjective and 2-valent symbol φ satisfying

$$\rho_\varphi(h) \leqslant m\big(\{z \in \mathbb{T};\ |\varphi^*(z)| > 1 - h\}\big) \leqslant \delta(h).$$

Now from (11.29) in Theorem 11.22, we know that $C_\varphi : H^\Psi \to H^\Psi$ is compact.

Take $\Phi = \varphi_a^2 \circ \varphi$ for some $a \in \mathbb{D} \setminus \{0\}$. From the argument given in Section 4.4, we know that $\Phi(\mathbb{D}) = \mathbb{D}$ and that Φ is 4-valent.

Moreover $C_\Phi : H^\Psi \to H^\Psi$ is compact since $C_\Phi = C_\varphi \circ C_{\varphi_a^2}$.

(ii) Since $\|\varphi\|_\infty = 1$, there exists a sequence (z_k) in $\mathbb{D}$ converging to a point of the boundary such that $N_\varphi(z_k) \neq 0$ for every $k \in \mathbb{N}$, equivalently $\rho_\varphi(h) > 0$ for every $h \in (0, 1)$. We are going to show that (11.28) is not satisfied for a suitable Ψ. It suffices to construct (by induction) a piecewise linear strictly increasing convex function $\Psi : \mathbb{R}^+ \to \mathbb{R}^+$ such that

$$\forall n \in \mathbb{N}, \qquad \Psi\big(2^{n+1}\big) \geqslant \frac{1}{\rho_\varphi\big(1/\Psi\big(2^n\big)\big)}.$$

Writing $h_n = \dfrac{1}{\Psi\big(2^n\big)}$, we have

$$\rho_\varphi(h_n) \geqslant \frac{1}{\Psi\big(2\Psi^{-1}(1/h_n)\big)},$$

and (11.28) is not satisfied, hence $C_\varphi : H^\Psi \to H^\Psi$ is not compact. □

We can wonder when the behavior of a composition operator will always look like the classical framework, from the viewpoint of compactness. The next result shows that the condition Δ_2 precisely reflects this.

Proposition 11.28 *Let Ψ be an Orlicz function. The following assertions are equivalent:*

(i) $\Psi \in \Delta_2$.
(ii) *For every symbol, $C_\varphi : H^\Psi \to H^\Psi$ is compact if and only if $C_\varphi : H^2 \to H^2$ is compact.*

Proof: The fact that (i) implies (ii) is clear from Theorem 11.22: see (11.28).

For the converse, we refer to [97, theorem 4.2] for a construction of a symbol not compact on H^Ψ but compact on H^2 as soon as $\Psi \notin \Delta_2$. □

Weak Compactness

We recall that we assumed $\Psi \in \nabla_2$, and it turns out that this plays a key role here.

Theorem 11.29 *Let $\varphi : \mathbb{D} \to \mathbb{D}$ be a symbol.*

(i) *The operator $C_\varphi : H^\Psi \to H^\Psi$ is a Dunford–Pettis operator if and only if it is compact.*
(ii) *Assume that the Orlicz function Ψ satisfies condition Δ^0. Then, the operator $C_\varphi : H^\Psi \to H^\Psi$ is weakly compact if and only if it is compact.*

It is not surprising that some specific condition appears in (ii) for Ψ: when $\Psi(x) = x^p$ with $p > 1$, the composition operator C_φ is always weakly compact.

Proof: (i) This is mainly the argument given in Theorem 11.22 to prove that C_φ is compact if and only if (11.31) is satisfied. More precisely, let us assume that $C_\varphi : H^\Psi \to H^\Psi$ is a Dunford–Pettis operator. Whatever the sequences (ξ_n) in $\mathbb{T}$ and (h_n) converging to 0, the sequence of analytic functions $g_n = \Psi^{-1}(1/h_n)u_{\xi_n, 1-h_n}$ converges to 0 in $\mathscr{H}(\mathbb{D})$: in other words, (g_n) is weak-star converging to 0 in H^Ψ; but g_n belongs to HM^Ψ; therefore (g_n) is actually weakly converging to 0 in HM^Ψ. But the restriction of C_φ to HM^Ψ is Dunford–Pettis as well, and we conclude that $C_\varphi(g_n)$ is norm-converging to 0, that is, we get (11.31).

(ii) Assume that $C_\varphi : H^\Psi \to H^\Psi$ is weakly compact, then $C_\varphi : HM^\Psi \to H^\Psi$ is weakly compact as well. We now use a result proved in [95, theorem 4] using that $\Psi \in \Delta^0$: for every $\varepsilon > 0$, we can find $C_\varepsilon > 0$ such that, for every $f \in HM^\Psi$, one has

$$\|C_\varphi(f)\|_\Psi \leqslant C_\varepsilon \|f\|_1 + \varepsilon \, \|f\|_\Psi.$$

Using our test functions $u_{\xi, 1-h}$, where $\xi \in \mathbb{T}$ and $h \in (0, 1)$, and Lemma 11.6, we get

$$\|C_\varphi(u_{\xi, 1-h})\|_{H^\Psi} \leqslant C_\varepsilon h + \varepsilon \, \frac{1}{\Psi^{-1}(\frac{1}{h})}.$$

But $\frac{\Psi(x)}{x} \underset{x \to +\infty}{\longrightarrow} +\infty$ so $h\Psi^{-1}(\frac{1}{h}) \underset{h \to 0}{\longrightarrow} 0$ and (11.31) is satisfied. □

11.3 Composition Operators on Bergman–Orlicz Spaces

11.3.1 Boundedness

As in the classical case, we can deduce the boundedness of every composition operator on Bergman–Orlicz spaces from that on Hardy–Orlicz spaces. The following theorem justifies this fact and we also give some upper and lower estimates of the norm of the composition operators on Bergman–Orlicz spaces.

Theorem 11.30 *Let φ be a symbol.*

(i) *Assume that φ fixes the origin. Then the composition operator C_φ is a contraction on $\mathscr{B}^\Psi$.*
(ii) *The composition operator C_φ is bounded on $\mathscr{B}^\Psi$ and*

$$\frac{1}{(1 + |\varphi(0)|)^4 \Psi^{-1}(1)} \Psi^{-1}\left(\left(\frac{1 + |\varphi(0)|}{1 - |\varphi(0)|}\right)^2\right) \leqslant \|C_\varphi\|_{B(\mathscr{B}^\Psi)} \leqslant \left(\frac{1 + |\varphi(0)|}{1 - |\varphi(0)|}\right)^2.$$

(iii) *C_φ also maps $\mathscr{B}M^\Psi$ into $\mathscr{B}M^\Psi$, and $C_\varphi\colon \mathscr{B}^\Psi \to \mathscr{B}^\Psi$ is the bi-adjoint of the composition operator $C_\varphi\colon \mathscr{B}M^\Psi \to \mathscr{B}M^\Psi$. Moreover $\|C_\varphi\|_{B(\mathscr{B}^\Psi)} = \|C_\varphi\|_{B(\mathscr{B}M^\Psi)}$.*

(iv) *When $\Psi \in \nabla_1$, we have*

$$\|C_\varphi\|_{B(\mathscr{B}^\Psi)} \approx \Psi^{-1}\left(\left(\frac{1+|\varphi(0)|}{1-|\varphi(0)|}\right)^2\right).$$

Note that the functions $\Psi(x) = x^p$ for $p \geqslant 1$ and the functions $\Psi = \Psi_r$, with $r > 0$ (see Appendix H) satisfy $\Psi \in \nabla_1$. In particular, from (iv), we have (recall Theorem 11.13)

$$\|C_\varphi\|_{B(\mathscr{B}^{\Psi_r})} \approx \|C_\varphi\|_{B(H^{\Psi_r})} \approx \left[\log\left(\frac{2}{1-|\varphi(0)|}\right)\right]^{\frac{1}{r}}.$$

The proof below actually gives the lower estimate:

$$\|C_\varphi\|_{B(\mathscr{B}^\Psi)} \geqslant \sup_{z\in\mathbb{D}} \frac{1}{(1+|\varphi(z)|)^4} \cdot \frac{\Psi^{-1}\left(\left(\frac{1+|\varphi(z)|}{1-|\varphi(z)|}\right)^2\right)}{\Psi^{-1}\left(\left(\frac{1+|z|}{1-|z|}\right)^2\right)}. \tag{11.34}$$

Proof: (i) We follow the strategy of Section 2.4.2, exploiting (ii) in Theorem 11.13.

Let f be in the unit ball of $\mathscr{B}^\Psi$. For every $r \in (0,1)$, we know from the Schwarz lemma that the analytic function defined by

$$\forall z \in \mathbb{D}, \qquad \chi_r(z) = \frac{1}{r}\varphi(rz),$$

is a symbol, also fixing the origin.

On the other hand, the function $f_r(z) = f(rz)$ belongs to $H^\infty \subset H^\Psi$ and we can apply (11.12) with the symbol χ_r, which fixes the origin:

$$\int_{\mathbb{T}} \Psi(|f\circ\varphi(r\xi)|)\,dm(\xi) = \int_{\mathbb{T}} \Psi(|f_r\circ\chi_r(\xi)|)\,dm(\xi) \leqslant \int_{\mathbb{T}} \Psi(|f(r\xi)|)\,dm(\xi).$$

Now integrating over r with respect to the measure $2r dr$, we get

$$\begin{aligned}\int_{\mathbb{D}} \Psi(|f\circ\varphi(z)|)\,d\sigma(z) &= \int_0^1 \int_{\mathbb{T}} \Psi(|f\circ\varphi(r\xi)|)\,dm(\xi)\,2r dr\\ &\leqslant \int_0^1 \int_{\mathbb{T}} \Psi(|f(r\xi)|)\,dm(\xi)\,2r dr\\ &= \int_{\mathbb{D}} \Psi(|f(z)|)\,d\sigma(z) \leqslant 1.\end{aligned}$$

In other words, $\|C_\varphi(f)\|_{\mathscr{B}^\Psi} \leqslant 1$ and C_φ is a contraction on $\mathscr{B}^\Psi$.

(ii) For the lower estimate, we follow the lines of the proof of (i) in Theorem 11.13: we apply (2.28) in Proposition 2.15:

$$\|\delta_{\varphi(z)}\|_{(\mathscr{B}^\Psi)^*} \leqslant \|C_\varphi\|_{B(\mathscr{B}^\Psi)} \|\delta_z\|_{(\mathscr{B}^\Psi)^*}. \tag{11.35}$$

We also know from (11.8) that, for every $a \in \mathbb{D}$,

$$\frac{1}{(1+|a|)^4}\Psi^{-1}\left(\left(\frac{1+|a|}{1-|a|}\right)^2\right) \leqslant \|\delta_a\|_{(\mathscr{B}^\Psi)^*} \leqslant \Psi^{-1}\left(\left(\frac{1+|a|}{1-|a|}\right)^2\right). \tag{11.36}$$

From relations (11.35) and (11.36), we get (11.34) and the specific case $z = 0$ gives the lower estimate.

For the upper estimate, we first focus on the case of a symbol given by an automorphism φ_a with $a \in \mathbb{D}$. In this case, we can perform a "true" change of variable and we have, for every f in the unit ball of $\mathscr{B}^\Psi$,

$$\begin{aligned} \int_{\mathbb{D}} \Psi(|f \circ \varphi_a(z)|)\, d\sigma(z) &= \int_{\mathbb{D}} \Psi(|f(z)|)\, |\varphi_a'(z)|^2\, d\sigma(z) \\ &= \int_{\mathbb{D}} \Psi(|f(z)|) \frac{(1-|a|^2)^2}{|1-\bar{a}z|^4}\, d\sigma(z) \\ &\leqslant \int_{\mathbb{D}} \Psi(|f(z)|) \left(\frac{1+|a|}{1-|a|}\right)^2 d\sigma(z) \leqslant \left(\frac{1+|a|}{1-|a|}\right)^2 \end{aligned}$$

since $\|f\|_{\mathscr{B}^\Psi} \leqslant 1$, and this proves that

$$\|C_{\varphi_a}\|_{B(\mathscr{B}^\Psi)} \leqslant \left(\frac{1+|a|}{1-|a|}\right)^2. \tag{11.37}$$

The general case follows once again from the Dévissage principle 2.3.2.

To justify (iii) and (iv), it suffices to mimick the proof of (iii) and (iv) in the proof of Theorem 11.13. □

11.3.2 Carleson Measures on $\mathscr{B}^\Psi$

As for Hardy–Orlicz spaces, we aim to provide here a full characterization of the compactness of composition operators on $\mathscr{B}^\Psi$ using the viewpoint of Carleson measures.

Nevertheless, in the present section, as in the case of the classical spaces (H^p and $\mathscr{B}^p$ spaces), we have to be aware that the associated objects – Carleson square and circular window – are not exactly the same. Actually, this is due to the fact that there are no radial limits in general for Bergman functions, hence *we always work here on* $\mathbb{D}$ *and not on* $\overline{\mathbb{D}}$.

We focus not only on the pullback measure σ_φ but more generally on a (positive finite) measure μ on $\mathbb{D}$.

Hence, given a positive finite measure μ on $\mathbb{D}$, we denote (without specifying Ψ for short) the "embedding" map by

$$J_\mu : \begin{array}{ccc} \mathscr{B}^\Psi & \longrightarrow & L^\Psi(\mathbb{D},\mu) \\ f & \longmapsto & f. \end{array} \tag{11.38}$$

Be aware that we use the same notation as for Hardy spaces. Nevertheless, in the context, there should be no ambiguity.

Let us mention here the following lemma:

Lemma 11.31 *Assume that $J_\mu : \mathscr{B}^\Psi \longrightarrow L^\Psi(\mathbb{D},\mu)$ is bounded.*

Then it is weak-star continuous.

Proof: The proof mimicks the one of Lemma 11.15, just replacing H^Ψ by $\mathscr{B}^\Psi$ and $\overline{\mathbb{D}}$ by $\mathbb{D}$, so we do not give the details. □

The main geometrical tools are again the Carleson windows, but, as already mentioned, we use here the versions involving $\mathbb{D}$, and not $\overline{\mathbb{D}}$. Nevertheless, we keep the same notations here (in accordance with established practice):

The "square" Carleson box

$$W(\xi,h) = \big\{z \in \mathbb{D} : 1 - |z| < h \text{ and } |\arg(\bar{\xi} z)| \leqslant \pi h\big\}$$

and the "circular" Carleson windows

$$S(\xi,h) = \{z \in \mathbb{D} :\ |z - \xi| < h\}.$$

We recall that, for $h \in (0,1)$ and $\xi \in \mathbb{T}$, $W(\xi,h/4) \subset S(\xi,h) \subset W(\xi,h)$.

About the fact that we now "forget" the boundary part, that is, the part lying in $\mathbb{T}$, we can also justify that keeping the same notations is safe: indeed, we focus in the sequel of this section on measures μ carried by $\mathbb{D}$, so that, anyway, even if we took some Borel set $A \subset \overline{\mathbb{D}}$, we could set $\mu(A) = \mu(A \cap \mathbb{D})$.

We still write

$$\rho_\mu(h) = \sup_{\xi \in \mathbb{T}} \mu\big(W(\xi,h)\big)$$

and

$$K_{\mu,2}(h) = \sup_{0<t\leqslant h} \frac{\rho_\mu(t)}{t^2}.$$

The measure μ is a Carleson measure for classical Bergman spaces (i.e., J_μ is bounded on $\mathscr{B}^p$ where $p \geqslant 1$) if and only if, for some constant $C > 0$, we have $\mu\big(S(\xi,h\big) \leqslant Ch^2$ for every $\xi \in \mathbb{T}$. In other words μ is a Carleson measure for classical Bergman spaces if and only if $K_{\mu,2}(h)$ is bounded on $(0,1]$, and this happens as soon as $K_{\mu,2}(h_0)$ is finite for some $h_0 \in (0,1)$.

In the same way, the results of Chapter 5 on compactness on Bergman spaces express that J_μ is compact on some $\mathscr{B}^p$ if and only if μ is a vanishing 2-Carleson measure, that is, $\rho_\mu(h) = o(h^2)$ when $h \to 0$. In other words, μ is vanishing 2-Carleson if and only if $K_{\mu,2}(h) \longrightarrow 0$ when $h \to 0$.

Before stating the main theorem of this section, let us mention the following result, which is the Bergman version of Proposition 11.16. Here also it emphasizes the role of the restriction operator

$$I_r(f) = f\mathbb{1}_{\mathbb{D}\setminus r\mathbb{D}}$$

where $r \in (0,1)$.

Proposition 11.32 *Let μ be a finite Borel measure on $\mathbb{D}$. The following assertions are equivalent:*

(i) *J_μ is bounded and moreover compact.*
(ii) *Every sequence in the unit ball of $\mathscr{B}^\Psi$, which is convergent to 0 in $\mathscr{H}(\mathbb{D})$, is norm-convergent to 0 in $L^\Psi(\mathbb{D},\mu)$.*
(iii) *J_μ is bounded and $\lim_{r\to 1^-} \|I_r\|_{B(\mathscr{B}^\Psi,\, L^\Psi(\mathbb{D},\mu))} = 0$.*

Proof: It suffices to reproduce the proof given for Proposition 11.16. □

Now we can state the main result of this section.

Theorem 11.33 *Let μ be a finite positive Borel measure on $\mathbb{D}$. Then:*

(i) (a) *If $J_\mu : \mathscr{B}^\Psi \to L^\Psi(\mathbb{D},\mu)$ is (defined and) bounded, then there exists some $A > 0$ such that*

$$\rho_\mu(h) \leqslant \frac{1}{\Psi\big(A\Psi^{-1}(1/h^2)\big)} \qquad \textit{for every } h \in (0,1]. \tag{11.39}$$

(b) *If there exists some $A > 0$ such that*

$$K_{\mu,2}(h) \leqslant \frac{1/h^2}{\Psi\big(A\Psi^{-1}(1/h^2)\big)} \qquad \textit{for every } h \in (0,1], \tag{11.40}$$

then $J_\mu : \mathscr{B}^\Psi \to L^\Psi(\mathbb{D},\mu)$ is bounded.

(ii) (a) *If $J_\mu : \mathscr{B}^\Psi \to L^\Psi(\mathbb{D},\mu)$ is compact, then*

$$\rho_\mu(h) = o\left(\frac{1}{\Psi\big(A\Psi^{-1}(1/h^2)\big)}\right) \qquad \textit{as } h \to 0, \textit{ for every } A > 0. \tag{11.41}$$

(b) *If, for every $A > 0$,*

$$K_{\mu,2}(h) = o\left(\frac{1/h^2}{\Psi\big(A\Psi^{-1}(1/h^2)\big)}\right) \qquad \text{as } h \to 0\,, \tag{11.42}$$

then $J_\mu : \mathscr{B}^\Psi \to L^\Psi(\mathbb{D}, \mu)$ is compact.

(c) *Condition (11.41) is equivalent to*

$$\lim_{h\to 0} \frac{\Psi^{-1}\left(\dfrac{1}{h^2}\right)}{\Psi^{-1}\left(\dfrac{1}{\rho_\mu(h)}\right)} = 0\,. \tag{11.43}$$

(iii) *Assume that $\Psi \in \nabla_0$. The compactness of $J_\mu : \mathscr{B}^\Psi \to L^\Psi(\mathbb{D}, \mu)$ is equivalent to the condition (11.41) (hence to (11.43)) and to (11.42).*

We mention that, in the same spirit as for Hardy–Orlicz spaces:

- In (ii)(c), if $\rho_\mu(h) = 0$ for h small enough (think of the case of $\mu = \sigma_\varphi$ with $\|\varphi\|_\infty < 1$), then the whole quotient expression must be understood as 0.
- When $x \mapsto \log\big(\Psi(\mathrm{e}^x)\big)$ is convex, conditions (11.41), (11.42), and (11.43) are equivalent and characterize compactness.
- For $\Psi(x) = x^p$ where $p \geqslant 1$, we recover the results of Chapter 5.
- In conditions (11.41) and (11.40), it is equivalent to consider a uniform big-oh condition. Moreover, we could also give a formula similar to (11.43), using $K_{\mu,2}(h)$, equivalent to (11.40).
- If μ is a 2-Carleson measure, then conditions (11.39) and (11.40) are fulfilled. In other words, if $J_\mu : \mathscr{B}^p \to L^p(\mathbb{D}, \mu)$ is bounded for some $p \geqslant 1$ then $J_\mu : \mathscr{B}^\Psi \to L^\Psi(\mathbb{D}, \mu)$ is bounded.
- Conditions (11.39) and (11.40) are not equivalent in full generality. Neither are (11.41) and (11.42). See [102] for counter-examples.

Even though the general strategy of the proof looks like that of Theorem 11.17, with quite similar intermediate results, the concrete arguments are quite different and we must give the details. We first focus on several required lemmas and we shall need the following dyadic maximal function:

$$\Lambda_f = \sum_{k=0}^{\infty} \Big(\sup_{\Delta_k} |f|\Big)\, \mathbb{1}_{\Delta_k}, \tag{11.44}$$

where (Δ_k) is an enumeration of the dyadic sets

$$\Delta_k = \left\{ z \in \mathbb{D}\,;\; 1 - \frac{1}{2^n} \leqslant |z| < 1 - \frac{1}{2^{n+1}} \text{ and } \frac{(2j-1)\pi}{2^n} \leqslant arg(z) < \frac{(2j+1)\pi}{2^n} \right\},$$

where $k = 2^n + j - 1$ with $n \geqslant 0$ and $0 \leqslant j \leqslant 2^n - 1$ (note that $\Delta_0 = D(0, 1/2)$).

Lemma 11.34 *Let Ψ be an Orlicz function and μ be a positive Borel measure on $\mathbb{D}$ such that $J_\mu : \mathscr{B}^1 \to L^1(\mathbb{D}, \mu)$ is bounded. Then:*

(i) *For every $\alpha > -1$, the embedding $J_{\mu_\alpha} : \mathscr{B}^1_\alpha \to L^1(\mathbb{D}, \mu_\alpha)$ is bounded (with a norm depending on the norm of J_μ on $\mathscr{B}^1$ and α only), where the measure μ_α is defined by $d\mu_\alpha(z) = (1 - |z|^2)^\alpha \, d\mu(z)$.*

(ii) *$J_\mu : \mathscr{B}^\Psi \to L^\Psi(\mathbb{D}, \mu)$ is bounded (with a norm depending on the norm of J_μ on $\mathscr{B}^1$ only).*

Proof: (i) This is just the Carleson embedding theorem for weighted Bergman spaces. Indeed, we just apply Theorem 5.15 since for any Carleson window W with opening $h > 0$, we have

$$\mu_\alpha(W) = O(h^\alpha \mu(W)) = O(h^{\alpha+2}).$$

(ii) Let $C \geqslant 1$ to be specified later and let f be in the unit ball of $\mathscr{B}^\Psi$. From (11.9), we know that, for every $z \in \mathbb{D}$,

$$\Psi\left(\frac{|f(z)|}{C}\right) \leqslant \int_{\mathbb{D}} \Psi\left(\frac{|f(w)|}{C}\right) . |\varphi_z'(w)|^2 \, d\sigma(w)$$

(this is just the Jensen inequality applied to the representation of f with the Berezin kernel).

Now integrating over $\mathbb{D}$ with respect to μ and using Fubini, we get

$$\begin{aligned}\int_{\mathbb{D}} \Psi\left(\frac{|f(z)|}{C}\right) d\mu(z) &\leqslant \int_{\mathbb{D}} \Psi\left(\frac{|f(w)|}{C}\right) \int_{\mathbb{D}} |\varphi_z'(w)|^2 \, d\mu(z) \, d\sigma(w) \\ &= \int_{\mathbb{D}} \Psi\left(\frac{|f(w)|}{C}\right) \int_{\mathbb{D}} \left|\frac{1}{(1 - \bar{w}z)^4}\right| (1 - |z|^2)^2 d\mu(z) \, d\sigma(w).\end{aligned}$$

Now we can use (i) with $\alpha = 2$ and we get a constant $C_2 > 0$ such that

$$\int_{\mathbb{D}} \left|\frac{1}{(1 - \bar{w}z)^4}\right| (1 - |z|^2)^2 d\mu(z) \leqslant C_2 \int_{\mathbb{D}} \frac{1}{|1 - \bar{w}z|^4} \, d\sigma_2(z) = 3C_2.$$

Choosing $C = \max\{1, 3C_2\}$, we obtain

$$\int_{\mathbb{D}} \Psi\left(\frac{|f(z)|}{C}\right) d\mu(z) \leqslant C \int_{\mathbb{D}} \Psi\left(\frac{|f(w)|}{C}\right) d\sigma(w) \leqslant \int_{\mathbb{D}} \Psi(|f(w)|) \, d\sigma(w) = 1$$

because Ψ is convex and $1/C \leqslant 1$. The result follows. □

Lemma 11.35 *For every Orlicz function Ψ, the map $f \in \mathscr{B}^\Psi \mapsto \Lambda_f \in L^\Psi(\mathbb{D}, \sigma)$ is bounded.*

Proof: Since $\Lambda_{tf} = |t|\Lambda_f$, it suffices to work with f in the unit ball of $\mathscr{B}^\Psi$. We set $c_k = \sup_{\Delta_k} |f|$ for any $k \geqslant 1$. There exists $\alpha_k \in \Delta_k$ such that

$$|f(\alpha_k)| \geqslant \frac{1}{2} \sup_{\Delta_k} |f| = \frac{c_k}{2} \cdot$$

Let μ be the measure $\mu = \sum_{k \geqslant 0} \sigma(\Delta_k)\, \delta_{\alpha_k}$. For every Carleson window W, we can write

$$\mu(W) = \sum_{\alpha_k \in W} \sigma(\Delta_k) \leqslant \sum_{\Delta_k \cap W \neq \emptyset} \sigma(\Delta_k) = \sigma\Big(\bigcup_{\Delta_k \cap W \neq \emptyset} \Delta_k \Big)$$

since the Δ_k's are disjoint.

Since $\bigcup_{\Delta_k \cap W \neq \emptyset} \Delta_k$ is contained in the window $\widetilde{W}$ with the same center as W, but with size two times that of W, one has $\mu(W) \leqslant \sigma(\widetilde{W}) = 4\sigma(W)$. Hence μ is a 2-Carleson measure; in other words, it is a Bergman–Carleson measure and we can use Lemma 11.34: there exists C_0, not depending on f (point out that the measure μ depends on f since the α_k's do) such that

$$\int_{\mathbb{D}} \Psi\left(\frac{|f|}{C_0}\right) d\mu \leqslant 1.$$

A fortiori,

$$\begin{aligned} \int_{\mathbb{D}} \Psi\left(\frac{\Lambda_f}{2C_0}\right) d\sigma &= \sum_{k \geqslant 0} \Psi\left(\frac{c_k}{2C_0}\right) \sigma(\Delta_k) \\ &\leqslant \sum_{k \geqslant 0} \Psi\left(\frac{|f(\alpha_k)|}{C_0}\right) \sigma(\Delta_k) \\ &= \int_{\mathbb{D}} \Psi\left(\frac{|f|}{C_0}\right) d\mu \leqslant 1, \end{aligned}$$

which exactly gives the conclusion of the lemma. □

Lemma 11.36 *For every $f \in \mathscr{B}^1$ and every finite positive Borel measure μ on $\mathbb{D}$, one has, for $h \in (0, 1/2)$ and $t > 0$,*

$$\mu(\{z \in \mathbb{D};\ |z| > 1 - h \text{ and } |f(z)| > t\}) \leqslant 4\, K_{\mu,2}(2h)\, \sigma(\{\Lambda_f > t\}).$$

Proof: Remark that when $z \in \Delta_k$ and $|z| > 1 - h$, we must have $h > 2^{-(n+1)}$ since $1 - 2^{-(n+1)} > |z| > 1 - h$. Moreover, since $k = 2^n + j - 1 \geqslant 2^{n-1}$, we also must have $k \geqslant 1/4h$.

Fix $t > 0$. Let us define the sets $I = \big\{k \geqslant 1;\ \sup_{\Delta_k} |f| > t\big\}$ and $I_h = \big\{k \geqslant 1/4h;\ \sup_{\Delta_k} |f| > t\big\}$. Let W_k be the smallest Carleson window containing Δ_k; its opening is less than $2h$ and we have

$$\begin{aligned}\mu\big(\{z \in \mathbb{D};\ |z| > 1-h \quad \text{and} \quad |f(z)| > t\}\big) &\leqslant \sum_{k\in I_h} \mu(\Delta_k) \leqslant \sum_{k\in I_h} \mu(W_k)\\ &\lesssim \sum_{k\in I_h} K_{\mu,2}(2h)\,\sigma(W_k) \leqslant 4\sum_{k\in I_h} K_{\mu,2}(2h)\,\sigma(\Delta_k)\\ &\leqslant 4K_{\mu,2}(2h)\sum_{k\in I}\sigma(\Delta_k) = 4K_{\mu,2}(2h)\,\sigma(\{\Lambda_f > t\}),\end{aligned}$$

and Lemma 11.36 is proved. □

Lemma 11.37 *Let μ be a finite Borel measure on $\mathbb{D}$ and Ψ be an Orlicz function. We assume that, for some positive constant A, there is $h_A \in (0,1]$ such that, for every $h \in (0, h_A)$,*

$$K_{\mu,2}(h) \leqslant \frac{1/h^2}{\Psi[A\Psi^{-1}(1/h^2)]}\cdot$$

Then, for every f in the unit ball of $\mathscr{B}^\Psi$ and every Borel subset E of $\mathbb{D}$, one has, with $x_A = \dfrac{A}{32}\Psi^{-1}(16/h_A^2)$,

$$\int_E \Psi\left(\frac{A\,|f|}{32}\right) d\mu \leqslant \mu(E)\Psi(x_A) + \frac{1}{16}\int_{\mathbb{D}} \Psi(\Lambda_f)\,d\sigma\,.$$

Proof: For every $s > 0$, the inequality $|f(z)| > s$ implies that the norm of the evaluation δ_z at z is greater than s. But we know from (11.8) that

$$\|\delta_z\|_{(\mathscr{B}^\Psi)^*} \leqslant \Psi^{-1}\left(\left(\frac{2}{1-|z|}\right)^2\right);$$

therefore,

$$|z| > 1 - \frac{2}{\sqrt{\Psi(s)}}\cdot$$

Lemma 11.36 gives, when $\Psi(s) \geqslant 4$,

$$\begin{aligned}\mu(\{|f(z)| > s\}) &\leqslant \mu\left(\left\{|z| > 1 - \frac{2}{\sqrt{\Psi(s)}} \text{ and } |f(z)| > s\right\}\right)\\ &\leqslant 4\,K_{\mu,2}\left(\frac{4}{\sqrt{\Psi(s)}}\right)\sigma(\{\Lambda_f > s\})\,.\end{aligned}$$

But, by our assumption, if $\Psi(s) \geqslant 16/h_A^2$,

$$K_{\mu,2}\left(\frac{4}{\sqrt{\Psi(s)}}\right) \leqslant \frac{\Psi(s)/16}{\Psi\big[A\Psi^{-1}\big(\Psi(s)/16\big)\big]} \leqslant \frac{\Psi(s)}{16\Psi\,(As/16)};$$

hence

$$\mu\left(\{|f(z)| > s\}\right) \leqslant \frac{\Psi(s)}{16\Psi(As/16)}\,\sigma(\{\Lambda_f > s\}).$$

We can now estimate

$$\begin{aligned}
\int_E \Psi(A|f|/32)\,d\mu &= \int_0^{+\infty} \Psi'(t)\,\mu(\{|f| > 32t/A\} \cap E)\,dt \\
&\leqslant \int_0^{x_A} \Psi'(t)\,\mu(E)\,dt \\
&\quad + \int_{x_A}^{+\infty} \Psi'(t)\,\frac{\Psi(32t/A)}{16\Psi(2t)}\,\sigma(\{\Lambda_f > 32t/A\})\,dt \\
&\leqslant \Psi(x_A)\,\mu(E) \\
&\quad + \int_{x_A}^{+\infty} \frac{\Psi'(t)}{16\Psi(2t)}\,\Psi(32t/A)\,\sigma(\{\Lambda_f > 32t/A\})\,dt\,.
\end{aligned}$$

But, as Ψ is an Orlicz function, one has for every $t > 0$, $t\Psi'(t) \leqslant \Psi(2t)$, and $\Psi(32t/A) \leqslant (32t/A)\Psi'(32t/A)$; hence

$$\begin{aligned}
\int_{x_A}^{+\infty} &\frac{\Psi'(t)}{16\Psi(2t)}\,\Psi(32t/A)\,\sigma(\{\Lambda_f > 32t/A\})\,dt \\
&\leqslant \frac{2}{A}\int_0^{+\infty} \Psi'(32t/A)\,\sigma(\{\Lambda_f > 32t/A\})\,dt \\
&= \frac{1}{16}\int_0^{+\infty} \Psi'(x)\,\sigma(\{\Lambda_f > x\})\,dx\,, \\
&= \frac{1}{16}\int_0^{+\infty} \Psi\big(\Lambda_f\big)\,d\sigma\,,
\end{aligned}$$

and the proof of Lemma 11.37 is finished. □

Proof of Theorem 11.33: (i) We first focus on boundedness.

(a) We assume that $J_\mu : \mathscr{B}^\Psi \to L^\Psi(\mathbb{D}, \mu)$ is bounded with norm less than C.

For $a \in \mathbb{D}$, we use the function $h_a(w) = \dfrac{\left(1 - |a|^2\right)^2}{\left(1 - \bar{a}w\right)^4}$ whose main properties are summarized in (11.10). In particular, we know that $h_a \in \mathscr{B}M^\Psi$ with

$$\|h_a\|_{\mathscr{B}M^\Psi} \leqslant \frac{\left(\dfrac{1 + |a|}{1 - |a|}\right)^2}{\Psi^{-1}\left(\left(\dfrac{1 + |a|}{1 - |a|}\right)^2\right)} \leqslant \frac{\dfrac{4}{(1 - |a|)^2}}{\Psi^{-1}\left(\dfrac{4}{(1 - |a|)^2}\right)}.$$

In other words, the following function f_a belongs to the unit ball of $\mathscr{B}M^\Psi$:

$$f_a(w) = \frac{(1-|a|)^2}{4}\,\Psi^{-1}\Big(\frac{4}{(1-|a|)^2}\Big)\,h_a(w) \qquad \text{where } w \in \mathbb{D}. \quad (11.45)$$

Now, let $\xi \in \mathbb{T}$ and $h \in (0,1)$ and denote $a = (1-h)\xi$ so that $1-|a| = h$.

When $z \in W(\xi,h)$, we have $|1-\bar{a}z| \leqslant 6h$ (see the proof of (i)(a) of Theorem 11.17), and it follows that

$$\inf_{z\in W(\xi,h)} |f_a(z)| \geqslant k_0\Psi^{-1}\Big(\frac{4}{h^2}\Big),$$

where $k_0 = \dfrac{1}{4\times 6^4} = \dfrac{1}{5184}\cdot$ We obtain

$$\begin{aligned} 1 &\geqslant \int_{\mathbb{D}} \Psi\Big(\frac{|f_a|}{C}\Big)\,d\mu \geqslant \int_{W(\xi,h)} \Psi\Big(\frac{|f_a|}{C}\Big)\,d\mu \\ &\geqslant \Psi\Big(\frac{k_0}{C}\Psi^{-1}(4/h^2)\Big)\,\mu\big(W(\xi,h)\big), \end{aligned}$$

which is (11.39).

(b) By Lemma 11.35, there exists $C \geqslant 1$ such that $\|\Lambda_f\|_{L^\Psi(\mu)} \leqslant C\,\|f\|_{\mathscr{B}^\Psi}$ for every $f \in \mathscr{B}^\Psi$. Let g be in the unit ball of $\mathscr{B}^\Psi$; we apply Lemma 11.37 with $f = \dfrac{1}{C}\,g$ (whose norm is also less than 1) and $E = \mathbb{D}$, and with $h_A = 1$ (thanks to (11.40) we are free to choose any h_A). We define

$$\tilde{C} = \max\Big\{(1\,;\,\mu(\mathbb{D})\Psi(x_A) + \frac{1}{16}\Big\},$$

and we can now estimate

$$\begin{aligned} \int_{\mathbb{D}} \Psi\Big(\frac{A}{32C\tilde{C}}\,|g|\Big)\,d\mu &\leqslant \frac{1}{\tilde{C}}\int_{\mathbb{D}} \Psi\Big(\frac{A}{32C}\,|g|\Big)\,d\mu \\ &\leqslant \frac{1}{\tilde{C}}\Big[\mu(\mathbb{D})\Psi(x_A) + \frac{1}{16}\int_{\mathbb{D}} \Psi(\Lambda_g/C)\,d\sigma\Big] \\ &\leqslant \frac{1}{\tilde{C}}\Big[\mu(\mathbb{D})\Psi(x_A) + \frac{1}{16}\Big] \leqslant 1, \end{aligned}$$

which means that $\|g\|_{L^\Psi(\mathbb{D},\mu)} \leqslant \dfrac{32C\tilde{C}}{A}\cdot$

(ii) (a) Suppose that J_μ is compact, but that condition (11.41) is not satisfied. In other words, there exist $\varepsilon_0 \in (0,1)$, $A > 0$, a sequence of positive numbers $(h_n)_n$ decreasing to 0, and a sequence of $\xi_n \in \mathbb{T}$, such that

$$\mu\big(W(\xi_n,h_n)\big) \geqslant \frac{\varepsilon_0}{\Psi\big(A\Psi^{-1}(1/h_n^2)\big)}\cdot$$

We now use the test functions like in (11.45), namely, with $a_n = (1-h_n)\xi_n$:

$$f_n(w) = F_{a_n}(w) = \Psi^{-1}\left(\frac{4}{(1-|a_n|)^2}\right)\frac{(1-|a_n|)^2(1-|a_n|^2)^2}{4(1-\overline{a_n}w)^4} \quad \text{where } w \in \mathbb{D}.$$

Again, f_n lies in the unit ball of $\mathscr{B}M^\Psi \subset \mathscr{B}^\Psi$. On the other hand, $(f_n)_n$ clearly converges to 0 in $\mathscr{H}(\mathbb{D})$.

By the compactness criterion (Proposition 3.2), $(f_n)_n$ is norm-converging to 0 in $L^\Psi(\mathbb{D},\mu)$.

But, using the estimates already made in the proof of (i), we also have, for every $n \geqslant 1$,

$$\forall z \in W(\xi_n, h_n), \qquad |f_n(z)| \geqslant k_0 \Psi^{-1}\left(\frac{1}{h_n^2}\right),$$

where $k_0 > 0$ is numerical.

$$\begin{aligned}\int_{\mathbb{D}} \Psi\left(\frac{A}{k_0\varepsilon_0}|f_n|\right) d\mu &\geqslant \Psi\left(\frac{A}{\varepsilon_0}\Psi^{-1}(1/h_n^2)\right) \mu\big(W(\xi_n,h_n)\big)\\ &\geqslant \Psi\left(\frac{A}{\varepsilon_0}\Psi^{-1}(1/h_n^2)\right) \frac{\varepsilon_0}{\Psi\left(A\Psi^{-1}(1/h_n^2)\right)} \geqslant 1,\end{aligned}$$

and this implies that $\|f_n\|_{L^\Psi(\mathbb{D},\mu)} \geqslant k_0\varepsilon_0/A$ and gives a contradiction.

(b) We already know that $J_\mu : \mathscr{B}^\Psi \to L^\Psi(\mathbb{D},\mu)$ is bounded thanks to (i)(b) since (11.40) is fulfilled as a particular case.

We now use Proposition 11.32: given $\varepsilon > 0$, we shall prove that, for $r \in (0,1)$ small enough, the norm of the injection $I_r : \mathscr{B}^\Psi \longrightarrow L^\Psi(\mathbb{D}\setminus r\mathbb{D},\mu)$ is smaller than ε.

Again $f \mapsto \Lambda_f$ is bounded: there exists a constant $C \geqslant 1$ such that $\|\Lambda_f\|_\Psi \leqslant C\|f\|_\Psi$ for every $f \in L^\Psi(\mathbb{T})$.

We know from our assumption (11.42), applied with $A = \dfrac{32C}{\varepsilon}$, that there exists $h_0 > 0$ such that

$$\forall h \in (0,h_0) \qquad K_{\mu,2}(h) \leqslant \frac{1}{h^2}\cdot\frac{1}{\Psi\left(A\Psi^{-1}(1/h^2)\right)}\cdot$$

Let f be in the unit ball of $\mathscr{B}^\Psi$ and $r \in (0,1)$. By Lemma 11.37,

$$\begin{aligned}\int_{\mathbb{D}\setminus r\mathbb{D}} \Psi\left(\frac{|f|}{\varepsilon}\right) d\mu &= \int_{\mathbb{D}\setminus r\mathbb{D}} \Psi\left(\frac{A}{32C}|f|\right) d\mu\\ &\leqslant \mu\big(\mathbb{D}\setminus r\mathbb{D}\big)\,\Psi(x_A) + \frac{1}{16}\int_{\mathbb{T}} \Psi\left(\frac{\Lambda_f}{C}\right) d\sigma\\ &\leqslant \mu\big(\mathbb{D}\setminus r\mathbb{D}\big)\,\Psi(x_A) + \frac{1}{16}\cdot\end{aligned}$$

But there exists some $r_0 \in (0,1)$ such that $\mu\big(\mathbb{D} \setminus r\mathbb{D}\big)\,\Psi(x_A) \leqslant 1$, for every $r \in (r_0, 1)$ and this implies that $\|I_r(f)\|_\Psi \leqslant \varepsilon$ for every f in the unit ball of $\mathscr{B}^\Psi$ and every $r \in (r_0, 1)$.

(c) The proof is similar to that of (iii) for Theorem 11.17, so we omit it.

(iii) Here again, this is similar to the case of Hardy–Orlicz spaces so we just give a sketchy proof. Assume that Ψ satisfies condition ∇_0 and that condition (11.41) is fulfilled: for every $\alpha > 0$ and every $h \in (0, h_\alpha)$ (h_α small enough), we have

$$\rho_\mu(h) \leqslant \frac{1}{\Psi\big[\alpha \Psi^{-1}(1/h^2)\big]} \cdot$$

This implies that

$$K_{\mu,2}(h) = \sup_{0<s\leqslant h} \frac{\rho_\mu(s)}{s^2} \leqslant \sup_{0<s\leqslant h} \frac{1/s^2}{\Psi\big[\alpha\Psi^{-1}(1/s^2)\big]} = \sup_{x\geqslant \Psi^{-1}(1/h^2)} \frac{\Psi(x)}{\Psi(\alpha x)} \cdot$$

Fix an arbitrary $A > 1$. From (H.6), we can choose some $\alpha > 1$ such that, for h small enough and $x \geqslant \Psi^{-1}(1/h^2)$,

$$\frac{\Psi(x)}{\Psi(\alpha\, x)} \leqslant \frac{\Psi\big[\Psi^{-1}(1/h^2)\big]}{\Psi\big[A\Psi^{-1}(1/h^2)\big]};$$

hence

$$K_{\mu,2}(h) \leqslant \frac{1/h^2}{\Psi\,\big[A\Psi^{-1}(1/h^2)\big]}$$

and condition (11.42) is fulfilled. □

11.3.3 Compactness of Composition Operators on $\mathscr{B}^\Psi$

We shall focus on the compactness issues for composition operators and specialize here the results of Section 11.3.2 to the case $\mu = \sigma_\varphi$ where φ is a symbol. We recall that, in the same spirit as for Hardy spaces, for every $f \in \mathscr{B}^\Psi$ and every symbol φ, we have

$$\|f \circ \varphi\|_{\mathscr{B}^\Psi} = \|f\|_{L^\Psi(\mathbb{D},\sigma_\varphi)};$$

therefore $C_\varphi : \mathscr{B}^\Psi \to \mathscr{B}^\Psi$ is compact if and only if $J_{\sigma_\varphi} : \mathscr{B}^\Psi \to L^\Psi(\mathbb{D}, \sigma_\varphi)$ is compact.

We first mention an easy necessary condition for compactness:

Proposition 11.38 *Let φ be a symbol and Ψ be an Orlicz function. Assume that $C_\varphi : \mathscr{B}^\Psi \to \mathscr{B}^\Psi$ is compact. Then we have*

$$\lim_{|z|\to 1^-} \frac{\Psi^{-1}\left(\dfrac{1}{(1-|\varphi(z)|)^2}\right)}{\Psi^{-1}\left(\dfrac{1}{(1-|z|)^2}\right)} = 0. \tag{11.46}$$

Proof: The proof follows the lines of the one of Theorem 11.21. The only specific ingredient is the use of Proposition 11.9. □

We recall (cf. (5.31)) the notation $\rho_{\varphi,2}(h) = \sup_{\xi\in\mathbb{T}} \sigma_\varphi\Big(W(\xi,h)\Big)$ and we shall now apply the results of Section 11.3.2 to characterize the compactness of $C_\varphi : \mathscr{B}^\Psi \to \mathscr{B}^\Psi$. In the statement, the Nevanlinna counting function of order 2 appears. There are two natural ways to define it and we refer (for instance) to Chapter 7 for more comments around this fact. Anyway the behavior near the boundary of these two functions is the same: they are equivalent when $|z| \to 1^-$. Let us say that we make here the choice

$$N_{\varphi,2}(z) = \sum_{\substack{\varphi(a)=z\\ a\in\mathbb{D}}} \left(\log\frac{1}{|a|}\right)^2 \qquad \text{for } z \in \varphi(\mathbb{D}) \setminus \{\varphi(0)\}.$$

Theorem 11.39 *Let φ be a symbol and Ψ be an Orlicz function.*

Then the following assertions are equivalent:

(i) *$C_\varphi : \mathscr{B}^\Psi \to \mathscr{B}^\Psi$ is compact.*

(ii) *For every $A > 0$,*

$$\rho_{\varphi,2}(h) = o\left(\frac{1}{\Psi\big(A\Psi^{-1}(1/h^2)\big)}\right) \qquad \text{as } h \to 0. \tag{11.47}$$

(iii) *We have*

$$\lim_{h\to 0} \frac{\Psi^{-1}\left(\dfrac{1}{h^2}\right)}{\Psi^{-1}\left(\dfrac{1}{\rho_{\varphi,2}(h)}\right)} = 0. \tag{11.48}$$

(iv) *We have*

$$\lim_{|z|\to 1^-} \frac{\Psi^{-1}\left(\dfrac{1}{(1-|z|)^2}\right)}{\Psi^{-1}\left(\dfrac{1}{N_{\varphi,2}(z)}\right)} = 0. \tag{11.49}$$

(v) *When $h \to 0$, we have*

$$\sup_{\xi \in \mathbb{T}} \left\| \frac{1}{\big(1 - \bar{\xi}(1-h)\,\varphi\big)^4} \right\|_{\mathscr{B}^\Psi} = o\left(\frac{1}{h^4 \Psi^{-1}(1/h^2)} \right). \tag{11.50}$$

Here also, when $\sigma_\varphi(h) = 0$ the whole quotient in (11.48) has to be understood as 0. In the same spirit, when $N_{\varphi,2}(z) = 0$, the whole quotient in (11.49) has to be understood as 0. This is the case when $z \notin \varphi(\mathbb{D})$. In particular, when $\|\varphi\|_\infty < 1$, this whole quotient is equal to 0 on an annulus near the boundary.

We slightly modified the functions f_a (see (11.45)) to make them a bit simpler, but the unmodified functions would do the job as well.

Proof: The equivalence between (i)–(ii)–(iii) is a direct consequence of Theorem 11.33, applied to the measure $\mu = \sigma_\varphi$, as soon as we have proved that (11.47) implies that, for every $A > 0$, we have

$$K_{\sigma_\varphi,2}(h) = o\left(\frac{1/h^2}{\Psi\big(A\Psi^{-1}(1/h^2)\big)} \right) \qquad \text{as } h \to 0,$$

but this latter fact is just a direct consequence of Proposition 7.12.

Now assume that $C_\varphi : \mathscr{B}^\Psi \to \mathscr{B}^\Psi$ is compact, or, equivalently, that (11.48) is satisfied. From Theorem 7.7 (see (i) with $\alpha = 0$), we know that, for some constants $C, k > 1$, we have $N_{\varphi,2}(z) \leqslant C\rho_{\varphi,2}(k(1-|z|))$ when $|z|$ is close enough to 1. Therefore, with $h = k(1-|z|)$ small enough, we have

$$\frac{\Psi^{-1}\left(\dfrac{1}{(1-|z|)^2}\right)}{\Psi^{-1}\left(\dfrac{1}{N_{\varphi,2}(z)}\right)} \leqslant \frac{\Psi^{-1}\left(\dfrac{k^2}{h^2}\right)}{\Psi^{-1}\left(\dfrac{1}{C\rho_{\varphi,2}(h)}\right)} \leqslant Ck^2 \frac{\Psi^{-1}\left(\dfrac{1}{h}\right)}{\Psi^{-1}\left(\dfrac{1}{\rho_{\varphi,2}(h)}\right)},$$

and we get (11.49).

Now assume that (11.49) is satisfied. From Theorem 7.7 (see (ii) with $\alpha = 0$), we know that, for some constants $C, k > 1$, we have

$$\rho_{\varphi,2}(h) \leqslant C \sup_{|w| \geqslant 1-kh} N_{\varphi,2}(w)$$

as soon as h is small enough and we also easily get (iv).

Assuming that $C_\varphi : \mathscr{B}^\Psi \to \mathscr{B}^\Psi$ is compact, point out that the family of functions

$$g_{\xi,h}(w) = \frac{h^4 \Psi^{-1}(1/h^2)}{\big(1 - \bar{\xi}(1-h)z\big)^4} \qquad \text{(where } w \in \mathbb{D}\text{)}$$

is bounded in $\mathscr{B}^\Psi$ and converging to 0 in $\mathscr{H}(\mathbb{D})$, as $h \to 0$, uniformly relatively to ξ, since $\Psi^{-1}(1/h^2)\, h^4 \to 0$. Thanks to the compactness criterion (Proposition 3.2), $(C_\varphi(g_{\xi,h}))$ is norm-converging to 0 in $\mathscr{B}^\Psi$, uniformly relatively to ξ, and we get (11.50).

Conversely, assume that condition (11.50) is satisfied. It means that we have for every $A > 0$, as soon as h is small enough,

$$1 \geqslant \int_{\mathbb{D}} \Psi\left(6^4 A |g_{\xi,h}|\right) d\sigma_\varphi$$

for every $\xi \in \mathbb{T}$. But, for every $z \in W(\xi, h)$, we have $|1 - \bar{\xi}(1-h)z| \leqslant 6h$ so $|g_{\xi,h}(z)| \geqslant 6^{-4}\Psi^{-1}(1/h^2)$, and this implies that

$$1 \geqslant \int_{W(\xi,h)} \Psi\left(A\Psi^{-1}(1/h^2)\right) d\sigma_\varphi = \sigma_\varphi(W(\xi,h))\, \Psi\left(A\Psi^{-1}(1/h^2)\right).$$

Therefore $\rho_{\varphi,2}(h) \leqslant \dfrac{1}{\Psi\big(A\Psi^{-1}(1/h^2)\big)}$ with $A > 0$ arbitrary. This implies (11.47). □

We mention an easy consequence (actually in terms of order-bounded operator into $M^\Psi(\mathbb{D})$).

Corollary 11.40 *Let Ψ be an Orlicz function.*
Assume $\Psi^{-1}\left(\dfrac{1}{(1-|\varphi|)^2}\right) \in M^\Psi(\mathbb{D}, \sigma)$. Then $C_\varphi : \mathscr{B}^\Psi \to \mathscr{B}^\Psi$ is compact.

Proof: The proof is easily adapted from the one of Theorem 11.23 and we leave it as an exercise. □

We also mention that for finitely valent symbols the characterization can be written in a simpler way with (11.46) (equivalently (11.51)). In the framework of *classical* Bergman spaces, this condition is actually a characterization for *any* symbol. It turns out to be true under some sufficient condition for Ψ.

Theorem 11.41 *Let Ψ be an Orlicz function.*

(i) *Let φ be a finitely valent symbol. The operator $C_\varphi : \mathscr{B}^\Psi \to \mathscr{B}^\Psi$ is compact if and only if the following condition is satisfied*

$$\Psi^{-1}\left(\frac{1}{(1-|\varphi(z)|)^2}\right) = o\left(\Psi^{-1}\left(\frac{1}{(1-|z|)^2}\right)\right) \qquad \text{when } |z| \to 1^-. \tag{11.51}$$

(ii) *Let φ be a symbol. Assume that $\Psi \in \nabla_0$.*
Then $C_\varphi : \mathscr{B}^\Psi \to \mathscr{B}^\Psi$ is compact if and only if (11.51) is satisfied.

Proof: (i) This is similar to the corresponding proof for H^{Ψ}, and we leave it as an exercise (see Theorem 11.24).

(ii) We already know that (11.51) is always a necessary condition by Proposition 11.38. Let us prove that it is sufficient.

Fix $A \geqslant 1$.

From (H.6), one can easily find some $\beta_A > 1$ such that, for $y \geqslant x$ large enough,

$$\frac{\Psi(Ax)}{\Psi(x)} \leqslant \frac{\Psi(\beta_A y)}{\Psi(y)}.$$

In particular we have, with $\alpha_A = A\beta_A$,

$$\Psi(Ax)^2 \leqslant \Psi(x)\Psi(\alpha_A x). \tag{11.52}$$

When $z \in \varphi(\mathbb{D})$ is close enough to the boundary (so that we may assume $z \neq \varphi(0)$),

$$N_{\varphi,2}(z) = \sum_{\substack{\varphi(a)=z \\ a\in\mathbb{D}}} \left(\log \frac{1}{|a|}\right)^2 \leqslant N_\varphi(z) \sup_{\substack{\varphi(a)=z \\ a\in\mathbb{D}}} \log \frac{1}{|a|} \leqslant C(1-|z|) \sup_{\substack{\varphi(a)=z \\ a\in\mathbb{D}}} (1-|a|)$$

for some constant $C > 1$ (depending on $\varphi(0)$ only). But, by hypothesis, we know that, when $a \in \mathbb{D}$ is close enough to the boundary,

$$1-|a| \leqslant \left(\Psi\left(\alpha_A \Psi^{-1}\left(\frac{1}{(1-|\varphi(a)|)^2}\right)\right)\right)^{-\frac{1}{2}}.$$

Therefore, having in mind (1.16), we can say that there exists some $\delta \in (0,1)$ such that

$$N_{\varphi,2}(z)^2 \leqslant C^2(1-|z|)^2 \left(\Psi\left(\alpha_A \Psi^{-1}\left(\frac{1}{(1-|z|)^2}\right)\right)\right)^{-1} = \frac{C^2}{\Psi(X)\Psi(\alpha_A X)}$$

for every $z \in \varphi(\mathbb{D})$ with $|z| \geqslant 1-\delta$, where

$$X = \Psi^{-1}\left(\frac{1}{(1-|z|)^2}\right).$$

From (11.52), we get $N_{\varphi,2}(z) \leqslant \dfrac{C}{\Psi(AX)} \leqslant \dfrac{1}{\Psi(AX/C)}$ since Ψ is convex.

In other words, for every $z \in \varphi(\mathbb{D})$ with $|z| \geqslant 1-\delta$, we have

$$\frac{1}{\Psi^{-1}\left(\dfrac{1}{N_{\varphi,2}(z)}\right)} \leqslant \frac{C}{AX} = \frac{C}{A}\,\frac{1}{\Psi^{-1}\left(\dfrac{1}{(1-|z|)^2}\right)},$$

and (11.49) is satisfied because A is arbitrary, that is, $C_\varphi : \mathscr{B}^{\Psi} \to \mathscr{B}^{\Psi}$ is compact. □

An interesting consequence of the preceding results concerns the specific case of very fast-growing Orlicz functions.

Theorem 11.42 *Let $\Psi \in \Delta^2$ and φ be a symbol.*
Then the following assertions are equivalent:

(i) $C_\varphi : \mathscr{B}^\Psi \to \mathscr{B}^\Psi$ *is compact.*
(ii) $C_\varphi : \mathscr{B}^\Psi \to \mathscr{B}^\Psi$ *is weakly compact.*
(iii) $\Psi^{-1}\left(\dfrac{1}{(1-|\varphi|)^2}\right) \in M^\Psi(\mathbb{D},\sigma)$.
(iv) $$\lim_{|z|\to 1^-} \frac{\Psi^{-1}\left(\dfrac{1}{(1-|\varphi(z)|)^2}\right)}{\Psi^{-1}\left(\dfrac{1}{(1-|z|)^2}\right)} = 0\,.$$

In that case, in particular, $\|\varphi^n\|_{\mathscr{B}^p} = o\left(n^{-s}\right)$ for every $p,s \geqslant 1$ when $n \to +\infty$. A fortiori $C_\varphi : \mathscr{B}^p \to \mathscr{B}^p$ is nuclear for any $p \geqslant 1$.

The proof of this theorem is straightforward using the former results; the fact that condition Δ^2 implies condition ∇_0 and ideas of the proof of the corresponding result for Hardy spaces. We do not give more details.

In particular, we have the following proposition:

Proposition 11.43 *Let $r > 0$ and φ be a symbol.*
Then the following assertions are equivalent:

(i) $C_\varphi : \mathscr{B}^{\Psi_r} \to \mathscr{B}^{\Psi_r}$ *is compact.*
(ii) *For every $\gamma \geqslant 1$, we have $\rho_{\varphi,2}(h) = o\left(h^\gamma\right)$ when $h \to 0$.*
(iii) *For every $\gamma \geqslant 1$, we have* $\displaystyle\int_{\mathbb{D}} \frac{1}{(1-|\varphi|)^\gamma}\, d\sigma < \infty$.

In that case, $C_\varphi : \mathscr{B}^2 \to \mathscr{B}^2$ belongs to all the Schatten classes.

Point out that the compactness of $C_\varphi : \mathscr{B}^{\Psi_r} \to \mathscr{B}^{\Psi_r}$ does not depend on r: this is a similar phenomenon as for the scale of classical Bergman spaces $\mathscr{B}^r$ (but the concrete characterization is very different).

Let us mention that there is a Bergman variant of Theorem 11.27 (see Section 11.4 to notice that the assertion (i) in the two propositions is not a consequence of each other in full generality). Nevertheless (ii) below implies (ii) in Theorem 11.27: indeed it suffices to choose a fast-growing Orlicz function, and if C_φ were compact on H^Ψ, it would be also on $\mathscr{B}^\Psi$.

Theorem 11.44 (i) *Given an Orlicz function Ψ, there exists a symbol φ such that φ is* onto *(a fortiori $\|\varphi\|_\infty = 1$) and $C_\varphi : \mathscr{B}^\Psi \to \mathscr{B}^\Psi$ is compact. Moreover φ is* 4*-valent.*

(ii) *Given a symbol φ such that $\|\varphi\|_\infty = 1$, there exists an Orlicz function Ψ such that $C_\varphi : \mathscr{B}^\Psi \to \mathscr{B}^\Psi$ is not compact.*

Proof: (i) Take the function $\Delta(h) = \sqrt{H(h^2/k^2)}$ where H is the function appearing in (11.32) and $k > 1$ is the constant appearing in (7.28).

Now apply Theorem 4.6 to produce an almost surjective (finitely valent) symbol φ satisfying, for h small enough,

$$\rho_{\varphi,2}(h) \lesssim \big(\rho_\varphi(kh)\big)^2 \leqslant \Delta(kh)^2 \leqslant H(h^2).$$

Now from (11.48) in Theorem 11.39, we know that $C_\varphi : \mathscr{B}^\Psi \to \mathscr{B}^\Psi$ is compact.

Take $\Phi = \varphi_a^2 \circ \varphi$ for some $a \in \mathbb{D} \setminus \{0\}$. From the argument given in Section 4.4, we know that $\Phi(\mathbb{D}) = \mathbb{D}$ and that Φ is 4-valent. Moreover $C_\Phi : \mathscr{B}^\Psi \to \mathscr{B}^\Psi$ is compact since $C_\Phi = C_\varphi \circ C_{\varphi_a^2}$.

(ii) Since $\|\varphi\|_\infty = 1$, we have $\rho_{\varphi,2}(h) > 0$ for every $h \in (0,1)$. We are going to show that (11.47) is not satisfied for a suitable Ψ. It suffices to construct (by induction) a piecewise linear strictly increasing convex function $\Psi : \mathbb{R}^+ \to \mathbb{R}^+$ such that

$$\forall n \in \mathbb{N}, \qquad \Psi\big(2^{n+1}\big) \geqslant \frac{1}{\rho_{\varphi,2}\Big(1/\sqrt{\Psi(2^n)}\Big)}.$$

Writing $h_n = \dfrac{1}{\sqrt{\Psi(2^n)}}$, we have $\rho_{\varphi,2}(h_n) \geqslant \dfrac{1}{\Psi\left(2\Psi^{-1}(1/h_n^2)\right)}$ and (11.47) is not satisfied; hence $C_\varphi : \mathscr{B}^\Psi \to \mathscr{B}^\Psi$ is not compact. □

Weak Compactness

We recall again that we assumed $\Psi \in \nabla_2$.

Theorem 11.45 *Let $\varphi : \mathbb{D} \to \mathbb{D}$ be a symbol.*

(i) *The operator $C_\varphi : \mathscr{B}^\Psi \to \mathscr{B}^\Psi$ is a Dunford–Pettis operator if and only if it is compact.*
(ii) *Assume that the Orlicz function Ψ satisfies condition Δ^0. Then the operator $C_\varphi : \mathscr{B}^\Psi \to \mathscr{B}^\Psi$ is weakly compact if and only if it is compact.*

It is not surprising that some specific condition appears in (ii) for Ψ: when $\Psi(x) = x^p$ with $p > 1$, the composition operator C_φ is always weakly compact.

Proof: The proof is similar to the one of Theorem 11.29, and we leave the details as an exercise. We just mention that we can use the function $g_{\xi,h}$ above instead of functions $u_{\xi,1-h}$. □

11.4 Comparison between Hardy–Orlicz and Bergman–Orlicz

In the classical framework (it means $\Psi(x) = x^p$), any compact composition operator on H^Ψ is compact on $\mathscr{B}^\Psi$; the converse is true for finitely valent symbols but not in general: see Theorem 3.22.

In the general framework of Orlicz spaces, this remains partially true.

Let us notice that the necessary conditions (11.27) for H^Ψ and (11.46) for $\mathscr{B}^\Psi$ are similar but relatively to different Orlicz functions. Actually (11.46) for Ψ^2 is (11.27) for Ψ. Even though we do not use this characterization for the proof of (i) in the next theorem, this is what gives the idea of the link.

Theorem 11.46 *Let Ψ be an Orlicz function and φ be a symbol.*

(i) *If $C_\varphi : H^\Psi \to H^\Psi$ is compact then $C_\varphi : \mathscr{B}^{\Psi^2} \to \mathscr{B}^{\Psi^2}$ is compact.*
(ii) *Let Ψ_1 and Ψ_2 be two Orlicz functions. There exists some inner symbol B such that $C_B : H^{\Psi_2} \to H^{\Psi_2}$ is an isometry (in particular not compact) and $C_B : \mathscr{B}^{\Psi_1} \to \mathscr{B}^{\Psi_1}$ is compact.*
(iii) *Assume that Ψ satisfies, for some $B \geqslant A > 0$,*

$$\Psi\left(A\Psi^{-1}(x^2)\right) \leqslant \left[\Psi\left(B\Psi^{-1}(x)\right)\right]^2 \qquad \text{where } x \text{ is large enough.} \tag{11.53}$$

If $C_\varphi : H^\Psi \to H^\Psi$ is compact then $C_\varphi : \mathscr{B}^\Psi \to \mathscr{B}^\Psi$ is compact.
(iv) *There exists some Orlicz function Ψ such that C_φ is compact on H^Ψ but not on $\mathscr{B}^\Psi$.*

Let us mention that (11.53) is "often" satisfied: for instance when $\Psi(x) = x^p$ or when ψ grows fast. Nevertheless (iv) shows that it is not always the case.

In (ii), the statement and the proof show that we can find a symbol φ such that C_φ is not compact on H^Ψ whereas C_φ is compact both on $\mathscr{B}^\Psi$ and $\mathscr{B}^{\Psi^2}$.

Proof: For (i), just point out that $N_{\varphi,2}(z) \leqslant \left(N_\varphi(z)\right)^2$ because the ℓ^1-norm is larger than the ℓ^2-norm. Now the proof follows directly from the characterizations of compactness.

(ii) We take the function $F(x) = \dfrac{1}{2\Psi_1\left[\left(\Psi_1^{-1}(1/x^2)\right)^2\right]}$, where $x > 0$, and Δ its inverse, which is well defined since we assumed in this chapter that our functions Ψ are increasing (in full generality we could consider the Lévy-pseudo-inverse).

In other words, we have $\Delta(1 - |z|) \leqslant h$ if and only if $F(h) \geqslant 1 - |z|$. Now, Theorem 4.11 provides us with a slow Blaschke product B_0 satisfying the inequality $1 - |B_0(z)| \geqslant \Delta(1 - |z|)$ and we define $B(z) = zB_0(z)$, which is a Blaschke product fixing the origin.

We know from Proposition 11.14 that C_B is an isometry on H^{Ψ_2}.

On the other hand, we have

$$\begin{aligned}\rho_{\varphi,2}(h) &= \sup_{\xi\in\mathbb{T}} \sigma\left(\{z \in \mathbb{D} : B(z) \in W(\xi,h)\}\right) \leqslant \sigma_B(\mathbb{D} \setminus (1-h)\mathbb{D}) \\ &\leqslant \sigma\left(\{z \in \mathbb{D} : 1 - |B_0(z)| \leqslant h\}\right) \leqslant \sigma\left(\{z \in \mathbb{D} : 1 - |z| \leqslant F(h)\}\right) \\ &\leqslant 1 - (1 - F(h))^2 \leqslant 2F(h) = \frac{1}{\Psi_1\left[\left(\Psi_1^{-1}(1/h^2)\right)^2\right]}\end{aligned}$$

since σ is the normalized area measure; and we easily conclude that condition (11.48) is satisfied, hence $C_B : \mathscr{B}^{\Psi_1} \to \mathscr{B}^{\Psi_1}$ is compact.

(iii) We have (11.30) from Theorem 11.22 and the conclusion comes from (11.49) in Theorem 11.39 as soon as we show that (11.30) implies (11.49). But on the one hand, $N_{\varphi,2}(z) \leqslant \left(N_\varphi(z)\right)^2$, and on the other hand the regularity hypothesis on Ψ exactly means that this is the case.

(iv) See [102, theorem 4.2] for a construction. □

12

Composition Operators on Weighted Hardy Spaces

12.1 Introduction

The aim of this chapter is twofold: we first give a full description of those weights $\beta = (\beta_n)_{n\geqslant 0}$ such that the composition operators C_φ are bounded on the weighted Hardy space $H^2(\beta)$ for *all* symbols φ (analytic self-maps of $\mathbb{D}$). In a second part, we compare the approximation numbers of a given C_φ acting on two different weighted spaces $H^2(\beta)$ and $H^2(\gamma)$.

Weighted Hardy Spaces

Let $\beta = (\beta_n)_{n\geqslant 0}$ be a sequence of positive numbers satisfying

$$\beta_0 = 1 \qquad \text{and} \qquad L := \varliminf_{n\to\infty} \beta_n^{1/n} \geqslant 1. \tag{12.1}$$

The associated weighted Hardy space $H^2(\beta)$ is the Hilbertian space of analytic functions $f(z) = \sum_{n=0}^{\infty} a_n z^n$ such that

$$\|f\|_\beta^2 := \sum_{n=0}^{\infty} |a_n|^2 \beta_n < \infty .$$

Condition (12.1) is equivalent to the inclusion $H^2(\beta) \subset \mathscr{H}(\mathbb{D})$. Indeed, if (12.1) holds, and if $f(z) = \sum_{n=0}^{\infty} a_n z^n \in H^2(\beta)$, we have $|a_n| = O(1/\sqrt{\beta_n})$; therefore

$$\varlimsup_n |a_n|^{1/n} \leqslant \varlimsup_n \beta_n^{-1/2n} = \frac{1}{\sqrt{L}} \leqslant 1,$$

and $f \in \mathscr{H}(\mathbb{D})$. Conversely, testing the inclusion $H^2(\beta) \subset \mathscr{H}(\mathbb{D})$ on the function $f(z) = \sum_{n=1}^{\infty} \frac{1}{n\sqrt{\beta(n)}} z^n \in H^2(\beta)$, we get (12.1) by the Cauchy–Hadamard formula.

We will assume *once and for all* throughout this chapter that (12.1) is satisfied.

The most interesting case is $L = 1$ (even $\lim_{n\to\infty} \beta_n^{1/n} = 1$) because if $L > 1$, any element of $H^2(\beta)$ is indeed analytic in the larger disk $D(0, \sqrt{L})$.

Let us mention some specific examples of spaces $H^2(\beta)$, most of them already met in the book:

(i) When $\beta_n \equiv 1$, we recover the usual Hardy space H^2.

(ii) The (classical) Bergman space corresponds to $\beta_n = 1/(n+1)$, and more generally, for $\alpha > -1$, weighted Bergman spaces $\mathscr{B}^2_\alpha$ are such examples with $\beta_n \sim \Gamma(\alpha+2)(n+1)^{-(\alpha+1)}$ (see (1.69)).

(iii) The Dirichlet space $\mathcal{D}^2$ for which $\beta_n = n+1$ and more generally, for $\alpha > -1$, weighted Dirichlet spaces $\mathcal{D}^2_\alpha$.

More generally, two classes of weights reveal quite interesting:

(iv) The polynomial weights $\beta_n = (n+1)^a$, for $a \in \mathbb{R}$ (see, for instance, spaces H_α in Chapter 9).

(v) The subexponential weights $\beta_n = \exp(n^a)$, for $0 < a < 1$. When $a \geqslant 1/2$, the latter space $H^2(\beta)$ is even contained in a quasi-analytic class (see Exercise I.28), thanks to Korenbljum's theorem [85].

The Hilbertian space $H^2(\beta)$ has the canonical orthonormal basis

$$e_n^\beta(z) = \frac{1}{\sqrt{\beta_n}}\, z^n\,, \quad \text{for } n \geqslant 0\,, \tag{12.2}$$

and the reproducing kernel $K_w \in H^2(\beta)$, meaning that

$$f(w) = \langle f, K_w\rangle \quad \text{for } f \in H^2(\beta) \text{ and } w \in \mathbb{D}. \tag{12.3}$$

The kernel (analytic in z, anti-analytic in w) is given for all $z, w \in \mathbb{D}$ by

$$K(z,w) = K_w(z) = \sum_{n=0}^{\infty} e_n^\beta(z)\, \overline{e_n^\beta(w)} = \sum_{n=0}^{\infty} \frac{1}{\beta_n}\, \overline{w}^n z^n\,. \tag{12.4}$$

In particular, polynomials are always dense in $H^2(\beta)$.

We abbreviate $\|.\|_\beta$ to $\|.\|$ for short.

There are now *two significantly distinct cases:*

(i) $\sum 1/\beta_n = \infty$. Then, the following important property holds:

$$\frac{K_a}{\|K_a\|} \xrightarrow{w} 0 \quad \text{as} \quad |a| \to 1^-. \tag{12.5}$$

Indeed, as $|a| \to 1^-$, $\dfrac{\langle K_a, f\rangle}{\|K_a\|} = \dfrac{\overline{f(a)}}{\|K_a\|} = O\left(\dfrac{1}{\|K_a\|}\right) \to 0$ when f is a polynomial.

(ii) $\sum 1/\beta_n < \infty$. Then, if $f = \sum_{n=0}^{\infty} a_n z^n \in H^2(\beta)$, we have $\sum_{n=0}^{\infty} |a_n| < \infty$ by the Cauchy–Schwarz inequality. In particular, $H^2(\beta) \subset A(\mathbb{D})$, the disk algebra. Moreover, (12.3) and (12.4) also make sense and hold for $w \in \partial\mathbb{D}$.

The difference between both cases will be apparent in some further statements or proofs. Let us also observe the following.

For $\theta \in \mathbb{R}$, denote by R_θ the rotation $z \mapsto e^{i\theta} z$. The following obvious fact needs to be outlined.

Fact: C_{R_θ} is unitary on $H^2(\beta)$ for any rotation R_θ and any weight β.

Actually (see exercises) the spaces $H^2(\beta)$ are exactly the rotation-invariant Hilbert spaces H of analytic functions on $\mathbb{D}$, meaning that

$$f \in H \Longrightarrow f \circ R_\theta \in H \text{ with } \|f \circ R_\theta\|_H = \|f\|_H \quad \text{for all } \theta \in \mathbb{R}.$$

The study of weighted Hardy spaces was initiated by Shields [167] and pursued by J. Shapiro [163] and Zorboska [185] in interesting works. See also [51].

12.2 The Boundedness Issue

One of our main purposes in this chapter will be to give a full description of those weights β such that C_φ is bounded on $H^2(\beta)$ (i.e., $C_\varphi \colon H^2(\beta) \to H^2(\beta)$) *for all symbols* φ. This may appear as a complete answer to a long-standing question posed by Zorboska [185] at the end of her thesis (problem 1); and the final answer is surprisingly neat.

Indeed, the situation so far appeared as intricate:

(i) If $\beta_n = 1$ or $\beta_n = 1/(n+1)$ (classical Hardy and Bergman spaces, respectively), all the C_φ are bounded on $H^2(\beta)$.

(ii) If $\beta_n = n + 1$ (classical Dirichlet space), all the C_φ are bounded when φ is an automorphism, but some with $\varphi(0) = 0$ are unbounded (take for φ an infinite Blaschke product vanishing at 0; then, $C_\varphi(z) = \varphi \notin H^2(\beta)$).

(iii) If β_n is non-increasing, all the C_φ with $\varphi(0) = 0$ are bounded. But

(iv) if, for example, β is the decreasing sequence $\beta_n = \exp(-\sqrt{n})$, all the C_φ with $\varphi(0) = 0$ are bounded, while no C_φ, with φ a non-trivial automorphism, is bounded.

We will be able here to unify the situation, thanks to the main result of this chapter: Theorem 12.30 (appearing in [106]).

It will turn out that there are two main issues, putting into play different properties of β:

(i) the case $\varphi(0) = 0$.
(ii) The case of automorphisms φ_a (with $a \in \mathbb{D}$) where $\varphi_a(z) = \dfrac{a - z}{1 - \bar{a}\, z}$, with $\varphi_a^{-1} = \varphi_a$.

But once those two cases are fixed, the general case follows thanks to the *dévissage principle* (see Chapter 2).

12.2.1 Conditions on the Weight

Several (new) definitions on weights will naturally appear in this chapter.

Definition 12.1 A sequence of positive numbers $\beta = (\beta_n)_{n \geqslant 0}$ is said to be *essentially decreasing* if, for some constant $C \geqslant 1$, we have, for all $m \geqslant n \geqslant 0$,

$$\beta_m \leqslant C\, \beta_n. \tag{12.6}$$

Note that the essentially decreasing character of β means that the forward shift operator on $H^2(\beta)$ is power-bounded.

If β is essentially decreasing, and if we set

$$\tilde{\beta}_n = \sup_{m \geqslant n} \beta_m\,,$$

the sequence $\tilde{\beta} = (\tilde{\beta}_n)$ is non-increasing and satisfies $\beta_n \leqslant \tilde{\beta}_n \leqslant C\, \beta_n$. In particular, the space $H^2(\beta)$ is canonically isomorphic to $H^2(\tilde{\beta})$ and H^2 is continuously embedded in $H^2(\beta)$.

A weaker related notion will come into play.

Definition 12.2 A sequence of positive numbers $\beta = (\beta_n)_{n \geqslant 0}$ is said to be *weakly decreasing* if, for every $\delta > 0$, there exists a positive constant $C = C(\delta)$ such that

$$\beta_m \leqslant C\, \beta_n \quad \text{whenever } m > (1 + \delta)\, n \text{ and } n \in \mathbb{N}_0. \tag{12.7}$$

Note that (12.7) automatically implies the boundedness of β, so that requiring to fulfill (12.7) for every n or for n large enough is equivalent.

Definition 12.3 We say that a sequence β is *slowly oscillating* if there is a function $\rho\colon (0, \infty) \to (0, \infty)$, bounded above on each compact subset of $(0, \infty)$, and satisfying

$$\frac{\beta_m}{\beta_n} \leqslant \rho\left(\frac{m}{n}\right) \qquad \text{for all } m, n \in \mathbb{N}\,. \tag{12.8}$$

We observe that (12.3) automatically implies the symmetric relation

$$\text{for all } m, n \in \mathbb{N}, \qquad \frac{1}{\rho\left(\dfrac{n}{m}\right)} \leqslant \frac{\beta_m}{\beta_n} \leqslant \rho\left(\frac{m}{n}\right).$$

This is also equivalent to an apparently weaker notion, reminiscent of an analog condition used for Orlicz functions (see Appendix H): the sequence of positive numbers $\beta = (\beta_n)$ is sometimes said to satisfy the Δ_2-*condition* if $\beta_m \geqslant \delta\, \beta_n$ whenever $n/2 \leqslant m \leqslant 2n$.

Definition 12.4 The sequence of positive numbers $\beta = (\beta_n)$ is said to have *polynomial decay* if there are positive constants c and γ such that, for all integers $n \geqslant 1$,

$$\beta_n \geqslant c\, n^{-\gamma}. \tag{12.9}$$

This means, with $\alpha = \gamma - 1 > -1$, that $H^2(\beta)$ is continuously embedded in the weighted Bergman space $\mathscr{B}^2_\alpha$ since $\mathscr{B}^2_\alpha = H^2(\tilde{\beta})$ with $\tilde{\beta}_n \approx n^{-(\alpha+1)}$.

Definition 12.5 The sequence of positive numbers $\beta = (\beta_n)$ is said to have *polynomial growth* if there are positive constants C and γ such that, for all integers $n \geqslant 1$,

$$\beta_n \leqslant C\, n^{\gamma}. \tag{12.10}$$

Of course, this means that $\mathscr{B}^2_\alpha$ is continuously embedded in $H^2(\beta)$ with $\alpha = \gamma - 1$. The following simple proposition links those notions.

Proposition 12.6 *The following properties hold:*

(i) *Every slowly oscillating sequence β has polynomial decay and growth.*
(ii) *There are sequences that are non-increasing, and have polynomial decay, but are not slowly oscillating.*
(iii) *There are bounded sequences that are slowly oscillating but not essentially decreasing.*

Proof: (i) This is clear, because if $2^j \leqslant n < 2^{j+1}$, then

$$\beta_n \geqslant c\, \beta_{2^j} \geqslant c^{j+1} \beta_1 \geqslant c\, \beta_1\, n^{-\alpha}$$

with $\alpha = (\log(1/c)) / \log 2$; and

$$\beta_n \leqslant C\, \beta_{2^j} \leqslant C^{j+1} \beta_1 \leqslant C\, n^{\gamma}$$

with $\gamma = \log C / \log 2$.

(ii) Let $\delta > 0$. We set $\beta_0 = \beta_1 = 1$ and for $n \geqslant 2$:

$$\beta_n = \frac{1}{(k!)^{\delta}} \quad \text{when } k! < n \leqslant (k+1)!$$

The sequence β is non-increasing.

For n and k as above, we have

$$\beta_n = \frac{1}{(k!)^\delta} \geqslant \frac{1}{n^\delta}\,;$$

hence β has arbitrarily slow polynomial decay. However we have, for $k \geqslant 2$,

$$\frac{\beta_{2(k!)}}{\beta_{k!}} = \frac{(k!)^{-\delta}}{[(k-1)!]^{-\delta}} = \frac{1}{k^\delta} \underset{k\to\infty}{\longrightarrow} 0;$$

so β fails to be slowly oscillating.

(iii) We define β_n as follows. Let (a_k) be an increasing sequence of positive square integers such that $\lim_{k\to\infty} a_{k+1}/a_k = \infty$, for example, $a_k = 4^{k^2}$, and let $b_k = \sqrt{a_k a_{k+1}}$; with our choice, this is an integer and we clearly have $a_k < b_k < a_{k+1}$. We set

$$\beta_n = \begin{cases} a_k/n & \text{for } a_k \leqslant n < b_k \\ \\ (a_k/b_k^2)\, n = (1/a_{k+1})\, n & \text{for } b_k \leqslant n < a_{k+1}. \end{cases}$$

This sequence (β_n) is slowly oscillating by construction. Indeed, it suffices to check that the quotient β_m/β_n remains bounded from below and above when $n/2 \leqslant m \leqslant n$ (if $n < m \leqslant 2n$, we permute the role of m and n). Let now k satisfy $a_k \leqslant n < a_{k+1}$. We separate three cases.

- $n > b_k$. If $m > b_k$, we have

$$\frac{\beta_m}{\beta_n} = \frac{m/a_{k+1}}{n/a_{k+1}} = \frac{m}{n} \in \big[1/2\,, 1\big].$$

  If $m \leqslant b_k$, then $m \geqslant b_k/2 \geqslant a_k$ and

$$\frac{\beta_m}{\beta_n} = \frac{a_k/m}{n/a_{k+1}} = \frac{a_k a_{k+1}}{mn} = \frac{b_k^2}{mn}\,,$$

  which is less than $2\, b_k^2/n^2 \leqslant 2$ and greater than $b_k^2/n^2 \geqslant (n/2)^2/n^2 = 1/4$.
- $2a_k \leqslant n < b_k$. Then again $\beta_m/\beta_n = m/n$.
- $a_k \leqslant n < 2a_k$. The same discussion as in the first case works. Now (β_n) is bounded, since $\beta_n \leqslant 1$ for $a_k \leqslant n < b_k$ and since, for $b_k \leqslant n < a_{k+1}$,

$$\beta_n \leqslant \beta_{a_{k+1}-1} = \frac{1}{a_{k+1}}\,(a_{k+1}-1) \leqslant 1.$$

  However, (β_n) is not essentially decreasing since

$$\frac{\beta_{a_{k+1}-1}}{\beta_{b_k}} = \frac{1}{\sqrt{a_k a_{k+1}}}\,(a_{k+1}-1) \sim \sqrt{\frac{a_{k+1}}{a_k}} \underset{k\to\infty}{\longrightarrow} \infty\,.$$

□

Now, we are going to recall some well-known facts about matrix representation of an operator T defined on a Hilbert space with an orthonormal basis $(\mathrm{e}_n)_{n\geqslant 0}$ and how this is translated in our framework.

The entry $a_{m,n}$ (where $m,n \geqslant 0$) is defined by the mth coordinate of $T(\mathrm{e}_n)$:

$$a_{m,n} = \mathrm{e}_m^*\big(T(\mathrm{e}_n)\big) = \langle T(\mathrm{e}_n), \mathrm{e}_m\rangle,$$

where $\mathrm{e}_k^*(x)$ stands for the kth coordinate of the vector x.

We shall use the notation $\widehat{f}(k)$ for the kth Fourier coefficient of a function $f \in L^1(\mathbb{T})$:

$$\widehat{f}(k) = \int_{\mathbb{T}} f(\xi)\,\bar{\xi}^k\,dm(\xi).$$

If $\widehat{f}(k) = 0$ for $k < 0$, then $\widehat{f}(k)$ is none other than the Taylor coefficient of the analytic function $f(z) = \sum_{k=0}^{\infty} \widehat{f}(k)z^k$ (extending f on $\mathbb{D}$).

Let us point out that when the operator is the composition operator C_φ associated to the symbol φ, viewed on $H^2(\beta)$, its matrix representation in the basis $\big(\mathrm{e}_n^\beta\big)_{n\geqslant 0}$ has an entry (m,n), which can be written as

$$(\mathrm{e}_m^\beta)^*\big(C_\varphi(\mathrm{e}_n^\beta)\big) = \sqrt{\frac{\beta_m}{\beta_n}}\,\widehat{\varphi^n}(m) \tag{12.11}$$

with (as above) the mth Taylor coefficient of φ^n equal to the mth Fourier coefficient of $\varphi^n \in H^\infty \subset L^1(\mathbb{T})$.

12.2.2 Difficulties Attached to General Weights

When $H^2(\beta)$ is the usual Hardy space H^2, the Littlewood subordination principle for H^2 tells that all symbols φ generate bounded composition operators. The case of automorphisms is claimed to be simple using an integral representation for the norm and some change of variable (see Section 2.3.1 of Chapter 2). The same trick occurs for Hilbertian classical weighted Bergman spaces. For Hilbertian weighted Bergman spaces with general radial weights, we still have an integral representation, and this is why finding some (sufficient) condition remains rather easy.

But this integral representation *is precisely the missing point* for general weights β, and as we already mentioned, the conditions on β for the boundedness of the composition operators C_φ on $H^2(\beta)$ are not the same, depending on whether we consider the class of all symbols such that $\varphi(0) = 0$, or the class of symbols $\varphi = \varphi_a$. Let us elaborate on this.

In many occurrences, the weight β is defined as

$$\beta_n = \int_0^1 t^n \, d\mu(t), \tag{12.12}$$

where μ is a positive measure on (0, 1); and when μ has a density ω, we recover the framework of weighted Bergman spaces $\mathscr{B}^2_\omega$ introduced in Section 1.11.6. Such weighted Bergman-type spaces are used, for instance, in [89], [82], and in [113]. We have $\mathscr{B}^2_\omega = H^2(\beta)$ with

$$\beta_n = 2\int_0^1 r^{2n+1}\omega(r)\,dr = \int_0^1 t^n\,\omega(\sqrt{t})\,dt, \tag{12.13}$$

and the sequence $\beta = (\beta_n)_n$ is *non-increasing* (actually the above representation (12.12) is equivalent, by the Hausdorff moment theorem, to a high regularity of the sequence β, namely its *complete monotony*). When the weight β is non-increasing (or, more generally, essentially decreasing), all the symbols vanishing at the origin induce a bounded composition operator, as we will see below.

For weights defined as in (12.12), we dispose of integral representations for the norm in $H^2(\beta)$, and, as we mentioned, this integral representation allows us rather easily to decide when the boundedness of C_φ on $H^2(\beta)$ occurs. The boundedness does not always occur, as shown in the following example due to Kriete and MacCluer in [89]. They consider the previous Bergman-type spaces $\mathscr{B}^2_\omega$ and prove, for example [89, theorem 3], that for

$$\omega(r) = \exp\left(-B\,\frac{1}{(1-r)^\alpha}\right), \qquad \text{with } B > 0,\ 0 < \alpha \leqslant 2,$$

and

$$\varphi(z) = z + t(1-z)^\gamma, \qquad \text{where } 1 < \gamma \leqslant 3, \text{ and } 0 < t < 2^{1-\gamma},$$

then φ is a symbol and C_φ is bounded on $\mathscr{B}^2_\omega$ if and only if $\gamma \geqslant \alpha + 1$. Here (by the Laplace method)

$$\beta_n = \int_0^1 t^n \mathrm{e}^{-B/(1-\sqrt{t})^\alpha}\,dt \approx \exp(-c\,n^{\alpha/(\alpha+1)}).$$

Even though β in non-increasing, we see that, when $\gamma < \alpha + 1$, not all C_{φ_a} are bounded. Actually, for all $\gamma \in (1, 3]$, no C_{φ_a} $(a \neq 0)$ is bounded because β has no polynomial decay (Proposition 12.18 to come).

In the spirit of the *dévissage principle*, we will accordingly organize the forthcoming sections as follows.

(i) We first discuss sufficient conditions when $\varphi(0) = 0$ (this is slightly easier).
(ii) We next discuss sufficient conditions for $\varphi = \varphi_a$.
(iii) We continue with necessary conditions on φ_a, which let the slow oscillation condition appear.
(iv) We are then able to provide a necessary and sufficient condition (slow oscillation and essential decay) in the general case.

12.3 Sufficient Conditions when $\varphi(0) = 0$

12.3.1 The Essentially Decreasing Case

We begin with this simple observation (rather a necessary condition!):

Proposition 12.7 *If all composition operators with symbol vanishing at* 0 *are bounded on $H^2(\beta)$, then the sequence β is bounded above.*

Proof: Let $f \in H^\infty$. Write $f = A\varphi + f(0)$ where A is a constant and φ a symbol vanishing at 0. We have $\varphi = C_\varphi(z) \in H^2(\beta)$ by hypothesis. So that $f \in H^2(\beta)$ and $H^\infty \subseteq H^2(\beta)$. It follows (by the closed graph theorem) that there exists a constant M such that $\|f\|_{H^2(\beta)} \leqslant M\,\|f\|_\infty$ for all $f \in H^\infty$. Testing that with $f(z) = z^n$, we get $\beta_n \leqslant M^2$. □

In Chapter 2, we pointed out a so-called "strong subordination principle." A nice corollary of this principle was observed by Goluzin (it will be reproved later as a corollary of Theorem 12.13).

Theorem 12.8 (Goluzin) *Let φ be a symbol fixing the origin.*

(i) *If β is non-increasing, the operator C_φ is a contraction:* $H^2(\beta) \to H^2(\beta)$.
(ii) *If β is* essentially decreasing *with constant C, then $\|C_\varphi\| \leqslant C^{1/2}$.*

Proof: Let $f(z) = \sum_{n=0}^\infty a_n z^n \in H^2(\beta)$ and $g(z) = f\big(\varphi(z)\big) = \sum_{n=0}^\infty b_n z^n$. We have to show that

$$\sum_{n=0}^\infty |b_n|^2 \beta_n \leqslant \sum_{n=0}^\infty |a_n|^2 \beta_n,$$

but this is a simple Abel summation by parts. Indeed, set

$$A_n = \sum_{k=0}^n |a_k|^2, \quad B_n = \sum_{k=0}^n |b_k|^2, \quad \text{and} \quad A_{-1} = B_{-1} = 0.$$

By the strong subordination principle and the non-increasing character of β,

$$\sum_{n=0}^{\infty} |b_n|^2 \beta_n = \sum_{n=0}^{\infty} (B_n - B_{n-1})\beta_n = \sum_{n=0}^{\infty} B_n(\beta_n - \beta_{n+1})$$
$$\leqslant \sum_{n=0}^{\infty} A_n(\beta_n - \beta_{n+1}) = \sum_{n=0}^{\infty} |a_n|^2 \beta_n.$$

Now, if $C^{-1}\gamma_n \leqslant \beta_n \leqslant \gamma_n$ with $\gamma = (\gamma_n)$ non-increasing, we get with the above notation $g = f \circ \varphi$

$$\|g\|_\beta^2 \leqslant \|g\|_\gamma^2 \leqslant \|f\|_\gamma^2 \leqslant C\|f\|_\beta^2,$$

and the essentially decreasing case follows. □

Theorem 12.8 has a restricted converse, namely the following:

Theorem 12.9 *The following assertions are equivalent:*

(i) *All symbols φ such that $\varphi(0) = 0$ induce bounded composition operators C_φ on $H^2(\beta)$ and moreover*

$$\sup_{\varphi(0)=0} \|C_\varphi\| < \infty\,; \tag{12.14}$$

(ii) *β is an essentially decreasing sequence.*

Proof: That (ii) implies (i) is the content of Theorem 12.8.

For the implication from (i) to (ii), set $M = \sup_{\varphi(0)=0} \|C_\varphi\|$. Let $m > n$ and take

$$\varphi(z) = \varphi_{m,n}(z) = z\left(\frac{1 + z^{m-n}}{2}\right)^{1/n}.$$

Then $\varphi(0) = 0$ and $[\varphi(z)]^n = \dfrac{z^n + z^m}{2}$; hence

$$\frac{1}{4}(\beta_n + \beta_m) = \|\varphi^n\|^2 = \|C_\varphi(z^n)\|^2 \leqslant \|C_\varphi\|^2 \|z^n\|^2 \leqslant M^2 \beta_n,$$

so $\beta_m \leqslant (4M^2 - 1)\beta_n$ and β is essentially decreasing. □

If $M = 1$ (in the proof above), we will see a better estimate in Exercise I.21. We mention in passing the following easy proposition:

Proposition 12.10 *If all C_φ with $\varphi(0) = 0$ are bounded on $H^2(\beta)$, then necessarily (with C_k depending only on k)*

$$\beta_{kn} \leqslant C_k\, \beta_n \qquad \textit{for all integers } k \geqslant 2,\ n \geqslant 1.$$

Proof: Just use the symbol $\varphi(z) = z^k$ to see that

$$\beta_{kn} = \|C_\varphi(z^n)\|^2 \leqslant C_k \|z^n\|^2 = C_k \beta_n. \qquad \square$$

Remarks:

(i) The boundedness of β_n is not sufficient to imply Theorem 12.8, and we used the essentially decreasing character of β in a crucial way. For example, let (β_n) be the sequence defined by the relations $\beta_{2k} = 1$ and $\beta_{2k+1} = 1/(k+1)$, which satisfies (12.1); with $\varphi(z) = z^2$, the operator C_φ is unbounded on $H^2(\beta)$, thanks to Proposition 12.10.
(ii) We shall see in Theorem 12.14 that there are weights β for which all composition operators C_φ with $\varphi(0) = 0$ are bounded, but $\sup_{\varphi(0)=0} \|C_\varphi\| = +\infty$.

Hence, Theorem 12.9 cannot be considered as a characterization of boundedness in the case of symbols fixing the origin.

12.3.2 A Less Stringent Sufficient Condition

In this subsection, we prove the following:

Theorem 12.11 *Let $\beta = (\beta_n)_{n=0}^\infty$ be a sequence of positive numbers satisfying (12.1). Assume that β is* weakly decreasing.

Then, for all symbols φ fixing the origin, the composition operator C_φ is bounded on $H^2(\beta)$.

Note that each φ with $\varphi(0) = 0$ will generate its own δ and $C(\delta)$ in (12.7).

Two lemmas are needed for the proof of Theorem 12.11.

Lemma 12.12 *Let φ be a symbol such that $\varphi(0) = 0$ and $|\varphi'(0)| < 1$. Set $r = \mathrm{e}^{-1/2}$. Then there exists $\rho > 0$ such that for all integers m, n,*

$$|\widehat{\varphi^n}(m)| \leqslant r^{(1+\rho)\,n-m}.$$

Proof: Since $\varphi(0) = 0$, we can write $\varphi(z) = z\,\psi(z)$. Since $|\varphi'(0)| < 1$, we have $\psi : \mathbb{D} \to \mathbb{D}$. Let $M(r) = \sup_{|z|=r} |\psi(z)|$. Cauchy's inequalities say that $|\widehat{\psi^n}(m)| \leqslant [M(r)]^n / r^m$. We have $M(r) < 1$, so there exists a positive number $\rho = \rho(r)$ such that $M(r) = r^\rho$. We get

$$|\widehat{\varphi^n}(m)| = |\widehat{\psi^n}(m-n)| \leqslant \frac{r^{\rho n}}{r^{m-n}} = r^{(1+\rho)\,n-m},$$

and the result follows. $\square$

The next theorem is proved in Appendix G (see Theorem G.23).

Theorem 12.13 *Let $A\colon \ell^2 \to \ell^2$ be a bounded operator represented by the matrix $(a_{m,n})_{m,n}$, that is, $a_{m,n} = \langle A\, e_n, e_m\rangle$, with $(e_n)_{n\geqslant 0}$ the canonical basis of ℓ^2.*

Let (d_n) be a sequence of positive numbers such that, for every m and n,

$$d_m < d_n \quad \Longrightarrow \quad a_{m,n} = 0. \tag{12.15}$$

Then, D being the (possibly unbounded) diagonal operator with entries d_n, we have

$$\|D^{-1}AD\| \leqslant \|A\|.$$

In matrix terms (Schur product underlying),

$$\left\| \left(a_{m,n}\frac{d_n}{d_m} \right) \right\| \leqslant \|(a_{m,n})\|\,.$$

Let us mention that assumption (12.15) is not so artificial as it might seem.

(i) It is naturally satisfied when $M_w C_\varphi$ is a weighted composition operator (with say $w \in H^\infty$) acting on some weighted Hardy space $H^2(\beta)$ equipped with its canonical basis

$$\mathrm{e}_n^\beta = \frac{z^n}{\sqrt{\beta_n}}\cdot$$

Then, if $\varphi(0) = 0$ and if $A = (\widehat{w\varphi^n}(m))$, the matrix A on the canonical basis of H^2 is clearly lower-triangular: $m < n \Longrightarrow a_{m,n} = 0$; and the matrix of B on the basis (e_n^β) of $H^2(\beta)$ is

$$(b_{m,n}) = \left(a_{m,n}\frac{d_n}{d_m} \right) \qquad \text{with } d_n = \frac{1}{\sqrt{\beta_n}}\cdot$$

Suppose now that β is non-increasing, that is, $m \geqslant n \Longrightarrow \beta_m \leqslant \beta_n$. Then

$$d_m < d_n \Longrightarrow \beta_m > \beta_n \Longrightarrow m < n \Longrightarrow a_{m,n} = 0.$$

We can therefore apply Theorem 12.13 to conclude that the norm of B acting on $H^2(\beta)$, or of the matrix B on ℓ^2, satisfies

$$\|B\|_{B(H^2(\beta))} \leqslant \|A\|_{B(H^2)} \leqslant \|w\|_\infty \|C_\varphi\|_{B(H^2)} \leqslant \|w\|_\infty,$$

where we used Littlewood's subordination principle for H^2, namely that $\varphi(0) = 0 \Longrightarrow \|C_\varphi\|_{B(H^2)} = 1$. In particular (taking $w = 1$), we recover Goluzin's result of Theorem 12.8: if $\varphi(0) = 0$ and if β is non-increasing, then the operator $C_\varphi : H^2(\beta) \to H^2(\beta)$ is a contraction.

(ii) We will *force* (12.15) to hold, through a change of weight, for the forthcoming proof of Theorem 12.11.

Proof of Theorem 12.11: First, if $|\varphi'(0)| = 1$, we have $\varphi(z) = \alpha z$ for some α with $|\alpha| = 1$, and the result is clear ($H^2(\beta)$ is rotationally invariant).

So, we assume that $|\varphi'(0)| < 1$. Then, by Lemma 12.12 (recall that $r = \mathrm{e}^{-1/2}$), there exists $\rho > 0$ such that, for all m, n,

$$|\widehat{\varphi^n}(m)| \leqslant r^{(1+\rho)\,n-m}. \tag{12.16}$$

Since $\varphi(0) = 0$, we also know that $\widehat{\varphi^n}(m) = 0$ if $m < n$.

Take $\delta = \rho/2$ and use our hypothesis (12.7) as well as (12.1): there exist $M > 0$ and $\eta_\delta > 0$ such that

$$\frac{\beta_m}{\beta_n} \leqslant M \qquad \text{whenever } m > (1{+}\delta)\,n \quad \text{and} \quad \beta_n \geqslant \eta_\delta \mathrm{e}^{-(\delta/2)n} \quad \text{for all } n.$$

Now define a new sequence $\gamma = (\gamma_n)$, designed to replace $C(\delta)$ by 1 in (12.7), as

$$\gamma_n = \max\Big\{\beta_n, \sup_{m>(1+\delta)\,n} \beta_m\Big\}.$$

This new sequence satisfies $\gamma_m \leqslant \gamma_n$ when $m > (1+\delta)\,n$.

It also satisfies $\beta_n \leqslant \gamma_n \leqslant M\,\beta_n$, so that $H^2(\gamma) = H^2(\beta)$, and we are reduced to prove that the operator $C_\varphi\colon H^2(\gamma) \to H^2(\gamma)$ is bounded.

Let $A = \big(a_{m,n}\big)_{m,n} = \big(\widehat{\varphi^n}(m)\big)_{m,n}$. We must show that

$$B = \big(\gamma_m^{1/2}\gamma_n^{-1/2}a_{m,n}\big)_{m,n}$$

represents a bounded operator on ℓ^2. Define for that the matrix

$$A_1 = \big(a_{m,n}\mathbb{1}_{\{(m,n)\,;\; m\leqslant(1+\delta)\,n\}}\big)_{m,n}$$

and set $A_2 = A - A_1$. Define analogously B_1 and $B_2 = B - B_1$.

The matrix A_1 defines a Hilbert–Schmidt operator, because

$$m \leqslant (1+\delta)n \Rightarrow (1+\rho)n - m \geqslant (1+2\delta-1-\delta)n = \delta n \Rightarrow |a_{m,n}|^2 \leqslant \mathrm{e}^{-\delta n}$$

by (12.16), and

$$\begin{aligned}\sum_{n=0}^{\infty}\sum_{m=0}^{(1+\delta)\,n}|a_{m,n}|^2 &\leqslant \sum_{n=0}^{\infty}\sum_{m=0}^{(1+\delta)\,n}\exp\big(-\delta n\big)\\ &\leqslant \sum_{n=0}^{\infty}\big[(1+\delta)\,n+1\big]\exp(-\delta\,n) < +\infty\,.\end{aligned}$$

Now, γ (as β) is bounded above and satisfies $\gamma_n \geqslant \beta_n \geqslant \eta_\delta e^{-(\delta/2)n}$, so that, for some positive constants M_δ and M'_δ, we have

$$\sum_{n=0}^{\infty} \sum_{m=0}^{(1+\delta)n} \frac{\gamma_m}{\gamma_n} |a_{m,n}|^2 \leqslant M_\delta \sum_{n=0}^{\infty} \sum_{m=0}^{(1+\delta)n} \exp\big((\delta/2 - \delta)\, n\big)$$
$$\leqslant M'_\delta \sum_{n=1}^{\infty} n \exp\big(-(\delta/2)\, n\big) < +\infty,$$

meaning that B_1 is also a Hilbert–Schmidt operator.

We now switch to A_2 and B_2. Since A is bounded, it follows that $A_2 = A - A_1$ is bounded. Remark that, writing $A_2 = \big(\alpha_{m,n}\big)_{m,n}$, we have, with $d_n = 1/\sqrt{\gamma_n}$,

$$d_m < d_n \quad\Longrightarrow\quad \gamma_m > \gamma_n \quad\Longrightarrow\quad m \leqslant (1+\delta)\, n \quad\Longrightarrow\quad \alpha_{m,n} = 0\,.$$

Hence we can apply Theorem 12.13 to A_2 to conclude that $B_2 = D^{-1} A_2 D$ is bounded, and therefore that $B = B_1 + B_2$ is bounded as well, as desired. □

As a corollary of Theorem 12.11, we can provide the following example, to be compared with Theorem 12.9.

Theorem 12.14 *There exists a bounded sequence β, with polynomial decay (hence satisfying* (12.1)*), but* not essentially decreasing, *and for which every composition operator with symbol vanishing at* 0 *is bounded on $H^2(\beta)$. Hence, we must have* $\sup_{\varphi(0)=0} \|C_\varphi\| = +\infty$.

It should be noted that for this weight, not all the composition operators are bounded, as we will see in Proposition 12.20.

Proof: Define $\beta_n = 1$ for $n \leqslant 3!$ and for $k \geqslant 3$:

$$\begin{cases} \beta_n = \dfrac{1}{k!} & \text{for } k! < n \leqslant (k+1)! - 2 \text{ and for } n = (k+1)! \\[2ex] \beta_n = \dfrac{1}{(k+1)!} & \text{for } n = (k+1)! - 1. \end{cases}$$

Note that, for $m > n$, we have $\beta_m > \beta_n$ only if $n = (k+1)! - 1$ and $m = (k+1)! = n+1$ for some $k \geqslant 3$.

However β is not essentially decreasing since, for every $k \geqslant 3$, we have $\beta_{n+1}/\beta_n = k+1$ if $n = (k+1)! - 1$.

The sequence β has polynomial decay because $\beta_n \geqslant 1/(2n)$ for all $n \geqslant 1$. In fact, for $k \geqslant 3$, we have $\beta_n \geqslant (k+1)/n \geqslant 1/n$ if $k! < n \leqslant (k+1)! - 2$ or if $n = (k+1)!$; and for $n = (k+1)! - 1$, we have $n\, \beta_n = [(k+1)! - 1]/(k+1)! \geqslant 1/2$. It is bounded above, by 1.

Now it remains to check (12.7) in order to apply Theorem 12.11 and finish the proof of Theorem 12.14. But this clearly follows from the fact that $\beta_m/\beta_n \leqslant 1$ if $m \geqslant n+2$. The last assertion follows from Theorem 12.9. □

Remark. We do not know if the sufficient condition of Theorem 12.11 is necessary. But it can be proved [106] that we cannot take a fixed $\delta > 0$ independent of φ, which indicates that Theorem 12.11 is fairly sharp.

12.4 Sufficiency of Slow Oscillation for φ_a

The following sufficient condition (slow oscillation) explains why all composition operators C_{φ_a} are bounded on the Dirichlet space, even though the corresponding weight $\beta = (n+1)_{n\geqslant 0}$ is unbounded. Nevertheless, this particular case can be checked directly thanks to the integral representation and a change of variable since φ_a is injective.

Theorem 12.15 *If β is* slowly oscillating, *then all symbols that extend analytically in a neighborhood of $\overline{\mathbb{D}}$ induce a bounded composition operator on $H^2(\beta)$.*

In particular, all the operators C_{φ_a}, with $a \in \mathbb{D}$, are bounded on $H^2(\beta)$.

To prove Theorem 12.15, we begin with an elementary fact.

Lemma 12.16 *Let $u : \mathbb{D} \to \mathbb{D}$ have an analytic extension to an open neighborhood Ω of $\overline{\mathbb{D}}$. Then there are a constant $b > 0$ and an integer $\lambda > 1$ such that*

$$|\widehat{u^n}(m)| \leqslant \begin{cases} \mathrm{e}^{-bn} & \text{if } n \geqslant \lambda m\,, \\ \mathrm{e}^{-bm} & \text{if } m \geqslant \lambda n. \end{cases}$$

In particular, given $\varepsilon > 0$, for every $N \geqslant 1$ large enough, we have

$$\forall q \geqslant 1, \qquad \sum_{j=Nq}^{\infty} |\widehat{u^q}(j)|^2 \leqslant \varepsilon. \tag{12.17}$$

Proof: Let $R > 1$ such that $\overline{D(0,R)} \subseteq \Omega$. For $0 < r \leqslant R$, we set

$$M(r) = \sup_{|z|=r} |u(z)|\,.$$

For the first inequality, take any $r \in (0,1)$, for instance, $r = \mathrm{e}^{-1}$. We have $M(r) < 1$, so we can write $M(r) = \mathrm{e}^{-\rho}$ for some positive ρ.

Cauchy's inequalities give

$$|\widehat{u^n}(m)| \leqslant \frac{[M(r)]^n}{r^m} = \mathrm{e}^{m-\rho n}.$$

Choose $\lambda_1 = \max(2; 2/\rho)$ and $b_1 = \rho - \lambda_1^{-1}$; then $|\widehat{u^n}(m)| \leqslant \mathrm{e}^{-b_1 n}$ if $n \geqslant \lambda_1 m$. For the second inequality, write $R =: \mathrm{e}^{\beta}$, with $\beta > 0$. Let $\alpha > 0$ with $M(R) \leqslant \mathrm{e}^{\alpha}$. Cauchy's inequalities again give

$$|\widehat{u^n}(m)| \leqslant \frac{[M(R)]^n}{R^m} \leqslant \mathrm{e}^{\alpha n-\beta m}.$$

Choose $\lambda_2 = \max(2; 2\alpha/\beta)$ and $b_2 = \beta - \alpha\lambda_2^{-1}$; then $|\widehat{u^n}(m)| \leqslant \mathrm{e}^{-b_2 m}$ if $m \geqslant \lambda_2 n$.

We get the conclusion for the first part, taking $b = \min(b_1; b_2)$ and an integer $\lambda \geqslant \max(\lambda_1; \lambda_2)$.

Now, for any integer $N \geqslant \lambda$, we have

$$\sum_{j=Nq}^{\infty} |\widehat{u^q}(j)|^2 \leqslant (1-\mathrm{e}^{-2b})^{-1}\mathrm{e}^{-2bNq} \leqslant (1-\mathrm{e}^{-2b})^{-1}\mathrm{e}^{-2bN} \leqslant \varepsilon$$

as soon as N is large enough. □

Lemma 12.17 will give the implication $(iii) \Rightarrow (ii)$ of Theorem 12.26 with

$$a_{m,n} = \widehat{\varphi_a^n}(m)$$

(by Littlewood's principle on H^2, the matrix $A = (a_{m,n})_{m,n}$ is bounded on ℓ^2).

Lemma 12.17 *Let (β_n) be a* slowly oscillating *sequence of positive numbers. Let $A = (a_{m,n})_{m,n}$ be the matrix of a bounded operator on ℓ^2. Assume that, for some positive integer λ and positive constants c, b, we have*

(i) $|a_{m,n}| \leqslant c\,\mathrm{e}^{-bn}$ *for $n \geqslant \lambda m$;*
(ii) $|a_{m,n}| \leqslant c\,\mathrm{e}^{-bm}$ *for $m \geqslant \lambda n$.*

Then the matrix $\tilde{A} = \left(a_{m,n}\sqrt{\dfrac{\beta_m}{\beta_n}}\right)_{m,n}$ *also defines a bounded operator on ℓ^2.*

Proof: We have to show that the hermitian form

$$\langle \tilde{A}x, y\rangle = \sum a_{m,n}\sqrt{\frac{\beta_m}{\beta_n}}x_n\overline{y_m} =: \sum \alpha_{m,n}x_n\overline{y_m}$$

is uniformly bounded on finitely supported vectors $x, y \in \ell^2$ of norm 1.

Since we are dealing with finite sums, we can collect terms as we wish, and will write

$$\langle \tilde{A}x, y\rangle = \sum_{(m,n)\in E} \alpha_{m,n} x_n \overline{y_m} + \sum_{(m,n)\in F} \alpha_{m,n} x_n \overline{y_m}\,,$$

where E and $F = E^c$ will be specified later.

First, since β_m/β_n remains bounded from above and below around the diagonal, the matrix $\tilde{A}$ behaves like A near the diagonal. More precisely, let the integer λ be as in Lemma 12.17. Set (with ρ as in (12.8))

$$C = \sup_{1/\lambda^2 \leqslant x \leqslant \lambda^2} \rho(x).$$

If I, J are blocks of integers such that $(m,n)\in I\times J$ implies that $\dfrac{n}{\lambda^2} \leqslant m \leqslant \lambda^2 n$, then, setting $y = (y_m)$, $P_I y = (y_m)_{m\in I}$, $x = (x_n)$, and $P_J x = (x_n)_{n\in J}$, the slow oscillation of β gives, by Cauchy-Schwarz,

$$\begin{aligned}\left| \sum_{(m,n)\in I\times J} a_{m,n}\, x_n \overline{y_m} \sqrt{\frac{\beta_m}{\beta_n}} \right| &\leqslant \|A\| \left(\sum_{m\in I} |y_m|^2 \beta_m\right)^{\frac12} \left(\sum_{n\in J} \frac{|x_n|^2}{\beta_n}\right)^{\frac12} \\ &= \|A\| \left(\sum_{(m,n)\in I\times J} |x_n|^2\, |y_m|^2\, \frac{\beta_m}{\beta_n}\right)^{\frac12} \\ &\leqslant C^{1/2} \|A\|\, \|P_J x\|_{\ell^2}\, \|P_I y\|_{\ell^2}.\end{aligned}$$

For $k = 0, 1, 2, \ldots$, let $J_k = [\lambda^k, \lambda^{k+1}[$ and, for $k = 1, 2, \ldots$, we define $I_k = [\lambda^{k-1}, \lambda^{k+2}[$. We also define $I_0 = [0, \lambda^2[$.

We recall that (e_n) denotes the canonical basis of ℓ^2; we set

$$E = \bigsqcup_{k\geqslant 0} I_k \times J_k,$$

and we define the matrix S whose entries are

$$S_{m,n} = \begin{cases} \alpha_{m,n} = \sqrt{\dfrac{\beta_m}{\beta_n}}\, a_{m,n} & \text{if } (m,n) \in E, \\ 0 & \text{if } (m,n) \notin E, \end{cases}$$

Let $H_k = span\,\{e_n : n \in I_k\}$, and $\tilde{H}_k = span\,\{e_n : n \in J_k\}$. Let P_k be (the matrix of) the orthogonal projection of ℓ^2 with range H_k and Q_k the one with range $\tilde{H}_k$. Then $S_k = P_k \tilde{A} Q_k$ is the matrix with entries $\alpha_{m,n}$ when $(m,n) \in I_k \times J_k$ and 0 elsewhere; and we have

$$\sum_{(m,n)\in E} \alpha_{m,n} x_n \overline{y_m} = \sum_{k\geqslant 0} \left(\sum_{(m,n)\in I_k\times J_k} \alpha_{m,n} x_n \overline{y_m} \right) = \sum_{k\geqslant 0} \langle S_k(x), y\rangle.$$

By the above discussion (observe that $(m,n) \in I_k \times J_k$ implies $n/\lambda^2 \leqslant m \leqslant \lambda^2 n$), we have

$$|\langle S_k(x), y\rangle| \leqslant C^{1/2}\|A\|\,\|Q_k x\|_{\ell^2}\,\|P_k y\|_{\ell^2}.$$

Point out that, for every $y \in \ell^2$, we have $\sum \|P_k y\|_{\ell^2}^2 \leq 3\|y\|_{\ell^2}^2$ since each integer belongs to at most three intervals I_k; and in the same way, for every $x \in \ell^2$, we have $\sum \|Q_k x\|_{\ell^2}^2 \leq \|x\|_{\ell^2}^2$ since the subspaces $\tilde{H}_k$ are orthogonal.

Summing up over k, using Cauchy–Schwarz again, we get

$$\left|\sum_{(m,n)\in E} \alpha_{m,n} x_n \overline{y_m}\right| \leqslant (3C)^{1/2}\|A\|\,\|x\|_{\ell^2}\,\|y\|_{\ell^2} \leqslant (3C)^{1/2}\|A\|. \qquad (12.18)$$

Next, the matrix $\tilde{A}$ is Hilbert–Schmidt far from the diagonal, namely on $F = E^c$. Let indeed $(m,n) \in F$, that is, (m,n) does not belong to the union of $I_k \times J_k$'s. When $n \geq 1$, it means that n belongs to some J_p but $m \notin I_p$: either $m < \lambda^{p-1}$ or $m \geq \lambda^{p+2}$, hence either $m/n < \lambda^{-1}$ or $m/n > \lambda$. Therefore, the non-zero entries (m,n) of $\tilde{A} - S$ satisfy either $n > \lambda m$ or $m > \lambda n$.

Moreover, since β is slowly oscillating, it has polynomial decay and growth: for some $\alpha > 0$ and $\delta \in (0,1)$, we have $\delta\, n^{-\alpha} \leqslant \beta_n \leqslant \delta^{-1} n^{\alpha}$. So that by Lemma 12.16

$$\sum_{n=1}^{\infty}\sum_{\lambda m<n} |\alpha_{m,n}|^2 \lesssim \sum_{n=1}^{\infty}\sum_{\lambda m<n} \delta^{-2} n^{2\alpha}|a_{m,n}|^2 \lesssim \sum_{n=1}^{\infty} n^{2\alpha+1}\mathrm{e}^{-2bn} < +\infty,$$

and similarly

$$\begin{aligned}\sum_{n=0}^{\infty}\sum_{m>\lambda n} |\alpha_{m,n}|^2 &\leqslant \sum_{n=0}^{\infty}\sum_{m>\lambda n} \delta^{-2} n^{2\alpha}|a_{m,n}|^2\\ &\lesssim \sum_{n=0}^{\infty} n^{2\alpha}\Big(\sum_{m>\lambda n} \mathrm{e}^{-2bm}\Big)\\ &\lesssim \sum_{n=0}^{\infty} n^{2\alpha}\mathrm{e}^{-2b\lambda n} < +\infty.\end{aligned}$$

Hence $\sum_{(m,n)\in F} |\alpha_{m,n}|^2 =: M^2 < \infty$, and the Cauchy–Schwarz inequality gives

$$\left|\sum_{(m,n)\in F} \alpha_{m,n} x_n \overline{y_m}\right| \leqslant M\left(\sum_{m,n\geqslant 0} |x_m|^2|y_n|^2\right)^{1/2} = M\,\|x\|_{\ell^2}\,\|y\|_{\ell^2} \leqslant M. \qquad (12.19)$$

Relations (12.18) and (12.19) end the proof. □

12.5 Necessity of Slow Oscillation for φ_a

We devote a whole section to this fairly long proof of the *necessity* of slow oscillation (implication $(ii) \Rightarrow (iii)$ in the main Theorem 12.26 to follow) for the boundedness of the operators C_{φ_a}.

12.5.1 A Simple Necessary Condition

We begin with an elementary necessary condition. It is implied by Theorem 12.19 to come, but its statement deserves to be pointed out. Moreover, its proof is simple and highlights the role of the reproducing kernel.

Proposition 12.18 *Let $a \in (0,1)$ and assume that φ_a induces a bounded composition operator on $H^2(\beta)$. Then β has polynomial decay.*

Equivalently, the reproducing kernels have a slow growth,

$$\|K_w\| \leqslant \frac{C}{(1-|w|)^s}, \tag{12.20}$$

for positive constants C and s.

Proof: Since

$$\|K_w\|^2 = \sum_{n=0}^{\infty} \frac{|w|^{2n}}{\beta_n} = \|K_{|w|}\|^2 ,$$

we have $\|K_x\| \leqslant \|K_y\|$ for $0 \leqslant x \leqslant y < 1$.

We set $T_a(z) = \varphi_a(-z)$, alternatively $T_a = -\varphi_{-a}$, and C_{T_a} is also bounded on $H^2(\beta)$. Define inductively a sequence $(u_n)_{n \geqslant 0}$ by

$$u_0 = 0 \qquad \text{and} \qquad u_{n+1} = T_a(u_n)\,.$$

Since $T_a(1) = 1$ (recall that $a \in (0,1)$), we have

$$1 - u_{n+1} = \int_{u_n}^{1} T_a'(t)\,dt = \int_{u_n}^{1} \frac{1-a^2}{(1+at)^2}\,dt;$$

hence

$$\frac{1-a}{1+a}\,(1-u_n) \leqslant 1 - u_{n+1} \leqslant (1-a^2)(1-u_n)\,.$$

Let $0 < x < 1$. We can find $N \geqslant 0$ such that $u_N \leqslant x < u_{N+1}$. Then

$$1 - x \leqslant 1 - u_N \leqslant (1-a^2)^N .$$

On the other hand, since $C_{T_a}^*(K_z) = K_{T_a(z)}$ for all $z \in \mathbb{D}$, it holds (recall $\beta_0 = 1$) that

$$\|K_x\| \leqslant \|K_{u_{N+1}}\| \leqslant \|C_{T_a}\|\,\|K_{u_N}\| \leqslant \|C_{T_a}\|^{N+1}\|K_{u_0}\| = \|C_{T_a}\|^{N+1}.$$

Let $s \geqslant 0$ be such that $(1-a^2)^{-s} = \|C_{T_a}\|$. We obtain

$$\|K_x\| \leqslant \|C_{T_a}\| \, \frac{1}{(1-x)^s} \tag{12.21}$$

and the reproducing kernel has a *slow growth.* But

$$\|K_x\|^2 = \sum_{k=0}^{\infty} \frac{x^{2k}}{\beta_k}\,;$$

so we get, for any $n \geqslant 2$, $\dfrac{x^{2n}}{\beta_n} \leqslant \|C_{T_a}\|^2 \dfrac{1}{(1-x)^{2s}}\cdot$
Taking $x = 1 - \frac{1}{n}$, we obtain $\beta_n \geqslant C\, n^{-2s}$ and β has polynomial decay.
For the converse, if $\beta_n \geqslant \delta n^{-\alpha}$, we just note that

$$\|K_w\|^2 = \frac{1}{\beta_0} + \sum_{n=1}^{\infty} \frac{|w|^{2n}}{\beta_n} \leqslant \frac{1}{\beta_0} + \delta^{-1} \sum_{n=1}^{\infty} n^{\alpha} |w|^{2n} \leqslant \frac{C}{(1-|w|^2)^{\alpha+1}}$$

and hence K_w has a slow growth. □

12.5.2 A More Precise Necessary Condition

The main result of this subsection is the following:

Theorem 12.19 *If, for all $r \in (0,1)$, the composition operator C_{φ_a} is uniformly bounded on $H^2(\beta)$ for all $a \in [0,r]$, then β is* slowly oscillating.

Let us give right now a corollary of this result.

Corollary 12.20 *For the weight β constructed in the proof of Theorem 12.14, no automorphism φ_a with $a \in \mathbb{D} \setminus \{0\}$ can be bounded.*

Proof: Indeed, it is clear that β is not slowly oscillating since

$$\beta_{(k+1)!-1} / \beta_{(k+1)!} = 1/(k+1) \underset{k\to\infty}{\longrightarrow} 0\,.$$ □

To show Theorem 12.19, we need estimates on the Fourier coefficients of φ_a^n. Such sharp estimates are given in the papers [171] and [172] for other purposes. However, we need minorations of $|\widehat{\varphi_a^n}(m)|$ when m is close to n, and the estimations of [171] and [172] show that this quantity oscillates and, for individual a, can be too small for our purpose. Hence, we will prove instead an estimate *in mean* for a in some interval of $(0,1)$. This estimate relies on the stationary phase method and the van der Corput lemmas (see Appendix D) and goes back at least to [175, pp. 77–80]. See also [60].

We begin by recalling a standard fact.

Lemma 12.21 *Let $a \in (0,1)$ and let*

$$P_a(x) = \frac{1-a^2}{1-2a\cos x + a^2}$$

be the Poisson kernel at point a. Then, for all $x \in [-\pi, \pi]$,

$$\varphi_a(e^{ix}) = -\exp\big[i\, V_a(x)\big], \tag{12.22}$$

where

$$V_a(x) = \int_0^x P_a(t)\, dt. \tag{12.23}$$

Proof: Let $\tau(x) = \varphi_a(e^{ix})$. We notably have, since $\tau(0) = -1$ and τ does not vanish,

$$\tau(x) = -\exp\left(\int_0^x \frac{\tau'(t)}{\tau(t)} dt\right) \quad \text{for all } x \in \mathbb{R}$$

(differentiate $\tau(x)\exp\big(-\int_0^x \frac{\tau'(t)}{\tau(t)}dt\big)$).

But $\dfrac{\tau'(t)}{\tau(t)} = \dfrac{a^2-1}{1-ae^{it}} \times \dfrac{ie^{it}}{a-e^{it}} = i\dfrac{1-a^2}{|1-ae^{it}|^2} = i P_a(t).$

The claimed formula (12.22) clearly follows. □

Let us note that, with V_a the function of Lemma 12.21, the Fourier formulas give, since $a_{m,n}$ is real or since $nV_a(x) - mx$ is odd,

$$2\pi a_{m,n} := 2\pi\widehat{\varphi_a^n}(m) = (-1)^n \int_{-\pi}^{\pi} \exp i[nV_a(x) - mx]\, dx = 2(-1)^n \operatorname{Re} I_{m,n}, \tag{12.24}$$

where

$$I_{m,n} = \int_0^{\pi} \exp i[nV_a(x) - mx]\, dx. \tag{12.25}$$

Now the main ingredient for proving Theorem 12.19 is the following. We shall first need its version for $s = 2$ in our Hilbertian framework, but we shall also exploit the L^s-version later when we treat the case of Beurling–Sobolev spaces with $p \neq 2$.

Theorem 12.22 *Let $I := \left[\frac{1}{3}, \frac{2}{3}\right]$ and $s \geqslant 1$. There exist constants $\alpha > 1$, for example, $\alpha = 5/4$, and $\delta \in (0, 1/2)$ such that, for n large enough ($n \geqslant n_0$), it holds*

$$\int_I |\widehat{\varphi_a^n}(m))|^s da \gtrsim \left(\int_I |\widehat{\varphi_a^n}(m)|\, da\right)^s \geqslant \delta n^{-s/2} \qquad \textit{for every } m \in [\alpha^{-1}n, \alpha\, n].$$

In order to prove this theorem, we need first to estimate several associated parameters. We will set once and for all

$$q = \frac{m}{n}, \tag{12.26}$$

so that $\alpha^{-1} \leqslant q \leqslant \alpha$ where $\alpha = 5/4$ (say). We will only consider pairs (a,q) satisfying

$$a \in I := \left[\frac{1}{3}, \frac{2}{3}\right], \qquad q \in J := \left[\frac{4}{5}, \frac{5}{4}\right]. \tag{12.27}$$

Such pairs will be called *admissible*.

With these notations, we set, for $0 \leqslant x \leqslant \pi$,

$$F_q(x) = n\left(\int_0^x P_a(t)\,dt - qx\right) = n\left(V_a(x) - qx\right).$$

We have

$$F_q'(x) = n\left(\frac{(1-a^2)}{1-2a\cos x + a^2} - q\right), \tag{12.28}$$

and the unique (if some) critical point $x_q = x_q(a)$ of F_q in $[0,\pi]$ is given by $P_a(x_q) = q$, that is,

$$\cos x_q = \frac{1+a^2}{2a} - \frac{1}{q}\left(\frac{1-a^2}{2a}\right) =: h_q(a).$$

We now proceed through a series of simple lemmas, estimating $h_q(a)$ and $x_q(a)$.

Lemma 12.23 *There are positive constants $C > 1$ and $\delta \in (0,1/2)$ such that, for every admissible pair (a,q), we have*

$$0 \leqslant h_q(a) \leqslant 1 - \delta \quad \textit{and} \quad |h_q'(a)| \leqslant C.$$

So there is one critical point $x_q(a)$ satisfying

$$\delta \leqslant x_q(a) \leqslant \frac{\pi}{2} \quad \textit{and} \quad \sin x_q(a) \geqslant \delta; \tag{12.29}$$

moreover

$$|x_q'(a)| \leqslant C \quad \textit{and} \quad \delta \leqslant |P_a'(x_q(a))| \leqslant C.$$

Proof: We have for (a,q) admissible, that is, $(a,q) \in I \times J$:

$$h_q(a) = \left(\frac{1+a^2}{2a}\right) - \frac{1}{q}\left(\frac{1-a^2}{2a}\right) \leqslant \left(\frac{1+a^2}{2a}\right) - \frac{4}{5}\left(\frac{1-a^2}{2a}\right) = \frac{9a}{10} + \frac{1}{10a},$$

which gives

$$h_q(a) \leqslant \frac{6}{10} + \frac{3}{10} = \frac{9}{10}.$$

Similarly

$$h_q(a) \geqslant \left(\frac{1+a^2}{2a}\right) - \frac{5}{4}\left(\frac{1-a^2}{2a}\right) = \frac{9a^2-1}{8a} \geqslant 0.$$

Since $\cos x_q(a) = h_q(a)$, we get the estimate (12.29). Next, we observe that $2h_q'(a) = \left(1+\frac{1}{q}\right) - \left(1-\frac{1}{q}\right)\frac{1}{a^2}$; hence $|h_q'(a)| \leqslant C$. So that, writing $x_q = x_q(a) = \arccos h_q(a)$, we get for another constant $C > 0$

$$|x_q'(a)| = \frac{|h_q'(a)|}{\sqrt{1-h_q(a)^2}} \leqslant C,$$

since $h_q(a)^2 \leqslant 1-\delta$. Finally, $\frac{1}{9} \leqslant (1-a)^2 \leqslant 1 - 2a\cos x_q + a^2 \leqslant 4$ and

$$P_a'(x_q) = -\frac{2a(1-a^2)\sin x_q}{(1-2a\cos x_q + a^2)^2}$$

can be now easily bounded. □

We saw in (12.24) that the value of $a_{m,n} := \widehat{\varphi_a^n}(m)$ is given by the formula

$$a_{m,n} = \frac{(-1)^n}{\pi}\,\mathrm{Re}\, I_{m,n}. \tag{12.30}$$

We also need the following crucial estimate.

Proposition 12.24 *We have*

$$I_{m,n} = \sqrt{2\pi}\, n^{-1/2}\, \frac{e^{i\,[F_q(x_q)-\pi/4]}}{\sqrt{|P_a'(x_q)|}} + O\,(n^{-3/5}), \tag{12.31}$$

where O only depends on a, and so is absolute as long as (a, q) is admissible.

Note that $3/5 > 1/2$. We hence have

$$a_{m,n} = (-1)^n \sqrt{\frac{2}{\pi}}\, n^{-1/2}\, \frac{\cos\,[F_q(x_q)-\pi/4]}{\sqrt{|P_a'(x_q)|}} + O\,(n^{-3/5}). \tag{12.32}$$

Proof: We use Theorem D.3 with $F = F_q$, $[A, B] = [0, \pi]$, $c = x_q$, $\lambda_2 = \kappa_0\, n$, $\lambda_3 = C_0 n$, and $\eta = (\lambda_2\lambda_3)^{-1/5}$ as detailed below.

Actually, there is a slight difficulty here: our function F_q has a second derivative vanishing at 0 and π (and nowhere else). We circumvent this difficulty by working first on $[\delta, \pi - \delta]$, obtaining then an estimate as in

Theorem D.3 of $\int_{A+\delta}^{B-\delta} e^{iF(x)}\,dx$ with uniform error terms independent of δ and then letting $\delta \to 0^+$. We omit the details.

The parameter η is chosen in order to make both error terms in Theorem D.3 equal: $\dfrac{1}{\eta\lambda_2} = \eta^4\lambda_3$; so we choose

$$\eta = \kappa\, n^{-2/5}$$

and we have

$$\frac{1}{\eta\lambda_2} + \eta^4\lambda_3 =: \tilde{\kappa}\, n^{-3/5} = O\,(n^{-3/5}). \tag{12.33}$$

We also know, using (12.28), that

$$F_q''(x) = n P_a'(x) = -\frac{2na(1-a^2)\,\sin x}{(1-2a\cos x + a^2)^2}.$$

From the estimates (and the constants involved there) of Lemma 12.23, there exists a positive (and absolute) constant κ_0 such that

$$F_q''(x) \leqslant -\kappa_0\, n = -\lambda_2 \qquad \text{for } x \in [\delta/2, \pi - \delta/2]. \tag{12.34}$$

Now (for n large enough), $[x_q - \eta, x_q + \eta] \subseteq [\delta/2, \pi - \delta/2]$. Hence the assumptions (1) and (2) of Proposition D.3 are satisfied with $\varepsilon = -1$.

Finally, since $F_q(x) = n[V_a(x) - qx]$, $F_q''' = nV_a''' = nP_a''$ we have, for all $x \in [0,\pi]$ and (a,q) admissible,

$$|F_q'''(x)| \leqslant C_0\, n = \lambda_3,$$

where C_0 is absolute and assertion (3) of Theorem D.3 holds.

With (12.33) this ends the proof of (12.31), once we remarked that $F_q''(x_q) = nV_a''(x_q) = nP_a'(x_q)$. □

For the proof of Proposition 12.22, it will be convenient to introduce a new function $\psi_q(a)$ (*this* function will meet the assumptions of van der Corput's lemma) by setting

$$F_q\big(x_q(a)\big) = n\,\psi_q(a). \tag{12.35}$$

To apply van der Corput's lemma, we will then need estimates on the derivatives of ψ_q.

Lemma 12.25 *If (a,q) is admissible, then ψ_q decreases on I and moreover*

(i) $|\psi_q'(a)| \geqslant \delta$;
(ii) $|\psi_q''(a)| \leqslant C$.

Proof: Note in passing that, with $x_q = x_q(a) \in [0,\pi]$,

$$\psi_q(a) = \int_0^{x_q} [P_a(t) - P_a(x_q)]\,dt \geqslant 0,$$

since the integrand is non-negative and $q = P_a(x_q)$. Next, if f and g are real C^1-functions and

$$\Phi(a) = \int_0^{f(a)} g(a,t)\,dt,$$

the chain rule gives

$$\Phi'(a) = f'(a)\,g\big(a, f(a)\big) + \int_0^{f(a)} \frac{\partial g}{\partial a}(a,t)\,dt.$$

With $g(a,t) = P_a(t)$ and $f(a) = x_q(a)$, we get, remembering $P_a(x_q) = q$,

$$\psi_q'(a) = x_q'(a)\big[P_a\big(x_q(a)\big) - q\big] + \int_0^{x_q(a)} \frac{\partial P_a}{\partial a}(a,t)\,dt = \int_0^{x_q(a)} \frac{\partial P_a}{\partial a}(a,t)\,dt.$$

But $P_a(t) = 1 + 2\sum_{k=1}^{\infty} a^k \cos kt$, so that

$$\psi_q'(a) = \int_0^{x_q(a)} \left(2\sum_{k=1}^{\infty} k\,a^{k-1}\cos kt\right) dt = \frac{2}{a}\sum_{k=0}^{\infty} a^k \sin\,[k x_q(a)],$$

that is (with an absolute constant $\delta > 0$),

$$\psi_q'(a) = \frac{2}{a}\,\mathrm{Im}\,\frac{1}{1 - a\,\mathrm{e}^{i x_q(a)}} = \frac{2\sin x_q(a)}{1 - 2a\cos x_q(a) + a^2} \geqslant \delta. \tag{12.36}$$

Now, the second estimate of Lemma 12.23 proves (i).

Since $|x_q'(a)| \leqslant C$ by Lemma 12.23, the chain rule and (12.36) clearly give the uniform boundedness of $|\psi_q''(a)|$ when (a,q) is admissible, and this ends the proof. □

Proof of Theorem 12.22: The first inequality is obvious using Hölder.

We first give a (short) argument explaining why the second term is greater than the third one in the specific case $s = 2$ (which, by the way, is sufficient in the Hilbertian framework $p = 2$).

We have $2\cos^2(\pi/4 - x) = 1 + \sin 2x$, and since $1/2 + 3/5 = 11/10$ and $|P_a'(x_q)| \geqslant \delta$ by Lemma 12.23, we have

$$a_{m,n}^2 = \frac{1}{\pi n}\,\frac{1 + \sin\,[2n\,\psi_q(a)]}{|P_a'(x_q)|} + O\left(n^{-11/10}\right).$$

Since $|P_a'(x_q)| \leqslant C$ by Lemma 12.23 again, for (a,q) admissible, this implies (changing δ if necessary)

$$a_{m,n}^2 \geqslant \frac{\delta}{n}(1 + \sin[2n\,\psi_q(a)]) + O\left(n^{-11/10}\right). \qquad (12.37)$$

We know that Lemma D.1 can be applied with $A = 1/3$, $B = 2/3$, $f = 2n\psi_q$, since Lemma 12.25 shows that this f meets the assumptions of Lemma D.1. This gives us, uniformly with respect to (a, q) admissible,

$$\left|\int_I \sin[2n\psi_q(a)]\,da\right| \leqslant \left|\int_I e^{2in\psi_q(a)}\,da\right| \leqslant \frac{C}{n}. \qquad (12.38)$$

Now, integrating (12.37) on I and using (12.38) give for some numerical $\delta \in (0, 1/2)$ (recall once again that $a_{m,n} = \widehat{\varphi_a^n}(m)$ and that $q = m/n$), we get

$$\int_I |\widehat{\varphi_a^n}(m)|^2\,da \geqslant \delta n^{-1} + O\left(n^{-2}\right) + O\left(n^{-11/10}\right) \geqslant (\delta/2)\,n^{-1}$$

for $n \geqslant n_0$ and $\alpha^{-1} \leqslant m/n \leqslant \alpha$. This ends the proof of Theorem 12.22 for $s = 2$.

Now, let us prove the full result, which amounts to treat the case $s = 1$.

We come back to (12.32) and write

$$|\widehat{\varphi_a^n}(m)| \geqslant \delta n^{-1/2}|\cos(n\psi_q(a) - \pi/4)| + O(n^{-3/5}). \qquad (12.39)$$

Moreover, we have the classical Fourier expansion

$$|\cos x| = c + \sum_{l=1}^{\infty} \delta_l \cos(lx) \quad \text{with } c > 0 \text{ and } \delta_l = O(l^{-2}).$$

So that (indeed, the obvious $\sum_{l=1}^{\infty} |\delta_l|^2 < \infty$ would suffice)

$$|\widehat{\varphi_a^n}(m)| \geqslant \delta n^{-1/2}\Big[c + \sum_{l=1}^{\infty} \delta_l \cos\left(l(n\psi_q(a) - \pi/4)\right)\Big] + O(n^{-3/5}). \quad (12.40)$$

We now apply Lemma D.1 with $f(a) = l(n\psi_q(a) - \pi/4)$. Here, for given l, we have $|f'| \geqslant \delta nl$ and $|f''| \leqslant Mnl$ on I, so that

$$\left|\int_I \cos(l(n\psi_q(a) - \pi/4))\,da\right| \leqslant C\left(\frac{1}{nl} + \frac{nl}{n^2 l^2}\right) = \frac{2C}{nl}.$$

It now follows from (12.40) that

$$\begin{aligned}\int_I \left|\widehat{\varphi_a^n}(m)\right| da &\geqslant \frac{\delta}{\sqrt{n}}\left[c + O\left(\sum_{l=1}^{\infty} \frac{|\delta_l|}{nl}\right)\right] + O\left(n^{-3/5}\right)\\ &\geqslant \frac{\delta}{\sqrt{n}} + O\left(n^{-3/5}\right) \geqslant \frac{\delta}{\sqrt{n}}.\end{aligned}$$

This ends the proof of Theorem 12.22. □

We are now ready to prove the main theorem of this section.

Proof of Theorem 12.19: By hypothesis, C_{φ_a} is bounded for all $a \in \mathbb{D}$, with moreover

$$K := \sup_{1/3 \leqslant a \leqslant 2/3} \|C_{\varphi_a}\| < +\infty .$$

Matricially, this writes, for all $a \in [1/3, 2/3]$, as

$$\left\| \left(\widehat{\varphi_a^n}(m) \sqrt{\frac{\beta_m}{\beta_n}} \right)_{m,n} \right\| \leqslant K.$$

In particular, we have, considering the columns and rows of the previous matrix, for every $n \geqslant 1$,

$$\sum_{m=1}^{\infty} \left|\widehat{\varphi_a^n}(m)\right|^2 \frac{\beta_m}{\beta_n} \leqslant K^2, \quad \text{that is,} \quad \sum_{m=1}^{\infty} \left|\widehat{\varphi_a^n}(m)\right|^2 \beta_m \leqslant K^2 \beta_n$$

and, for all $m \geqslant 1$,

$$\sum_{n=1}^{\infty} \left|\widehat{\varphi_a^n}(m)\right|^2 \frac{\beta_m}{\beta_n} \leqslant K^2, \quad \text{that is} \quad \sum_{n=1}^{\infty} \left|\widehat{\varphi_a^n}(m)\right|^2 \frac{1}{\beta_n} \leqslant \frac{K^2}{\beta_m}.$$

This implies, for every $n \geqslant 1$ (recall that $\alpha = 5/4$),

$$\sum_{\alpha^{-1} n \leqslant j \leqslant \alpha n} |\widehat{\varphi_a^n}(j)|^2 \beta_j \leqslant K^2 \beta_n \tag{12.41}$$

and, for every $m \geqslant 1$,

$$\sum_{\alpha^{-1} m \leqslant k \leqslant \alpha m} |\widehat{\varphi_a^k}(m)|^2 \frac{1}{\beta_k} \leqslant \frac{K^2}{\beta_m} \cdot \tag{12.42}$$

Integrating on $a \in (1/3, 2/3)$ and using Theorem 12.22, we get, from (12.41) and (12.42), with some constant $C \geqslant 1$ and all integers $m, n \geqslant 1$,

$$\frac{1}{n} \sum_{\alpha^{-1} n \leqslant j \leqslant \alpha n} \beta_j \leqslant C \beta_n \qquad \text{and} \qquad \frac{1}{m} \sum_{\alpha^{-1} m \leqslant k \leqslant \alpha m} \frac{1}{\beta_k} \leqslant \frac{C}{\beta_m} \cdot \tag{12.43}$$

Let now $r = \sqrt{\alpha}$. Assume that $r^{-1} n \leqslant m \leqslant rn$ and let E be the interval of integers between $r^{-1} m$ and rm. Observe that $E \subset [\alpha^{-1} n, \alpha n] \cap [\alpha^{-1} m, \alpha m]$ and that $|E| \geqslant \eta \max(m, n)$ where η is a positive numerical constant. Therefore, we obtain from (12.43) that

$$\frac{\eta}{|E|} \sum_{j \in E} \beta_j \leqslant C \beta_n \qquad \text{and} \qquad \frac{\eta}{|E|} \sum_{k \in E} \frac{1}{\beta_k} \leqslant \frac{C}{\beta_m} \cdot \tag{12.44}$$

Now denote by M^{ha} and M^{ar} the respective harmonic and arithmetic means of the sequence β on E. We know that $M^{ha} \leqslant M^{ar}$; this implies that (using (12.44))

$$\beta_m \leqslant \frac{C}{\eta} M^{ha} \leqslant \frac{C}{\eta} M^{ar} \leqslant \frac{C^2}{\eta^2} \beta_n =: C' \beta_n .$$

All in all, we proved that $r^{-1}n \leqslant m \leqslant rn \Longrightarrow \beta_m \leqslant C'\beta_n$. Since $r > 1$ and since m, n play symmetric roles, this proves that β is slowly oscillating. □

12.5.3 A Necessary and Sufficient Condition for φ_a

We now have the following theorem [106], which summarizes and precises what precedes. The equivalence of (i) and (ii) is already in [185, proposition 3.6].

We recall, for a and z in $\mathbb{D}$, the notation $\varphi_a(z) = \dfrac{a - z}{1 - \bar{a}\, z}$.

Theorem 12.26 *Let β be a sequence of positive numbers. The following assertions are equivalent:*

(i) *For some $a \in \mathbb{D}$, $a \neq 0$, the map φ_a induces a bounded composition operator C_{φ_a} on $H^2(\beta)$.*
(ii) *For all $a \in \mathbb{D}$, the maps φ_a induce bounded composition operators C_{φ_a} on $H^2(\beta)$; moreover, $a \mapsto \|C_{\varphi_a}\|$ is bounded on compact subsets of $\mathbb{D}$.*
(iii) *β is* slowly oscillating.

Of course, if β satisfies either of the equivalent conditions of this theorem, $C_\varphi : H^2(\beta) \to H^2(\beta)$ is bounded for every automorphism φ of the disk. We then say, with Zorboska [185], that $H^2(\beta)$ is *disc-automorphism invariant.*

Condition (iii) is essential in this chapter, because we will rely on it to obtain the full characterization on the weight β alluded to in the introduction.

The proof of (i) implies (ii) requires simple lemmas.

Lemma 12.27 *Let U, V denote two automorphisms of $\mathbb{D}$.*

Then $|V(0)| = |U(0)|$ if and only if $V = R_{\theta_0} U R_\theta$ for some rotations R_{θ_0}, R_θ.

In particular, the composition operator C_{φ_b} is bounded if and only if $C_{\varphi_{|b|}}$ is bounded, with equal norms (see also [56, proposition 2.1]).

Proof: The right-to-left implication is obvious.

Suppose now that $V(0) = U(0)$. Then $U^{-1}V(0) = 0$. By the Schwarz lemma (automorphisms fixing the origin are rotations), we get $U^{-1} \circ V = R_\theta$ for some θ and $V = U \circ R_\theta$.

Finally, if $|V(0)| = |U(0)|$, let θ_0 be such that $(R_{-\theta_0} V)(0) = U(0)$. By what precedes, $R_{-\theta_0} V = U R_\theta$ and $V = R_{\theta_0} U R_\theta$ as desired. □

Lemma 12.28 *Set $r = |a|$. For $|b| \leqslant \frac{2r}{1+r^2}$, we have*

$$\|C_{\varphi_b}\| \leqslant \|C_{\varphi_r}\|^2 = \|C_{\varphi_a}\|^2 .$$

Proof: Let S be the circle $C(0,r)$ and $u\colon S \to \mathbb{R}_+$ be the continuous function defined by

$$u(s) = \left| \frac{s-r}{1-\overline{s}r} \right| . \tag{12.45}$$

By connectedness, $u(S)$ contains the segment $\left[0, \frac{2r}{1+r^2}\right] = [u(r), u(-r)]$. Let now $b \in \overline{D}\left(0, \frac{2r}{1+r^2}\right)$. By the above, there exists $s \in S$ such that $|b| = u(s)$. In other terms, we have $|\varphi_b(0)| = |\varphi_s(r)| = |(\varphi_s \circ \varphi_r)(0)|$. Now, Lemma 12.27 implies that $\varphi_b = R_{\theta'} \circ \varphi_s \circ \varphi_r \circ R_\theta$, so that $C_{\varphi_b} = C_{R_\theta} C_{\varphi_r} C_{\varphi_s} C_{R_{\theta'}}$. Since $C_{R_\theta}, C_{R_{\theta'}}$ are isometries, we get using Lemma 12.27 for C_{φ_s} (recall $|s| = r = |a|$)

$$\|C_{\varphi_b}\| = \|C_{\varphi_r} C_{\varphi_s}\| \leqslant \|C_{\varphi_r}\| \, \|C_{\varphi_s}\| = \|C_{\varphi_r}\|^2 .$$ □

Proof of Theorem 12.26: (ii) obviously implies (i)

We must show here that (i) automatically implies (ii), which was an assumption in Theorem 12.19 where, based on this, the fact that (ii) implies (iii) has been proved; and that (iii) implies (ii) is Theorem 12.15.

Now, it suffices to use Lemmas 12.27 and 12.28 and perform an iteration, in noting that if $r_0 = |a| > 0$ and $r_{n+1} = \frac{2r_n}{1+r_n^2}$, then $(r_n)_{n \geqslant 0}$ increases to 1 (at super-exponential speed: $1 - r_{n+1} \leqslant (1 - r_n)^2$). □

12.6 Boundedness of All Composition Operators

In this section, we are finally able to characterize the sequences β for which *all composition operators* are bounded on $H^2(\beta)$. The main remaining step is the following theorem.

Theorem 12.29 *Let β be a* slowly oscillating *sequence, and assume that all composition operators C_φ are bounded on $H^2(\beta)$. Then, β is* essentially decreasing.

Once this theorem is proved, we have our complete answer, the main theorem of this chapter.

Theorem 12.30 *Let $\beta = (\beta_n)$ be a weight.*

A necessary and sufficient condition for all composition operators C_φ to be bounded on $H^2(\beta)$ is that β is both slowly oscillating *and* essentially decreasing.

In other words, a necessary and sufficient condition for all composition operators C_φ to be bounded on $H^2(\beta)$ is the following:

$$\exists C > 0\,, \forall m, n \in \mathbb{N}_0, \quad m \geqslant \frac{n}{2} \Longrightarrow \beta_m \leqslant C\beta_n\,. \tag{12.46}$$

The fact that *slowly oscillating and essentially decreasing* can be translated as (12.46) is clear.

Proof: We first show the sufficiency.

Since β is essentially decreasing, all composition operators C_φ with $\varphi(0) = 0$ are bounded on $H^2(\beta)$ (Theorem 12.8). Since β is slowly oscillating, all composition operators C_{φ_a}, with $a \in \mathbb{D}$, are bounded (Theorem 12.15). The case of a general symbol follows by the dévissage principle.

The necessity goes as follows.

Since all composition operators are bounded on $H^2(\beta)$, we know that the operators C_{φ_a} are all bounded; hence β must be slowly oscillating, by Theorem 12.19. It then follows from Theorem 12.29 that β is essentially decreasing. □

We now tackle the proof of Theorem 12.29.

Proof of Theorem 12.29: Assume by contradiction that the slowly oscillating sequence β is not essentially decreasing. We are going to construct a symbol φ such that the composition operator C_φ is unbounded on $H^2(\beta)$. We prefer to use here the automorphisms

$$T_a(z) = \varphi_a(-z), \quad \text{that is,} \quad T_a(z) = \frac{z+a}{1+az} = a + (1-a^2)z + \cdots \quad \text{for } a \in (0,1).$$

This function φ will be an infinite product of the form

$$\varphi(z) = \prod_{k=1}^{\infty} T_{a_k}(z^{n_k}) = \prod_{k=1}^{\infty} \frac{z^{n_k} + a_k}{1 + a_k z^{n_k}} =: \prod_{k=1}^{\infty} \psi_k(z)$$

for a sequence of numbers $a_k \in (0,1)$ such that $\sum_{k\geqslant 1}(1-a_k) < \infty$ and a sequence (n_k) of positive integers increasing to infinity.

Then, $|T_{a_k}(z^{n_k}) - 1| \leqslant \frac{2(1-a_k)}{1-|z|}$ and φ is a locally uniformly convergent infinite product. Observe that φ is indeed a convergent Blaschke product, with n_k zeroes of modulus a_k^{1/n_k}, $k = 1, 2, \ldots$, and (setting $a_k = \mathrm{e}^{-\varepsilon_k}$):

$$\sum_{k\geqslant 1} n_k(1 - a_k^{1/n_k}) \leqslant \sum_{k\geqslant 1} n_k(\varepsilon_k/n_k) = \sum_{k\geqslant 1} \varepsilon_k < \infty.$$

Those sequences $(a_k), (n_k)$ will be constructed together with two other sequences of positive integers $(N_k)_{k\geqslant 0}$ and $(m_k)_{k\geqslant 0}$, $m_k \geqslant 2$. Indeed, we shall always put $a_k = 1 - 1/m_k$ and our three sequences $(N_k), (m_k), (n_k)$ will be defined inductively as follows:

We start with $n_0 = m_0 = 2$ and $a_0 = 1 - \dfrac{1}{m_0} = \dfrac{1}{2}\cdot$

Using Lemma 12.16 (see (12.17)) with $u = u_0 = T_{a_0}$, we get $N_0 \geqslant 1$ such that

$$\sum_{j=N_0 m}^{\infty} |\widehat{u_0^m}(j)|^2 = \sum_{j=N_0 m}^{\infty} |\widehat{T_{a_0}^m}(j)|^2 \leqslant \rho\,, \quad \text{for all } m \geqslant 1,$$

where ρ is a positive numerical constant to be adjusted in the end.

Assume now that we have constructed increasing sequences of integers

$$m_0, \ldots, m_k, \quad n_0, \ldots, n_k, \quad N_0, \ldots, N_k$$

such that, for $0 \leqslant l \leqslant k-1$, we have

$$m_{l+1} \geqslant 2m_l \quad \text{and} \quad n_{l+1} \geqslant 2n_l$$

and, for $1 \leqslant l \leqslant k$,

$$n_l > N_{l-1} m_l \quad \text{and} \quad \beta_{n_l} \geqslant 2^l \beta_{m_l}$$

and, for $0 \leqslant l \leqslant k$,

$$\sum_{j=N_l m}^{\infty} |\widehat{u_l^m}(j)|^2 \leqslant \rho\,, \quad \text{for all } m \geqslant 1,$$

where $a_l = 1 - \dfrac{1}{m_l}$ and $u_l = \psi_1 \cdots \psi_l$ with

$$\psi_l(z) = T_{a_l}(z^{n_l}) = a_l + (1 - a_l^2) z^{n_l} + \cdots$$

Since β is not essentially decreasing *but is slowly oscillating,* we claim that there exist integers $m_{k+1} \geqslant 2m_k$ and $n_{k+1} \geqslant 2n_k$ such that $n_{k+1} > N_k m_{k+1}$ and $\beta_{n_{k+1}} \geqslant 2^{k+1} \beta_{m_{k+1}}$, but let us give some words of explanation about this.

Since β is slowly oscillating the quotients $\dfrac{\beta_n}{\beta_m}$ stay bounded for all pairs (n,m) such that $1 \leqslant \dfrac{n}{m} \leqslant N_k' := \max(N_k, 2n_k)$, say, bounded by a constant C_k (recall (12.8)). On the other hand, our hypothesis is that β is not essentially decreasing, so that the family $\left(\dfrac{\beta_n}{\beta_m}\right)_{n\geqslant m\geqslant 2m_k}$ is not bounded, and in particular there exists a pair (n_{k+1}, m_{k+1}) satisfying $n_{k+1} \geqslant m_{k+1} \geqslant 2m_k$ and

$$\frac{\beta_{n_{k+1}}}{\beta_{m_{k+1}}} \geqslant \max\{C_k + 1, 2^{k+1}\}$$

so that $\beta_{n_{k+1}} \geqslant 2^{k+1}\beta_{m_{k+1}}$; but also (n_{k+1}, m_{k+1}) cannot satisfy $\dfrac{n_{k+1}}{m_{k+1}} \leqslant N'_k$, so $n_{k+1} > N_k m_{k+1}$ and $n_{k+1} \geqslant 2n_k$. The claim is proved.

We set

$$a_{k+1} = 1 - \frac{1}{m_{k+1}} \quad \text{and} \quad \psi_{k+1}(z) = T_{a_{k+1}}(z^{n_{k+1}}),$$

and we apply Lemma 12.16 again to the function $u = u_{k+1} = \psi_1 \cdots \psi_{k+1}$. We get some $N_{k+1} > N_k$ such that

$$\sum_{j=N_{k+1}m}^{\infty} |\widehat{u_{k+1}^m}(j)|^2 \leqslant \rho \quad \text{for all } m \geqslant 1. \tag{12.47}$$

That ends the induction.

Clearly $\sum_{k=1}^{\infty}(1 - a_k) = \sum_{k=1}^{\infty}\dfrac{1}{m_k} \leqslant \sum_{k=1}^{\infty} 2^{-k} = 1$; therefore, the infinite product $\varphi = \prod_{k\geqslant 1}\psi_k$ converges uniformly on compact subsets of $\mathbb{D}$.

To show that the composition operator C_φ is unbounded on $H^2(\beta)$, *it suffices to show that,* for some constant $c > 0$, we have, for all $k \geqslant 2$,

$$\sum_{j=n_k}^{2n_k} |\widehat{\varphi^{m_k}}(j)|^2 \geqslant c\,. \tag{12.48}$$

Indeed, *since β is slowly oscillating*, there is a positive constant $\delta < 1$ such that

$$\beta_j \geqslant \delta\,\beta_{n_k} \quad \text{for} \quad j = n_k, n_k+1, \ldots, 2n_k.$$

Then, we have, since $C_\varphi(z^{m_k}) = \varphi^{m_k}$

$$\frac{\|C_\varphi(z^{m_k})\|^2_{H^2(\beta)}}{\|z^{m_k}\|^2_{H^2(\beta)}} \geqslant \frac{\sum_{j=n_k}^{2n_k}|\widehat{\varphi^{m_k}}(j)|^2\beta_j}{\beta_{m_k}} \geqslant \frac{c\delta\,\beta_{n_k}}{\beta_{m_k}} \geqslant 2^k c\delta \underset{k\to\infty}{\longrightarrow} +\infty\,,$$

so that C_φ is unbounded on $H^2(\beta)$.

Let us agree to write formally, for an analytic function $f(z) = \sum_{k=0}^{\infty} f_k z^k$ and an arbitrary positive integer p, $f(z) = \sum_{k=0}^{p} f_k z^k + O(z^{p+1})$.

We now have to show (12.48). We have, for $k \geqslant 2$,

$$\varphi(z) = v_k(z)\,\psi_k(z)\,G_k(z),$$

where we put $v_k = \psi_1 \cdots \psi_{k-1} = u_{k-1}$ (the function defined before) and where

$$G_k(z) = \prod_{l=k+1}^{\infty} \psi_l(z) = \prod_{l=k+1}^{\infty} a_l + O\left(z^{n_{k+1}}\right).$$

We observe that

$$\varphi^{m_k} = v_k^{m_k} \psi_k^{m_k} G_k^{m_k} =: v_k^{m_k} R_k,$$

and we will proceed in three steps. The first two are simple.

Step 1. We claim that $\widehat{\psi_k^{m_k}}(n_k)$ is bounded below: more precisely (by the binomial formula)

$$[\psi_k(z)]^{m_k} = \left[a_k + (1 - a_k^2) z^{n_k} + + O(z^{2n_k})\right]^{m_k} =: \lambda_k + \mu_k z^{n_k} + O(z^{2n_k}) \tag{12.49}$$

with (using $1 - x \geqslant \mathrm{e}^{-2x}$ for $0 \leqslant x \leqslant 1/2$)

$$\lambda_k = a_k^{m_k} = \left(1 - \frac{1}{m_k}\right)^{m_k} \geqslant \mathrm{e}^{-2};$$

$$\mu_k = \widehat{\psi_k^{m_k}}(n_k) = \left(1 - a_k^2\right) m_k a_k^{m_k - 1} \geqslant (1 - a_k) m_k a_k^{m_k} = a_k^{m_k} \geqslant \mathrm{e}^{-2}.$$

Step 2. $\widehat{\psi_k^{m_k} G_k^{m_k}}(n_k) =: \widehat{R_k}(n_k)$ is again bounded below. Indeed

$$[G_k(z)]^{m_k} =: \nu_k + O\left(z^{n_{k+1}}\right) \tag{12.50}$$

with

$$\nu_k = \left(\prod_{l=k+1}^{\infty} a_l\right)^{m_k} \geqslant \exp\left(-2 m_k \left(\sum_{l=k+1}^{\infty} \frac{1}{m_l}\right)\right) \geqslant \exp\left(-2 \sum_{l=1}^{\infty} 2^{-l}\right) = \mathrm{e}^{-2}. \tag{12.51}$$

Now, (12.49), (12.50), and (12.51) imply (note that $\mu_k \nu_k = \widehat{R_k}(n_k)$)

$$R_k(z) = \lambda_k \nu_k + \mu_k \nu_k z^{n_k} + O(z^{2n_k}) \quad \text{with } \mu_k \nu_k \geqslant \mathrm{e}^{-4}. \tag{12.52}$$

Step 3. This is the most delicate one. With $\varphi^{m_k} = v_k^{m_k} R_k$, a lower bound for $\widehat{\varphi^{m_k}}(n_k)$ seems out of reach. But a lower bound for a long sum (but not too long! since $N_{k-1} m_k < n_k$),

$$\sum_{0 \leqslant l \leqslant N_{k-1} m_k} |\widehat{\varphi^{m_k}}(n_k + l)|^2 = \sum_{0 \leqslant l \leqslant N_{k-1} m_k} |\widehat{v_k^{m_k} R_k}(n_k + l)|^2,$$

will be available, and this will show (12.48). To that effect, write $v_k^{m_k} = g_1 + g_2$, with

$$\begin{cases} g_1(z) &= \sum_{j=0}^{N_{k-1}m_k} \widehat{v_k^{m_k}}(j)\, z^j, \\ g_2(z) &= \sum_{j>N_{k-1}m_k} \widehat{v_k^{m_k}}(j)\, z^j. \end{cases}$$

The functions g_1 and g_2 are orthogonal in $L^2(\mathbb{T})$, and g_2 will be "negligible." Indeed, with $\|\cdot\|_2 = \|\cdot\|_{L^2(\mathbb{T})}$, (12.47) gives us, since $v_k = u_{k-1}$,

$$\|g_2\|_2^2 = \sum_{j>N_{k-1}m_k}^{\infty} |\widehat{v_k^{m_k}}(j)|^2 \leqslant \rho\,. \tag{12.53}$$

Besides, $v_k^{m_k}$ is inner, so that

$$\|g_1\|_2^2 = \|v_k^{m_k}\|_2^2 - \|g_2\|_2^2 = 1 - \|g_2\|_2^2 \geqslant 1 - \rho\,.$$

Now, write $\varphi^{m_k} = v_k^{m_k} R_k = F_1 + F_2$, with $F_1 = g_1 R_k$ and $F_2 = g_2 R_k$. We now show that F_1 and F_2 are again "large" and "small." To that effect, we first note that

$$\widehat{F_1}(n_k + l) = \widehat{R_k}(n_k)\, \widehat{g_1}(l) \qquad \text{for } 0 \leqslant l \leqslant N_{k-1}m_k. \tag{12.54}$$

Indeed, we can write

$$F_1(z) = \sum_{l=0}^{N_{k-1}m_k} \widehat{g_1}(l) z^l \left(\widehat{R_k}(0) + \widehat{R_k}(n_k) z^{n_k} + O(z^{2n_k})\right)$$

or else

$$F_1(z) = \sum_{l=0}^{N_{k-1}m_k} \widehat{g_1}(l) \widehat{R_k}(0) z^l + \sum_{l=0}^{N_{k-1}m_k} \widehat{g_1}(l) \widehat{R_k}(n_k) z^{n_k+l} + O(z^{2n_k}),$$

and (12.54) follows since $n_k + N_{k-1}m_k < 2n_k$.

It follows that

$$\sum_{j=n_k}^{n_k+N_{k-1}m_k} |\hat{F}_1(j)|^2 = \mu_k^2 \nu_k^2 \sum_{l=0}^{N_{k-1}m_k} |\hat{g}_1(l)|^2 \geqslant (1-\rho)\, \mathrm{e}^{-4}.$$

Finally, since the inner function R_k satisfies $\|R_k\|_\infty = 1$,

$$\|F_2\|_2^2 \leqslant \|g_2\|_2^2\, \|R_k\|_\infty^2 = \|g_2\|_2^2 \leqslant \rho\,,$$

and we get, using the inequality $|a+b|^2 \geqslant \frac{1}{2}|a|^2 - |b|^2$ between complex numbers,

$$\begin{aligned}\sum_{j=n_k}^{2n_k} |\widehat{\varphi^{m_k}}(j)|^2 &\geqslant \sum_{j=n_k}^{n_k+N_{k-1}m_k} |\widehat{F_1}(j) + \widehat{F_2}(j)|^2 \\ &\geqslant \frac{1}{2}\sum_{j=n_k}^{n_k+N_{k-1}m_k} |\widehat{F_1}(j)|^2 - \sum_{j=n_k}^{n_k+N_{k-1}m_k} |\widehat{F_2}(j)|^2 \\ &\geqslant \frac{1}{2}(1-\rho)\,\mathrm{e}^{-4} - \rho =: c > 0\end{aligned}$$

if we now adjust $0 < \rho < \dfrac{1}{2\mathrm{e}^4 + 1}$. The proof is complete. □

12.7 The Non-Hilbertian Case

According to the chess champion J. R. Capablanca, "A good player is always lucky."

The example of Littlewood's subordination principle (with Littlewood in the role of the good player) seems to us illuminating in this respect: the average player could try to extend it to the non-Hilbertian case. But as soon as $p \neq 2$, this principle dramatically fails. We shall prove this as a by-product of the tools elaborated in this chapter, as in [107], but it can also be obtained using a different path as did Charpentier–Espoullier–Zarouf [26] using the results of the paper [172].

For $1 \leqslant p \leqslant \infty$, let $\ell^p_A = h^p$ be the Banach space (a copy of ℓ^p, called the Beurling–Sobolev space in [172]) of analytic functions $f(z) = \sum_{n=0}^{\infty} x_n z^n$ on the unit disk $\mathbb{D}$ such that

$$\|f\|_p^p := \sum_{n=0}^{\infty} |x_n|^p < \infty.$$

If $\beta = (\beta_n)_{n\geqslant 0}$ is a sequence of positive numbers, satisfying again (12.1), the weighted Beurling–Sobolev space $h^p(\beta)$, sometimes also denoted $\ell^p_A(\beta)$, is defined by the finiteness of

$$\|f\|_\beta^p := \sum_{n=0}^{\infty} |x_n|^p \beta_n.$$

Identifying sequences and their generating functions, $\ell_A^p(\beta)$ is none other than the space of sequences $x = \sum_{n\geqslant 0} x_n z^n$ such that

$$\|x\| := \left(\sum_{n\geqslant 0} |x_n|^p \beta_n\right)^{1/p} < \infty.$$

Note that (12.1) amounts to the inclusion $h^p(\beta) \subset \mathscr{H}(\mathbb{D})$ by testing this inclusion on the function

$$f(z) = \sum_{n=1}^{\infty} \frac{1}{(n^2 \beta_n)^{1/p}} z^n \in h^p(\beta).$$

When p is finite, the dual space $h^p(\beta)^*$ is (isometrically) the space of sequences $y = \sum_{n\geqslant 0} y_n z^n$ such that (with the obvious change when $p = 1$)

$$\|y\| := \left(\sum_{n\geqslant 0} |y_n|^{p'} \beta_n^{-p'/p}\right)^{1/p'} < \infty$$

with the duality

$$\langle x, y\rangle := \sum_{n\geqslant 0} x_n y_n.$$

A normalized Schauder basis of $h^p(\beta)$ is

$$e_n = \frac{z^n}{\beta_n^{1/p}}, \qquad \text{for } n \in \mathbb{N}_0,$$

and any $x = \sum_{n\geqslant 0} x_n z^n \in h^p(\beta)$ writes as

$$x = \sum_{n\geqslant 0} \tilde{x}_n e_n, \qquad \|x\| = \left(\sum_{n\geqslant 0} |\tilde{x}_n|^p\right)^{1/p}.$$

The dual basis (e_n^*) of (e_n), equally normalized, is the sequence of monomials

$$e_n^* = \beta_n^{1/p} z^n,$$

and any $y = \sum_{n\geqslant 0} y_n z^n \in h^p(\beta)^*$ writes as

$$y = \sum_{n\geqslant 0} \tilde{y}_n e_n^*, \qquad \|y\| = \left(\sum_{n\geqslant 0} |\tilde{y}_n|^{p'}\right)^{1/p'}.$$

If an operator $T : h^p(\beta) \to h^p(\beta)$ has the matrix $(t_{m,n})$ on the Schauder basis (e_n), meaning that $T(e_n) = \sum_{m\geqslant 0} t_{m,n} e_m$, the transposed operator $T^* : h^p(\beta)^* \to h^p(\beta)^*$ has the matrix $(t_{n,m})$ on the dual Schauder basis (e_n^*).

Note that h^2 is none other than the usual Hardy space H^2. Also note that h^p, $p \neq 2$, has little to see with the usual Hardy space H^p, even if, for example, $H^p \subset h^\infty$. But in the spirit of what precedes, the case of h^p seems a natural one to consider. For $p = 2$, we know (Littlewood's principle) that all composition operators C_φ, in particular the operators C_{φ_a} with $a \in \mathbb{D}$, generated by automorphisms

$$\varphi_a(z) = \frac{a - z}{1 - \overline{a}z}$$

are bounded on h^2. We have just studied the weighted version $h^2(\beta)$ of this space and gave a complete characterization of those weights $\beta = (\beta_n)$ for which either C_φ is bounded on $h^2(\beta)$ for all automorphisms, or for all symbols.

In this section, we show that this is never the case for $p \neq 2$, whatever the choice of the weight β. We mention in passing that, in the non-weighted case, this boundedness issue was considered by Blyudze and Shimorin [15] in the case when the initial space is h^1 and the target space h^p. In that case, the authors show that $C_{\varphi_a} : h^1 \to h^p$ is bounded if and only if $p \geqslant 2$. This has been made much more precise in [172].

Here is now the main result of this section.

Theorem 12.31 *Let $1 \leqslant p \leqslant \infty$ with $p \neq 2$, and $a \in \mathbb{D}$ with $a \neq 0$ fixed. Then C_{φ_a} is unbounded on h^p.*

Actually C_{φ_a} is never bounded on $h^p(\beta)$, whatever the choice of β satisfying (12.1).

Proof: Set $I = [1/3, 2/3]$ and $J = [\alpha^{-1}, \alpha]$ with $\alpha = 5/4$.

The rotation invariance of $h^p(\beta)$ gives the following: if C_{φ_a} is bounded on $h^p(\beta)$ for some $a \in \mathbb{D}, a \neq 0$, then there exists a constant K such that, abbreviating $\|C_{\varphi_a}\|_{h^p(\beta)\to h^p(\beta)}$ to $\|C_{\varphi_a}\|$,

$$\|C_{\varphi_a}\| \leqslant K \quad \text{for all } a \in I. \tag{12.55}$$

We use an argument, directly relying on the beginning of this chapter, which works for all finite $p \neq 2$ and for all weights β. Actually, everything will rely on the matrix

$$A = \left(\widehat{\varphi_a^n}(m) \left(\frac{\beta_m}{\beta_n} \right)^{1/p} \right),$$

which represents C_{φ_a} on the canonical (Schauder) basis (e_n) of $h^p(\beta)$.

If this matrix defines a bounded operator, its columns and rows (the columns of the transposed operator on the dual basis) are, respectively, uniformly bounded on ℓ^p and on the dual space $h^p(\beta)^* \equiv \ell^{p'}$, as we saw at the beginning of this section. That implies (with some R independent of $a \in I$)

$$C_n = \sum_{m=0}^{\infty} |\widehat{\varphi_a^n}(m)|^p \left(\frac{\beta_m}{\beta_n}\right) \quad \text{satisfies} \quad \sup_n C_n \leqslant R < \infty. \tag{12.56}$$

$$L_m = \sum_{n=0}^{\infty} |\widehat{\varphi_a^n}(m)|^{p'} \left(\frac{\beta_m}{\beta_n}\right)^{p'/p} \quad \text{satisfies} \quad \sup_m L_m \leqslant R < \infty \tag{12.57}$$

(with the obvious adaptation of (12.57) when $p = 1$ but anyway we use only (12.56) in this case).

We also set $J_l = [\alpha^{-1/2}l, \alpha^{1/2}l]$ in the sequel (so that $m, n \in J_l \Rightarrow m/n \in J$).

We will show that, for $p \neq 2$, one of the necessary conditions, (12.56) or (12.57), fails. We will hence separate two cases.

The case $p < 2$

It follows from (12.56) that, for $n \in J_l$,

$$\sum_{m \in J_l} |\widehat{\varphi_a^n}(m)|^p \left(\frac{\beta_m}{\beta_n}\right) \leqslant R.$$

Integrating on I and using Theorem 12.22 gives, since $m, n \in J_l$ have the same growth as l,

$$l^{-p/2} \sum_{m \in J_l} \frac{\beta_m}{\beta_n} \lesssim R.$$

Summing over $n \in J_l$ further gives

$$l^{-p/2} \left(\sum_{k \in J_l} \beta_k\right) \left(\sum_{k \in J_l} \beta_k^{-1}\right) \lesssim Rl.$$

But, by Cauchy–Schwarz,

$$l^2 \lesssim |J_l|^2 \leqslant \left(\sum_{k \in J_l} \beta_k\right) \left(\sum_{k \in J_l} \beta_k^{-1}\right),$$

and we get $l^{2-p/2} \lesssim Rl$ or $R \gtrsim l^{1-p/2}$. Since $1 - p/2 > 0$, we have a contradiction for large l.

The case $p \in (2, \infty)$. The proof is nearly the same. It now follows from (12.57) that, for $m \in J_l$,

$$\sum_{n \in J_l} |\widehat{\varphi_a^n}(m)|^{p'} \left(\frac{\beta_m}{\beta_n}\right)^{p'/p} \leqslant R.$$

Set $\gamma_k = \beta_k^{p'/p}$ and proceed as in the case $p < 2$ to get

$$l^{-p'/2}\Big(\sum_{k\in J_l}\gamma_k\Big)\Big(\sum_{k\in J_l}\gamma_k^{-1}\Big) \lesssim Rl,$$

or again, by Cauchy–Schwarz, $l^{2-p'/2} \lesssim Rl$ and $R \gtrsim l^{1-p'/2}$. But since this time $1 - p'/2 > 0$, we have again a contradiction for large l.

When $p = \infty$. The "β_k" naturally disappear and the proof is even shorter. □

12.8 Composition Operators between Different $H^2(\beta)$

We now switch to another topic of this chapter: we compare the action of a composition operator with a *given* symbol φ, but acting on various weighted Hardy spaces $H^2(\beta)$.

It turns out that the properties (compactness, membership in a Schatten class, even boundedness) can be very different from one space to another. The (extended) Kacnelson comparison principle discussed in Theorem G.23, and already used in the first half of this chapter, will again play a quite important role. In certain situations, it will reveal unsufficient, and we will have to appeal to specific properties of composition operators (see Chapter 9).

We first present a general result, valid not only for composition operators.

Let $\beta = (\beta_k)_{k\geqslant 0}$ and $\gamma = (\gamma_k)_{k\geqslant 0}$ be two weights with $d_k := \sqrt{\frac{\beta_k}{\gamma_k}}$ *increasing*, and, as before, $H^2(\beta)$ and $H^2(\gamma)$ the corresponding weighted Hilbert spaces, with respective orthonormal bases

$$e_k^\beta = \frac{1}{\sqrt{\beta_k}} z^k \quad \text{and} \quad e_k^\gamma = \frac{1}{\sqrt{\gamma_k}} z^k = d_k e_k^\beta .$$

Theorem 12.32 *Let* $T : \mathscr{H}(\mathbb{D}) \to \mathscr{H}(\mathbb{D})$ *be a linear operator whose restriction* T_β *to* $H^2(\beta)$ *satisfies* $T(H^2(\beta)) \subset H^2(\beta)$ *and is bounded on* $H^2(\beta)$*, and represented by a lower triangular matrix* $(a_{i,j})$ *on the basis* (e_i^β)*. Then:*

(i) T_γ *is also bounded on* $H^2(\gamma)$*, and* $\|T_\gamma\|_{B(H^2(\gamma))} \leqslant \|T_\beta\|_{B(H^2(\beta))}$
(ii) *if* T_β *is compact on* $H^2(\beta)$*, so is* T_γ *on* $H^2(\gamma)$*;*
(iii) *the sequence* $s^{(\gamma)}$ *of singular numbers of* T_γ *on* $H^2(\gamma)$ *is log-subordinate to the analog sequence* $s^{(\beta)}$*, so that*

- $s_{2n}^{(\gamma)} \leqslant \sqrt{s_1^{(\beta)} s_n^{(\beta)}}$, *for all* $n \geqslant 1$;
- $T_\beta \in \mathscr{S}^p(H^2(\beta)) \Rightarrow T_\gamma \in \mathscr{S}^p(H^2(\gamma))$, *for any* $p > 0$;
- $\mathfrak{e}_n(T_\gamma) \lesssim \mathfrak{e}_n(T_\beta)$ *for all* $n \geqslant 1$.

This clearly applies perfectly well to a weighted composition operator when the symbol fixes the origin, and that is why we reproduce this theorem in the specific case of composition operators.

The first two items (and the third one when $p \geqslant 1$) of this corollary were previously obtained by Chalendar and Partington in [20] and [21].

Theorem 12.33 *Let $H^2(\beta)$ and $H^2(\gamma)$ be two weighted Hilbert spaces. Assume that γ is dominated by β in the sense that the sequence (β_k/γ_k) is* increasing, *so that the continuous inclusion $H^2(\beta) \subseteq H^2(\gamma)$ holds.*

Then, for any symbol φ with $\varphi(0) = 0$:

(i) *if C_φ is bounded on $H^2(\beta)$, C_φ is also bounded on $H^2(\gamma)$, and moreover $\|C_\varphi\|_{B(H^2(\gamma))} \leqslant \|C_\varphi\|_{B(H^2(\beta))}$;*
(ii) *if C_φ is compact on $H^2(\beta)$, so is C_φ on $H^2(\gamma)$;*
(iii) *the sequence $s^{(\gamma)}$ of singular numbers of C_φ on $H^2(\gamma)$ is log-subordinate to the analog sequence $s^{(\beta)}$, so that*
   - $s_{2n}^{(\gamma)} \leqslant \sqrt{s_1^{(\beta)} s_n^{(\beta)}}$, *for all* $n \geqslant 1$;
   - $C_\varphi \in \mathscr{S}^p(H^2(\beta)) \Rightarrow C_\varphi \in \mathscr{S}^p(H^2(\gamma))$, *for any* $p > 0$;
   - $\mathfrak{e}_n(C_\varphi : H^2(\gamma) \to H^2(\gamma)) \lesssim \mathfrak{e}_n(C_\varphi : H^2(\beta) \to H^2(\beta))$ *for all* $n \geqslant 1$.

Remark. Let us mention that we can apply the previous statement in the framework of weighted Bergman spaces. Indeed, let $\alpha' > \alpha > -1$ and consider the two weights

$$\beta_k = \frac{k!\,\Gamma(\alpha+2)}{\Gamma(k+\alpha+2)} \qquad \text{and} \qquad \gamma_k = \frac{k!\,\Gamma(\alpha'+2)}{\Gamma(k+\alpha'+2)}$$

associated with the weighted Bergman spaces $\mathfrak{B}_\alpha^2$ and $\mathfrak{B}_{\alpha'}^2$, respectively, so that $\mathfrak{B}_\alpha^2 \subset \mathfrak{B}_{\alpha'}^2$. In order to apply our comparison to Theorem 12.33, we have to show that the sequence (β_k/γ_k) increases. But

$$\frac{\beta_k}{\gamma_k} = \frac{\Gamma(\alpha+2)}{\Gamma(\alpha'+2)}\,\frac{\Gamma(\alpha'+k+2)}{\Gamma(\alpha+k+2)} =: \frac{\Gamma(\alpha+2)}{\Gamma(\alpha'+2)} A_k.$$

Setting $h = \alpha' - \alpha > 0$ and $x_k = \alpha + k + 2$, we see that

$$A_k = \frac{\Gamma(x_k+h)}{\Gamma(x_k)} \text{ increases}$$

Indeed, the function $x \mapsto \frac{\Gamma(x+h)}{\Gamma(x)}$ increases on $(0,\infty)$ since Γ is log-convex.

Of course, in this framework, we already proved that composition operators are always bounded on spaces $\mathscr{B}^2_\alpha$ (and are contractions when the symbol fixes the origin), and that compactness does not depend on α, so that the real interest of Theorem 12.33 here is in the comparison of singular and entropy numbers, for $C_\varphi : \mathscr{B}^2_\alpha \to \mathscr{B}^2_\alpha$ for different α's, which is not trivial.

Actually the same kind of result holds for weighted Dirichlet spaces: for $\alpha' > \alpha$, we have $\mathcal{D}^2_\alpha \subset \mathcal{D}^2_{\alpha'}$.

Proof of Theorem 12.32: Let $J : H^2(\beta) \to H^2(\gamma)$ be the unitary operator defined by $J(e_k^\beta) = e_k^\gamma$, so that

$$\langle Jf, Jg\rangle_\gamma = \langle f, g\rangle_\beta \quad \text{for all } f, g \in H^2(\beta).$$

Let then $D : H^2(\beta) \to H^2(\beta)$ be the diagonal operator on the basis (e_k^β) with $D(e_k^\beta) = d_k e_k^\beta$.

If we set $a_{i,j} = \langle T_\beta(e_j^\beta), e_i^\beta\rangle_\beta$, we observe that

$$T_\gamma(e_j^\gamma) = T(e_j^\gamma) = d_j T(e_j^\beta) = d_j T_\beta(e_j^\beta) = \sum_i a_{i,j} d_j e_i^\beta = \sum_i a_{i,j} \frac{d_j}{d_i} e_i^\gamma;$$

hence

$$\langle T_\gamma(e_j^\gamma), e_i^\gamma\rangle_\gamma = a_{i,j} \frac{d_j}{d_i} \cdot$$

In other words, for all indices i, j,

$$\langle T_\gamma(e_j^\gamma), e_i^\gamma\rangle_\gamma = \langle D^{-1} T_\beta D(e_j^\beta), e_i^\beta\rangle_\beta, \tag{12.58}$$

implying that $J^* T_\gamma J = D^{-1} T_\beta D$, that is, T_γ is unitarily equivalent to $D^{-1} T_\beta D$ and, from Theorem G.23, we obtain

$$\|T_\gamma\|_{B(H^2(\gamma))} = \|D^{-1} T_\beta D\|_{B(H^2(\beta))} \leqslant \|T_\beta\|_{B(H^2(\beta))},$$

proving (i).

Actually, from Theorem G.23, we have $\log s(D^{-1} T_\beta D) \prec \log s(T_\beta)$, but $s(D^{-1} T_\beta D) = s(T_\gamma)$ by unitary equivalence, and this proves (ii) and (iii) using also Proposition F.4. □

12.9 Some Results on Multipliers

In this section, we give some results on the multipliers on $H^2(\beta)$, which show how the different notions of regularity for β intervene.

The set $\mathscr{M}\big(H^2(\beta)\big)$ of multipliers of $H^2(\beta)$ is by definition the vector space of functions h analytic on $\mathbb{D}$ and such that $hf \in H^2(\beta)$ for all $f \in H^2(\beta)$.

When $h \in \mathscr{M}(H^2(\beta))$, the operator M_h of multiplication by h is bounded on $H^2(\beta)$ by the closed graph theorem. The space $\mathscr{M}(H^2(\beta))$ equipped with the operator norm is a Banach space. We note the obvious property

$$\mathscr{M}(H^2(\beta)) \hookrightarrow H^\infty \quad \text{contractively.} \tag{12.59}$$

Indeed, if $h \in \mathscr{M}(H^2(\beta))$, we get, for $w \in \mathbb{D}$, the weighted mapping equation

$$M_h^*(K_w) = \overline{h(w)} K_w$$

so that taking norms and simplifying we are left with $|h(w)| \leqslant \|M_h\|$, showing that $h \in H^\infty$ with $\|h\|_\infty \leqslant \|M_h\|$.

Proposition 12.34 *We have $\mathcal{M}(H^2(\beta)) = H^\infty$ isomorphically if and only if β is essentially decreasing.*

Proof: The sufficient condition follows from Kacnelson's principle (see Appendix G), since, if $h \in H^\infty$, the operator M_h of multiplication by h is bounded on H^2 and clearly lower triangular.

For the necessity, we then have $\|M_h\| \approx \|h\|_\infty$ for every $h \in H^\infty$ by the Banach isomorphism theorem. Now, for $m > n$ (recall that $e_n(z) = z^n$),

$$e_m(z) = z^{m-n} z^n = (M_{e_{m-n}} e_n)(z)\,;$$

so, since $\|M_{e_{m-n}}\| \leqslant C\, \|e_{m-n}\|_\infty = C$ for some positive constant C,

$$\beta_m = \|e_m\|^2 \leqslant C^2\, \|e_n\|^2 = C^2\, \beta_n. \qquad \square$$

We coin the following notion of H^2-*compatible* Hilbert space of analytic functions.

Definition 12.35 A Hilbert space H of analytic functions on $\mathbb{D}$, containing the constants, with reproducing kernels K_a, $a \in \mathbb{D}$, is said to be H^2-*compatible* if the following conditions hold:

(i) H^2 is continuously embedded in H, that is, $\beta_n \leqslant C$;
(ii) $\mathscr{M}(H) = H^\infty$;
(iii) the automorphisms of $\mathbb{D}$ induce bounded composition operators on H;
(iv) the reproducing kernels are slowly oscillating, meaning that

$$\frac{\|K_a\|_H}{\|K_b\|_H} \leqslant h\Big(\frac{1-|b|}{1-|a|}\Big)$$

for $a, b \in \mathbb{D}$, where $h\colon \mathbb{R}^+ \to \mathbb{R}^+$ is a non-decreasing function.

In view of Theorem 12.15, we get the following result.

Proposition 12.36 *Let β be a weight.*

(i) *If β is essentially decreasing, we have (i), (ii), (iii) in Definition 12.35.*
(ii) *If β is slowly oscillating, then we have (iv) in Definition 12.35.*

Proof: (i). Because β is essentially decreasing, item (i) holds, as well as item (ii), by Proposition 12.34. Item (iii) is Theorem 12.15.
(ii). We have to show (iv), which simply expresses the following: the slow oscillation of the sequence β implies that of the reproducing kernels. Indeed, let $0 < s < r < 1$. Without loss of generality, we may assume that $r, s \geqslant 1/2$. It is enough to prove that

$$\|K_r\|^2 \leqslant C\,\|K_{r^2}\|^2 \tag{12.60}$$

for some constant $C > 1$. Because iteration of (12.60) then gives

$$\|K_r\|^2 \leqslant C^k\,\|K_{r^{2^k}}\|^2$$

and if k is the smallest integer such that $r^{2^k} \leqslant s$, we have $2^{k-1}\log r > \log s$ and $2^k \leqslant D\,\frac{1-s}{1-r}$ where D is a numerical constant. Writing $C = 2^\alpha$ with $\alpha > 1$, we obtain

$$\left(\frac{\|K_r\|}{\|K_s\|}\right)^2 \leqslant C^k = (2^k)^\alpha \leqslant D^\alpha\left(\frac{1-s}{1-r}\right)^\alpha =: h\left(\frac{1-s}{1-r}\right).$$

To prove (12.60), we pick some $M > 1$ such that $\min(\beta_{2n}, \beta_{2n+1}) \geqslant M^{-1}\beta_n$ for all $n \geqslant 1$ (this is possible since β is slowly oscillating) and write $t = r^2$. We have

$$\|K_r\|^2 = \sum_{n=0}^{\infty}\frac{t^n}{\beta_n} = \sum_{n=0}^{\infty}\frac{t^{2n}}{\beta_{2n}} + \sum_{n=0}^{\infty}\frac{t^{2n+1}}{\beta_{2n+1}} \leqslant 2M\sum_{n=0}^{\infty}\frac{t^{2n}}{\beta_n} = 2M\,\|K_t\|^2,$$

proving (12.60). □

The notion of H^2-compatible Hilbert space H is useful to study the set of *conditional multipliers*

$$\mathscr{M}(H,\varphi) = \big\{w \in H \,:\, w\,(f\circ\varphi) \in H \text{ for all } f \in H\big\}.$$

One obviously has $H^\infty \subset \mathscr{M}(H,\varphi) \subset H.$
The extreme cases were characterized by Attele [6] when $H = H^2$.

Theorem 12.37 (Attele) *Let $\varphi : \mathbb{D} \to \mathbb{D}$ be a symbol. Then:*

(i) $\mathscr{M}(H^2,\varphi) = H^2 \iff \|\varphi\|_\infty < 1$.
(ii) $\mathscr{M}(H^2,\varphi) = H^\infty \iff \varphi$ *is a finite Blaschke product.*

Proof: We just prove the left-to-right part of the second item. The rest will be covered by the next theorem.

The key point, and indeed $\mathscr{M}(H,\varphi) \subset H^p$ for some exponent $p > 2$ instead of $\mathscr{M}(H,\varphi) \subset H^\infty$ would be enough, is the following: if $\mathscr{M}(H^2,\varphi) = H^\infty$, we have for some constant C' and for any $a \in \mathbb{D}$

$$\frac{1 - |\varphi(a)|}{1 - |a|} \leqslant C'. \tag{12.61}$$

Indeed, denote $\|w\|_{op}$ the norm $\|M_w C_\varphi\|_{B(H^2)}$. By Banach's isomorphism theorem, there exists a constant C such that $\|w\|_\infty \leqslant C\|w\|_{op}$ for all $w \in H^\infty$. To exploit this with $w = K_a$, the reproducing kernel of H^2 at $a \in \mathbb{D}$, first note that

$$\|K_a\|_{op} \leqslant \sqrt{\frac{1 + |\varphi(a)|}{1 - |\varphi(a)|}} \times \frac{1}{\sqrt{1 - |a|^2}}.$$

For that, set $b = \varphi(a)$. Noting the important relation $(1 - |a|^2)|K_a|^2 = |\varphi_a'|$, we have for any $f \in H^2$

$$\begin{aligned}
(1 - |a|^2) \int_{\mathbb{T}} |K_a|^2 |f \circ \varphi|^2 dm &= \int_{\mathbb{T}} |\varphi_a'| \, |f \circ \varphi \circ \varphi_a \circ \varphi_a|^2 \, dm \\
&= \int_{\mathbb{T}} |f \circ \varphi \circ \varphi_a|^2 \, dm \\
&= \int_{\mathbb{T}} |f \circ \varphi_b \circ \varphi_b \circ \varphi \circ \varphi_a|^2 \, dm \\
&\leqslant \int_{\mathbb{T}} |f \circ \varphi_b|^2 \, dm \\
&\leqslant \frac{1 + |b|}{1 - |b|} \int_{\mathbb{T}} |f|^2 \, dm
\end{aligned}$$

by Littlewood's subordination principle (note that $\varphi_b \circ \varphi \circ \varphi_a(0) = 0$), giving the above inequality.

Now, since $\|K_a\|_\infty = \dfrac{1}{1 - |a|}$, the relation $\|K_a\|_\infty \leqslant C\|K_a\|_{op}$ gives (12.61) after simplification, with $C' = 2C^2$. But then (see Theorem 1.65), φ must be a finite Blaschke product. □

We no longer have precise estimates on $\|K_a\|_{op}$ for H^2-compatible spaces, but we have (i) to follow!! We have the following extension of Attele's result:

Theorem 12.38 *Let β be essentially decreasing and slowly oscillating, so that $H^2(\beta)$ is H^2-compatible. Then:*

(i) $\mathscr{M}(H^2,\varphi) \subseteq \mathscr{M}\big(H^2(\beta),\varphi\big)$;
(ii) $\mathscr{M}\big(H^2(\beta),\varphi\big) = H^2(\beta)$ *if and only if* $\|\varphi\|_\infty < 1$;
(iii) $\mathscr{M}\big(H^2(\beta),\varphi\big) = H^\infty$ *if and only if* φ *is a finite Blaschke product.*

Proof: Without loss of generality we can assume that β is non-increasing.

(i) Assume first that $\varphi(0) = 0$. Let $w \in \mathscr{M}(H^2,\varphi)$ and let T be the weighted composition operator $T = M_w C_\varphi$. Then, T is bounded on $H^2 = H^2(\mathbb{1})$, and moreover lower triangular on the canonical basis (z^k) of $H^2(\mathbb{1})$ since

$$\langle T(z^j), z^i\rangle = \langle w\varphi^j, z^i\rangle$$

and $w\varphi^j = \sum_{k\geqslant j} c_k z^k$. Applying Theorem 12.32 (which is licit since $1/\beta_n$ is non-decreasing), we get that T is bounded on $H^2(\beta)$ as well, meaning that $w \in \mathscr{M}(H^2(\beta),\varphi)$.

In the general case, let $\varphi(0) = a$, so that $(\varphi_a \circ \varphi)(0) = 0$. Property (iii) of H^2-compatible spaces implies that

$$\mathscr{M}(H^2,\varphi) = \mathscr{M}(H^2,\varphi_a\circ\varphi) \subset \mathscr{M}(H^2(\beta),\varphi_a\circ\varphi) = \mathscr{M}(H^2(\beta),\varphi)$$

since $f \in H^2 \iff f\circ\varphi_a \in H^2$ and $f \in H^2(\beta) \iff f\circ\varphi_a \in H^2(\beta)$.

(ii) The left-to-right implication is proved as follows (see also [6] for H^2); we start from the (already used) mapping equation

$$(M_w C_\varphi)^*(K_z) = \overline{w(z)}\, K_{\varphi(z)}. \tag{12.62}$$

The assumption implies the existence of a constant C such that

$$\|M_w C_\varphi\|_{B(H^2(\beta))} \leqslant C\|w\|_\beta \quad \text{for all } w \in H^2(\beta).$$

As a consequence, for given $z \in \mathbb{D}$,

$$\|(M_w C_\varphi)^*(K_z)\|_\beta \leqslant C\|w\|_\beta \|K_z\|_\beta.$$

That is, in view of (12.62),

$$|w(z)|\, \|K_{\varphi(z)}\|_\beta \leqslant C\|w\|_\beta\|K_z\|_\beta.$$

Testing this inequality with $w = K_z$ and simplifying by $\|K_z\|_\beta^2$, we get that $\|K_{\varphi(z)}\|_\beta \leqslant C$. Since $\lim_{|a|\to 1^-}\|K_a\|_\beta = \infty$ (because $\sum 1/\beta_n = +\infty$), this implies $\|\varphi\|_\infty < 1$.

The right-to-left implication is clear, since then $C_\varphi : H^2(\beta) \to H^\infty$ and $\mathscr{M}(H^2(\beta)) = H^\infty$ by (ii) in Definition 12.35.

(iii) The implication from right to left goes as follows: finite Blaschke products φ are characterized by (see Theorem 1.65)

$$\sup_{z\in\mathbb{D}} \frac{1-|\varphi(z)|}{1-|z|} =: R_\varphi < \infty.$$

Let now $w \in \mathscr{M}(H^2(\beta), \varphi)$, so that $\|M_w C_\varphi\| = C < \infty$. We can assume that $\|w\|_\beta \leqslant 1$. The mapping equation (12.62) gives, for $z \in \mathbb{D}$,

$$|w(z)|\, \|K_{\varphi(z)}\|_\beta \leqslant C \|K_z\|_\beta.$$

This implies (meaning that $w \in H^\infty$)

$$|w(z)| \leqslant C \frac{\|K_z\|_\beta}{\|K_{\varphi(z)}\|_\beta} \leqslant Ch\left(\frac{1 - |\varphi(z)|}{1 - |z|}\right) \leqslant Ch(R_\varphi).$$

Finally, for the implication from left to right, assume that $\mathscr{M}(H^2(\beta), \varphi) = H^\infty$; then $\mathscr{M}(H^2, \varphi) = H^\infty$ by (i) and φ is a finite Blaschke product by Theorem 12.37 proved above. □

Here is a corollary, a partial answer to a question of Zorboska [185, problem 5], which extends the Dirichlet space case $\beta_n = n + 1$.

Corollary 12.39 *Assume that $H^2(\beta)$ is "disc-automorphism invariant" (i.e., automorphisms induce bounded composition operators on $H^2(\beta)$) and that $C_\varphi : H^2(\beta) \to H^2(\beta)$ is compact. Then, the Denjoy–Wolff point of φ lies in $\mathbb{D}$ (alternatively, φ has a fixed point in $\mathbb{D}$).*

This statement is a complete answer to a question raised in problem 5 of [185]. It will be easily proved now that we know (Theorem 12.19) that disc-automorphism invariance of $H^2(\beta)$ implies the slow oscillation of β.

Proof: We separate two cases.

(i) $\sum 1/\beta_n < \infty$. Then, the even stronger conclusion $\|\varphi\|_\infty < 1$ holds (under the disc-automorphism invariance assumption): this is a result of Shapiro [163, theorem 2.1]. Moreover, the symbol φ has exactly one fixed point in $\overline{\mathbb{D}}$ (and slow oscillation is not required here). This is a result of Zorboska [185, corollary 4.5]. The proofs are sketched in Exercises I.29 and I.30.

(ii) $\sum 1/\beta_n = \infty$. We know by the above proof of Proposition 12.36 that condition (iv) for H^2-compatibility of $H^2(\beta)$ holds, namely with $h : \mathbb{R}^+ \to \mathbb{R}^+$ an increasing function

$$\frac{\|K_z\|^2}{\|K_{\varphi(z)}\|^2} \leqslant h\left(\frac{1 - |\varphi(z)|}{1 - |z|}\right). \tag{12.63}$$

But by (12.5) of this chapter, we also have that $\dfrac{K_z}{\|K_z\|} \xrightarrow{w} 0$ as $|z| \to 1^-$. Since C_φ^* is compact and $C_\varphi^*(K_z) = K_{\varphi(z)}$, this implies in turn $\dfrac{\|K_{\varphi(z)}\|}{\|K_z\|} \to 0$, that is, $\dfrac{\|K_z\|}{\|K_{\varphi(z)}\|} \to \infty$. So that the RHS of (12.63) tends to ∞ as $|z| \to 1^-$. Hence $\dfrac{1 - |\varphi(z)|}{1 - |z|} \to \infty$ since h is non-decreasing, and φ has an angular derivative nowhere on $\partial\mathbb{D}$. By the Denjoy–Wolff theorem, the Denjoy–Wolff point of φ lies in $\mathbb{D}$.

We could also argue as for H^2: there exists $0 < r < 1$ such that $1 - |\varphi(z)| > 1 - |z|$, or else $|\varphi(z)| < |z|$, for all z of modulus r; and Rouché's theorem shows that $\varphi(z) - z$ has a zero $z_0 \in D(0, r)$, which is the Denjoy–Wolff point of φ.

□

Remark. In the case $\sum 1/\beta_n = \infty$, under the assumptions of Corollary 12.39, we cannot in general obtain the stronger statement $\|\varphi\|_\infty < 1$. Indeed, if φ is the *cusp map* and $\beta_n = n + 1$, the assumptions of Corollary 12.39 are met and $\sum 1/\beta_n = \infty$. We saw in Chapter 10 that $C_\varphi : H^2(\beta) \to H^2(\beta)$ is "very" compact: its approximation numbers satisfy

$$a' \exp(-b' \sqrt{n}) \leqslant a_n(C_\varphi) \leqslant a \exp(-b \sqrt{n})$$

for some positive constants a, b, a', b'. The Denjoy–Wolff point of φ is 0. But we know that $\varphi(1) = 1$, and $\|\varphi\|_\infty = 1$.

Appendix A Schur Test Lemma

We present here an elementary version of the Schur Test Lemma. This version is sufficient for the framework of our book.

Theorem A.1 (Schur Test Lemma) *Let $p \geqslant 1$ and p' its conjugate exponent. Let (Ω, ν) be a measurable space, with ν positive and σ-finite, and K be a measurable kernel on $\Omega \times \Omega$ satisfying, for two positive constants A, B,*

(i) $\displaystyle\int_{\Omega} |K(z,w)|\, d\nu(w) \leqslant A$ *for ν-almost all $z \in \Omega$;*

(ii) $\displaystyle\int_{\Omega} |K(z,w)|\, d\nu(z) \leqslant B$ *for ν-almost all $w \in \Omega$.*

Then, the operator $T = T_K$, formally given by the kernel K,

$$Tf(z) = \int_{\mathbb{T}} K(z,w) f(w)\, d\nu(w) \quad \textit{for } z \in \Omega,$$

is well defined and bounded from $L^p(\Omega)$ to $L^p(\Omega)$ with norm less than $A^{\frac{1}{p'}} B^{\frac{1}{p}}$.

In particular, T is bounded on $L^2(\Omega)$ with norm $\|T\| \leqslant \sqrt{AB}$.

Proof: The case $p = 1$ is clear from (ii) and we assume from now on that $p > 1$; moreover, thanks to Fubini–Tonelli, we also get that $T_K(f)(z)$ is defined for a.e. $z \in \Omega$.

By duality, it suffices to control $\displaystyle\int_{\Omega \times \Omega} |K(z,w) f(w) g(z)|\, d\nu(w)\, d\nu(z)$ for every f in the unit ball of $L^p(\Omega)$ and every g in the unit ball of $L^{p'}(\Omega)$.

We make use of the basic Young inequality: for every $x, y \geqslant 0$ and $\lambda > 0$, we have

$$xy \leqslant \frac{\lambda^p}{p} x^p + \frac{\lambda^{-p'}}{p'} y^{p'}.$$

Therefore, for any $\lambda > 0$ (we specify our choice later), we have, for almost every $z, w \in \Omega$,

$$|K(z,w) f(w) g(z)| \leqslant \frac{\lambda^p}{p} |K(z,w)|.|f(w)|^p + \frac{\lambda^{-p'}}{p'} |K(z,w)|.|g(z)|^{p'}; \qquad \text{(A.1)}$$

hence

$$\begin{aligned}\int_{\Omega\times\Omega} |K(z,w)f(w)g(z)|\,d\nu(w)\,d\nu(z) &\leqslant \frac{\lambda^p}{p}\int_\Omega\left(\int_\Omega |K(z,w)|\,d\nu(z)\right)|f(w)|^p\,d\nu(w)\\ &\quad+\frac{\lambda^{-p'}}{p'}\int_\Omega\left(\int_\Omega |K(z,w)|\,d\nu(w)\right)|g(z)|^{p'}\,d\nu(z)\\ &\leqslant \frac{\lambda^p}{p}B+\frac{\lambda^{-p'}}{p'}A.\end{aligned}$$

Then searching to the minimum with respect to λ (equivalently when the two terms $\lambda^p B$ and $\lambda^{-p'} A$ are equal), we choose $\lambda = \left(\frac{A}{B}\right)^{\frac{1}{p+p'}}$.

Rearranging, we get the result. □

Appendix B Multiplier

Let X be a (Banach) space of analytic functions on the unit disk $\mathbb{D}$. We assume that the point evaluation $\delta_a : f \in X \mapsto f(a)$ is a (continuous) functional for any $a \in \mathbb{D}$, which is non-trivial: there exists some $f \in X$ such that $f(a) \neq 0$.

Given an analytic function u on $\mathbb{D}$, the multiplier $M_u : \mathscr{H}(\mathbb{D}) \to \mathscr{H}(\mathbb{D})$ is defined by

$$M_u(f) = u.f.$$

We first recall the following proposition:

Proposition B.1 *Any bounded multiplier M_u on such a space X must be associated to some u belonging to H^∞. Moreover*

$$\|u\|_\infty \leqslant \|M_u\|_{B(X)}.$$

Proof: For every $a \in \mathbb{D}$ and every $f \in X$ we have

$$|u(a)|.|f(a)| = |\delta_a(M_u(f))| \leqslant \|\delta_a\|_{X^*}\|M_u\|_{B(X)}\|f\|_X.$$

Taking the supremum over the unit ball of X we get

$$|u(a)|.\|\delta_a\|_{X^*} \leqslant \|\delta_a\|_{X^*}\|M_u\|_{B(X)},$$

and since $\|\delta_a\|_{X^*} \neq 0$ we get the result. □

Actually for the main spaces involved in this book, this is a characterization.

Proposition B.2 *Let X be the Hardy space H^p or the Bergman space $\mathscr{B}^p_\alpha$, where $p \geqslant 1$, $\alpha > -1$. Let u be an analytic function.*

Then:

(i) *The multiplier M_u is bounded on X if and only if $u \in H^\infty$.*
Moreover $\|M_u\|_{B(X)} = \|u\|_\infty$.

(ii) *The multiplier M_u is compact on X if and only if $u = 0$.*

Proof: (i) If M_u is bounded, we already know from the previous proposition that $u \in H^\infty$ and $\|u\|_\infty \leqslant \|M_u\|_{B(X)}$.

Assume that $u \in H^\infty$. For every $f \in H^p$, we have

$$\|M_u(f)\|_{H^p}^p = \int_{\mathbb{T}} |u.f|^p \, dm \leqslant \|u\|_\infty^p \int_{\mathbb{T}} |f|^p \, dm,$$

so that $\|M_u\|_{B(H^p)} \leqslant \|u\|_\infty$.

For every $f \in \mathscr{B}_\alpha^p$, we have

$$\|M_u(f)\|_{\mathscr{B}_\alpha^p}^p = \int_{\mathbb{D}} |u.f|^p \, d\sigma_\alpha \leqslant \|u\|_\infty^p \int_{\mathbb{D}} |f|^p \, d\sigma_\alpha,$$

so that $\|M_u\|_{B(\mathscr{B}_\alpha^p)} \leqslant \|u\|_\infty$. Hence (i) is proved.

Now assume that M_u is compact on H^p. The sequence $\left(z^n\right)_{n\in\mathbb{N}}$ is weakly converging to 0 by the Riemann–Lebesgue theorem. By compactness, we get

$$\int_{\mathbb{T}} |u|^p \, dm = \int_{\mathbb{T}} |u.z^n|^p \, dm = \|M_u(z^n)\|_{H^p}^p \longrightarrow 0$$

and $u = 0$.

Assume now that M_u is compact on $\mathscr{B}_\alpha^p$. Since $u \in H^\infty$, there exists some $E \subset \mathbb{T}$ with full m-measure such that, for every $\xi \in E$, there is a radial limit for u at point ξ. Consider the functions (where $\rho_n \in (0, 1)$ converges to 1)

$$f_n(z) = \frac{(1-\rho_n^2)^{\frac{2+\alpha}{p}}}{(1-\rho_n\bar{\xi}z)^{\frac{4+2\alpha}{p}}}.$$

Clearly this sequence satisfies $\|f_n\|_{\mathscr{B}_\alpha^p} = 1$ and converges to 0 in $\mathscr{H}(\mathbb{D})$, so does $(u.f_n)$ (in other words these two sequences are weak star converging to 0). By compactness of M_u,

$$\|M_u(f_n)\|_{\mathscr{B}_\alpha^p}^p \longrightarrow 0.$$

It means that

$$\int_{\mathbb{D}} \frac{(1-\rho_n^2)^{2+\alpha}}{|1-\rho_n\bar{\xi}z|^{4+2\alpha}} |u(z)|^p \, d\sigma_\alpha(z) \longrightarrow 0,$$

but, on the other hand, it is plain to see that, by an involutive change of variable and subharmonicity,

$$\int_{\mathbb{D}} \frac{(1-\rho_n^2)^{2+\alpha}}{|1-\rho_n\bar{\xi}z|^{4+2\alpha}} |u(z)|^p \, d\sigma_\alpha = \int_{\mathbb{D}} |u \circ \theta_n(w)|^p \, d\sigma_\alpha \geqslant |u \circ \theta_n(0)|^p = |u(\rho_n\xi)|^p,$$

where $\theta_n(z) = \dfrac{\rho_n\xi - z}{1-\rho_n\bar{\xi}z}$. We conclude that $u(\xi) = 0$. Therefore $u = 0$ a.e. on $\mathbb{T}$. □

Appendix C Toeplitz Operators

In this section, we aim to prove that a (classical) Toeplitz operator with symbol $\psi \in L^2(\mathbb{T})$ is bounded on H^2 if and only if $\psi \in L^\infty(\mathbb{T})$. Moreover, its norm is equal to $\|T_\psi\| = \|\psi\|_\infty$.

Let us recall that, given $\psi \in L^2(\mathbb{T})$, the operator is defined in the following way:

$$T_\psi : \begin{array}{ccc} H^\infty & \longrightarrow & H^2 \\ f & \longmapsto & P_+(\psi f), \end{array}$$

where $P_+ : L^2(\mathbb{T}) \to H^2$ is the natural Riesz projection; to each $g \in L^2(\mathbb{T})$, with Fourier coefficient $\hat{g}(n)$ (with $n \in \mathbb{Z}$), is associated $P_+(g) = \sum_{n \geqslant 0} \hat{g}(n) z^n$, which is clearly well defined and belongs to H^2.

We point out that at this moment we defined T_ψ only on $H^\infty \subset H^2$ to be sure that $\psi f \in L^2$. At this stage, it is not clear when we can extend the operator on H^2. For the sake of lightness, we shall keep the same notation for the operator viewed on H^2. The following result clarifies the situation:

Theorem C.1 *The Toeplitz operator* T_ψ,

$$\begin{array}{ccc} H^2 & \longrightarrow & H^2 \\ f & \longmapsto & P_+(\psi f), \end{array}$$

is defined and bounded from H^2 *to* H^2 *if and only if* $\psi \in L^\infty(\mathbb{T})$.

Moreover

$$\|T_\psi\|_{B(H^2)} = \|\psi\|_\infty.$$

Proof: First it is obvious that when $\psi \in L^\infty(\mathbb{T})$, $\psi f \in L^2$ for every $f \in H^2$, and $\|P_+(\psi f)\|_2 \leqslant \|\psi\|_\infty \|f\|_2$. We conclude that T_ψ is bounded on H^2 with norm less than $\|\psi\|_\infty$.

From now on, we assume that T_ψ is defined, and so bounded (by the closed graph theorem) from H^2 to H^2.

Fix $f \in L^2(\mathbb{T})$ and an integer $N \geqslant 1$. The function $F_N = P_+\big(z^N f\big)$ belongs to H^2. For every $a \in \mathbb{D}$, we consider the reproducing kernel K_a at point a and we have

$$\langle z^N K_a, T_\psi(F_N)\rangle = \langle z^N K_a, \psi . F_N\rangle = \int_{\mathbb{T}} \frac{\psi(w)}{1 - a\overline{w}} \overline{w}^N P_+(z^N f)(w)\, dm(w).$$

But the L^2-norm of $w \mapsto \overline{w}^N P_+(z^N f)(w)$ is less than $\|f\|_2$ and

$$\overline{z}^N P_+(z^N f) = \sum_{k \geqslant -N} \hat{f}(k) z^k \longrightarrow f$$

in $L^2(\mathbb{T})$ when $N \to +\infty$. Therefore

$$\langle z^N K_a, T_\psi(F_N) \rangle \longrightarrow \int_{\mathbb{T}} \frac{\psi(w)}{1 - a\overline{w}} f(w)\, dm(w).$$

On the other hand,

$$\begin{aligned} \left| \langle z^N K_a, T_\psi(F_N) \rangle \right| &\leqslant \|z^N K_a\|_2 . \|T_\psi(F_N)\|_2 \leqslant \|K_a\|_2 . \|T_\psi\| . \|F_N\|_2 \\ &\leqslant \frac{1}{\sqrt{1 - |a|^2}} \|T_\psi\| . \|f\|_2 . \end{aligned}$$

Finally we get

$$\sup_{\|f\|_2 = 1} \left| \int_{\mathbb{T}} \frac{\sqrt{1 - |a|^2}}{1 - a\overline{w}} \psi(w) f(w)\, dm(w) \right| \leqslant \|T_\psi\|;$$

hence

$$\int_{\mathbb{T}} P_a(w) |\psi(w)|^2\, dm(w) = \int_{\mathbb{T}} \frac{1 - |a|^2}{|1 - a\overline{w}|^2} |\psi(w)|^2\, dm(w) \leqslant \|T_\psi\|^2.$$

Since $|\psi|^2 \in L^1(\mathbb{T})$, and writing $a = r\xi$ where $\xi \in \mathbb{T}$ and $r \to 1^-$, we have

$$\int_{\mathbb{T}} P_a(w) |\psi(w)|^2\, dm(w) \longrightarrow |\psi(\xi)|^2$$

for almost every ξ.

We obtain $\psi \in L^\infty(\mathbb{T})$ with $\|\psi\|_\infty \leqslant \|T_\psi\|_{B(H^2)}$. □

Appendix D Van der Corput and the Stationary Phase

We present here several (very) classical results around van der Corput–type inequalities and versions of the stationary phase.

We begin with the following easy (yet extremely useful) lemma attributed to van der Corput.

Lemma D.1 *Let $f\colon [A,B] \to \mathbb{R}$, with $A < B$, be a $\mathscr{C}^2$-function satisfying $|f'| \geqslant \delta > 0$ and $|f''| \leqslant C$. Then*

$$\left| \int_A^B e^{if(x)}\, dx \right| \leqslant \frac{2}{\delta} + \frac{C(B-A)}{\delta^2}\,.$$

Proof: Write $\mathrm{e}^{if} = \dfrac{(\mathrm{e}^{if})'}{if'}$ and integrate by parts to get

$$\int_A^B \mathrm{e}^{if(x)}\, dx = \left[\frac{\mathrm{e}^{if}}{if'} \right]_A^B - i \int_A^B \mathrm{e}^{if(x)} \frac{f''(x)}{[f'(x)]^2}\, dx =: M_1 + M_2\,, \tag{D.1}$$

with $|M_1| \leqslant \dfrac{2}{\delta}$ and $|M_2| \leqslant (B-A)\, \dfrac{C}{\delta^2}$. This gives the result. □

A variant is the following:

Lemma D.2 *Let $f\colon [u,v] \to \mathbb{R}$, with $u < v$, be a $\mathscr{C}^2$-function with $f'' > 0$, and f' not vanishing on $[u,v]$. Let*

$$J = \int_u^v e^{if(x)}\, dx\,.$$

We have:

(i) *If $f' > 0$ on $[u,v]$, then $|J| \leqslant \dfrac{2}{f'(u)}$;*

(ii) *If $f' < 0$ on $[u,v]$, then $|J| \leqslant \dfrac{2}{|f'(v)|}$.*

Proof: The starting point is exactly the same and we write (D.1) (replacing (A,B) by (u,v)). Now the standard triangular inequality allows to conclude easily after integration, splitting the discussion according to the sign of f'. □

We present now a variant of [175, lemma 4.6] of the van der Corput's version of the stationary phase method. A careful reading of the proof gives the version below, which only needs local estimates on the second derivative F'', as occurs in the situations we encounter in the present book.

Theorem D.3 (Stationary phase) *Let F be a real function on the interval $[A, B]$, with continuous derivatives up to the third order and $F'' \neq 0$ throughout $[A, B]$. Assume that there is a (unique) point c in $]A, B[$ such that $F'(c) = 0$, and that, for some positive numbers λ_2, λ_3, and η, the following assertions hold:*

1) $[c-\eta, c+\eta] \subseteq]A, B[$;
2) $|F''(x)| \geqslant \lambda_2$ *for all* $x \in [c-\eta, c+\eta]$;
3) $|F'''(x)| \leqslant \lambda_3$ *for all* $x \in [A, B]$.

Then

$$I := \int_A^B e^{iF(x)}\,dx = \sqrt{2\pi}\,\frac{e^{i[F(c)+\varepsilon\pi/4]}}{|F''(c)|^{1/2}} + O\left(\frac{1}{\eta\lambda_2} + \eta^4\lambda_3\right), \tag{D.2}$$

with $\varepsilon = \pm 1$ depending on whether $F'' > 0$ or $F'' < 0$ on $[c-\eta, c+\eta]$; and where O involves an absolute constant.

It is good to have in mind the imaginary gaussian case whose Theorem D.3 is a generalization:

$$I_n = \int_{-1}^{1} e^{inx^2}\,dx = \frac{1}{\sqrt{n}}\int_{-\sqrt{n}}^{\sqrt{n}} e^{iy^2}\,dy = \frac{1}{\sqrt{n}}\left(\sqrt{\pi}e^{i\pi/4} + O(1/\sqrt{n})\right),$$

corresponding to $F(x) = nx^2$, $c = 0$, $\eta = 1$, $\lambda_2 = 2n$, $\lambda_3 = 0$.

Proof of Theorem D.3: We can assume $F''(x) \geqslant \lambda_2$ for all $x \in [c-\eta, c+\eta]$ (and hence $F'' > 0$ on $[A, B]$): for the case $F''(x) \leqslant -\lambda_2$ (i.e., $\varepsilon = -1$), just write

$$\int_A^B e^{iF(x)}dx = \overline{\int_A^B e^{-iF(x)}dx}.$$

Now write the integral I of Theorem D.3 as $I = I_1 + I_2 + I_3$ with

$$I_1 = \int_A^{c-\eta} e^{iF(x)}\,dx, \quad I_2 = \int_{c-\eta}^{c+\eta} e^{iF(x)}\,dx, \quad I_3 = \int_{c+\eta}^{B} e^{iF(x)}\,dx.$$

Lemma D.2 with $u = A$ and $v = c - \eta$ implies

$$|I_1| \leqslant \frac{2}{|F'(c-\eta)|} \leqslant \frac{2}{\eta\lambda_2}, \tag{D.3}$$

where, for the last inequality, we just have to write, for some $\xi \in [c-\eta, c]$, $|F'(c-\eta)| = F'(c) - F'(c-\eta) = \eta\, F''(\xi)$, where $F''(\xi) \geqslant \lambda_2$.

Similarly, Lemma D.2 with $u = c + \eta$ and $v = B$ implies

$$|I_3| \leqslant \frac{2}{F'(c+\eta)} \leqslant \frac{2}{\eta\lambda_2}. \tag{D.4}$$

We now estimate I_2. We know that $F(x) = F(c) + \dfrac{(x-c)^2}{2} F''(c) + R$ with $|R| \leqslant \dfrac{|x-c|^3}{6} \lambda_3$, thanks to the Taylor formula. Hence

$$I_2 = e^{iF(c)} \int_0^\eta 2 \exp\left(\frac{i}{2} x^2 F''(c)\right) dx + S$$

with $|S| \leqslant \lambda_3 \displaystyle\int_0^\eta \frac{x^3}{3} dx = \frac{\eta^4}{12} \lambda_3$. Finally, set $K = \displaystyle\int_0^\eta 2 \exp\left(\frac{i}{2} x^2 F''(c)\right) dx$.

We make the change of variable $x = \sqrt{\dfrac{2t}{F''(c)}}$. Recall the value of the classical Fresnel integral $\displaystyle\int_0^{+\infty} \frac{e^{it}}{\sqrt{t}} dt = \sqrt{\pi}\, e^{i\pi/4}$; an integration by parts gives, for $m > 0$,

$$\left| \int_m^{+\infty} \frac{e^{it}}{\sqrt{t}} dt \right| \leqslant \frac{2}{\sqrt{m}}.$$

Therefore, with $m = \dfrac{\eta^2}{2} F''(c)$,

$$K = \sqrt{\frac{2}{F''(c)}} \int_0^m \frac{e^{it}}{\sqrt{t}} dt = \sqrt{\frac{2\pi}{F''(c)}}\, e^{i\pi/4} + R_m,$$

where (C denoting an absolute constant)

$$|R_m| \leqslant C \sqrt{\frac{1}{F''(c)}} \frac{1}{\sqrt{m}} \leqslant \frac{C}{\eta \lambda_2}.$$

All in all, we proved that

$$I_2 = \sqrt{\frac{2\pi}{F''(c)}} \exp\left[i(F(c) + \pi/4)\right] + O\left(\frac{1}{\eta\lambda_2} + \eta^4 \lambda_3\right), \tag{D.5}$$

and the same estimate holds for I, thanks to (D.3) and (D.4). □

Appendix E Several Variables

E.1 Introduction

In this appendix, we fix an integer $d \geqslant 2$; we denote by $\mathbb{D}^d$ the d-dimensional polydisk, and by $H = H^2(\mathbb{D}^d)$ the Hardy space of $\mathbb{D}^d$, namely the Hilbertian space of analytic functions $f(z) = \sum_\alpha a_\alpha z^\alpha$ such that

$$\sum_\alpha |a_\alpha|^2 =: \|f\|_2^2 < \infty, \text{ where } \alpha = (\alpha_j)_{1\leqslant j\leqslant d} \in \mathbb{N}_0^d.$$

Here, if $z = (z_j)_{1\leqslant j\leqslant d}$ and $\alpha = (\alpha_j)_{1\leqslant j\leqslant d} \in \mathbb{N}_0^d$, we denote $z^\alpha := \prod_{1\leqslant j\leqslant d} z_j^{\alpha_j}$ and $|\alpha| = \sum_j \alpha_j$. The monomials $z \mapsto z^\alpha$ clearly form an ONB of H. Hence, as for $d = 1$, we have if $r = (r_j)_{1\leqslant j\leqslant d}$ with $0 < r_j < 1$

$$\sum_\alpha |a_\alpha|^2 r^{2|\alpha|} = \int_{\mathbb{T}^d} |f(ru)|^2\, dm_d(z), \tag{E.1}$$

where m_d denotes the Haar measure of $\mathbb{T}^d$, the "distinguished" boundary of $\mathbb{D}^d$. The multiplier space of $H^2(\mathbb{D}^d)$ is the space $H^\infty(\mathbb{D}^d)$ of bounded holomorphic functions on $\mathbb{D}^d$, equipped with the norm

$$\|f\|_\infty = \sup_{z\in\mathbb{D}^d} |f(z)| \geqslant \|f\|_2, \tag{E.2}$$

the latter inequality following obviously from (E.1). If $\Phi = (\Phi_j)_{1\leqslant j\leqslant d}$ is an analytic self-map of $\mathbb{D}^d$, the corresponding composition operator

$$C_\Phi : \mathscr{H}(\mathbb{D}^d) \to \mathscr{H}(\mathbb{D}^d)$$

is naturally defined by

$$C_\Phi(f) = f \circ \Phi. \tag{E.3}$$

We also set

$$\|\Phi\|_\infty = \sup_{z\in\mathbb{D}^d} \left(\max_{1\leqslant j\leqslant d} |\Phi_j(z)| \right) = \max_{1\leqslant j\leqslant d} \|\Phi_j\|_\infty.$$

Here, a major difference with the one-dimensional case appears: there is *no Littlewood subordination principle*, and in general, C_Φ does not map H to itself. An easy example in that respect is given below. Indeed, the theory of composition operators in several variables is a rather little explored domain in comparison to the one-variable case, and certainly deserves further investigation.

Theorem E.1 *We have the following:*

(i) *If* $\|\Phi\|_\infty < 1$, *then* $C_\Phi : H \to H$ *boundedly.*
(ii) *If* $\Phi(z_1, z_2) = (z_1, z_1)$, *then* C_Φ *is unbounded on* $H^2(\mathbb{D}^2)$.

Proof: By (E.2), we have

$$\|\Phi^\alpha\|_2 \leqslant \|\Phi^\alpha\|_\infty \leqslant \|\Phi\|_\infty^{|\alpha|}.$$

Let now $f = \sum_\alpha a_\alpha z^\alpha \in H$. Since $C_\Phi(f) = \sum_\alpha a_\alpha \Phi^\alpha$, the result follows by Cauchy–Schwarz inequality, which provides

$$\|C_\Phi(f)\|_2 \leqslant \|f\|_2 \times (1 - \|\Phi\|_\infty^2)^{-d/2}.$$

(ii) Let $f_n(z) = (z_1 + z_2)^n$, where n is a positive integer. By Stirling's formula, it holds that

$$\|f_n\|_2^2 = \sum_{k=0}^{n} \binom{n}{k}^2 = \binom{2n}{n} \sim \frac{4^n}{\sqrt{\pi n}}$$

while $C_\Phi(f_n)(z) = (2z_1)^n$, so that $\|C_\Phi(f_n)\|_2^2 = 4^n$ and that C_Φ is unbounded. □

Now, for $d \geqslant 1$, we define, for an operator T,

$$\beta_d^-(T) = \varliminf \left[a_{n^d}(T)\right]^{1/n} \qquad \text{and} \qquad \beta_d^+(T) = \varlimsup \left[a_{n^d}(T)\right]^{1/n}.$$

We have

$$0 \leqslant \beta_d^-(T) \leqslant \beta_d^+(T) \leqslant 1,$$

and we simply write $\beta_d(T)$ in case of equality.

It may well happen in general (consider diagonal operators) that $\beta_d^-(T) = 0$ and $\beta_d^+(T) = 1$. If $d = 1$, we proved in Chapter 10 that $\beta_1(C_\Phi)$ always exists and is given by a "spectral radius formula." As a consequence, we have

$$\|\Phi\|_\infty = 1 \Longrightarrow \beta_1(C_\Phi) = 1. \tag{E.4}$$

We will see that such a consequence no longer holds for $d \geqslant 2$. For simplicity, we only consider the case $d = 2$. We begin by showing the relevance of the parameters $\beta_d^-(C_\Phi)$, $\beta_d^+(C_\Phi)$ in a slightly more general context, but the reader can think of Ω as the polydisk $\mathbb{D}^d$. In what follows, a domain $\Omega \subset \mathbb{C}^d$ is called symmetric [41] if it is (up to a biholomorphism) balanced and convex, such as the ball or the polydisk in $\mathbb{C}^d$.

Theorem E.2 *Let* Ω *be a bounded and symmetric domain of the Hermitian space* $\mathbb{C}^d$ *and let* $\Phi\colon \Omega \to \Omega$ *be a non-degenerate analytic map inducing a compact composition operator* $C_\Phi\colon H^2(\Omega) \to H^2(\Omega)$. *Then, for some constants* $0 < c \leqslant C < \infty$, *independent of n, we have*

$$a_{n^d}(C_\Phi) \geqslant c\, e^{-Cn}, \qquad \forall n \geqslant 1,$$

that is, $\beta_d^-(C_\Phi) > 0$.

In order to prove this theorem, we shall use the following results, whose first is due to D. Clahane [30, theorem 2.1] (and to B. MacCluer [124] in the particular case of the unit ball B_d).

Theorem E.3 (D. Clahane) *Let Ω be a bounded, symmetric domain of $\mathbb{C}^d$ and $\Phi\colon \Omega \to \Omega$ be a holomorphic map inducing a compact composition operator $C_\Phi\colon H^2(\Omega) \to H^2(\Omega)$. Then Φ has a unique fixed point $z_0 \in \Omega$ and the spectrum of C_Φ consists of 0 and all possible products of eigenvalues of the derivative $\Phi'(z_0)$.*

When Φ is non-degenerate, 0 cannot be an eigenvalue of C_Φ since if $f \circ \Phi = 0$, then f vanishes on $\Phi(\Omega)$, which has a non-void interior, and hence $f \equiv 0$. Note that 1 is an eigenvalue by taking an empty product of eigenvalues of $\Phi'(z_0)$.

In fact, in our case, we will not need the existence of z_0, for we will force 0 to be a fixed point by a harmless change of the symbol Φ. By the multi-dimensional version of Schwarz's lemma ([41], or [153, theorem 8.1.2]), $\Phi'(0)$ has norm bounded by 1 if $\mathbb{C}^d$ is normed by taking Ω as its open unit ball; this implies that the eigenvalues of $\Phi'(0)$ have modulus bounded by 1. Moreover, when C_Φ is compact, its eigenvalues tend to 0, and it follows from Theorem E.3 that the eigenvalues of $\Phi'(0)$ have modulus (strictly) less than 1.

One may also add that $\Phi'(0)$ has a spectral radius always bounded by 1 (for Ω bounded and connected) and has at least an eigenvalue of modulus (strictly) less than 1 if and only if Φ is not an automorphism (see [86, Cartan–Caratheodory–Kaup-Wu Theorem, p. 374]).

We next need the following lemma:

Lemma E.4 *Let N_p be the number of multi-indices $\alpha = (\alpha_1, \ldots, \alpha_d)$ such that $|\alpha| = \alpha_1 + \cdots + \alpha_d \leqslant p$. Then, as p goes to infinity,*

$$N_p \sim \frac{p^d}{d!}\,. \tag{E.5}$$

Proof: Actually N_p is also the number of multi-indices $(\alpha_1, \ldots, \alpha_d, \alpha_{d+1})$ such that $\alpha_1 + \cdots + \alpha_d + \alpha_{d+1} = p$. We have, classically, for $|t| < 1$

$$\sum_{p=0}^{\infty} N_p t^p = \Big(\sum_{\alpha_1=0}^{\infty} t^{\alpha_1}\Big) \cdots \Big(\sum_{\alpha_{d+1}=0}^{\infty} t^{\alpha_{d+1}}\Big) = \Big(\sum_{k=0}^{\infty} t^k\Big)^{d+1} = \frac{1}{(1-t)^{d+1}}$$

so that $N_p = \binom{d+p}{p} = \dfrac{(p+d)\cdots(p+1)}{d!} \sim \dfrac{p^d}{d!}$, as claimed. □

Proof of Theorem E.2: We may assume that $\Phi(0) = 0$ and $\Phi'(0)$ is invertible. Indeed, since Φ is non-degenerate, there exists $a \in \Omega$ such that $\Phi'(a)$ is invertible. Since Ω is homogeneous, there exist two automorphisms Ψ_a and $\Psi_{\Phi(a)}$ of Ω such that $\Psi_a(0) = a$ and $\Psi_{\Phi(a)}[\Phi(a)] = 0$. Set $\chi = \Psi_{\Phi(a)} \circ \Phi \circ \Psi_a$. Then $\chi(0) = 0$. Moreover, using the chain rule, we see that $\chi'(0)$ is invertible, since $\Phi'(a)$ is.

Now every analytic automorphism Ψ of Ω induces a bounded composition operator on $H^2(\Omega)$ and $C_\Psi^{-1} = C_{\Psi^{-1}}$ ([30, theorem 3.1]).We can write $C_\chi = C_{\Psi_a} \circ C_\Phi \circ C_{\Psi_{\Phi(a)}}$

and it follows that C_χ, as well as C_Φ, is compact. The ideal property of approximation numbers implies that, for all $n = 1, 2, \ldots$, one has

$$\left(\|C_{\Psi_a}\| \, \|C_{\Psi_{\Phi(a)}}\|\right)^{-1} a_{n^d}(C_\Phi) \leqslant a_{n^d}(C_\chi) \leqslant \|C_{\Psi_a}\| \, \|C_{\Psi_{\Phi(a)}}\| \, a_{n^d}(C_\Phi),$$

so $\beta_d^-(C_\chi) = \beta_d^-(C_\Phi)$.

Let now $\mu_1, \ldots, \mu_d$ be the eigenvalues of $\chi'(0)$, with $|\mu_1| \geqslant \cdots \geqslant |\mu_d|$. As mentioned above, we have $|\mu_1| < 1$. Set $|\mu_d| = \mathrm{e}^{-A}$, with $A > 0$. By Theorem E.3, the eigenvalues $\lambda_1, \ldots, \lambda_n, \ldots$ of C_χ are the numbers $z_\alpha = \mu_1^{\alpha_1} \cdots \mu_d^{\alpha_d}$ rearranged in descending order (of modulus). We next claim that

$$|\lambda_{N_p}| \geqslant |\mu_d|^p = \mathrm{e}^{-Ap}.$$

Indeed, there are N_p numbers $|z_\alpha|$ with $|\alpha| \leqslant p$ (cf. Lemma E.4), and all of them are greater than $|\mu_d|^p$ since $|z_\alpha| \geqslant |\mu_d|^{|\alpha|} \geqslant |\mu_d|^p$ (because $|\mu_d| \leqslant 1$). This proves the claim.

If $M_p = \lfloor N_p/2 \rfloor$ where $\lfloor . \rfloor$ stands for the integer part, equation (G.11) (see Lemma G.11) gives

$$\mathrm{e}^{-2Ap} \leqslant |\lambda_{N_p}|^2 \leqslant |\lambda_{2M_p}|^2 \leqslant a_1 \, a_{M_p}.$$

Since $M_p \sim C_d \, p^d$ in view of Lemma E.4, inverting this relation and using the monotonicity of the a_n's clearly gives the claimed result. □

E.2 A Counterexample

The purpose here is to show that the spectral radius formula of Chapter 10 no longer holds in dimension at least 2. More precisely, the main theorem of this section is the following [116]:

Theorem E.5 *There exists an analytic self-map Φ of $\mathbb{D}^2$ satisfying the following properties:*

(i) *the composition operator $C_\Phi \colon H^2(\mathbb{D}^2) \to H^2(\mathbb{D}^2)$ is bounded and compact;*
(ii) *we have $\|\Phi\|_\infty = 1$ and Φ is* non-degenerate, *so that $\beta_2^-(C_\Phi) > 0$;*
(iii) *the approximation numbers of C_Φ satisfy $a_n(C_\Phi) \leqslant \alpha \, e^{-\beta \sqrt{n}}$ for some positive constants α, β; in other terms $\beta_2^+(C_\Phi) < 1$.*

The fact that Φ is *non-degenerate* means that its jacobian is not identically 0.

Proof: Our symbol Φ will be of the form

$$\Phi(z_1, z_2) = \big(\varphi(z_1), h(z_2)\psi(z_1)\big), \tag{E.6}$$

where φ, ψ are analytic self-maps of $\mathbb{D}$ chosen as follows:

$$\begin{cases} \varphi &= \dfrac{1 + \Lambda_\theta}{2} \\ S(z) &= \left(\dfrac{1+z}{1-z}\right)^\theta \\ w &= \mathrm{e}^{-S} \\ \psi &= w \circ \varphi. \end{cases}$$

Recall that Λ_θ is the lens map of parameter $0 < \theta < 1$; note that we have $\|\varphi\|_\infty = \varphi(1) = 1$, $\overline{\varphi(\mathbb{D})} \cap \mathbb{T} = \{1\}$, and $\varphi(0) = 1/2$ so that $\|C_\varphi\|_{B(H^2)} \leqslant \sqrt{3}$.

Next, let h be a *Rudin function,* namely $h \in H^\infty$ with $\|h\|_\infty = 1$, $h(0) = 0$, and moreover the powers of h are orthogonal in $H^2(\mathbb{D})$:

$$m \neq n \Longrightarrow h^m \perp h^n.$$

For our example, any such function h is enough, for example, $h(z_2) = z_2$! Or any inner function with $h(0) = 0$. But there are others by a beautiful theorem of Bishop (see Theorem 5.45) that provides a wealth of examples in Theorem E.5. We now set

$$\Phi(z_1, z_2) = \big(\varphi(z_1), h(z_2)\psi(z_1)\big). \tag{E.7}$$

This Φ is non-degenerate since its jacobian is $J_\Phi = \psi(z_1)\,\varphi'(z_1)\,h'(z_2)$.

A preliminary observation is that the function ψ satisfies, for every $z \in \mathbb{D}$,

$$|\psi(z)| = |w(\varphi(z))| \leqslant \exp\left(-\frac{\delta}{|1-\varphi(z)|^\theta}\right) \leqslant \exp\left(-\frac{\delta'}{|1-z|^{\theta^2}}\right), \tag{E.8}$$

where $\delta, \delta' > 0$ (and depending on θ).

In particular, since $|1 - \varphi(z)| \leqslant 1$, it holds that

$$\|\psi\|_\infty \leqslant \exp(-\delta) < 1.$$

To prove (E.8), setting $\delta = \cos(\theta\pi/2) > 0$, we have

$$\operatorname{Re}\left(\frac{1+\varphi(z)}{1-\varphi(z)}\right)^\theta \geqslant \delta\left|\frac{1+\varphi(z)}{1-\varphi(z)}\right|^\theta = \delta\frac{|1+\varphi(z)|^\theta}{|1-\varphi(z)|^\theta}$$

with $|1+\varphi(z)| \geqslant \operatorname{Re}(1+\varphi(z)) \geqslant 1$, which gives the first inequality in (E.8). Moreover, the second inequality comes from the fact that, for some $C_\theta > 0$,

$$|1-\varphi(z)| = \frac{1}{2}|1-\Lambda_\theta(z)| \leqslant C_\theta|1-z|^\theta \quad \text{for all } z \in \mathbb{D}.$$

See (4.1) in Chapter 4.

Now the proof consists of three steps, whose first indicates that the *two-dimensional* composition operator $C_\Phi : H^2(\mathbb{D}^2) \to H^2(\mathbb{D}^2)$, with Φ non-degenerate, behaves as a *direct Hilbertian sum of one-dimensional* weighted composition operators: $H^2(\mathbb{D}) \to H^2(\mathbb{D})$.

Step 1.

$$C_\Phi \sim \bigoplus_{k=0}^{\infty} T_k \qquad \text{with } T_k = M_{\psi^k} C_\varphi, \tag{E.9}$$

where $\sim$ indicates unitary equivalence and T_k acts as a weighted composition operator on a copy H_k of $H^2(\mathbb{D})$.

Indeed, let $f(z_1, z_2)$ be a polynomial, written as

$$f(z_1, z_2) = \sum_{j,k} a_{j,k} z_1^j z_2^k =: \sum_k z_2^k f_k(z_1), \text{ with } \|f\|^2 = \sum_k \|f_k\|^2,$$

where $f_k(z_1) = \sum_j a_{j,k} z_1^j$. We see that

$$C_\Phi f(z_1, z_2) = \sum_k h^k(z_2) \left(\psi^k(z_1) \sum_j a_{j,k} \varphi^j(z_1) \right) = \sum_k h^k(z_2) M_{\psi^k} C_\varphi f_k(z_1).$$

Since h is a Rudin function, it ensues that

$$\|C_\Phi(f)\|^2 = \sum_k \|M_{\psi^k} C_\varphi(f_k)\|^2 = \sum_k \|T_k(f_k)\|^2,$$

clearly implying (E.9).

Step 2.

We have the following inequality, in which $T = \oplus_{k=0}^\infty T_k$ acts on a direct Hilbertian sum $H = \oplus_{k=0}^\infty H_k$; $n_0, \ldots, n_L$ are positive integers and $N = \sum_{k=0}^N n_k$:

$$a_N(T) \leqslant \max\big(\max_{0 \leqslant k \leqslant L} a_{n_k}(T_k), \ \sup_{k > L} \|T_k\|\big). \tag{E.10}$$

Indeed, denote by μ the RHS of (E.10). Let R_k with $0 \leqslant k \leqslant L$ be operators on H_k of respective ranks $< n_k$ such that $\|T_k - R_k\| = a_{n_k}(T_k)$ and let $R = \bigoplus_{k=0}^L R_k$ an operator of rank $\leqslant n_0 + \cdots + n_L - L - 1 < N$. If $f = \sum_{k=0}^\infty f_k \in H$, we see that

$$\begin{aligned}\|Tf - Rf\|^2 &= \sum_{k=0}^L \|T_k f_k - R_k f_k\|^2 + \sum_{k>L} \|T_k f_k\|^2 \\ &\leqslant \sum_{k=0}^L \big(a_{n_k}(T_k)\big)^2 \|f_k\|^2 + \sum_{k>L} \|T_k f_k\|^2 \leqslant \mu^2 \sum_{k=0}^\infty \|f_k\|^2 = \mu^2 \|f\|^2.\end{aligned}$$

Step 3.

Taking $m_k = c\lfloor \log(b\frac{n}{k})\rfloor$, for $1 \leqslant k \leqslant n$, where a, b, c are positive numerical constants, we claim that

$$a_{nm_k+1}(T_k) \lesssim e^{-an}. \tag{E.11}$$

This is the most delicate point and will be only sketched. We refer to [116, theorems 2.3 and 5.12] for a more detailed exposition. Things go as follows:

Fact 1. Let $T = M_{w\circ\varphi} C_\varphi$ with w and φ as above, and B be a Blaschke product of length $< N$. Then

$$a_N(T) \lesssim \sup_{\substack{|z-1| \leqslant 1 \\ z \in \varphi(\mathbb{D})}} |B(z)|\,|w(z)|. \tag{E.12}$$

An appropriate choice of B, with $N = m$, now gives the following:

Fact 2. Let φ be as in Theorem E.5, and w some weight satisfying

$$|w(z)| \leqslant \exp\left(-\frac{R}{|1-z|^\theta} \right) \qquad \text{for all } z \in \mathbb{D}.$$

Then, the approximation numbers of $T = M_{w\circ\varphi} C_\varphi$ satisfy, for all $m, n \in \mathbb{N}$,

$$a_{nm+1}(T) \lesssim \max\left[\exp(-an), \ \exp(-R2^{m\theta})\right]. \tag{E.13}$$

We get the existence of positive constants $a,\ b,\ c$ depending only on θ such that, since

$$m_k = \big[c\ \log(bn/k)\big],$$

the approximation numbers of $T_k = M_{\psi^k} C_\varphi$ satisfy, for all $n, k \in \mathbb{N}$,

$$a_{nm_k+1}(T_k) \lesssim \exp(-an).$$

Just apply (E.8) and (E.13) to φ and the weight $\psi^k = w^k \circ \varphi$, which satisfy their assumptions with $R = k\delta^2$ and θ changed into θ^2. We get the claim (E.11).

Let us now finish the proof of Theorem E.5. We first observe that

$$N := \sum_{k=0}^{n} nm_k + 1 \approx \sum_{k=1}^{n} n \log(n/k) \leqslant c^2 n^2 \int_0^1 \log(1/x)\, dx = c^2 n^2$$

(with $c \in \mathbb{N}$) using Riemann sums of the non-increasing, positive function $\log(1/x)$.

Next note that $\|T_k\| \leqslant \|\psi^k\|_\infty \|C_\varphi\|_{B(H^2)} \leqslant \sqrt{3}\,\mathrm{e}^{-\delta k}$ so that (E.10) and (E.11) give

$$a_{c^2n^2}(C_\Phi) \leqslant a_N(C_\Phi) \lesssim \max\big(\mathrm{e}^{-an}, \mathrm{e}^{-\delta n}\big),$$

implying $\overline{\lim}\,\big[a_{(cn)^2}(C_\Phi)\big]^{1/cn} < 1$, equivalently $\beta_2^+(C_\Phi) < 1$ whereas we have $\|\Phi\|_\infty \geqslant \|\varphi\|_\infty = 1$. □

Observe that $\Phi(\mathbb{D}^2)$ does not "touch" the distinguished boundary $\mathbb{T}^2$ of $\mathbb{D}^2$. A more sophisticated example is given in [117] with Φ non-degenerate,

$$\Phi(\overline{\mathbb{D}}^2) \cap \mathbb{T}^2 \neq \emptyset,$$

and yet $\beta_2^-(C_\Phi) < 1$.

It would be interesting to exhibit a reasonably large class of symbols Φ for which

$$\|\Phi\|_\infty = 1 \Longrightarrow \beta_2^-(C_\Phi) = 1.$$

Appendix F Reminders on Subordination of Sequences

Let $\mathcal{S}$ (resp. $\mathcal{S}^+$) be the set of non-increasing sequences $u = (u_j)_{j\geqslant 1}$ of real (resp. of positive) numbers. If $u, v \in \mathcal{S}$, the sequence u is said to be *subordinate* to the sequence v, and we write $u \prec v$ if

$$\sum_{j=1}^{n} u_j \leqslant \sum_{j=1}^{n} v_j \qquad \text{for all } n \geqslant 1. \tag{F.1}$$

For example, if $u = (1, 1, 0, 0, \ldots)$ and $v = (2, 0, 0, \ldots)$, we have $u \prec v$.

We have a basic stability property of this notion (see [168, theorem 1.16]).

Proposition F.1 *Let I be an interval of $\mathbb{R}$, let $h\colon I \to \mathbb{R}$ be increasing and convex, and $u, v \in \mathcal{S}$ with values in I. Then*

$$u \prec v \quad \Longrightarrow \quad h(u) \prec h(v).$$

Proof: We may assume that h is $\mathscr{C}^2$. We set $a = \min(u_n, v_n)$ if the positive integer n is given. Then, for $x > a$, with $x \in I$,

$$h(x) = h(a) + (x-a)h'(a) + \int_a^{+\infty} (x-t)^+ h''(t)\, dt.$$

One easily checks, using (F.1), that $\sum_{j=1}^{n}(u_j - t)^+ \leqslant \sum_{j=1}^{n}(v_j - t)^+$ for all $t \geqslant a$. Hence, thanks to the positivity of $h'(a)$ and h'',

$$\begin{aligned}
\sum_{j=1}^{n} h(u_j) &= nh(a) + h'(a)\sum_{j=1}^{n}(u_j - a) + \int_a^{+\infty} \sum_{j=1}^{n}(u_j - t)^+ h''(t)\, dt \\
&\leqslant nh(a) + h'(a)\sum_{j=1}^{n}(v_j - a) + \int_a^{+\infty} \sum_{j=1}^{n}(v_j - t)^+ h''(t)\, dt \\
&= \sum_{j=1}^{n} h(v_j).
\end{aligned}$$

□

A stronger notion is that of log-subordination.

Definition F.2 We say that the sequence $u \in \mathcal{S}^+$ is *log-subordinate* to the sequence $v \in \mathcal{S}^+$ if $\log u \prec \log v$. In other terms,

$$\prod_{j=1}^{n} u_j \leqslant \prod_{j=1}^{n} v_j \qquad \text{for all } n \geqslant 1.$$

The following result will be useful:

Proposition F.3 *For $u, v \in \mathcal{S}^+$, the following two conditions are equivalent:*

(i) $\log u \prec \log v$;
(ii) $u^p \prec v^p$ *for all* $p > 0$.

Proof: If $\log u \prec \log v$, it suffices to apply Proposition F.1 to the function $h(x) = \mathrm{e}^{px}$ to get $u^p \prec v^p$.

Conversely, if $u^p \prec v^p$ for all $p > 0$, we have

$$\left(\frac{1}{n}\sum_{j=1}^{n} u_j^p\right)^{1/p} \leqslant \left(\frac{1}{n}\sum_{j=1}^{n} v_j^p\right)^{1/p},$$

and by letting p go to 0, we get

$$\left(\prod_{j=1}^{n} u_j\right)^{1/n} \leqslant \left(\prod_{j=1}^{n} v_j\right)^{1/n},$$

that is, $\log u \prec \log v$. □

A related result is the following proposition:

Proposition F.4 *Let $u, v \in \mathcal{S}^+$ be two sequences such that u is log-subordinate to v. Then for $N \geqslant n$,*

$$u_N \leqslant v_1^{n/N} v_n^{1-n/N}. \tag{F.2}$$

In particular, for any $n \geqslant 1$

$$u_{2n} \leqslant \sqrt{v_1 v_n}\,. \tag{F.3}$$

Proof: We have

$$u_N^N \leqslant \prod_{j=1}^{N} u_j \leqslant \prod_{j=1}^{N} v_j = \prod_{j=1}^{n} v_j \prod_{j=n+1}^{N} v_j \leqslant v_1^n v_n^{N-n},$$

and (F.2) follows. Now the choice $N = 2n$ gives (F.3). □

We prove below a basic result in the theory of Schatten classes whose proof is worth mentioning here as an application of the results of the present appendix.

We will say that the matrix $(\alpha_{m,n})$ with non-negative entries is *doubly substochastic* if

$$\sup_n \left(\sum_m \alpha_{m,n}\right) \leqslant 1 \quad \text{and} \quad \sup_m \left(\sum_n \alpha_{m,n}\right) \leqslant 1.$$

Lemma F.5 *Let* (x_n), (y_m) *be two sequences of non-negative numbers such that*

$$y_m = \sum_n \alpha_{m,n} x_n \qquad \text{for all } m,$$

where $(\alpha_{m,n})$ *is doubly substochastic. Then, for any function* $\phi : \mathbb{R}^+ \to \mathbb{R}^+$ *with* $\phi(0) = 0$, *one has:*

(i) *when* ϕ *is convex,*

$$\sum_m \phi(y_m) \leqslant \sum_n \phi(x_n).$$

(ii) *when* ϕ *is concave and moreover*

$$\sum_m \alpha_{m,n} = 1, \tag{F.4}$$

the conclusion is reversed: $\sum_m \phi(y_m) \geqslant \sum_n \phi(x_n)$.

Proof: (i) Recalling that $y_m = \sum_n \alpha_{m,n} x_n$, we have for each m, since ϕ is convex, vanishes at 0, and $\sum_n \alpha_{m,n} \leqslant 1$,

$$\phi(y_m) \leqslant \sum_n \alpha_{m,n}\, \phi(x_n).$$

Then by Fubini and double substochasticity

$$\sum_m \phi(y_m) \leqslant \sum_n \left(\sum_m \alpha_{m,n} \right) \phi(x_n) \leqslant \sum_n \phi(x_n),$$

and (i) is proved.

When ϕ is concave, the same strategy works taking into account (F.4) and the inequalities are reversed:

$$\sum_m \phi(y_m) \geqslant \sum_n \phi(x_n). \qquad \square$$

An interesting application to Schatten classes is the following proof of Theorem 1.111 (we recall the statement for convenience).

Lemma F.6 *Let* $T : H \to H$ *be a compact operator, and* $(e_m)_{m\geqslant 0}$ *an ONB of* H. *Then:*

(i) *If* $0 < p < 2$ *and* $\sum_m \|T(e_m)\|^p < \infty$, *we have* $T \in \mathscr{S}^p$, *and*

$$\big(\mathfrak{s}_p(T)\big)^p \leqslant \sum_m \|T(e_m)\|^p.$$

(ii) *If* $p \geqslant 2$ *and* $T \in \mathscr{S}^p$, *we have*

$$\sum_m \|T(e_m)\|^p \leqslant \big(\mathfrak{s}_p(T)\big)^p.$$

(iii) *For every $p \geqslant 1$,*

$$\mathfrak{s}_p(T) = \sup \left(\sum_{n \geqslant 1} |\langle T(h_n), k_n \rangle|^p \right)^{\frac{1}{p}},$$

where the supremum runs over all the orthonormal sequences $(h_n)_{n \geqslant 1}$ and $(k_n)_{n \geqslant 1}$ in H.

Proof: Let $T(x) = \sum_{n \geqslant 1} s_n \langle x, a_n \rangle b_n$ be the Schmidt decomposition of T, with

$$\|T(e_m)\|^2 = \sum_{n \geqslant 1} |\langle e_m, a_n \rangle|^2 s_n^2 =: \sum_{n \geqslant 1} \alpha_{m,n} s_n^2.$$

We observe that the matrix $(\alpha_{m,n})$ is doubly substochastic, and satisfies (F.4) since (e_m) is an ONB.

We now apply Lemma F.5 to $(\alpha_{m,n})$, to the function $\phi(t) = t^{p/2}$ (concave or convex according to whether $p \leqslant 2$ or $p \geqslant 2$), and the two sequences $x_n = s_n^2$, $y_m = \|T(e_m)\|^2$, to infer the result for (i) and (ii). Let $T(x) = \sum_{n \geqslant 1} s_n \langle x, a_n \rangle\, b_n$ be the Schmidt decomposition of T.

Just apply the preceding to the doubly substochastic matrix

$$(\alpha_{m,n}) = (|\langle e_m, a_n \rangle|^2),$$

which satisfies (F.4) since (e_m) is an ONB, to $\|T(e_m)\|^2 = \sum_{n \geqslant 1} s_n^2 |\langle e_m, a_n \rangle|^2$, and to the function $\phi(t) = t^{p/2}$ (according to whether $p \leqslant 2$ or $p \geqslant 2$, this function is concave or convex), to the sequences $x_n = s_n^2$ and $y_m = \|T(e_m)\|^2$, to infer the result for (i) and (ii).

We now prove item (iii). Assume first that $T \in \mathscr{S}^p$, under Schmidt form

$$T = \sum_{n \geqslant 1} s_n\, (b_n \otimes a_n) \qquad \text{or else} \qquad T(x) = \sum_{n \geqslant 1} s_n\, \langle x, a_n \rangle b_n.$$

Let $(h_m)_{m \geqslant 1}, (k_m)_{m \geqslant 1}$ be two orthonormal sequences in H. We have, for every $m \geqslant 1$,

$$\langle T h_m, k_m \rangle = \sum_{n \geqslant 1} s_n \langle h_m, a_n \rangle . \langle b_n, k_m \rangle$$

so that

$$|\langle T h_m, k_m \rangle| \leqslant \sum_{n \geqslant 1} \alpha_{m,n}\, s_n =: t_m \qquad \text{with} \quad \alpha_{m,n} = |\langle h_m, a_n \rangle . \langle k_m, b_n \rangle|$$

Now, the inequalities of Cauchy–Schwarz and Bessel show that the matrix $(\alpha_{m,n})$ is doubly substochastic. The subordination Lemma F.5 applied to the convex function $\phi(t) = t^p$ and the sequences (s_n) and (t_m) now gives us

$$\sum_{m \geqslant 1} |\langle T h_m, k_m \rangle|^p \leqslant \sum_{m \geqslant 1} t_m^p \leqslant \sum_{n \geqslant 1} s_n^p,$$

and the RHS term in (iii) is finite.

Moreover, we have the equality $\langle Ta_m, b_m\rangle = s_m$ for the special orthonormal sequences (a_m) and (b_m).

Actually, this proves that, as soon as $T \in \mathscr{K}(H)$, the RHS is finite if and only if $T \in \mathscr{S}^p$. □

We mention that if the RHS is finite, then T must be compact (hence the equality in (iii) holds): we leave this part as an exercise.

The next lemma is not directly related to subordination, but we include it here because it is needed for the slow decay issues in Chapter 10.

Lemma F.7 *Let $(u_n)_{n\geqslant 0}$ be a sequence of non-negative numbers with limit* 0.

There exists a log-convex sequence $(v_n)_{n\geqslant 0}$ (a fortiori a convex one) with limit 0 *such that*

$$v_n \geqslant u_n \qquad \text{for all } n \geqslant 0.$$

Proof: Replacing u_n by $u'_n = (\sup_{k\geqslant n} u_k) + \dfrac{1}{n+1}$, we can assume that (u_n) is decreasing and positive.

We then define (v_n) by $v_0 = u_0, v_1 = u_1$; and for $n \geqslant 1$,

$$v_{n+1} = \max\,(u_{n+1}, v_n^2/v_{n-1}).$$

It is easily seen by induction that (v_n) is itself decreasing; it is log-convex by construction, and positive since $v_n \geqslant u_n > 0$. Hence, it has a limit $\ell \geqslant 0$. In case $\ell > 0$, one must have for some large integer n_0

$$\log v_{n+1} = 2\log v_n - \log v_{n-1}, \qquad \text{for } n \geqslant n_0,$$

implying $\log v_n = an + b$ for $n \geqslant n_0$, with a, b constants. But since $\log v_n$ converges to $\log \ell$, we have $a = 0$, which is absurd. This proves that indeed $\ell = 0$. □

Appendix G Singular Numbers and Their Comparisons

G.1 Definitions

This appendix is partly an exposition of some results of [113], [114], [115], and [110], as well as [145]. Let $T : X \to Y$ be a compact operator between Banach spaces (mostly Hilbert spaces, but not always). We wish to study, in a quantitative way, the "degree of compactness" of T. In most (but not all) cases, such an operator is a uniform limit of finite rank operators, and the speed at which those operators approximate T, namely the decay rate of

$$a_n(T) = \inf_{\text{rank} R < n} \|T - R\|, \tag{G.1}$$

the so-called sequence of approximation numbers of T, is a good indicator of this degree of compactness (the non-commutative analog of Bernstein's theory of polynomial approximation of functions). To make this more precise, we first need to coin several definitions and compare them. But first, it is often convenient to restrict the set of approximating operators of rank $< n$, especially in the Banach space case (this is more or less automatic in the Hilbertian case). In a recent paper, Pouliasis ([140]) uses the following result, which he attributes to T. Ransford:

Lemma G.1 *Let $T : X \to X$ and let*

$$\widetilde{a_n}(T) = \inf\{\|T - R\| : R(X) \subset T(X),\ \textit{rank}\ (R) < n)\}.$$

Then

$$\widetilde{a}_n(T) \leqslant n\, a_n(T). \tag{G.2}$$

The proof consists in a clever application of Auerbach's lemma. Here is a slight improvement, using instead the Kadec–Snobar theorem on projections.

Lemma G.2 *Let $T : X \to X$ and let*

$$\widehat{a_n}(T) = \inf\{\|T - TP\| : P \textit{ is projection of rank } < n\}.$$

Then

$$\widehat{a_n}(T) \leqslant \left(\sqrt{n-1} + 3\right) a_n(T). \tag{G.3}$$

This result is better than the previous one in two respects: the power for n, and the class of approximating operators.

Proof: Fix $\varepsilon > 0$. Let R be an operator of rank $m < n$. Let $Q : X \to \operatorname{Ker} R$ be a projection of norm $\leqslant \sqrt{m} + 1 + \varepsilon$ (see [180, p. 117]) and $P = I - Q$. Then P has rank less than $n - 1$ and norm less than $\sqrt{n-1} + 2 + \varepsilon$. Let $R_0 = TP$. We see that, if x belongs to the unit ball of X and $x = Q(x) + P(x)$, we have $R(x) = R(P(x))$ and hence

$$\|R(x) - R_0(x)\| \leqslant \|T - R\| \, \|P(x)\| \leqslant (\sqrt{n-1} + 2 + \varepsilon) \, \|T - R\|$$

so that

$$\widehat{a_n}(T) \leqslant \|T - R_0\| \leqslant \|T - R\| + \|R - R_0\| \leqslant (\sqrt{n-1} + 3 + \varepsilon)\|T - R\|.$$

Letting ε tend to zero, and taking the infimum on R, we get the result. □

Before proceeding, we coin some notations. For $n = 1, 2, \ldots$, we shall denote by B_E (resp. S_E) the unit ball (resp. sphere) of a subspace E of X. And if F is a subspace of Y, we shall denote by $\gamma_E(T)$, $\delta_F(T)$ and $\delta'_E(T)$ the quantities

$$\gamma_E(T) = \inf_{x \in S_E} \|Tx\|, \qquad \delta_F(T) = \sup_{x \in B_X} d(Tx, F), \qquad \delta'_E(T) = \sup_{x \in B_X} d(Tx, T(E)). \tag{G.4}$$

We now attach to T six sequences of numbers (often called sequences of s-numbers, and which share many similarities but can also have subtle differences; see [16]) defined below.

Definition G.3 (s-numbers) We define here the most renowned s-numbers:

(i) $a_n(T) = \inf_{\operatorname{rank} R < n} \|T - R\|$, the nth *approximation number* of T.
(ii) $b_n(T) = \sup_{\dim E = n} [\gamma_E(T)]$, the nth *Bernstein number* of T.
(iii) $c_n(T) = \inf_{\operatorname{codim} G < n} \|T_{|G}\|$, the nth *Gelfand number* of T.
(iv) $d_n(T) = \inf_{\dim F < n} \delta_F(T)$, the nth *Kolmogorov number* of T.
(v) $d'_n(T) = \inf_{\dim E < n} \delta'_E(T) \geqslant d_n(T)$, the nth *modified Kolmogorov number* of T.
(vi) $\mathfrak{e}_n(T) = \inf\{\varepsilon > 0 \; : \; N(TB_X, \varepsilon B_Y) \leqslant 2^{n-1}\}$, the nth *entropy number* of T.

Here we denote by $N(A, B) \leqslant \infty$ the smallest number of translates of B needed to cover A. Note that all those numbers are defined as minimax, except the Bernstein numbers that are defined as maximin. Even if we will not make strong use of them, we have the feeling that entropy numbers are the most natural in this context, since by definition T is compact if and only if $T(B_X)$ is precompact, that is,

$$T \text{ compact} \iff \mathfrak{e}_n(T) \to 0.$$

For approximation numbers, we only have (due to the existence of Banach spaces X without the approximation property [108, chapter 9]

$$a_n(T) \to 0 \Longrightarrow T \text{ compact}.$$

A fundamental property shared by those six sequences (say $(\alpha_n(T))$ is the easily checked two-sided Ideal Property, which says the following:

Proposition G.4 (Ideal property) *For operators* $B : E \to X$, $T : X \to Y$ *and* $A : Y \to F$, *one always has*

$$\alpha_n(ATB) \leqslant \|A\|\, \alpha_n(T)\, \|B\|.$$

G.2 The Banach Case

The relations between various "singular" numbers are well known to experts, but sparse in the literature, and we think it worthwhile including a short proof of some of those relations. Let $T : X \to Y$ be a compact operator. When $Y = X$, we denote by $(\lambda_n(T))_{n\geqslant 1}$ the sequence of non-zero eigenvalues of T rearranged in decreasing order (of modulus). We have the following:

Theorem G.5 *The sequences* $a_n, b_n, c_n, d_n, d'_n, \mathfrak{e}_n, \lambda_n$ *are related by the following inequalities:*

(i) $b_n(T) \leqslant d_n(T) \leqslant a_n(T)$.
(ii) $c_n(T) \leqslant a_n(T) \leqslant 2\sqrt{n}\, c_n(T)$.
(iii) $\sup_{k\leqslant n} k^{\alpha} \mathfrak{e}_k(T) \leqslant C_\alpha \sup_{k\leqslant n} k^{\alpha} a_k(T)$ *for each* $\alpha > 0$.
(iv) $|\lambda_n(T)| \leqslant 4\mathfrak{e}_n(T)$.
(v) $b_n(T) \leqslant 4\mathfrak{e}_n(T)$.
(vi) $a_n(T) \geqslant a_n(T^*)$.
(vii) $d'_n(T) \leqslant \widehat{a_n}(T)$.
(viii) *One has for all* S, T *and all* $p, q \geqslant 1$: $a_{p+q-1}(S+T) \leqslant a_p(S) + a_q(T)$ *and* $a_{p+q-1}(ST) \leqslant a_p(S)a_q(T)$.
(ix) *Finally,* $|a_n(S) - a_n(T)| \leqslant \|S - T\|$ *for all* n, S, T.

In the sequel, for short, we often write a_n instead of $a_n(T), \ldots, \mathfrak{e}_n$ for $\mathfrak{e}_n(T)$.

Proof: (i) Let E and F be subspaces of X and Y of respective dimensions n and p with $p < n$. Let us show (cf. (G.4)) that $\gamma_E(T) \leqslant \delta_F(T)$. If T is not injective on E, this is obvious. Otherwise, $T(E)$ has dimension $n > \dim F$. By a celebrated consequence of the Borsuk–Ulam antipodal theorem ([120, p. 137]), there exists $y_0 \in T(E)$, $y_0 \neq 0$, such that $\|y_0\| = d(y_0, F)$. By homogeneity, we can assume that $y_0 = Tx_0$ with $x_0 \in S_E$, implying

$$\gamma_E(T) \leqslant \|Tx_0\| = d(Tx_0, F) \leqslant \delta_F(T),$$

and we are done for the bound $b_n \leqslant d_n$ (see also [137, p. 13]). Moreover, if R has rank $< n$ and $F = R(X)$, then $\dim F < n$ and

$$\sup_{x\in B_X} d(Tx, F) \leqslant \sup_{x\in B_X} \|Tx - Rx\| = \|T - R\|,$$

implying the inequality $d_n \leqslant a_n$.

(ii) Let R be an operator of rank $< n$. Then the kernel G of R has codimension $< n$ and clearly

$$\|T_{|G}\| \leqslant \|T - R\|$$

so that $c_n \leqslant a_n$. The reverse inequality needs an extra factor: let $\varepsilon > 0$ and G such that $\operatorname{codim} G < n$ and $\|T_{|G}\| \leqslant c_n + \varepsilon$. By ([16, proposition 2.4.3]), there exists a projection $P : X \to G$ such that $\|P\| \leqslant 2\sqrt{n}$. Then $R = T(I - P)$ has rank $< n$ and moreover

$$a_n \leqslant \|T - R\| = \|TP\| \leqslant \|T_{|G}\| \, \|P\| \leqslant 2\sqrt{n}(c_n + \varepsilon),$$

which gives the second inequality.

(iii) We will not make strong use of that result, and refer to [138, p. 64] for a detailed proof due to B. Carl.

(iv) For simplicity, let $\mathbb{R}$ be the scalar field. Fix n, let m be the Lebesgue measure on $\mathbb{R}^n$, and $\delta > 0$. By localization, we may assume that $X = \mathbb{R}^n$ (see [138, p. 67] for some more details). One has by definition

$$TB_X \subset \bigcup_{1 \leqslant j \leqslant 2^{n-1}} B_j, \tag{G.5}$$

where B_j's are balls of radius $\mathfrak{e}_n + \delta$. Taking volumes in that inclusion gives

$$\Big|\prod_{j=1}^{n} \lambda_j\Big| m(B_X) = |\det T| \, m(B_X) = m(T(B_X)) \leqslant 2^{n-1}\, (\mathfrak{e}_n + \delta)^n\, m(B_X);$$

hence taking nth roots and letting δ tend to zero, we have

$$|\lambda_n| \leqslant \prod_{j=1}^{n} |\lambda_j|^{1/n} \leqslant 2\mathfrak{e}_n.$$

(v) Let E be a n-dimensional subspace of X and $N(\varepsilon)$ (resp. $P(\varepsilon)$) the minimum number of closed balls of radius ε needed to cover B_E (resp. the maximum number of points in B_E mutually ε apart). One notoriously has ([138, p. 49])

$$\left(\frac{1}{\varepsilon}\right)^n \leqslant N(\varepsilon) \leqslant P(\varepsilon). \tag{G.6}$$

Taking $\varepsilon = 1/2$, one gets a finite set $A \subset B_E$, with cardinality at least 2^n, such that

$$a, b \in A \quad \text{and} \quad a \neq b \Rightarrow \|a - b\| > 1/2. \tag{G.7}$$

We now use (G.5). As there are $2^n > 2^{n-1}$ points Ta with $a \in A$, one can find two distinct points $a, b \in A$ with images under T in a common ball B_j, which implies

$$\|Ta - Tb\| \leqslant \operatorname{diam} B_j \leqslant 2(\mathfrak{e}_n + \delta). \tag{G.8}$$

It follows from (G.8) and (G.7) that

$$\gamma_E(T) \leqslant \frac{\|Ta - Tb\|}{\|a - b\|} \leqslant 4(\mathfrak{e}_n + \delta).$$

We then obtain $\gamma_E(T) \leqslant 4\mathfrak{e}_n$ and $b_n \leqslant 4\mathfrak{e}_n$. In particular, a compact operator ($\mathfrak{e}_n \to 0$) is a so-called finitely strictly singular (FSS for short) operator ($b_n \to 0$).

(vi) If $R : X \to Y$ has rank $< n$, then $R^* : Y^* \to X^*$ also has rank $< n$, so that

$$a_n(T^*) \leqslant \|T^* - R^*\| = \|T - R\|.$$

This gives the result. Note that the converse fails in general ([75]), but holds if T is compact ([16, p. 78]) or if X and Y are reflexive spaces.

(vii) Let $P : X \to X$ be a projection of rank $< n$ and $E = P(X)$, as well as $x \in B_X$. Just note that $d(Tx, TE) \leqslant \|Tx - TPx\| \leqslant \|T - TP\|$, clearly giving the inequality.

(viii) We refer to [16, p. 42] for this easily verified but quite useful property.

(ix) With $p = n$, $q = 1$ in the above, we get

$$a_n(S) \leqslant a_n(T) + a_1(S - T) = a_n(T) + \|S - T\|,$$

and this gives the result. □

We saw that compact implies FSS: there is no converse in full generality, as shown by the canonical injection of $L^\infty(0,1)$ to $L^2(0,1)$, which is FSS ([154, pp. 114–115]) by a result of Grothendieck and non-compact.

G.3 The Hilbertian Case

In the Hilbertian case, we can be more precise and will also use the numbers $s_j = s_j(T)$ (by definition the eigenvalues of $|T| := \sqrt{T^*T}$), called the *singular values (or singular numbers)* of T.

Before that we mention a useful tool for the study of s-numbers, since they share the *ideal property*: the following factorization result, often referred to as *Douglas lemma.*

Lemma G.6 (Douglas) *Let H be a Hilbert space, and $A, B \in B(H)$ such that*

$$\|A(x)\| \leqslant \|B(x)\| \qquad \textit{for all } x \in H.$$

Then there exists a linear contraction $C \in B(H)$ such that $A = CB$.

Proof: Let M be the range of B and define C_0 on M by

$$C_0(y) = A(x) \quad \text{if } y = B(x) \in M.$$

This does not depend on the choice of x since $Ker B \subset Ker A$ and hence C_0 is linear with $\|C_0\|_{B(M)} \leqslant 1$ by our assumption. Extend C_0 by continuity to the closure $\overline{M}$ and set $C = C_0 P$ where $P : H \to \overline{M}$ is the orthogonal projection. This C does the job: it remains a contraction since $\|P\| \leqslant 1$ and

$$x \in H \Longrightarrow B(x) \in M \Longrightarrow C[B(x)] = C_0 P[B(x)] = C_0[B(x)] = A(x). \qquad \square$$

The numbers s_n can be estimated thanks to the *minimax formula* (see [168, p. 4]; this is an easy consequence of the polar decomposition, or orthonormal diagonalization of A): for every integer $n \geqslant 1$,

(i) one has $s_n(A) = \inf_{\dim E < n} \sup_{\substack{\|x\|=1 \\ x \in E^\perp}} \|A(x)\|.$

(ii) If A is moreover positive,

$$s_n(A) = \inf_{\dim E < n} \sup_{\substack{\|x\|=1 \\ x \in E^\perp}} \langle A(x), x \rangle. \tag{G.9}$$

As a consequence, $0 \leqslant A \leqslant B$ implies $s_n(A) \leqslant s_n(B)$ for all $n \in \mathbb{N}$.

(iii) If $\dim H = N < \infty$, and if $A \in B(H)$ is hermitian with eigenvalues in decreasing order $\lambda_1 \geqslant \lambda_2 \geqslant \cdots \geqslant \lambda_N$ and with an orthonormal diagonalized form written $A = \sum_{j=1}^{N} \lambda_j u_j \otimes u_j$, then, for all $1 \leqslant n \leqslant N$,

$$\lambda_n = \inf_{\dim E < n} \sup_{\substack{\|x\|=1 \\ x \in E^{\perp}}} \langle A(x), x \rangle.$$

Concerning other s-numbers, we have the following:

Theorem G.7 *Let $T : H \to H$ be compact, with Schmidt decomposition $T = \sum_{j \geqslant 1} s_j (u_j \otimes v_j)$. Then, for all $n \geqslant 1$,*

$$s_n(T) = a_n(T) = b_n(T) = c_n(T) = d_n(T) = d'_n(T).$$

Moreover

$$a_n(T) \leqslant 4\mathfrak{e}_n(T).$$

Proof: Once more, we omit the dependence in T. We first prove that

$$a_n = b_n = d_n = s_n. \tag{G.10}$$

Indeed let R_{n-1} be defined by $R_{n-1}(x) = \sum_{j=1}^{n-1} s_j \langle x, v_j \rangle u_j$, so that $(T - R_{n-1})(x) = \sum_{j=n}^{\infty} s_j \langle x, v_j \rangle u_j$, and

$$\|(T - R_{n-1})(x)\|^2 = \sum_{j=n}^{\infty} s_j^2 |\langle x, v_j \rangle|^2 \leqslant s_n^2 \sum_{j=n}^{\infty} |\langle x, v_j \rangle|^2 \leqslant s_n^2 \|x\|^2, \text{ implying}$$

$$a_n \leqslant \|T - R_{n-1}\| \leqslant s_n.$$

Next take for E the n-dimensional subspace generated by $v_1, \ldots, v_n$. For $x \in S_E$, $T(x) = \sum_{j=1}^{n} s_j \langle x, v_j \rangle u_j$ so that

$$\|T(x)\|^2 = \sum_{j=1}^{n} s_j^2 \, |\langle x, v_j \rangle|^2 \geqslant s_n^2 \sum_{j=1}^{n} |\langle x, v_j \rangle|^2 = s_n^2 \, \|x\|^2 = s_n^2.$$

It follows that $b_n \geqslant \gamma_E(T) \geqslant s_n$. Since we already know from Theorem G.5 that $b_n \leqslant d_n \leqslant a_n$, we arrive at (G.10). Similarly, if E is the $(n-1)$-dimensional subspace generated by $v_1, \ldots, v_{n-1}$, since $T(v_j) = s_j u_j$, we see that $F := T(E)$ is the space generated by $u_1, \ldots, u_{n-1}$ so that, for $x \in B_H$,

$$(d'_n)^2 \leqslant \sup_{x \in B_H} d(Tx, F)^2 = \sup_{x \in B_H} \|(T - R_{n-1})(x)\|^2 = \|T - R_{n-1}\|^2 = a_n^2$$

giving $d'_n \leqslant a_n$ and $d'_n = d_n$.

Finally, let $\varepsilon > 0$ and $G \subset H$ of codimension $< n$ verifying $\|T_{|G}\| \leqslant c_n + \varepsilon$. If P is the orthogonal projection on $G^{\perp}$, then $R = TP$ has rank $< n$ and verifies

$$\|T - R\| \leqslant \|T_{|G}\| \, \|I - P\| = \|T_{|G}\| \leqslant c_n + \varepsilon.$$

This implies $a_n \leqslant c_n$, and so $a_n = c_n$.

We now claim that $\mathfrak{e}_n(T) \leqslant 4a_n(T)$. Indeed, using a polar decomposition $T = UP$, $P = U^*T$, where P is a positive operator and U a partial isometry, as well as Theorem G.5 and the Ideal Property, we get

$$\begin{aligned} a_n(T) &= a_n(UP) \leqslant \|U\| a_n(P) \leqslant s_n(P) = \lambda_n(P) \\ &\leqslant 4\mathfrak{e}_n(P) = 4\mathfrak{e}_n(U^*T) \leqslant 4\mathfrak{e}_n(T). \end{aligned}$$

This finishes the proof. □

Remarks:

(1) One might wonder what is the point of defining six times the same object a_n. It will (surprisingly) turn out that, according to the involved situation, using one definition rather than another can be illuminating. Of course, Theorem G.7 is fairly soft, and we will make strong use of additional results borrowed from function theory: interpolation sequences, Carleson measures, rational approximation "à la Widom," and so on.

(2) The following obvious fact still deserves to be outlined.

Proposition G.8 *The sequence $(a_n(T))$ of approximation numbers of a compact operator T on a Hilbert space H is non-increasing with limit zero and otherwise* arbitrary.

Proof: Let $(\varepsilon_n)_{n\geqslant 1}$ be a non-increasing sequence with limit zero. Let (u_n) be an orthonormal basis of H and T be the diagonal operator defined by

$$T(u_n) = \varepsilon_n u_n.$$

This operator T obviously does the job. □

The following simple proposition also deserves to be mentioned.

Proposition G.9 *Let $T : H \to H$ be a Hilbert–Schmidt operator. Then*

$$a_n(T) \leqslant \frac{1}{\sqrt{n}} \|T\|_{HS} \textit{ for all } n \geqslant 1.$$

Proof: It suffices to write

$$n[a_n(T)]^2 \leqslant \sum_{j=1}^{n} [a_j(T)]^2 \leqslant \sum_{j=1}^{\infty} [a_j(T)]^2 = \|T\|_{HS}^2. \qquad \square$$

Whereas it is possible to have access to eigenvalues for classes of compact operators, this is much more difficult for singular numbers. In this respect, and in the Hilbertian case, the following beautiful inequality of H. Weyl (for short $\log|\lambda| \prec \log a$; see Appendix F) will reveal quite useful.

Theorem G.10 (Weyl) *Let $T : H \to H$ be a compact linear operator with eigenvalues $\lambda_j(T)$ and singular numbers $a_j(T)$. Then we have for every $n \in \mathbb{N}$*

$$\prod_{j=1}^{n} |\lambda_j(T)| \leqslant \prod_{j=1}^{n} a_j(T).$$

This is useful when we have information on the eigenvalues of T, which will often be the case for weighted composition operators.

Corollary G.11 *Let H be a complex Hilbert space and $T\colon H \to H$ be a compact operator with eigenvalues $\lambda_1, \dots, \lambda_n, \dots$ arranged so that $(|\lambda_n|)_{n\geqslant 1}$ is non-increasing. Then, for any $n \geqslant 1$,*

$$|\lambda_{2n}|^2 \leqslant a_1\, a_n\,. \tag{G.11}$$

Proof: It follows from Weyl's inequalities that $|\lambda_{2n}| \leqslant (a_1 \cdots a_{2n})^{1/2n}$.
Just square to get $|\lambda_{2n}|^2 \leqslant (a_1 \cdots a_{2n})^{1/n} \leqslant (a_1^n\, a_n^n)^{1/n} = a_1\, a_n$. □

We know from Theorems G.5(iii) and G.7 that

$$\sup_{k\leqslant n} k^\alpha \mathfrak{e}_k(T) \leqslant C_\alpha \sup_{k\leqslant n} k^\alpha a_k(T) \quad \text{and} \quad a_n(T) \leqslant 4\mathfrak{e}_n(T).$$

Those inequalities indicate that entropy numbers are always bigger than singular numbers, up to a constant, and that, as far as the scale of powers n^α is implied, they are dominated by approximation numbers in a weak sense. But it turns out that, individually, they can be much bigger than the latter for composition operators, as we shall see.

We will for that purpose rely on the following estimate ([16, p. 17]) in which ℓ^2 denotes the space of square-summable sequences $x = (x_k)_{k\geqslant 1}$ of complex numbers. This estimate is given for the sequence (ε_n) of *covering numbers* (not coined in this book). But $\mathfrak{e}_n = \varepsilon_{2^{n-1}}$, by definition, and we can as well formulate the Carl–Stephani theorem below in terms of entropy numbers as follows:

Theorem G.12 *There exist absolute constants $0 < a < b$ such that, for any diagonal compact operator $\Delta\colon \ell^2 \to \ell^2$ with positive and non-increasing eigenvalues $(\sigma_k)_{k\geqslant 1}$, namely $\Delta\big((x_k)_k\big) = (\sigma_k x_k)_k$, we have, for all $n \geqslant 1$,*

$$a \sup_{k\geqslant 1} \bigg[2^{-n/2k}\bigg(\prod_{j=1}^k \sigma_j\bigg)^{1/k}\bigg] \leqslant \mathfrak{e}_n(\Delta) \leqslant b \sup_{k\geqslant 1} \bigg[2^{-n/2k}\bigg(\prod_{j=1}^k \sigma_j\bigg)^{1/k}\bigg]. \tag{G.12}$$

The following useful corollary of Theorem G.12 indicates that a pointwise control on singular numbers of an operator implies a pointwise control on its entropy numbers as well.

Theorem G.13 *Let $T\colon H_1 \to H_2$ be a compact operator between Hilbert spaces, and let $(a_n)_{n\geqslant 1}$ be its sequence of approximation numbers. Then, for all $n \geqslant 1$,*

$$\alpha \sup_{k\geqslant 1} \bigg[2^{-n/2k}\bigg(\prod_{j=1}^k a_j(T)\bigg)^{1/k}\bigg] \leqslant \mathfrak{e}_n(T) \leqslant \beta \sup_{k\geqslant 1} \bigg[2^{-n/2k}\bigg(\prod_{j=1}^k a_j(T)\bigg)^{1/k}\bigg],$$

where α and β are positive numerical constants.

Proof: Let $Tx = \sum_{n=1}^\infty s_n \langle x, v_n\rangle\, u_n$ be the Schmidt decomposition of T, where $(v_n)_n$ and $(u_n)_n$ are orthonormal sequences of H_1 and H_2, respectively, and $(s_n)_n$ the sequence of singular numbers of T. Let $\Delta\colon \ell_2 \to \ell_2$ be the diagonal operator with diagonal values s_n, $n \geqslant 1$. Then $T = V_1 \Delta U_1$ and $\Delta = V_1^{-1} T U_1^{-1}$, where

$$U_1 x = \big(\langle x, v_n\rangle\big)_n \quad \text{and} \quad V_1\big((t_n)_n\big) = \sum_n t_n u_n$$

are unitary operators.

Hence the result follows from Theorem G.12 with $\sigma_k = a_k$ and from the ideal property. □

Weyl's inequality was generalized much later to Banach spaces as follows:

Theorem G.14 (Johnson–König–Maurey–Retherford) *Let $T : X \to X$ be a bounded and power-compact linear operator from a Banach space X to itself and $0 < r < \infty$. Then, for some numerical constant $c_r > 0$,*

$$\|(\lambda_j(T))\|_{\ell^r} \leqslant c_r \|(a_j(T))\|_{\ell^r}. \tag{G.13}$$

Here $\lambda_j(T)$ and $a_j(T)$ stand, respectively, for the eigenvalues and approximation numbers of T. Soon after this result was established, Pietsch found the following multiplicative improvement (see [136, p. 156] or [136, lemma 13]), which we will use as well.

Theorem G.15 (Pietsch) *Let $T : X \to X$ be a bounded and power-compact linear operator from a Banach space X to itself. Then*

$$|\lambda_{2n}(T)| \leqslant \mathrm{e} \left(\prod_{j=1}^{n} a_j(T) \right)^{1/n}. \tag{G.14}$$

G.4 Kacnel'son Comparison Principle

G.5 Schur Products

In the theory of weighted Hardy spaces, when one switches, for example, from H^2 to $H^2(\beta)$, equivalently from ℓ^2 to $\ell^2(\beta) = \{(a_n)_{n\in\mathbb{N}_0} : \sum |a_n|^2 \beta_n < \infty\}$, equipped with their respective canonical orthonormal bases (e_n) and (e_n^β), when an operator T has the matrix $A = (a_{m,n})$ on (e_n), it has the matrix $B = (b_{m,n})$ on (e_n^β) with

$$b_{m,n} = a_{m,n} \sqrt{\frac{\beta_m}{\beta_n}} \quad \text{on } (e_n^\beta).$$

In other terms, with $p_{m,n} = \sqrt{\beta_m/\beta_n}$, we have $B =: P.\, A$, the so-called Schur (or Hadamard) product (in ℓ^2) of A and the matrix $P = (p_{m,n})$, namely the pointwise product

$$b_{m,n} = a_{m,n} \cdot p_{m,n}.$$

This fact, already pointed out by C. Cowen [33], justifies that we devote a few lines to the notion. As our Hilbert spaces (even if separable) will not always have a canonical orthonormal basis naturally indexed by the integers, we adopt a slightly more general notation.

Definition. Let H be a Hilbert space with a privileged orthonormal basis $(e_\alpha)_{\alpha\in I}$. We will then identify a bounded linear operator T on H ($T \in B(H)$) with its matrix on this basis

$$A = (a_{\alpha,\beta}) \quad \text{with} \quad a_{\alpha,\beta} = \langle T(e_\beta), e_\alpha \rangle.$$

A function $p : I \times I \to \mathbb{C}$ is called a *Schur multiplier* if the pointwise product $P.\,T = (a_{\alpha,\beta}\, p(\alpha,\beta)) \in B(H)$ for all $T \in B(H)$. Then the operator $M_p : B(H) \to B(H)$ defined by $M_p(T) = P.\,T$ is continuous and its norm is called the Schur norm of p. A famous theorem of Grothendieck [139, theorem 5.1] characterizes all those continuous Schur multipliers.

Theorem G.16 (Grothendieck) *Let $p : I \times I \to \mathbb{C}$. The following are equivalent:*

(i) *p is a Schur multiplier of $H = \ell^2(I)$ and $\|M_p\| \leqslant C$.*
(ii) *There exist two families $(u_\alpha), (v_\beta), \alpha, \beta \in I$, of vectors of some Hilbert space K such that for all $(\alpha,\beta) \in I \times I$, it holds*

$$p(\alpha,\beta) = \langle u_\alpha, v_\beta \rangle$$

with $\sup \|u_\alpha\| \times \sup \|v_\beta\| \leqslant C$.

It is not always easy to test the assumptions of this theorem. We will yet see an example in what follows (even if a direct proof will be available).

G.6 Tensor and Exterior Products of Hilbert Spaces

Here we briefly recall the basics on (exterior) tensor products of Hilbert spaces and which provide a very neat proof of Weyl's inequality (G.10). Good, even if sketchy, references for this are [138, chapter 15] and [168, pp. 9–11].

Let H be an infinite-dimensional separable complex Hilbert space, and let us fix a positive integer n. Let AL be the complex Banach space of (bounded) *n-antilinear* forms h on H, equipped with the norm

$$\|h\| = \sup_{\|x_i\| \leqslant 1} |h(x_1,\ldots,x_n)|.$$

If $x_1,\ldots,x_n \in H$, denote by $x_1 \otimes x_2 \cdots \otimes x_n$ the element of AL defined by

$$(z_1,\ldots,z_n) \longmapsto \prod_{j=1}^{n} \langle x_j, z_j \rangle$$

and denote by $H^{\otimes n}$ the closure in AL of the linear subspace $H_0^{\otimes n}$ of AL generated by the $x_1 \otimes x_2 \cdots \otimes x_n$. If

$$s = \sum_p a_p (x_1^p \otimes x_2^p \cdots \otimes x_n^p) \text{ and } t = \sum_q b_q (y_1^q \otimes y_2^q \cdots \otimes y_n^q) \text{ belong to } H_0^{\otimes n},$$

we define the scalar product of s and t by

$$\langle s,t \rangle = \sum_{p,q} a_p \overline{b_q} \prod_{j=1}^{n} \langle x_j^p, y_j^q \rangle. \tag{G.15}$$

The following simple proposition justifies and precises this definition, basic for the study of this tensor product $H^{\otimes n}$.

Proposition G.17 (i) *The value of* $\langle s,t\rangle$ *does not depend on the decomposition of* s *and* t.
(ii) *This scalar product is positive-definite:* $s \neq 0 \Longrightarrow \langle s,s\rangle > 0$.

Proof: (i) If s,t are as above, it suffices to observe that

$$\sum_{p,q} a_p\overline{b_q}\prod_{j=1}^{n}\langle x_j^p, y_j^q\rangle = \sum_q \overline{b_q}\, s(y_1^q, y_2^q, \ldots, y_n^q) = \sum_p a_p\, \overline{t(x_1^p, x_2^p, \ldots, x_n^p)}.$$

So that $\langle s,t\rangle$ does not depend on the special decomposition of s or t.

(ii) Thanks to (i), we can choose for s a decomposition

$$s = \sum_{1\leqslant p\leqslant N} a_p(x_1^p \otimes x_2^p \cdots \otimes x_n^p)$$

in which the family (x_1^p) is orthonormal and all the families (x_j^p), $1 \leqslant j \leqslant n$, $1 \leqslant p \leqslant N$, are formed by unit vectors. This then gives for $s \neq 0$

$$\langle s,s\rangle = \sum_{1\leqslant p,q\leqslant N} a_p\overline{a_q}\prod_{j=1}^{n}\langle x_j^p, x_j^q\rangle = \sum_{1\leqslant p\leqslant N} |a_p|^2 > 0. \qquad \square$$

This allows us to equip $H_0^{\otimes n}$ with the prehilbertian norm associated to this scalar product and to consider its completion, the Hilbert space $H^{\otimes n}$. We note in passing that, for every $s \in H^{\otimes n}$,

$$\|s\|_{H^{\otimes n}} = \|s\|_{AL}. \tag{G.16}$$

Indeed, we can assume that $s = \sum_{1\leqslant p\leqslant N} a_p(x_1^p \otimes x_2^p \cdots \otimes x_n^p) \in H_0^{\otimes n}$, with a decomposition as above. If $(y_1, \ldots, y_n) \in H^n$, the Cauchy–Schwarz inequality gives

$$\begin{aligned}
|s(y_1,\ldots,y_n)|^2 &\leqslant \left(\sum_p |a_p|^2\right) \times \left(\sum_p |\langle x_1^p, y_1\rangle|^2 \cdots |\langle x_n^p, y_n\rangle|^2\right)\\
&\leqslant \|s\|_{H^{\otimes n}}^2 \|y_2\|^2 \cdots \|y_n\|^2 \times \sum_p \left|\langle x_1^p, y_1\rangle\right|^2\\
&\leqslant \|s\|_{H^{\otimes n}}^2 \|y_1\|^2\|y_2\|^2 \cdots \|y_n\|^2,
\end{aligned}$$

where we used Bessel's inequality. This shows that $\|s\|_{AL} \leqslant \|s\|_{H^{\otimes n}}$ and the reverse inequality is obvious. We will then simply write $\|s\|$.

The Hilbert space $H^{\otimes n}$ is called the nth tensor power (product) of H. The following proposition is important.

Proposition G.18 (i) *Let* $(e_i)_{i\geqslant 0}$ *be an orthonormal basis of* H. *Then the set*

$$(e_{i_1} \otimes e_{i_2} \otimes \cdots e_{i_n})_{i_1,\ldots,i_n\geqslant 0}$$

is an orthonormal basis of $H^{\otimes n}$.
(ii) *If* $A \in B(H)$, *there exists a unique continuous operator* $A^{\otimes n} \in B(H^{\otimes n})$, *called the* n*th tensor power of* A, *with* $\|A^{\otimes n}\| = \|A\|^n$, *such that*

$$A^{\otimes n}(x_1 \otimes \cdots \otimes x_n) = A(x_1) \otimes \cdots \otimes A(x_n) \quad \textit{for all } x_1, \ldots, x_n \in H.$$

(iii) *Moreover, for all* $A, B \in B(H)$,

$$(AB)^{\otimes n} = A^{\otimes n} B^{\otimes n}.$$

Proof: The first statement is easy. For the second one, set formally $B = A^{\otimes n}$ and let $s = \sum_{1 \leqslant p \leqslant N} a_p (x_1^p \otimes x_2^p \cdots \otimes x_n^p) \in H_0^{\otimes n}$. We have, for all $y_1, \ldots, y_n \in H$,

$$B(s)(y_1, \ldots, y_n) = s(A^* y_1, \ldots, A^* y_n).$$

This shows that

$$|B(s)(y_1, \ldots, y_n)| \leqslant \|s\| \times \prod_{j=1}^{n} \|A^* y_j\| \leqslant \|s\| \times \|A\|^n \prod_{j=1}^{n} \|y_j\|.$$

By (G.16), this implies that $\|B(s)\| \leqslant \|A\|^n \|s\|$. Hence, B extends to a continous linear map $A^{\otimes n}$ on the whole of $H^{\otimes n}$, with norm less than $\|A\|^n$, and the reverse inequality is clear since

$$\|A^{\otimes n}(x \otimes x \otimes \cdots \otimes x)\| = \|Ax \otimes Ax \cdots \otimes Ax\| = \|Ax\|^n.$$

For the third statement, just write

$$\begin{aligned} AB^{\otimes n}(x_1 \otimes \cdots \otimes x_n) &= ABx_1 \otimes \cdots \otimes ABx_n = A^{\otimes n}(Bx_1 \otimes \cdots \otimes Bx_n) \\ &= A^{\otimes n} B^{\otimes n}(x_1 \otimes \cdots \otimes x_n). \end{aligned}$$

□

We now define the wedge (or skew, exterior) nth power of H as follows: first we define the anti-symmetrization of the tensor $x_1 \otimes x_2 \cdots \otimes x_n$ by the formula

$$x_1 \wedge x_2 \cdots \wedge x_n = \frac{1}{\sqrt{n!}} \sum_{\sigma \in \mathfrak{S}_n} \varepsilon_\sigma x_{\sigma(1)} \otimes x_{\sigma(2)} \cdots \otimes x_{\sigma(n)} \in H^{\otimes n}, \tag{G.17}$$

where σ runs over the group $\mathfrak{S}_n$ of all permutations of $\{1, 2, \ldots, n\}$ and ε_σ is the signature of the permutation σ.

The anti-symmetric character of the operation can be seen on the obvious formula

$$x_{\sigma(1)} \wedge x_{\sigma(2)} \cdots \wedge x_{\sigma(n)} = \varepsilon_\sigma (x_1 \wedge x_2 \cdots \wedge x_n) \tag{G.18}$$

(where $\sigma \in \mathfrak{S}_n$). Let now $H^{\wedge n}$ be the closed subspace of $H^{\otimes n}$ generated by $x_1 \wedge x_2 \cdots \wedge x_n$. This new Hilbertian subspace is called the nth skew (or exterior) power of H. Now, if $A \in B(H)$, and $x_1, \ldots, x_n \in H$, it is readily verified that

$$A^{\otimes n}(x_1 \wedge x_2 \cdots \wedge x_n) = A(x_1) \wedge A(x_2) \cdots \wedge A(x_n). \tag{G.19}$$

So that $A^{\otimes n}$ leaves the subspace $H^{\wedge n}$ invariant. The restriction of $A^{\otimes n}$ to $H^{\wedge n}$ is called the nth skew power of A and is denoted by $A^{\wedge n}$. We first note a simple fact (connection with determinants) on this skew product.

Proposition G.19 *Let* $x_1, \ldots, x_n, y_1, \ldots, y_n \in H$. *Then we have*

$$\langle x_1 \wedge \cdots \wedge x_n, y_1 \wedge \cdots \wedge y_n \rangle = (\det \langle x_i, y_j \rangle)_{1 \leqslant i, j \leqslant n}.$$

Proof: The LHS, say v, of the equality writes

$$v = \frac{1}{n!} \sum_{\sigma, \tau \in \mathfrak{S}_n} \varepsilon_\sigma \varepsilon_\tau \times \prod_{j=1}^{n} \langle x_{\sigma(j)}, y_{\tau(j)} \rangle,$$

and a simple change of variable gives

$$v = \sum_{\chi \in \mathfrak{S}_n} \varepsilon_\chi \prod_{k=1}^{n} \langle x_k, y_{\chi(k)} \rangle = (\det \langle x_i, y_j \rangle)_{1 \leqslant i, j \leqslant n}$$

by definition. □

We now have the analogue of Proposition G.18, in which I denotes the set of all increasing sequences $\alpha = (i_1 < i_2 < \cdots < i_n)$ with length n of non-negative integers.

Proposition G.20 (i) *Let $(e_i)_{i \geqslant 0}$ be an orthonormal basis of H. Then the set*

$$(e_\alpha)_{\alpha \in I} := (e_{i_1} \wedge e_{i_2} \cdots \wedge e_{i_n})_{(i_1, \ldots, i_n) \in I}$$

is an orthonormal basis of $H^{\wedge n}$.

(ii) *If $A \in B(H)$, there exists a unique continuous operator $A^{\wedge n} \in B(H^{\wedge n})$, called the nth exterior power of A, with $\|A^{\wedge n}\| = \|A\|^n$, such that*

$$A^{\wedge n}(x_1 \wedge \cdots \wedge x_n) = A(x_1) \wedge \cdots \wedge A(x_n) \quad \textit{for all } x_1, \ldots, x_n \in H.$$

(iii) *Moreover, for all $A, B \in B(H)$*

$$(AB)^{\wedge n} = A^{\wedge n} B^{\wedge n}.$$

(iv) *For all $A \in B(H)$, we have $|A^{\wedge n}| = |A|^{\wedge n}$.*

Proof: (i) If $\alpha = (i_1 < \cdots < i_n)$ and $\beta = (j_1 < \cdots < j_n)$ are distinct, let M be the $n \times n$ matrix $m_{p,q} = \langle e_{i_p}, e_{j_q} \rangle$ and r be the first index such that $i_r \neq j_r$. If $i_r < j_r$, the rth row of M is zero. If $i_r > j_r$, the rth column of M is zero. So that $\det M = 0$ and $e_\alpha \perp e_\beta$ by Proposition G.19.

The generating character of (e_α) follows from the fact that $H^{\wedge n}$ is by definition generated by vectors of the form $x_1 \wedge \cdots \wedge x_n$ with $x_j \in H$, and each of those vectors can be approximated by finite combinations of the vectors e_α, since (e_n) is a Hilbertian basis of H.

(ii) and (iii) obviously follow from Proposition G.18, and (iv) from the uniqueness of the positive square root of a positive operator. □

Let now J be the set of increasing sequences $\alpha = (j_1 < j_2 < \cdots < j_n)$ of n positive integers. An essential fact for our study of approximation numbers is the following:

Theorem G.21 *Let $A \in K(H)$ be a compact operator with singular values $s_1 \geqslant s_2 \cdots \geqslant s_n \geqslant \cdots$ and Schmidt decomposition*

$$A(x) = \sum_{j=1}^{\infty} s_j \langle x, v_j \rangle u_j.$$

(i) *Then the Schmidt decomposition of* $A^{\wedge n}$ *is*

$$A^{\wedge n}(X) = \sum_{\alpha \in J} s_\alpha \langle X, v_\alpha \rangle u_\alpha,$$

where $v_\alpha = v_{j_1} \wedge \cdots \wedge v_{j_n}$, $u_\alpha = u_{j_1} \wedge \cdots \wedge u_{j_n}$, *and* $s_\alpha = s_{j_1} \cdots s_{j_n}$.

(ii) *In particular, for all positive integers n, we have*

$$\|A^{\wedge n}\| = \prod_{j=1}^{n} s_j.$$

Proof: Observe first that all families of vectors in this proof are summable, due to the orthonormality of the sequences (u_n) and (v_n), so there are no problems of convergence. Now start from the Schmidt decomposition of A,

$$A(x) = \sum_{j=1}^{\infty} s_j \langle x, v_j \rangle u_j,$$

and let $X = x_1 \wedge \cdots \wedge x_n \in H^{\wedge n}$. By (G.19), we have

$$A^{\wedge n}(X) = Ax_1 \wedge \cdots \wedge Ax_n = \sum_{j_1 \geqslant 1, \ldots, j_n \geqslant 1} s_{j_1} \cdots s_{j_n} \langle x_1, v_{j_1} \rangle \cdots \langle x_n, v_{j_n} \rangle u_{j_1} \wedge \cdots \wedge u_{j_n}.$$

Rearranging the indices in increasing order and using (G.18) gives, noting that $u_{j_1} \wedge \cdots \wedge u_{j_n} = 0$ as soon as two j_k coincide, and with $\alpha = (j_1 < \cdots < j_n)$,

$$A^{\wedge n}(X) = \sum_{\alpha \in J} s_\alpha \left(\sum_{\sigma \in \mathfrak{S}_n} \varepsilon_\sigma \prod_{i=1}^{n} \langle x_i, v_{j_{\sigma(i)}} \rangle \right) u_\alpha.$$

By Proposition G.19, the quantity between brackets (a determinant) equals $\langle X, v_\alpha \rangle$. This proves (i) for X of the form $x_1 \wedge \cdots \wedge x_n$. Since those X generate $H^{\wedge n}$, (i) is fully proved.

Now, for (ii), the norm of $A^{\wedge n}$ is the largest value of the numbers s_α, $\alpha \in J$, namely $s_1 \cdots s_n$. □

Remark. Let $(\lambda_n)_{n \geqslant 1}$ be the sequence of eigenvalues of A in decreasing order of modulus. We could prove exactly in the same way that $\lambda_\alpha := \lambda_{j_1} \cdots \lambda_{j_n}$, where $\alpha = (j_1 < \cdots < j_n) \in J$ are the eigenvalues of $A^{\wedge n}$. In particular

$$\prod_{j=1}^{n} |\lambda_j| \leqslant \|A^{\wedge n}\| = \prod_{j=1}^{n} s_j.$$

This is nothing but Weyl's inequality.

G.7 A Comparison Principle

In the Hilbertian case, here is a useful comparison principle initially due to V. Kacnel'son ([76]). This comparison will have several applications to (weighted) composition operators, approximation numbers, and also multipliers issues. For a good comprehension of those applications, it is important to have in mind the following relation between the weight β_n, and d_n to come.

$$d_n = \frac{1}{\sqrt{\beta_n}} \cdot \tag{G.20}$$

In particular, (d_n) is increasing whenever (β_n) is decreasing.

We indeed enunciate directly a generalization of Kacnel'son's result, which initially concerned only the case $I = \mathbb{N}_0$ and (d_n) increasing. Here, in order to exploit that result for approximation numbers or weighted Hardy spaces, we need a general set of indices, denoted I, and non-monotonic sequences (d_n). The original result is due to V. È. Kacnel'son and the following variant is due to Lefèvre–Li–Queffélec–Rodríguez–Piazza.

Theorem G.22 *Let H be a complex Hilbert space and $(e_\alpha)_{\alpha\in I}$ a fixed orthonormal basis of H. Let $T\colon H \to H$ be a bounded linear operator with matrix $A = (a_{\alpha,\,\beta})$ on the basis (e_α), that is, $a_{\alpha,\beta} = \langle T(e_\beta), e_\alpha\rangle$. Let also $(d_\alpha)_{\alpha\in I}$ be a family of positive real numbers. We assume that the matrix of T on this basis is lower-triangular with respect to the order induced by (d_α), that is,*

$$d_\alpha < d_\beta \Longrightarrow \langle Te_\beta, e_\alpha\rangle = 0. \tag{G.21}$$

Let D be the (generally unbounded) diagonal operator formally defined by $D(e_\alpha) = d_\alpha e_\alpha$, $\alpha \in I$. Then the operator $D^{-1}TD\colon H \to H$ is bounded and moreover

$$\|D^{-1}TD\| \leqslant \|T\|\,. \tag{G.22}$$

In other terms, the Hermitian form $A = \left(a_{\alpha,\,\beta}\frac{d_\beta}{d_\alpha}\right)$ is bounded on H, with a norm less than $\|A\| = \|T\|$.

Proof 1: We begin with a real variable proof, more in the spirit of Hermitian forms and Hadamard–Schur product. We take the liberty of confusing operators and their matrix on the fixed orthonormal basis (e_α). The idea is the following: given (G.21), the action on A of the (mostly unbounded) multiplier d_β/d_α coincides with the action on A of a notably *bounded* Schur multiplier

$$p(\alpha,\beta) = \widehat{f}(\rho_\alpha - \rho_\beta), \quad \|M_p\| \leqslant \|f\|_1,$$

where $f \in L^1(\mathbb{R})$, $\|f\|_1 = 1$, $\rho_\alpha \in \mathbb{R}$, and $\widehat{f}$ denotes the Fourier transform of f. And this is all.

For details, since $d_\alpha > 0$, write $d_\alpha = e^{-\rho_\alpha}$ with $\rho_\alpha \in \mathbb{R}$ and observe that, if $x = (x_\beta)$ and $y = (y_\alpha)$ are two finitely supported vectors of ℓ^2_I, we have

$$\sum_{\alpha,\,\beta} a_{\alpha,\,\beta}\frac{d_\beta}{d_\alpha}x_\beta\overline{y_\alpha} = \sum_{\alpha,\,\beta} a_{\alpha,\,\beta}e^{-|\rho_\alpha-\rho_\beta|}x_\beta\overline{y_\alpha}. \tag{G.23}$$

Indeed, both sums bear by hypothesis on pairs (α,β) such that $d_\alpha \geqslant d_\beta$, and we have term-by-term equality for those pairs.

Now, we observe that

$$\mathrm{e}^{-|x|} = \widehat{f}(x) = \int_{\mathbb{R}} f(t)\mathrm{e}^{-ixt}dt, \quad \text{with } f(t) = \frac{1}{\pi(1+t^2)},$$

and the matrix $p := p(\alpha, \beta) = \widehat{f}(\rho_\alpha - \rho_\beta)$ is a Schur multiplier of all bounded matrices on ℓ^2_I with Schur norm $\|M_p\| \leqslant \|f\|_1 = 1$.

More generally, if we have $p(\alpha, \beta) = \widehat{\mu}(\rho_\alpha - \rho_\beta)$ where μ is a finite complex Borel measure on $\mathbb{R}$, we get $\|M_p\| \leqslant \|\mu\|$. Indeed, write $\mu = h|\mu|$ where h is a unimodular function and $|\mu|$ the total variation of μ. Then, in the Hilbert space $L^2(|\mu|)$, we can write

$$\widehat{\mu}(\rho_\alpha - \rho_\beta) = \langle u_\alpha, v_\beta \rangle \text{ with } u_\alpha = h e^{i\rho_\alpha t} \text{ and } v_\beta = e^{i\rho_\beta t}.$$

The previously mentioned theorem of Grothendieck ([139]) would then ensure that M_h is a Schur multiplier with

$$\|M_h\| \leqslant \sup \|u_\alpha\| \times \sup \|v_\beta\| = (\|\mu\|^{1/2})^2 = \|\mu\|.$$

We can give here a much more direct and elementary proof. We write for that

$$\begin{aligned}\sum_{\alpha,\beta} a_{\alpha,\beta} x_\beta \overline{y_\alpha}\, \widehat{\mu}(\rho_\alpha - \rho_\beta) &= \int_{\mathbb{R}} \sum a_{\alpha,\beta} x_\beta e^{i\rho_\beta t} \overline{y_\alpha e^{i\rho_\alpha t}} h(t)\, d|\mu|(t) \\ &= \int_{\mathbb{R}} \langle Ax(t), y(t)\rangle\, d|\mu|(t)\end{aligned}$$

with $x(t) = h(t)(x_\beta e^{i\rho_\beta t})$, $y(t) = (y_\alpha e^{i\rho_\alpha t})$, $\|x(t)\| = \|x\|$, and $\|y(t)\| = \|y\|$.

We get $|\langle Ax(t), y(t)\rangle| \leqslant \|A\|\, \|x\|\, \|y\|$ for all t, and the triangle inequality in $L^1(|\mu|)$ now gives

$$|\langle D^{-1}ADx, y\rangle| \leqslant \|A\|\, \|\mu\|\, \|x\|\, \|y\|$$

and the result. □

Proof 2: The second, complex-variable-style proof goes as follows.

Let us first assume that H is finite-dimensional, so that all forthcoming operators are well defined and bounded. Observe that

$$D^{-1}AD = \left(a_{m,n}\frac{d_n}{d_m}\right).$$

Let $\mathbb{C}_0 = \{z \in \mathbb{C} : \operatorname{Re} z > 0\}$ be the right half-plane; denote by D^z the operator defined by $D^z(e_n) = d_n^z e_n$ where $z \in \mathbb{C}_0$ and consider the function $u : \overline{\mathbb{C}_0} \to \mathbb{R}^+$ defined by

$$u(z) = \|D^{-z}AD^z\| = \left\| \left(a_{m,n}\left(\frac{d_n}{d_m}\right)^z\right) \right\|.$$

This function is subharmonic on $\mathbb{C}_0$ and continuous on $\overline{\mathbb{C}_0}$. Moreover, given (G.21), it is bounded; hence, the maximum principle applies and gives

$$\|D^{-1}TD\| = u(1) \leqslant \sup_{y\in\mathbb{R}} u(iy) = \|T\|,$$

the latter equality holding because $D^{\pm iy}$ is unitary. And this is all.

Here are now some details in the general case, written for $I = \mathbb{N}_0$ for simplicity of notations. Let $H_N = \operatorname{span}\{e_j\,; j \leqslant N\}$ where N is a positive integer. We set

$$t_{i,j} = \langle Te_j, e_i\rangle$$

and denote by $T_N : H_N \to H_N$ the operator $T_N = P_N T J_N$ where J_N is the canonical injection of H_N into H and P_N the orthogonal projection from H onto H_N. We consider, for $z \in \overline{\mathbb{C}_0}$, the operator

$$T_N(z) = D^{-z} T_N D^z : H_N \to H_N,$$

where once again $D^z(e_n) = d_n^z e_n$. If $\left(t_{i,j}^{(N)}(z)\right)_{i,j}$ is the matrix of $T_N(z)$ on the orthonormal basis $\{e_j, 0 \leqslant j \leqslant N\}$ of H_N, we clearly have

$$t_{i,j}^{(N)}(z) = t_{i,j}\,(d_j/d_i)^z, \quad \text{for } 0 \leqslant i, j \leqslant N.$$

In particular, since $t_{i,j}^{(N)}(z) = 0$ for $d_i < d_j$ by hypothesis

$$|t_{i,j}^{(N)}(z)| \leqslant \sup_{k,l} |t_{k,l}| =: M \qquad \text{for all } z \in \overline{\mathbb{C}_0}.$$

We next observe that

$$\|T_N(z)\| \leqslant (N+1) \times M \qquad \text{for all } z \in \overline{\mathbb{C}_0}.$$

Indeed, using the Hilbert–Schmidt norm,

$$\|T_N(z)\|^2 \leqslant \|T_N(z)\|_{HS}^2 = \sum_{0 \leqslant i, j \leqslant N} |t_{i,j}^{(N)}(z)|^2 \leqslant (N+1)^2 M^2.$$

Let us consider the function $u : \overline{\mathbb{C}_0} \to \overline{\mathbb{C}_0}$ defined by

$$u(z) = \left\| T_N(z) \right\|. \tag{G.24}$$

This function u is continuous on $\overline{\mathbb{C}_0}$. It is moreover bounded by $(N+1) \times M$ as we just saw. It is subharmonic on $\mathbb{C}_0$ as the norm of the operator-valued analytic function $z \mapsto D^{-z} A D^z$. Alternatively, $u = \sup_h u_h$ where h runs over the unit ball of the dual of $B(H)$ and $u_h(z) = |h(D^{-z} A D^z)|$ is subharmonic as the modulus of a holomorphic function. We can therefore apply to u the maximum principle (see Chapter 1) to get as before

$$\|D^{-1} T_N D\| \leqslant \|T_N\| \leqslant \|T\|.$$

Letting N tend to infinity, we get $\|D^{-1} T D\| \leqslant \|T\|$. Indeed, for each finitely supported $x = \sum_{k=0}^{l} x_k e_k$ with norm 1, we have $D^{-1} T_N D(x) = D^{-1} T D(x)$ as soon as $N \geqslant l$, so that $\|D^{-1} T D(x)\| \leqslant \|D^{-1} T_N D\| \leqslant \|T\|$. □

Here is an important corollary of Theorem G.22.

Theorem G.23 *Take $I = \mathbb{N}_0$ in Theorem G.22. Assume again that*

$$d_i < d_j \Longrightarrow \langle T(e_j), e_i \rangle = 0.$$

Then:

(i) *If T is bounded (resp. compact), $D^{-1} T D$ is also bounded (resp. compact), with $\|D^{-1} T D\| \leqslant \|T\|$.*

(ii) *Moreover, if T is compact, we have for every $n \geqslant 1$*

$$\prod_{j=1}^{n} s_j(D^{-1}TD) \leqslant \prod_{j=1}^{n} s_j(T). \tag{G.25}$$

In other words, the sequence $\big(s_j(D^{-1}TD)\big)_j$ is log-subordinate to $\big(s_j(T)\big)_j$. As a consequence, in terms of Schatten norms,

$$\mathfrak{s}_p(D^{-1}TD) \leqslant \mathfrak{s}_p(T) \quad \textit{for all} \quad p > 0.$$

(iii) *The entropy numbers satisfy:* $\mathfrak{e}_n(D^{-1}TD) \leqslant C\mathfrak{e}_n(T)$.

In [20], part (i) of this theorem was extended to the framework of Banach spaces with 1-unconditional basis and used for the study of composition operators and in [21], [105] to compare the Schatten-class norms of weighted Hilbert spaces of analytic functions.

Proof: (i) The inequality has already been proved. If T is compact, $T_N \to T$ in norm as $N \to \infty$. Moreover, $T - T_N$ is also "triangular" so that the inequality gives

$$\|D^{-1}TD - D^{-1}T_N D\| = \|D^{-1}(T - T_N)D\| \leqslant \|T - T_N\|.$$

This shows that $D^{-1}TD$ is compact, as the limit in norm of the finite rank operators $D^{-1}T_N D$.

For (ii), we use anti-symmetric tensor products to show that Theorem G.22 implies the result. Things go as follows: let once again I denote the set of all increasing n-tuples $\alpha = (i_1 < i_2 < \cdots < i_n)$ of non-negative integers. Let $(e_\alpha)_{\alpha \in I}$ be the orthonormal basis of $H^{\wedge n}$ defined as before by

$$e_\alpha = e_{i_1} \wedge e_{i_2} \wedge \ldots \wedge e_{i_n}, \quad \alpha \in I. \tag{G.26}$$

Now we use the result of Theorem G.21:

$$\prod_{j=1}^{n} s_j(D^{-1}TD) = \|(D^{-1}TD)^{\wedge n}\|,$$

where $\wedge^n$ denotes the nth skew power. But $(UV)^{\wedge n} = U^{\wedge n}V^{\wedge n}$ by Proposition G.20; therefore

$$(D^{-1}TD)^{\wedge n} = (D^{-1})^{\wedge n}T^{\wedge n}D^{\wedge n} =: \Delta^{-1}T\,\Delta,$$

where Δ is the diagonal operator on the basis (e_α) with diagonal elements

$$\delta_\alpha = d_{i_1}d_{i_2}\cdots d_{i_n} \qquad \text{if } \alpha = (i_1 < i_2 < \cdots < i_n).$$

We then claim that $T^{\wedge n}$ is *lower triangular* in the sense of Theorem G.22. Let for that $\alpha = (i_1 < i_2 < \cdots < i_n)$ and $\beta = (j_1 < j_2 < \cdots < j_n)$ in I. We have

$$\delta_\alpha < \delta_\beta \Rightarrow \langle T^{\wedge n}(e_\beta), e_\alpha\rangle = 0. \tag{G.27}$$

Indeed, write (using Proposition G.19)

$$\begin{aligned} \langle T^{\wedge n}(e_\beta), e_\alpha\rangle &= \langle Te_{j_1} \wedge Te_{j_2} \cdots \wedge Te_{j_n},\, e_{i_1} \wedge e_{i_2} \cdots \wedge e_{i_n}\rangle \\ &= \det\big(\langle Te_{j_p}, e_{i_q}\rangle\big)_{1 \leqslant p,q \leqslant n} =: X. \end{aligned}$$

If $X \neq 0$, by definition of a determinant, there exists a permutation σ of $\{1,2,\dots,n\}$ such that

$$\prod_{1\leqslant k\leqslant n} (\langle Te_{j_k}, e_{i_{\sigma(k)}}\rangle) \neq 0, \text{ implying } d_{i_{\sigma(k)}} \geqslant d_{j_k} \text{ for each } k.$$

But then

$$\delta_\alpha = \prod_{1\leqslant k\leqslant n} d_{i_{\sigma(k)}} \geqslant \prod_{1\leqslant k\leqslant n} d_{j_k} = \delta_\beta,$$

contradicting our hypothesis $\delta_\alpha < \delta_\beta$. This proves (G.27), and then Theorem G.22 (and Theorem G.21 again) gives

$$\prod_{1\leqslant j\leqslant n} s_j(D^{-1}TD) = \|\Delta^{-1}T^{\wedge n}\Delta\| \leqslant \|T^{\wedge n}\| = \prod_{1\leqslant j\leqslant n} s_j(T).$$

(iii) Abbreviate $a_k(T)$ and $a_k(D^{-1}TD)$, respectively, to a_k and a'_k. We use the double inequality

$$\alpha \sup_{k\geqslant 1}\left[2^{-n/2k}\left(\prod_{j=1}^{k} a_j\right)^{1/k}\right] \leqslant \mathfrak{e}_n(T) \leqslant \beta \sup_{k\geqslant 1}\left[2^{-n/2k}\left(\prod_{j=1}^{k} a_j\right)^{1/k}\right]$$

of Theorem G.13 and (ii) to get

$$\begin{aligned}\mathfrak{e}_n(D^{-1}TD) &\leqslant \beta \sup_{k\geqslant 1}\left[2^{-n/2k}\left(\prod_{j=1}^{k} a'_j\right)^{1/k}\right] \leqslant \beta \sup_{k\geqslant 1}\left[2^{-n/2k}\left(\prod_{j=1}^{k} a_j\right)^{1/k}\right]\\ &\leqslant \frac{\beta}{\alpha}\,\mathfrak{e}_n(T).\end{aligned}$$

□

We cannot expect having the pointwise domination $s_j(D^{-1}TD) \leqslant s_j(T)$ as shown by the following example in dimension 2, where $a > 0$ and $0 < d_1 < d_2$:

$$T_a = \begin{pmatrix} 1 & 0 \\ a & 1 \end{pmatrix}$$

as well as

$$D = \begin{pmatrix} d_1 & 0 \\ 0 & d_2 \end{pmatrix}$$

and then

$$D^{-1}T_aD = \begin{pmatrix} 1 & 0 \\ a\frac{d_1}{d_2} & 1 \end{pmatrix} = T_{a\frac{d_1}{d_2}}.$$

We observe that

$$\|T_a\|^2 = \frac{a^2+2+a\sqrt{a^2+4}}{2}$$

(strictly) increases with a. Set $T'_a = D^{-1}TD_a$, with

$$s_1(T'_a) = \|T'_a\| < \|T_a\| = s_1(T_a).$$

But then, since

$$s_1(T_a')s_2(T_a') = \det T_a' = 1 = \det T_a = s_1(T_a)s_2(T_a),$$

we have $s_2(T_a') > s_2(T_a)$.

G.8 Rotfel'd's Inequality

We recall here and prove the Rotfel'd theorem used in Chapter 9: this is an old but not-so-well-known and quite interesting inequality, which trivially holds for ℓ^p-spaces, but requires more work in the non-commutative setting of operators. This inequality as stated below is due to Rotfel'd. It is indicated (for $h(x) = x^p, 0 < p < 1$) in Peller's book [134, p. 706] where a proof due to Alexandrov is presented, and another one due to C. A. McCarthy is mentioned. We thank N. Nikolski for quite useful information in this respect.

We present below the proof of Rotfel'd for general concave functions, since it is not given in Peller's book (moreover strong use of it is made, as we saw, in the El-Fallah and El Ibbaoui approach to Schattenness).

Theorem G.24 *Let $h : \mathbb{R}^+ \to \mathbb{R}^+$ be a concave increasing function with $h(0) = h(0^+) = 0$. Let $A, B \in \mathscr{K}(H)$. Then, for any $n \in \mathbb{N}$,*

$$\sum_{k=1}^{n} h\big(s_k(A+B)\big) \leqslant \sum_{k=1}^{n} h\big(s_k(A)\big) + \sum_{k=1}^{n} h\big(s_k(B)\big). \tag{G.28}$$

Equivalently

$$Tr\, h(|A+B|) \leqslant Tr\, h(|A|) + Tr\, h(|B|). \tag{G.29}$$

In particular, for $0 < p < 1$, the functional $A \mapsto \mathfrak{s}_p(A)^p$ is sub-additive on $\mathscr{S}^p(H)$:

$$\mathfrak{s}_p(A+B)^p \leqslant \mathfrak{s}_p(A)^p + \mathfrak{s}_p(B)^p. \tag{G.30}$$

Proof: First, the equivalence of forms (G.28) and (G.29) is seen as follows: if the operator $T \in \mathscr{K}(H)$ is positive with Schmidt form

$$T = \sum_{k=1}^{\infty} s_k(T)\, e_k \otimes e_k,$$

the functional calculus gives

$$h(T) = \sum_{k=1}^{\infty} h\big(s_k(T)\big)\, e_k \otimes e_k$$

so that, when U is unitary,

$$\operatorname{Tr} h(T) = \sum_{k=1}^{\infty} h\big(s_k(T)\big) \quad \text{and} \quad \operatorname{Tr} h(U^*TU) = \operatorname{Tr} h(T), \tag{G.31}$$

the first equality clearly giving the equivalence.

Let $C = A + B$. Note the following: let

$$u_k = s_k(A),\ v_k = s_k(B),\ w_k = s_k(C),\ u = (u_k),\ v = (v_k),\ w = (w_k).$$

By Weyl's theory, we know that $w \prec u + v$, so that Theorem G.24 amounts to prove that $h(w) \prec h(u) + h(v)$. The stronger inequality $h(w) \prec h(u+v)$ fails in general. We now perform successive reductions to prove either form.

(i) We can assume that $\dim H < \infty$.

Indeed, in the general case, consider $C = W|C|$ the polar (Schmidt) decomposition of C and write

$$|C| = W^*C = \sum_{k=1}^{\infty} s_k(C)\,(\chi_k \otimes \chi_k).$$

Let E_n be the span of $\chi_1, \dots, \chi_n$; let $P_n : H \to E_n$ the orthogonal projection, and

$$A_n = P_n W^* A P_n,\ B_n = P_n W^* B P_n,\ C_n = P_n W^* C P_n = P_n |C| P_n,$$

which satisfy, for $1 \leqslant k \leqslant n$,

$$C_n = A_n + B_n,\ s_k(C_n) = s_k(C),\ s_k(A_n) \leqslant s_k(A),\ s_k(B_n) \leqslant s_k(B),$$

where we used the ideal property of singular numbers.

Now, assuming that the finite dimensional case is proved, we apply this case to A_n, B_n, C_n and the space E_n; this gives, since h increases,

$$\begin{aligned}\sum_{k=1}^{n} h\big(s_k(C)\big) = \sum_{k=1}^{n} h(s_k(C_n)) &\leqslant \sum_{k=1}^{n} h\big(s_k(A_n)\big) + \sum_{k=1}^{n} h\big(s_k(B_n)\big)\\ &\leqslant \sum_{k=1}^{n} h\big(s_k(A)\big) + \sum_{k=1}^{n} h\big(s_k(B)\big).\end{aligned}$$

This is exactly (G.28).

(ii) We can assume that A, B are *Hermitian* matrices of order n. We need for that a lemma. Of course, given an operator $T : \mathbb{C}^n \to \mathbb{C}^n$, we identify T and its matrix on the canonical basis of $\mathbb{C}^n$.

Lemma G.25 *Let A be a square matrix of order n and $\widehat{A}$ the hermitian matrix of order $2n$ defined as*

$$\widehat{A} = \begin{pmatrix} 0 & A \\ A^* & 0 \end{pmatrix}.$$

Then

$$s_{2k-1}(\widehat{A}) = s_{2k}(\widehat{A}) = s_k(A), \quad 1 \leqslant k \leqslant n.$$

Proof of Lemma G.25: If the Schmidt decomposition of A in $\mathbb{C}^n$ is

$$Ax = \sum_{k=1}^{n} s_k \langle x, v_k \rangle u_k \quad \text{with} \quad A^*x = \sum_{k=1}^{n} s_k \langle x, u_k \rangle v_k,$$

we have for $z = (x, y) \in \mathbb{C}^n \oplus \mathbb{C}^n$

$$\widehat{A}z = (Ay, A^*x) = \Big(\sum_{k=1}^{n} s_k \langle y, v_k \rangle u_k, \sum_{k=1}^{n} s_k \langle x, u_k \rangle v_k \Big),$$

or else

$$\widehat{A}z = \sum_{k=1}^{n} s_k \langle z, w_k' \rangle w_k + \sum_{k=1}^{n} s_k \langle z, w_k \rangle w_k' \quad \text{with} \quad w_k = (u_k, 0),\ w_k' = (0, v_k).$$

Once we have noticed that $(w_1, \dots, w_n, w_1', \dots, w_n')$ is an orthonormal system in the direct sum $\mathbb{C}^n \oplus \mathbb{C}^n$, this Schmidt decomposition of $\widehat{A}$ makes the lemma clear. □

Now, assuming that the hermitian case is fixed, the inequality

$$\sum_{k=1}^{2n} h\big(s_k(\widehat{C})\big) \leqslant \sum_{k=1}^{2n} h\big(s_k(\widehat{A})\big) + \sum_{k=1}^{2n} h\big(s_k(\widehat{B})\big)$$

reads

$$2\Big[\sum_{k=1}^{n} h\big(s_k(C)\big) \Big] \leqslant 2\Big[\sum_{k=1}^{n} h\big(s_k(A)\big) + \sum_{k=1}^{n} h\big(s_k(B)\big) \Big],$$

which proves Step 2.

(iii) We can assume that A, B are Hermitian and *positive.* We need for that a new lemma, a substitute for the wrong inequality $|A + B| \leqslant |A| + |B|$.

Lemma G.26 *If A and B are Hermitian matrices of order n, and B* has rank 1, *there are unitary matrices U and V of order n such that*

$$|A + B| \leqslant U^{-1}|A|U + V^{-1}|B|V.$$

Proof: First we can take B (of rank one) positive, changing if necessary B into $-B$ and then A to $-A$. Next write $A, B, C = A + B$ under Schmidt form

$$A = \sum_{j=1}^{n} \alpha_j (\varphi_j \otimes \varphi_j), \quad B = \beta_1 (\psi_1 \otimes \psi_1), \quad C = \sum_{j=1}^{n} \gamma_j (\chi_j \otimes \chi_j)$$

with $\beta_1 > 0$ and

$$\alpha_1 \geqslant \alpha_2 \geqslant \cdots \geqslant \alpha_l \geqslant 0 > \alpha_{l+1} \geqslant \cdots \geqslant \alpha_n,$$
$$\gamma_1 \geqslant \gamma_2 \geqslant \cdots \geqslant \gamma_k \geqslant 0 > \gamma_{k+1} \geqslant \cdots \geqslant \gamma_n.$$

We first observe that

$$\gamma_1 \leqslant \alpha_1 + \beta_1. \tag{G.32}$$

Indeed, setting $\mathbb{C}^n = H$ and recalling that S_X denotes the unit sphere of the normed space X, we have

$$\gamma_1 = \sup_{x \in S_H} \langle Cx, x \rangle \leqslant \sup_{x \in S_H} \langle Ax, x \rangle + \sup_{x \in S_H} \langle Bx, x \rangle = \alpha_1 + \beta_1.$$

We next show that

$$\gamma_{p+1} \leqslant \alpha_p \leqslant \gamma_p \qquad \text{for } 1 \leqslant p < n. \tag{G.33}$$

Use for that the minimax principle: the RHS inequality (also valid for $p = n$) is obvious since $A \leqslant C$. Next there is a $(p-1)$-dimensional subspace E such that

$$\alpha_p = \sup_{x \in S_{E^\perp}} \langle Ax, x\rangle = \delta_E(A).$$

If $F = \operatorname{span}(E, \psi_1)$, we have $F \subset L$ for some L with $\dim L = p < n$, so that $L^\perp \subset F^\perp$ and

$$\gamma_{p+1} \leqslant \delta_L(C) \leqslant \delta_F(C) = \alpha_p$$

since $\langle Bx, x\rangle = \beta_1 |\langle x, \psi_1\rangle|^2 = 0$ for all $x \in F^\perp$, proving (G.33).

Note in passing two obvious consequences of (G.33), the second one crucial for what follows:

$$l \leqslant k \leqslant l + 1. \tag{G.34}$$

Indeed, $\gamma_l \geqslant \alpha_l \geqslant 0$ and $\gamma_{l+2} \leqslant \alpha_{l+1} < 0$.

Let us proceed. We clearly have $|C| = C_+ + C_-$ with

$$C_+ = \sum_{j=1}^{k} \gamma_j (\chi_j \otimes \chi_j), \quad C_- = -\sum_{j=k+1}^{n} \gamma_j (\chi_j \otimes \chi_j)$$

and similarly $|A| = A_+ + A_-$ with

$$A_+ = \sum_{j=1}^{l} \alpha_j (\varphi_j \otimes \varphi_j), \quad A_- = -\sum_{j=l+1}^{n} \alpha_j (\varphi_j \otimes \varphi_j).$$

Let G be the span of $\chi_1, \ldots, \chi_k$ and Q the orthogonal projection on G.

Let us introduce the operators

$$\tilde{A} = QAQ \text{ and } \tilde{B} = QBQ, \text{ so } \tilde{A} + \tilde{B} = QCQ = C_+.$$

We write $\tilde{A}$ and $\tilde{B}$ under Schmidt form (with $\tilde{\varphi}_j, \tilde{\psi}_j \in G$):

$$\tilde{A} = \sum_{j=1}^{k} \tilde{\alpha}_j (\tilde{\varphi}_j \otimes \tilde{\varphi}_j), \quad \tilde{B} = \sum_{j=1}^{k} \tilde{\beta}_j (\tilde{\psi}_j \otimes \tilde{\psi}_j). \tag{G.35}$$

The minimax principle implies, for $1 \leqslant j \leqslant k$,

$$\tilde{\alpha}_j \leqslant \alpha_j \quad \text{and} \quad \tilde{\beta}_1 \leqslant \beta_1, \quad \tilde{\beta}_2 = \cdots = \tilde{\beta}_k = 0. \tag{G.36}$$

Indeed, let X be a $(j-1)$-dimensional subspace such that

$$\alpha_j = \sup_{x \in S_{X^\perp}} \langle Ax, x\rangle = \delta_X(A).$$

Let $Y = Q(X)$, which has dimension $< j$ and let $u \in Y^\perp$ with norm 1. Then $v = Qu \in X^\perp$ with norm at most 1, so that

$$\langle \tilde{A}u, u\rangle = \langle Av, v\rangle \leqslant \alpha_j$$

and that $\tilde{\alpha}_j \leqslant \delta_Y(\tilde{A}) \leqslant \alpha_j$ as claimed.

The second item of (G.36) is obvious.

We next consider unitary operators U, V such that $V^{-1}(\psi_1) = \tilde{\psi}_1$ and

$$U^{-1}(\varphi_j) = \begin{cases} \tilde{\varphi}_j & \text{if } 1 \leqslant j \leqslant k \\ \chi_j & \text{if } k+1 \leqslant j \leqslant n. \end{cases}$$

Using (G.36), we see that

$$\tilde{B} \leqslant V^{-1}BV \quad \text{and} \quad \tilde{A} \leqslant U^{-1}A_+U. \tag{G.37}$$

The first item is clear: for every x, we have

$$\begin{aligned} \langle \tilde{B}x, x\rangle &= \sum_{j=1}^{k} \tilde{\beta}_j |\langle \tilde{\psi}_j, x\rangle|^2 = \tilde{\beta}_1 |\langle \tilde{\psi}_1, x\rangle|^2 \\ &\leqslant \beta_1 |\langle \tilde{\psi}_1, x\rangle|^2 = \langle V^{-1}BVx, x\rangle. \end{aligned} \tag{G.38}$$

For the second one, if $x = \sum_{j=1}^{n} x_j \varphi_j \in \mathbb{C}^n$, it holds $\tilde{A}U^{-1}x = \sum_{j=1}^{k} \tilde{\alpha_j}\, x_j\, \tilde{\varphi_j}$ and

$$\langle U\tilde{A}U^{-1}x, x\rangle = \langle \tilde{A}U^{-1}x, U^{-1}x\rangle = \sum_{j=1}^{k} \tilde{\alpha_j} |x_j|^2.$$

On the other hand,

$$\langle A_+x, x\rangle = \sum_{k=1}^{l} \alpha_j |x_j|^2. \tag{G.39}$$

We now separate two cases and will only detail the second one, since both cases are similar:

(a) $k = l + 1$.
(b) $k = l$. Then, (G.36) and (G.39) give $U\tilde{A}U^{-1} \leqslant A_+$ or else $\tilde{A} \leqslant U^{-1}A_+U$, which is the first item.

This proves (G.37).

Now, since $C_+ = \tilde{A} + \tilde{B}$, the above inequalities yield (G.33) and (G.38):

$$C_+ \leqslant U^{-1}A_+U + V^{-1}BV \quad \text{as well as} \quad C_- \leqslant U^{-1}A_-U.$$

And by addition, we obtain (since $B \geqslant 0$)

$$|C| = C_+ + C_- \leqslant U^{-1}(A_+ + A_-)U + V^{-1}BV = U^{-1}|A|U + V^{-1}|B|V.$$

□

Now, assuming that the positive hermitian case is fixed, and since $U^{-1}|A|U$ and $V^{-1}|B|V$ are positive, we have

$$\begin{aligned} \operatorname{Tr} h(|C|) &\leqslant \operatorname{Tr} h(U^{-1}|A|U + V^{-1}|B|V) \leqslant \operatorname{Tr} h(U^{-1}|A|U) + \operatorname{Tr} h(V^{-1}|B|V) \\ &= \operatorname{Tr} h(|A|) + \operatorname{Tr} h(|B|), \end{aligned}$$

which proves step 3.

(iv) Finally, the result holds if A, B are positive. We can moreover assume that B has rank 1. Indeed, suppose that case settled. In the general case, write $B = \sum_{j=1}^{p} B_j$ with each operator B_j positive, self-adjoint, of rank 1, and moreover $s_1(B_j) = s_j(B), 1 \leqslant j \leqslant p$ (this is possible by the Schmidt decomposition of B). By induction on p, we get

$$\begin{aligned}\operatorname{Tr} h(A+B) &\leqslant \operatorname{Tr} h(A) + \sum_{j=1}^{p} \operatorname{Tr} h(B_j) = \operatorname{Tr} h(A) + \sum_{j=1}^{p} h\big(s_1(B_j)\big)\\ &= \operatorname{Tr} h(A) + \sum_{j=1}^{p} h\big(s_j(B)\big) = \operatorname{Tr} h(A) + \operatorname{Tr} h(B).\end{aligned}$$

Let us finally deal with the crucial case when B has rank 1.

Keeping in mind relation (G.33), we have

$$Tr\, h(C) = Tr\, h(A) + \sum_{p=1}^{n} \big[h(\gamma_p) - h(\alpha_p)\big].$$

Since h is concave, we know that, for every $\delta \in [0, \alpha_p]$,

$$h(\gamma_p) - h(\alpha_p) \leqslant h(\gamma_p - \delta) - h(\alpha_p - \delta).$$

Applying this with $\delta = \sum_{j=p}^{n} \alpha_j - \sum_{j=p+1}^{n} \gamma_j \in [0, \alpha_p]$ by (G.33), we get

$$h(\gamma_p) - h(\alpha_p) \leqslant h(r_p) - h(r_{p+1}),$$

where $r_p = \sum_{j=p}^{n} (\gamma_j - \alpha_j)$, with $r_{n+1} = 0$, and $r_1 = Tr\,(A+B) - Tr\, A = \beta_1$.

Now summing over p, we obtain

$$Tr\, h(C) \leqslant Tr\, h(A) + h(r_1) = Tr\, h(A) + h(\beta_1) = Tr\, h(A) + Tr\, h(B),$$

and from (iii), this implies the statement of Theorem G.24. □

An alternative proof (and extension to the convex case) of Theorem G.24, obvious from this theorem in the concave case, can be found in Thompson [174], and is used in El-Fallah and El Ibbaoui's approach to Schattenness (see Section 9.6.3), as follows.

Corollary G.27 *Let H be a Hilbert space and $(T_n)_{n\geqslant 1}$ a sequence of positive compact operators on H such that $T = \sum_{n\geqslant 1} T_n$ exists (with operator-norm convergence). Let also $h : \mathbb{R}^+ \to \mathbb{R}^+$ be an increasing function satisfying $h(0) = h(0^+) = 0$. Then:*

(i) *If h is convex, $Tr\,[h(T)] \geqslant \sum_{n\geqslant 1} Tr\,[h(T_n)]$.*

(ii) *If h is concave, $Tr\,[h(T)] \leqslant \sum_{n\geqslant 1} Tr\,[h(T_n)]$.*

A word of comment here: a convex (resp. concave) function $h : \mathbb{R}^+ \to \mathbb{R}^+$ vanishing at 0 is notably super-additive (resp. sub-additive). The Rotfel'd–Thompson result can be qualified as *non-commutative concavity or convexity* and is one more manifestation of the fact that the trace *erases* non-commutativity.

Appendix H Orlicz Spaces

Chapter 11 (obviously) makes a crucial use of the concept of Orlicz spaces. We recall here just a tiny part of this wide area of analysis, focusing on the concepts and tools required to understand the construction of Hardy–Orlicz spaces and Bergman–Orlicz spaces, and the action of composition operators on these spaces.

Most of the results in the present appendix are not proved, but we mention references. In particular, monographs [88, 148] are classics.

H.1 Framework

We consider here Orlicz spaces defined on a (separable) probability space $(\Omega, \mathbb{P})$. Concretely, we shall mainly work on $(\mathbb{T}, m)$ (resp. $(\mathbb{D}, \sigma)$) in the framework of Hardy–Orlicz spaces (resp. Bergman–Orlicz spaces). We shall also sometimes adopt the viewpoint of *Carleson measures* in an Orlicz framework, and in this case the space can be $(\overline{\mathbb{D}}, m_\varphi)$ or $(\mathbb{D}, \sigma_\varphi)$ when studying composition operators.

By an Orlicz function, we understand that $\Psi \colon [0, +\infty] \to [0, +\infty]$ is a nondecreasing convex function such that $\Psi(0) = 0$ and $\lim_{+\infty} \Psi = +\infty$. In particular Ψ is *continuous on* $[0, +\infty]$.

To avoid pathologies, and because, in the spirit of the book, we are interested in a better understanding of the difference between the behavior of composition operators on H^p spaces and the one on space H^∞, we always assume that we work with an Orlicz function Ψ having the following additional properties: *strictly convex (hence increasing) and such that*

$$\frac{\Psi(x)}{x} \underset{x \to +\infty}{\longrightarrow} +\infty.$$

In particular $\Psi^{-1} \colon [0, +\infty] \to [0, +\infty]$ is well defined.

This is essentially to exclude the case of $\Psi(x) = ax$, but once again we are mainly interested here in "small" spaces "between" H^p $(p > 1)$ and H^∞.

Definition H.1 The *Orlicz space* $L^\Psi(\Omega) = L^\Psi(\Omega, \mathbb{P})$ is the space of all (equivalence classes of) measurable functions $f : \Omega \to \mathbb{C}$ for which there is a constant $C > 0$ such that

$$\int_\Omega \Psi\left(\frac{|f(t)|}{C}\right) d\mathbb{P}(t) \tag{H.1}$$

is finite and then $\|f\|_{L^\Psi(\Omega)} = \|f\|_\Psi$ (the *Luxemburg norm*) is the infimum of all possible constants C such that this integral is less than 1.

Point out that a measurable function f belongs to the unit ball of $L^\Psi(\Omega)$ if and only if $\int_\Omega \Psi(|f|)\, d\mathbb{P} \leqslant 1$. In the same spirit, assume that $\int_\Omega \Psi(|f|)\, d\mathbb{P} \leqslant C$ for some $C \geqslant 0$, then use that Ψ is convex to get $\|f\|_\Psi \leqslant \max(1, C)$.

Another easy remark is that $L^\infty(\Omega) \subset L^\Psi(\Omega) \subset L^1(\Omega)$.

Moreover, we have for every $f \in L^\infty(\Omega)$ with $f \neq 0$

$$\|f\|_\Psi \leqslant \frac{\|f\|_\infty}{\Psi^{-1}(\|f\|_\infty / \|f\|_1)} \leqslant \frac{1}{\Psi^{-1}(1)} \|f\|_\infty. \tag{H.2}$$

Indeed take $f \in L^\Psi(\Omega)$ with $f \neq 0$. For some $A > 0$ and every $x \geqslant A$, we have $\Psi(x) \geqslant x$. Therefore,

$$\begin{aligned}\int_\Omega |f|\, d\mathbb{P} &\leqslant A\|f\|_\Psi + \|f\|_\Psi \int_{|f| \geqslant A\|f\|_\Psi} \frac{|f|}{\|f\|_\Psi}\, d\mathbb{P} \\ &\leqslant A\|f\|_\Psi + \|f\|_\Psi \int_{|f| \geqslant A\|f\|_\Psi} \Psi\left(\frac{|f|}{\|f\|_\Psi}\right) d\mathbb{P} \leqslant (A+1)\|f\|_\Psi,\end{aligned}$$

and we get $f \in L^1(\Omega)$.

For the comparison between $L^\infty(\Omega)$ and $L^\Psi(\Omega)$, take $f \in L^\infty(\Omega)$ with $f \neq 0$. Thanks to the convexity of Ψ, we have for every $C > 0$

$$\int_\Omega \Psi\left(\frac{|f|}{C}\right) d\mathbb{P} \leqslant \int_\Omega \frac{|f|}{\|f\|_\infty} \Psi\left(\frac{\|f\|_\infty}{C}\right) d\mathbb{P} = \frac{\|f\|_1}{\|f\|_\infty} \Psi\left(\frac{\|f\|_\infty}{C}\right),$$

which is 1 if we choose $C = \|f\|_\infty \left(\Psi^{-1}(\|f\|_\infty / \|f\|_1)\right)^{-1}$. The last inequality is clear since Ψ (as well as Ψ^{-1}) is increasing; (H.2) is proved.

For general Orlicz functions, there is no reason that each integral (H.1) may be finite (think of the case of the gaussian function Ψ_2 for instance; see below). Therefore, the following space is also natural.

Definition H.2 The *Morse–Transue space* $M^\Psi(\Omega)$ is the subspace of $L^\Psi(\Omega)$ generated by $L^\infty(\Omega)$, or, equivalently, the subspace of all functions f for which the integral (H.1) is finite for *every* $C > 0$. It is a separable space.

We shall often write L^Ψ instead of $L^\Psi(\Omega)$ and M^Ψ instead of $M^\Psi(\Omega)$.

In full generality, $L^\Psi \neq M^\Psi$ and actually (see Definition H.5):

$$L^\Psi = M^\Psi \qquad \Longleftrightarrow \qquad \Psi \in \Delta_2. \tag{H.3}$$

In the Morse–Transue space, we still have a dominated convergence theorem:

Theorem H.3 (Dominated convergence theorem in M^Ψ) *Let $\{f_n\}_n$ be a sequence of measurable functions converging pointwise a.e. to f. Assume there exists $g \in M^\Psi$ such that $|f_n| \leqslant g$ a.e. for every n.*

Then $f_n \xrightarrow{\|\cdot\|_\Psi} f$.

We leave the proof as an (easy) exercise. It is also easy to see that this statement is false in L^Ψ as soon as $L^\Psi \neq M^\Psi$.

To every Orlicz function is associated the *complementary Orlicz function* $\Phi = \Psi^*: [0, +\infty] \to [0, +\infty]$ defined by

$$\Phi(x) = \sup_{y \geqslant 0} \big(xy - \Psi(y)\big).$$

With this definition, we get obviously *Young's inequality* $xy \leqslant \Phi(x) + \Psi(y)$ for every $x, y \geqslant 0$.

We also get an Orlicz version of Hölder's inequality: for $f \in L^\Psi$ and $g \in L^\Phi$,

$$\left| \int_\Omega fg \, d\mathbb{P} \right| \leqslant 2\|f\|_\Psi . \|g\|_\Phi . \tag{H.4}$$

Indeed, integrating Young's inequality over Ω, we obtain

$$\left| \int_\Omega \frac{f}{\|f\|_\Psi} \cdot \frac{g}{\|g\|_\Phi} \, d\mathbb{P} \right| \leqslant \int_\Omega \Psi\left(\frac{|f|}{\|f\|_\Psi}\right) + \Phi\left(\frac{|g|}{\|g\|_\Phi}\right) d\mathbb{P} \leqslant 2.$$

Let us mention also the following inequalities

$$\forall x \geqslant 0, \qquad x \leqslant \Phi^{-1}(x) . \Psi^{-1}(x) \leqslant 2x \,. \tag{H.5}$$

The RHS inequality is obvious with Young's inequality. For the LHS inequality, first observe that for every $u, y > 0$, we have $\dfrac{y}{u}\Psi(u) \leqslant \Psi(u) + \Psi(y)$ (just using the convexity when $y > u$ and this is trivial when $u \geqslant y$). Then, from the definition of Φ we get $\Phi\big(\Psi(u)/u\big) \leqslant \Psi(u)$. Finally appyling this inequality with $u = \Psi^{-1}(x)$ we get (H.5) for every $x > 0$.

Important examples:

- The most classical one is the function $\Psi(x) = x^p$, and in this case we recover the space $L^p(\Omega) = L^p(\Omega, \mathbb{P})$.
  Our viewpoint imposes $p > 1$ (we recover L^1 as well with the former definition but, in the sequel, some of the properties of L^p with $p > 1$ are not valid anymore for L^1). We have $L^p(\Omega) = M^p(\Omega)$.
- The fast-growing functions $\Psi_r(x) = \mathrm{e}^{x^r} - 1$ with $r \geqslant 1$ also play an important role in this book. This includes the gaussian function Ψ_2. We can also consider functions defined by $x \mapsto \Psi_r(x) = \mathrm{e}^{x^r} - 1$, for $r \in (0, 1)$ and x large enough ($x > \sqrt{(1 - r)/r}$ say). We have $L^{\Psi_r}(\Omega) \neq M^{\Psi_r}(\Omega)$.

H.2 Growth Conditions

The behavior of composition operators on Orlicz-type spaces may be (very) different according to the underlying Orlicz function, and in a natural way various growth conditions for the Orlicz function Ψ come into play. These conditions are usually denoted as Δ-conditions. As we already mentioned, there is a particular interest in Orlicz functions that have a somewhat fast growth. Some of these conditions are defined through a moderate growth condition on the complementary function Φ of Ψ, and the condition Δ for the Orlicz function is translated as a ∇-condition for the complementary function. Therefore, we have to distinguish between *moderate growth* conditions, which we shall define for the complementary Orlicz function, and *fast growth* conditions.

Moderate Growth Conditions

Definition H.4 The Orlicz function Ψ satisfies the Δ_1*-condition* ($\Psi \in \Delta_1$) if, for some constant $c > 0$, one has (submultiplicativity)

$$\Psi(xy) \leqslant c\,\Psi(x)\Psi(y) \qquad \text{for } x, y \text{ large enough.}$$

This is equivalent to saying that $\Psi(axy) \leqslant \Psi(x)\Psi(y)$ for some constant $a > 0$ and x, y large enough.

To know more on this condition, see [148, condition Δ', p. 28].

Definition H.5 Ψ satisfies the Δ_2*-condition* ($\Psi \in \Delta_2$) if, for some constant $K > 1$, one has

$$\Psi(2x) \leqslant K\,\Psi(x) \qquad \text{for } x \text{ large enough.}$$

We see with (H.3) the key role played by this condition in the theory of Orlicz spaces. Another result closely retated to (H.3) is that L^Ψ is separable if and only if $\Psi \in \Delta_2$.

It is easy to check that $\Psi \in \Delta_1 \implies \Psi \in \Delta_2$.

Fast Growth Conditions

Definition H.6 The Orlicz function Ψ satisfies the Δ^0*-condition* ($\Psi \in \Delta^0$) if, for some $\beta > 1$,

$$\lim_{x \to +\infty} \frac{\Psi(\beta x)}{\Psi(x)} = +\infty.$$

An example is $\Psi(x) = \exp\big[\log(x + \mathrm{e}) \log\log(x + \mathrm{e})\big] - 1$.

Definition H.7 The Orlicz function Ψ satisfies the Δ^1*-condition* ($\Psi \in \Delta^1$) if there is some $\beta > 1$ such that

$$\Psi(\beta x) \geqslant x\Psi(x) \qquad \text{for } x \text{ large enough.}$$

When $\Psi \in \Delta^1$, we have $\Psi(x) \geqslant \exp\big(\alpha\,(\log x)^2\big)$ for some $\alpha > 0$ and x large enough [148, proposition 2, p. 37]. An example is $\Psi(x) = \mathrm{e}^{(\log(x+1))^2} - 1$. To know more about this condition, see [148, Δ_3-condition, §2.5].

Definition H.8 The Orlicz function $\Psi\colon [0, +\infty) \to [0, +\infty)$ is said to satisfy the Δ^2*-condition* ($\Psi \in \Delta^2$) if there exists some $\alpha > 1$ such that

$$\Psi(\alpha x) \geqslant \Psi(x)^2 \qquad \text{for } x \text{ large enough.}$$

A function $\Psi \in \Delta^2$ must satisfy $\Psi(x) \geqslant \exp(x^\alpha)$ for some $\alpha > 0$ and x large enough [148, proposition 6, p. 40]). Typical examples are the functions Ψ_r.

We mention here a lemma useful to compare M^Ψ and the growth in $L^1(\Omega)$.

Lemma H.9 *Assume that $\Psi \in \Delta^2$ and let f be a (mesurable) positive function, bounded by* 1, *such that $\Psi^{-1}(1/(1 - f)) \in M^\Psi(\Omega, \mathbb{P})$.*

Then, for every $q \geqslant 1$, we have $\displaystyle\int_\Omega f^n \, d\mathbb{P} = o\left(n^{-q}\right)$ *when $n \to +\infty$.*

Proof: It suffices to treat the case of a dyadic integer q. Let $s \in \mathbb{N}$; we know that $\Psi(x)^{2^s} \leqslant \Psi(\alpha^s x)$ for x large enough (say $x \geqslant B > 1$).

On the other hand, for every integer $N \geqslant 1$ and every $t \in [0, 1)$, we have, just looking at the Taylor expansion,

$$\frac{1}{1-t} \geqslant (N+1)t^N .$$

We get, when $1 > f \geqslant 1 - \dfrac{1}{B}$,

$$\Psi\left(\alpha^s \Psi^{-1}\left(\frac{1}{1-f}\right)\right) \geqslant (N+1)^{2^s} f^{N2^s} .$$

By hypothesis, for some $C > 0$, we have $C \geqslant \displaystyle\int_{\Omega_B} \Psi\left(\alpha^s \Psi^{-1}\left(\frac{1}{1-f}\right)\right) d\mathbb{P}$, where $\Omega_B = \left\{1 > f \geqslant 1 - \dfrac{1}{B}\right\}$. We get, for n integer, as soon as $n \geqslant 2^s$,

$$C \geqslant \int_{\Omega_B} (N+1)^{2s} f^{N2^s} \, d\mathbb{P} \geqslant \frac{n^{2^s}}{2^{s2^s}} \int_{\Omega_B} f^n \, d\mathbb{P},$$

choosing N such that $N + 1 > 2^{-s}n \geqslant N$. The upper estimate of the integral over $\Omega \setminus \Omega_B$ is obvious: it is then geometrically decreasing. □

Conditions of Regularity

Definition H.10 The Orlicz function Ψ satisfies the ∇_2-*condition* ($\Psi \in \nabla_2$) if its complementary function Φ satisfies the Δ_2-condition.

This is equivalent to saying that for some constant $\beta > 1$, one has

$$\Psi(\beta x) \geqslant 2\beta\, \Psi(x) \qquad \text{for } x \text{ large enough.}$$

This condition implies that $\dfrac{\Psi(x)}{x} \underset{x\to+\infty}{\longrightarrow} +\infty$.

Definition H.11 The Orlicz function Ψ satisfies the ∇_1-*condition* ($\Psi \in \nabla_1$) if its complementary function Φ satisfies the Δ_1-condition.

This is equivalent to saying that, for some constant $b > 0$,

$$\Psi(x)\Psi(y) \leqslant \Psi(bxy) \qquad \text{for } x, y \text{ large enough.}$$

All power functions $\Psi(x) = x^p$ satisfy ∇_1, but $\Psi(x) = x^p \log(x+1)$ does not.

Another regularity condition is particularly adapted when studying Carleson embeddings in the Orlicz framework.

Definition H.12 We say that Ψ satisfies the ∇_0-*condition* if, for some $x_0 > 0$, $C \geqslant 1$, and every $x_0 \leqslant x \leqslant y$, one has

$$\frac{\Psi(2x)}{\Psi(x)} \leqslant \frac{\Psi(2Cy)}{\Psi(y)}. \tag{H.6}$$

This is satisfied as soon as Ψ has some homegeneity property, for instance $\Psi(x) = x^p$, but it is also satisfied by functions Ψ_r.

Let us mention the following result: the proof is quite straightforward and we omit the details (see [98]).

Proposition H.13 *The function $\kappa(x) = \log \Psi(e^x)$ is convex on some interval $(x_0, +\infty)$ if and only if $\Psi \in \nabla_0$ with constant $C = 1$.*

We summarize here the link between all these conditions [148, p. 43]:

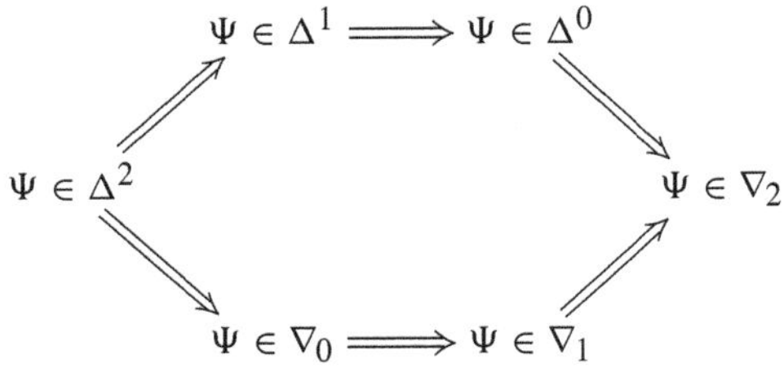

Duality

In the most general framework, it is natural to define the duality bracket by $\langle f, g\rangle_{L^\Psi, M^\Phi} = \int_\Omega f(z)g(z)\, d\mathbb{P}(z)$. Nevertheless, in the framework of the book, Ω is stable under conjugation (think of $\mathbb{T}$, $\mathbb{D}$, or $\overline{\mathbb{D}}$) and the duality seems easier to describe with convolution (especially in classical Hardy spaces); that is why in this book our choice of the duality bracket will be in the case of $\mathbb{T}$ with $\mathbb{P} = m$, and in the case $\mathbb{D}$ or $\overline{\mathbb{D}}$ with $\mathbb{P} = \sigma$:

$$\langle f, g\rangle_{L^\Psi, M^\Phi} = \int_\Omega f(z)g(\bar{z})\, d\mathbb{P}(z).$$

Proposition H.14 *The space L^Ψ is the dual of M^Φ.*

As an immediate corollary, we get that, when Φ satisfies the Δ_2-condition (i.e., $\Psi \in \nabla_2$), the space L^Ψ is the dual space of L^Φ. Hence we have

$$\Psi \in \nabla_2 \quad \Longrightarrow \quad L^\Psi = \big(M^\Psi\big)^{**}.$$

Maximal Function

Our framework here is $(\Omega, \mathbb{P}) = (\mathbb{T}, m)$. Recall that, for $f \in L^1(\mathbb{T})$, the maximal non-tangential function $N_\alpha(f)$ is defined by

$$N_\alpha(f)(\theta) = \sup_{\substack{r\xi \in S_\theta \\ \xi \in \mathbb{T}, r \in (0,1)}} |(f * P_r)(\xi)|,$$

where S_θ is the Stolz domain at $\theta \in \mathbb{T}$ with opening α (recall Sections 1.3.1 and 1.3.2). Here, for $z = r\xi$, the formula $f(z) = f * P_r(\xi)$ defines a harmonic function in $\mathbb{D}$.

Theorem H.15 (Marcinkiewicz) *Assume that $\Psi \in \nabla_2$. Every linear or sublinear operator on $L^0(\Omega)$ that is of weak type $(1,1)$ and (strong) type (∞,∞) is bounded on $L^\Psi(\Omega)$.*

In particular, for every $f \in L^\Psi(\mathbb{T})$, the function $N_\alpha(f)$ belongs to $L^\Psi(\mathbb{T})$.

See [98, proposition 3.5] for a proof.

Appendix I Exercises

Exercise I.1 Let $a \in \mathbb{D}$ and $r > 0$ with $|a| + r < 1$, N an integer greater than 1, $\omega = e^{2i\pi/N}$, and $Z = (z_1, \ldots, z_N)$ with $z_n = a + r\omega^n$, where $1 \leqslant n \leqslant N$.

(a) Show that the Carleson constant $\delta = \delta(Z)$ of Z satisfies

$$\delta \geqslant N \frac{r^N}{2^N}.$$

Hint: Use the identity $\dfrac{z^N - 1}{z - \omega^j} = \prod_{\substack{1 \leqslant k \leqslant N, \\ k \neq j}} (z - \omega^k)$ to show that

$$\left| \prod_{\substack{1 \leqslant k \leqslant N, \\ k \neq j}} (\omega^j - \omega^k) \right| = N.$$

(b) Show that the interpolation constant I_Z of Z satisfies $I_Z \leqslant C^N$ with $C = C(a, r) > 0$.

Exercise I.2 Prove that if a symbol is inner and univalent then it is an automophism.

Exercise I.3 Let φ be a symbol. Show that φ is inner if and only if we have equality almost everywhere in the Littlewood inequality (see Theorem 1.37).

Exercise I.4 Let (z_n) be a sequence of distinct points of $\mathbb{D}$ and (K_{z_n}) be the associated sequence of reproducing kernels in H^2.

Show that (K_{z_n}) is total in H^2 if and only if we have $\sum(1 - |z_n|) = +\infty$ (Müntz theorem for H^2).

Exercise I.5 We consider the Bloch space, a priori equipped with its usual norm $|f(0)| + \sup_{z\in\mathbb{D}}(1 - |z|)|f'(z)|$.

Using the Littlewood–Paley formula, show that an equivalent norm on the Bloch space is $|f(0)| + \sup\limits_{a\in\mathbb{D}} \|f \circ \varphi_a - f(a)\|_{\mathscr{B}^2}$.

Exercise I.6 Show that every symbol induces a bounded composition operator on the Bloch space.

Find a necessary and sufficient condition on a symbol, characterizing the compactness of the associated composition operator (acting on the Bloch space).

Hint: Show that when $\lim_{\rho \to \to 1^-} \sup_{|\varphi(z)| > \rho} \frac{1-|z|}{1-|\varphi(z)|} |\varphi'(z)| = 0$, the operator C_φ is compact on the Bloch space. For the converse, use the family of functions $z \in \mathbb{D} \mapsto \log(1/(1-\overline{w}z))$ where $w \in \mathbb{D}$. See [127] for details.

Exercise I.7 This exercise proposes an alternative presentation of Proposition 1.86.

Let $(\varepsilon_n)_{n \geqslant 0}$ be a Rademacher sequence, and $f_\omega(z) = \sum_{n \geqslant 0} \varepsilon_n(\omega) z^{2^n}$. We recall that f_ω belongs to the Bloch space (see the proof of Proposition 1.86) for each ω, and note δ a "varying" positive constant.

(a) Fix $\theta \in \mathbb{T}$ and let

$$X_r = |f_\omega(re^{i\theta})|, \quad 1/2 \leqslant r < 1.$$

Show that

$$\mathbb{E}(X_r) \geqslant \delta \sqrt{\log 1/(1-r)}.$$

Infer that

$$\mathbb{P}(X_r \geqslant \delta \sqrt{\log 1/(1-r)}) \geqslant \delta.$$

Hint: Use the Khintchine and Paley–Zygmund inequalities.

(b) Show that, almost surely,

$$\limsup_{r \to 1^-} \left(|f_\omega(re^{i\theta})| / (\sqrt{\log 1/(1-r)}) \right) \geqslant \delta.$$

Hint: Use (a) and the zero-one law.

(c) Show that, almost surely, f_ω is unbounded on almost every ray $[0, e^{i\theta})$.
Hint: Use Fubini's theorem.

Exercise I.8 Recall that $\mathcal{F}$ denotes the set of finite Blaschke products. Let f, g be two symbols.

(a) Show that $fg \in \mathcal{F} \Longrightarrow f$ and $g \in \mathcal{F}$.
(b) Assume that $f(0) = 0$. Show that $f \circ g \in \mathcal{F} \Longrightarrow g \in \mathcal{F}$.
Hint: Use the angular derivative characterization of finite Blaschke products.

Exercise I.9 (a) Let $B \in \mathcal{F}$ be a Blaschke product of length n. Show that, for any $w \in \mathbb{D}$, the equation $B(z) = w$ has exactly n roots, counted with multiplicity.

(b) Prove that there exists Blaschke products B such that $B(\mathbb{D})$ is strictly contained in $\mathbb{D}$.

Exercise I.10 Hilbert–Schmidt classes can be separated on weighted Bergman spaces. Namely, let $\alpha' > \alpha > -1$.

Show that if C_φ is Hilbert–Schmidt on $\mathscr{B}^2_\alpha$ then C_φ is Hilbert–Schmidt on $\mathscr{B}^2_{\alpha'}$.

Show that there exist symbols φ such that C_φ is Hilbert–Schmidt on $\mathscr{B}^2_{\alpha'}$ but not on $\mathscr{B}^2_\alpha$.

Hint: See Theorem 3.18. Use Shapiro–Taylor maps and estimates in Section 9.4.

Exercise I.11 Let φ be a symbol and $\alpha > -1$.

Show that C_φ is nuclear on H^1 if and only if $\int_{\mathbb{T}} \frac{1}{1-|\varphi|}\,dm < \infty$.

Show that C_φ is nuclear on $\mathscr{B}^1_\alpha$ if and only if $\int_{\mathbb{D}} \frac{d\sigma_\alpha}{(1-|\varphi|)^{\alpha+2}} < \infty$.

Hint: For H^1 (resp. $\mathscr{B}^1_\alpha$), show that when C_φ is nuclear, then $\|\delta_z\|$ must belong to $L^1(\mathbb{T}, m)$ (resp. $L^1(\mathbb{D}, \sigma_\alpha)$). See [94] for details.

Exercise I.12 Let $\alpha > -1$ and $a, b \in \mathbb{D}$ with $|a| + |b| = 1$ and $a \neq 0$.

Show that the norm of the composition operator $\varphi(z) = az + b$ viewed on the Bergman space $\mathscr{B}^2_\alpha$ is

$$\|C_\varphi\|_{B(\mathscr{B}^2_\alpha)} = |a|^{-\frac{\alpha+2}{2}}.$$

Hint: Adapt the H^2 case.

Exercise I.13 Let $f \in A(\mathbb{D})$, the disk algebra, satisfying

$$|f(u) - f(v)| \leqslant M|u - v|^\theta \qquad \text{for all } u, v \in \mathbb{T},$$

where $M > 0$ and $0 < \theta \leqslant 1$. We wish to show that

$$|f(u) - f(v)| \leqslant M|u - v|^\theta \qquad \text{for all } u, v \in \overline{\mathbb{D}}. \tag{I.1}$$

(a) Show that we can assume f to be a polynomial, by Fejér approximation ($f * K_n$ satisfies the same assumption as f). Finish the proof when $\theta = 1$. We now assume that $\theta < 1$.

(b) Let $v \in \mathbb{T}$. Show that the function defined by $g_v(z) = \frac{|f(v) - f(z)|}{|v - z|^\theta}$ and $g_v(v) = 0$ is subharmonic on $\mathbb{D}$ and continuous on $\overline{\mathbb{D}}$. Deduce that (I.1) holds for all $z \in \overline{\mathbb{D}}$ (you can begin with $v = 1$).

(c) Let $v \in \mathbb{D}$. Show that g_v is locally subharmonic (i.e., locally satisfies the sub-mean property) at each point z_0 of $\mathbb{D}$ (separate the cases $z_0 = v, z_0 \neq v$), and $g_v \leqslant M$ on $\partial\mathbb{D}$. Conclude.

Exercise I.14 Let φ be the symbol $\varphi(z) = \frac{b}{1 - az}$ for $z \in \mathbb{D}$, where $a, b \in (0, 1)$ with $a + b = 1$. Let $C_\varphi : H^2 \to H^2$ be the associated composition operator.

(a) Is C_φ compact?

(b) Show that the adjoint of C_φ is given by

$$C_\varphi^* = w.C_\psi + \delta_0 \qquad \text{with } w(z) = \frac{bz}{1 - bz} \text{ and } \psi(z) = \frac{a}{1 - bz}.$$

Exercise I.15 Let Λ_θ be the lens map.

(a) Show that Λ_θ belongs to the Wiener algebra W^+ of absolutely convergent Taylor series.
Hint: The nth Taylor coefficient of $(1 \pm z)^\theta$ is $O(n^{-1-\theta})$ and $(1+z)^\theta + (1-z)^\theta$ does not vanish on $\overline{\mathbb{D}}$.

(b) Is the associated composition operator C_{Λ_θ} bounded on W^+?

Exercise I.16 Using the ideas of the proof of Theorem 4.6, write a proof of Theorem 4.7.

Exercise I.17 In Chapter 5, prove (5.5) and (5.6): compute $m\big(S(\xi,h)\big)$ and $\sigma\big(S(\xi,h)\big)$.

Hint: Use polar coordinates centered at the point $\xi = 1$. First compute the variation of the angle, then deduce the variation of the associated angle centered at 0 to get easily $m\big(S(1,h)\big)$. For the area, compute the integral to get $\int_0^h 2\rho \arccos(\rho/2)\, d\rho$. Then make a change of variable.

Exercise I.18 In Chapter 5, prove (5.17) and (5.18).

Exercise I.19 Prove the estimates left as exercises in Section 6.4 of Chapter 6: (6.18) and (6.19).

Hint for (6.18): For the lower estimate, you can consider an appropriate power of the normalized reproducing kernel at a point $a \in \mathbb{D}$. For the upper estimate, replace the partial sum operator in the proof in the Hilbertian case by a convolution-type operator with a Fejér kernel and use the Green formula in H^p.

Hint for (6.19): Use the same ideas, starting from the Bergman version of the Green formula.

Exercise I.20 Let H be a rotation-invariant Hilbert space of analytic functions on $\mathbb{D}$, containing the polynomials as a dense subset.

Show that $H = H^2(\beta)$ for some weight $\beta = (\beta_n)_{n \geqslant 0}$.

Hint: Use that

$$\langle z^j, z^k \rangle = e^{i(j-k)\theta} \langle z^j, z^k \rangle \qquad \text{for all } j,k,\theta.$$

Then show that $\beta_n = \|z^n\|^2$ works.

Exercise I.21 Let $H^2(\beta)$ be a weighted Hardy space, and suppose that, for all symbols φ such that $\varphi(0) = 0$, we have $\|C_\varphi\| \leqslant 1$.

Fix integers m,n with $m \geqslant n$.

(a) Test the inequality $\|\varphi^n\| \leqslant \|z^n\|$ on the symbol

$$\varphi(z) = z\big[\psi(z^{m-n})\big]^{1/n}$$

with ψ another *zero-free* symbol, to get $|\psi(0)|^2\beta_n + |\psi'(0)|^2\beta_m \leqslant \beta_n$.

(b) Set $I(\psi) = \dfrac{1 - |\psi(0)|^2}{|\psi'(0)|^2}$. Show that

$$\beta_m \leqslant I\beta_n \text{ where } I = \inf_\psi I(\psi).$$

(c) Show that the value of I is

$$I = \inf_{x>0} \frac{e^x - 1}{x^2} =: J \approx 1.5467. \tag{I.2}$$

Hint: We can write $\psi = e^{-g}$ with Re $g > 0$ and hence

$$|g'(0)| \leqslant 2\mathrm{Re}\, g(0). \tag{I.3}$$

To see that, let $h = \dfrac{1-g}{1+g}$ with $h : \mathbb{D} \to \mathbb{D}$. Using the Schwarz–Pick inequality $|h'(0)| \leqslant 1 - |h(0)|^2$, show that

$$\frac{|g'(0)|}{|1+g(0)|^2} \leqslant \frac{2\mathrm{Re}\, g(0)}{|1+g(0)|^2},$$

which gives (I.3). Next, setting $2\mathrm{Re}\, g(0) = a$, and using $\psi'(0) = -g'(0)e^{-g(0)}$, show that

$$I(\psi) = \frac{1-e^{-a}}{|\psi'(0)|^2} \geqslant \frac{1-e^{-a}}{a^2 e^{-a}} = \frac{e^a - 1}{a^2} \geqslant J.$$

On the other hand, taking $g(z) = a\dfrac{1+z}{1-z}$, with $|g'(0)| = 2\mathrm{Re}\, g(0) = 2a$, show that we get exactly $I(\psi) = \dfrac{e^{2a}-1}{4a^2}$ and finally $I \leqslant J$, proving (I.2).

(d) Conclude that $\beta_m \leqslant I\beta_n$, with $I \approx 1.5467$ and the method cannot give more.

(e) Can you decide whether $\beta_m \leqslant \beta_n$ when $m \geqslant n$?

Exercise I.22 Let $\alpha > 0$ and $I_n = \displaystyle\int_0^1 (1-x)^n \exp(-x^{-\alpha})\, dx$.

(a) Writing $I_n = \int_0^\delta + \int_\delta^1$ with $\delta = n^{-1/(\alpha+1)}$, show that $I_n \leqslant e^{-n^{\alpha/(\alpha+1)}}$.

(b) Perform a sharper study of I_n.

Hint: First analyze the integral $J_n = \displaystyle\int_0^1 \exp(-nx)\exp(-x^{-\alpha})\, dx$ by making the change of variable $x = n^{-1/(\alpha+1)}y$ and applying the Laplace asymptotic formula for integrals [40, p. 125] (see also Exercise I.27 to follow).

Exercise I.23 Show that the weight sequence defined by $\beta_0 = 1$ and $\beta_n = n|\sin n|$, for $n \geqslant 1$ has both polynomial growth and decay.
Hint: Use that π is diophantine (i.e., "badly approximable by rationals" in an extended sense compared to the usual terminology).
Is β slowly oscillating?

Exercise I.24 Recall that $T_a(z) = \dfrac{z+a}{1+\bar{a}z}$. Suppose a and $b \in \mathbb{D}$ have the same argument mod. π. Show that $T_a \circ T_b = T_{\frac{a+b}{1+\bar{a}b}}$.

In particular, $(T_a)_{0 \leqslant a < 1}$ forms a unital semi-group for composition, and $(T_a)_{-1<a<1}$ a group for the same law.

Exercise I.25 Let φ be a symbol and $n_\varphi(z)$ the number of solutions of $\varphi(w) = z$ (where $w, z \in \mathbb{D}$). Show that C_φ is bounded on $\mathcal{D}$ (the Dirichlet space) if and only

$$\sup_{\xi \in \mathbb{T}} \int_{W(\xi, h)} n_\varphi(z)\, d\sigma(z) = O(h^2).$$

Hint: Show that $C_\varphi : \mathcal{D} \to \mathcal{D}$ is bounded if and only if the Carleson embedding $\mathscr{B}^2 \to L^2(\mathbb{D}, n_\varphi.d\sigma)$ is bounded.

When is $C_\varphi : \mathcal{D} \to \mathcal{D}$ compact?

Answer the same questions for $\mathcal{D}_\alpha$, where $\alpha > -1$.

Exercise I.26 Show that there are symbols $\varphi \in \mathcal{D}$, the Dirichlet space, such that C_φ is unbounded on $\mathcal{D}$.

Hint: The difficulty is that (why?) all powers of φ lie in $\mathcal{D}$.

But try to exhibit such a φ such that

$$\sup_{h \in (0,1)} \sup_{\xi \in \mathbb{T}} h^{-2} \int_{W(\xi, h)} n_\varphi(z)\, d\sigma(z) = +\infty.$$

See also [114].

Exercise I.27 Here is an exotic example in the framework $H = H^2(\beta)$. Take $\beta_n = \exp(-\sqrt{n})$.

(a) Show that, for $0 < r < 1$, $r = \mathrm{e}^{-\varepsilon}$,

$$\|K_r\|^2 = \sum_{n=0}^{\infty} r^{2n} \exp(\sqrt{n}) = \sum_{n=0}^{\infty} \mathrm{e}^{-2n\varepsilon} \exp(\sqrt{n}) =: S(\varepsilon).$$

(b) Using Euler–McLaurin summation formula, show that, when $\varepsilon \to 0^+$,

$$S(\varepsilon) \sim I(\varepsilon) = \int_0^\infty \exp(\sqrt{t} - 2\varepsilon t)dt = (4\varepsilon^2)^{-1} \int_0^\infty \exp\left[\frac{\sqrt{x} - x}{2\varepsilon}\right] dx. \tag{I.4}$$

(c) Using the Laplace theorem [40, p. 125] on the equivalent of integrals, which tells that

$$\int_0^\infty \mathrm{e}^{A\varphi(x)} dx \sim \sqrt{2\pi(|\varphi''(x_0)|)^{-1}}\, A^{-1/2} \mathrm{e}^{A\varphi(x_0)}$$

as $A \to \infty$, taking $A = \frac{1}{2\varepsilon}$ and the function $\varphi(x) = \sqrt{x} - x$, which takes its maximum at $x_0 = 1/4$, with $\varphi(x_0) = 1/4$, show that

$$\|K_r\|^2 = S(\varepsilon) \sim C\varepsilon^{-3/2} \exp\left(\frac{1}{8\varepsilon}\right) \sim C(1-r)^{-3/2} \exp\left[\frac{1}{8(1-r)}\right], \tag{I.5}$$

where C is a numerical constant (note that $\frac{1}{1-r} - \frac{1}{\varepsilon} \to \frac{1}{2}$).

(d) Now consider the hyperbolic automorphism of $\mathbb{D}$ given by

$$T(z) = \frac{2z+1}{z+2}.$$

For $r < 1$, set $s = T(r)$ and note that $1 - s \sim (1-r)/3$.

Show that

$$\frac{\|K_s\|^2}{\|K_r\|^2} \sim C \exp\left[\frac{1}{8}\left(\frac{1}{1-s} - \frac{1}{1-r}\right)\right] = C \exp\left[\frac{s-r}{8(1-r)(1-s)}\right] \overset{r \to 1^-}{\longrightarrow} \infty. \tag{I.6}$$

Hint: Using again $1 - s \sim (1-r)/3$ as well as $s - r \sim \frac{2}{3}(1-r)$, note that

$$\frac{s-r}{(1-r)(1-s)} \sim \frac{2}{1-r} \to \infty.$$

(e) Show that (I.6) implies the non-boundedness of C_T on $H^2(\beta)$.
Hint: Use the mapping equation $C_T^*(K_r) = K_{T(r)}$.

Exercise I.28 We remind here a theorem of Korenbljum [85]: let $(M_k)_{k\geqslant 0}$ be a logarithmically convex ($M_k^2 \leqslant M_{k-1}M_{k+1}$) sequence of positive numbers and $C\{M_k\}$ be the associated class of C^∞ functions of "Taylor type" defined as

$$C\{M_k\} = \left\{ F(x) = \sum_{n=0}^{\infty} a_n e^{inx} : \|F^{(k)}\|_\infty \leqslant AB^k M_k \quad \forall k \geqslant 0 \right\}$$

(where A, B depend on F). If $\sum_{k\geqslant 1} \sqrt{\frac{M_{k-1}}{M_k}} = \infty$ or $\sum_{k\geqslant 1} M_k^{-1/2k} = \infty$, then $C\{M_k\}$ is *quasi-analytic*, that is, $F \in C\{M_k\}$ and $F^{(k)}(0) \equiv 0$ implies $F = 0$.

(a) Let $\beta_n = \exp(\sqrt{n})$. Show that $H^2(\beta) \subset C\{M_k\}$ with $M_k = \sum_{n=1}^{\infty} n^{2k} e^{-\sqrt{n}}$.
(b) Show that $M_k \leqslant Ck^{2k}$.
(c) Show that $H^2(\beta)$ is quasi-analytic.

Exercise I.29 Let φ be a symbol such that the associated composition operator $C_\varphi\colon H^2(\beta) \to H^2(\beta)$ is compact (hence φ is not an automorphism), where $\beta = (\beta_n)$ satisfies the condition $\sum 1/\beta_n < \infty$, so that $H^2(\beta) \subset A(\mathbb{D})$ continuously. Denote $\varphi^{[n]}$ the nth iterate of φ.

(a) Let $\alpha \in \overline{\mathbb{D}}$ be the Denjoy–Wolff point of φ and $E_\alpha = \{f \in H^2(\beta) : f(\alpha) = 0\}$.
Show that E_α is a closed, C_φ-invariant, subspace of $H^2(\beta)$.
(b) Let T be the restriction of C_φ to E_α. Let λ be an eigenvalue of T and $f \in E_\alpha$ with $Tf = \lambda f$ and $f(z_0) \neq 0$ for some $z_0 \in \mathbb{D}$. Show that $\lambda^n f(z_0) \to 0$; infer that the spectral radius $r(T) < 1$.
(c) Let $e_1(z) = z$ and $e_0(z) = 1$. Observing that $e_1 - \alpha\, e_0 \in E_\alpha$, show that

$$\|\varphi^{[n]} - \alpha\, e_0\|_\infty = \|T^n(e_1 - \alpha\, e_0)\|_\infty \to 0.$$

(d) Conclude that the only fixed point of φ in $\overline{\mathbb{D}}$ is α. Give an example of a non-trivial symbol $\varphi \in A(\mathbb{D})$ with two fixed points or more in $\overline{\mathbb{D}}$.
Hint: Take $\varphi = zf$ where $f \in A(\mathbb{D})$ peaks on $E \subset \mathbb{T}$, with $m(E) = 0$.

Exercise I.30 Under the same assumptions as in Exercise I.29, suppose that, for some $z_0 = e^{i\theta_1} \in \partial\mathbb{D}$, we also have $\varphi(z_0) = e^{i\theta_2} \in \partial\mathbb{D}$.

(a) Let $\psi = R_{-\theta_2} \circ \varphi \circ R_{\theta_1}$ and $a = \psi(0)$. Show that $\psi(1) = 1$ and that there exists an automorphism χ of $\mathbb{D}$ such that $\chi(1) = 1$ and $\chi(a) = 0$.
(b) Let $\Phi = \chi \circ \psi$. Show that $C_\Phi \colon H^2(\beta) \to H^2(\beta)$ is compact and that Φ fixes 0 and 1.
(c) Conclude that, under the assumptions of Exercise I.29, we indeed have $\|\varphi\|_\infty < 1$.

Exercise I.31 Prove that, in Theorem 12.22, the estimate

$$\int_I |\widehat{T_a^n}(m))|^s \, da \gtrsim \delta n^{-s/2} \qquad \text{for every } m \in [\alpha^{-1} n, \alpha\, n]$$

still holds for any $s > 0$.

Hint: $|\cos x|^s = c + \sum_{l=1}^{\infty} \delta_l \cos(lx)$ with $c > 0$ and $\delta_l = O(l^{-(s+1)})$.

Appendix J A Small List of Open Problems

The questions below are just a very tiny sample among all the problems still open involving composition operators.

Problems

Questions J.1 Can one give a (simple?) explicit formula for the norm of $C_\varphi : H^2 \to H^2$ depending on the symbol φ?

The same for $C_\varphi : \mathscr{B}^2_\alpha \to \mathscr{B}^2_\alpha$, or for $C_\varphi : \mathscr{B}^p_\alpha \to \mathscr{B}^p_\alpha$.

More generally the same question holds for $C_\varphi : H^2(\beta) \to H^2(\beta)$.

For H^2, this has been done for fractional linear symbols with a very complicated expression.

Questions J.2 Compute the exact value of the essential norm of the composition operator $C_\varphi : H^p \to H^p$ with the help of the associated Nevanlinna counting function, or with the help of the Carleson function.

The same question holds for $C_\varphi : \mathscr{B}^p_\alpha \to \mathscr{B}^p_\alpha$ (here we can also ask the question in term of the hyperbolic version of the Carleson function).

Questions J.3 Characterize nuclear (weighted) composition operators on Hardy spaces H^p (the problem is already solved for $p = 2$ since the membership in $\mathscr{S}^1(H^2)$ is characterized by Luecking's criterion; and for $p = 1$, see Exercise I.11).

For Bergman spaces, the question also holds for weighted composition operators (see [42] in the unweighted case).

More generally, characterize nuclear composition operators in the framework of Orlicz spaces.

Questions J.4 Assume that any symbol φ fixing the origin induces a contraction on $H^2(\beta)$. Does this imply that β is non-increasing?

We saw that $m \geqslant n \Longrightarrow \beta_m \leqslant C\beta_n$ with $C \leqslant 1.5467\ldots$ (see I.21, in particular (I.2)).

What is the best possible value of C? Can we take $C = 1$?

Questions J.5 The sufficient condition β *essentially decreasing* for the boundedness of all symbols φ vanishing at 0 is not necessary. The weaker condition β *weakly decreasing* is sufficient (see Theorems 12.11 and 12.14): Is this also a necessary condition ? Otherwise, is there a fairly simple and explicit necessary and sufficient condition for that boundedness?

Questions J.6 Beyond the results mentioned in Chapter 8, there are many other results in the literature characterizing the spectrum of a composition operator. Nevertheless there is still a substantial work to do: given an arbitrary symbol φ and a space X of analytic functions on $\mathbb{D}$, characterize $\sigma(C_\varphi : X \to X)$.

Bibliography

[1] J. R. Akeroyd. On Shapiro's compactness criterion for composition operators. *J. Math. Anal. Appl.* **379** (1) (2011), 1–7.

[2] E. Amar, A. Lederer. Points exposés de la boule unité de $H^\infty(\mathbb{D})$ (French). *C. R. Acad. Sci. Paris Sér. A-B*, **272** (1971), A1449–A1452.

[3] E. Amar, Suites d'interpolation pour les classes de Bergman de la boule et du polydisque. *Canadian J. Math.* **XXX** (4) (1978), 711–737.

[4] J. Agler, J. McCarthy. Pick interpolation and Hilbert function spaces, vol. 44, Graduate Studies in Mathematics. American Mathematical Society, 2002.

[5] W. Arendt, E. Bernard, B. Celaries, I. Chalendar. Denjoy-Wolff theory and spectral properties of weighted composition operators on Hol $(\mathbb{D})$. *Illinois Math. Journ.* **66** (4) (2022), 463–489.

[6] K. Attele. Multipliers of the range of composition operators. *Tokyo. J. Math.* **15** (1) (1992), 185–198.

[7] F. Bayart. Compact composition operators on a Hilbert space of Dirichlet series. *Illinois. J. Math.* **47** (2003), 725–743.

[8] C. J. Bishop. Orthogonal functions in H^∞. *Pacific J. Math.* **220** (1) (2005), 1–31.

[9] R. P. Boas Jr. A general moment problem. *Amer. J. Math.* **63** (1941), 361–370.

[10] S. V. Bočkarev. Existence of a basis in the space of functions analytic in the disk, and some properties of Franklin's system (in Russian). *Mat. Sb. (n.s.)* **95** (137) (1974), 3-18, 159. Translated in *Math. USSR-Sb.* 24 (1974), 1–16.

[11] J. Bonet, P. Domański, M. Lindström. Essential norm and weak compactness of composition operators on weighted Banach spaces of analytic functions. *Canadian Math. Bull.* **42** (2) (1999), 139–148.

[12] A. A. Borichev, K. Fouchet, R. Zarouf. On the Fourier coefficients of powers of a Blaschke factor and strongly annular functions. *Construct. Approx.* **60** (1) (2024), 33–86.

[13] P. Bourdon, J. Shapiro. Mean growth of Koenigs eigenfunctions. *J. Amer. Math. Soc.* **10** (2)(1997), 299–325.

[14] P. S. Bourdon, E. E. Fry, C. Hammond, C. H. Spofford. Norms of linear-fractional composition operators. *Trans. Amer. Math. Soc.* **356** (2004), 2459–2480.

[15] M. Blyudze, S. Shimorin. Estimates of the norms of powers of functions in certain Banach spaces. *J. Math. Sciences* **80** (4), (1996), 1880–1891.

[16] B. Carl, M. Stephani. *Entropy, Compactness and the Approximation of Operators.* Cambridge University Press, 1990.

[17] L. Carleson. An interpolation problem for bounded analytic functions. *Amer. J. Math.* **80** (1958), 921–930.

[18] T. Carroll, C. C. Cowen, Compact composition operators not in the Schatten classes. *J. Operator Theory* **26** (1991), 109–120.

[19] J. Caughran, H. Schwartz. Spectra of composition operators. *Proc. Amer. Math. Soc.* **51** (1975), 127–130.

[20] I. Chalendar, J. R. Partington. Norm estimates for weighted composition operators on spaces of holomorphic functions. *Complex Anal. Oper. Theory* **8** (5) (2014), 1087–1095.

[21] I. Chalendar, J. R. Partington. Compactness and norm estimates for weighted composition operators on spaces of holomorphic functions. *Theta Ser. Adv. Math.* **19** (2017), 81–89.

[22] I. Chalendar, L. Oger, J. R. Partington. Linear isometries on the annulus: Description and spectral properties. Banach J. Math. Anal. **19** (2) (2025), 18 pp.

[23] S. Charpentier. Essential norm of composition operators on the Hardy space H^1 and the weighted Bergman spaces A^p_α on the ball. *Arch. Math. (Basel)* **98** (2012), 327–340.

[24] S. Charpentier. Compact composition operators on the Hardy-Orlicz and weighted Bergman-Orlicz spaces on the ball. *J. Operator Theory* **69** (2) (2013), 463–481.

[25] S. Charpentier. Composition operators on weighted Bergman-Orlicz spaces on the ball. *Complex Anal. Oper. Theory* **7** (1) (2013), 43–68.

[26] S. Charpentier, N. Espoullier, R. Zarouf. Composition operators on Beurling-Sobolev spaces. In preparation.

[27] S. Charpentier, S. Sehba. Carleson measure theorems for large Hardy-Orlicz and Bergman-Orlicz spaces. *J. Funct. Spaces Appl.* (2012), n.pag.

[28] J. A. Cima, A. L. Matheson. Completely continuous composition operators. *Trans. Amer. Math. Soc.* **344** (1994), 849–856.

[29] J. A. Cima, A. L. Matheson. Essential norms of composition operators and Aleksandrov measures. *Pacific J. Math.* **179** (1) (1997), 59–64.

[30] D. D. Clahane. Spectra of compact composition operators over bounded symmetric domains. *Integr. Equ. Oper. Theory.* **51** (1) (2005), 41–56.

[31] M. Contreras, S. Diaz-Madrigal, C. Pommerenke. Iteration in the unit disk: The parabolic zoo. *Proceedings of Complex and Harmonic Analysis.* DEStech Publications, 2007, pp. 63–91.

[32] C. Cowen. Linear fractional composition operators on H^2. *Integr. Equ. Oper. Theory.* **11** (1988), 151–160.

[33] C. Cowen. An application of Hadamard multiplication to operators on weighted Hardy spaces. *Linear Algebra Appl.* **133** (1990), 21–32.

[34] C. Cowen, B. McCluer. *Composition Operators on Spaces of Analytic Functions.* Studies in Advanced Mathematics. CRC Press, 1994.

[35] D. Choimet, H. Queffélec. *Twelve Landmarks of Twentieth-Century Analysis.* Cambridge University Press, 2015.

[36] Ž. Čučković, R. Zhao. Weighted composition operators between different weighted Bergman spaces and different Hardy spaces. *Illinois J. Math.* **51** (2) (2007), 479–498.

[37] A. Danielyan. Fatou's interpolation theorem implies the Rudin-Carleson theorem *J. Fourier Analysis App.* **23** (3) (2017), 656–659.

[38] R. Demazeux. Essential norms of weighted composition operators between Hardy spaces H^p and H^q. *Studia Math.* **206** (3) (2011), 191–209.

[39] J. Diestel, H. Jarchow, A. Tonge. Absolutely summing operators. *Cambridge Studies in Advanced Mathematics*, 43. Cambridge University Press, 1995.

[40] J. Dieudonné. *Calcul infinitésimal.* Deuxième édition. Hermann, 1980.

[41] S. Dineen. *The Schwarz Lemma.* Clarendon Press, 1989.

[42] T. Domenig. Composition operators belonging to operator ideals. *J. Math. Anal. Appl.*, **237**(1) (1999), 327–349.

[43] R.G. Douglas. On majorization, factorization, and range inclusion of operators on Hilbert space, *Proc. Amer. Math. Soc.* **17** (1966), 413–415.

[44] J. Dugundji. *Topology.* Allyn and Bacon, 1966.

[45] P. Duren. *Theory of H^p-spaces.* Dover, 1970.

[46] P. Duren, A. Schuster. Bergman spaces. *Mathematical Surveys and Monographs* 100. American Mathematical Society, 2004.

[47] J. Earl. On the interpolation of bounded sequences by bounded functions. *J. London Math. Soc.* **2** (1970), 544–548.

[48] O. El-Fallah, M. El Ibbaoui. Trace estimates of Toeplitz operators on Bergman spaces and applications to composition operators, *Rev. Mat. Iberoamericana* **6**, 1723–1762.

[49] O. El-Fallah, K. Kellay. Nevanlinna counting function and pull-back measure. *Proc. Amer. Math. Soc.* **144** (6) (2016), 2559–2564.

[50] O. El-Fallah, K. Kellay, J. Mashreghi, T. Ransford. A primer on the Dirichlet space. *Cambridge Tracts in Mathematics*, 203. Cambridge University Press, 2014.

[51] O. El-Fallah, N. Nikolski, M. Zarrabi. Resolvant estimates in Beurling-Sobolev algebras. *Algebra i Analiz* **10** (6) (1998), 1–92; translation in *St Petersburg Math. J.* **10** (6) (1999), 901–964.

[52] T. Farès, P. Lefèvre. Nuclear composition operators on Bloch spaces. *Proc. Amer. Math. Soc.* **148** (6) (2020), 2487–2496.

[53] T. Farès, P. Lefèvre. Absolutely summing composition operators on Bloch type spaces. *J. Operator Theory* **88** (2) (2022), 407–443.

[54] H. Federer. *Geometric Measure Theory.* Band 153. Springer-Verlag, 1969.

[55] E. Fricain, J. Mashreghi. *The Theory of $\mathcal{H}(b)$-spaces.* Volume 1–2. Cambridge University Press, 2016.

[56] E. A. Gallardo-Gutiérrez, J. R. Partington. Norms of composition operators on weighted Hardy spaces. *Israel J. Math.* **196** (1) (2013), 273–283.

[57] S. Garcia, J. Mashreghi, W. Ross. *Finite Blaschke Products and Their Connections.* Springer, 2018.

[58] J. Garnett. *Bounded Analytic Functions.* Revised first edition. Graduate Texts in Mathematics, **236**. Springer, 2007.

[59] J. Garnett, D. Marshall. Harmonic measure. *New Math Monographs.* Cambridge, 2005.

[60] D. M. Girard. The behavior of the norm of an automorphism of the unit disk. *Pacific J. Math.* **47** (2) (1973), 443–456.

[61] E. Gluskin, M. Meyer, A. Pajor. Zeros of analytic functions and norms of inverse matrices. *Israel J. Math.* **87** (1994), 225–242.

[62] I. C. Gohberg, M. G. Krein, Introduction to the Theory of Lineat Non-selfadjoint Operators. Nauka, 1965; English translation: Transl. Math. Monographs, 18. American Mathematical Society, 1969.

[63] F. González-Dona. *Operadores de Composición y Clases de Schatten.* Universidad de Sevilla, 2015.

[64] J. Gordon, H. Hedenmalm. The composition operators on the space of Dirichlet series with square-summable coefficients. *Michigan. Math. J.* **46** (1999), 313–329.

[65] P. Gorkin, B. MacCluer. Essential norms of composition operators. *Integral Equations Operator Theory* **48** (1) (2004), 27–40.

[66] S. Goroliha (a.k.a. I. Alam), I. Chalendar, F. El Chami, E. Fricain, P. Lefèvre. Eventual ideal properties of the Riemann-Liouville analytic semigroup. *Preprint* (2024), n.pag.

[67] L. Grafakos. *Classical Fourier Analysis.* Second edition. Springer, 2008.

[68] G. Gunatillake. Spectrum of a compact weighted composition operator. *Proc. Amer. Math. Soc.* **135** (2) (2007), 461–467.

[69] C. Hammond. On the norm of a composition operator with linear fractional symbol. *Acta Sci. Math. (Szeged)* **69** (2003), 813–829.

[70] W. W. Hastings. A Carleson measure theorem for Bergman spaces. *Proc. Amer. Math. Soc.* **52** (1975), 237–241.

[71] W. Hayman, P. Kennedy, W. Ross. *Subharmonic Functions,* vol.I. London Mathematical Society Monographs, vol.9. Academic Press, 1976.

[72] H. Hedenmalm, P. Lindqvist, K. Seip. A Hilbert space of Dirichlet series and systems of dilated functions. *Duke. Math. J.* **86** (1997), 1–37.

[73] H. Hedenmalm, B. Korenblum, K. Zhu. Theory of Bergman spaces. *Graduate Texts in Mathematics*, 199. Springer-Verlag, 2000.

[74] K. Hoffman. *Banach Spaces of Analytic Functions.* Dover, 1988.

[75] N. Hutton. On the approximation numbers of an operator and its adjoint. *Math. Annalen* **210** (1974), 277–280.

[76] V. È Kacnel'son. A remark on canonical factorization in certain spaces of analytic functions. *J. Soviet Math.* **4** (2) (1976), 444–445.

[77] J. P. Kahane. Some random series of functions. *Cambridge Studies in Adv. Math.* 5. Cambridge University Press, 1985.

[78] J.P. Kahane, Y. Katznelson. Sur le comportement radial des fonctions analytiques. *C. R. Acad. Sci. Paris Sér. A-B* **272** (1971), A718–A719.

[79] N. Kalton, C. Lemerdy. Solution of a problem of Peller concerning similarity. *J. Operator Th.* **47** (2) (2002), 379–387.

[80] H. Kamowitz. The spectra of composition operators on H^p. *J. Funct. Anal.* **(18)** (1975), 32–150.

[81] S. Karlin, J. MacGregor. Embedding iterates of analytic functions. *Trans. Amer. Math. Soc.* **132** (1968), 137–145.

[82] K. Kellay, P. Lefèvre. Compact composition operators on weighted Hilbert spaces of analytic functions. *J. Math. Anal. Appl.* **386** (2012), 718–727.

[83] M. Klimek. *Pluripotential Theory*. London Math. Soc. Monographs, New Series 6. Oxford University Press, 1991.

[84] H. König. Eigenvalue distribution of compact operators. *Operator Theory, Advances and Applications,* vol. 86. Birkhäuser, 1986.

[85] B. Korenbljum. Quasianalytic classes of functions in a circle. *Soviet Math. Doklady* (6) (1965), 1155–1158.

[86] S. G. Krantz. Function theory of several complex variables. *Pure and Applied Mathematics*. Wiley-Interscience, 1982.

[87] S. G. Krantz. Complex analysis: The Geometric Viewpoint. *Caradus Mathematical Monographs* (1990), n.pag.

[88] M. A. Krasnosel'skiĭ, Ja. B. Rutickiĭ. *Convex Functions and Orlicz Spaces*. P. Noordhoff, 1961.

[89] T. L. Kriete, B. D. MacCluer. A rigidity theorem for composition operators on certain Bergman spaces. *Michigan Math. J.* **42** (1995), 379–386.

[90] N. S. Landkof. *Foundations of Modern Potential Theory*. Springer, 1972.

[91] C. Le Merdy. Oral communication in the OT Conference 2008 in Timisoara.

[92] G. Lechner, D. Li, H. Queffélec, L. Rodríguez-Piazza. Approximation numbers of weighted composition operators. *J. Funct. Anal.* **274** (7) (2018), 1928–1958.

[93] P. Lefèvre. Generalized essential norm of weighted composition operators on some uniform algebras of analytic functions. *Integral Equations Operator Theory* **63** (4) (2009), 557–569.

[94] P. Lefèvre. Nuclear Carleson embeddings on Hardy and Bergman spaces with exponent 1. Preprint.

[95] P. Lefèvre, D. Li, H. Queffélec, L. Rodríguez-Piazza, A criterion of weak compactness for operators on subspaces of Orlicz spaces. *J. Funct. Spaces Appl.* **6** (3) (2008), 277–292.

[96] P. Lefèvre, D. Li, H. Queffélec, L. Rodríguez-Piazza, Some examples of compact composition operators on H^2. *J. Funct. Anal.* **255** (11) (2008), 3098–3124.

[97] P. Lefèvre, D. Li, H. Queffélec, L. Rodríguez-Piazza. Compact composition operators on H^2 and Hardy-Orlicz spaces. *J. Math. Anal. Appl.* **354** (1) (2009), 360–371.

[98] P. Lefèvre, D. Li, H. Queffélec, L. Rodríguez-Piazza. Composition operators on Hardy-Orlicz spaces. *Memoirs Amer. Math. Soc.* **207** (974) (2010), n.pag.

[99] P. Lefèvre, D. Li, H. Queffélec, L. Rodríguez-Piazza. Nevanlinna counting function and Carleson function of analytic maps. *Math. Ann.* **351** (2) (2011), 305–326.

[100] P. Lefèvre, D. Li, H. Queffélec, L. Rodríguez-Piazza. Some revisited results about composition operators on Hardy spaces. *Revista Mat. Iberoamer.* **28** (1) (2012), 57–76.

[101] P. Lefèvre, D. Li, H. Queffélec, L. Rodríguez-Piazza. Compact composition operators on the Dirichlet space and capacity of sets of contact points. *J. Funct. Anal.* **264** (4) (2013), 895–919.

[102] P. Lefèvre, D. Li, H. Queffélec, L. Rodríguez-Piazza. Compact composition operators on Bergman-Orlicz spaces. *Trans. Amer. Math. Soc.* **365** (8) (2013), 3943–3970.

[103] P. Lefèvre, D. Li, H. Queffélec, L. Rodríguez-Piazza. Some new properties of composition operators associated with lens maps. *Israel J. Math.* **195** (2) (2013), 801–824.

[104] P. Lefèvre, D. Li, H. Queffélec, L. Rodríguez-Piazza. Approximation numbers of composition operators on the Dirichlet space. *Ark. Mat.* **53** (1) (2015), 155–175.

[105] P. Lefèvre, D. Li, H. Queffélec, L. Rodríguez-Piazza. Comparison of singular numbers of composition operators on different Hilbert spaces of analytic functions. *J. Funct. Anal.* **280** (3) (2021), 1–47.

[106] P. Lefèvre, D. Li, H. Queffélec, L. Rodríguez-Piazza. Characterization of weighted Hardy spaces on which all composition operators are bounded. *Analysis & PDE.* **18** (8) (2025), 1921–1954.

[107] P. Lefèvre, D. Li, H. Queffélec, L. Rodríguez-Piazza. On some questions about composition operators on weighted Hardy spaces. *Pure and Applied Functional Analysis* **10** (4) (2025), 859–874.

[108] D. Li, H. Queffélec. *Introduction to Banach Spaces. Analysis and Probability.* Cambridge University Press, 2018.

[109] D. Li. Espaces de Hardy et interpolation. *Ellipses* (2022), n. pag.

[110] D. Li, H. Queffélec, L. Rodríguez-Piazza. On approximation numbers of composition operators. *Journ. of Approx. Theory* **164** (4) (2012), 431–459.

[111] D. Li, H. Queffélec, L. Rodríguez-Piazza. Estimates for approximation numbers of some classes of composition operators on the Hardy space. *Annales Academiae Scientiarum Fennicae* **38** (2013), 547–564.

[112] D. Li, H. Queffélec, L. Rodríguez-Piazza. Infinitesimal Carleson property for weighted measures induced by analytic self-maps of the unit disk. *Complex Anal. Op. Theory* **7** (4) (2013), 1371–1387.

[113] D. Li, H. Queffélec, L. Rodríguez-Piazza. A spectral radius-type formula for approximation numbers of composition operators. *J. Funct. Anal.* **267** (2014), 4753–4774.

[114] D. Li, H. Queffélec, L. Rodríguez-Piazza. Two results on composition operators on the Dirichlet space. *J. Math. Anal. Appl.* **426** (2) (2015), 734–746.

[115] D. Li, H. Queffélec, L. Rodríguez-Piazza. Approximation numbers of composition operators on H^p. *Concr. Oper.* **2** (1) (2015), 98–109.

[116] D. Li, H. Queffélec, L. Rodríguez-Piazza. Some examples of composition operators and their approximation numbers on the Hardy space of the bidisk. *Trans. Amer. Math. Soc.* **372** (2019), 2631–2658.

[117] D. Li, H. Queffélec, L. Rodríguez-Piazza. An extremal composition operators on the Hardy space of the bidisk with small approximation numbers. *Journal of Approximation Theory* **252** (2020) 1–14.

[118] J. Lindenstrauss, A. Pełczyński. Contribution to the theory of classical Banach spaces. *J. Funct. Anal.* **8** (1971), 225–249.

[119] J.E. Littlewood. On inequalities in the theory of functions. *Proc. London Math. Soc.* **23** (7) (1925), 481–519.

[120] G. G. Lorentz. *Approximation of Functions*. Second edition AMS Chelsea Publishing, 1986.

[121] D. H. Luecking, Inequalities in Bergman spaces. *Illinois. J. Math.* (25) (1981), 1–11.

[122] D. H. Luecking. Trace Ideal Criteria fo composition operators. *J. Funct. Anal.*, **73** (2) (1987), 345–368.

[123] D. H. Luecking, K. H. Zhu. Composition operators belonging to the Schatten ideals. *Amer. J. Math.* **114** (2) (1992), 1127–1145.

[124] B. MacCluer. Spectra of compact composition operators on $H^p(B_N)$. *Analysis 4* (1984), 87–103.

[125] B. MacCluer. Compact composition operators on $H^p(B_N)$. *Michigan Math. J.* **32** (2) (1985), 237–248.

[126] B. MacCluer, J. H. Shapiro. Angular derivatives and compact composition operators on the Hardy and Bergman spaces. *Canadian J. Math.* **38** (4) (1986), 878–906.

[127] K. Madigan, A. L. Matheson. Compact composition operators on the Bloch space. *Trans. Amer. Math. Soc.* **347** (7) (1995), 2679–2687.

[128] A. V. Megretskii, V. Peller, S. Treil. The inverse spectral problem for self-adjoint Hankel Operators. *Acta Mathematica* **174** (1999), 241–309.

[129] H. L. Montgomery. Ten lectures on the interface between analytic number theory and harmonic analysis. *CBMS Regional Conference Series in Mathematics* 84. American Mathematical Society, 1994.

[130] D. Newman, H. Shapiro. The Taylor coefficients of inner functions. *Michigan Math. J.* **9** (1962), 249–255.

[131] E. Nordgren. Composition operators. *Canadian J. Math.* **20** (1968), 442–449.

[132] V. L. Oleinik, B. S. Pavlov. Embedding theorems for weighted classes of harmonic and analytic functions. *J. Soviet Math.* **2** (1974), 135–142; a translation of Zap. Nauch. Sem. LOMI Steklov, 22 (1971).

[133] O. G. Parfenov. Estimates of the singular numbers of the Carleson embedding operator. *Math. USSR Sbornik* **59** (2) (1988), 497–511.

[134] V. Peller. *Hankel Operators and Their Applications*. *Springer*, 2003.

[135] H. J. Petzsche. On E. Borel's theorem. *Math. Ann.* **282** (2) (1988), 299–313.

[136] A. Pietsch. Weyl numbers and eigenvalues of operators in Banach spaces. *Math. Ann.*, **247** (1980), 149–168.

[137] A. Pinkus. *n-Widths in Approximation Theory.* Springer, 1985.

[138] G. Pisier. *The Volume of Convex Bodies and Banach Space Geometry.* Cambridge University Press, 1989.

[139] G. Pisier. *Similarity Problems and Completely Bounded Maps*. Second edition. Springer, 1996.

[140] S. Pouliasis. Approximation numbers of composition operators on weighted Besov spaces of analytic functions. *Proc. Edinb. Math. Soc.* **65** (2) (2022), 311–325.

[141] S. C. Power. Vanishing Carleson measures. *Bull. London Math. Soc.* **12** (3) (1980), 207–210.

[142] H. Queffélec. Norms of composition operators with affine symbols. *J. Anal.* **20** (2012), 47–58.

[143] H. Queffélec. Approximation numbers of classes of symbolic operators. *Recent trends in Analysis* (2013), 159–175.

[144] H. Queffélec. *Topologie,* 7 ieme edition. Dunod, 2025.

[145] H. Queffélec, K. Seip. Approximation numbers of composition operators on the H^2 space of Dirichlet series. *J. Funct. Anal.* **268** (2015), 1612–1648.

[146] H. Queffélec, K. Seip. Decay rates for approximation numbers of composition operators. *J. Anal. Math.* **125** (2015), 371–399.

[147] H. Queffélec, C. Zuily. *Analyse pour l'agrégation* Fifth edition. Dunod, 2020.

[148] M. M. Rao, Z. D. Ren. Theory of Orlicz spaces. *Pure and Applied Mathzematics* 146. *Marcel Dekker*, 1991.

[149] F. Riesz. Sur une inégalite de M. Littlewood dans la théorie des fonctions. *Proc. London Math. Soc.* **23** (1925), 36–39.

[150] W. Ross. Lens lectures on Aleksandrov-Clark measure. Université d'Artois, 2013.

[151] S. Rotfel'd. The singular numbers of the sum of completely continuous operators. *Funkcional.Anal. i Prilozen.* **1** (3) (1967), 95–96.

[152] W. Rudin. *Function Theory in the Unit Ball of* $\mathbb{C}^n$. Springer, 1980.

[153] W. Rudin. *Real and Complex Analysis.* Third edition. McGraw-Hill, 1987.

[154] W. Rudin. *Analyse fonctionnelle.* Ediscience International, 1995.

[155] J.V. Ryff. Subordinate H^p functions. *Duke Math.* **J.33** (1966), 347–354.

[156] E. Saksman. An elementary introduction to Clark measures. *Topics in Complex Analysis and Operator Theory,* 2007, 85–136.

[157] E. Saksman, oral communication, Oberwolfach, 2015.

[158] H. Schwartz. Composition operators on H^p. Thesis (Ph.D.), The University of Toledo, 1969.

[159] J. Schäffer. Norms and determinants of linear mappings. *Math. Z.* **118** (1970), 331–339.

[160] K. Seip. *Interpolation and Sampling in Spaces of Analytic Functions.* Amer. Math. Soc., 2004.

[161] K. Seip. Zeros of functions in Hilbert spaces of Dirichlet series. *Math. Z.* **274** (2013), 1327–1339.

[162] H. Shapiro, A. Shields. On some interpolation problems for analytic functions. *Amer. J. Math.* **83** (1961), 513–532.

[163] J. H. Shapiro. Compact composition operators on spaces of boundary regular holomorphic functions. *Proc. Amer. Math. Soc.* **100** (1987), 49–57.

[164] J. H. Shapiro. The essential norm of a composition operator. *Ann. of Math.* **125** (2) (1987), 375–404.

[165] J. H. Shapiro. *Composition Operators and Classical Function Theory.* Springer-Verlag, 1993.

[166] J. H. Shapiro, P. D. Taylor. Compact, nuclear, and Hilbert-Schmidt composition operators on H^2. *Indiana Univ. Math. J.* **23** (1973/74), 471–496.

[167] A. Shields. Weighted shift operators and analytic function theory. *Topics in Operator Theory. Math. Surveys*, (1974), 49–128.

[168] B. Simon. Trace ideal classes and their applications. *London Mathematical Society Lecture Notes Series* **35** (1979), n.pag.

[169] D. A. Stegenga. Multipliers of the Dirichlet space. *Illinois J. Math.* **24** (1) (1980), 113–139.

[170] E. M. Stein. *Singular Integrals and Differentiability Properties of Functions.* Princeton University Press, 1970.

[171] O. Szehr, R. Zarouf. l_p-norms of Fourier coefficients of powers of a Blaschke factor. *J. Anal. Math.* **140** (2020), 1–30.

[172] O. Szehr, R. Zarouf. Explicit counterexamples to Schäeffer's conjecture. *J. Math. Pures Appl.* **146** (2021), 1–30.

[173] G. D. Taylor. Multipliers on D_α. *Trans. Amer. Math. Soc.* **123** (1966), 229–240.

[174] R.C. Thompson. Convex and concave functions of singular values of matrix sums. *Pacific J. Math.* **66** (1) (1976), 285–290.

[175] E. C. Titchmarsch. *The Theory of the Riemann Zeta-Function.* Second edition revised by D. R. Heath-Brown. Oxford Science, 1986.

[176] N. Tomczak-Jaegermann. Banach-Mazur distances and finite-dimensional operator ideals. *Longman Scientific & Technical* (1987), n.pag.

[177] G. Valiron. Sur l'itération des fonctions holomorphes dans un demi-plan. *Bull. Sci. Math.* **2** (1931), 105–128.

[178] V. I. Videnskii. An analog of the hyperbolic metric generated by a Hilbert space kernel with the Schwarz-Pick kernel. *Journ. of Math. Sciences.* **118** (2016), 331–339.

[179] H. Widom. Rational approximation and n-dimensional diameter. *J. Approx. Theory*, **5** (1972), 342–361.

[180] P. Wojtaszczyk. Banach spaces for analysts. *Cambridge Studies in Advanced Mathematics* **25** (1991), n.pag.

[181] R. Young. *An Introduction to Nonharmonic Fourier Series.* Academic Press, 1980.

[182] L. Zheng. The essential norms and spectra of composition operators on H^∞. *Pac. J. Math.* **203** (2) (2002), 503–510.

[183] K. Zhu. Schatten class composition operators on weighted Bergman spaces of the disk. *J. Operator Theory* **46** (1) (2001), 173–181.

[184] K. Zhu. *Operator Theory in Function Spaces and Finite-Dimensional Operator Ideals.* American Math. Soc., 2007.

[185] N. Zorboska. Composition operators on weighted Hardy spaces. Thesis (Ph.D.), University of Toronto, 1988.

[186] N. Zorboska. Composition operators with closed range. *Trans. Amer. Math. Soc.* **344** (2) (1994), 791–801.

[187] N. Zorboska. Composition operators on weighted Dirichlet spaces. *Proc. Amer. Math. Soc.* **126** (7) (1998), 2013–2023.

[188] A. Zygmund. *Trigonometric Series, Vol.I and II.* Second edition. Cambridge Mathematical Library, 1977.

Index

For EU product safety concerns, contact us at Calle de José Abascal, 56–1°, 28003 Madrid, Spain or eugpsr@cambridge.org.

www.ingramcontent.com/pod-product-compliance
Ingram Content Group UK Ltd.
Pitfield, Milton Keynes, MK11 3LW, UK
UKHW020223290926
15470UKWH00001B/342
* 9 7 8 1 0 0 9 3 8 8 8 9 4 *